Two-Phase Heat Transfer

Wiley-ASME Press Series

Advanced Multifunctional Lightweight Aerostructures: Design, Development, and Implementation
Kamran Behdinan and Rasool Moradi-Dastjerdi

Vibration Assisted Machining: Theory, Modelling and Applications
Lu Zheng, Wanqun Chen, Dehong Huo

Two-Phase Heat Transfer
Mirza Mohammed Shah

Computer Vision for Structural Dynamics and Health Monitoring
Dongming Feng, Maria Q Feng

Theory of Solid-Propellant Nonsteady Combustion
Vasily B. Novozhilov, Boris V. Novozhilov

Introduction to Plastics Engineering
Vijay K. Stokes

Fundamentals of Heat Engines: Reciprocating and Gas Turbine Internal Combustion Engines
Jamil Ghojel

Offshore Compliant Platforms: Analysis, Design, and Experimental Studies
Srinivasan Chandrasekaran, R. Nagavinothini

Computer Aided Design and Manufacturing
Zhuming Bi, Xiaoqin Wang

Pumps and Compressors
Marc Borremans

Corrosion and Materials in Hydrocarbon Production: A Compendium of Operational and Engineering Aspects
Bijan Kermani and Don Harrop

Design and Analysis of Centrifugal Compressors
Rene Van den Braembussche

Case Studies in Fluid Mechanics with Sensitivities to Governing Variables
M. Kemal Atesmen

The Monte Carlo Ray-Trace Method in Radiation Heat Transfer and Applied Optics
J. Robert Mahan

Dynamics of Particles and Rigid Bodies: A Self-Learning Approach
Mohammed F. Daqaq

Primer on Engineering Standards, Expanded Textbook Edition
Maan H. Jawad and Owen R. Greulich

Engineering Optimization: Applications, Methods and Analysis
R. Russell Rhinehart

Compact Heat Exchangers: Analysis, Design and Optimization using FEM and CFD Approach
C. Ranganayakulu and Kankanhalli N. Seetharamu

Robust Adaptive Control for Fractional-Order Systems with Disturbance and Saturation
Mou Chen, Shuyi Shao, and Peng Shi

Robot Manipulator Redundancy Resolution
Yunong Zhang and Long Jin

Stress in ASME Pressure Vessels, Boilers, and Nuclear Components
Maan H. Jawad

Combined Cooling, Heating, and Power Systems: Modeling, Optimization, and Operation
Yang Shi, Mingxi Liu, and Fang Fang

Applications of Mathematical Heat Transfer and Fluid Flow Models in Engineering and Medicine
Abram S. Dorfman

Bioprocessing Piping and Equipment Design: A Companion Guide for the ASME BPE Standard
William M. (Bill) Huitt

Nonlinear Regression Modeling for Engineering Applications: Modeling, Model Validation, and Enabling Design of Experiments
R. Russell Rhinehart

Geothermal Heat Pump and Heat Engine Systems: Theory and Practice
Andrew D. Chiasson

Fundamentals of Mechanical Vibrations
Liang-Wu Cai

Introduction to Dynamics and Control in Mechanical Engineering Systems
Cho W.S. To

Two-Phase Heat Transfer

Mirza Mohammed Shah

This Work is a co-publication between John Wiley & Sons Ltd and ASME Press.

WILEY

This Work is a co-publication between John Wiley & Sons Ltd and ASME Press.

Registered Office(s)
John Wiley & Sons, Inc., 111 River Street, Hoboken, NJ 07030, USA
John Wiley & Sons Ltd, The Atrium, Southern Gate, Chichester, West Sussex, PO19 8SQ, UK

Editorial Office
The Atrium, Southern Gate, Chichester, West Sussex, PO19 8SQ, UK

For details of our global editorial offices, customer services, and more information about Wiley products visit us at www.wiley.com.

Wiley also publishes its books in a variety of electronic formats and by print-on-demand. Some content that appears in standard print versions of this book may not be available in other formats.

Library of Congress Cataloging-in-Publication Data Applied for:
9781119618614 (Hardback)

Cover Design: Wiley
Cover Image: © Jub Rubjob/Getty Images

Set in 9.5/12.5pt STIXTwoText by SPi Global, Chennai, India

Printed and bound by CPI Group (UK) Ltd, Croydon, CR0 4YY

10 9 8 7 6 5 4 3 2 1

Contents

Preface *xvii*

1 Introduction *1*
1.1 Scope and Objectives of the Book *1*
1.2 Basic Definitions *1*
1.3 Various Models *2*
1.3.1 Homogeneous Model *2*
1.3.2 Separated Flow Models *2*
1.3.3 Flow Pattern-Based Models *3*
1.4 Classification of Channels *3*
1.4.1 Based on Physical Dimensions *3*
1.4.2 Based on Condensation Studies *3*
1.4.3 Based on Boiling Flow Studies *4*
1.4.4 Based on Two-Component Flow *4*
1.4.5 Discussion *5*
1.4.6 Recommendation *5*
1.5 Flow Patterns in Channels *5*
1.5.1 Horizontal Channels *5*
1.5.1.1 Description of Flow Patterns *5*
1.5.1.2 Flow Pattern Maps *6*
1.5.2 Vertical Channels *7*
1.5.3 Inclined Channels *7*
1.5.4 Annuli *8*
1.5.5 Minichannels *8*
1.5.6 Horizontal Tube Bundles with Crossflow *9*
1.5.7 Vertical Tube Bundles *10*
1.5.8 Effect of Low Gravity *10*
1.5.9 Recommendations *12*
1.6 Heat Transfer in Single-Phase Flow *12*
1.6.1 Flow Inside Channels *12*
1.6.2 Vertical Tube/Rod Bundles with Axial Flow *13*
1.6.3 Various Geometries *14*
1.6.4 Liquid Metals *14*
1.7 Calculation of Pressure Drop *14*
1.7.1 Single-Phase Pressure Drop in Pipes *14*
1.7.2 Two-Phase Pressure Drop in Pipes *15*
1.7.3 Annuli and Vertical Tube Bundles *17*
1.7.4 Horizontal Tube Bundles *17*
1.7.5 Recommendations *17*

1.8 Calculation of Void Fraction *17*
1.8.1 Flow Inside Pipes *17*
1.8.2 Flow in Tube Bundles *18*
1.8.3 Recommendations *18*
1.9 CFD Simulation *18*
1.10 General Information *19*
Nomenclature *19*
References *20*

2 Heat Transfer During Condensation *25*
2.1 Introduction *25*
2.2 Condensation on Plates *25*
2.2.1 Nusselt Equations *25*
2.2.2 Modifications to the Nusselt Equations *26*
2.2.3 Condensation with Turbulent Film *27*
2.2.4 Condensation on Underside of a Plate *27*
2.2.5 Recommendations *28*
2.3 Condensation Inside Plain Channels *28*
2.3.1 Laminar Condensation in Vertical Tubes *28*
2.3.2 The Onset of Turbulence *28*
2.3.3 Prediction of Heat Transfer in Turbulent Flow *29*
2.3.3.1 Analytical Models *29*
2.3.3.2 CFD Models *30*
2.3.3.3 Empirical Correlations *30*
2.3.3.4 Correlations Applicable to Both Macro and Minichannels *34*
2.3.4 Recommendation *41*
2.4 Condensation Outside Tubes *41*
2.4.1 Single Tube *41*
2.4.1.1 Stagnant Vapor *41*
2.4.1.2 Moving Vapor *42*
2.4.2 Bundles of Horizontal Tubes *42*
2.4.2.1 Vapor Entry from Top *42*
2.4.2.2 Vapor Entry from Side *44*
2.4.3 Recommendations *44*
2.5 Condensation with Enhanced Tubes *44*
2.5.1 Condensation on Outside Surface *44*
2.5.1.1 Single Tubes *44*
2.5.1.2 Tube Bundles *46*
2.5.2 Condensation Inside Enhanced Tubes *47*
2.5.3 Recommendations *49*
2.6 Condensation of Superheated Vapors *49*
2.6.1 Stagnant Vapor on External Surfaces *49*
2.6.2 Forced Flow on External Surfaces *49*
2.6.3 Flow inside Tubes *50*
2.6.4 Plate-Type Heat Exchangers *50*
2.6.5 Recommendations *51*
2.7 Miscellaneous Condensation Problems *51*
2.7.1 Condensation on Stationary Cone *51*
2.7.2 Condensation on a Rotating Disk *51*
2.7.3 Condensation on Rotating Vertical Cone *52*
2.7.4 Condensation on Rotating Tubes *52*
2.7.5 Plate-Type Condensers *53*

2.7.5.1 Recommendation *54*
2.7.6 Effect of Oil in Refrigerants *54*
2.7.6.1 Recommendation *55*
2.7.7 Effect of Gravity *55*
2.7.7.1 Some Formulas for Zero Gravity *55*
2.7.7.2 Experimental Studies *55*
2.7.7.3 Conclusion *55*
2.7.8 Effect of Non-condensable Gases *56*
2.7.8.1 Prediction Methods *56*
2.7.8.2 Recommendation *57*
2.7.9 Flooding in Upflow *57*
2.7.10 Condensation in Thermosiphons *58*
2.7.11 Condensation in Helical Coils *58*
2.8 Condensation of Vapor Mixtures *59*
2.8.1 Physical Phenomena *59*
2.8.2 Prediction Methods *60*
2.8.3 Recommendation *61*
2.9 Liquid Metals *61*
2.9.1 Stagnant Vapors *61*
2.9.2 Interfacial Resistance *62*
2.9.3 Moving Vapors *62*
2.9.4 Recommendation *62*
2.10 Dropwise Condensation *63*
2.10.1 Prediction of Mode of Condensation *63*
2.10.2 Theories of Dropwise Condensation *63*
2.10.3 Methods to Get Dropwise Condensations *63*
2.10.4 Some Experimental Studies *64*
2.10.5 Prediction of Heat Transfer *64*
2.10.6 Recommendations *66*
Nomenclature *66*
References *67*

3 Pool Boiling *77*
3.1 Introduction *77*
3.2 Nucleate Boiling *77*
3.2.1 Mechanisms of Nucleate Boiling *77*
3.2.1.1 Bubble Agitation *77*
3.2.1.2 Vapor–Liquid Exchange *77*
3.2.1.3 Evaporative Mechanism *78*
3.2.2 Bubble Nucleation *78*
3.2.2.1 Inception of Boiling *78*
3.2.2.2 Bubble Nucleation Cycle *79*
3.2.2.3 Active Nucleation Site Density *81*
3.2.2.4 Recommendations *81*
3.2.3 Correlations for Heat Transfer *81*
3.2.3.1 Conclusion and Recommendation *83*
3.2.4 Multicomponent Mixtures *83*
3.2.4.1 Physical Phenomena *83*
3.2.4.2 Prediction of Heat Transfer *84*
3.2.4.3 Recommendation *86*
3.2.5 Liquid Metals *86*

3.2.5.1 Physical Phenomena *86*
3.2.5.2 Prediction of Heat Transfer *87*
3.2.5.3 Recommendations *88*
3.3 Critical Heat Flux *90*
3.3.1 Models of Mechanisms *90*
3.3.1.1 Bubble Interference Model *90*
3.3.1.2 Hydrodynamic Instability Model *90*
3.3.1.3 Macrolayer Dryout Model *91*
3.3.1.4 Dry Spot Model *91*
3.3.1.5 Interfacial Lift-off Model *92*
3.3.2 Correlations for Inclined Surfaces *92*
3.3.3 Various Correlations *93*
3.3.4 Effect of Subcooling *93*
3.3.5 Various Other Factors Affecting CHF *94*
3.3.6 Evaluation of CHF Prediction Methods *94*
3.3.7 Recommendations *94*
3.3.8 Multicomponent Mixtures *95*
3.3.8.1 Physical Phenomena and Prediction Methods *95*
3.3.8.2 Recommendation *95*
3.3.9 Liquid Metals *95*
3.3.9.1 Physical Phenomena *97*
3.3.9.2 Prediction of CHF *98*
3.3.9.3 Recommendations *102*
3.4 Transition Boiling *102*
3.5 Minimum Film Boiling Temperature *104*
3.5.1 Prediction Methods *104*
3.5.1.1 Analytical Models *104*
3.5.1.2 Empirical Correlations *105*
3.5.2 Recommendations *106*
3.6 Film Boiling *106*
3.6.1 Methods for Predicting Heat Transfer *106*
3.6.1.1 Vertical Plates *106*
3.6.1.2 Horizontal Cylinders *107*
3.6.1.3 Horizontal Plates *108*
3.6.1.4 Inclined Plates *108*
3.6.1.5 Spheres *109*
3.6.2 Liquid Metals *109*
3.6.3 Recommendations *110*
3.7 Various Topics *110*
3.7.1 Effect of Gravity *110*
3.7.1.1 Scaling Method of Raj et al. *110*
3.7.1.2 Scaling for Hydrogen *112*
3.7.1.3 Some Other Studies *112*
3.7.1.4 Recommendations *113*
3.7.2 Effect of Oil in Refrigerants *113*
3.7.2.1 Mechanisms *114*
3.7.2.2 Correlations *114*
3.7.2.3 Recommendation *115*
3.7.3 Thermosiphons *115*
3.7.4 Effect of Some Organic Additives *115*
Nomenclature *115*
References *116*

4 Forced Convection Subcooled Boiling *123*
4.1 Introduction *123*
4.2 Inception of Boiling in Channels *123*
4.2.1 Analytical Models and Correlations *123*
4.2.2 Minichannels *125*
4.2.3 Effect of Dissolved Gases *126*
4.2.4 Recommendations *126*
4.3 Prediction of Subcooled Boiling Regimes in Channels *126*
4.3.1 Recommendation *127*
4.4 Prediction of Void Fraction in Channels *127*
4.4.1 Recommendations *129*
4.5 Heat Transfer in Channels *129*
4.5.1 Visual Observations and Mechanisms *129*
4.5.2 Prediction of Heat Transfer *130*
4.5.2.1 Some Dimensional Correlations *130*
4.5.2.2 The Shah Correlation *130*
4.5.2.3 Various Correlations *132*
4.5.2.4 Recommendations *135*
4.6 Single Cylinder with Crossflow *135*
4.6.1 Experimental Studies *135*
4.6.2 Prediction of Heat Transfer *135*
4.6.2.1 Shah Correlation *135*
4.6.2.2 Other Correlations *137*
4.6.3 Recommendation *138*
4.7 Various Geometries *138*
4.7.1 Tube Bundles with Axial Flow *138*
4.7.2 Tube Bundles with Crossflow *138*
4.7.3 Flow Parallel to a Flat Plate *138*
4.7.4 Helical Coils *138*
4.7.5 Bends *139*
4.7.6 Rotating Tube *139*
4.7.7 Jets Impinging on Hot Surfaces *141*
4.7.7.1 Experimental Studies and Correlations *142*
4.7.7.2 Recommendations *145*
4.7.8 Spray Cooling *145*
Nomenclature *146*
References *146*

5 Saturated Boiling with Forced Flow *151*
5.1 Introduction *151*
5.2 Boiling in Channels *151*
5.2.1 Effect of Various Parameters *151*
5.2.2 Prediction of Heat Transfer *152*
5.2.2.1 Correlations for Macro Channels *152*
5.2.2.2 Correlations for Minichannels *158*
5.2.2.3 Correlations for Both Minichannels and Macrochannels *159*
5.2.2.4 Recommendations *162*
5.3 Plate-Type Heat Exchangers *162*
5.3.1 Herringbone Plate Type *162*
5.3.1.1 Longo et al. Correlation *163*
5.3.1.2 Almalfi et al. Correlation *163*
5.3.1.3 Ayub et al. Correlation *164*

5.3.1.4 Recommendation *164*
5.3.2 Plane Plate Heat Exchangers *164*
5.3.3 Serrated Fin Plate Heat Exchangers *164*
5.3.4 Plate Fin Heat Exchangers *165*
5.4 Boiling in Various Geometries *166*
5.4.1 Helical Coils *166*
5.4.1.1 Correlations for Heat Transfer *166*
5.4.1.2 Evaluation of Correlations *167*
5.4.1.3 Discussion *167*
5.4.1.4 Recommendation *167*
5.4.2 Rotating Disk *168*
5.4.3 Cylinder Rotating in a Liquid Pool *169*
5.4.3.1 Recommendation *169*
5.4.4 Bends *170*
5.4.5 Spiral Wound Heat Exchangers (SWHE) *170*
5.4.6 Falling Thin Film on Vertical Surfaces *171*
5.4.6.1 Various Studies and Correlations *171*
5.4.6.2 Recommendation *171*
5.4.7 Vertical Tube/Rod Bundles with Axial Flow *172*
5.4.8 Spiral Plate Heat Exchangers *172*
5.5 Horizontal Tube Bundles with Upward Crossflow *172*
5.5.1 Physical Phenomena *172*
5.5.2 Prediction Methods for Heat Transfer *173*
5.5.2.1 Shah Correlation *175*
5.5.3 Conclusion and Recommendation *176*
5.6 Horizontal Tube Bundles with Falling Film Evaporation *177*
5.6.1 Flow Patterns/Modes *177*
5.6.2 Heat Transfer *178*
5.6.3 Conclusion and Recommendation *180*
5.7 Boiling of Multicomponent Mixtures *180*
5.7.1 Boiling in Tubes *180*
5.7.2 Boiling in Various Geometries *182*
5.7.3 Conclusions and Recommendations *182*
5.8 Liquid Metals *182*
5.8.1 Inception of Boiling *182*
5.8.2 Heat Transfer *184*
5.8.2.1 Sodium *184*
5.8.2.2 Potassium *184*
5.8.2.3 Mercury *186*
5.8.2.4 Cesium and Rubidium *186*
5.8.2.5 Mixtures of Liquid Metals *187*
5.8.3 Conclusions and Recommendations *187*
5.9 Effect of Gravity *187*
5.9.1 Experimental Studies *188*
5.9.2 Conclusions and Recommendation *189*
5.9.3 Effect of Oil in Refrigerants *189*
5.9.3.1 Heat Transfer with Immiscible Oils *189*
5.9.3.2 Heat Transfer with Miscible Oils *190*
5.9.3.3 Conclusions and Recommendations *190*
Nomenclature *191*
References *192*

6 Critical Heat Flux in Flow Boiling *201*
6.1 Introduction *201*
6.2 CHF in Tubes *201*
6.2.1 Types of Boiling Crisis and Mechanisms *201*
6.2.2 Prediction Methods *201*
6.2.2.1 Analytical Models *201*
6.2.2.2 Lookup Tables of CHF *202*
6.2.2.3 Dimensional Correlations for Water *203*
6.2.2.4 General Correlations *203*
6.2.2.5 Fluid-to-Fluid Modeling *213*
6.2.2.6 Non-uniform Heat Flux *214*
6.2.3 Recommendations *216*
6.3 CHF in Annuli *216*
6.3.1 Vertical Annuli with Upflow *216*
6.3.1.1 Dimensional Correlations for Water *216*
6.3.1.2 General Correlations *217*
6.3.1.3 Recommendations *220*
6.3.2 Horizontal Annuli *221*
6.3.3 Eccentric Annuli *221*
6.4 CHF in Various Geometries *222*
6.4.1 Single Cylinder with Crossflow *222*
6.4.2 Horizontal Tube Bundles *224*
6.4.2.1 Recommendation *226*
6.4.3 Vertical Tube/Rod Bundles *227*
6.4.3.1 Mixed Flow Analyses *227*
6.4.3.2 Subchannel Analysis *228*
6.4.3.3 Phenomenological Analyses *228*
6.4.4 Falling Films on Vertical Surfaces *229*
6.4.5 Flow Parallel to a Flat Plate *230*
6.4.6 Helical Coils *230*
6.4.6.1 Recommendation *232*
6.4.7 Spiral Wound Heat Exchangers (SWHE) *232*
6.4.8 Rotating Liquid Film *232*
6.4.9 Bends *233*
6.4.10 Jets Impinging on Hot Surfaces *234*
6.4.10.1 Correlations for CHF in Free Stream Jets *234*
6.4.10.2 Effect of Contact Angle *235*
6.4.10.3 Multiple Jets *236*
6.4.10.4 Effect of Heater Thickness *236*
6.4.10.5 Confined Jets *236*
6.4.10.6 Submerged Jets *236*
6.4.10.7 Recommendations *236*
6.4.11 Spray Cooling *236*
6.4.12 Effect of Gravity *237*
6.4.12.1 Terrestrial Studies *237*
6.4.12.2 Experimental Studies at Low Gravities *238*
6.4.12.3 CHF Prediction Methods *239*
6.4.12.4 Recommendation *239*
Nomenclature *239*
References *240*

7 Post-CHF Heat Transfer in Flow Boiling *247*
7.1 Introduction *247*
7.2 Film Boiling in Vertical Tubes *247*
7.2.1 Physical Phenomena *247*
7.2.2 Prediction of Dispersed Flow Film Boiling in Upflow *248*
7.2.2.1 Empirical Correlations *248*
7.2.2.2 Mechanistic Analyses *249*
7.2.2.3 Phenomenological Correlations *249*
7.2.2.4 Lookup Tables *254*
7.2.2.5 Recommendations *256*
7.2.3 Prediction of Inverted Annular Film Boiling in Upflow *256*
7.2.3.1 Recommendations *257*
7.2.4 Film Boiling in Downflow *257*
7.3 Film Boiling in Horizontal Tubes *257*
7.3.1 Prediction Methods *258*
7.3.2 Recommendations *259*
7.4 Film Boiling in Various Geometries *259*
7.4.1 Annuli *259*
7.4.2 Vertical Tube Bundles *260*
7.4.3 Single Horizontal Cylinder *261*
7.4.3.1 Recommendation *262*
7.4.4 Spheres *262*
7.4.5 Jets Impinging on Hot Surfaces *264*
7.4.6 Bends *265*
7.4.7 Helical Coils *265*
7.4.8 Chilldown of Cryogenic Pipelines *266*
7.4.9 Flow Parallel to a Plate *267*
7.4.10 Spray Cooling *267*
7.5 Minimum Film Boiling Temperature and Heat Flux *268*
7.5.1 Flow in Channels *268*
7.5.2 Jets Impinging on Hot Surfaces *268*
7.5.3 Chilldown of Cryogenic Lines *269*
7.5.4 Spheres *269*
7.5.5 Spray Cooling *270*
7.6 Transition Boiling *270*
7.6.1 Flow in Channels *270*
7.6.2 Jets on Hot Surfaces *271*
7.6.3 Spheres *272*
7.6.4 Spray Cooling *272*
Nomenclature *273*
References *274*

8 Two-Component Gas–Liquid Heat Transfer *279*
8.1 Introduction *279*
8.2 Pre-mixed Mixtures in Channels *279*
8.2.1 Flow Pattern-Based Prediction Methods *279*
8.2.1.1 Bubbly Flow *279*
8.2.1.2 Slug Flow *281*
8.2.1.3 Annular Flow *282*
8.2.1.4 Post-dryout Dispersed Flow *283*
8.2.2 General Correlations *283*
8.2.2.1 Horizontal Channels *283*

8.2.2.2 Vertical Channels *286*
8.2.2.3 Horizontal and Vertical Channels *288*
8.2.2.4 Inclined Channels *289*
8.2.3 Recommendations *289*
8.3 Gas Flow through Channel Walls *290*
8.3.1 Experimental Studies *290*
8.3.2 Heat Transfer Prediction *292*
8.3.3 Conclusions *292*
8.4 Cooling by Air–Water Mist *292*
8.4.1 Single Cylinders in Crossflow *292*
8.4.2 Flow over Tube Banks *294*
8.4.3 Flow Parallel to Plates *294*
8.4.4 Wedges *295*
8.4.5 Jets *295*
8.4.6 Sphere *297*
8.5 Evaporation from Water Pools *297*
8.5.1 Introduction *297*
8.5.2 Empirical Correlations *297*
8.5.3 Analytical Models *298*
8.5.3.1 Shah Model *298*
8.5.3.2 Other Models *300*
8.5.4 CFD Models *301*
8.5.5 Occupied Swimming Pools *301*
8.5.6 Conclusions and Recommendations *301*
8.6 Various Topics *301*
8.6.1 Jets Impinging on Hot Surfaces *301*
8.6.2 Vertical Tube Bundle *302*
8.6.3 Effect of Gravity *302*
8.7 Liquid Metal–Gas in Channels *303*
8.7.1 Mercury *303*
8.7.2 Various Liquid Metals *304*
8.7.3 Discussion *305*
Nomenclature *305*
References *306*

9 Gas-Fluidized Beds *311*
9.1 Introduction *311*
9.2 Regimes of Fluidization *311*
9.2.1 Regime Transition Velocities *312*
9.2.1.1 Minimum Fluidization Velocity *312*
9.2.1.2 Various Regime Transition Velocities *312*
9.2.2 Void Fraction and Bed Expansion *313*
9.3 Properties of Solid Particles *313*
9.3.1 Density *313*
9.3.2 Particle Diameter *313*
9.3.3 Particle Shape Factor *314*
9.3.4 Classification of Particles *314*
9.4 Parameters Affecting Heat Transfer to Surfaces *315*
9.4.1 Gas Velocity *315*
9.4.2 Particle Size and Shape *315*
9.4.3 Pressure and Temperature *316*
9.4.4 Heat Transfer Surface Diameter *317*

9.4.5 Properties of Gas and Solid *317*
9.4.6 Gas Distribution *317*
9.4.7 Length and Location of Tube *317*
9.4.8 Bed Diameter and Height *318*
9.4.9 Tube Inclination *318*
9.5 Theories of Heat Transfer *318*
9.5.1 Film Theory *318*
9.5.2 Penetration Theory *318*
9.5.2.1 Particle Theory *319*
9.5.2.2 Packet Theory *319*
9.6 Prediction Methods for Single Tubes and Spheres *319*
9.6.1 Analytical Models *319*
9.6.1.1 Particle Models *319*
9.6.1.2 Packet Models *320*
9.6.2 Empirical Correlations *321*
9.6.2.1 Maximum Heat Transfer *321*
9.6.2.2 Correlations for the Entire Range *324*
9.6.3 Recommendations *325*
9.7 Tube Bundles *326*
9.7.1 Horizontal Tube Bundles *326*
9.7.2 Vertical Tube Bundles *328*
9.7.3 Recommendations *328*
9.8 Radiation Heat Transfer *329*
9.8.1 Radiation Heat Transfer Coefficient and Effective Emissivity *329*
9.8.2 Temperature for Significant Radiation Contribution *329*
9.8.3 Conclusions and Recommendations *330*
9.9 Heat Transfer to Bed Walls *330*
9.9.1 Prediction Methods *330*
9.9.2 Conclusions and Recommendations *331*
9.10 Heat Transfer in Freeboard Region *331*
9.10.1 Experimental Studies and Prediction Methods *332*
9.10.2 Recommendation *332*
9.11 Heat Transfer Between Gas and Particles *332*
9.12 Gas–Solid Flow in Pipes *333*
9.12.1 Regimes of Gas–Solid Flow *333*
9.12.2 Experimental Studies of Heat Transfer *334*
9.12.3 Prediction of Heat Transfer *334*
9.12.3.1 Various Methods *334*
9.12.3.2 Shah Correlation *336*
9.12.4 Recommendation *337*
9.13 Solar Collectors with Particle Suspensions *337*
Nomenclature *338*
References *340*

Appendix *347*
Index *357*

Preface

The two-phase systems covered in this book include boiling, condensation, gas–liquid mixtures, and gas–solid mixtures. While there are many books on these topics, most of them are concerned mainly with theoretical aspects while information of practical use is addressed only briefly. The very few books that were intended to help the practicing engineers are greatly out of date. I therefore felt that there was a need for an up-to-date book that emphasized the practical aspects while also addressing the theoretical bases. This book is intended to fulfil this need.

The emphasis in this book is on information that is of practical use. For this reason, theories and methods that do not provide useable and adequately verified solutions are dealt only briefly though sufficient references are provided for more information about them. Effort has been made to provide a review of the state-of-art and the best available information for the design of a wide variety of heat exchangers in a clear and concise manner. This information includes experimental data, theoretical solutions, and empirical correlations. Accuracy and range of applicability of formulas/correlations presented is stated. Clear recommendations are made for application of the methods presented. A very wide variety of heat exchangers and applications is covered. These include boiling and condensation of pure fluids and their mixtures in tubes and tube bundles, plate heat exchangers of various types, falling film heat exchangers, coils, bends, heat pipes, cryogenic pipelines, surfaces cooled by jets, mist cooling, rotating surfaces, spheres, disks, cones, etc. Boiling and condensation of metallic fluids is also discussed. Also included are heat exchangers with two-component gas–liquid mixtures, fluidized beds, and flowing gas–solid mixtures. As space travel and colonization are of much current interest, available information on effects of low gravity has been addressed.

While this book is primarily intended to assist practicing engineers and researchers, it may also be used as textbook for courses on two-phase heat transfer.

Finally, I thank Dr. Milaz Darzi for his help in getting some of the publications studied during the preparation of this book.

Redding, CT
11 April 2020

Mirza Mohammed Shah

1

Introduction

1.1 Scope and Objectives of the Book

The two-phase systems covered in this book include boiling, condensation, gas–liquid mixtures, and gas–solid mixtures.

Two-phase heat transfer is involved in numerous applications. These include heat exchangers in refrigeration and air conditioning, conventional and nuclear power generation, solar power plants, aeronautics, chemical processes, petroleum industry, etc. In recent years, there has been increasing use of miniature heat exchangers for computers and other electronic intensive products.

The emphasis in this book is on information that is of practical use. For this reason, theories and methods that do not provide useable and adequately verified solutions are dealt only briefly though sufficient references are provided for more information about them. Effort is made to provide the best available information for the design of a wide variety of heat exchangers in a clear and concise manner. This information includes experimental data, theoretical solutions, and empirical correlations. Accuracy and range of applicability of formulas/correlations presented is stated. Clear recommendations are made for application of the methods presented. A very wide variety of heat exchangers is covered. These include boiling and condensation in tubes and tube bundles, plate heat exchangers of various types, falling film heat exchangers, coils, surfaces cooled by jets, mist cooling, rotating surfaces (tubes, disks, cones, etc.), spheres, etc. Boiling and condensation of metallic fluids is discussed besides those of non-metallic fluids. Also included are heat exchangers with two-component gas–liquid mixtures, fluidized beds, and flowing gas–solid mixtures.

In this chapter, information is provided that is needed for understanding and using the material in other chapters as well as in other publications. This includes explanation of commonly used terms, various models used in solving two-phase flow and heat transfer problems, distinction between minichannels and conventional channels, flow patterns and their prediction, etc. While the focus of this book is on two-phase heat transfer, methods for calculation of single-phase heat transfer, void fraction and pressure drop have also been briefly discussed as these are needed in the design of heat exchangers. References to sources for more information on these topics have been provided.

Only Newtonian fluids are considered in this book. All discussions pertain to non-metallic fluids except where stated otherwise.

1.2 Basic Definitions

Some commonly used terms are explained in the following.

Mass flux or mass velocity is the mass flow rate per unit area. It is usually designated as G. If W be the mass flow rate $\mathrm{kg\,s^{-1}}$ in a tube of cross-sectional area A_c ($\mathrm{m^2}$), $G = W/A_c$ ($\mathrm{kg\,m^{-2}\,s^{-1}}$).

Void fraction is the part of the total volume occupied by the gas phase. Consider a gas–liquid mixture flowing in a pipe. If A_L is the flow area occupied by liquid and A_G is the flow area occupied by gas, void fraction α is

$$\alpha = \frac{A_G}{A_L + A_G} = \frac{A_G}{A_c} \tag{1.2.1}$$

Liquid holdup R_L is the part of flow area occupied by liquid phase.

$$R_L = 1 - \alpha \tag{1.2.2}$$

Quality, usually given the symbol x, is mass flow rate of vapor divided by the total flow rate. With W_L as the flow rate of liquid and W_G that of gas,

$$x = \frac{W_G}{W_L + W_G} \tag{1.2.3}$$

Two types of phase velocities are used, actual, and superficial. The actual velocity of gas phase u_G is that in the area occupied by the gas phase:

$$u_G = \frac{W_G}{\rho_g A_c \alpha} = \frac{Gx}{\rho_g \alpha} \tag{1.2.4}$$

Two-Phase Heat Transfer, First Edition. Mirza Mohammed Shah.

where ρ_g is the density of gas. The actual liquid velocity is similarly defined and is given by

$$u_L = \frac{W_L}{\rho_L A_c(1-\alpha)} = \frac{G(1-x)}{\rho_f(1-\alpha)} \tag{1.2.5}$$

Superficial gas velocity u_{GS} is the velocity assuming that gas alone is flowing through the entire flow area. In other words, liquid is assumed to be absent. Then,

$$u_{GS} = \frac{W_G}{\rho_g A_c} = \frac{Gx}{\rho_g} \tag{1.2.6}$$

Similarly, superficial liquid velocity u_{LS} is defined as

$$u_{LS} = \frac{W_L}{\rho_L A_c} = \frac{G(1-x)}{\rho_f} \tag{1.2.7}$$

The superficial gas and liquid velocities are also called volumetric gas and liquid flux represented by the symbols j_G and j_L, respectively.

Gas and liquid velocities are often not equal. The difference in phase velocities $(u_G - u_L)$ is called the slip velocity, while u_G/u_L is known as slip ratio. The latter is expressed by the following relation obtained using Eqs. (1.2.4) and (1.2.5):

$$\frac{u_G}{u_L} = \left(\frac{x}{1-x}\right)\left(\frac{1-\alpha}{\alpha}\right)\left(\frac{\rho_f}{\rho_g}\right) \tag{1.2.8}$$

The relative velocity between phases u_{GL} can be written as

$$u_{GL} = (u_G - u_L) = \frac{j_G}{\alpha} - \frac{j_L}{(1-\alpha)} \tag{1.2.9}$$

The drift flux j_{GL} is defined as

$$j_{GL} = u_{GL}\alpha(1-\alpha) = j_G - \alpha j \tag{1.2.10}$$

where

$$j = j_{GS} + j_{LS} \tag{1.2.11}$$

The drift velocity of gas u_{Gj} with respect to a plane moving at a velocity j is defined as

$$u_{Gj} = u_G - j \tag{1.2.12}$$

The drift velocity of the liquid phase is

$$u_{Lj} = u_L - j \tag{1.2.13}$$

Heat flux, usually represented as q, is defined as the heat applied to a surface per unit area per unit time. If Q Watts are applied to a tube of diameter D and length L,

$$q = \frac{Q}{\pi DL} \tag{1.2.14}$$

In boiling systems, quality is usually defined assuming thermodynamic equilibrium between vapor and liquid phases, i.e. all the heat applied is used to evaporate the liquid. Thus, if W kg s^{-1} of saturated liquid enters a tube of length L with heat flux q, quality at exit from tube is

$$x = \frac{\pi DLq/i_{fg}}{W} \tag{1.2.15}$$

where i_{fg} is the latent heat of vaporization. Equilibrium quality during condensation is defined in a similar way; all heat removed is used to condense the vapor. Unless stated otherwise, the quality used in equations and given in test data is the equilibrium quality.

If T_w be the wall temperature and T_{SAT} the saturation temperature during boiling, $(T_w - T_{SAT}) = \Delta T_{SAT}$ is known as the wall superheat. In condensation, $(T_{SAT} - T_w)$ is called wall subcooling. If a liquid is at a temperature T that is lower than the saturation temperature, $(T_{SAT} - T) = \Delta T_{SC}$ is called subcooling.

The term "film temperature" is frequently used. It means the mean of wall and fluid temperature. Unless stated otherwise, it is the arithmetic mean. Thus,

$$T_{film} = \frac{T_{wall} + T_{fluid}}{2} \tag{1.2.16}$$

1.3 Various Models

Some basic models used in the analysis of two-phase systems are discussed herein.

1.3.1 Homogeneous Model

It is assumed that gas and liquid are flowing at the same velocity and form a homogeneous mixture. By putting $u_G = u_L$ in Eq. (1.2.8) and rearranging, the following expression for void fraction α is obtained:

$$\alpha = \left[1 + \left(\frac{1-x}{x}\right)\left(\frac{\rho_g}{\rho_f}\right)\right]^{-1} \tag{1.3.1}$$

For use in calculation of heat transfer and pressure drop with this model, the properties of the mixture are considered to be the mean of those of gas and liquid. Various methods of calculating the mean values have been proposed, for example, weighted according to the mass fractions of gas and liquid in the mixture.

Homogeneous model works fairly well for bubble flow and mist flow though it has been used in some empirical correlations without regard to the flow pattern.

1.3.2 Separated Flow Models

In the separated flow model, the gas and liquid phases are considered to be separated. Separate equations can then be written for each phase. Additional equations are needed for determining areas occupied by the two phases and interfacial shear. These can be empirical or semi-theoretical correlations or sophisticated analyses such as the two-fluid models in which momentum, energy, and continuity equations are written separately for each phase together with equations for interaction between phases. Closed-form solutions of these equations are rarely possible and hence have to be solved numerically on computers. The two-fluid models are difficult to use and not necessarily more accurate than the simpler models. Empirical and semi-theoretical models are generally used in practical designs.

1.3.3 Flow Pattern-Based Models

In these models, the gas and liquid are considered to be arranged according to the expected flow pattern, and prediction methods are developed specific to particular flow patterns. The prediction methods are most often empirical correlations. Analytical solutions have also been developed notably for stratified, slug, and annular flow patterns. Such analytical solutions use idealized geometry of the flow patterns. For example, annular flow is usually assumed to have uniform liquid layer, no interfacial waves, and no liquid entrainment. These assumptions are usually not correct. Still, the analytical solutions are useful as they provide understanding of the physical phenomena.

The accuracy of flow pattern-based models is further limited by the accuracy of flow pattern prediction methods. One of the most verified flow pattern correlation is that of Mandhane et al. (1974). They report an accuracy of 68% in prediction. Researchers often report that their observed flow patterns do not agree with well-known flow pattern correlations. For example, Kim (2000) found large differences between his own flow pattern observation in air–water flow and the predictions of the Taitel and Dukler (1976) map.

Due to the previously mentioned factors, the accuracy of flow pattern-based prediction methods is not good.

1.4 Classification of Channels

In recent years, there has been increasing use of small diameter channels known as mini- or microchannels as they offer more compact and economical heat exchangers. Most of the methods for predicting heat transfer were developed with data for larger tubes known as conventional or macro channels.

The generally held view is that there is no effect of surface tension on heat transfer in tubes of larger diameter, while in tubes of small diameter, surface tension affects heat transfer. The implication is that methods for predicting heat transfer in macro channels are not applicable to mini-/microchannels. It is therefore necessary to demarcate the boundary between macro channels and minichannels to ensure use of macro channel correlations only within their applicable range.

Many classifications of channels have been proposed. These have most recently been discussed by Shah (2018).

1.4.1 Based on Physical Dimensions

According to Shah (1986), the heat exchangers with area-to-volume ratio more than 700 $m^2\,m^{-3}$ are compact. This results in 6 mm diameter being the boundary between minichannels and macro channels.

Mehendale et al. (2000) proposed the following:

$D > 6$ mm, macro channels
$D = 1$–6 mm, compact channel
$D = 100$ μm to 1 mm, meso channel
$D = 1$–100 μm, microchannel

A widely used one is by Kandlikar (2002), according to which

Conventional channels: $D > 3$ mm
Minichannels: $3\text{ mm} \geq D > 0.2$ mm
Microchannels: $0.2\text{ mm} \geq D > 0.01$ mm

This classification was based mainly on single-phase flow of gases, but for uniformity, he also recommended it for boiling and condensing flows. This is the most widely used classification.

1.4.2 Based on Condensation Studies

Li and Wang (2003) studied condensation in minichannels. They observed the transition of flow patterns from symmetrical to asymmetrical and noted that these depend on the capillary length L_{cap} (also known as Laplace constant) defined as

$$L_{cap} = \left[\frac{\sigma}{g(\rho_f - \rho_g)}\right]^{0.5} \quad (1.4.1)$$

Their conclusions were as follows:

- $D < 0.224L_{cap}$: Gravity forces are negligible compared with surface tension forces. Flow regimes are symmetrical.
- $0.224L_{cap} < D < 1.75L_{cap}$: Gravity and surface tension forces are comparable. Flow distribution is slightly stratified.
- $1.75L_{cap} < D$: Gravity forces dominate surface tension forces and the flow regimes are similar to macro channels.

Cheng and Wu (2006) rearranged the preceding results of Li and Wang in terms of Bond number as follows:

Microchannel, if $Bd < 0.5$ (negligible effect of gravity)
Minichannel, if $0.5 < Bd < 3.0$ (both gravity and surface tension have significant effect)
Macro channel, if $Bd > 3.0$ (surface tension has negligible effect).

Bond number is the ratio of surface tension and gravitational forces and is defined as

$$Bd = \frac{gD^2(\rho_f - \rho_g)}{\sigma} \tag{1.4.2}$$

It is also the ratio of channel diameter to capillary length.

Based on the comparison of his general correlation for condensation in tubes, Shah (2009, 2013), with a wide-ranging database, Shah (2016) gave the following criterion. It is minichannel if

$$We_{GT} < 100 \tag{1.4.3}$$

where

$$We_{GT} = \frac{G^2 D}{\rho_G \sigma} \tag{1.4.4}$$

The data for $We_{GT} < 100$ included Bond numbers up to 105. Hence the criteria based on Bond number were found to be incorrect as they consider effect of surface tension to occur at Bond numbers between 1 and 4.

1.4.3 Based on Boiling Flow Studies

The growth of bubbles during boiling in small channels may be restricted due to the limitation of tube diameter. This has led several authors to use the confinement number Co defined as

$$Co = \frac{1}{D}\left[\frac{\sigma}{g(\rho_f - \rho_g)}\right]^{0.5} \tag{1.4.5}$$

Kew and Cornwell (1997) compared the data from their tests on heat transfer during boiling in tubes of diameter 1.39, 2.87, and 3.69 mm, and a square channel 2 mm × 2 mm, to several correlations based on macro channel data. They found that these failed when the confinement number Co is less than 0.5. Accordingly, they gave the following classification:

Micro-/minichannel: $Co > 0.5$
Macro channel: $Co < 0.5$

According to Ong and Thome (2011a), the lower threshold of macroscale flow is Co = 0.3–0.4, while the upper threshold of symmetric microscale flow is Co = 1 with a transition (or mesoscale) region in between. This was based on the experimental two-phase flow pattern transition data together with a top/bottom liquid film thickness comparison for refrigerants R-134a, R-236fa, and R-245fa during flow boiling in channels of 1.03, 2.20, and 3.04 mm diameter.

Li and Wu (2010a,b) have given a transition criterion based on their analysis of data for boiling heat transfer in a variety of channels. According to it, it is minichannel if

$$Bd\, Re_{LS}^{0.5} \le 200 \tag{1.4.6}$$

Shah (2017b) compared a very wide-ranging database for saturated boiling prior to CHF with several correlations for macro channels including Shah (1982). He concluded that it is minichannel if

$$F = (2.1 - 0.008 We_{GT} - 110 Bo) > 1 \tag{1.4.7}$$

Bo is the boiling number. For horizontal channels, $F = 1$ if $Fr_{LT} < 0.01$. If $F \le 1$, it is macro channel. The data for $F > 1$ (minichannel) included diameters up to 6.4 mm and Bd up to 13.7. The data for $F \le 1$ (macro channels) include diameters down to 0.38 mm and Bond numbers down to 0.15. Hence the criteria based on Bond number and diameter are not satisfactory

Shah (2017a) compared his general correlation for subcooled boiling in tubes and annuli with a wide-ranging database that included diameters as small as 0.176 mm and Bond number down to 0.025. Data over the entire range were satisfactorily predicted. This correlation did not include any factor for surface tension effects. No effect of diameter or Bond number was found.

Shah (2017c) compared his correlation for dispersed flow film boiling in horizontal and vertical tubes with a wide-ranging database. This correlation did not have any factors for surface tension effects. Data over the entire range were satisfactorily predicted. These included tube diameters as low as 0.98 mm and Bond numbers down to 2. The minimum We_{GT} in these data was 32. This shows that the Shah criterion for saturated boiling, Eq. (1.4.7), does not apply to film boiling.

Shah (2015, 2017d) compared his correlation for CHF in vertical and horizontal tubes with a very wide range of data. These correlations had no factors for the effect of surface tension. The data included diameters down to 0.13 mm and Bond numbers down to 0.026. The minimum We_{GT} in the data was 6. Hence the criterion of Eq. (1.4.7) for saturated boiling heat transfer is not applicable to CHF.

1.4.4 Based on Two-Component Flow

Triplett et al. (1999) studied gas–liquid flow in small diameter channels. They proposed that mini-/microchannels are those with diameter less than capillary length L_{cap}. This is equivalent to $Bd < 1$.

Ullmann and Brauner (2007) studied flow pattern transitions in gas–liquid flow in channels, and based on their analyses, they proposed that the transition between minichannels and macro channels depends on the Eotvos number Eo, which is the ratio of buoyancy force to surface tension force. It is written as

$$Eo = \frac{g(\rho_f - \rho_g)D^2}{8\sigma} \tag{1.4.8}$$

They proposed that minichannels are those with $Eo < 0.2$.

1.4.5 Discussion

The criteria given earlier are summarized in Table 1.4.1. To make the comparison easy, the criteria using Eo and Co have been given in terms of Bond number. These are related by the following equation:

$$Eo = \frac{Bd}{8} = \frac{1}{8Co^2} \tag{1.4.9}$$

It is seen that the value of the transition Bond number in various criteria varies from 1 to 4. As seen in the discussion earlier, many data for condensation and saturated boiling show effect of surface tension at much higher Bond numbers, while some data at lower Bond numbers do not show effect of surface tension. Hence the criteria based on Bond number are inaccurate. Similarly, many data for tube diameters smaller than 3 mm showed satisfactory agreement with macro channel correlations for saturated boiling and condensation, while many data for larger diameters showed effect of surface tension. Hence the limit of applicability of macro channel correlations to minichannels cannot be based on criteria based on tube diameter or Bond number. For saturated boiling and condensation, the criteria given by Shah are well verified. For subcooled boiling, film boiling, and CHF, limits of applicability of macrochannel correlations are as yet unknown.

1.4.6 Recommendation

Distinction has to be made between naming convention and the actual boundary according to the limit of applicability of macro channel correlations.

In most literature, channels with diameter > 3 mm are called conventional or macro channels, while those with diameter ≤ 3 mm are called minichannels. Hence this naming convention is also followed in this book. However, this is not the limit of applicability of macro channel correlations. The following are the recommendations for this limit:

- For condensation heat transfer, Shah's criterion $We_{GT} > 100$.
- For saturated boiling heat transfer, Shah's Eq. (1.4.7).
- For subcooled boiling, film boiling, and CHF, the limit is undefined. Use macro channel correlations within the range of data in the analyses of such data performed by Shah (2017a,c,d).

1.5 Flow Patterns in Channels

1.5.1 Horizontal Channels

1.5.1.1 Description of Flow Patterns

There is a great deal of variation in the description and names of flow patterns used by different authors. Rouhani

Table 1.4.1 Criteria for macro to mini transition by various authors.

Author	Criterion for minichannel	Basis
Shah (1986)	$D < 6$ mm	Surface-area-to-volume ratio > 700 m^2 m^{-3}
Mehendale et al. (2000)	$D < 6$ mm	Same as above
Kandlikar (2002)	$D \leq 3$ mm	Based on mean free path of common gases
Kew and Cornwell (1997)	$Bd < 4$	Bubble growth confinement during boiling in channels
Triplett et al. (1999)	$Bd < 1$	Flow pattern transitions in gas–liquid flows
Ullman and Brauner (2007)	$Bd < 1.6$	Flow pattern transitions in adiabatic gas–liquid flows
Cheng and Wu (2006)	$Bd < 3$	Flow pattern transitions during condensation in tubes
Ong and Thome (2011a)	$Bd < 1$	Flow pattern transitions and top-bottom liquid film thickness during boiling in channels
Li and Wu (2010aa,b)	$Bd \cdot Re_{LS}^{0.5} \leq 200$	Correlation of heat transfer coefficients during saturated boiling in channels
Shah (2016)	$We_{GT} < 100$	Comparison of test data with correlation for condensation heat transfer in macro channels
Shah (2017b)	$F = (2.1 - 0.008We_{GT} - 110Bo) > 1$	Comparison of test data with correlation for saturated boiling heat transfer in macro channels

Source: From Shah (2018).

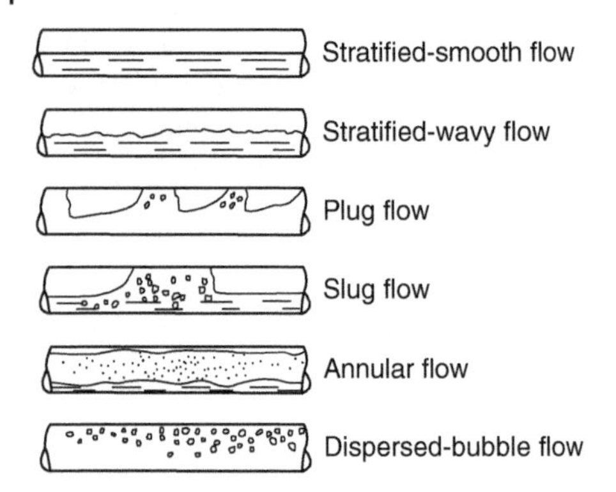

Figure 1.5.1 Flow patterns during co-current gas–liquid flow in horizontal tubes. Source: From Rouhani and Sohal (1983). © 1983 Elsevier.

and Sohal (1983) note that there are 84 different flow pattern labels in literature, 60 of them being for horizontal co-current flow. The most common names are used herein.

Typical flow patterns occurring during flow of gas–liquid mixtures in horizontal channels are shown in Figure 1.5.1. Flow in horizontal channels is subjected to gravity, inertia, and surface tension forces, and the flow patterns result from the balance of these forces. Gravity force tends to pull the heavier liquid phase to the bottom, while the inertia force tends to keep the flow symmetrical. At low flow rates, stratified flow occurs in which the liquid flows at the bottom, while gas flows at the top. The interface is smooth at the lowest flow rates. As flow rates increase, gas–liquid interface becomes rough with appearance of ripples and waves. This is usually called the stratified-wavy pattern. If the waves are of significant height, many authors call it the wavy pattern. With further increase in flow rates, the wave amplitude increases and they reach the top of tube. Gas pockets/plugs then get trapped between the liquid crests, resulting in the plug and slug flow regimes. The difference between plug and slug patterns is mainly that gas pockets are larger in slug flow. Slug flow is often called Taylor flow. Plug and slug flow are also called intermittent flow. Considerable pressure fluctuations occur during intermittent flow. As gas and liquid flow rates increase further, the annular flow pattern occurs. The churn flow pattern may occur in transition from slug flow to annular flow as the slugs begin to disintegrate. In annular flow, liquid is in the form of a layer around the tube circumference, and gas flows in the middle of the tube. Considerable amounts of liquid drops may be entrained in the vapor core and interfacial waves occur. If the liquid layer is thin or non-existent at the upper part of tube, some authors call it semi-annular or crescent pattern. This pattern often occurs during evaporation in tubes. At high gas/vapor velocities, large amount of liquid is torn off the liquid film, and the gas core carries large amounts of liquid droplets. This is called the mist-annular flow; it is called mist flow if there is no liquid film. During dispersed bubble flow (also called bubble flow), vapor bubbles are carried in the continuous liquid stream. It occurs at high liquid flow rates together with low gas flow rate.

Figure 1.5.2 shows the flow patterns during evaporation in horizontal tubes under two conditions common in refrigeration evaporators.

1.5.1.2 Flow Pattern Maps

Many maps for prediction of flow patterns have been proposed. One of the best known is that of Baker (1954). The original map was in terms of dimensional parameters. Figure 1.5.3 is an essentially dimensionless version. This map was developed by analysis of mostly air–water data in pipes of diameters up to 50 mm. G_l and G_g are the superficial mass velocities of liquid and gas. The other terms are

$$\lambda = \left(\frac{\rho_g}{\rho_{air}} \frac{\rho_f}{\rho_{water}} \right)^{0.5} \tag{1.5.1}$$

$$\phi = \left(\frac{\sigma_{water}}{\sigma} \right) \left[\frac{\mu_f}{\mu_{water}} \left(\frac{\rho_{water}}{\rho_f} \right)^2 \right]^{1/3} \tag{1.5.2}$$

The subscripts "air" and "water" indicate air and water at room temperature and pressure. While this map was based entirely on adiabatic flow, several authors have reported its agreement with boiling and condensation data. For example, Shah (1975) reported it to be in fairly good agreement with his data for ammonia evaporating in a 26.2 mm diameter pipe.

A well-verified correlation is by Mandhane et al. (1974) shown in Figure 1.5.4. It was developed using adiabatic gas–liquid data for many gas–liquid combinations. The range of those data is given in Table 1.5.1. Its success in correctly predicting the flow patterns was 67.1%. For the same data, Baker map was able to correctly predict only 41.5% of them. Other researchers have generally found it to be fairly good. Mandhane et al. have also given a version that includes fluid properties but its accuracy was about the same as that of this simple version. They have given a computer subroutine for this correlation.

Another widely quoted flow pattern map is that of Taitel and Dukler (1976). It was developed analytically. Kim (2000) found large differences between his own flow pattern observation in air–water flow and the predictions of the Taitel and Dukler (1976) map.

A number of maps have been developed specifically for boiling and condensation. Among them are those of El Hajal et al. (2003) for condensation and Kattan et al. (1998) for boiling. These were verified with data for halocarbon refrigerants.

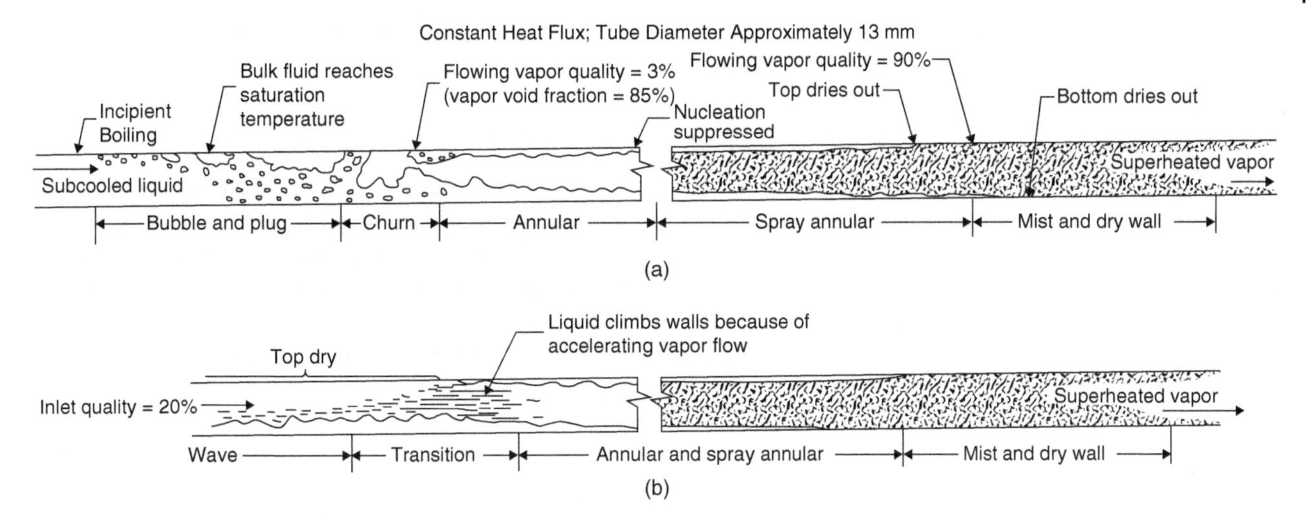

Figure 1.5.2 Flow patterns during evaporation in horizontal tubes. (a) High mass velocity (400 kg s^{-1} m^{-2}), subcooled liquid at inlet. (b) Low mass velocity (200 kg s^{-1} m^{-2}), 20% flash gas at inlet. Source: From ASHRAE (2017).

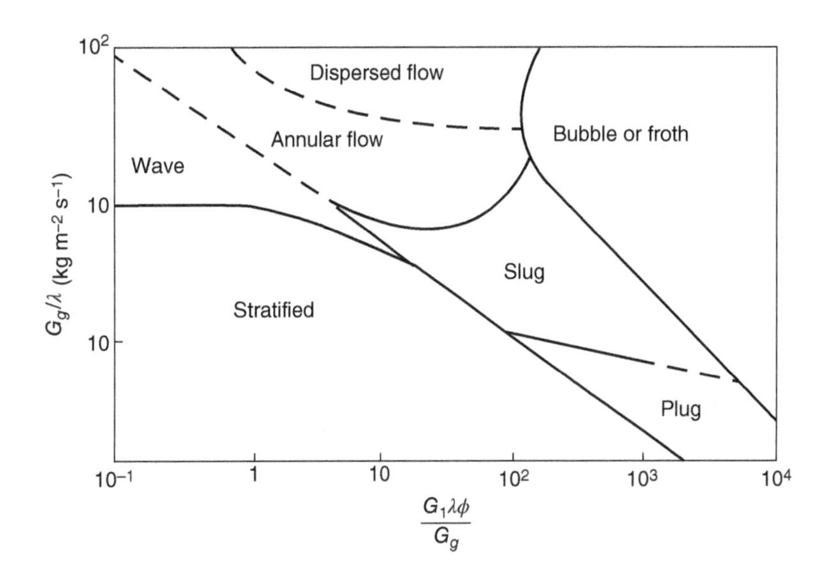

Figure 1.5.3 Baker flow pattern map for co-current gas–liquid flow in horizontal pipes. Source: From Rouhani and Sohal (1983). © 1983 Elsevier.

1.5.2 Vertical Channels

Figure 1.5.5 shows the most common flow patterns in vertical upward co-current flow. These are mostly similar to the horizontal flow patterns except that they are more axisymmetric and the stratified pattern does not occur. This is because gravity force is parallel to the flow direction.

An early flow pattern map is by Hewitt and Roberts (1969), which was based on steam–water flow. Mishima and Ishii (1984) analytically developed criteria for transitions between flow patterns. These were compared to data for air–water and boiling water in round and rectangular channels from several sources and found to be in fair agreement with them. Flow pattern maps can be drawn for any conditions using these criteria. Figure 1.5.6 shows their predicted map for air–water flow at room conditions in a 25.4 mm pipe. In the Region A shown in it, it is difficult to distinguish between churn and annular flow as it is highly agitated.

McQuillan and Whalley (1985) analytically derived criteria for transitions between flow patterns in co-current upflow. Figure 1.5.7 is an example of their predictions. They compared their map with data for air–water as well as for boiling water and refrigerants. Agreement was generally good with 84.1% of the data points predicted correctly.

1.5.3 Inclined Channels

By inclined channels is meant channels with flow directions other than horizontal and vertical up.

The flow patterns in different inclinations change due to the changing relative direction of gravitational force. Analytical expressions for transition criteria have been developed by several authors for particular flow directions.

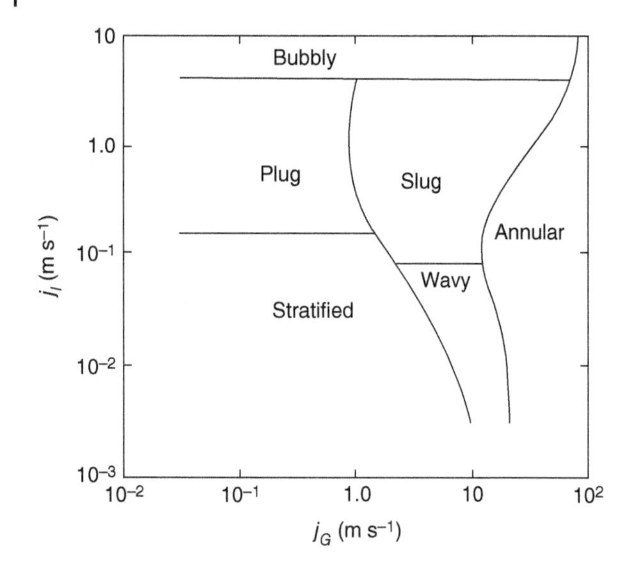

Figure 1.5.4 The Mandhane et al. flow pattern map for co-current flow in horizontal pipes. Source: From Ghiaasiaan et al. and Cambridge University Press. © 1974 Elsevier.

Table 1.5.1 Range of data with which flow pattern map of Mandhane et al. was verified.

	Range
Pipe diameter (mm)	12.7–165.1
Liquid density (kg m^{-3})	705–1009
Gas density (kg m^{-3})	0.8–50.5
Liquid viscosity (Pa s)	3×10^{-4}–9×10^{-2}
Gas viscosity (Pa s)	10^{-5}–2.2×10^{-5}
Surface tension (N m^{-1})	0.024–0.103
Superficial liquid velocity (m s^{-1})	0.9×10^{-3}–7.3
Superficial gas velocity (m s^{-1})	0.04–171

Source: Modified from Mandhane et al. (1974).

Barnea (1987) developed a comprehensive model that is applicable to all flow directions from vertical up to vertical down. It consists of equations for transitions between flow patterns. Figure 1.5.8 shows their predicted flow patterns over the entire range of inclinations from vertical up to vertical down. Good agreement with data from one source is seen. Data from the same source for a 25 mm pipe also showed good agreement. Comparison of this model with data from many sources covering a wide range of parameters is needed.

Mehta and Banerjee (2014) observed flow patterns during air–water flow in a 2.1 mm diameter tube whose orientation was varied at various angle from vertical up to vertical down. They compared their observations with several flow pattern maps, but none was found to agree with their data. They developed maps for horizontal, vertical upflow, and vertical downflows, which were verified only with their own data.

1.5.4 Annuli

Kelessidis and Dukler (1989) investigated flow patterns in vertical upward gas–liquid flow in a concentric and an eccentric annulus (eccentricity 50%). Flow patterns observed were essentially the same as in tubes. Eccentricity was found to have only minor effect on flow patterns. They derived expressions for transitions between flow patterns. These were found to be in agreement with their own data.

Das et al. (1999a,b) observed flow patterns during adiabatic gas–liquid upflow in vertical annuli and developed a mechanistic model of the flow pattern transitions that agreed with their data.

Julia and Hibiki (2011) developed criteria for transitions between flow patterns during upflow in annuli. They compared their map with adiabatic data mentioned earlier as well as boiling water data of Hernandez et al. (2010). Satisfactory agreement was found.

1.5.5 Minichannels

All the foregoing discussions were on flow patterns in macro/conventional channels. Those in minichannels are addressed in this section.

Numerous experimental studies have been done on flow patterns in minichannels, and many flow pattern maps have been proposed. Cheng et al. (2008) reviewed many of them. Experimental studies show that flow patterns in minichannels are the same as in macro channels, but criteria for transitions between flow patterns are usually different.

Triplett et al. (1999) studied flow patterns during air–water flow in horizontal tubes of diameter 1.09 and 1.49 mm. They compared their data to some macro channel maps and found them unsatisfactory.

Akbar et al. (2002, 2003) studied data for air–water in horizontal and vertical minichannels from six sources and proposed a new flow pattern map.

Chen et al. (2006) performed tests with R-134 boiling with upward flow in vertical tubes of diameter 1.10, 2.01, 2.88, and 4.26 mm. They found the Akbar et al. (2003) map unsatisfactory for their data. They noticed subtle differences between the flow patterns of the two larger tubes and the two smaller tubes. The smaller tubes had slimmer vapor slugs and thinner liquid films around the vapor slugs, suggesting greater influence of surface tension. They therefore considered $D = 2$ mm as the boundary between minichannels and macro channels for the conditions of their tests.

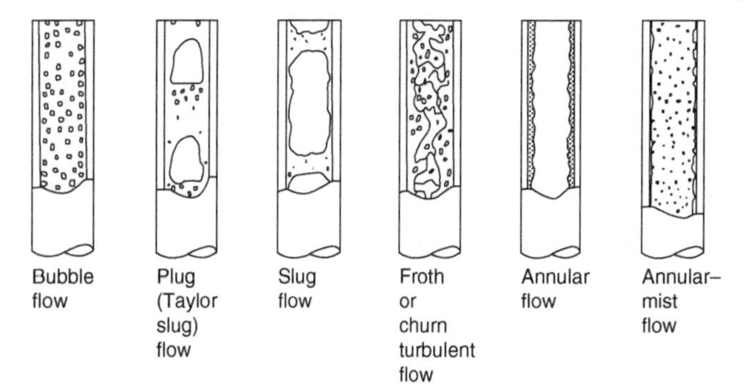

Figure 1.5.5 Flow patterns during upflow in vertical pipes. Source: From Rouhani and Sohal (1983). © 1983 Elsevier.

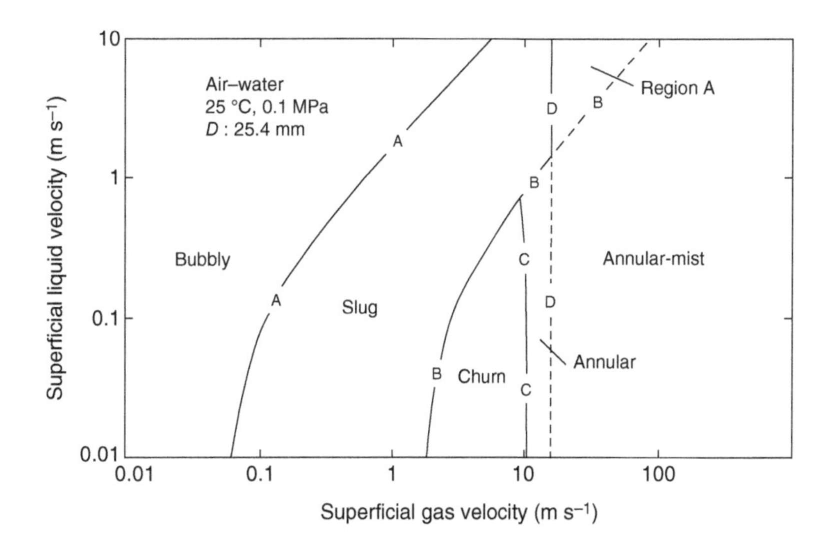

Figure 1.5.6 Example of flow patterns according to the transition criteria of Mishima and Ishii. Source: From Mishima and Ishii (1984). © 1984 Elsevier.

Ullmann and Brauner (2007) analytically developed criteria for transition between flow patterns. They concluded that for flow patterns, transition between minichannel and macro channel occurs at Eotvos number of 0.2. They compared their map with the flow patterns observed by Triplett et al. (1999) in a 1 mm diameter tube with air–water flow. Satisfactory agreement was found.

Ong and Thome (2011b) studied boiling of three refrigerants in tubes of diameter 1.03, 2.20, and 3.04 mm. They gave a new flow pattern map that agreed with their data. Saisorn et al. (2018) found satisfactory agreement with this map of their data for R-134 boiling in a 1 mm diameter tube in horizontal, vertical upflow, and vertical downflow.

Jige et al. (2018) experimentally investigated R-32 boiling in horizontal multiport rectangular minichannels with hydraulic diameters of 0.5 and 1.0 mm. Mass velocity range was 30–400 kg m^{-2} s^{-1} at a saturation temperature of 15 °C. They compared their observations with flow pattern transition criteria of Garimella et al. (2002) and Enoki et al. (2013), both for minichannels. Agreement was not good. They developed their own map.

As is evident from the earlier discussions, many flow pattern maps for minichannels have been proposed, but none of them has been verified with a wide range of data.

1.5.6 Horizontal Tube Bundles with Crossflow

A number of experimental studies have been done on upflow across horizontal tube bundles. Most of them were done with air–water, while a few were with boiling and condensation. Flow patterns observed included bubble, slug, churn, and annular. Xu et al. (1998) studied both downflow and upflow of air–water. In downflow, they also noticed a falling film flow pattern. This occurred at low superficial velocities of gas and liquid. The liquid formed a film around tube wall and flowed down on the tube below. Their observations during upflow and downflow are shown in Figure 1.5.9.

Flow pattern maps have been proposed by Grant and Chisholm (1979), Pettigrew et al. (1989), Ulbrich and Mewes (1994), Xu et al. (1998), Aprin et al. (2007), and Kanizawa and Ribatski (2016). All of them are based on air–water data except that of Aprin et al., which was based on their own boiling data.

Xu et al. (1998) compared their upflow data with the maps of Ulbrich and Mewes (1994) and Grant and Chisholm (1979). Significant differences were found.

Kanizawa and Ribatski (2016) performed tests with air–water flowing up across a bundle of 19 mm tubes on an

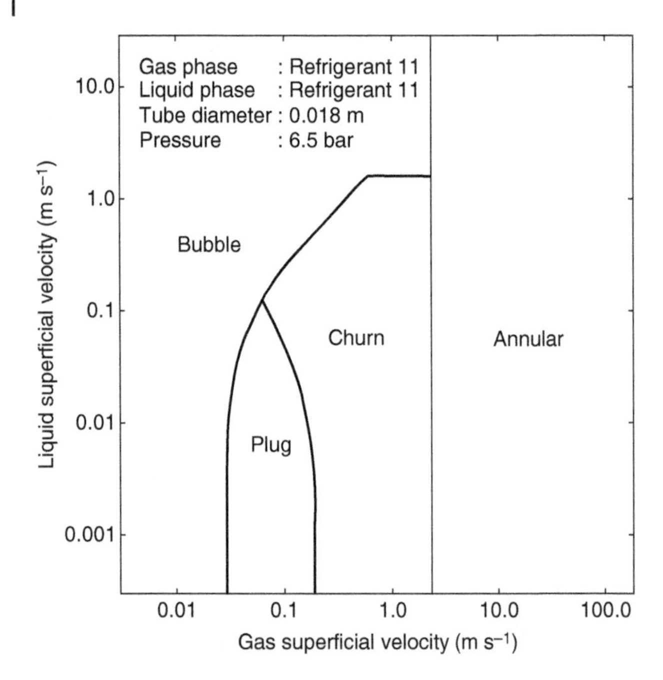

Figure 1.5.7 Flow patterns predictions of McQuillan and Whalley for evaporating R-11 during upflow in a vertical tube under the conditions shown. Source: From McQuillan and Whalley (1985). © 1985 Elsevier.

equilateral triangular arrangement. They compared their data with the maps of Xu et al. (1998), Ulbrich and Mewes (1994), and Grant and Chisholm (1979). These were able to correctly predict 48%, 69%, and 58% of the flow patterns, respectively. They developed their own flow pattern map, which agrees well with their own data.

1.5.7 Vertical Tube Bundles

Vertical tube/rod bundles are especially of interest due to their use in light water nuclear reactors in normal and post-accident conditions. A number of studies on flow patterns in such bundles have been done.

Williams and Peterson (1978) studied upflow of high pressure boiling water in a bundle consisting of a single row of four 6.35 mm rods. The observed two-phase flow patterns were bubble flow, froth flow, slug flow, and annular flow.

Venkateswararao et al. (1982) performed an experiment of vertical adiabatic air–water flow in a rod bundle under atmospheric pressure. There were 24 rods arranged on a square pitch in a cylindrical shell with 12.7 mm outside diameter and 17.5 mm pitch. They identified five flow patterns. These are bubbly, finely dispersed bubbly, slug, churn, and annular. They proposed an analytically based flow pattern map that agreed with their data.

Mizutani et al. (2007) also studied air–water upflow in a 4×4 bundle of 12 mm rods with pitch of 16 mm. They identified the following flow patterns: bubbly, bubbly–churn, churn, churn–annular, and annular flows. They developed a map that agreed well with their own data.

Paranjape et al. (2011) had air–water flowing up an 8×8 bundle of 12.7 mm diameter rods with square arrangement and a pitch of 16.7 mm. They observed four flow patterns, namely, bubbly, cap-bubbly, cap-turbulent, and churn–turbulent flows. Cap-bubbly indicates that the bubbles were cap shaped. A map of their flow patterns was presented.

Zhou et al. (2015) studied vertical boiling steam–water flow in a 3×3 heated rod bundle at atmospheric pressure. The rods were 10 mm diameter at 15 mm square pitch. The flow patterns observed were bubbly, bubbly–churn, churn, and annular. They also proposed a map that agreed with their data.

Liu and Hibiki (2017) showed that the flow pattern maps mentioned earlier do not agree well with data other than their own. Liu and Hibiki analytically developed their own flow pattern map that identifies six flow patterns. It was shown to be in fair agreement with data of Zhou et al. (2015), Paranjape et al. (2011), Mizutani et al. (2007), and Venkateswararao et al. (1982). They did not compare it with the data of Williams and Peterson (1978).

1.5.8 Effect of Low Gravity

All the foregoing discussions were for systems operating under Earth gravity. Flow patterns under micro gravity (<0.03 Earth gravity) condition are addressed herein.

Experimental studies show that flow patterns in microgravity are the same as under Earth gravity but the transitions between flow patterns are different.

The earliest experimental study at near-zero gravity was by Heppner et al. (1975). They used air–water in 25.4 mm diameter tube. They compared the observed flow pattern transitions to those at Earth gravity and found large differences.

Dukler et al. (1988) performed tests under microgravity conditions in a drop tower as well as in parabolic flights with air–water flowing in horizontal tubes of diameter 9.5 and 12.7 mm. Study of their data and analysis led them to the following criteria for transitions between flow patterns:

$$\text{Bubble to slug,} \quad u_{LS} = 1.2\, u_{GS} \tag{1.5.3}$$

$$\text{Slug to annular,} \quad \frac{u_{GS}}{u_{LS} + u_{GS}} = C_0 \tag{1.5.4}$$

Study of their data showed C_0 between 1.15 and 1.3. They tentatively chose a value of 1.25. Rezkallah (1990) found these criteria to be in fair agreement with data from several sources as seen in Figure 1.5.10. The data of Hill et al. (1987) were for Freon 114 boiling in a 15.8 mm diameter tube. In

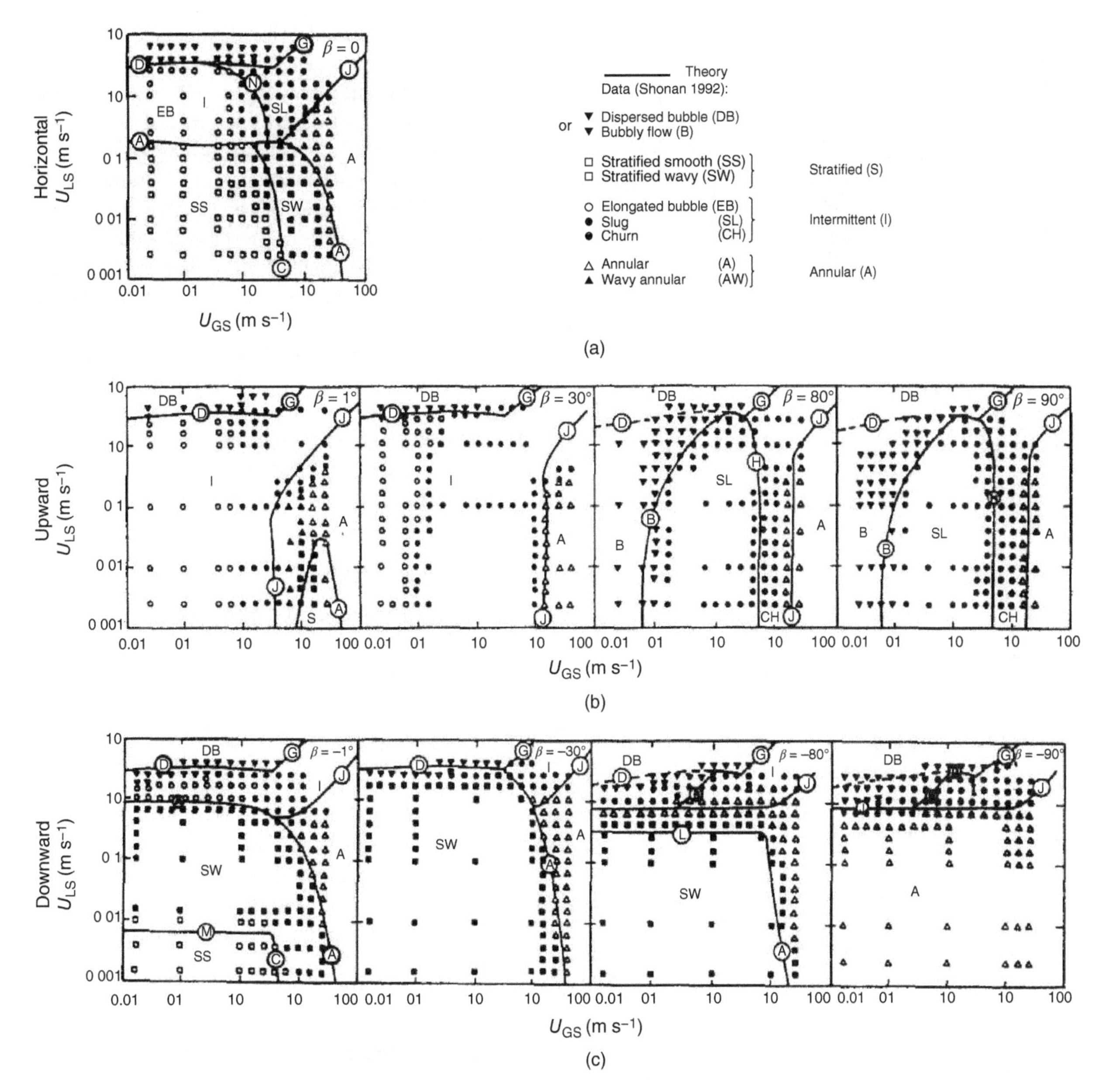

Figure 1.5.8 Effect of pipe inclination β on flow patterns during air–water flow in 51 mm diameter pipe at room temperature and pressure. The predictions are by Barnea (1987) and the data are of Shoham (1982). Source: From Barnea (1987). © 1987 Elsevier.

the tests of Karri and Mathur (1988), oil–water was used in a 25.4 mm tube. Lee et al. (1987) data are from a theoretical study. The only data showing large differences are the annular flow data of Karri and Mathur, which mostly fall in the adjacent slug flow regime. These were the only data for liquid–liquid flow. All others are for gas–liquid flow.

Zhao and Rezkallah (1993) performed microgravity tests on air–water flowing up a vertical 9.5 mm tube. The liquid superficial velocity was 0.09–3.73 m s^{-1}, and the gas superficial velocity was 0.2–29.9 m s^{-1}. Flow patterns observed were bubble, slug, frothy slug–annular, and annular. Considering the balance of forces involved (inertia, gravity, viscous, and surface tension), they concluded that annular flow occurs when inertia force dominates and slug flow occurs when surface tension force dominates. Frothy annular–slug occurs in between these two. They derived the following criteria for flow pattern transitions:

Slug flow and frothy slug–annular flow boundary, $We_{GS} = 1$.
Frothy slug–annular flow to annular flow, $We_{GS} = 20$.

These criteria were found to be in agreement with their own data as well as horizontal flow air-water data from Dukler et al. (1988) in a 9.1 mm tube and Colin et al. (1991) in a 40 mm tube.

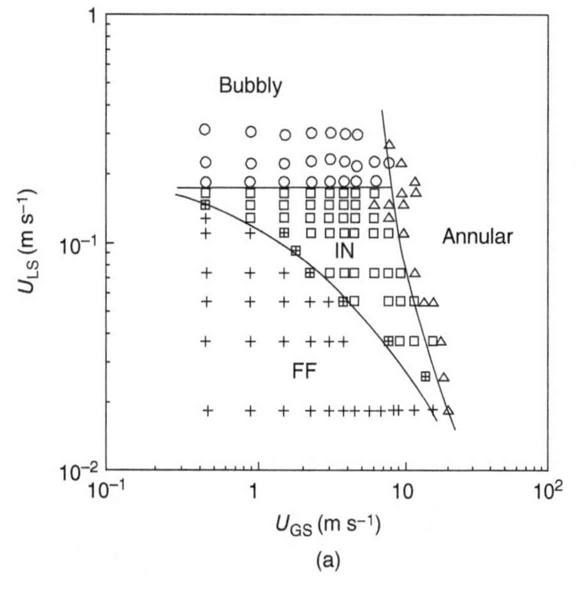

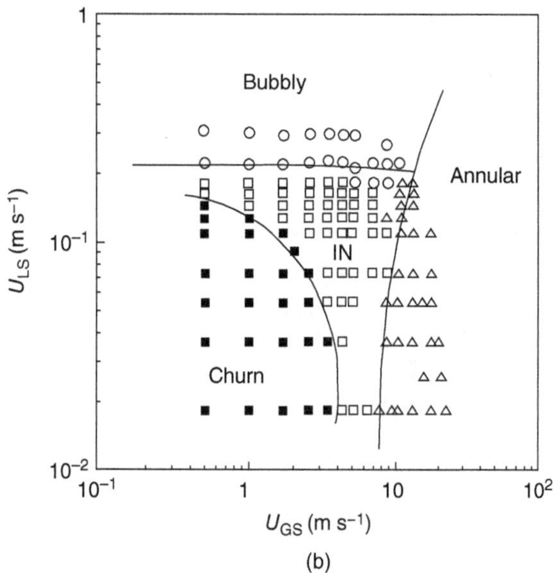

Figure 1.5.9 Flow patterns observed during flow of air–water across a horizontal tube bundle: (a) downflow and (b) upflow. Source: From Xu et al. (1998). © 1998 Elsevier.

Rezkallah and Zhao (1995) studied air–glycerin data under microgravity conditions. The transition between slug and froth slug–annular flow occurred at We_{GS} of 2–3 as Re_{SL} increased from 70 to 700.

Witte et al. (1996) performed tests on a tube during parabolic flights in aircraft with gravity less than 0.1% of Earth gravity. The tube was horizontal and mixtures were air–water and 50% glycerol solution in water. For all data flow pattern was either annular or slug. Slug-to-annular transition occurred at $We_{SG} \approx 10$. Shah (2018) found that most of their annular flow data were predicted to be in the slug or wave flow regimes by the Mandhane et al. (1974) map, which is based on data at Earth gravity.

Narcy et al. (2014) studied boiling of HFE-7000 in a 6 mm diameter tube under Earth gravity and microgravity. Mass flux was 100–400 kg m^{-2} s^{-1} and quality was 0.0–0.7. Flow patterns observed in both conditions were the same except that bubbles under microgravity were smaller at low flow rates. There was little difference in plots of flow patterns under Earth and micro gravities. According to Narcy and Colin (2015), these data are in agreement with the Dukler et al. (1988) map.

The preceding discussions indicate that the Dukler et al. (1988) map is the most verified, but much more verification is needed.

Additional information may be found in the review papers by Rezkallah (1988, 1990), Cheng et al. (2008), and Narcy and Colin (2015).

1.5.9 Recommendations

For horizontal flow in pipes, the most verified map is that of Mandhane et al. (1974). It is very simple and empirical. There is no evidence that the more complex and theoretical maps are more reliable. This should be preferred unless another map has been verified for the conditions of interest.

For vertical and inclined tubes, annuli, and tube bundles, none of the proposed maps has been validated with a wide range of data. Their application is advisable only in the range of data with which they were verified.

For flow under microgravity, the most verified map is that of Dukler et al. (1988). However, range of those data is limited. It can be applied with confidence only in the range of those data.

1.6 Heat Transfer in Single-Phase Flow

1.6.1 Flow Inside Channels

For fully developed laminar flow, perhaps the most commonly used correlation is the following by Sieder and Tate (1936):

$$Nu = \frac{hD}{k} = 1.86(Re \cdot Pr)^{1/3}\left(\frac{D}{L}\right)^{1/3}\left(\frac{\mu}{\mu_w}\right)^{0.14} \tag{1.6.1}$$

All properties are evaluated at bulk fluid temperature except μ_w, which is evaluated at the wall temperature.

For turbulent flow, Sieder and Tate (1936) gave the following correlation:

$$Nu = 0.027Re^{0.8}Pr^{1/3}\left(\frac{\mu}{\mu_w}\right)^{0.14} \tag{1.6.2}$$

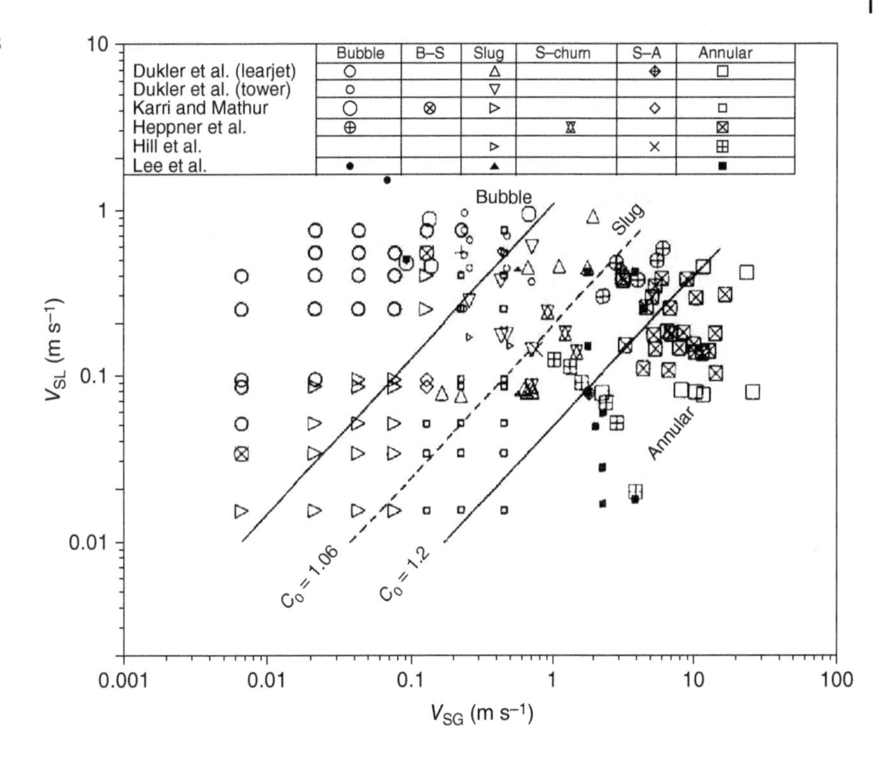

Figure 1.5.10 Comparison of flow patterns under microgravity from several studies with the map by Dukler et al. (1988). Source: From Rezkallah (1990). © 1990 Elsevier.

Colburn (1933) gave the following correlation for turbulent flow:

$$St = \frac{Nu}{Re \cdot Pr} = \frac{h}{GC_p} = \frac{f}{2} Pr_f^{-2/3} \tag{1.6.3}$$

All properties are at bulk fluid temperature except Pr_f, which is at the film temperature, the mean of bulk and wall temperatures. f is the Fanning friction factor.

The most widely used correlation for turbulent flow is the following by Dittus and Boelter (1930):

$$Nu = 0.023 Re^{0.8} Pr^n \tag{1.6.4}$$

All properties are evaluated at bulk fluid temperature. Dittus and Boelter gave $n = 0.3$ for cooling and $n = 0.4$ for heating. McAdams (1954) gave $n = 0.4$ for both heating and cooling. This is the form that is called Dittus–Boelter correlation by most authors. In view of this widespread usage, this has also been done in this book.

Pethukov and Kirillov (1958) have given the following formula that is usually considered to be the most accurate correlation:

$$Nu = \frac{(f/2) Re \cdot Pr}{C + 12.7 (f/2)^{1/2} (Pr^{2/3} - 1)} \tag{1.6.5}$$

where C is given by

$$C = 1.07 + \frac{900}{Re} - [0.63/(1 + 10Pr)] \tag{1.6.6}$$

In a simplified version given by these authors, $C = 1.07$. This correlation is applicable to $Re > 4000$. It is often referred to simply as the Pethukov correlation.

Gnielinski (1976) has modified the Pethukov and Kirillov correlation to the following form to extend it to Re down to 2300:

$$Nu = \frac{\left(\frac{f}{2}\right)(Re - 1000)Pr}{1 + 12.7 (f/2)^{1/2} (Pr^{2/3} - 1)} \tag{1.6.7}$$

For application to noncircular channels and annuli, D in the above equations is usually replaced by the hydraulic equivalent diameter D_{HYD} that is defined as

$$D_{HYD} = \frac{4 \times \text{flow area}}{\text{wetted perimeter}} \tag{1.6.8}$$

In some cases, a heated equivalent diameter D_{HP} is used. It is defined as

$$D_{HP} = \frac{4 \times \text{flow area}}{\text{perimeter through which heat transfer occurs}} \tag{1.6.9}$$

1.6.2 Vertical Tube/Rod Bundles with Axial Flow

Heat transfer coefficients for ordinary fluids can be calculated with equations for flow in tubes using the hydraulic diameter;

For square arrays,

$$\frac{D_{HYD}}{D} = \frac{4}{\pi}\left(\frac{P}{D}\right)^2 - 1 \tag{1.6.10}$$

For triangular arrays,

$$\frac{D_{HYD}}{D} = \frac{2\sqrt{3}}{\pi}\left(\frac{P}{D}\right)^2 - 1 \tag{1.6.11}$$

P is the pitch and D the diameter of rods.

Rehme (1987) reports good agreement of this method for turbulent flow when $P/D > 1.2$. For more information, see Rehme (1987) and Kays and Perkins (1973).

1.6.3 Various Geometries

For laminar flow parallel to a plate, Pletcher (1987) recommends the following equations for local heat transfer coefficient:

For $Pr > 0.6$,

$$Nu_z = \frac{hz}{k} = 0.332 Re_z^{1/2} Pr^{1/3} \tag{1.6.12}$$

$$\text{where } Re_z = \frac{u \rho z}{\mu} \tag{1.6.13}$$

For turbulent flow over a plate, $Pr = 0.6\text{–}60$,

$$Nu_z = 0.0296 Re_z^{4/5} Pr^{1/3} \tag{1.6.14}$$

Typical laminar to turbulent transition Reynolds number is 5×10^5.

For single cylinders in crossflow, Zukauskas (1987) recommends the following correlations:

$$Re\ 10^3 - 2 \times 10^5, \quad Nu = 0.26 Re^{0.6} (Pr/Pr_w)^n \tag{1.6.15}$$

$$Re\ 2 \times 10^5 - 10^7, \quad Nu = 0.023 Re^{0.8} (Pr/Pr_w)^n \tag{1.6.16}$$

Exponent n is 0.25 for fluid heating and 0.2 for cooling.

For single spheres in crossflow, Zukauskas (1987) recommends the following equations:

For $Re < 7 \times 10^4$,

$$Nu = 2 + (0.4 Re^{0.5} + 0.06 Re^{0.7}) Pr^{0.4} (Pr/Pr_w)^{0.25} \tag{1.6.17}$$

For $4 \times 10^5 < Re < 5 \times 10^6$,

$$Nu = (496 + 5.767 \times 10^{-4} Re + 0.288 \times 10^{-9} Re^2 - 3.58 \times 10^{-17} Re^3) Pr^{0.4} \tag{1.6.18}$$

1.6.4 Liquid Metals

For laminar flow parallel to a flat plate, Pletcher (1987) recommends

$$Nu_z = 0.453 Re_z^{1/2} Pr^{1/2} \tag{1.6.19}$$

For flow in uniformly heated tubes, Reed (1987) recommends the following equation:

$$Nu = 5.0 + 0.025 Pe_m^{0.8} \tag{1.6.20}$$

For flow in tubes with uniform wall temperature, Reed recommends

$$Nu = 3.3 + 0.002 Pe_m^{0.8} \tag{1.6.21}$$

The subscript m means that properties are to be calculated at the mean of inlet and outlet temperatures. These equations apply to $Pe > 100$.

For liquid metals parallel to the axis of vertical tube bundles, Mikityuk (2009) developed the following formula:

$$Nu = (1 - e^{-3.8(P/D-1)})(Pe^{0.77} + 250) \tag{1.6.22}$$

This correlation was developed with data from four sources. Data included triangular and square arrays. It gave better agreement with data than four other correlations. Mikityuk recommends this correlation for Pe 30–5000 and P/D 1.1–1.95. This correlation gives an asymptotic value as P/D approaches infinity. Wu et al. (2018) have discussed several correlations for liquid metal flow in annuli and rod bundles.

1.7 Calculation of Pressure Drop

1.7.1 Single-Phase Pressure Drop in Pipes

Pressure drop of single-phase fluids flowing in tubes is given by the following relation:

$$\Delta p = f \left(\frac{2 \rho u^2 L}{D} \right) \tag{1.7.1}$$

u is the average fluid velocity and f is the Fanning friction factor. Another friction factor used by some authors is the Darcy friction factor that is four times the Fanning friction factor. If the Darcy friction factor is used, right-hand side of Eq. (1.7.1) has to be divided by 4. Fanning friction factor is used throughout this book unless stated otherwise:

For laminar flow,

$$f = \frac{16}{Re} \tag{1.7.2}$$

In turbulent flow, f depends also on surface roughness. It is given by the formula of Colebrook (1939) as follows:

$$\frac{1}{\sqrt{f}} = 1.48 - 1.7372 \ln \left(\frac{2\varepsilon}{D} + \frac{9.35}{Re\sqrt{f}} \right) \tag{1.7.3}$$

where ε is the pipe roughness.

The Colebrook equation requires iterations to find f. A number of explicit formulas for f approximating to the Colebrook equation have been proposed. The following by Zigrang and Sylvester (1982) provides results very close to the Colebrook equation:

$$\frac{1}{\sqrt{f}} = 3.4769 - 1.7372 \ln \left[\frac{2\varepsilon}{D} - \frac{16.1332}{Re} \ln A_3 \right] \tag{1.7.4}$$

A_3 is given by the following equation:

$$A_3 = \frac{\varepsilon/D}{3.7} + \frac{13}{Re} \quad (1.7.5)$$

This equation is applicable for $4000 < Re < 10^8$.

1.7.2 Two-Phase Pressure Drop in Pipes

Total pressure drop in two-phase flow, $(\Delta p)_{tot}$, is the sum of those due to change in static head, acceleration, and friction:

$$(\Delta p)_{tot} = (\Delta p)_{static} + (\Delta p)_a + (\Delta p)_{fric} \quad (1.7.6)$$

The static head loss $(\Delta p)_{static}$ due to change in height Δz is expressed as

$$\left(\frac{\Delta p}{\Delta z}\right)_{static} = [\alpha \rho_g + (1 - \alpha)\rho_f] g \sin\theta \quad (1.7.7)$$

θ is the inclination to horizontal.

When liquid evaporates in the pipe, the increased volume accelerates both the liquid and vapor. This acceleration causes the acceleration pressure drop $(\Delta p)_a$. The acceleration pressure drop, also known as momentum pressure drop, between points 1 and 2 in a pipe is expressed by

$$(\Delta p)_a = G^2 \left(\frac{1}{\rho_{mix,2}} - \frac{1}{\rho_{mix,1}}\right) \quad (1.7.8)$$

where $\rho_{mix,1}$ and $\rho_{mix,2}$ are the densities of vapor–liquid mixtures at points 1 and 2, respectively. The mixture density is given by

$$\rho_{mix} = [\alpha \rho_g + (1 - \alpha)\rho_f] \quad (1.7.9)$$

During condensation, acceleration pressure drop is negative.

To calculate the frictional pressure drop, there are two basic models. These are the homogeneous model and the separated flow model. In the homogeneous model, gas and liquid have the same velocity and have mixture mean properties. Calculation of pressure drop can then be done using the single-phase relations given in Section 1.7.1. This approach is most suitable for bubble flow and mist flow. The separated flow model has been used much more commonly.

A very well-known separated flow model is that of Lockhart and Martinelli (1949). They alternatively used the following two-phase pressure drop multipliers:

$$\phi_G^2 = \frac{(\Delta p)_{TP,fric}}{(\Delta p)_{GS}} \quad (1.7.10)$$

$$\phi_L^2 = \frac{(\Delta p)_{TP,fric}}{(\Delta p)_{LS}} \quad (1.7.11)$$

A parameter X is defined as

$$X^2 = \frac{(\Delta p)_{LS}}{(\Delta p)_{GS}} \quad (1.7.12)$$

X is known as the Lockhart–Martinelli parameter or simply Martinelli parameter.

Lockhart and Martinelli gave graphical correlations in the form of ${\phi_G}^2$ and ${\phi_L}^2$ vs. X^2. Four flow regimes were identified:

Laminar–laminar when both Re_{LS} and Re_{GS} are <1000.
Laminar–turbulent when $Re_{LS} < 1000$ and $Re_{GS} > 2000$.
Turbulent–laminar when $Re_{LS} > 2000$ and $Re_{GS} < 1000$.
Turbulent–turbulent when $Re_{LS} > 2000$ and $Re_{GS} > 2000$.

X_{tt} is X for the turbulent–turbulent regime and from Eq. (1.7.12) is obtained as

$$X_{tt} = \left(\frac{\rho_g}{\rho_f}\right)^{0.5} \left(\frac{\mu_f}{\mu_g}\right)^{0.1} \left(\frac{1-x}{x}\right)^{0.9} \quad (1.7.13)$$

X_{tt} has been used in numerous correlations for heat transfer.

Chisolm (1967) has given the following equations that give results very close to the Lockhart–Martinelli curves:

$$\phi_L^2 = 1 + \frac{C}{X} + \frac{1}{X^2} \quad (1.7.14)$$

$$\phi_G^2 = 1 + CX + X^2 \quad (1.7.15)$$

The values of C are as follows:

Turbulent–turbulent: $C = 20$
Laminar–turbulent: $C = 12$
Turbulent--laminar: $C = 10$
Laminar–laminar: $C = 5$

The Lockhart–Martinelli correlation has been found to agree with data from many sources for adiabatic as well as boiling/condensing flow. On the other hand, disagreement with many other data sets has also been reported.

Among other pressure drop correlations for macro channels that have had considerable verification are those of Friedel (1979), Müller-Steinhagen and Heck (1986), and Shannak (2008).

The Shannak correlation uses a two-phase Reynolds number Re_{TP} defined as

$$Re_{TP} = \frac{x^2 + (1-x)^2(\rho_g/\rho_f)}{(x^2/Re_{GS}) + [(1-x)^2/Re_{LS}](\rho_g/\rho_f)} \quad (1.7.16)$$

Friction factor f_{TP} is calculated using single-phase correlations such as the Colebrook equation or its explicit equivalents with Re_{TP} in place of Re. Frictional pressure drop is then given by

$$(\Delta p)_{TP,fric} = f_{TP}\left(\frac{2G^2 L}{\rho_{mix} D}\right) \quad (1.7.17)$$

G is the total mass flux (gas + liquid). The mixture density is calculated by the homogeneous model as

$$\frac{1}{\rho_{mix}} = \frac{x}{\rho_g} + \frac{1-x}{\rho_f} \quad (1.7.18)$$

This correlation was compared to Shannak's own data for air–water flow as well as from other sources. The entire database had about 17 000 data points that included boiling, condensing, and gas–liquid flows in channels of diameter 0.98–257 mm, mass flux 15–10 330 kg m^{-2} s^{-1}, and pressure from 0.6 to 220 bar. Horizontal and vertical round and rectangular channels were included. The Shannak correlation had a mean absolute deviation (MAD) of 25%, while those of Friedel and Mueller-Steinhagen were 30% and 34%, respectively.

The correlation of Müller-Steinhagen and Heck (1986) interpolates between the pressure drops of all liquid flow and all gas flow. It is as follows:

$$(\Delta p)_{\mathrm{TP,fric}} = Y(1-x)^{1/3} + Bx^3 \tag{1.7.19}$$

where

$$Y = A + 2(B-A)x \tag{1.7.20}$$

A and B are pressure drops assuming all mass to be flowing as liquid and gas, respectively, calculated by the single-phase correlations given in Section 1.7.1. Tribbe and Müller-Steinhagen (2000) have shown it to agree with a very wide range of data that included gas–liquid mixtures as well as boiling/condensing mixtures.

Xu et al. (2012) compared a database of 3480 points from many sources to many correlations and found that Mueller-Steinhagen correlation had a MAD of 27%, while the Shannak correlation had a MAD of 40%. The majority of data in Xu et al. database were for minichannels, the largest diameter was 14 mm, and it did not include boiling or condensing water. The Shannak database included diameters up to 220 mm and included water with phase change.

Many correlations have been proposed specifically for mini-/microchannels. Among these, the most verified are the correlations of Kim and Mudawar (2012, 2013). In Kim and Mudawar (2012), a correlation for adiabatic and condensing flows was presented. In Kim and Mudawar (2013) a modification was provided to make it applicable to boiling flow. The final correlation is given as follows.

This correlation has the same four regimes as the Lockhart and Martinelli correlation, except that the transitions are all at $Re = 2000$, while it is $Re = 1000$ in the Lockhart and Martinelli correlation in some cases:

Pressure drops for each phase flowing alone is given by

$$(\Delta p)_{\mathrm{SL}} = f_f\left(\frac{2\rho u_{\mathrm{SL}}^2 L}{D}\right) \tag{1.7.21}$$

$$(\Delta p)_{\mathrm{SG}} = f_g\left(\frac{2\rho u_{\mathrm{SG}}^2 L}{D}\right) \tag{1.7.22}$$

The friction factors f_f and f_g are given by

$$Re_k < 2000, \quad f_k = 16Re_k^{-1} \tag{1.7.23}$$

For $2000 \le Re_k < 20\,000$,

$$f_k = 0.079Re_k^{-0.25} \tag{1.7.24}$$

$$Re_k \ge 20\,000, \quad f_k = 0.046Re_k^{-0.2} \tag{1.7.25}$$

For laminar flow in rectangular channels,

$$f_k Re_k = 24(1 - 1.3553\beta + 1.9467\beta^2 - 1.7012\beta^3 + 0.9564\beta^4 - 0.2537\beta^5) \tag{1.7.26}$$

Subscript k denotes f for liquid and g for liquid phase. β is the channel aspect ratio, less than 1.

For non-boiling flow, C in Eq. (1.7.14) is replaced by C_{NB} given by the following relations:

$$Re_{\mathrm{LS}} > 2000 \text{ and } Re_{\mathrm{GS}} > 2000,\quad C_{\mathrm{NB}} = 0.39Re_{\mathrm{LT}}^{0.03} Su_g^{0.1}(\rho_f/\rho_g)^{0.35} \tag{1.7.27}$$

$$Re_{\mathrm{LS}} > 2000 \text{ and } Re_{\mathrm{GS}} < 2000,\quad C_{\mathrm{NB}} = 8.7\times10^{-4}Re_{\mathrm{LT}}^{0.17} Su_g^{0.5}(\rho_f/\rho_g)^{0.14} \tag{1.7.28}$$

$$Re_{\mathrm{LS}} < 2000 \text{ and } Re_{\mathrm{GS}} > 2000,\quad C_{\mathrm{NB}} = 0.0015Re_{\mathrm{LT}}^{0.59} Su_g^{0.19}(\rho_f/\rho_g)^{0.36} \tag{1.7.29}$$

$$Re_{\mathrm{LS}} < 2000 \text{ and } Re_{\mathrm{GS}} < 2000,\quad C_{\mathrm{NB}} = 3.5\times10^{-5}Re_{\mathrm{LT}}^{0.0.44} Su_g^{0.50}(\rho_f/\rho_g)^{0.48} \tag{1.7.30}$$

For boiling flow, C in Eq. (1.7.14) is as given as follows:

$$Re_{\mathrm{LS}} \ge 2000, \quad C = C_{\mathrm{NB}}\left[1 + 60We_{\mathrm{LT}}^{0.32}\left(Bo\frac{\rho_f}{\rho_g}\right)^{0.78}\right] \tag{1.7.31}$$

$$Re_{\mathrm{LS}} < 2000, \quad C = C_{\mathrm{NB}}\left[1 + 530We_{\mathrm{LT}}^{0.52}\left(Bo\frac{\rho_f}{\rho_g}\right)^{1.09}\right] \tag{1.7.32}$$

Su_g is the Suratman number defined as

$$Su_g = \frac{\rho_g \sigma D}{\mu_g^2} \tag{1.7.33}$$

We_{LT} is the Weber number for all mass flowing as liquid.

$$We_{\mathrm{LT}} = \frac{G^2 D}{\rho_f \sigma} \tag{1.7.34}$$

G is the total mass flux including both liquid and vapor. Bo is the boiling number.

X is calculated by Eq. (1.7.12); $\phi_L{}^2$ is calculated by Eq. (1.7.14); and then $(\Delta p)_{\mathrm{TP,fric}}$ is calculated by Eq. (1.7.11). For noncircular channels, D was replaced by D_{HYD}.

This correlation was compared to a very wide range of data from many sources. The database for adiabatic and condensing flow included 17 working fluids, hydraulic diameters from 0.0695 to 6.22 mm, mass velocities from 4.0 to 8528 kg m^{-2} s^{-1}, liquid-only Reynolds numbers from 3.9 to 89 798, flow qualities from 0 to 1, and reduced pressures from 0.0052 to 0.91. The database for boiling flow consisted of nine fluids, hydraulic diameters from 0.349 to 5.35 mm, mass velocities from 33 to 2738 kg m^{-2} s^{-1}, liquid-only Reynolds numbers from 156 to 28 010, qualities from 0 to 1, reduced pressures from 0.005 to 0.78, and both single-channel and multichannel data. The MAD for boiling data was 17.2%. MAD for adiabatic data was about 24%, and MAD for condensing data was 17.5%. Many other correlations were compared to the same database. These included Lockhart and Martinelli, Friedel (1979), Müller-Steinhagen and Heck (1986), Mishima and Hibiki (1996), and Li and Wu (2010a,b). The first three were based on macro channel data, while the last two were specifically for minichannels. All had much higher deviations than the Kim and Mudawar correlation.

1.7.3 Annuli and Vertical Tube Bundles

The methods given for pressure drop in pipes can be applied to annuli and vertical tube bundles with flow parallel to their axis by using the hydraulic equivalent diameter:

For annuli,

$$D_{\mathrm{HYD}} = (D_{\mathrm{out}} - D_{\mathrm{in}}) \tag{1.7.35}$$

For rod bundles, the formulas for D_{HYD} are in Section 1.6.2.

Many theoretical and empirical formulas have been developed for single-phase pressure drop in tube bundles. Many of them are discussed in Rehme (1987) and Kays and Perkins (1973).

Qiu et al. (2015) and Wu et al. (2018) have listed and discussed many correlations for single-phase and two-phase pressure drop for liquid metals. It is interesting that the Lockhart–Martinelli correlation gives fairly good agreement with liquid metal data, while it was based entirely on data for ordinary fluids.

1.7.4 Horizontal Tube Bundles

Kanizawa and Ribatski (2017) compared data for upflow across horizontal tube bundles from seven sources with seven published prediction methods. The data included air–water flow as well as condensation of hydrocarbons. Tube diameter was 7.9–19.1 mm, P/D 1.08–1.47, and mass flux 200–959 kg m^{-2} s^{-1}. All the published correlations gave poor results with MAD from 46% to 1211%. They therefore developed a new correlation in the form of Eq. (1.7.11). Their expression for $\phi_L{}^2$ is

$$\phi_L^2 = 1 + \left[\left(98.9\frac{\alpha^{1.75}}{We^{0.76}}\right)^2 + \left\{7.76\times10^5\frac{\alpha^{1.95}}{Re_{\mathrm{LS}}^{1.26}}\frac{1}{1+(6500/Re_{\mathrm{LS}}^2)^2}\right\}^2\right]^{1/2} \tag{1.7.36}$$

where

$$We = \frac{G^2 D}{(\rho_f - \rho_g)\sigma} \tag{1.7.37}$$

Void fraction was calculated with their own correlation, while single-phase liquid pressure drop was calculated by the method of Zukauskas and Ulinskas (1983). The entire database was predicted with a MAD of 21.6% with 94% data within ±30%.

1.7.5 Recommendations

The recommendations for inside channels are as follows:

$D \le 6$ mm, Kim and Mudawar correlation is recommended.
$6 < D < 14$ mm, Mueller-Steinhagen correlation is recommended.
$D > 14$ mm, Shannok correlation is recommended.

For horizontal tube bundles, correlation of Kanizawa and Ribatski is recommended.

1.8 Calculation of Void Fraction

If the velocities of the liquid and gas phase are considered equal (homogeneous model), void fraction is obtained from Eq. (1.3.1). This assumption is a fair approximation for bubble and mist flow. In other flow patterns, the velocities of the two phases usually differ considerably. Some of the correlations that take into consideration the velocity slip are discussed in the following text.

1.8.1 Flow Inside Pipes

Lockhart and Martinelli (1949) gave a graphical correlation in which void fraction is a function of the Martinelli parameter X. For the turbulent–turbulent regime, their correlation is represented well by the following equation, Hewitt (1982):

$$\alpha = \frac{\phi_L - 1}{\phi_L} \tag{1.8.1}$$

Rouhani and Axelsson (1970) used the drift flux model to develop a correlation for void fraction that is given in Section 4.4. This correlation has been widely used.

Godbole et al. (2011) evaluated 52 correlations against a wide-ranging database for gas–liquid flowing up in vertical pipes of diameter 12.7–76 mm. Only a few of them performed well. The Rouhani and Axelsson correlation was among the three best performing correlations. In using the Rouhani–Axelsson correlation, Godbole et al. (2011) used the following formulas to calculate the distribution factor C_o:

$$\text{For } \alpha < 0.25, \quad C_o = 1 + 0.2(1 - x)\left(\frac{gD\rho_f^2}{G^2}\right)^{0.25} \tag{1.8.2}$$

$$\text{For } \alpha > 0.25, \quad C_o = 1 + 0.2(1 - x) \tag{1.8.3}$$

Melkamu et al. (2007) evaluated 68 correlations against data from many sources, which included gas–liquid flow in horizontal, inclined, and vertical tubes. The data included tube diameters from 12.7 to 102.26 mm. Most of them performed poorly. The Rouhani and Axelsson correlation with C_o by Eq. (1.8.3) was among the two-best.

Godbole et al. (2011) and Melkamu et al. (2007) have listed all the correlations evaluated by them; these may be consulted for more information about them.

1.8.2 Flow in Tube Bundles

A large number of correlations have been proposed for void fraction in vertical and horizontal tube bundles. These were reviewed by Hibiki et al. (2017) and compared with test data from many sources. They recommend the following correlation of Smith (1969):

$$\alpha = \left[1 + S\frac{\rho_g}{\rho_f}\left(\frac{1 - X}{X}\right)\right]^{-1} \tag{1.8.4}$$

The velocity slip ratio S is given by

$$S = E + (1 - E)\left[\frac{\frac{\rho_f}{\rho_g} + \left(\frac{1-x}{x}\right)E}{1 + \left(\frac{1-x}{x}\right)E}\right]^{1/2} \tag{1.8.5}$$

E is the entrainment factor defined as the ratio of the mass of liquid droplets entrained in the gas core to the total mass of liquid. They compared this and other correlation to data from many sources for vertical bundles with vertical flow and horizontal bundles with crossflow. Bundles included square and triangular configurations. They recommend that the Smith correlation with $E = 0.5$ can be used for both horizontal and vertical bundles.

Kanizawa and Ribatski (2017) have given the following correlation for void fraction in horizontal bundles with crossflow:

$$\alpha = \left[1 + \left(\frac{K_f}{K_g}\right)^{1/3}\left(\frac{\rho_g}{\rho_f}\right)^{1/3}\left(\frac{1 - x}{x}\right)^{2/3}\right]^{-1} \tag{1.8.6}$$

The factor K is related to non-uniformities of velocity profile along the flow cross section. They developed the following expression for it:

$$\left(\frac{K_f}{K_g}\right)^{1/3} = \left[\left(\frac{87.7}{We^{1.33}}\left(\frac{Re_{LS}}{10\,000}\right)^{2.24}\frac{1}{1 + (Re_{LS}/6500)^4}\right)^2 + \left(\frac{0.26}{Re_{LS}^{0.75}x^{0.92}}\right)^2\right]^{1/2} \tag{1.8.7}$$

We is defined by Eq. (1.7.34).

They compared this correlation with data from six sources. These included square and triangular tube configurations, tube diameter 6.17–25 mm, P/D 1.08–1.75, and quality up to 0.68. Fluids were air–water, R-11, and R-113. This correlation predicted 83% of the experimental data within ±30%, with an absolute deviation of 20%, The same data were compared to seven published correlations. Their MAD were from 32% to 210%.

1.8.3 Recommendations

The recommended prediction methods are as follows:

For flow inside pipe, the Rouhani–Axelsson correlation is recommended.

For vertical tube bundles with flow parallel to axis, Hibiki et al. correlation is recommended.

For horizontal bundles with crossflow, Kanizawa and Ribatski correlation is recommended.

1.9 CFD Simulation

Computational fluid dynamics (CFD) methods, are being increasingly used to simulate a wide range of problems in two-phase heat transfer and fluid flow. This is firstly because of the limitations of theoretical and empirical methods. It is hoped that CFD simulation will provide accurate fundamentally based design methods. Secondly, it offers the possibility of solving complex problems for which experimentation is difficult or impossible. Finally, such simulations offer the possibility of obtaining solutions without expensive experimentation.

Most common methods solve the conservation equations for mass, momentum, and energy, for each phase, treating them as continuum. The methods for solving these equations have been reviewed, among others, by Kharangate and Mudawar (2017) and Ilic et al. (2019).

Kharangate and Mudawar (2017) reviewed the numerous published CFD studies on boiling and condensation. They concluded that though considerable progress has been made, capabilities of CFD are presently very limited. Much more needs to be done before investigators can achieve the long-term objective of developing a more unified, physically based, accurate, and computationally efficient methodology.

Ilic et al. (2019) reviewed research on application of CFD to boiling systems. They note that significant progress has been made in CFD techniques and methods. Nevertheless, due to the complexity of the boiling phenomenon, all the approaches used have serious deficiencies and cannot provide reliable and accurate computational predictions of boiling heat transfer.

It is thus clear that while CFD is a promising research field, it is as yet not suitable for use in practical design. Therefore, CFD studies and methodology have not been discussed in this book.

1.10 General Information

The terms gas and vapor have been used interchangeably in this book. Same is the case for tube and pipe.

All equations given in this book are dimensionless except where noted otherwise. Where dimensional equations are given, the units to be used are stated there. Only those units should be used in them. If units for dimensional equations are not given, the units in Nomenclature are to be used.

Attempt was made to keep the Nomenclature uniform throughout the book but this was not possible even within a single chapter due to the wide range of topics, the limited number of available symbols, and the desirability of using symbols typically used. List of symbols has been provided in each chapter. Those are the symbols used in most of the chapter, but in some cases some of those symbols have also been used for some other quantity. The definitions of units stated for the particular equation should be used in such cases.

Since about 40 years, almost all research publications have been using the SI units. Earlier publications used a variety of units such as metric, MKS, and British inch-pound. To help in the use of publications in those units, conversion tables have been provided in the Appendix. A point to note is that in some older publications, dimensionless equations written using units other than SI often have a conversion factor g_c. This factor equals 1 when using SI units.

For evaluating the accuracy of prediction methods, a variety of criteria are used in publications. Most common are the average deviation (AVD) and the mean absolute deviation (MAD). These are defined as follows, using y for the parameter. Subscript p is for predicted and m is for measured;

$$\mathrm{MAD} = \frac{1}{N}\sum_{1}^{N} \mathrm{ABS}((y_{p,i} - y_{m,i})/y_{m,i}) \tag{1.10.1}$$

$$\mathrm{AVD} = \frac{1}{N}\sum_{1}^{N} ((y_{p,i} - y_{m,i})/y_{m,i}) \tag{1.10.2}$$

Nomenclature

A	area (m^2)
A_c	cross-sectional area (m^2)
Bd	Bond number (–)
Bo	boiling number = $qG^{-1}i_{fg}^{-1}$ (–)
C_p	specific heat at constant pressure (J kg^{-1} K^{-1})
Co	Confinement number (–)
D	diameter of tube (m)
D_{HP}	equivalent diameter based on heated perimeter (m)
D_{HYD}	hydraulic equivalent diameter (m)
f	Fanning pipe friction factor (–)
g	acceleration due to gravity at Earth (m s^{-2})
h	heat transfer coefficient (W m^{-2} K^{-1})
i_{fg}	latent heat of vaporization (J kg^{-1})
j	volumetric flux, superficial velocity (ms^{-1})
k	thermal conductivity (W m^{-1} K^{-1})
L	length (m)
N	number of data points (–)
Nu	Nusselt number (–)
p	pressure (Pa)
p_c	critical pressure (Pa)
p_r	reduced pressure = p/p_c (–)
P	pitch of tube bundles (m)
Pr	Prandtl number (–)
q	heat flux (W m^{-2})
Re	Reynolds number = (GD/μ) (–)
T	temperature (K)
ΔT_{SAT}	$(T_w - T_{SAT})$ (K)
ΔT_{SC}	$(T_{SAT} - T_f)$ (K)
u	velocity (m s^{-1})
W	mass flow rate (kg s^{-1})
We	Weber number (–)
X	Lockhart–Martinelli parameter (–)
x	equilibrium quality (–)
z	length or distance (m)

Greek letters

α Void fraction (–)
μ dynamic viscosity (Pa s)
ϕ_L two-phase pressure drop multiplier (–)
ρ density (kg m^{-3})
σ surface tension (N m^{-1})

Subscripts

f liquid
G vapor, gas
GS superficial gas, meaning gas phase flowing alone in pipe
GT total gas, all mass flowing as gas/vapor
g vapor
L liquid
LS liquid superficial, meaning liquid flowing alone in pipe
LT liquid total, all mass flowing as liquid
l liquid
SAT saturation
SC subcooled
w wall

References

Akbar, M.K., Plummer, D.A., and Ghiaasiaan, S.M. (2002, 2002). Gas-liquid two-phase flow regimes in microchannels. In: *Proceedings of the ASME 2002 International Mechanical Engineering Congress and Exposition. New Orleans, Louisiana, USA. 17–22 November*, Heat Transfer, vol. **7**, 527–534.

Akbar, M.K., Plummer, D.A., and Ghiaasiaan, S.M. (2003). On gas–liquid two-phase flow regimes in microchannels. *Int. J. Multiphase Flow* **29**: 855–865.

Aprin, L., Mercier, P., and Tadrist, L. (2007). Experimental analysis of local void fractions measurements for boiling hydrocarbons in complex geometry. *Int. J. Multiphase Flow* **33**: 371–393.

ASHRAE (2017). *ASHRAE Handbook Fundamentals*. Atlanta, GA: ASHRAE.

Baker, O. (1954). Simultaneous flow of oil and gas. *Oil Gas J.* **53**: 185–195.

Barnea, D. (1987). A unified model for predicting flow-pattern transitions for the whole range of pipe inclinations. *Int. J. Multiphase Flow* **13** (1): 1–12.

Chen, L., Tian, Y.S., and Karayiannis, T.G. (2006). The effect of tube diameter on vertical two-phase flow regimes in small tubes. *Int. J. Heat Mass Transfer* **49**: 4220–4230.

Cheng, P. and Wu, H.Y. (2006). Macro and microscale phase-change heat transfer. *Adv. Heat Transfer* **39**: 461–563.

Cheng, L., Ribatski, G., and Thome, J.R. (2008). Two-phase flow patterns and flow-pattern maps: fundamentals and applications. *Appl. Mech. Rev.* **61**: 050802-1–050802-28.

Chisolm, D. (1967). A theoretical basis for the Lockhart-Martinelli correlation for two-phase flow. NEL Report No. 310.,

Colburn, A.P. (1933). A method for correlating forced convection heat transfer data and a comparison with fluid friction. *Trans. AIChE* **19**: 174–210. Reprinted in: Int. J. Heat Mass Transfer, 7, 1359–1384, 1964.

Colebrook, C.F. (1939). Turbulent flow in pipes with particular reference to the transition region between the smooth and rough pipe laws. *J. Inst. Civil Eng.* **11**: 133–156.

Colin, C., Fabre, J.A., and Dukler, A.E. (1991). Gas-liquid flow at microgravity conditions – I. Dispersed bubble and slug flow. *Int. J. Multiphase Flow* **17**: 533–544.

Das, G., Das, P.K., Purohit, N.K., and Mitra, A.K. (1999a). Flow pattern transition during gas liquid upflow through vertical concentric annuli – I. Experimental investigations. *J. Fluids Eng.-Trans. ASME* **121**: 895–901.

Das, G., Das, P.K., Purohit, N.K., and Mitra, A.K. (1999b). Flow pattern transition during gas liquid upflow through vertical concentric annuli – Part II: Mechanistic models. *Trans. ASME J. Fluids Eng.* **121**: 902–907.

Dittus, P.W. and Boelter, L.M.K. (1930). Heat transfer in the radiators of the tubular type. *Univ. Calif. Eng. Pub.* **2** (13): 443–461.

Dukler, A.E., Fabre, J.A., McQuillen, J.B., and Vernon, R. (1988). Gas-liquid flow at microgravity conditions: flow pattern and their transitions. *Int. J. Multiphase Flow* **14**: 389–400.

El Hajal, J., Thome, J.R., and Cavallini, A. (2003). Condensation in horizontal tubes. Part 1: two-phase flow pattern map. *Int. J. Heat Mass Transf.* **46**: 3349–3363.

Enoki, K., Miyata, K., Mori, H. et al. (2013). Boiling heat transfer and pressure drop of a refrigerant flowing vertically upward in small rectangular and triangular tubes. *Heat Transfer Eng.* **34** (11–12): 966–975.

Friedel, L. (1979). Improved friction pressure drop correlations for horizontal and vertical two-phase pipe flow. *European Two-phase Group Meeting*, Paper E2 June, Ispra, Italy.

Garimella, S., Killion, J.D., and Coleman, J.W. (2002). An experimentally validated model for two-phase pressure drop in the intermittent flow regime for circular microchannels. *J. Fluids Eng.* **124** (1): 205–214.

Ghiaasiaan, S.M. (2008). *Two-Phase Flow, Boiling, and Condensation*. New York, NY: Cambridge University Press.

Gnielinski, V. (1976). New equations for heat and mass transfer in turbulent pipe and channel flow. *Int. Chem. Eng.* **16**: 359–368.

Godbole, P.V., Tang, C.C., and Ghajar, A.J. (2011). Comparison of void fraction correlations for different flow patterns in upward vertical two-phase flow. *Heat Transfer Eng.* **32** (10): 843–860.

Grant, I.D.R. and Chisholm, D. (1979). Two-phase flow on the shell-side of a segmentally baffled shell-and-tube heat exchanger. *J. Heat Transfer* **101**: 38–42.

Heppner, D. B., King, C. D. & Littles, J. W. (1975) Zero-gravity experiments in two-phase fluids flow patterns. ASME Paper No. TS-ENAs-24.

Hernandez, L., Julia, J.E., Ozar, B., et al. (2010). Flow regime identification in boiling two-phase flow in a vertical annulus. *7th International Conference on Multiphase Flow*, Tampa, FL, USA. Quoted in Julia and Hibiki (2011).

Hewitt, G.F. (1982). Void fraction. In: *Handbook of Multiphase Systems* (ed. G. Hetsroni), 2-76–2-94. Washington, DC: Hemisphere Publishing Corporation.

Hewitt, G.F. and Roberts, D.N. (1969). Studies of two-phase flow patterns by simultaneous X-ray and flash photography. UKAE Report AERE-M 2159.

Hibiki, T., Mao, K., and Ozaki, T. (2017). Development of void fraction-quality correlation for two-phase flow in horizontal and vertical tube bundles. *Prog. Nucl. Energy* **97**: 38–52.

Hill, N., Downing, S., Rogers, D., et al. (1987). A study of two-phase flow in a reduced gravity environment. Final Report DRL No. T-1884, Sunstrand Energy Systems, Rockford, IL.

Ilic, M.M., Petrovic, M.M., and Stevanvic, V.D. (2019). Boiling heat transfer modelling a review and future prospectus. *Therm. Sci.* **23** (1): 87–107.

Jige, D., Kikuchi, S., Eda, H. et al. (2018). Two-phase flow characteristics of R32 in horizontal multiport minichannels: flow visualization and development of flow regime map. *Int. J. Refrig* **95**: 156–164.

Julia, J.E. and Hibiki, T. (2011). Flow regime transition criteria for two-phase flow in a vertical annulus. *Int. J. Heat Fluid Flow* **32**: 993–1004.

Kandlikar, S.G. (2002). Fundamental issues related to flow boiling in minichannels and microchannels. *Exp. Therm Fluid Sci.* **26** (2–4): 389–407.

Kanizawa, F.T. and Ribatski, G. (2016). Two-phase flow patterns across triangular tube bundles for air–water upward flow. *Int. J. Multiphase Flow* **80**: 43–56.

Kanizawa, F.T. and Ribatski, G. (2017). Void fraction and pressure drop during external upward two-phase cross flow in tube bundles – Part II: Predictive methods. *Int. J. Heat Fluid Flow* **65**: 210–219.

Karri, S.B.R. and Mathur, V.K. (1988). Two-phase flow pattern map predictions under microgravity. *AlChE J.* **34**: 137–139. Quoted in Rezkallah (1990).

Kattan, N., Thome, J.R., and Farvat, D. (1998). Flow boiling in horizontal tubes: Part 1 – Development of a diabatic two-phase flow pattern map. *J. Heat Transfer* **120**: 140–147.

Kays, W.H. and Perkins, H.C. (1973). Forced convection, internal flow in ducts. In: *Handbook of Heat Transfer* (eds. W.M. Rohsenow and J.P. Hartnett), 7-1–7-193. New York, NY: McGraw-Hill.

Kelessidis, V.C. and Dukler, A.E. (1989). Modeling flow pattern transitions for upward gas–liquid flow in vertical concentric and eccentric annuli. *Int. J. Multiphase Flow* **15**: 173–191.

Kew, P.A. and Cornwell, K. (1997). Correlations for prediction of flow boiling heat transfer in small-diameter channels. *Appl. Therm. Eng.* **17**: 705–715.

Kharangate, C.R. and Mudawar, I. (2017). Review of computational studies on boiling and condensation. *Int. J. Heat Mass Transfer* **108**: 1164–1196.

Kim, D. (2000). An experimental and empirical investigation of convective heat transfer for gas–liquid two-phase flow in vertical and horizontal pipes. PhD thesis. Oklahoma State University, Stillwater, OK.

Kim, S.M. and Mudawar, I. (2012). Universal approach to predicting two-phase frictional pressure drop for adiabatic and condensing mini/micro-channel flows. *Int. J. Heat Mass Transfer* **55**: 3246–3261.

Kim, S.M. and Mudawar, I. (2013). Universal approach to predicting two-phase frictional pressure drop for mini/micro-channel saturated flow boiling. *Int. J. Heat Mass Transfer* **58**: 718–734.

Lee, D., Best, F.R., and McGraw, N. (1987). Microgravity two phase flow pattern modeling. In: *Proceedings of the American Nuclear Society Winter Meeting*, 94–100. Los Angeles, CA: American Nuclear Society. Quoted in Rezkallah (1990).

Li, J.-M. and Wang, B.-X. (2003). Size effect on two-phase regime for condensation in micro/mini tubes. *Heat Transfer Asian Res.* **32** (1): 65–71.

Li, W. and Wu, Z. (2010a). A general criterion for evaporative heat transfer in micro/mini-channels. *Int. J. Heat Mass Transfer* **53**: 1967–1976.

Li, W. and Wu, Z. (2010b). A general correlation for adiabatic two-phase pressure drop in micro/mini-channels. *Int. J. Heat Mass Transfer* **53**: 2732–2739.

Liu, H. and Hibiki, T. (2017). Flow regime transition criteria for upward two-phase flow in vertical rod bundles. *Int. J. Heat Mass Transfer* **108**: 423–433.

Lockhart, R.W. and Martinelli, R.C. (1949). Proposed correlation of data for isothermal two-phase, two-component flow in pipes. *Chem. Eng. Prog.* **45** (1): 39–48.

Mandhane, J.M., Gregory, G.A., and Aziz, K. (1974). Flow pattern map for gas–liquid flow in horizontal pipes. *Int. J. Multiphase Flow* **1**: 537–553.

McAdams, W.H. (1954). *Heat Transmission*, 3e. New York, NY: McGraw-Hill.

McQuillan, K.W. and Whalley, P.B. (1985). Flow patterns in vertical two-phase flow. *Int. J. Muhiphase Flow* **11** (2): 161–175.

Mehendale, S.S., Jacobi, A.M., and Shah, R.K. (2000). Fluid flow and heat transfer at micro- and meso-scales with application to heat exchanger design. *Appl. Mech. Rev.* **53** (7): 175–193.

Mehta, H.B. and Banerjee, J. (2014). An investigation of flow orientation on air–water two-phase flow in circular minichannel. *Heat Mass Transfer* **50**: 1353–1364. https://doi.org/10.1007/s00231-014-1332-2.

Melkamu, A., Woldesmayat, M.A., and Ghajar, A.J. (2007). Comparison of void fraction correlations for different flow patterns in horizontal and upward inclined pipes. *Int. J. Multiphase Flow*, **33**, 347–370.

Mikityuk, K. (2009). Heat transfer to liquid metal: review of data and correlations for tube bundles. *Nucl. Eng. Des.* **239**: 680–687.

Mishima, K. and Hibiki, T. (1996). Some characteristics of air–water two-phase flow in small diameter vertical tubes. *Int. J. Multiphase Flow* **22**: 703–712.

Mishima, K. and Ishii, K. (1984). Flow regime transition criteria for upward two-phase flow in vertical tubes. *Int. J. Heat Mass Transfer* **27** (5): 723–731.

Mizutani, Y., Tomiyama, A., Hosokawa, S. et al. (2007). Two-phase flow patterns in a four by four rod bundle. *J. Nucl. Sci. Technol.* **44**: 894–901.

Müller-Steinhagen, H. and Heck, K. (1986). A simple friction pressure drop correlation for two-phase flow in pipes. *Chem. Eng. Process.* **20**: 297–308. Quoted in Thome, J.R. (2006). *The Heat Exchanger Data Book III*. Ulm, Germany: Wieland-Werke AG.

Narcy, M. and Colin, C. (2015). Two-phase pipe flow in microgravity with and without phase change: recent progress and future prospects. *Interfacial Phenom. Heat Transfer* **3** (1): 1–17. ISSN 2169-2785.

Narcy, M., Malmazet, E., and Colin, C. (2014). Flow boiling in tube under normal gravity and microgravity conditions. *Int. J. Multiphase Flow* **60**: 50–63.

Ong, C.L. and Thome, J.R. (2011a). Macro-to-microchannel transition in two-phase flow: Part 1 – Two-phase flow patterns and film thickness measurements. *Exp. Therm Fluid Sci.* **35** (1): 37–47.

Ong, C.L. and Thome, J.R. (2011b). Macro-to-Microchannel transitionin two-phase flow: part 2 – flow boiling heat transfer and critical heat flux. *Exp. Therm. Fluid Sci.* **35**: 873–886.

Paranjape, S., Chen, S.W., Hibiki, T., and Ishii, M. (2011). Flow regime identification under adiabatic upward two-phase flow in a vertical rod bundle geometry. *J. Fluids Eng.* **133** (9): 091302.

Pethukov, B.S. and Kirillov, V.V. (1958). The problem of heat exchange in the turbulent flow of liquids in tubes. *Teploenergetika* **4** (4): 63–68. Quoted in Bhatti, M.S. & Shah, R. K. (1987) Turbulent and transition flow convective heat transfer in ducts. In: Kacak, S., Shah, R.K., & Aung, W. (eds.) Handbook of Single-Phase Heat Transfer, New York, NY, Wiley, pp. 4-1–4-166.

Pettigrew, M.J., Taylor, C.E., and Kim, B.S. (1989). Vibration of tube bundles in two-phase cross-flow: Part 1 – Hydrodynamic mass and damping. *J. Pressure Vessel Technol.* **111**: 466–477.

Pletcher, R.H. (1987). External flow forced convection. In: *Handbook of Single-Phase Heat Transfer* (eds. S. Kacac, R.K. Shah and W. Aung), 2-1–2-67. New York, NY: Wiley.

Qiu, Z.C., Ma, Z.Y., Qiu, S.Z. et al. (2015). Experimental research on the thermal hydraulic characteristics of sodium boiling in an annulus. *Exp. Therm Fluid Sci.* **60**: 263–274.

Reed, C.B. (1987). Convective heat transfer in liquid metals. In: *Handbook of Single-Phase Heat Transfer* (eds. S. Kacac, R.K. Shah and W. Aung), 8-1–8-30. New York, NY: Wiley.

Rehme, K. (1987). Convective heat transfer over rod bundles. In: *Handbook of Single-Phase Heat Transfer* (eds. S. Kakac, R.K. Shah and W. Aung), 7-1–7-62. New York, NY: Wiley.

Rezkallah, K.S. (1988). Two-phase flow and heat transfer at reduced gravity: a literature survey. *Proc. Am. Nucl. Soc.* **3**: 435–444.

Rezkallah, K.S. (1990). A comparison of existing flow-pattern predictions during forced-convective two-phase flow under microgravity conditions. *Int. J. Multiphase Flow* **16** (2): 243–259.

Rezkallah, K.S. and Zhao, L. (1995). A flow pattern map for two-phase liquid-gas flows under reduced gravity conditions. *Adv. Space Res.* **16** (7): 133–136.

Rouhani, S.Z. and Axelsson, E. (1970). Calculation of void volume fraction in the subcooled and quality boiling regions. *Int. J. Heat Mass Transfer* **13** (2): 383–393.

Rouhani, S.Z. and Sohal, M.S. (1983). Two-phase flow patterns: a review of research results. *Prog. Nucl. Energy* **11** (3): 219–259.

Saisorn, S., Wongprom, P., and Wongwises, S. (2018). The difference in flow pattern, heat transfer and pressure drop characteristics of mini-channel flow boiling in horizontal and vertical orientations. *Int. J. Multiphase Flow* **101**: 97–112.

Shah, M.M. (1975). Visual observations in an ammonia evaporator. *ASHRAE Trans.* **82** (1).

Shah, M.M. (1982). Chart correlation for saturated boiling heat transfer: equations and further study. *ASHRAE Trans.* **88** (1): 185–196.

Shah, R.K. (1986). Classification of heat exchangers. In: *Heat Exchangers: Thermal Hydraulic Fundamentals and Design* (eds. S. Kakac, A.E. Bergles and F. Mayinger). Washington, DC: Hemisphere Publishing Corporation.

Shah, M.M. (2009). An improved and extended general correlation for heat transfer during condensation in plain tubes. *HVAC&R Res.* **15** (5): 889–913.

Shah, M.M. (2013). General correlation for heat transfer during condensation in plain tubes: further development and verification. *ASHRAE Trans.* **119** (2).

Shah, M.M. (2015). A general correlation for CHF in horizontal channels. *Int. J. Refrig* **59**: 37–52.

Shah, M.M. (2016). Comprehensive correlations for heat transfer during condensation in conventional and mini/micro channels in all orientations. *Int. J. Refrig* **67**: 22–41.

Shah, M.M. (2017a). New correlation for heat transfer during subcooled boiling in plain channels and annuli. *Int. J. Therm. Sci.* **112**: 358–370.

Shah, M.M. (2017b). Unified correlation for heat transfer during boiling in plain mini/micro and conventional channels. *Int. J. Refrig* **74**: 604–624.

Shah, M.M. (2017c). Comprehensive correlation for dispersed flow film boiling heat transfer in mini/macro tubes. *Int. J. Refrig* **78**: 32–46.

Shah, M.M. (2017d). Applicability of general correlations for CHF in conventional tubes to mini/micro channels. *Heat Transfer Eng.* **38** (1): 1–10.

Shah, M.M. (2018). Applicability of correlations for boiling/condensing in macrochannels to minichannels. *Heat Mass Transfer Res. J.* **2** (1): 20–32.

Shannak, B.A. (2008). Frictional pressure drop of gas liquid two-phase flow in pipes. *Nucl. Eng. Des.* **238**: 3277–3284.

Shoham, O. (1982). Flow pattern transitions and characterization in gas–liquid two phase flow in inclined pipes. PhD dissertation. Tel-Aviv University, Ramat-Aviv, Israel. Quoted in Barnea (1987).

Sieder, E.N. and Tate, C.E. (1936). Heat transfer and pressure drop of liquids in tubes. *Ind. Eng. Chem.* **28**: 1429–1435.

Smith, S.L. (1969). Void fractions in two-phase flow: a correlation based upon an equal velocity head model. *Proc. Inst. Mech. Eng.* **184**: 647–664. Quoted in Hibiki et al. (2017).

Taitel, Y. and Dukler, A.E. (1976). A model for predicting flow regime transitions in horizontal and near horizontal gas–liquid flow. *AlChE J.* **22** (1): 47–54.

Tribbe, C. and Müller-Steinhagen, H.M. (2000). An evaluation of the performance of phenomenological models for predicting pressure gradient during gas–liquid flow in horizontal pipelines. *Int. J. Multiphase Flow* **26**: 1019–1036.

Triplett, K.A., Ghiaasiaan, S.M., Abdel-Khalik, S.I., and Sadowski, D.L. (1999). Gas liquid two-phase flow in microchannels: Part 1 – Two-phase flow patterns. *Int. J. Multiphase Flow* **25**: 377–394.

Ulbrich, R. and Mewes, D. (1994). Vertical, upward gas–liquid two-phase flow across a tube bundle. *Int. J. Multiphase Flow* **20**: 249–272.

Ullmann, A. and Brauner, N. (2007). The prediction of flow pattern maps in minichannels. *Multiphase Sci. Technol.* **19** (1): 49–73.

Venkateswararao, P., Semiat, R., and Dukler, A.E. (1982). Flow pattern transition for gas-liquid flow in a vertical rod bundle. *Int. J. Multiphase Flow* **8**: 509–524.

Williams, C.L. and Peterson, A.C. Jr., (1978). Two-phase flow patterns with high-pressure water in a heated four-rod bundle. *Nucl. Sci. Eng.* **68**: 155–169.

Witte, L.C., Bousman, W.S., and Fore, L.B. (1996). Studies of two-phase flow dynamics and heat transfer at reduced gravity conditions. NASA Contractor Report No 198459.

Wu, Y., Luo, S., Wang, L. et al. (2018). Review on heat transfer and flow characteristics of liquid sodium (2): two phase. *Prog. Nucl. Energy* **103**: 151–164.

Xu, G.P., Tso, C.P., and Tou, K.W. (1998). Hydrodynamics of two-phase flow in vertical up and down-flow across a horizontal tube bundle. *Int. J. Multiphase Flow* **24**: 1317–1342.

Xu, Y., Fang, X., Su, X. et al. (2012). Evaluation of frictional pressure drop correlations for two-phase flow in pipes. *Nucl. Eng. Des.* **253**: 86–97.

Zhao, L. and Rezkallah, K.S. (1993). Gas–liquid flow patterns at microgravity conditions. *Int. J. Multiphase Flow* **19** (5): 751–763.

Zhou, Y., Hou, Y., Li, H. et al. (2015). Flow pattern map and multi-scale entropy analysis in 3 × 3 rod bundle channel. *Ann. Nucl. Energy* **80**: 144–150.

Zigrang, D.J. and Sylvester, N.D. (1982). Explicit approximation to the solution of Colebrook's friction factor equation. *AlChE J.* **28**: 514–516. Quoted in Bhatti, M.S. & Shah, R. K. (1987) Turbulent and transition flow convective heat transfer in ducts. In: Kacak, S., Shah, R.K., & Aung, W. (eds.) Handbook of Single-Phase Heat Transfer, New York, NY, Wiley, pp. 4-1–4-166.

Zukauskas, A. (1987). Convective heat transfer in cross flow. In: *Handbook of Single-Phase Heat Transfer* (eds. S. Kacac, R.K. Shah and W. Aung), 6-1–6-45. New York, NY: Wiley.

Zukauskas, A. and Ulinskas, R. (1983). Section 2.2.4 Banks of plain and finned tubes. In: *Heat Exchanger Design Handbook. Heat Exchanger Design Handbook 2 – Fluid Mechanics and Heat Transfer* (ed. E.U. Schlunder). Washington, DC. USA: Hemisphere Publishing Corporation.

2

Heat Transfer During Condensation

2.1 Introduction

Condensation is the phenomenon of vapor changing to liquid. There are two basic modes of condensation, namely, film condensation and dropwise condensation. During film condensation, vapor condenses in the form of a continuous liquid film over the condensing surface. During dropwise condensation, vapor condenses in the form of drops covering parts of the surface. Heat transfer coefficients during dropwise condensation are much higher than those during film condensation. Hence dropwise condensation is desirable. However, most common fluids condense in the film mode over ordinary metallic surfaces. Most of the condensers used in the industry today operate with film condensation. The mechanism of film condensation is well understood, and the rate of heat transfer during film condensation can be predicted with reasonable accuracy in most cases. Much research has been and is being done to develop durable heat exchangers with dropwise condensers but has not yet achieved success.

Most of this chapter is concerned with film condensation. Dropwise condensation is discussed at the end of the chapter in Section 2.10. All material up to and including Section 2.8 is applicable only to non-metallic fluids unless otherwise noted. Section 2.8 deals with condensation of mixtures; sections 2.2–2.7 are for single component vapors. Section 2.9 is devoted to liquid metals.

2.2 Condensation on Plates

2.2.1 Nusselt Equations

Nusselt (1916) presented analytical solutions to several problems of laminar film condensation. His analysis for a vertical plate is presented in the following.

Consider a vertical flat plate as shown in Figure 2.2.1. The wall surface is at a temperature T_w while the pure saturated vapor is at the temperature T_{SAT}, which is higher than the wall temperature. As the wall is below its saturation temperature, the vapor condenses, and the condensate flows down in a film under the influence of gravity, opposed by viscous and buoyancy forces. Considering the liquid film to be laminar and taking force balance on an element of thickness $(\delta - y)$,

$$\rho_f g(\delta - y)dz = \mu_f \frac{du}{dy}dz + \rho_g g(\delta - y)dz \tag{2.2.1}$$

Integrating with the boundary condition $u = 0$ at $z = 0$,

$$u = \frac{(\rho_f - \rho_g)g}{\mu_f}\left(\delta y - \frac{y^2}{2}\right) \tag{2.2.2}$$

Integrating over the film thickness,

$$\Gamma = \rho_f \int_0^\delta u dy = \frac{\rho_f(\rho_f - \varrho_g)g\delta^3}{3\mu_f} \tag{2.2.3}$$

where Γ is the mass flow rate of condensate per unit width of plate. The rate of increase of mass flow rate with film thickness is obtained by differentiating with the boundary condition $\delta = 0$ at $z = 0$.

$$\frac{d\Gamma}{d\delta} = \frac{\rho_f(\rho_f - \rho_g)g\delta^2}{\mu_f} \tag{2.2.4}$$

Neglecting subcooling of condensate, the latent heat of the condensate added over a length dz. must equal the heat conducted through the liquid film:

$$\frac{\rho_f(\rho_f - \rho_g)g\delta^2 d\delta}{\mu_f} i_{fg} = -k_f dz \frac{T_{SAT} - T_w}{\delta} \tag{2.2.5}$$

Rearranging and integrating with $\delta = 0$ at $z = 0$ gives

$$\delta = \left[\frac{4\mu_f k_f z(T_{SAT} - T_w)}{g i_{fg} \varrho_f(\rho_f - \rho_g)}\right]^4 \tag{2.2.6}$$

The local heat transfer coefficient h_z at distance z from the top is then

$$h_z = \frac{k_f}{\delta} = \left[\frac{\rho_f(\rho_f - \rho_g)g i_{fg} k_f^3}{4\mu_f z(T_{SAT} - T_w)}\right]^{1/4} \tag{2.2.7}$$

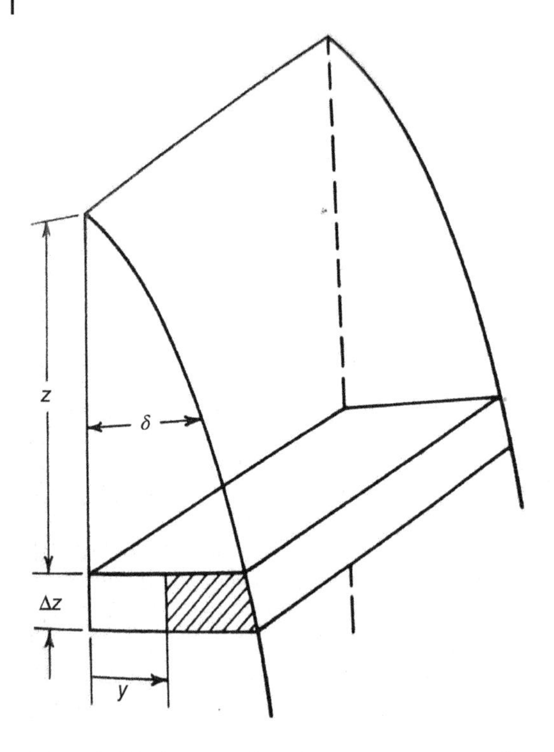

Figure 2.2.1 Condensation of stagnant vapor on a flat plate. Source: Rohsenow (1973a). © 1973 McGraw-Hill.

Integrating from $z = 0$ to $z = L$, the mean heat transfer coefficient h_m for the length L is

$$h_m = \frac{1}{L}\int_0^L h_z dz = 0.943\left[\frac{\rho_f(\rho_f - \rho_g)g i_{fg} k_f^3}{\mu_f L(T_{SAT} - T_w)}\right]^{1/4} \tag{2.2.8}$$

Equations (2.2.6) and (2.2.7) can be expressed in terms of Reynolds number defined as

$$Re_z = \frac{4\Gamma_z}{\mu_f} \tag{2.2.9}$$

The mean heat transfer coefficient can also be defined as

$$h_m = \frac{i_{fg}\Gamma}{L(T_{SAT} - T_w)} \tag{2.2.10}$$

Using Eq. (2.2.10), the alternative form of Nusselt equations for mean heat transfer is

$$h_m\left[\frac{\mu_f^2}{k_f^3\rho_f(\rho_f - \rho_g)g}\right]^{1/3} = 1.47 Re_{z=L}^{-1/3} \tag{2.2.11}$$

Similarly, the local heat transfer coefficient equation is

$$h_z\left[\frac{\mu_f^2}{k_f^3\rho_f(\rho_f - \rho_g)g}\right]^{1/3} = 1.1 Re_z^{-1/3} \tag{2.2.12}$$

If the plate is inclined at angle θ to the vertical, g is replaced by (g sin θ) in the above expressions.

Nusselt equations have been extensively tested against experimental data with generally good results.

2.2.2 Modifications to the Nusselt Equations

McAdams (1954) reports a derivation by T.B. Drew, and according to which liquid viscosity should be calculated at the temperature T given by

$$T = T_{SAT} - 0.75(T_{SAT} - T_w) \tag{2.2.13}$$

The above derivation was made assuming linear variation of viscosity with temperature.

Bromley (1952) noted that the condensing vapor is subcooled from the saturation temperature to the film temperature. He showed that to account for this effect, i_{fg} in the Nusselt analysis should be replaced by i_{fg}^* defined as

$$i_{fg}^* = i_{fg}\left[1 + 0.4\frac{C_{pf}(T_{SAT} - T_w)}{i_{fg}}\right] \tag{2.2.14}$$

Rohsenow (1956) has given an empirical equation for the effective latent heat. Based on their theoretical analysis, Minkowycz and Sparrow (1966) recommended that liquid properties, except latent heat, should be calculated at a reference temperature T_{ref} given by

$$T_{ref} = T_w + 0.31(T_{SAT} - T_w) \tag{2.2.15}$$

As the temperature drop across the liquid film is generally small, the above modifications usually have negligible effect.

Based on comparison with experimental data, McAdams (1954) recommended increasing the heat transfer coefficients from Nusselt equations by 20%. He attributed the increase to thinning of liquid film due to vapor shear besides the mixing action of ripples that become noticeable at quite low Reynolds numbers. Stephan (1992) recommends increasing Nusselt predictions by 15%.

Comprehensive analyses have been done by Koh et al. (1961) and Chen (1961a). These authors considered several factors that had been neglected in Nusselt's derivation. Among them are the effect of vapor shear due to the relative motion of liquid with respect to stagnant vapor, inertia effects, and the effect of energy convection. The results of Koh et al. and Chen are in fairly good agreement. Both found that for Prandtl number of the order of 1 or larger, Nusselt was justified in neglecting these factors. At low Prandtl numbers, which are characteristic of liquid metals, these factors become important, and then heat transfer coefficients are much lower than Nusselt predictions. Heat transfer coefficients decrease with increasing subcooling and are also reduced by vapor drag. Chen (1961a) has given the following formula to represent the results of his analytical solution:

$$\frac{h}{h_{Nu}} = \left(\frac{1 + 0.68\beta + 0.02\beta\gamma}{1 + 0.85\gamma - 0.15\beta\gamma}\right) \tag{2.2.16}$$

where

$$\beta = \frac{C_{pf}\Delta T_{SAT}}{i_{fg}} \tag{2.2.17}$$

$$\gamma = \frac{k_f \Delta T_{SAT}}{\mu_f i_{fg}} \tag{2.2.18}$$

Equation (2.2.16) is applicable to liquid Prandtl numbers less than 0.05 or greater than 1, $\beta \le 2, \gamma \le 20$.

2.2.3 Condensation with Turbulent Film

Ripples begin to form on condensate film at fairly low Reynolds number, about 30 according to Ghiaasiaan (2008). The condensate becomes turbulent when Reynolds number exceeds a critical value, for which various values have been given. According to McAdams (1954), it occurs at 1800, while Isachenko et al. (1977) state it to be at 1600.

The mean heat transfer coefficient for a vertical plate or on the outer surface of a vertical tube with both laminar and turbulent film may be calculated by the following correlation given by Kirkbride (1934):

$$h_m\left(\frac{\mu_f^2}{g\rho_f^2 k_f^3}\right)^{1/2} = 0.0077\left(\frac{4\Gamma}{\mu_f}\right)^{0.4} \tag{2.2.19}$$

This correlation was based on condensation of diphenyl oxide and Dowtherm A on the outside of vertical tubes.

Chen et al. (1987) presented the following equation for vertical surfaces when $Re_z > 30$:

$$\frac{h_m}{k_f}\left(\frac{\nu_f^2}{g}\right)^{1/3} = (Re_z^{-0.44} + 5.8 \times 10^{-6} Re_z^{0.8} \mathrm{Pr}_f^{1/3})^{1/2} \tag{2.2.20}$$

The range of applicability includes both laminar and turbulent films.

Isachenko et al. (1977) recommend the following equation for mean heat transfer when both laminar and turbulent regimes are involved:

$$\frac{h_m}{k_f}\left(\frac{\nu_f^2}{g}\right)^{1/3} = \frac{0.25Re_z}{2300 + 41\,\mathrm{Pr}_f^{-0.5}[(0.25Re_z)^{3/4} - 89](\mathrm{Pr}_f/\mathrm{Pr}_w)^{0.25}} \tag{2.2.21}$$

This equation was shown to be in agreement with data from several sources for Reynolds numbers up to 50 000.

2.2.4 Condensation on Underside of a Plate

This problem was theoretically and experimentally investigated by Gerstmann and Griffith (1967). They condensed R-113 at atmospheric pressure on the underside of plates whose inclination to horizontal varied from 0° to 90°. For horizontal surface, the condensate leaves the plate in the form of drops. For R-113, there was approximately one drop location per square centimeter. For the plates inclined between 3° and 7°, liquid ridges were formed from which the drops fell. As angle of inclination was increased, ridges became less pronounced. At inclinations greater than about 20°, the liquid drained from the trailing edge of the plate. Data for inclination of 21° and higher were in good agreement with the Nusselt theory. For lower inclinations, measurements are higher than the Nusselt theory.

For inclined surfaces, their analysis of a ridge model yielded the following expression for mean Nusselt number:

$$Nu_m = 0.9Ra^{1/6}[1 + 1.1Ra^{-1/6}] \tag{2.2.22}$$

where

$$Nu_m = \frac{h}{k_f}\left[\frac{\sigma}{g(\rho_f - \rho_g)\cos\theta}\right]^{1/2} \tag{2.2.23}$$

and

$$Ra = \frac{g\cos\theta\rho_f(\rho_f - \rho_g)i_{fg}}{k_f\mu_f\Delta T_{SAT}}\left[\frac{\sigma}{g(\rho_f - \rho_g)\cos\theta}\right]^{3/2} \tag{2.2.24}$$

where θ is the inclination of plate to the horizontal. Equation (2.2.22) is applicable only for $Ra > 10^6$.

For horizontal surfaces, they developed the following expressions through analysis:

For $10^{10} > Ra > 10^8$,

$$Nu_m = 0.81Ra^{0.193} \tag{2.2.25}$$

For $10^8 > Ra > 10^6$,

$$Nu_m = 0.69Ra^{0.2} \tag{2.2.26}$$

All data were within about 15% of these equations.

Howarth et al. (1978) presented a theoretical solution of condensation on underside of a plate, but they did not provide comparison with any data.

Stein et al. (1985) studied condensation of steam on the underside of a horizontal plate enclosed in a vessel. Pressure ranged from 0.31 to 1.24 MPa. Steam contained air from 9 to 500 ppm. Heat transfer coefficients were compared to the Gerstmann and Griffith correlation given earlier. The data for very low air content were lower than this correlation at 0.31 MPa and higher than the correlation at 1.24 MPa. Hence there appeared to be an effect of pressure. Their data were in the range of Eq. (2.2.26). They modified it to the following:

For $10^8 > Ra > 10^6$,

$$Nu_m = 0.787 p^{0.464} Ra^{0.2} \tag{2.2.27}$$

where p is in MPa.

2.2.5 Recommendations

For vertical plates and inclined plates with condensation on the upper side, use Nusselt equations for $Re_z < 1700$, and then increase their predictions by 20%.

For vertical plates for $Re_z > 1700$, use Eq. (2.2.21), the correlation given by Isachenko et al.

For plates with condensation on underside, use the following:

For inclinations to horizontal ≥21°, use Nusselt equations.

For inclinations <21°, use the correlations of Gerstmann and Griffith and Stein et al. in the range of their data.

2.3 Condensation Inside Plain Channels

2.3.1 Laminar Condensation in Vertical Tubes

As the condensate film thickness is small compared to curvature of all but the smallest diameter tubes, Nusselt equations presented in Section 2.2 are also applicable to vertical tubes. The Reynolds number for tubes is defined as

$$Re = \frac{4w}{\pi D \mu_L} = \frac{G(1-x)D}{\mu_f} \tag{2.3.1}$$

where w is the mass flow rate of condensate and G is the total mass flux of vapor and liquid. If condensation occurs on outside surface of the tube, D is the outside diameter. If it occurs on the inside of tube, D is the inside diameter.

Nusselt equations have been tested against a wide range of data by many researchers. McAdams (1954) summarized the results of many early studies. The measured heat transfer coefficients were found to be 0.78–1.7 times the Nusselt predictions. In most cases, the measurements were higher. Based on the data for Re_{LS} from 80 to 2000, McAdams recommended that the mean heat transfer coefficient given by Nusselt formulas be multiplied by 1.2. Further, he recommended that the Nusselt equations be used only for Reynolds numbers less than 1800.

Walt and Kroger (1972) studied condensation of R-12 inside a vertical tube. They found that the ratio of measured to Nusselt predicted heat transfer coefficients increased from 1.03 to 1.15 as Re_{LS} increased from 5 to 28.

Cavallini and Zecchin (1966) compared their data for R-11 at very low velocities in a vertical tube at Re_{LS} between 180 and 700. They found the data to be about 20% higher than the Nusselt equation.

Hayakawa et al. (1975) condensed ethanol, trichloroethylene, and water on the outer surface of a 10 mm outside diameter (OD) vertical tube. All local heat transfer coefficients were within ±20% of the Nusselt equation. They point out that wall temperature is often not constant. In such cases Nusselt equations for mean heat transfer coefficient are not valid as these were arrived at by assuming $(T_W - T_{SAT})$ to be constant. They appear to suggest that the larger deviations reported by experimenters may be due to the wall temperature not being constant.

The Nusselt equations presented till now are based on the assumption that vapor shear is negligible. This assumption is valid only for very low velocities. Nusselt (1916) carried out another analysis in which vapor shear was considered but liquid film was assumed to be laminar. This solution has been outlined by Jakob (1949). Calculations with this method are very tedious. Very few reports of comparison of this solution with data have come to this author's notice.

Dobran and Thorsen (1980) carried out an analysis for the case of laminar liquid film with vapor shear in a vertical tube. They found that the heat transfer coefficients become increasingly smaller than Nusselt equations for stagnant vapor as Fr/Re of vapor at entrance to tube decreases. The effect is very strong at low liquid Prandtl numbers of the order of 0.005. Their solution shows reasonable agreement with some low pressure data for steam with vapor velocities between 20 and 48 m s^{-1}. Their solution is too cumbersome for practical use.

2.3.2 The Onset of Turbulence

All solutions discussed till now assumed that the liquid film is laminar. For single-phase flow, it is known that liquid generally remains laminar till Reynolds number exceeds 2100–2300. Hence in the absence of vapor shear, one would except the same behavior for condensate film. Actually, transition to turbulent flow without vapor shear occurs at Reynolds number of 1600–1800 though ripples in condensate film are seen at quite low Reynolds numbers, and Kirkbride (1934) found that friction factor for liquid films exceeded those predicted by laminar film theory at Reynolds numbers exceeding 8.

In the presence of significant vapor velocities, the liquid film becomes turbulent at very low Reynolds numbers. Carpenter and Colburn (1951) compared their data for ethanol condensing in a vertical tube with the Nusselt solution that includes vapor shear. The measured heat transfer coefficients were found to be three times higher than the theoretical at vapor velocity of 27 m s^{-1} and seven times higher at vapor velocity of 71 m s^{-1}. Such wide discrepancies could be explained only by turbulence in liquid film. They concluded that in the presence of vapor shear, liquid film becomes turbulent at Reynolds number of 240.

Rohsenow et al. (1956) analyzed condensation on a vertical plate in the presence of vapor shear. They found that the Reynolds number for transition to turbulent flow decreases with increasing interfacial shear and can be as low as 50.

Kim and Mudawar (2012) analyzed condensation in a rectangular channel. They found the transition to turbulent film to occur at Reynolds number of 25.

Shah (2018a, 2018b) analyzed data for heat transfer to gas–liquid flow in vertical and horizontal channels. He found that the transition to turbulence occurred at $Re_{LS} = 170–175$.

Thus it is clear that laminar to turbulent transition during two-phase flow occurs at much lower Reynolds number than in single-phase flow, even as low as 25.

2.3.3 Prediction of Heat Transfer in Turbulent Flow

Numerous predictive methods, theoretical and empirical, have been published. These are discussed in the following text.

2.3.3.1 Analytical Models

In many condensers, annular flow prevails over much of the length of the condenser. This has prompted many attempts at analytical solutions for condensation in the annular regime. The first to do so were Carpenter and Colburn (1951). However, the agreement of their model with their own data was marginal, and it was found by other researchers to have large deviations from their data, for example, Altman et al. (1959). Soliman et al. (1968) presented an improved version of the Carpenter–Colburn model and showed it to agree with some data. No independent report of its evaluation has come to this author's notice. Calculations with this model are very tedious.

Many analyses based on a different approach have been made by many authors including Dukler (1960), Altman et al. (1959), Kosky and Staub (1971), Cavallini and Zecchin (1974), Azer et al. (1972), Razavi and Damle (1978), Traviss et al. (1973), Bae et al. (1968), etc. The starting point for all these analyses is the following two equations:

$$\frac{q}{C_{pf}\rho_f} = -(\alpha + \varepsilon_H)\frac{dT}{dy} \tag{2.3.2}$$

$$\frac{\tau}{\rho_f} = (\nu + \varepsilon_M)\frac{du}{dy} \tag{2.3.3}$$

where α is the laminar thermal diffusivity, ν is the kinematic viscosity, ε_M and ε_H are the turbulent diffusivities for heat and momentum, τ is the shear stress, and q is the heat flux. These equations were obtained by Prandtl through application of his mixing length theory. Their derivation can be seen in many textbooks including that by Knudsen and Katz (1958).

All the analyses mentioned previously apply these equations to the annular liquid film, which is assumed to be uniform around the circumference of the tube. Liquid entrainment is assumed zero, and the analogy between heat and momentum transfer is assumed to be applicable. Thus a fixed ratio between ε_M and ε_H, usually 1, is assumed. Indeed, this assumption was implicit in Prandtl's derivation of these equations as he assumed the mixing lengths for heat and momentum transfer to be equal. However, experiments have shown that they are generally not equal (Knudsen and Katz 1958).

In order to solve for dT/dy, either the velocity distribution or the eddy diffusivity must be known. Some authors assumed that the velocity distribution in the liquid film is the same as in single-phase flow and used Von Karman's universal velocity profile for smooth tubes. Others used eddy diffusivity expressions for single-phase flow. For calculating the shear stress, frictional pressure drop is calculated using some empirical correlation. Void fraction is also calculated with some empirical correlation.

With either of these two approaches, a closed form solution is not possible without further simplifications and assumptions. Most researchers proceeded to solve them numerically with computers. Dukler (1960) presented his numerical results in the form of graphs, which could be used for manual calculations. Azer et al. (1972) fitted an equation to their computer output, which fitted their own test data. However, most researchers did not try to develop easily useable equations.

Among those who have presented closed form solutions with the help of further assumptions and simplifications are Kosky and Staub (1971) and Traviss et al. (1973). The latter is easier to use.

An obvious difficulty with annular flow model analyses as well as those for other flow pattern-based correlations is that prediction of flow patterns can have significant inaccuracies as was discussed in Chapter 1. Many assumptions used in the development of these models are of limited accuracy. They assume uniform liquid layer, smooth vapor–liquid interface, and no entrainment of liquid. None of these is generally valid. Further they use empirical correlations for pressure drop and void fraction that can be in considerable error. Hence these analyses are not based on secure theoretical or physical grounds. Their reported accuracy is generally not good. For example, Razavi and Damle (1978) compared their own solution as well as several other solutions with data for water and other fluids from several sources. Their own solution came out best with mean absolute deviation (MAD) of 48%, while the Dukler solution had a MAD of 57%.

A similar though more sophisticated model was presented by Kim and Mudawar (2012). They analyzed annular flow in a rectangular minichannel. The model was solved numerically. The results were in good agreement with their data for F-72 condensing in the same channel. It was not compared with any other data.

In conclusion, the analytical solutions considered in this section are mainly of academic interest. These are not suitable for practical design and hence none of them has been given here in detail.

2.3.3.2 CFD Models

There is presently much interest in computational fluid dynamics (CFD) modeling. An example of the application of this method is the work of Da Riva et al. (2012) who performed a numerical simulation of condensation in a 1 mm diameter using the volume of fluid (VOF) method to track the interface between liquid and vapor. There are many variations of the VOF method as well as of other methods. Further there are many schemes for modeling turbulence. Del Col et al. (2015) and Kharangate and Mudawar (2017) have provided a detailed review of various CFD methods for condensation.

As was noted in Section 1.9, the CFD methodology is as yet not suitable for practical design. Therefore, a detailed discussion of this topic has not been done. Those interested can find more information in the papers referenced herein.

2.3.3.3 Empirical Correlations

Numerous empirical correlations have been published. Most of them are based only on the author's own data. Such correlations fail when compared with other data and hence cannot be relied upon. However, there are a number of correlations that have been shown to agree with data from many sources. Those are the ones that are discussed herein.

Correlations for Conventional (Macro)Channels

Cavallini et al. Correlation Cavallini et al. (2006) have given a correlation that was verified with data for refrigerants and hydrocarbons in horizontal tubes from many sources. In this correlation, there are two regimes, a ΔT-independent (or heat flux independent) regime and a ΔT-dependent (or heat flux dependent) regime. The distinction between the two regimes is made based on a dimensionless vapor velocity J_g defined as

$$J_g = \frac{xG}{[gD\rho_g(\rho_f - \rho_g)]^{0.5}} \tag{2.3.4}$$

The transition between the regimes occurs when $J_g = J_{g,t}$ given by

$$J_{g,t} = \left\{ \left[\frac{7.5}{4.3X_{tt}^{1.1} + 1} \right]^{-3} + C_T^{-3} \right\}^{-1/3} \tag{2.3.5}$$

If $J_g > J_{g,t}$, the regime is ΔT independent, else it is ΔT-dependent regime.

$C_T = 1.6$ for hydrocarbons, $=2.6$ for other fluids. X_{tt} is the Martinelli parameter defined as

$$X_{tt} = \left(\frac{\mu_f}{\mu_g}\right)^{0.1} \left(\frac{\rho_g}{\rho_f}\right)^{0.5} \left(\frac{1}{x} - 1\right)^{0.9} \tag{2.3.6}$$

In ΔT-independent regime, heat transfer coefficient is h_A given by

$$h_A = h_{LT}\left(1 + 1.28x^{0.817}\left(\frac{\rho_f}{\rho_g}\right)^{0.3685}\left(\frac{\mu_f}{\mu_g}\right)^{0.2363} \times \left(1 - \frac{\mu_f}{\mu_g}\right)^{2.144} \mathrm{Pr}_f^{-0.1}\right) \tag{2.3.7}$$

In ΔT-dependent regime,

$$h_{TP} = \left[h_A\left(\frac{J_{g,t}}{J_g}\right)^{0.8} - h_{strat}\right]\left(\frac{J_g}{J_{g,t}}\right) + h_{strat} \tag{2.3.8}$$

$$h_{strat} = \left\{1 + 0.741\left[\left(\frac{1-x}{x}\right)^{0.3321}\right]\right\}^{-1} h_{Nu} \tag{2.3.9}$$

h_{Nu} is the heat transfer coefficient for condensation on a single horizontal tube calculated by the following equation given by Nusselt (1916):

$$h_{Nu} = 0.725\left[\frac{k_f^3\rho_f(\rho_f - \rho_g)g i_{fg}}{\mu_f D \Delta T}\right]^{0.25} \tag{2.3.10}$$

h_{LT} is the heat transfer coefficient for all mass flowing as liquid. It may be calculated by the following equation given by McAdams (1954), generally known as the Dittus–Boelter equation:

$$h_{LT} = 0.023\left(\frac{GD}{\mu_f}\right)^{0.8} \mathrm{Pr}_f^{0.4}\frac{k_f}{D} \tag{2.3.11}$$

This correlation was validated with data that included tube diameters from 3.1 to 17 mm, several fluids, and mass velocity from 18 to 1002 $\mathrm{kg\,m^{-2}s^{-1}}$. As ΔT appears on the right-hand side of Eq. (2.3.10), iterative calculations are required with assumed ΔT. Kondo and Hrnjak (2012a) found that this correlation overpredicted their data for R-410A and carbon dioxide at $p_r > 0.7$, the deviations increasing as the critical pressure was approached. They proposed a modification of Cavallini et al. correlation that showed good agreement with their own data up to the critical pressure.

Shah Correlations A widely used correlation is Shah (1979), which is expressed by the following equations:

$$\psi = \frac{h_{\mathrm{TP}}}{h_{\mathrm{LS}}} = 1 + \frac{3.8}{Z^{0.95}} \tag{2.3.12}$$

where

$$Z = \left(\frac{1}{x} - 1\right)^{0.8} p_r^{0.4} \tag{2.3.13}$$

where p_r is the reduced pressure. h_{LO} is the heat transfer coefficient of the liquid phase flowing alone. It is calculated by the following equation:

$$h_{\mathrm{LS}} = h_{\mathrm{LT}}(1 - x)^{0.8} \tag{2.3.14}$$

h_{LT} is calculated with Eq. (2.3.11). Equations (2.3.11)–(2.3.14) may be combined to give the correlation in the following compact form:

$$\frac{h_{\mathrm{TP}}}{h_{\mathrm{LT}}} = (1 - x)^{0.8} + \frac{3.8x^{0.76}(1 - x)^{0.04}}{p_r^{0.38}} \tag{2.3.15}$$

This correlation was verified with data for water, refrigerants, and many chemicals in horizontal and vertical tubes with downflow. The range of data included tube diameters 7–40 mm, reduced pressures up to 0.44, and mass flux 11–1600 kg m^{-2} s^{-1}. It was realized that this correlation is likely to fail at low flow rates and higher reduced pressures. Shah (1981) recommended its use only if all the following three conditions are met: $U_{\mathrm{GT}} > 3\ \mathrm{m\,s^{-1}}$, $Re_{\mathrm{LT}} > 350$, and $Re_{\mathrm{GT}} > 35\,000$.

This correlation has been widely used and continues to be used even though the author has published improved versions to extend its range of applicability. Caution should be exercised to not use it beyond its recommended range, especially not for $p_r > 0.44$.

Shah (2009, 2013) presented a modified version of this correlation to extend it to higher reduced pressures and lower flow rates. In this correlation, there are three heat transfer regimes called Regime I, Regime II, and Regime III. The following two equations for heat transfer are used:

$$h_I = h_{\mathrm{LS}}\left(1 + \frac{3.8}{Z^{0.95}}\right)\left(\frac{\mu_L}{14\mu_G}\right)^{(0.0058+0.557p_r)} \tag{2.3.16}$$

$$h_{Nu} = 1.32Re_{\mathrm{LO}}^{-1/3}\left[\frac{\rho_L(\rho_L - \rho_G)gk_L{}^3}{\mu_L{}^2}\right]^{1/3} \tag{2.3.17}$$

Equation (2.3.16) is a modification of the Shah (1979) formula, done to extend it to high reduced pressures. At $p_r \leq 0.4$, it essentially equals the 1979 formula. It represents heat transfer exclusively by forced convection. Equation (2.3.17) is Nusselt's equation for condensation in a vertical tube, Eq. (2.2.11), multiplied by 1.2 as recommended by McAdams (1954). This equation represents heat transfer controlled by heat flux alone.

Equations (2.3.16) and (2.3.17) are used to calculate heat transfer coefficients in the three regimes as follows:

In Regime I,

$$h_{\mathrm{TP}} = h_{\mathrm{I}} \tag{2.3.18}$$

In Regime II,

$$h_{\mathrm{TP}} = h_I + h_{Nu} \tag{2.3.19}$$

In Regime III,

$$h_{\mathrm{TP}} = h_{Nu} \tag{2.3.20}$$

The boundaries of these three heat transfer regimes are defined as follows:

Horizontal tubes:

Regime I occurs when

$$J_g \geq 0.98(Z + 0.263)^{-0.62} \tag{2.3.21}$$

Regime III occurs when

$$J_g \leq 0.95(1.254 + 2.27Z^{1.249})^{-1} \tag{2.3.22}$$

If neither of the above conditions is satisfied, it is Regime II.

J_g is the dimensionless vapor velocity defined by Eq. (2.3.4).

Equation (2.3.22) for the boundary of Regime III was given in Shah (2013), and the rest of the correlation was given in Shah (2009).

Vertical downflow:

Regime I occurs when

$$J_g \geq \frac{1}{2.4Z + 0.73} \tag{2.3.23}$$

Regime III prevails when

$$J_g \leq 0.89 - 0.93\exp(-0.087Z^{-1.17}) \tag{2.3.24}$$

If the Regime is not determined to be I or III by Eqs. (2.3.23) and (2.3.24), it is Regime II.

Figure 2.3.1 shows the boundaries of the regimes given by Eqs. (2.3.21) to (2.3.24). It is seen that for most values of Z and J_g, the area between the curves for Regimes I and III is much smaller for vertical tubes than for horizontal tubes. This is the area in which Regime II occurs. Heat transfer coefficients in Regime II are always higher than those in Regime I or III as it is obtained by adding the equations for those two regimes. Thus, the Shah correlation indicates that heat transfer in vertical tubes will be usually lower than in horizontal tubes.

This correlation was verified with a very wide range of data. The 1736 data points for 24 fluids from 51 studies were predicted with MAD of 16.1%. The data included both horizontal and vertical tubes, while the correlations mentioned earlier were validated only with horizontal tube data. The reduced pressure range is 0.0008–0.945. This is the most

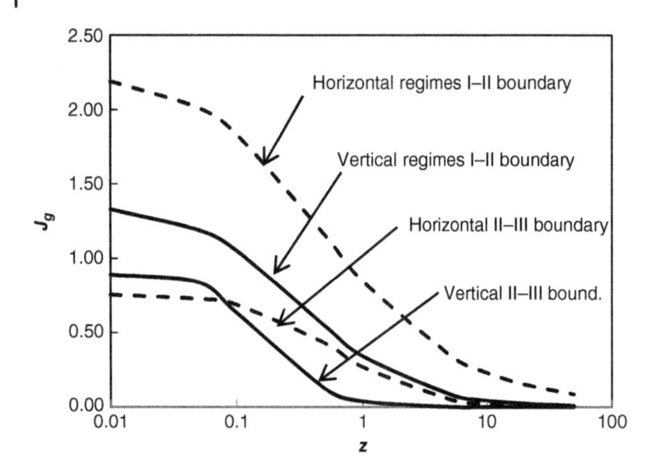

Figure 2.3.1 Boundaries of heat transfer regimes for vertical and horizontal tubes in the Shah correlation (2009, 2013). Source: Reprinted from Shah (2016a). © 2016, with permission from Elsevier.

verified correlation for macro channels and the only one that distinguishes between horizontal and vertical tubes. It shows that heat transfer coefficients for the two orientations can be different for the same pressure and flow rate. This is clear from Figure 2.3.1 as the regimes for horizontal and vertical orientations are seen to differ considerably.

For calculations in the heat flux-dependent range, this correlation uses the Nusselt equation for flow inside vertical tubes in terms of liquid Reynolds number. This has the advantage that heat flux (or ΔT) are taken into account implicitly and hence iterations with assumed values of heat flux or ΔT are not required as is the case with the Cavallini et al. correlation and others using the Nusselt formula for condensation outside tube in terms of ΔT. This makes design calculations easier and also enables analysis of test data for which ΔT is not given, as is usually the case.

It is of interest to know whether the heat transfer regimes in the Shah correlation have any physical interpretation. Shah (2014) compared the heat transfer regimes predicted by Shah (2009, 2013) for his database with the flow pattern map given by El Hajal et al. (2003). The only data for water was that of Varma (1977) for water in a 49 mm diameter tube. The El Hajal et al. map is recommended only for diameters ≤21.4 mm, and it was not validated with water data. The map of Baker, discussed in Chapter 1, was validated for diameters from 25 to 100 mm. It was therefore used for Varma data. The comparison showed the following:

- Heat transfer Regime I corresponded to the intermittent, annular, and mist flow patterns.
- Heat transfer Regime II corresponded to stratified-wavy flow pattern.
- Heat transfer Regime III corresponded to stratified flow pattern.

In view of these findings, Shah (2014) gave the following flow pattern-based correlation:

If flow pattern is intermittent, annular, or mist,

$$h_{\mathrm{TP}} = h_I \tag{2.3.25}$$

If flow pattern is stratified-wavy,

$$h_{\mathrm{TP}} = h_I + h_{Nu} \tag{2.3.26}$$

If flow pattern is stratified,

$$h_{\mathrm{TP}} = h_{Nu} \tag{2.3.27}$$

To determine flow patterns, it is recommended to use the Baker map for low pressure water and the El Hajal et al. map for other fluids. Applicability of other flow pattern maps is unknown. The MAD of the database with this correlation was only slightly higher than of the Shah (2013) correlation. Figure 2.3.2 shows the comparison of this and some other correlations with test data.

Some Other Correlations A few other well-known correlations are discussed herein.

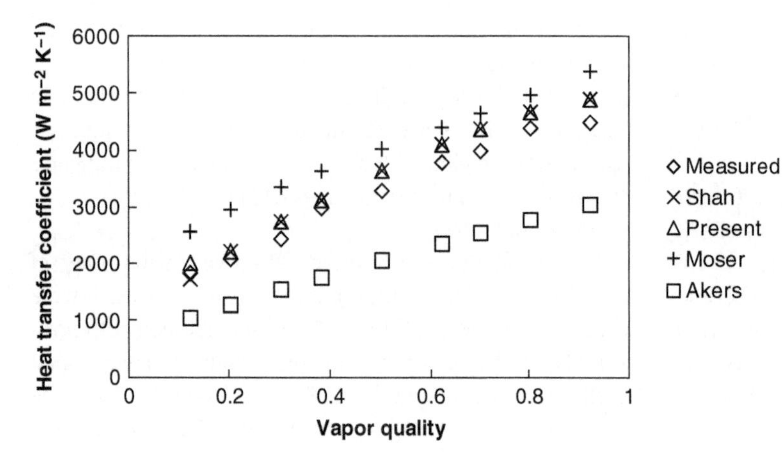

Figure 2.3.2 Comparison of the data of Jung et al. (2003) for R-142b with various correlations. $D = 8$ mm, $T_{\mathrm{SAT}} = 40\,°\mathrm{C}$, and $G = 300\ \mathrm{kg\,m^{-2}s^{-1}}$. "Present" is Shah (2014) flow pattern-based correlation and "Shah" is Shah (2013) correlation. Source: Reprinted from Shah (2014). ©2014, with permission from Begell House, Inc.

A well-known correlation is the following by Ananiev et al. (1961):

$$h_{TP} = h_{LT}\left(\frac{\rho_f}{\rho_m}\right)^{0.5} \tag{2.3.28}$$

where ρ_m is the density of vapor–liquid mixture calculated by the homogeneous model as

$$\rho_m = \frac{\rho_f \rho_g}{\rho_g + x(\rho_f - \rho_g)} \tag{2.3.29}$$

This correlation was originally based on water data but has been found by other researchers to work well with other fluids also. As it does not include heat flux, it fails in heat flux-dependent regime. Hence it should be used only at higher flow rates.

Thome et al. (2003) and Dobson and Chato (1998) have given flow pattern-based correlations using data for horizontal tubes from many sources. These also use the Nusselt formula Eq. (2.3.10) for condensation on horizontal tube for predicting heat transfer in heat flux affected regimes and hence iterative calculations are required. As they require prediction of flow patterns, their accuracy is limited by the accuracy of flow pattern prediction that is far from perfect as discussed in Chapter 1.

Moser et al. (1998) developed an equivalent Reynolds number-based semi-theoretical correlation that was shown to agree with data from several sources. Shah (2016c) found it to be generally inaccurate.

A correlation that is widely quoted in literature is that of Akers et al. (1959). Most authors, such as Shah (2016c), have found it to be very inaccurate.

Correlations for Minichannels There has been a tremendous amount of research on condensation in minichannels in recent years. Many empirical or semiempirical correlations have been published. Correlations for mini/micro channels were reviewed by Awad et al. (2014) and Del Col et al. (2015). Comparatively recent correlations for minichannels include those by Bohdal et al. (2011), Huang et al. (2010), Park et al. (2011), Jige et al. (2016), Rahman et al. (2018), and Keinath and Garimella (2018). These are based on limited databases and therefore cannot be expected to have general applicability. The one that has had extensive verification is that of Kim and Mudawar (2013). It is presented as follows:

For annular flow (smooth annular, wavy-annular, transition) where $We^* > 7X_{tt}^2$,

$$\frac{h_{TP}D}{k_f} = 0.048Re_{LO}^{0.69}\mathrm{Pr}_f^{0.34}\frac{\phi_g}{X_{tt}} \tag{2.3.30}$$

For slug and bubbly flow where $We^* < 7X_{tt}^2$,

$$\frac{h_{TP}D}{k_f} = \left[\left(0.048Re_{LO}^{0.69}\mathrm{Pr}_f^{0.34}\frac{\phi_g}{X_{tt}}\right)^2 + (3.2\times10^{-7}Re_{LO}^{-0.38}Su_g^{1.39})^2\right]^{0.5} \tag{2.3.31}$$

$$\phi_g^2 = 1 + CX + +X^2 \tag{2.3.32}$$

$$X^2 = \frac{(dp/dz)_f}{(dp/dz)_g} \tag{2.3.33}$$

$$-\left(\frac{dp}{dz}\right)_f = \frac{2f_f G^2(1-x)^2}{\rho_f D} \tag{2.3.34}$$

$$-\left(\frac{dp}{dz}\right)_g = \frac{2f_g G^2 x^2}{\rho_{fg} D} \tag{2.3.35}$$

For $Re_k < 2000$,

$$f_k = 16Re_k^{-1} \tag{2.3.36}$$

For $Re_k = 2000$–$20\,000$,

$$f_k = 0.079Re_k^{-0.25} \tag{2.3.37}$$

For $Re_k > 20\,000$,

$$f_k = 0.046Re_k^{-0.2} \tag{2.3.38}$$

For laminar flow in rectangular channel,

$$f_k Re_k = 24(1 - 1.3553\beta + 1.9467\beta^2 - 1.7012\beta^3 + 0.9564\beta^4 - 0.2537\beta^5) \tag{2.3.39}$$

β is the aspect ratio of channel; use its reciprocal if $\beta > 1$.

The subscript k denotes for liquid or vapor.

For $Re_{LO} \geq 2000$, $Re_{GO} \geq 2000$,

$$C = 0.39Re_{LT}^{0.03}Su_g^{0.1}\left(\frac{\rho_f}{\rho_g}\right)^{0.35} \tag{2.3.40}$$

For $Re_{LO} \geq 2000$, $Re_{GO} < 2000$,

$$C = 0.000\,87Re_{LT}^{0.17}Su_g^{0.5}\left(\frac{\rho_f}{\rho_g}\right)^{0.14} \tag{2.3.41}$$

For $Re_{LO} < 2000$, $Re_{GO} \geq 2000$,

$$C = 0.0015Re_{LT}^{0.59}Su_g^{0.19}\left(\frac{\rho_f}{\rho_g}\right)^{0.36} \tag{2.3.42}$$

For $Re_{LO} < 2000$, $Re_{GO} < 2000$,

$$C = 0.000\,035Re_{LT}^{0.44}Su_g^{0.5}\left(\frac{\rho_f}{\rho_g}\right)^{0.48} \tag{2.3.43}$$

In Eqs. (2.3.30)–(2.3.43), D_{HYD} is used in place of D for noncircular channels. The modified Weber number We^* is according to Soliman (1986) given as follows:

For $Re_{LO} < 1250$,

$$We^* = 2.45 \frac{Re_{GO}^{0.64}}{Su_g^{0.3}(1 + 1.09X_{tt}^{0.039})^{0.4}} \tag{2.3.44}$$

For $Re_{LO} > 1250$,

$$We^* = 0.85 \frac{Re_{GO}^{0.79} X_{tt}^{0.157}}{Su_g^{0.3}(1 + 1.09X_{tt}^{0.039})^{0.4}} \left[\left(\frac{\mu_g}{\mu_f} \right)^2 \left(\frac{\rho_f}{\rho_g} \right) \right]^{0.084} \tag{2.3.45}$$

Su_g is the Sugomel number defined as

$$Su_g = \frac{\rho_g \sigma D}{\mu_g^2} \tag{2.3.46}$$

For channels heated only on three sides, heat transfer coefficient calculated by the foregoing is corrected as follows:

$$h_{TP} = \left(\frac{Nu_3}{Nu_4} \right) h_{TP,circular} \tag{2.3.47}$$

where $h_{TP,circular}$ is the heat transfer coefficient for a circular tube calculated by the foregoing equations:

$$Nu_3 = (8.235(1 - 1.833\beta + 3.767\beta^2 - 5.814\beta^3 + 5.361\beta^4 - 2.0\beta^5)) \tag{2.3.48}$$

$$Nu_4 = (8.235(1 - 2.042\beta + 33.085\beta^2 - 2.477\beta^3 + 1.058\beta^4 - 0.186\beta^5)) \tag{2.3.49}$$

This correlation was compared to a database that consisted of single-channel and multi-channel data, 17 different working fluids (refrigerants, chemicals, CO_2), hydraulic diameters from 0.424 to 6.22 mm, mass velocities from 53 to 1403 kg m^{-2} s^{-1}, Re_{LO} from 276 to 89 798, qualities from 0 to 1, and reduced pressures from 0.04 to 0.91, and the MAD was 16.1%. However, Garimella et al. (2014) found it to be in poor agreement with their database for minichannels.

Garimella et al. (2014) developed a flow pattern-based predictive model that gave good agreement with their own data collected over many years, which included many fluids over a wide range of pressures and flow rate in channels of hydraulic diameters from 0.1 to 15 mm.

2.3.3.4 Correlations Applicable to Both Macro and Minichannels

Dorao and Fernandino (2018) have given the following correlation:

$$h_1 = 0.023(Re_{LO} + Re_{GO})^{0.8} Pr_{TP}^{0.3} k_f / D \tag{2.3.50}$$

$$h_2 = 41.5D^{0.6}(Re_{LO} + Re_{GO})^{0.4} Pr_{TP}^{0.3} k_f / D \tag{2.3.51}$$

$$h_{TP} = (h_1^9 + h_2^9)^{1/9} \tag{2.3.52}$$

$$Pr_{TP} = Pr_f(1 - x) + Pr_g x \tag{2.3.53}$$

Equation (2.3.51) is dimensional; D used in it should be in meter. For noncircular channels, D is to be replaced by hydraulic equivalent diameter. They compared this correlation with a very wide range of data from many sources that included many fluids, hydraulic diameters 0.067–20 mm, and round and noncircular shapes. Good agreement with data was reported. The simplicity of this correlation is remarkable.

Shah Correlation Shah (2016b) compared his correlation for macro channels, Shah (2009, 2013), with data for minichannels ($D \leq 3$ mm). It was found that while most data were correctly predicted, some of the data at low mass flux were being considerably underpredicted. Examination of results showed that in such cases, the correlation predicted Regime I while the data showed that good agreement will be obtained if Regime II were used. Visual studies show that the flow patterns are affected as flow passage becomes smaller. For example, Garimella et al. (2001) found that the boundary between intermittent and non-intermittent flow changes as hydraulic diameter varies from 4 mm down to 0.5 mm. This suggested that the boundary between the two regimes at low flow rates was being modified by surface tension effects. At high mass flow rates, inertia force dominates the surface tension force, while at low flow rates, the surface tension force dominates. Weber number is the ratio between surface tension force and inertia force. It was therefore felt that this behavior may be quantified by its use. For all mass flowing as vapor, it is defined as

$$We_{GT} = \frac{G^2 D}{\rho_G \sigma} \tag{2.3.54}$$

It was found that the change from Regime I to II occurs when $We_{GT} < 100$. With this modification, good agreement with data was obtained. Further, it was found that deviations become smaller if Eq. (2.3.16) in the Shah (2009) correlation is replaced by Eq. (2.3.7), which is the formula of Cavallini et al. (2006) for heat flux-independent regime.

In Shah (2016c), the correlation was compared to an extensive database that included both mini and macro channels. It was found that the use of Eq. (2.3.7) improved the agreement for horizontal minichannels, but for vertical minichannels and for macro channels in both orientations, Eq. (2.3.16) gave better agreement. Further data analysis and modifications were done in Shah (2019a). The modifications included the effects of Froude number and behavior of hydrocarbons. It was found that heat transfer regimes for hydrocarbons are different from those for other fluids. As noted earlier, this is also the case in the Cavallini et al. (2006) correlation. The reason for this behavior is not known. Data analysis showed that when $Fr_{LT} < 0.012$, Regimes I and III become Regime II, resulting in higher heat transfer coefficients. This appears to be due to stratification in which condensate flows at the bottom tube, and the liquid film in the upper circumference is very thin, resulting in higher heat transfer. In developing his correlation for boiling heat transfer, Shah (1976) found stratification to occur at $Fr_{LT} < 0.04$. Shah (2019a) reports that several data sets for condensation also indicated 0.04 as the transition value but deviations were minimized with transition at $Fr_{LT} = 0.012$.

The final version of the correlation is as follows.

Heat transfer regimes are the same as in Shah (2013) correlation for both horizontal and vertical downflows if any of the following conditions are applicable:

- Fluid is a hydrocarbon; Regime is I and $p_r < 0.4$.
- Fluid is hydrocarbon and Regime is III.
- $Re_{LT} < 100$.

If any of the mentioned conditions is fulfilled, the Shah (2013) correlation is to be used.

If none of the mentioned conditions is applicable, heat transfer regimes are determined as follows:

For horizontal flow:

Regime I occurs when $We_{GT} > 100$ and $Fr_{LT} > 0.012$ and

$$J_g \geq 0.98(Z + 0.263)^{-0.62} \tag{2.3.55}$$

Regime III occurs if $Fr_{LT} > 0.012$ and

$$J_g \leq 0.95(1.254 + 2.27Z^{1.249})^{-1} \tag{2.3.56}$$

If it is not Regime I or III, it is Regime II.

For vertical downflow:

Regime I occurs when $We_{GT} > 100$ and

$$J_g \geq \frac{1}{2.4Z + 0.73} \tag{2.3.57}$$

Regime III occurs when

$$J_g \leq 0.89 - 0.93 \exp(-0.087Z^{-1.17}) \tag{2.3.58}$$

If it is not Regime I or III according to Eqs. (2.3.57) and (2.3.58), it is Regime II.

There are two alternative correlations. Correlation 1 uses Eq. (2.3.7) for Regime I. Correlation 2 uses Eq. (2.3.16) for Regime I. The two are the same in all other respects. Equations (2.3.17)–(2.3.20) remain applicable. Correlation 1 is recommended for horizontal flow when $D \leq 3$ mm. Correlation 2 is recommended for horizontal channels with $D > 3$ mm and for vertical channels of all diameters. For noncircular channels, D_{HYD} is to be used as equivalent diameter in Weber and Froude numbers and D_{HP} in all other places. D_{HP} is defined as

$$D_{HP} = \frac{4 \times \text{flow area}}{\text{perimeter with heat transfer}} \tag{2.3.59}$$

Table 2.3.1 Range of data with which Shah (2019a) was verified.

Parameter	**Data range**
Fluids	Water, R-11, R-12, R-22, R-32, R-41, R-113, R-123, R-125, R-134a, R-141b, R-142b, R-152a, R-161, R-236ea, R-245fa, R-404A, R-410A, R-502, R-507, R-1234fa, R-1234yf, R-1234ze(E), DME, butane, propane, carbon dioxide, methane, FC-72, isobutane, propylene, benzene, ethanol, methanol, toluene, Dowtherm 209, HFE-7100, ethane, pentane, HFE-7000, Novec 649, and ammonia (43 fluids)
Geometry	Round, square, rectangle, semi-circle, triangle, and barrel shaped, single, and multi channels. All sides cooled or one side insulated
Orientation	Horizontal, vertical down
Aspect ratio, width/height	0.14–13.9
D_{HYD} (mm)	0.08–49.0
Reduced pressure	0.0008–0.946
G (kg m^{-2} s^{-1})	1.1–1400
x	0.01–0.99
We_{GT}	0.15–79 060
Fr_{LT} for horizontal channels	0.00 035–133
Bond number	0.033–2392
Number of data sources	88 (75 Horizontal, 13 vertical down)
Number of data sets	182 (161 horizontal, 21 vertical down)

Source: From Shah (2019a). © 2019 Elsevier.

This correlation was compared to an extensive database, whose range of parameters is given in Table 2.3.1. It includes 43 fluids, diameters 0.0.08–49.0 mm, reduced pressures 0.0008–0.946, mass flux from 1.1 to 1400 kg m^{-2} s^{-1}, channels of various shapes (round, rectangular, triangular, etc.) partially or fully cooled, and horizontal and vertical downflow. There are 182 data sets from 88 sources. The same data were also compared to the following correlations: Kim and Mudawar (2013), Ananiev et al. (1961), Moser et al. (1998), Akers et al. (1959), and Dorao and Fernandino (2018).

Considering all data, the Shah correlations 1 and 2 have MAD of 19.0% and 18.6%, respectively. Using the two in their recommended ranges, the MAD of the Shah correlation is 18.2%. The other correlations have MAD from 26.4% to 68.0%.

The deviations of various correlations in different ranges are listed in Table 2.3.2. It is seen that the deviations of all correlations except Shah (2019a) are much higher for

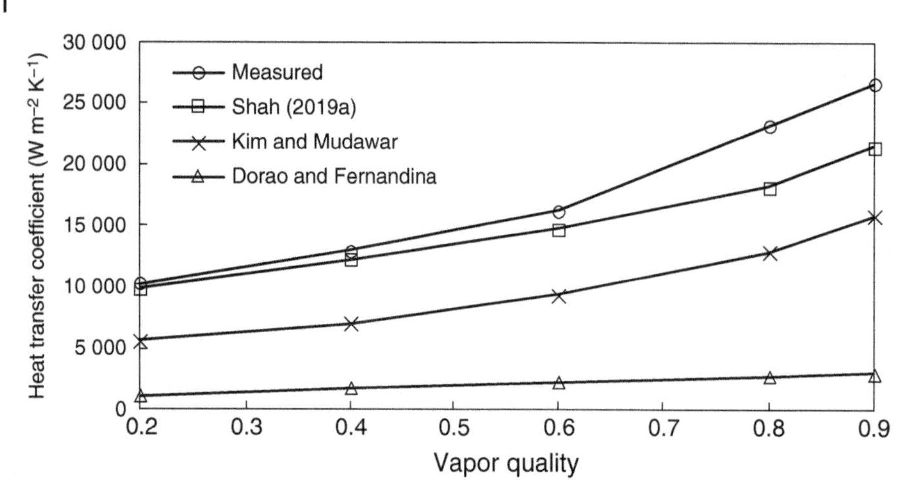

Figure 2.3.3 Comparison of correlations with the data of Wang and Du (2000) for water in a horizontal tube. $D = 4.98$ mm, $T_{SAT} = 105\,°C$, $G = 14.1\ kg\,m^{-2}\,s^{-1}$, and $We_{GT} = 24$. Source: From Shah (2019a). © 2019 Elsevier.

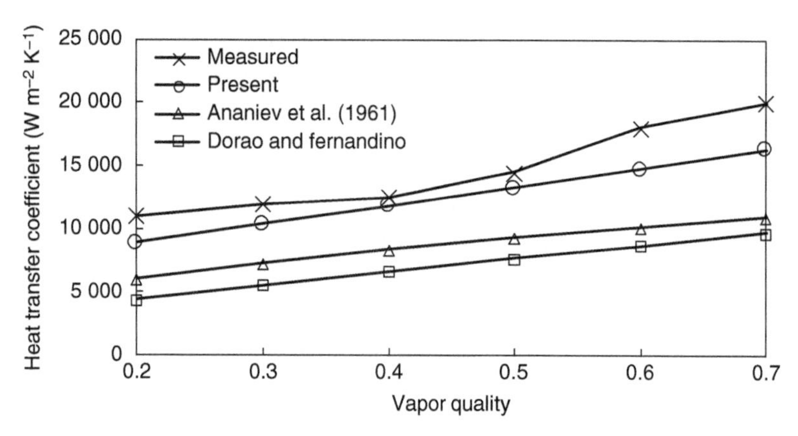

Figure 2.3.4 Data of Fronk and Garimella (2016) for ammonia in a 1.44 mm diameter horizontal tube compared to various correlations. $T_{SAT} = 40\,°C$, $G = 75\ kg\,m^{-2}\,s^{-1}$, and $We_{GT} = 39$. "Present" is Shah (2019a) correlation. Source: From Shah (2019a). © 2019 Elsevier.

$We_{GT} < 100$ than for $We_{GT} > 100$. This is true for $D \leq 3$ mm as well as $D > 3$ mm. For example, the MAD of Dorao and Fernandino correlation for $D \leq 3$ mm is 39.3% at $We_{GT} < 100$ and 19.9% for $We_{GT} > 100$. For $D > 3$ mm, the deviations of Dorao and Fernadino correlation are 56.5% and 24.1% for $We_{GT} < 100$ and $We_{GT} > 100$, respectively. The data at the lower Weber number are underpredicted. Results with other correlations are similar. This indicates that the effects of surface tension become important at $We_{GT} < 100$ for all tube diameters, not only for $D \leq 3$ mm that are usually considered to be the one affected by surface tension.

Figures 2.3.3–2.3.5 show the comparison of data for various fluids and parameters with various correlations. These show the values and trends of heat transfer coefficients for many conditions as well as the accuracy of various correlations.

While the Kim and Mudawar (2013) correlation was stated to be valid up to 6.2 mm diameter, it was found to be inaccurate when diameter exceeded 3 mm.

In view of the results presented earlier, it is clear that the Shah (2019a) is the only correlation that correctly predicts data at all Weber numbers. The Dorao and Fernandino and Kim and Mudawar correlations can only be used at $We_{GT} > 100$. The latter is further limited to minichannels.

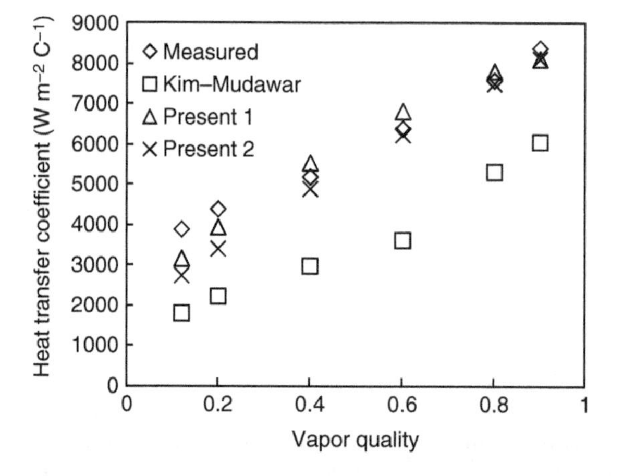

Figure 2.3.5 Comparison of the data of Son and Lee (2009) for R-134a with correlations. Present 1 and 2 are correlations of Shah (2016c). $D = 3.36$ mm, $T_{SAT} = 40\,°C$, $G = 400\ kg\,m^{-2}\,s^{-1}$, and $We_{GT} = 1757$. Source: From Shah (2016c). © 2016 Elsevier.

The Akers et al. correlation shows poor agreement with most data. The Ananiev correlation is suitable only for macro channels in the heat flux independent regime.

As seen in Table 2.3.1, the Shah correlation is in good agreement with data for noncircular channels of many

Table 2.3.2 Deviations of various correlations.

Orientation	Diameter (mm)	We_{GT}	Re_{LT}	Fr_{LT}	Fluid	N	Deviation, % Mean Absolute Average						
							Shah (2009, 2013)	Shah (2019a)	Kim and Mudawar (2013)	Ananiev et al. (1961)	Moser et al. (1998)	Akers et al. (1959)	Dorao and Fernandez
Horizontal	≤3	<100	All	All	All	479	33.8	22.3	29.6	38.0	39.4	191.3	39.3
							−23.6	3.0	−13.2	−33.2	5.6	190.6	−35.5
	≤3	>100				1450	22.6	20.3	19.5	19.7	39.4	111.9	19.9
							9.6	5.3	−6.3	−5.4	31.9	110.1	−4.4
	≤3	All				1929	25.3	20.5	22.0	24.2	39.4	131.6	24.7
							1.4	4.7	−8.1	−12.1	25.4	130.4	−12.1
	>3	<100				184	30.1	26.4	72.6	54.2	38.1	34.2	56.5
							−7.8	−6.0	25.6	−54.2	−35.2	12.9	−56.5
	>3	>100				2517	16.7	16.1	28.3	24.1	32.7	27.8	24.1
							2.3	3.2	−22.6	−13.8	17.3	−5.6	−16.0
	>3	All				2701	17.6	16.4	31.3	26.2	33.1	28.3	26.3
							1.6	4.1	−19.23	−16.6	13.7	−4.3	−18.8
	All	All		≤0.012		39	50.1	18.2	83.1	65.6	56.2	39.5	69.1
							−50.1	−13.5	−14.4	−65.6	−56.2	39.3	−69.1
	All	All		All		4630	20.8	18.5	27.4	25.4	35,7	71.3	25.6
							1.5	3.1	−14.5	−14.7	18.6	51.8	−15.9
Vertical downflow	≤3	<100	All	All		67	31.0	19.6	32.1	39.3	21.6	115.7	36.1
							−24.9	1.5	−24.7	−36.8	−9.5	115.7	−33.9
	≤3	>100				51	16.4	19.0	24.3	28.6	14.2	76.4	24.9
							−14.6	−18.5	−24.7	−28.6	−2.7	76.4	−24.9
	<3	All				118	24.7	19.5	28.7	34.6	18.4	98.7	31.3
							−20.4	−6.0	−24.2	−33.2	−6.6	98.7	−30.0
	>3	<100				56	20.7	20.5	52.9	51.1	57.9	42.2	76.8
							0.1	2.5	−12.7	−50.0	−54.3	−34.2	−76.8
	>3	>100				296	14.3	16.0	30.4	34.2	25.8	24.5	39.0
							−0.5	1.0	6.7	24.7	15.6	−16.1	−36.9
	>3	All				352	15.3	16.4	34.0	36.9	30.9	27.3	45.0
							−0.4	2.2	3.6	10.3	4.5	−19.0	−43.2
	All	All				470	17.7	17.5	32.7	36.3	27.8	45.2	41.6
							−5.5	−0.9	−3.4	−0.6	1.7	10.5	−39.9

(Continued)

Table 2.3.2 (Continued)

Orientation	Diameter (mm)	We_{GT}	Re_{LT}	Fr_{LT}	Fluid	N	Deviation, % Mean Absolute Average						
							Shah (2009, 2013)	Shah (2019a)	Kim and Mudawar (2013)	Ananiev et al. (1961)	Moser et al. (1998)	Akers et al. (1959)	Dorao and Fernandez
Horizontal and Vertical	All	<100		All		786	31.8	23.1	41.5	42.8	38.9	137.4	45.7
							−18.3	0.7	−5.1	−39.0	−9.5	126.6	−43.2
		<100	All	All	HCN	56	24.8	18.5	22.5	20.9	41.5	229.8	24.0
							14.9	8.3	9.8	−13.0	25.8	227.2	−9.4
		>100			All	4314	18.5	18.0	25.4	23.4	34.3	56.4	23.7
							4.4	2.3	−15.1	−8.2	21.9	33.7	−13.7
	All	All	≤100	All	All	52	23.9	20.1	39.2	29.8	92.2	568.2	24.7
							6.2	7.5	23.4	9.1	85.8	562.8	−13.2
	All	All	All	All	All	5100	20.5	18.2	27.9	26.4	35.0	68.9	27.1
							0.9	1.7	−13.6	−13.4	17.0	4.8	−18.2

Source: From Shah (2019a). © 2019 Elsevier.

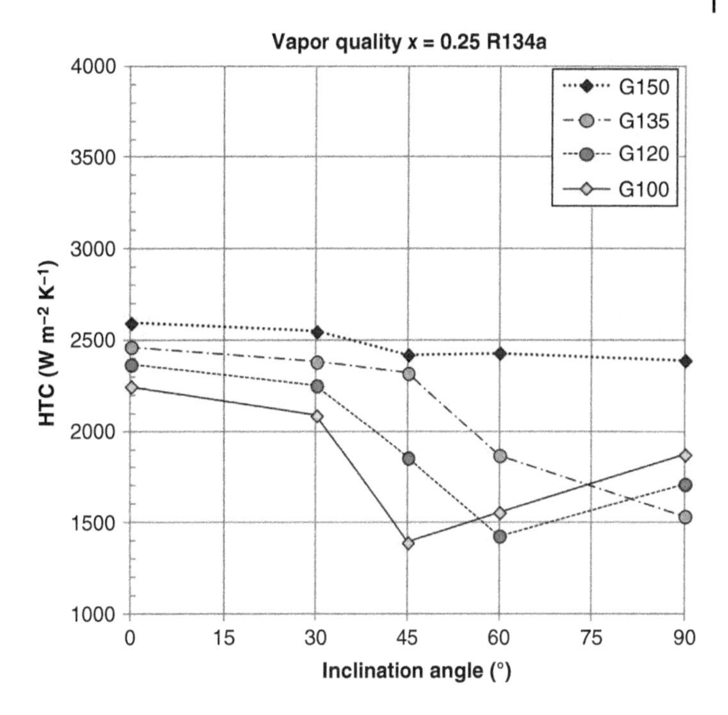

Figure 2.3.6 Effect of inclination on heat transfer in a square channel with $D_{HYD} = 1.25$ mm. R-134a with $T_{SAT} = 40$ °C. Source: From Del Col et al. (2014). © 2014 Elsevier.

shapes. The theoretical analyses by Wang and Rose (2005, 2006, 2011) showed heat transfer in noncircular channels to be higher than in round tubes. This was attributed to condensate accumulating in corners, thus thinning the liquid film on the channel surface, which reduced its thermal resistance. These analyses assumed laminar film. As discussed in Section 2.3.2, the liquid film becomes turbulent due to the effect of vapor shear. The liquid Reynolds number at which this occurs can be as low as 25. Hence the assumption of laminar film is applicable to only a very small length of the channel. This explains why correlations for round tubes also give good agreement with data for noncircular channels. This has been further discussed in Shah (2019b).

Inclined Channels Many applications involve condensation inside inclined tubes. By inclined tubes is meant tubes at angles other than 0° (horizontal) and − 90° (vertical downwards). In the foregoing, the flow orientations considered have been horizontal and vertical downwards. The possible range of angles of inclination is from −90° (vertical downwards) to +90° (vertical upwards). Examples are air-cooled condensers for air conditioning and refrigeration, air-cooled condensers for power plants, condensers for accident mitigation in advanced nuclear reactors, and thermosiphons. Thermosiphons have liquid and vapor flowing in opposite directions; these are not considered here but in Section 2.7.10.

Research on condensation in inclined channels was reviewed by Lips and Meyer (2011) and Shah (2016a). Table 2.3.3 lists the salient features and range of parameters in various studies on inclined channels. These are discussed in the following.

All studies showed that effect of inclination decreased with increasing mass flow rates. Figure 2.3.6 shows a typical case. It is seen that at G of 150 kg m^{-2} s^{-1}, there is negligible effect of inclination while it has considerable effect at lower mass flux. The values of mass flux at which the effect of inclination ends found by different researchers differ. In the studies in which measurements in vertical downflow were included, heat transfer coefficient in this orientation were found to be lowest. The only exception is Wurfel et al. (2003) who found it to increase continuously from 0° to −90°. The results for other orientations vary greatly as seen in Table 2.3.3. While some studies show increase in heat transfer coefficient with increasing downward inclination, others show it to decrease. Similar is the situation for upward inclinations. Most remarkable is the difference between the results of Mohseni et al. (2013) and Meyer et al. (2014). The two studies had the same tube diameter, the same fluid, and overlapping range of saturation temperature and mass flux. While Lips and Meyer report the inclination for highest heat transfer as −30°, Mohseni et al. had highest heat transfer at +30° inclination. This discrepancy cannot be explained by any mechanism.

Many prediction methods, theoretical and empirical have been published for calculating the effect of inclination on heat transfer coefficients. These are discussed in the following text.

Chato (1960) developed an analytical model that requires numerical solution of three differential equations. It agreed

Table 2.3.3 Salient features of studies on inclined tubes.

Source	Diameter (mm)	Inclination (°)	Fluid	T_{SAT} (°C)	p_r	G (kg m^{-2} s^{-1})	Re_{LT}	Re_{GT}	Inclination for maximum h	Maximum G for effect of inclination, kg m^{-2} s^{-1}	Data analyzable?
Chato (1960)	14.5	−3	R-113						−10	Not known	No
		0									
Meyer et al. (2014)	8.34	−90	R-134a	30	0.189	100	5156	6.7E4	−30 to −15	300	Yes
		+90		50	0.325	400	11 742	1.3E5			
Mohseni et al. (2013)	8.34	−90	R-134a	35	0.218	53	2577	3.7E4	+30	Not known	Yes
		+90				170	8267	1.2E5			
Del Col et al. (2014)	1.23	−90	R-134a	40	0.2494	100	760	9947	+60	150	Yes
		+90				390	2966	3.9E4			
		−90	R-32	40	0.427	100	1292	8902	0	200	Yes
		0				390	5041	34 717			
Wang and Du (2000)	1.94,	−45	Water	105	0.0065	28.8	208	4500	−45	Not known	Yes
		0				65.1	680	15 000			
	2.8					19.9	208	4500	−45	Not known	Yes
							680	15 000			
	3.95					14.1	208	4500	0	Not known	Yes
							680	15 000			
	4.98					11.2	208	4500	0	Not known	Yes
							680	15 000			
Lyulin et al. (2011)	4.8	−90	Ethanol	58	0.068	0.28	2.2	137	−30	Not known	No
		0				2.44	19.4	1195	−40		
Tepe and Mueller (1947)	18.5	−15	Methyl alcohol	64.4	0.0122	16.3	920	27 000	Not known	Not known	Yes
						30.5	1726	52 000			
			Benzene	80.8	0.021	24.9	1527	52 000	Not known	Not known	Yes
						65.9	3444	120 000			
Xing et al. (2015)	14.81	−90	R-245fa	55.4	0.113	199	11 000	260 000	+90	199	Yes
		+90				699	38 000	900 000			
Wurfel et al. (2003)	20.0	−90	n-Heptane						−90	Not known	No
		0									
Cho et al. (2013)	44.8	−3	Water	290	0.335				Not known	Not known	No
		+3									
Yunxiao and Li (2015)	8.0	+90	R-410A	31	0.0065	103	7559	52 274	NA	NA	No
				48	0.427	390	36 809	278 786			

Source: From Shah (2016a). © 2016 with permission from Elsevier.

with his own data for inclinations of −3° to 0°. Rufer and Kezios (1966) pointed out that the assumption made by Chato that depth of liquid decreases toward the outlet holds only for open channel flow. Inside tubes, liquid depth increases from inlet to outlet as more condensate is formed. They gave their own correlation.

Wang and Du (2000) noted that gravity forces tend to stratify the flow while surface tension and inertia forces tend to spread the liquid uniformly around the circumference. Considering these forces, they developed an analytical model that agreed with their own data.

Lips and Meyer (2012) developed a theoretical model of stratified flow. It agreed well with their own data for R-134a at 200 kg m^{-2} s^{-1}.

Saffari and Naziri (2010) developed a model of stratified downward flow. It consists of a system of differential equations that are to be solved numerically. Their model predicts maximum heat transfer to occur at −30° and lowest heat transfer occurs in the horizontal position, with that in vertical downflow in between the two.

Xing et al. (2015) and Mohseni et al. (2013) have also given correlations of their own data.

Wurfel et al. (2003) fitted the following formula to their own data for inclinations −90° to 0°:

$$\frac{h_{TP,\theta}}{h_{TP,0}} = (1 + \sin\,\theta)^{0.214} \tag{2.3.60}$$

According to this formula, heat transfer is highest in vertical down orientation.

Shah (2016a) developed the following correlation:

For inclinations of −30° to +90°,

$$h_{TP,\theta} = h_{TP,0} \tag{2.3.61}$$

For inclinations −90° to −30°,

$$h_{TP,\theta} = h_{TP,0} + (h_{TP,0} - h_{TP,-90})(\theta + 30)/60 \tag{2.3.62}$$

where $h_{TP,\theta}$ is the heat transfer coefficient at angle θ to the horizontal. In Shah (2016c), this correlation was used together with Shah (2016b) version of the correlation. The 550 data points from eight sources were predicted with MAD of 17%.

While this correlation does not reconcile all the differences in the trends reported by various researchers, it allows prediction of heat transfer in all orientations with reasonable accuracy. It could be that some of the data are incorrect. More experimental work is needed to resolve the differences.

2.3.4 Recommendation

For condensation of stagnant vapor inside vertical channels, calculate as for vertical plates. See Section 2.2.5.

For condensation of vapor with forced flow in channels:

- For horizontal or vertical downflow, use the Shah (2019a) correlation.
- For other orientations, use the Shah (2016a) correlation for inclined tubes. Use the Shah (2019a) correlation for calculating $h_{TP,0}$ and $h_{TP,-90°}$ required by it.
- Dorao and Fernadino (2018), Cavallini et al. (2006), and Kim and Mudawar (2013) correlations may be used as alternatives to Shah correlation for horizontal and vertical downflow when $We_{GT} > 100$. The last mentioned is further restricted to $D \le 3$ mm. The Cavallini et al. correlation should be restricted to $p_r < 0.7$.

2.4 Condensation Outside Tubes

2.4.1 Single Tube

2.4.1.1 Stagnant Vapor

For condensation outside vertical tubes, the analyses and correlations for vertical plates are applicable (see Section 2.2).

For condensation on the outer surface of a single horizontal tube, Nusselt performed an analysis similar to that for vertical plates. For the local heat transfer coefficient h_η at angle η to the vertical, he derived the following equation:

$$h_\eta = 0.693\left[\frac{\rho_f(\rho_f - \rho_g)g\,\sin\,\eta k_f^3}{\Gamma'_\eta}\right]^{1/3} \tag{2.4.1}$$

Γ_η' is the mass flow rate per unit length at that angular position. The mean heat transfer coefficient for the entire perimeter is

$$h_m = 0.725\left[\frac{\rho_f(\rho_f - \rho_g)gi_{fg}k_f^3}{D\mu_f\Delta T_{SAT}}\right]^{1/4} \tag{2.4.2}$$

These can be expressed in terms of Reynolds number as

$$h_m\left[\frac{\mu_f^2}{\rho_f(\rho_f - \rho_g)gk_f^3}\right]^{1/3} = 1.51Re_f^{-1/3} \tag{2.4.3}$$

The Reynolds number is defined as

$$Re_f = \frac{4\Gamma'}{\mu_f} \tag{2.4.4}$$

Γ' is the mass flow rate of condensate per unit length of tube.

In his analysis, Nusselt showed that 59.4% of condensate is formed on the upper half of the tube. The reason is that the liquid film is thicker at the lower half as the condensate from the upper half flows to the lower part from where it drains down.

According to Collier and Thome (1994), Eq. (2.4.3) is valid up to $Re_f = 3200$. According to Ghiaasiaan (2008), transition to turbulent flow occurs at $Re_f = 3600$.

Comparing Eq. (2.2.8) for vertical plate/tube with Eq. (2.4.2) for horizontal tube at the same ΔT, it is seen that the horizontal tube will have the same heat transfer coefficient as the vertical tube if length of tube is 2.87 times the diameter. For longer lengths, heat transfer coefficient of the horizontal tube will be higher. The reason is that the film thickness on vertical tube keeps on increasing with length while the condensate continuously drains down from the horizontal tube.

McAdams (1954) examined the data of early researchers and found that the average deviation of data from Nusselt equation was 1.23 for steam and 0.94 for various organic vapors. Berman (1969) examined data for steam from nine studies. The data for $\Delta T > 10\,°C$ were within +30% and −20%; deviations increased as ΔT decreased. Collier and Thome (1994) state that most data are within 15% of Nusselt equation. It may be concluded that the Nusselt equation is reliable.

Many analyses have been performed to determine the effect of factors neglected in Nusselt's analysis. A comprehensive analysis was performed by Chen (1961b). He took into consideration the effects of subcooling of condensate, inertial forces, and shear on moving liquid film from stagnant vapor. He found that for ordinary fluids with Prandl numbers about 1 or greater, there is little departure from the Nusselt equations. For low Prandl numbers typical of liquid metals, heat transfer coefficients are much lower than given by Nusselt equation. The results of this analysis are closely represented by Eq. (2.2.16) for flat plates, with h_{Nu} calculated by Nusselt equation for condensation on tubes. The results of the analysis by Sparrow and Gregg (1959a) are close to those of Chen. Berman (1969) has reviewed several other analyses besides these two.

2.4.1.2 Moving Vapor

Moving vapor disturbs the liquid film and causes significant increase in heat transfer coefficient compared with that of stagnant vapor. Berman and Tumanov (1962) performed tests with steam flowing downwards on tubes. Heat transfer coefficients were up to 1.6 times the predictions of Nusselt equation as vapor Reynolds number reached 800.

Shekriladze and Gomelauri (1966) developed an analytical predictive formula by analyzing the case of downward flow of vapor on a horizontal tube. They assumed that the condensate is laminar, and there is no separation of vapor boundary layer. Their formula has been put in the following more convenient form by Butterworth (1977):

$$h_m = [0.5h_{FC}^2 + (0.25h_{FC}^2 + h_{Nu}^4)^{1/2}]^{1/2} \tag{2.4.5}$$

where h_{Nu} is obtained from Nusselt equation for stagnant vapor, Eq. (2.4.2), and h_{FC} is the heat transfer coefficient due to forced convection given by

$$h_{FC} = 0.9k_f(\rho_f u_g/\mu_f D)^{1/2} \tag{2.4.6}$$

At high velocities, contribution of h_{Nu} becomes very small, and at very low velocities, Eq. (2.4.5) approximates to the case of stagnant vapor.

Shekriladze and Gomelauri (1966) also analyzed the case of vapor boundary layer separation. They argued that with boundary layer separation, there will be very little condensation on the area of tube beyond 82° from top and therefore recommended multiplying h_m from Eq. (2.4.5) by 0.66. Butterworth (1977) recommended multiplying only h_{FC} by that factor. With Butterworth modification, Eq. (2.4.6) becomes

$$h_{FC} = 0.59k_f(\rho_f u_g/\mu_f D)^{1/2} \tag{2.4.7}$$

Butterworth (1977) analyzed data for steam flowing down on a tube bank. He found that generally heat transfer coefficients were in between predictions with and without boundary layer separation. Hence use of Eq. (2.4.7) gives conservative predictions.

In the foregoing discussions, vapor was considered to be flowing vertically downwards. Honda and Fujii (1974) analyzed the case of vapor approaching the tube at different angles. Their results for horizontal flow are essentially the same as for downwards flow.

2.4.2 Bundles of Horizontal Tubes

2.4.2.1 Vapor Entry from Top

Nusselt (1916) assumed that condensate from tubes leaves in the form of sheet and falls on the tube below without splashing. Accordingly, he developed the following formula for a vertical column of horizontal tubes. The mean heat transfer coefficient h_n for the nth tube in the column is given by

$$\frac{h_n}{h_1} = n^{0.75} - (n-1)^{0.75} \tag{2.4.8}$$

where h_1 is the heat transfer coefficient on the top tube. The mean heat transfer coefficient h_N for a column of N tubes is given by

$$\frac{h_N}{h_1} = N^{-1/4} \tag{2.4.9}$$

Kern (1958) gave the following correlation:

$$\frac{h_N}{h_1} = N^{-1/6} \tag{2.4.10}$$

$$\frac{h_n}{h_1} = n^{5/6} - (n-1)^{-5/6} \tag{2.4.11}$$

Eisenberg (1972) has given the following correlation based on his own data for steam condensation:

$$\frac{h_N}{h_1} = 0.6 + 0.42N^{-1/4} \tag{2.4.12}$$

Belghazi et al. (2001) condensed R-134a at 40 °C in a 13 × 3 staggered bundle of 16.8 mm OD tubes arranged at 60°. Their data are compared to three correlation in Figure 2.4.1. The Nusselt theory underpredicts all data that are seen to be in between the predictions of Kern and Eisenberg formulas.

Berman (1969) studied a wide range of data. He found that a lot of data exceeded predictions of Nusselt equation. He points out two major reasons for the failure of the Nusselt theory. Firstly, condensate does not flow down from tubes as a continuous sheet but as widely separated drops and columnar streams. Figure 2.4.2a shows the condensate flow assumed by Nusselt, while Figure 2.4.2b shows the actual flow as described by Berman. The distance between the liquid columns may be as large as 80–100 mm. Thus, the thickening of liquid film at the bottom of tube assumed by Nusselt does not occur. The liquid where the drops and streams fall get agitated and hence its thermal resistance is much reduced. Berman also points out that in a large tube bank, the vapor velocity cannot be considered negligible except at the lowest tubes. The vapor velocity is highest at the top tubes as the entire vapor condensed in the bundle has to pass over them. Then it progressively decreases as it passes through the bundle. Hence effect of vapor velocity has to be taken into consideration. He recommends step-by-step calculations from top to bottom.

Figure 2.4.3 shows the data examined by Berman and the predictions of Nusselt and Kern models. Neither is seen to be satisfactory.

Honda et al. (1988) have given a correlation that was verified with data for steam and refrigerants from several sources.

Eckels (2005) condensing R134a downwards at 35 and 40 °C saturation temperatures on a bundle of 25 water-cooled smooth tubes and 10 dummy half tubes. The

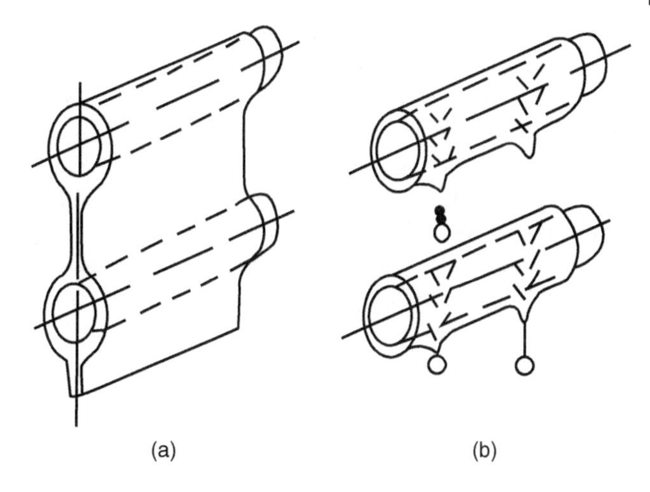

Figure 2.4.2 Downward flow of condensate: (a) in Nusselt model and (b) actual. Source: From Butterworth (1977). © 1977 American Society of Mechanical Engineers.

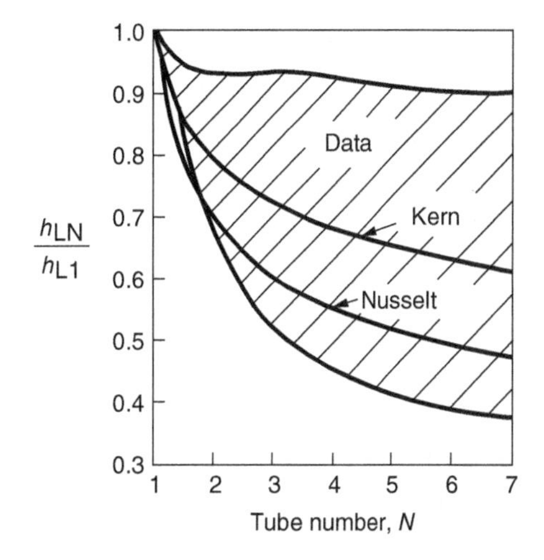

Figure 2.4.3 Comparison of heat transfer coefficients for lower tubes with those for the top tube in the bundle. Source: From Butterworth (1977). © 1977 American Society of Mechanical Engineers.

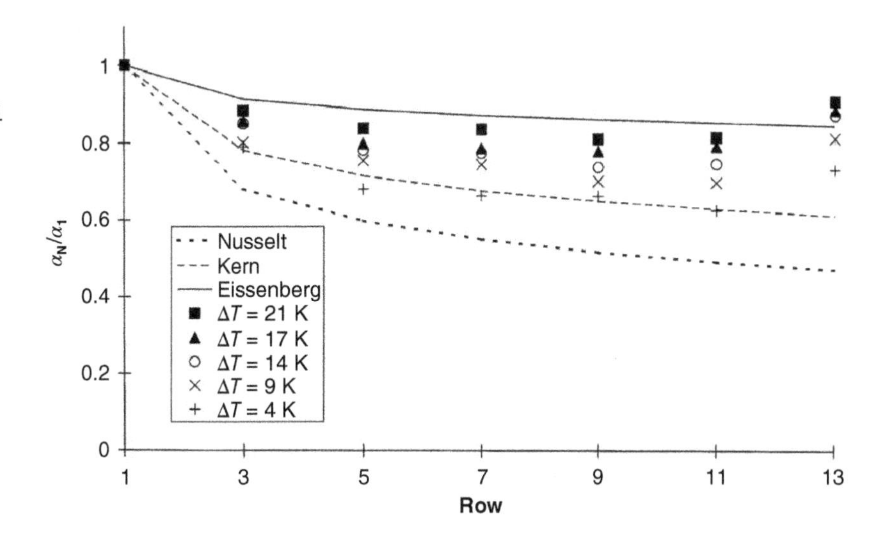

Figure 2.4.1 Ratio of heat transfer coefficients of first and nth row of a tube bundle compared to three correlations. Source: From Belghazi et al. (2001). © 2001 Elsevier.

tubes have 19.1 mm outer diameter and are arranged in a staggered pattern with a horizontal and a vertical pitch of 22.9 mm. He compared the data with the formulas of Nusselt, Kern, and Honda et al. (1988). Best agreement was found with Honda et al. method. However, even better agreement was found with the Nusselt formula if the exponent of N was changed to −0.08 form −0.25.

Zeinelabdeen et al. (2018) compared a number of empirical models for calculating the heat transfer coefficients for condensation on banks of tubes with the experimental data set obtained by previous investigators consisting of more than 4000 data points for 6 different condensing fluids and 13 different tube bank configurations. The configurations included inline and staggered bundles. The fluids were steam, R-11, R-21, R-113, methanol, and iso-propanol. Vapor velocity approaching the bundle varied from 0 to 21 m s^{-1}. They found the Fujii and Oda model (1986) to be the most accurate model with MAD of about 21.5%. It accounts for the effects of the shear stress on the surface of the film condensate and the inundation within the bank. It is given by the following equations:

$$Nu = \frac{h_m D}{k_f} = (Nu_{gr}^4 + Nu_{sh}^4)^{1/4} \tag{2.4.13}$$

$$Nu_{gr} = Nu_{Nu} N^{-s} \tag{2.4.14}$$

$$Nu_{sh} = 0.9(1 + \zeta^{-1})^{1/3} Re_{TP,mv}^{1/2} N^{-0.14} \tag{2.4.15}$$

$$\zeta = \frac{k_f \Delta T}{\mu_f i_{fg}} \left[\frac{\mu_f \rho_f}{\mu_g \rho_g} \right]^{1/2} \tag{2.4.16}$$

$$Re_{TP,mv} = \frac{u_{mv} \rho_g D}{\mu_g} \tag{2.4.17}$$

The exponent s is 0.16 and 0.08, respectively, for the inline and staggered bundles. The subscripts "gr" and "sh" indicate gravity dominated and shear dominated. u_{mv} is the mean vapor velocity through the area not occupied by the tubes. The other models that were compared to the same data included those of McNaught (1982) and Honda et al. (1986).

2.4.2.2 Vapor Entry from Side

In the foregoing, vapor was considered as flowing vertically downwards on the bundle. Vapor entry horizontally from the side is herein discussed.

Fuks and Zernova (1970) performed experiments in which measurements were done with both downwards and side entry horizontal flow of vapor. The bundle had eight staggered rows of 19 mm diameter tubes. Steam was saturated at 36–80 °C. While there were some differences between the results for top and side entry of vapor, there was little difference between the bundle average heat transfer coefficient.

Fujii et al. (1972) carried out experiments on condensation of saturated steam at 30 °C on a bundle with 14 mm diameter tubes. Vapor entered from the side. Their results show little difference from those with downflow.

Butterworth (1977) has quoted some unpublished data for side flow in which the measured heat transfer coefficients with approach velocity of 30 m s^{-1} were about half of those from Nusselt theory. Butterworth noted that in those tests, the velocity was higher and temperature difference was lower than in any published data for downflow or side flow. Hence the reason for low heat transfer coefficients may not necessarily be side flow.

The conclusion is that at low to moderate velocities. Heat transfer in side flow may be calculated by the same methods as for downflow. The situation at high velocities is unclear. It will be advisable to avoid high velocities.

2.4.3 Recommendations

Stagnant vapor on vertical tube: Use the recommendations for vertical plate in Section 2.2.5

Stagnant vapor on horizontal tube: Use the Nusselt formulas, Eq. (2.4.2) or Eq. (2.4.3).

Moving vapor on single tube: Eq. (2.4.5) with Eq. (2.4.7) for h_{FC}.

Tube bundles: Use the Fujii and Oda model, Eq. (2.4.13).

2.5 Condensation with Enhanced Tubes

2.5.1 Condensation on Outside Surface

A wide variety of enhanced tubes are in use. A commonly used type is the integral low finned tube. It has fins of rectangular or trapezoidal shape. With higher surface tension fluids, these tend to trap condensate between fins in the lower part of the tube that reduces heat transfer. A wide variety of tubes with two-dimensional and three-dimensional fins have been developed to minimize liquid trapping.

2.5.1.1 Single Tubes

Beatty and Katz (1948) were the first to present a model for condensation on horizontal low finned tubes. They assumed that in the space between fins, there is no effect of fins, and condensation can be calculated by Nusselt equation for single tube. Heat transfer on the fin is calculated by the Nusselt equation for vertical plate. An effective length of fin L_{fin} was defined as

$$L_{fin} = \frac{\pi(D_o^2 - D_r^2)}{4D_o} \tag{2.5.1}$$

where D_o is the diameter at the tip of fin and D_r is diameter at the root of fin (see Figure 2.5.1).

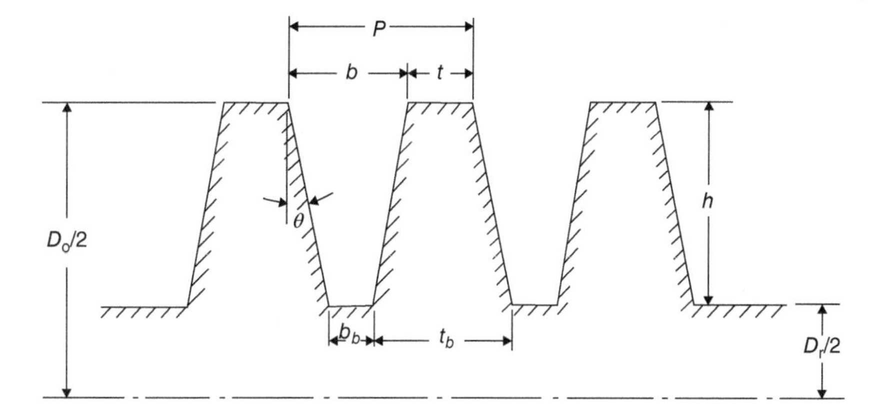

Figure 2.5.1 Definition of geometric variables for tube with trapezoidal fins. Source: From Marto (1988). © 1988 American Society of Mechanical Engineers.

The heat transfer coefficient of fin is calculated with Nusselt equation for vertical plate using L_{fin} from Eq. (2.5.1) as the characteristic length. Heat transfer coefficient for the tube surface is calculated using the Nusselt formula for condensation on horizontal tubes. The final relation is

$$h = 0.689\left[\frac{\rho_f(\rho_f - \rho_g)gi_{fg}k_f^3}{\mu_f(T_{\text{SAT}} - T_w)}\right]^{1/4} \times \left[\frac{A_{\text{tube}}}{A_{\text{total}}}D_{\text{tube}}^{-0.25} + 1.3\frac{\eta_{\text{fin}}A_{\text{fin}}}{A_{\text{total}}}L_{\text{fin}}^{-0.25}\right] \tag{2.5.2}$$

The constant 0.689 is empirical. A_{tube} is the area of tube surface, A_{fin} is the area of fin surface, and A_{total} is the sum of the two. The fin efficiency η_{fin} is calculated considering conduction from the tube surface into the fin; formulas for it may be found in most books. This equation gave satisfactory agreement with data for six low surface tension fluids (methyl chloride, SO_2, R-22, n-pentane, propane, and n-butane). The tubes had 433–633 fins m^{-1}.

Note that the heat transfer coefficient given by Eq. (2.5.2) is based on an effective area A_{eff} given by

$$A_{\text{eff}} = A_{\text{tube}} + \eta_{\text{fin}}A_{\text{fin}} \tag{2.5.3}$$

Beatty and Katz assumed that condensate flows down freely as for bare tubes. This assumption is valid only for low finned tubes and low surface tension liquids. In other cases, surface tension causes condensate to be retained between the fins, making the lower part of tube ineffective. This phenomenon is known as condensate retention or flooding. Many expressions have been proposed for predicting the flooding angle ϕ_f measured from the top; see Figure 2.5.2. Honda et al. (1983) performed a mechanistic analysis to develop the following expression for flooding in trapezoidal and rectangular fins with $h > b/2$:

$$\phi_f = \cos^{-1}\left[\left(\frac{4\sigma\cos\theta}{\rho_f gbD_o}\right) - 1\right] \tag{2.5.4}$$

where b is the space between fins at the tip and θ is one-half of the apex angle at the tip of fins; it is zero for rectangular fins. See Figure 2.5.1 for definition of geometrical variables. Figure 2.5.2 shows predictions of Eq. (2.5.4) for rectangular fins of thickness 0.5 mm on a tube with outside diameter of 21.05 mm. It is seen that there is a critical fin spacing below which flooding angle decreases sharply, which causes deterioration in heat transfer. The dependence of critical spacing on (σ/ρ) is also shown. The larger this ratio, the greater the critical fin spacing.

See Ali (2017) for various other methods for calculating the flooding angle.

Many predictive methods have been proposed for condensation on finned tubes taking condensate retention into consideration. The most conservative approach is to assume no condensation in the flooded part of surface. This will give the following formula:

$$\text{h} = h_{\text{bk}}\frac{\phi_f}{\pi} \tag{2.5.5}$$

where h_{bk} is the heat transfer coefficient assuming no flooding calculated by the Beatty and Katz model. This model underpredicts heat transfer.

Kumar et al. (2002) measured heat transfer coefficients during condensation of steam (p = 101–110 kPa) and R134a at 39 °C on five different single horizontal circular integral-fin tubes and on two spine integral-fin tubes. Fin tip diameter of the copper integral fin tubes ranged from 23.82 to 24.98 mm, fin height ranged from 0.6 to 1.11 mm. Longitudinal fin pitch varied from 0.53 to 1.07 mm for R134a and was 2.57 mm for steam. They presented a new model that is stated to be able to predict 90% of their own experimental data within ±15% and data from 13 investigators within 30%. Their model is described by the following equations:

$$h = \frac{q}{\pi D_o L\Delta T_{\text{SAT}}}0.024Re_f^{-1/3}We^{0.3}Y^{1.4}\left(\frac{\rho_f^2k_f^3g}{\mu_f^2}\right) \tag{2.5.6}$$

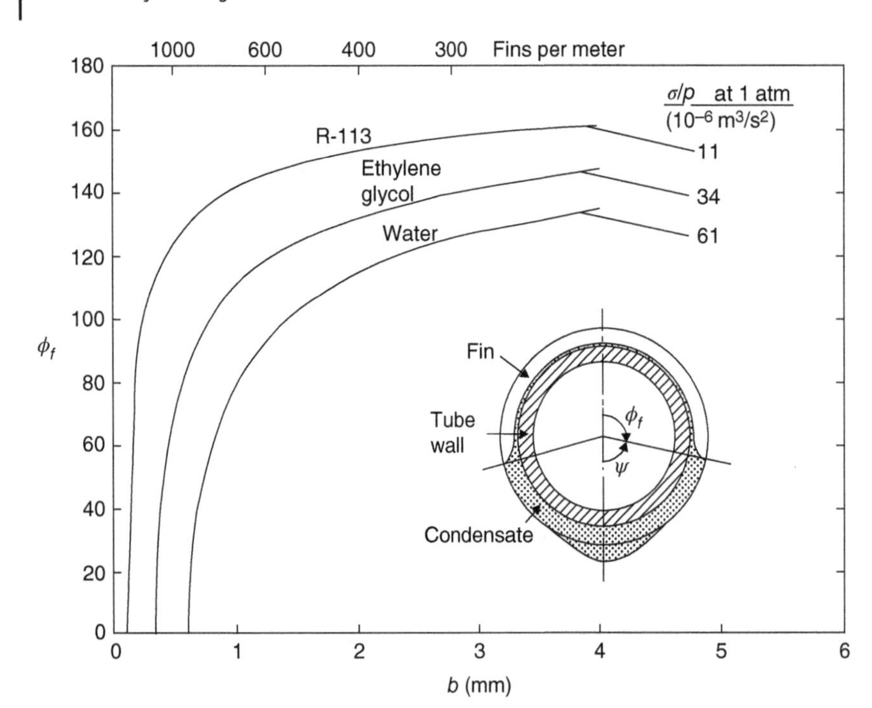

Figure 2.5.2 Finned tube with condensate retention. Source: From Marto (1988). © 1988 American Society of Mechanical Engineers.

$$Re_f = \frac{4w}{\mu_f p_{\text{fin}}} \tag{2.5.7}$$

$$We = \frac{4\sigma\left[\frac{1}{t_o} + \frac{1}{(p_{\text{fin}} - t_r)}\right]}{H_{\text{fin}}\rho_f g} \tag{2.5.8}$$

$$Y = \frac{4A}{D_r - p_{\text{fin}}} = \frac{4\pi\left[\frac{(D_o^2 - D_r^2)}{2} + D_o t_o + D_r(p_{\text{fin}} - t_r)\right]}{D_r p_{\text{fin}}} \tag{2.5.9}$$

A is the true area of the tube surface for one pitch length, t_r and t_o are the fin thickness at root and top, D_o and D_r are diameters of the fin at top and root, H_{fin} is the height of fin, p_{fin} is the fin pitch, and w is mass flow rate of condensate. Kumar et al. compared the same data with the model of Honda and Nozu (1987) and found it to have deviations up to 50%.

Cavallini et al. (2003) compared this model with data from several sources for steam and refrigerants. Large deviations were found with many data sets. They concluded that the Kumar et al. equation is applicable only to condensation on single copper tubes with fin height less than 1.11 mm and fin pitches between 0.5 and 1 mm for halogenated refrigerants and larger than 2.0 mm for steam.

Sajjan et al. (2015) measured condensation heat transfer coefficients of refrigerant R-600a (iso-butane) over a horizontal smooth tube of outer diameter 19 mm and five integral fin tubes of different fin-densities (945, 1024, 1102, 1181, and 1260 fpm). They compared their fin tube data with several models. Rudy and Webb (1985) and Webb et al. (1985) models overpredicted their data, while Beatty and Katz (1948) and Owen et al. (1983) models underpredicted.

It is concluded that none of the predictive methods can be considered generally applicable. Their use has to be limited to the database for which they have been validated.

2.5.1.2 Tube Bundles

Eckels (2005) tested two bundles with R-134a at 35 and 40 °C. One of the bundles had circular integral fins, while the other had three-dimensional integral fins. Nominal diameter of both was 19.1 mm. There were 1575 fins m^{-1}, and there were 10 rows of tubes arranged in staggered pattern. By pumping liquid to the top, simulation was made for up to 30 rows. The measurements at the top row were in good agreement with the Beatty and Katz model. The bundle average and tube-by-tube heat transfer coefficients were in satisfactory agreement if the Beatty and Katz prediction was adjusted as follows:

For bundle average,

$$\frac{h_N}{h_1} = N^{-m} \tag{2.5.10}$$

For the nth row,

$$\frac{h_n}{h_1} = n^{-m} - (n-1)^{(1-m)} \tag{2.5.11}$$

where h_1 is the heat transfer coefficient of the first tube calculated by Beatty and Katz model. They used the exponent m = 0.08. This is half that in the Kern formula and 1/3rd that in the Nusselt formula. This shows that the effect of inundation was small. This is in agreement with the results of most other researchers as stated by Browne and Bansal (1999) after review of data for integral low finned tubes from several sources.

Eckels also compared his data with the models of Sardesai et al. (1983), Briggs and Rose (1994), Sreepathi et al. (1996), and Murata and Hashizume (1992). These models are more complex and take into considerations the effects of surface tension. Deviations of Sardesai et al. model were comparable to those with Beatty and Katz model, while those of the others were higher.

Their tests on the bundle with three-dimensional tubes showed progressively lower heat transfer coefficients at the lower rows. Figure 2.5.3 is an example. They have given equations to fit their row-by-row data. After reviewing data for bundles of three-dimensional tubes, Browne and Bansal (1999) noted that most studies show that there is significant row effect in such bundles. Thus, the results of Eckels are in line with those of other researchers.

The difference in row effect between integral finned tubes and three-dimensional enhanced tubes can be explained by different condensate flow in the two. Honda et al. (1987) studied condensate flow in a bundle of integral fin tubes. They found four flow patterns: drop, column, column and sheet, and sheet. Columns occur at constant pitch. At high condensate rates, sheet mode occurs. Webb and Murawski (1990) studied condensation on three-dimensional tubes. They found that the drainage pattern was very unstable and that there was a lateral flow of condensate not seen with integral fin tubes. Condensate falling on a tube drains down not directly below there but at some distance away along the tube. There was no lateral flow of condensate in integral fins because condensate flows down through the fins at column locations, and thus the length between the columns remains largely condensate free. Therefore, heat transfer in lower tubes is not reduced much and the row effect is small if any. Due to lateral movement of condensate in three-dimensional enhanced tubes, no part of the tube is completely free of condensate and therefore heat transfer on the lower tubes gets lowered, resulting in the row effect.

For other studies and prediction methods, see the review papers of Browne and Bansal (1999) and Cavallini et al. (2003).

2.5.2 Condensation Inside Enhanced Tubes

Many types of internally enhanced tubes are in use. More common are those with helical or herringbone fins shown in Figure 2.5.4.

Condensation in enhanced tubes has been reviewed by Cavallini et al. (2003) and Dalkilic and Wongwises (2009).

Cavallini et al. (1995, 1999) presented a correlation for heat transfer during condensation in microfin and cross-grooved tubes by modifying the correlation of Cavallini and Zecchin (1974) for smooth tubes. It was based on data for refrigerants from 15 sources. Wang and Honda (2003) compared data for one-and two-dimensional integral finned tubes with several predictive techniques. The Cavallini et al. (1999) correlation was found to underpredict data for R-11 and R-123, which was at $p_r < 0.1$. They found best results with their own theoretical model that is difficult to use as it involves numerical solution of a set of equations. The next best results were with the correlation of Yu and Koyama (1998), which predicted the low pressure data well. This correlation is applicable only to microfin tubes. Cavallini et al. (2003) found it to give unsatisfactory agreement with data. They also report unsatisfactory results with the correlation of Kedzierski and Goncalves (1997).

Cavallini et al. (2009) developed a correlation for integral fin tubes by modifying their correlation for plain tubes, Cavallini et al. (2006). It is given by the following equations:

$$h_{TP} = [h_D^3 + h_A^3]^{1/3} \quad (2.5.12)$$

$$h_A = A\, Ch_{AS} \quad (2.5.13)$$

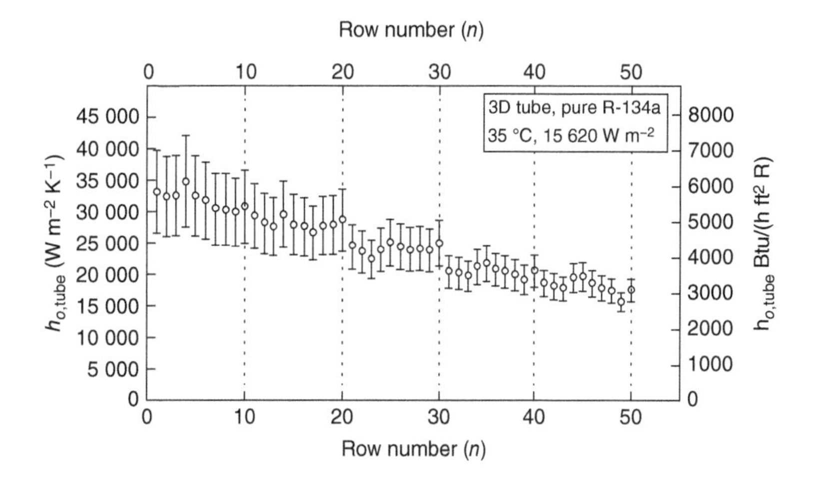

Figure 2.5.3 Variation of heat transfer with tube rows in a bundle of three-dimensional tubes. Source: From Eckels (2005). © 2005 ASHRAE.

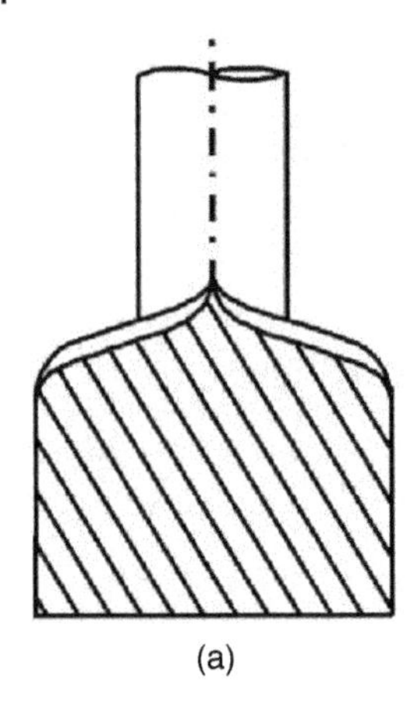

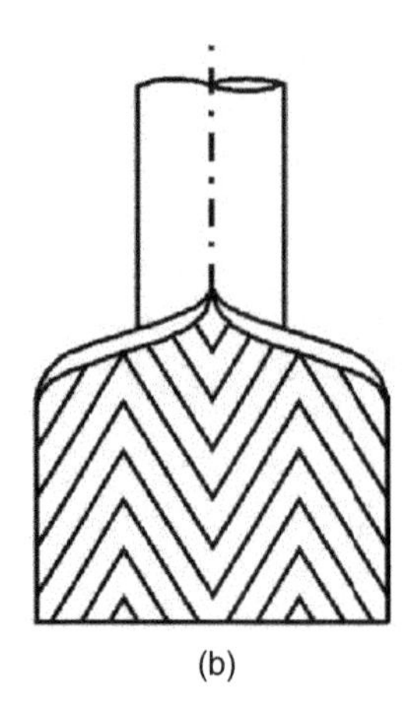

Figure 2.5.4 Internally enhanced tubes: (a) helical microfin tube and (b) herringbone tube. Source: From Cavallini et al. (2003). © 2003 Elsevier.

$$h_{AS} = h_{LT}\left(1 + 1.28x^{0.817}\left(\frac{\rho_f}{\rho_g}\right)^{0.3685}\left(\frac{\mu_f}{\mu_g}\right)^{0.2363} \times \left(1 - \frac{\mu_f}{\mu_g}\right)^{2.144} \mathrm{Pr}_f^{-0.1}\right) \tag{2.5.14}$$

$$h_D = C[2.4x^{0.1206}(R_x - 1)^{1.466}C_1^{0.6875} + 1]h_{D,S} + C(1 - x^{0.087})R_x h_{D,S} \tag{2.5.15}$$

$$h_{D,S} = \left\{1 + 0.741\left[\left(\frac{1-x}{x}\right)^{0.3321}\right]\right\}^{-1} h_{Nu} \tag{2.5.16}$$

where h_{Nu} is calculated with Eq. (2.4.2), the Nusselt formula for condensation on horizontal tubes. h_{LT} is calculated by the McAdams/Dittus–Boelter equation for single-phase flow, Eq. (2.3.11).

$$A = 1 + 1.119Fr^{-0.3821}(R_x - 1)^{-0.3821} \tag{2.5.17}$$

$$Fr = \frac{G^2}{gD(\rho_f - \rho_g)^2} \tag{2.5.18}$$

$$R_x = \left\{\frac{\left[2h_{\text{fin}}n_g\left(1 - \sin\left(\frac{\gamma}{2}\right)\right)\right]}{\left[\pi D\cos\left(\frac{\gamma}{2}\right)\right]} + 1\right\} / \cos\beta \tag{2.5.19}$$

$$n_{\text{opt}} = 4064.4D + 23.257D \tag{2.5.20}$$

with D in meter.

$C = 1$ if $(n_{\text{opt}}/n_g) \geq 0.8$

$C = (n_{\text{opt}}/n_g)^{1.904}$ if $(n_{\text{opt}}/n_g) < 0.8$

$C_1 = 1$ if $J_g \geq J_{g,*}$

$C_1 = (J_g/J_{g,*})$ if $J_g < J_{g,*}$

$$J_{g,*} = 0.6\left\{\left[\frac{7.5}{4.3X_{tt}^{1.111} + 1}\right]^{-3} + 2.5^{-3}\right\}^{-0.3333} \tag{2.5.21}$$

Figure 2.5.5 shows the geometrical parameters in these equations. D in all expressions above is that at the fin tip. h_{fin} is the fin height, n_g is the number of grooves, and n_{opt} is the optimum number of grooves. The parameter R_x accounts for the effect of the heat transfer area increase. In Eq. (2.5.11), the heat transfer coefficient is defined with reference to the cylindrical envelope surface area of the finned surface at the fin tip. This correlation was verified with 3500 data points from many sources covering the range: 6 mm $< D <$ 14 mm, 21 $< n_g <$ 82, 0.12 mm $< h_{\text{fin}} <$ 0.35 mm, $h_{\text{fin}}/D < 0.04$, 25° $< \gamma <$ 64°, 0° $< \beta <$ 30°, 0.1 $< p_r <$ 0.67, and 90 $< G <$ 900 kg m^{-2}s^{-1}. The experimental data used for the validation of the model includes CO_2 and halogenated refrigerants. It is not to be used for hydrocarbons, ammonia, and water. While

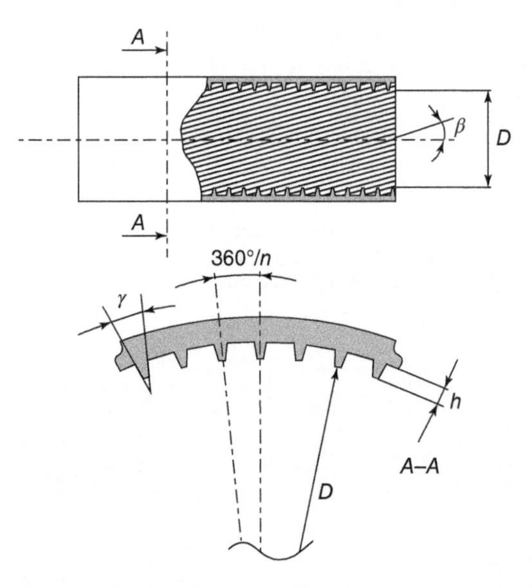

Figure 2.5.5 Geometrical parameters in the correlation of Cavallini et al. (2009). Source: From Cavallini et al. (2009). © 2009 Elsevier.

it showed good agreement with its database, a number of other correlations were also compared to it and found less accurate. Correlations for Kedzierski and Goncalves (1997), Han and Lee (2005), Chamra and Mago (2006), and Chamra et al. (2005) were found to be fairly good.

Miyara et al. (2000), Wellsandt and Vamling (2005), and Olivier et al. (2007) have given correlations for herringbone tubes. Aroonrat and Wongwises (2018) have given a correlation for dimpled tubes to fit their own test data. Data and limited correlations for many other types of enhanced tubes have been published.

2.5.3 Recommendations

Outside single tubes: Use the Beatty and Katz model for low surface tension fluids for low finned tubes. Use the Kumar et al. correlation for integral circular and spine fins in the range of their database. This includes fin height less than 1.11 mm and fin pitches between 0.5 and 1 mm for halogenated refrigerants and larger than 2.0 mm for steam.

Tube bundles: For integral low finned tubes, use Eqs. (2.5.10) and (2.5.11) with heat transfer coefficient of the first row calculated with a method suited for the tube geometry and fluid. For other enhanced tube, use test data and correlations suited for the design conditions.

Condensation inside tubes: For condensation in microfin tubes, use the Cavallini et al. correlation (2009) in the range recommended by these authors. Outside this range and for various other types of tubes and fluids, use the correlations proposed for those geometries in the range of their database.

2.6 Condensation of Superheated Vapors

2.6.1 Stagnant Vapor on External Surfaces

As long as the surface temperature is above the saturation temperature of vapor, heat transfer occurs by natural convection, and no condensation occurs. Once the wall temperature drops below the saturation temperature, condensation starts. A thermal boundary layer is created in which the vapor temperature drops to the interface temperature. If the effects of natural convection in the vapor boundary layer are neglected, the Nusselt analysis may be extended by including the de-superheating of vapor. Thus i_{fg} is replaced by $[i_{fg} + C_{pg}(T_g - T_{SAT})]$. This gives

$$\frac{h_{sh}}{h_{SAT}} = \left[\frac{i_{fg} + C_{pg}(T_g - T_{SAT})}{i_{fg}}\right] \tag{2.6.1}$$

where h_{sh} is the heat transfer coefficient of superheated vapor and h_{SAT} is the heat transfer coefficient of saturated vapor. As the sensible heat is usually very small compared with the latent heat, the increase predicted by Eq. (2.6.1) does not exceed a few percent.

Minkowycz and Sparrow (1966) performed a detailed analysis of condensation of steam on a vertical plate in which they considered the effect of superheat. Saturation temperature varied from 27 to 100 °C. Their result was that increase in heat transfer due to superheat does not exceed 5%. They also analyzed the case of steam containing air and found that superheat causes considerably more increase in heat transfer compared with that for pure vapor.

Many experimental studies have confirmed that the effect of superheat is small. Spencer and Ibele (1966) found increases of 4–12%. Walt and Kroger (1972) condensed R-12 with superheats up to 128 °C on a vertical surface and found that increase in heat transfer coefficients compared with saturated vapor by Nusselt equation did not exceed 8%.

It is concluded that effect of superheat on condensation heat transfer coefficient can be predicted with sufficient accuracy by Eq. (2.6.1).

The total heat flux is that which passes from liquid film to the wall and is calculated by

$$q = h_{sh}(T_w - T_{SAT}) \tag{2.6.2}$$

Heat transfer from the superheated vapor to the interface is

$$q_{sensible} = h_g(T_g - T_{SAT}) \tag{2.6.3}$$

where h_g is heat transfer coefficient between vapor and the interface. It may be calculated by correlations for single-phase convection. The condensate rate w from a plate of height L and unit width is

$$w = \frac{L(q - q_{sensible})}{i_{fg}} \tag{2.6.4}$$

2.6.2 Forced Flow on External Surfaces

Minkowycz and Sparrow (1969) analyzed laminar condensation of superheated steam with forced flow on a horizontal plate. Saturation temperatures of 27–100 °C and superheats up to 204 °C were considered. For saturation temperature of 27 °C, there was virtually no effect of superheat for $\Delta T_{SAT} > 2$ °C and rose to about 5% above that for saturated vapor when superheat was 204 °C. At 100 °C saturation temperature, the increase at 204 °C reached about 10%. It is interesting to note that Eq. (2.6.1) predicts a 17% increase for this condition. The increase was negligible at superheat of 60 °C. They also analyzed the case of steam containing air and found that superheat causes considerable increase in heat transfer. This is discussed further in Section 2.7.8

Michael et al. (1989) condensed steam at atmospheric pressure flowing down on a tube of 14 mm diameter. Vapor velocity was 5–81 m s^{-1} and superheat was up to 40 °C. No effect of superheat on heat transfer was found.

2.6.3 Flow inside Tubes

When superheated vapor enters a tube, heat transfer occurs by single-phase convection as long as the wall temperature is above the saturation temperature of vapor; heat transfer in this region may be calculated by single-phase equations. Once the wall temperature falls below saturation temperature, condensation can start even while the core of vapor is superheated. McAdams (1954) studied a number of experimental studies on whose basis he recommends that heat transfer coefficient h in this region be calculated as for saturated vapor and the total heat flux is then calculated as

$$q = h(T_w - T_{SAT}) \tag{2.6.5}$$

Based on their tests on condensation of steam at 8–180 bar, Miropolskiy et al. (1974) state that even if the wall temperature is below the saturation temperature, condensation will not start until the vapor temperature is below a border temperature T_{br} and quality is below a quality x_{br} defined by

$$T_{br} = T_{SAT} + \frac{q}{h_{GT}} \tag{2.6.6}$$

$$x_{br} = 1 + \frac{qC_{pg}}{h_g i_{fg}} \tag{2.6.7}$$

Specific heat C_{pg} is calculated at the mean of bulk and saturation temperatures. h_{GT} is the heat transfer coefficient of vapor calculated with a suitable forced convection equation such as Eq. (2.3.11) but its constant doubled. This was attributed to the variation of properties across the vapor core. They have given a graphical correlation for calculation of heat transfer for qualities between x_{br} and 1.

Mizoshina et al. (1977) condensed superheated benzene in vertical tubes at vapor Reynolds number between 3000 and 35 000. They concluded that heat transfer coefficient can be calculated by Eq. (2.6.1).

Kondo and Hrnjak (2011, 2012a, 2012b) condensed R-410A and carbon dioxide in a 6.1 mm diameter plain tube. Reduced pressure varied from 0.55 to 0.95 and superheats from 0 to 40 °C. They found that condensation started when the wall temperature reached the saturation temperature. This is in agreement with generally accepted view and in contrast to that of Miropoloskiy et al. mentioned earlier. They correlated their data by modifying the Cavallini et al. (2006) correlation. The modification included using the actual vapor quality instead of the thermodynamic vapor quality and defining the heat transfer coefficient based on $(T_{bulk} - T_w)$. Agarwal and Hrnjak (2015) condensed R134a, R1234ze(E), and R32 at superheats up to 50 °C in a 6.1 mm tube. They found satisfactory agreement with the modified Cavallini et al. correlation mentioned earlier. Xio and Hrnjak (2018) have given a flow pattern map for superheated vapor.

Webb (1998) have given a theoretically based methods for calculation of heat transfer during condensation of superheated vapor that is expressed by the following equation:

$$h_{sup} = h_{SAT} + F\left(h_{s\text{-ph}} + \frac{C_{pg}q_{lat}}{h_{fg}}\right) \tag{2.6.8}$$

where h_{sup} is the heat transfer coefficient of superheated vapor, h_{SAT} is the heat transfer coefficient of saturated vapor, $h_{s\text{-ph}}$ is the local single-phase heat transfer coefficient between super-heated vapor and the condensate interface, q_{lat} is the local heat flux due only to phase change, and F is a factor defined as

$$F = \frac{T_g - T_{SAT}}{T_{SAT} - T_w} \tag{2.6.9}$$

According to Webb, this model may be applied to different types of condensers by using the appropriate correlations to compute the saturated vapor condensation heat transfer coefficient h_{SAT} and the single-phase heat transfer coefficient $h_{s\text{-ph}}$.

Kondo and Hrnjak (2012c) measured heat transfer of superheated CO_2 and R410A flow in horizontal microfin tubes of 6 mm inner diameter at reduced pressures of 0.55–0.95. They found that condensation of superheated vapor starts when wall temperature reaches saturation temperature. They proposed a correlation that combines the correlations of Carnavos (1980) for single-phase flow and Cavallini et al. (2009) for condensation, both for microfin tubes. It evaluates the liquid properties at film temperature and was validated with experimental results at the reduced pressure ranging from 0.55 to 0.95.

2.6.4 Plate-Type Heat Exchangers

Sarraf et al. (2016) studied condensation of pentane in a brazed plate heat exchanger (BPHE) with chevron angle of 55° and a hydraulic diameter equal to 4.4 mm. Mass flux varied from 9 to 30 kg m^{-2}s^{-1}, and the vapor superheat 5–25 K, at a mean constant saturation temperature of 36.5 °C (1.029 bar). Flow inside the channels was simultaneously studied by an infrared camera. Condensation was found to be occurring while the vapor was still superheated. It was found that heat transfer coefficient

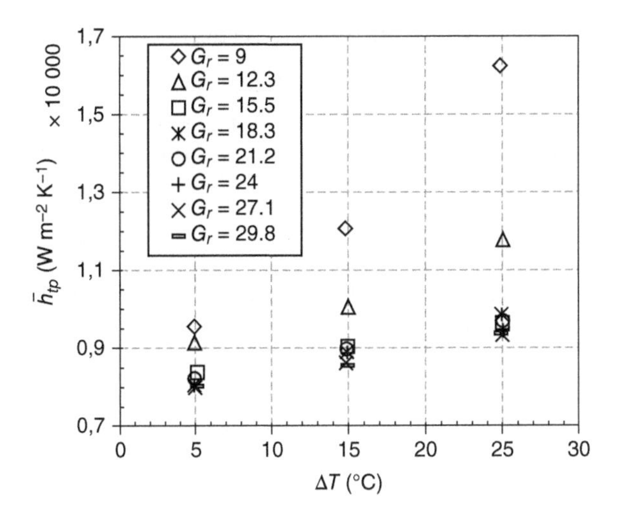

Figure 2.6.1 Effect of superheat on mean heat transfer coefficient in a brazed plate heat exchanger. G_r is the mass flux, kg m^{-2} s^{-1}, of condensing fluid pentane. Source: From Sarraf et al. (2016). © 2016 Elsevier.

increased with superheat, up to 70% at 9 kg m^{-2}s^{-1} flow and 25 K superheat. The effect of superheat decreased with increasing flow rate. Figure 2.6.1 shows their measurements. They fitted the following equation to their data:

$$h_{\text{TP}} = [(300891G^{-3.1})^2 + 73.3^2]^{0.5}(T_g - T_{\text{SAT}}) + 8491G^{-0.03} \tag{2.6.10}$$

Based on study of infrared images, it was concluded that the increase in heat transfer with superheat is caused by its effect on fluid distribution near the inlet. They found the method of Webb (1998) for calculating the effect of superheat unsatisfactory for their data.

Other researchers have also reported increase of heat transfer due to superheat. These include Longo (2009, 2011) and Mancin et al. (2012, 2013) with a variety of fluids. Increase due to superheat in mean heat transfer coefficients up to 20% were reported.

Longo et al. (2015) have given a computational model for calculating heat transfer during condensation in plate heat exchangers. It includes the effect of superheat by using the method of Webb (1998). It was verified with data from several sources. It is discussed in Section 2.7.5.

2.6.5 Recommendations

Stagnant vapor on external surfaces: Use Eq. (2.6.1).

Forced flow on external surfaces: Neglect effect of superheat.

Condensation inside plain tubes: Condensation to be considered to start when wall temperature reaches saturation temperature. For condensation of superheated vapor, use the Kondo and Hrnjak (2011) correlation in its verified range as given in this paper and in Agarwal and Hrnjak (2015). For other conditions, use the following simple method. Calculate heat transfer coefficient at $x = 0.995$ with a correlation for saturated vapor such as Shah (2019a). Quality slightly less than 1 is suggested as many correlations crash at $x = 1$. Then calculate total heat flux by Eq. (2.6.2). Calculate sensible heat flux by Eq. (2.6.3) between vapor and interface with h_g calculated by a single-phase correlation such as that of Dittus–Boelter, Eq. (2.3.11), using vapor properties in it.

Condensation inside microfin tubes: Use the correlation of Kondo and Hrnjak (2012c) in its verified range. Outside its range, use the same method as for plain tubes.

Condensation in plate heat exchangers: Use the method of Webb (1998), Eq. (2.6.8).

2.7 Miscellaneous Condensation Problems

2.7.1 Condensation on Stationary Cone

For a vertical cone, Rohsenow (1973a) analyzed condensation of stagnant vapor on a vertical cone in a way similar to the Nusselt analysis for an inclined plate. His expression for the local heat transfer coefficient for a cone with apex angle (2θ) is

$$h_z = 0.875\left[\frac{g\cos\theta\rho_f(\rho_f-\rho_g)k_f^3 i_{fg}}{z\mu_f\Delta T_{\text{SAT}}}\right]^{1/4} \tag{2.7.1}$$

where z is the distance from top along the slope. The expression for mean heat transfer coefficient for a cone of length $z = L$ is

$$h_m = \left[\frac{g\cos\theta\rho_f(\rho_f-\rho_g)k_f^3 i_{fg}}{L\mu_f\Delta T_{\text{SAT}}}\right]^{1/4} \tag{2.7.2}$$

2.7.2 Condensation on a Rotating Disk

Rohsenow (1973a) analyzed condensation of stagnant vapor on the upper side of a rotating horizontal disk. By equating shear force on the liquid film to the centrifugal force on it, he arrived with the following equation:

$$h_m = \left[\frac{2\omega^2\rho_f^2 k_f^3 i_{fg}}{3\mu_f\Delta T_{\text{SAT}}}\right]^{1/4} \tag{2.7.3}$$

where ω is angular velocity in radians per second. He reports that the data of Beatty and Nandkapur (1959) for condensation of methanol, ethanol, and R-113, at atmospheric pressure were 25% below this equation.

Sparrow and Gregg (1959a) performed a comprehensive analysis of condensation on a rotating disk taking into consideration shear and centrifugal forces and also taking into consideration momentum and convection terms. The model was solved numerically and the results presented graphically. For Prandtl number ≥ 1, these results are represented at $(C_{pf}\ \Delta T_{SAT}/i_{fg}) < 0.1$ by the following equation:

$$\frac{h(\nu_f/\omega)^{1/2}}{k_f} = 0.094\left(\frac{Pr_f}{C_{pf}\Delta T_{SAT}/i_{fg}}\right)^{1/4} \tag{2.7.4}$$

Yanniotis and Kolokotsa (1996) performed tests on condensation of low-pressure steam on the underside of a rotating horizontal disk. Saturation temperature was 40–50 °C. Speeds were up to 1000 rpm. Heat transfer coefficients were independent of the radial distance from the center. The experimental heat transfer coefficients were only 50–66% of those predicted by Sparrow and Gregg's formula, Eq. (2.7.4). They report that data from two other studies were 25–55% lower than Eq. (2.7.4), besides the data of Beatty and Nandkapur (1959) that were 25% below this equation. Hence despite its rigorous derivation, the Sparrow and Gregg formula does not seem reliable.

It would be interesting to compare the data of Yanniotis and Kolokotsa for stationary disk with the formulas of Gerstmann and Griffith (1967) presented in Section 2.2.4.

Visual observation by Yanniotis and Kolokotsa (1996) during operation showed the following. Condensation on the underside of the disk at 0 rpm forms a stationary film. As the thickness of the film increases, drops are formed and fall down. When the disk is put in rotation, centrifugal forces act on the film and throw the liquid out toward the periphery. The higher the rotational speed, the thinner the film becomes. At 1000 rpm observation using a stroboscopic light showed that the condensate forms rivulets on the outer area of the disc. Under such conditions it is doubtful that the disk surface will be fully wetted. The low visibility of the Plexiglas wall did not allow more definite observation.

Alasadi et al. (2013) studied condensation of steam at saturation temperature of 40–50 °C on a rotating disk. They found heat transfer coefficients to increase with rotational speed. They developed an analytical formula that agrees fairly well with their own data. They did not compare their data with the Sparrow–Gregg formula, nor did they compare their formula with any other data.

Rifert et al. (2018) have reviewed the literature on boiling and condensation on rotating disks. However, they have not offered any resolution of the differences between test data and theory.

2.7.3 Condensation on Rotating Vertical Cone

Sparrow and Hartnett (1961) analyzed condensation of a vertical cone rotating about its axis in stagnant vapor. They arrived at the following result for a cone with apex angle (2θ):

$$\frac{h_{cone}}{h_{disk}} = (\sin\ \theta)^{0.5} \tag{2.7.5}$$

where h_{disk} is the heat transfer coefficient of a rotating horizontal disk given graphically in Sparrow and Gregg (1959b) as well as by Eq. (2.7.4). This analysis took into consideration momentum and convection effects that were neglected in Rohsenow's analysis in arriving at Eq. (2.7.3).

Howe (1972) performed theoretical and experimental studies on condensation of steam on rotating cones.

2.7.4 Condensation on Rotating Tubes

Mathews (1962) measured condensation of steam on horizontal cylinders of diameter 102, 204, and 254 mm, rotating around their own axis. For centrifugal accelerations up to 2 g, the heat transfer coefficients were similar to the stationary-cylinder values but increased above 2 g accelerations.

Gacesa (1967) and Nicol and Gacesa (1970) condensed steam on a 25 mm diameter vertical tube, rotating on its axis. The condensing heat transfer coefficients increased with speed of rotation, and for the maximum rotational speed of 2700 rpm investigated, heat transfer coefficients were found to be four or five times the stationary value. However, for speeds up to about 600 rpm, heat transfer coefficients remained unchanged from the stationary value.

Chandran and Watson (1976) condensed methanol, isopropanol, ethyl acetate, n-butanol, and water, on horizontal 19 mm tubes. Some of the tubes were plain, some had pins on their surface, and some were coated with polytetrafluoroethylene (PTFE). Tests were made with tubes stationary and rotating. Heat transfer coefficients were found to initially decrease with increasing speed and then rose with further increase in speed. At a speed of 200 rad s^{-1}, a threefold increase in heat transfer was obtained. Pins and PTFE coatings increased heat transfer, and it was attributed to better drainage of condensate. A dimensionless correlation of their own data was developed but there was large scatter. They did not compare their correlation with data other than their own. They compared the predictions of their correlation with those proposed by others. The predictions of the various correlations differed greatly.

In conclusion, there is no prediction method that has been verified with data other than the authors' own. The data and correlations in these references may be used for guidance in design.

2.7.5 Plate-Type Condensers

The most widely used plate heat exchangers have herringbone pattern plates. Their geometry is shown in Figure 2.7.1. To ensure leak-tightness, condensers have brazed plates, and these are therefore known as BPHE. Many correlations have been proposed for condensation heat transfer in BPHE based on limited data. Among these are Han et al. (2003), Yan et al. (1999), and Kuo et al. (2005). A correlation based on data from many sources has been presented by Longo et al. (2015). In this correlation, there are two heat transfer regimes, a gravity dominated regime and a forced convection regime. The boundary between the two regimes is at $Re_{\text{eq}} = 1600$, where Re_{eq} is an equivalent Reynolds number defined as

$$Re_{\text{eq}} = G[(1-x) + x(\rho_f/\rho_g)^{1/2}]D_{\text{HYD}}/\mu_f \tag{2.7.6}$$

The mean of the inlet and outlet quality is used for x. $D_{\text{HYD}} = 2b$, where b is the height of corrugation.

The heat transfer coefficient calculated in this correlation are based on the projected area S of active plates given by

$$S = NWL_p \tag{2.7.7}$$

where N is the number of effective plates. At $Re_{\text{eq}} < 1600$, heat transfer coefficient $h_{\text{Nu,m}}$ is calculated with the Nusselt formula for condensation on vertical plates, Eq. (2.2.8), with L_p replacing L. This heat transfer coefficient $h_{Nu,m}$ is multiplied by the enlargement factor ϕ to get the heat transfer coefficient based on projected area S in the gravity controlled regime, $h_{\text{grav},m}$. Thus,

$$h_{\text{grav},m} = \phi h_{Nu,m} \tag{2.7.8}$$

For $Re_{\text{eq}} > 1600$, regime is of forced convection and the mean heat transfer coefficient h_{FC} is

$$h_{\text{FC}} = 1.875\phi(k_f/D_{\text{HYD}})Re_{\text{eq}}^{0.445}\text{Pr}_f^{1/3} \tag{2.7.9}$$

As was discussed in Section 2.6.5, superheated vapor begins to condense if wall temperature falls below the saturation temperature and that Webb (1998) gave a model for heat transfer in this region. That model is expressed by Eq. (2.6.8) given in Section 2.6.3. Longo et al. have incorporated the Webb model into the above correlation. Local single-phase heat transfer coefficient of vapor $h_{s\text{-ph}}$ is calculated by the following equation given by Thonon (1995):

$$h_{s\text{-ph}} = 0.2267(k_g/D_{\text{HYD}})Re_{\text{GS}}^{0.631}Pr_G^{1/3} \tag{2.7.10}$$

They compared this correlation with their own data as well as those from six other sources. The data included several fluids (halogenated refrigerants, propane, and carbon dioxide), chevron angles 30°–65°, (measured from the vertical axis) and mass flux 2.6–42 kg m^{-2} s^{-1}. The MAD for all data was 16%.

Zhang et al. (2019) condensed R1234ze(E) and R1233zd(E) in a plate heat exchanger with chevron angle of 65°. Saturation temperatures ranged from 30 to 70 °C, mass flux from 16 to 90 kg m^{-2} s^{-1}. They found satisfactory agreement

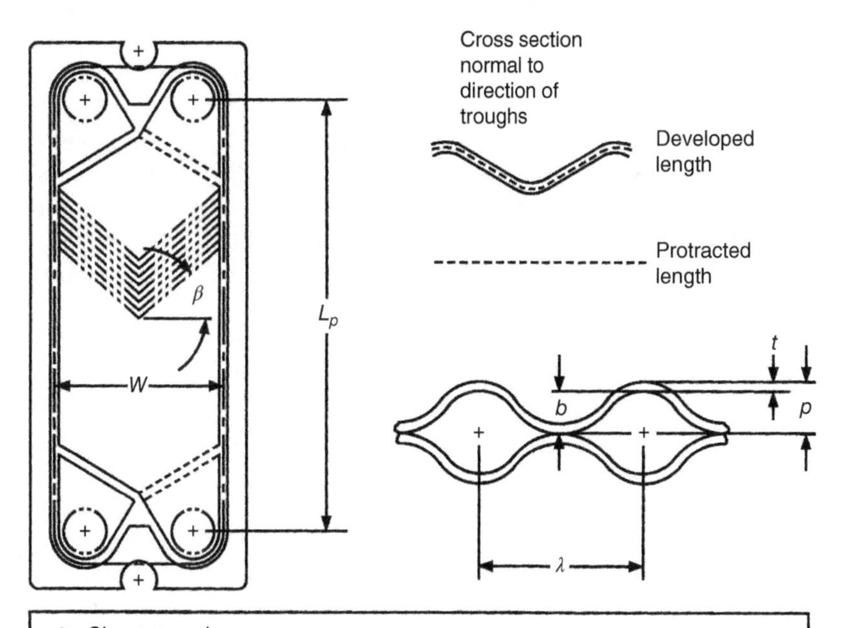

Figure 2.7.1 Corrugated plate heat exchanger. Note that in many publications, chevron angle β is measured from the vertical axis. Source: From ASHRAE (2017). © 2017 ASHRAE.

with the Longo et al. correlation, while several others had large deviations. They also gave a correlation of their own data.

Shon et al. (2018) condensed R-1233zd(E) in a BPHE with chevron angle 60°. They compared their data with correlations of Kuo et al. (2005), Yan et al. (1999), and Han et al. (2003). These were found unsatisfactory so they gave a correlation of their own data.

Eldeeb et al. (2016) have reviewed experimental studies and predictive techniques for plate condensers.

2.7.5.1 Recommendation

Longo et al. correlation is recommended in its verified range. Note that it has not been verified for water or ammonia. Their properties differ considerably from the fluids for which Longo et al. correlation was verified.

2.7.6 Effect of Oil in Refrigerants

In most compressors used in refrigeration and air conditioning, some lubricating oil gets mixed with refrigerant during compression and is carried into the system. While the larger systems use oil separators, they are not 100% effective. Small systems do not have an oil separator at all. Oil is thus carried into the condenser where it can affect its performance.

Depending on the type of oil, refrigerants may be miscible or immiscible with them. Immiscible oils were exclusively used in ammonia systems in the past and are also presently used in ammonia systems with flooded evaporators. Mirmov and Yemelyanov (1976) state that in the USSR, it was generally believed that oil films of 0.05–0.08 mm were formed on condensing surface and this resistance was added in the design of condensers. However, Abdukmanov and Mirmov (1971) state that they found in their experiments on horizontal tubes that oil increased the heat transfer coefficient by 30%. Mazukewitch (1952) found that oil reduced heat transfer in a vertical annulus by 30%. Short (2017) states that normal naphthenic or paraffinic lubricants and synthetic hydrocarbon oils have low solubility and miscibility in ammonia. These oils are heavier than ammonia and tend to form an oil film on the heat transfer surfaces. Shah (1975) studied evaporation of ammonia in a horizontal tube. Oil films were visually observed in both single-phase liquid flow and during evaporation. Single-phase heat transfer coefficients were lower than predicted by the Dittus–Boelter equation, Eq. (2.3.11). The difference in heat transfer was equivalent to the resistance oil films 0.04–0.1 mm. So, most of the evidence indicates that insulting oil films can form with immiscible oils in ammonia systems and it will be advisable to take it into account when sizing the condensers.

The most commonly used refrigerants in the past and at present are halogenated type. These are mostly miscible with and soluble in oil. Shen and Groll (2005) reviewed the research on condensation of such refrigerants. They concluded that most studies show that condensation heat transfer decreases with increasing oil content though its extent varies. Tichy et al. (1984) condensed R-12 with oil in a horizontal tube. Oil caused deterioration of heat transfer, which they expressed by the following formula:

$$\frac{h_{\text{with oil}}}{h_{\text{no oil}}} = \exp(-5x_{\text{oil}}) \tag{2.7.11}$$

where x_{oil} is the mass fraction of in the mixture.

Sur and Azer (1991) condensed refrigerant–oil mixtures in smooth and finned tubes. They found reductions in heat transfer coefficients of 7%, 12%, and 16%, with oil concentrations of 1.2%, 2.8%, and 4.0% in R-113, respectively.

Some studies also showed no effect of oil. Williams and Sauer (1981) condensed R-11 on a horizontal tube. Two types of oil were used: SUS 150 and SUS 500. Oil concentrations up to 7% had no effect on heat transfer but higher concentrations caused deterioration.

Eckels (2005) studied the effect on condensation of R-134a on horizontal banks of smooth and enhanced tubes. Two types of oils were used, ISO-68 and ISO-120. Oil concentrations up to 12.4% were used. It was found that ISO-120 oil had essentially no effect on heat transfer, while ISO-68 oil reduced heat transfer by up to 25%. The only notable difference between the two oils is that the viscosity of ISO-120 oil is about twice that of ISO-68 oil. Eckels had no explanation for the different behavior with the two oils.

Some prediction methods for miscible refrigerant–oil mixtures are now discussed. Many authors have given empirical correlations based on their own data. Equation (2.7.11) is an example of such correlations. Huang et al. (2010) compared many of these correlations with their data for R-410A in 4.18 and 1.6 mm diameter tubes. All of them gave large errors and so they gave their own correlation of their data.

A number of authors have taken the approach that the effect of oil on heat transfer is due to change in fluid properties. Thome (1998) recommended the use of pure fluid correlations but using the viscosity of oil–refrigerant mixture in place of the pure refrigerant viscosity. Several authors have recommended use of pure refrigerant condensing correlations but using mixture properties. Among them are Sur and Azer (1991), Schlager et al. (1990), and Shao and Granryd (1995). These authors were able to correlate their own data by this method. No comprehensive comparison of any method with data from many sources could be found.

It should be noted that the composition of mixture as well as the saturation temperature change as condensation progresses. Many methods for prediction of mixture properties have been proposed. Shen and Groll (2005) have discussed them and made recommendations for their choice. Calculations with this method are quite complex. Considering that the actual amount of oil is unknown and that no comprehensive verification of this method has been done, such calculations may be worthwhile only in exceptional situations, perhaps to get an idea of trends. As the oil content in normally operating systems is small, assumption of a reduction of 5–10% should be adequate for design purposes.

2.7.6.1 Recommendation

For immiscible oils, assume an oil film 0.04–0.08 mm thick. Add its resistance to that of pure fluid to calculate the effective heat transfer.

When refrigerants and oils are miscible, assume a 5–10% decrease in heat transfer due to the effect of oil.

2.7.7 Effect of Gravity

In recent years there has been much interest in space travel and space colonization. Low to zero gravity occurs under those conditions, and it is important to know their impact on heat transfer.

It is easy to see that the absence of gravity will have a huge impact and many of the predictive techniques used under terrestrial gravity will be inapplicable. For example, Nusselt analysis for condensation on horizontal and vertical surfaces is based on drainage of condensate by gravity. In the absence of gravity, condensate will not flow and hence Nusselt formulas will be inapplicable. During condensation in horizontal tubes, stratification occurs due to gravity. In the absence of gravity, there will be no stratification, and heat transfer will be the same in all orientations.

2.7.7.1 Some Formulas for Zero Gravity

Rohsenow (1973a) analyzed condensation from stagnant vapor on a tube at zero gravity. Condensate does not drain away. Instead, it forms a uniform layer around the tube whose thickness grows with time. Heat transfer in liquid occurs by conduction. The thickness of condensate $\delta = 0$ at time $t = 0$. By taking an energy balance and integrating, the following expression was obtained for value of δ at time t:

$$\left(\frac{R+\delta}{R}\right)^2 \left(\ln \frac{R+\delta}{R} - \frac{1}{2}\right) + \frac{1}{2} = \frac{2k_f \Delta T}{\varrho_f i_{fg} R^2} t \quad (2.7.12)$$

R is the radius of tube. Heat transfer coefficient h is given by

$$h = \frac{k_f}{R \ln\left(\frac{R+\delta}{R}\right)} \quad (2.7.13)$$

A similar analysis can be done for condensation of stagnant vapor on a flat plate as follows. Let y be the thickness of condensate layer at time t. Taking heat balance,

$$\frac{k_f}{\delta} = \rho_f i_{fg} \frac{dy}{dt} \quad (2.7.14)$$

Integrating with $y = 0$ at $t = 0$ and $y = \delta$ at time t,

$$\delta = \left(\frac{2k_f t}{\rho_f i_{fg}}\right)^{0.5} \quad (2.7.15)$$

The heat transfer coefficient h is

$$h = \frac{k_f}{\delta} = \left(\frac{k_f \rho_f i_{fg}}{2t}\right) \quad (2.7.16)$$

2.7.7.2 Experimental Studies

Azzolin et al. (2018) measured heat transfer during condensation of HFE-7000 in a horizontal 3.4 mm diameter tube. Tests were done in normal gravity as well as in microgravity achieved in parabolic flights. The result was that at 170 kg m^{-2} s^{-1} flow, heat transfer coefficients were the same in both gravity conditions. At lower flow rates, heat transfer coefficients in microgravity were lower, the difference increasing with decreasing mass flow rates. At 70 kg m^{-2} s^{-1}, heat transfer coefficients were 10–24% lower at various qualities under microgravity compared with normal gravity. Flow patterns were observed simultaneously. It was seen that at 170 kg m^{-2} s^{-1}, there was uniform liquid layer around the tube circumference at both gravity levels. At lower flow rates, liquid film was thicker at the bottom and thinner at top at normal gravity. At microgravity, liquid film remained uniform around the circumference at all flow rates. The thinner liquid film at top under normal gravity results in less thermal resistance and hence higher heat transfer. It is gravity that causes stratification and there is no stratification in its absence.

Lee et al. (2013) studied condensation of FC-72 at microgravity as well as gravity at Mars and the moon. Heat transfer measurements were in a horizontal tube while flow observations were made in a horizontal annulus. Liquid film thickness was uniform at microgravity while at Lunar and Martian gravities, liquid film was thicker at the bottom for lower flow rates. They did not compare their heat transfer measurements with any correlation. They developed a control volume-based model that had MAD of 25.7% with their data.

Mudawar (2017) has reviewed the literature on microgravity condensation.

2.7.7.3 Conclusion

Amount of available experimental data for condensation in microgravity conditions is very limited and there is no well-verified method for calculating heat transfer.

2.7.8 Effect of Non-condensable Gases

Presence of non-condensable gases reduces heat transfer. In the pioneering study by Othmer (1929), presence of 0.5% air in steam was found to reduce heat transfer by 50%. Similar results have been found in numerous subsequent experimental studies as well as theoretical analyses. The reason for this is explained by considering condensation of steam–air mixture on a vertical plate as shown in Figure 2.7.2. As steam condenses on the plate surface, air is left behind. A layer of air–vapor mixture of thickness δ is formed on the condensate layer. Steam vapor has now to diffuse through this air-rich layer to reach the interface. The partial pressure of steam p_{go} drops through this layer, while that of air p_{ao} rises through this layer. The total pressure at the condensate surface remains the same as in the region outside this layer. The temperature of the liquid–vapor interface is at the saturation temperature T_{gi} corresponding to the partial pressure of vapor at the interface. Heat transfer is reduced due to the mass diffusion resistance and also because air has to be cooled from the bulk temperature to the interface temperature.

Comprehensive laminar boundary layer analyses were done by Minkowycz and Sparrow (1966) for condensation from stagnant vapor and Sparrow et al. (1967) for forced convection on plates. They found that presence of non-condensables greatly reduces heat transfer. The effect is much greater in stagnant vapor than in forced flow. For steam at 100 °C, 2% mass content of air reduces heat transfer by over 70% for stagnant vapor and by 10% for forced convection. Minkowycz and Sparrow (1966) also studied the effect of superheat. For pure steam, superheat had essentially negligible effect. For steam–air mixtures, there was considerable effect of superheat. For stagnant steam at 100 °C saturation and 2% air, heat transfer was about 40% higher at 222 °C superheat.

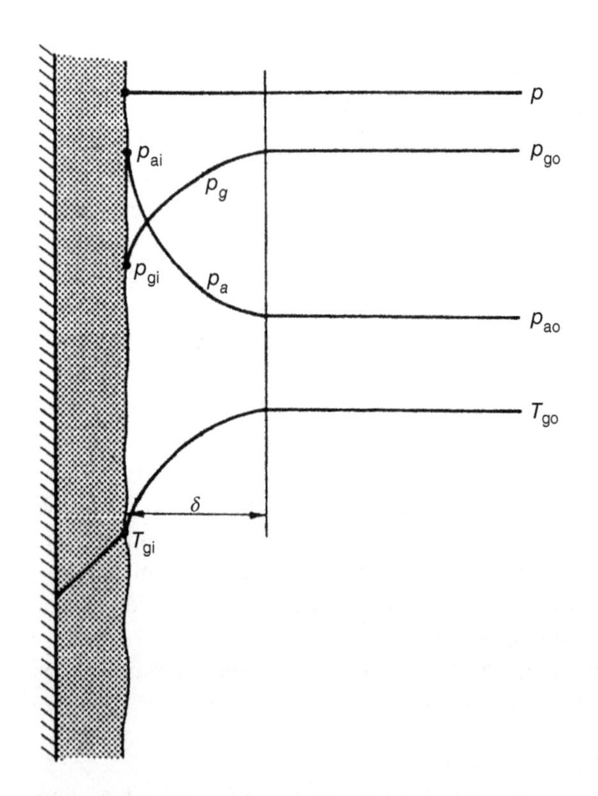

Figure 2.7.2 Effect of non-condensable gas on condensation of vapor on a vertical surface. Source: From Collier and Thome (1994). © 1994 Oxford University Press.

2.7.8.1 Prediction Methods

Colburn and Hougen (1934) presented a general method to calculate condensation in the presence of non-condensable that was further developed by later researchers. It may be written as

$$h_c(T_i - T_w) = \varsigma h_g(T_g - T_i) + G_c i_{fg} \tag{2.7.17}$$

T_i is the interface temperature and T_g is the bulk vapor–gas mixture temperature. ς is the Ackerman (1937) constant that corrects for mass transfer effects. h_c is the heat transfer coefficient for condensation of pure vapor, and h_g is the single-phase convective heat transfer coefficient between mixture and interface. ς is given by

$$\varsigma = \lambda/(1 - e^{-\lambda}) \tag{2.7.18}$$

$$\lambda = G_c C_{pg}/h_g \tag{2.7.19}$$

The mass flux of condensate G_c is given by

$$G_c = K_M \rho_g \ln\left[\frac{p - p_{g,i}}{p - p_{g,b}}\right] \tag{2.7.20}$$

where p is the total pressure, $p_{g,b}$ is the partial pressure of vapor in the bulk mixture, and $p_{g,I}$ is the partial pressure of vapor at the liquid–gas interface. The mass transfer coefficient K_M may be calculated by the analogy between heat and mass transfer, empirical correlations, or by other methods. Iterative solution of the above listed equations together with an equation for heat transfer between wall and coolant will yield the interface temperature T_i. This method can be used for external as well as internal condensation with appropriate choice of for h_c and h_g.

Many empirical correlations have been proposed. An early one is by Meisenburg et al. (1935) for condensation of steam–air outside a vertical tube. Ge et al. (2016) obtained data on a vertical plate, with the average vapor velocity of 1.2 m s^{-1}, CO_2 mass fraction in steam of 20–94%, and a pressure of 1 atm. They found the Meisenberg et al. correlation and several others to greatly underpredict their data. They fitted the following equation to their own data:

$$\frac{h_m}{h_{Nu}} = \frac{1.9864 - 0.498\ \ln(C + 16.2724)}{(T_g - T_w)^{0.18}} \tag{2.7.21}$$

where C is the mass percentage of carbon dioxide in the mixture.

A number of authors have given correlations in terms of a degradation factor defined as the ratio of heat transfer of mixture to that of pure vapor. Vierow and Shrock (1991), Kuhn et al. (1997), and Siddique et al. (1993) gave correlations based on their data for steam–gas mixtures in vertical tubes. Lee and Kim (2008) found them to be unsatisfactory on comparison of their own data for steam–nitrogen mixtures in a 13 mm diameter vertical tube. They gave the following correlation that fitted their own data as well as those of Siddiqui et al. and Kuhn et al.:

$$\frac{h_{\text{mixture}}}{h_{Nu}} = \tau_{\text{mix}}^{0.3124}(1 - 0.964C_{\text{nc}}^{0.402}) \tag{2.7.22}$$

For $0.06 < \tau_{\text{mix}} < 46.65$, $0.038 < C_{\text{nc}} < 0.814$. C_{nc} is the mass fraction of non-condensable gas and τ_{mix} is the dimensionless shear stress calculated as follows:

$$\tau_{\text{mix}} = \frac{0.5\rho_{\text{mix}}u_{\text{mix}}^2 f}{g\rho_f L} \tag{2.7.23}$$

$$u_{\text{mix}} = \frac{Re_{\text{mix}}\mu_{\text{mix}}}{\rho_{\text{mix}}D} \tag{2.7.24}$$

$$L = \left(\frac{v_f^2}{g}\right)^{1/3} \tag{2.7.25}$$

The friction factor f is given by

For $\text{Re}_{\text{mix}} > 2300$,

$$f = \frac{0.079}{Re_{\text{mix}}^{0.25}} \tag{2.7.26}$$

For $\text{Re}_{\text{mix}} < 2300$,

$$f = \frac{16}{Re_{\text{mix}}} \tag{2.7.27}$$

The range of data covered by this correlation includes tube diameters 13–49.5 mm, inlet non-condensable gas mass fraction up to 40%, pressure 0.03–0.5 MPa, and steam flow 6–62 kg m^{-2} s^{-1}.

Caruso et al. (2013) performed tests with steam–air mixtures in a tube inclined at 7° to the horizontal and gave an empirical correlation of their data. Xu et al. (2016) performed experiments with condensation of steam–air mixture in horizontal 25 mm diameter tube. The correlation of Caruso was found to overpredict their data. They have given a correlation to fit their own data.

Various correlations and theoretical methods have been reviewed by Huang et al. (2015).

2.7.8.2 Recommendation

The Lee and Kim correlation may be used for vertical tubes in its verified range. Colburn and Haugen method is recommended for general use.

2.7.9 Flooding in Upflow

The correlations for heat transfer presented in Section 2.3 are applicable only past the flooding point. Hence prediction of flooding point is important.

To understand the flooding phenomenon, consider a vapor entering at very low velocity into the bottom of a vertical tube at wall temperature below the saturation temperature of vapor. Condensate forms on the wall and flows down while the uncondensed vapor flows upwards. As velocity of vapor is increased, pressure drop rises, liquid continues flowing down though some droplets are carried away by the vapor. As vapor velocity continues to increase, a point is reached where pressure drop reaches a maximum, downwards flow of condensate ends, and condensate begins to move upwards. This is the flooding point and the vapor velocity at this point is called flooding velocity. With further increase in vapor velocity, pressure drop falls though liquid keeps moving upwards.

There are several possible mechanisms that may be the cause of flooding. These are interfacial vapor shear, entrainment of liquid in vapor stream, formation of large waves in liquid near the bottom of tube that may prevent downflow of liquid, and Bernoulli effect at tube entrance caused by high vapor velocity that causes liquid drops to be sucked upwards.

Many prediction methods, theoretical and empirical, have been proposed. Wallis (1969) considered flooding to occur due to vapor shear and gave the following correlation:

$$u_{\text{LS}}^{*1/2} + u_{\text{GS}}^{*1/2} = C \tag{2.7.28}$$

Wallis gave $C = 0.775$. Other values of C (0.88–1) have been proposed by other researchers as noted by McQuillan and Whalley (1985). In this equation u_{LS}^* and u_{GS}^* are dimensionless velocities for liquid and gas flowing alone, defined as

$$u_{\text{LS}}^* = u_{\text{LS}}\left(\frac{\rho_f}{gD(\rho_f - \rho_g)}\right)^{1/2} \tag{2.7.29}$$

$$u_{\text{GS}}^* = u_{\text{GS}}\left(\frac{\rho_g}{gD(\rho_f - \rho_g)}\right)^{1/2} \tag{2.7.30}$$

McQuillan and Whalley (1985) compared data for flooding in gas–liquid and steam-water systems from many source to a number of correlations. Best agreement was found with the correlation of Alekseev et al. (1972). They modified it to the following equation, which gives even better agreement:

$$K_g^* = 0.286Bd^{0.26}Fr^{*-0.22}\left(\frac{\mu_f}{\mu_{\text{water}}}\right)^{-0.18} \tag{2.7.31}$$

μ_{water} is the viscosity of water, taken to be 0.001 N s m^{-2} K_g^* is known as Kutateladze number defined as

$$K_g^* = J_g \frac{\rho_g^{0.5}}{[\sigma g(\rho_f - \rho_g)]^{0.25}} \tag{2.7.32}$$

Bd is the Bond number defined by

$$Bd = \frac{D^2 g(\rho_f - \rho_g)}{\sigma} \tag{2.7.33}$$

Fr^* is a dimensionless Froude number defined as

$$Fr^* = \left(\frac{w}{\pi D \rho_f}\right) \frac{g^{0.25}(\rho_f - \rho_g)^{0.25}}{\sigma^{0.75}} \tag{2.7.34}$$

where w is the mass flow rate of liquid.

Palen and Yang (2001) reviewed a number of flooding correlations. They suggested a different form of correlation but did not give its details, nor its validation.

Berrichon et al. (2015) compared their data for air–water in a 54 mm diameter tube with several correlations. The Wallis correlation with $C = 0.88$ gave the best agreement. The McQuillan and Whalley correlation also gave satisfactory agreement. They also did tests with low pressure steam.

The McQuillan and Whalley correlation was validated with a very wide-ranging database that included condensing steam. It is therefore recommended.

2.7.10 Condensation in Thermosiphons

Closed vertical thermosiphons are considered in this section. In such thermosiphons, evaporation occurs at the bottom. The vapor rises up to the section with cold wall where it condenses. The condensate flows down back into the evaporation section and thus the cycle continues.

Shiraishi et al. (1981) tested a closed vertical thermosiphon formed by a 37 mm diameter tube with water, ethanol, and R-113. Condensation heat transfer agreed fairly well with Nusselt equation.

Jouhara and Robinson (2010) studied a 6 mm diameter vertical thermosiphon with water, FC-84, FC-77, and FC-3283. In the condensation region, agreement was good with Nusselt equation at $Re_{LS} > 10$, but it over predicted at lower Re_{LS}. Hashimoto and Kaminaga (2002) had also found similar behavior in their tests on a thermosiphon. They attributed it to entrainment of liquid that thinned out liquid film and gave a formula to fit their own data. Jouhara and Robinson found it to somewhat underpredict their data and modified it to the following formula:

$$h_m = 0.85 Re_{LS}^{0.1} \exp\left(-0.000\,067\frac{\rho_f}{\rho_g} - 0.6\right) h_{Nu} \tag{2.7.35}$$

where h_{Nu} is the prediction of Nusselt formula, Eq. (2.2.8).

It may be concluded that Nusselt equations can be used with confidence for $Re_{LS} > 10$. For lower Reynolds numbers, Eq. (2.7.35) may be used but with caution as it has not had validation with other data.

2.7.11 Condensation in Helical Coils

Wongwises and Polsongkram (2006) condensed R-134a in a coil with vertical axis. Tube inside diameter was 8.3 mm and coil diameter 305 mm and coil pitch was 35 mm. Saturation condensing temperatures were from 40 to 50 °C and mass flux was 400–800 kg m^{-2} s^{-1}. They found that coil average heat transfer coefficients were 33–53% higher than for horizontal tubes. They correlated their data by the following equation:

$$\frac{h_{\text{TP}}}{D_{\text{tube}} k_f} = 0.1325 De_{\text{eq}}^{0.7654} \text{Pr}_f^{0.8144} X_{\text{tt}}^{0.0432} p_r^{-0.3356} (Bo \times 10^4)^{0.112} \tag{2.7.36}$$

De_{eq} is equivalent Dean number defined as

$$De_{\text{eq}} = \left[Re_{\text{LS}} + Re_{\text{GO}} \left(\frac{\mu_g}{\mu_f}\right) \left(\frac{\rho_f}{\rho_g}\right)^{0.5} \left(\frac{D_{\text{tube}}}{D_{\text{coil}}}\right)^{0.5} \right] \tag{2.7.37}$$

Bo is the boiling number defined as $[q/(Gi_{\text{fg}})]$, where q is the heat flux.

Gupta et al. (2014) measured heat transfer coefficients during condensation of R-134a in a horizontal coil. Tube diameter was 8.33 mm and coil diameter was 90.48 mm. Mass flux was 100–350 kg m^{-2} s^{-1} and evaporation temperature was 35–40 °C. They found satisfactory agreement with the correlation of Wongwises and Polsongkram (2006) given earlier. They also compared their data with the Shah (1979) correlation for straight tubes. Measurements were mostly 1.5–2.0 times the predictions. Deviations increased with decreasing mass flux. As was discussed in Section 2.3.3.2, the Shah (1979) correlation is inaccurate at low mass flux and has therefore been replaced by later versions, the latest being Shah (2019a).

Salimpour et al. (2017) condensed R-404A in horizontal coils. Tube diameter was 7.52 mm, while coil diameter was from 87 to 153 mm. Coil pitch varied from 15 to 35 mm. Mass flux was from 125 to 188 kg m^{-2} s^{-1} and saturation temperature was 29–39 °C. They fitted the following equation to their data:

$$h_{\text{TP}} = 0.086 De_{\text{eq}}^{0.8934} \text{Pr}_f^{0.9744} X_{\text{tt}}^{0.0832} \left(\frac{P_{\text{coil}}}{\pi D_{\text{coil}}}\right)^{-0.0355} \left(\frac{k_f}{D_{\text{tube}}}\right) \tag{2.7.38}$$

P_{coil} is the pitch of the coil.

Mozafari et al. (2015) condensed R-600a in horizontal and inclined coils. Tube diameter was 8.3 mm. The coil diameter, pitch, height, and the number of coil turns were 305 mm, 35 mm, 210 mm, and 6, respectively. Experiments were also done in a straight tube. Tests were performed at average saturation temperatures ranging between 38.5 and 47 °C. Refrigerant mass fluxes varied in the range of 155–265 kg m^{-2} s^{-1}. Inclinations of coil axis to horizontal were 0°, 30°, 60°, and 90°. The highest and lowest values of heat transfer coefficient occurred at 30° and 90° inclination angles, respectively. Heat transfer coefficients in the coils were 1.24–2.65 times those in the straight horizontal tube.

Al-Hajeri et al. (2007) have reported data for condensation of R-134a in a coil but have not compared it to any correlation.

Yu et al. (2018) condensed propane in a tube bent into a horizontal half coil with helix angle of 10°. Tube diameter was 32 mm, while the coil diameter was 2 m. Shah (2019a) found their data to be in good agreement with his correlation. The larger the coil diameter compared to tube diameter, the more the coil resembles a straight tube. The coil diameter to tube diameter ratio in these tests was 62. It indicates that for this ratio and higher ratios, straight tube correlations can be used for coils.

The preceding discussions show that heat transfer coefficients can be considerably higher than in straight tubes. Ratio of pitch to coil diameter and ratio of tube diameter to coil diameter affect heat transfer. While several correlations for heat transfer have been proposed, they have not been verified beyond the data on which they were based. Therefore, these correlations should be used only for the data range on which they were based. For $D_{\text{coil}}/D_{\text{tube}} \geq 62$, straight tube correlation may be used.

2.8 Condensation of Vapor Mixtures

Condensation of vapor mixtures is required in many industries that include chemical, petroleum, and refrigeration. In recent years, mixtures are being increasingly used as refrigerants because the traditional refrigerants such as R-12 were found to harm the ozone layer and cause global warming.

2.8.1 Physical Phenomena

Mixtures of vapors may be azeotropic or zeotropic (also called non-azeotropic). Azeotropic mixtures behave like single-component fluids and their heat transfer may be calculated by the same methods as for pure vapors using the mixture properties. In zeotropic mixtures, the components condense at different rates and their heat transfer cannot be calculated by methods for pure vapor. Their heat transfer coefficients are lower than that of the components. Figure 2.8.1 shows an example. Condensation of zeotropic mixtures of vapors of miscible fluids is discussed herein.

The boiling point of a mixture of a particular composition at a particular pressure is different from the condensation point of a mixture of the same composition. The former is known as the bubble point, while the latter is known as the dew point. The dew point is higher than the bubble point. The difference between the two is known as temperature glide or simply glide.

As condensation proceeds, the less volatile (that is with higher boiling point) component condenses at a more rapid rate. This results in the vapor mixture getting richer in the more volatile (that is with lower boiling point) component. As a result, the dew point temperature falls. With more and more condensation, dew point temperature keeps falling. When condensation is completed, the composition of the liquid is the same as of the original vapor. Sensible heat has to be removed from both vapor and liquid as their temperature falls from the initial dew point temperature to the bubble point temperature. As the composition of vapor

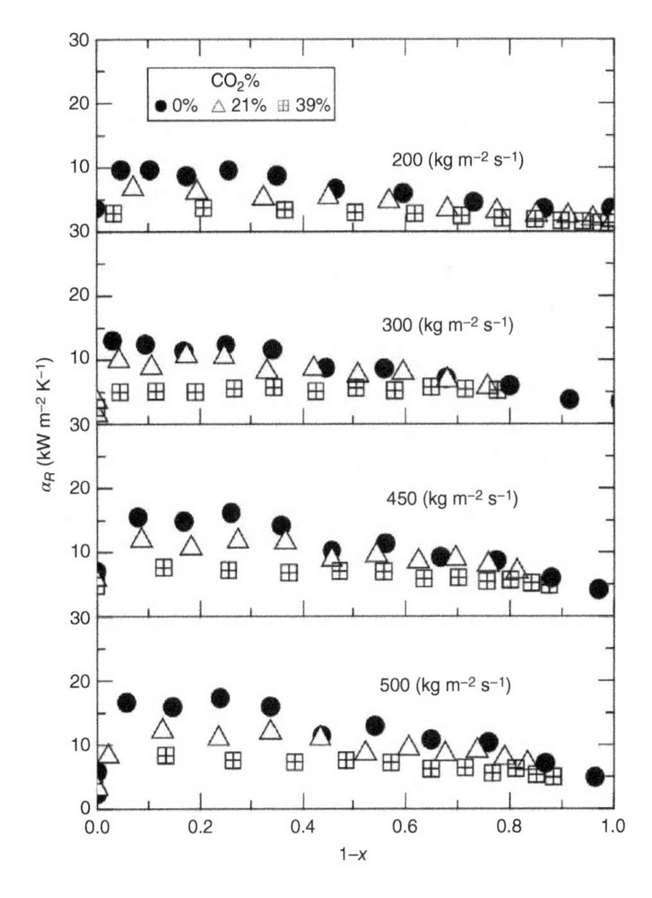

Figure 2.8.1 Effect of composition of mixture on heat transfer. CO_2/DME mixture condensing in a 4.35 mm diameter horizontal tube. Source: From Afroz et al. (2008). © 2008 Elsevier.

and liquid is varying, mass transfer resistance occurs in both vapor and liquid phases. The resistance in the vapor phase has much more effect. Heat transfer of mixtures is lowered compared to pure vapor because of these sensible heat transfer and mass transfer effects.

2.8.2 Prediction Methods

The classical method used for correcting the performance of binary mixtures is the film theory method proposed by Colburn and Drew (1937), which subsequently has been extended to multicomponent mixtures. Such methods are described in Taylor and Krishna (1993). These methods are quite tedious, especially when there are more than two components. Bell and Ghaly (1973) presented a simple method for calculating heat transfer during condensation of mixtures that has been widely used.

Bell and Ghaly assumed that the mass transfer resistance is approximately equal to the sensible heat transfer resistance of the vapor phase. Sensible heat is transferred from the vapor to the interface by single-phase convection. The sum of sensible and latent heats is transferred through the liquid film to the wall. Sensible heat flux from vapor to liquid interface is

$$q_{sens} = h_g(T_g - T_i) \tag{2.8.1}$$

where T_i is the interface temperature and h_g is the single-phase heat transfer coefficient of vapor calculated by single-phase correlations suitable for the geometry and conditions ignoring the presence of liquid. The total heat flux is

$$q_{tot} = h_c(T_i - T_w) \tag{2.8.2}$$

where h_c is the condensing heat transfer coefficient to be calculated with a pure vapor correlation. Eliminating T_i between these two equations, the following expression for the effective heat transfer coefficient h_{eff} is obtained:

$$\frac{1}{h_{eff}} = \frac{T_g - T_w}{q_{tot}} = \frac{1}{h_c} + \frac{q_{sens}/q_{tot}}{h_g} \tag{2.8.3}$$

Note that h_g and h_c are calculated using properties of mixture. h_g is calculated assuming liquid is not present. Thus for condensation in tubes, h_g is the superficial heat transfer coefficient assuming that the gas phase occupies the entire cross section. The ratio of sensible to total heat transfer may be written as

$$\frac{q_{sens}}{q_{tot}} = \frac{xC_{pg}\Delta T_{DP}}{\Delta H} \tag{2.8.4}$$

where ΔH is the change in enthalpy of mixture and ΔT_{DP} is the change in dew point temperature.

Several authors have proposed modifications of the Bell–Ghaly method. For condensation inside horizontal tubes, Del Col et al. (2005) have presented two methods applicable to different flow pattern groups. For the intermittent, annular, and mist flow patterns, they reasoned that the heat transfer coefficient of the vapor phase is increased because of interfacial waves and interfacial shear. To take these factors into account, they modified Eq. (2.8.3) to the following:

$$\frac{1}{h_{eff}} = \frac{1}{h_c} + \frac{q_{sens}/q_{tot}}{f_i h_{ga}} \tag{2.8.5}$$

where h_{ga} is the heat transfer coefficient of vapor phase in the tube cross section occupied by vapor and is calculated using the vapor velocity through the area occupied by the vapor. The interfacial friction factor, f_i, is calculated using a number of equations that include those for the calculation of flow patterns and void fraction. Their correction factor for the stratified and stratified-wavy flow patterns requires that heat flux be known.

According to McNaught (1979), the mass transfer resistance can be significantly higher than that calculated by the Bell and Ghaly method (1973). McNaught replaced h_g in Eq. (2.8.3) by $h_{g,mod}$ given by the following equation:

$$h_{g,mod} = \frac{h_g\phi}{(exp\phi - 1)} \tag{2.8.6}$$

where ϕ is given by

$$\phi = \sum_{i=1}^{n} \frac{m_i C_{pg,i}}{h_g} \tag{2.8.7}$$

where m is the individual mass flux of a component. The subscript i indicates for component i. The summation is carried out for all n components of the mixture. The predicted heat transfer coefficient by the McNaught method (1979) is always equal or lower than that by the Bell and Ghaly method (1973).

Shah et al. (2013) analyzed a database for condensation of inside tubes that included 529 data points from 22 studies, for 36 fluid mixtures, and temperature glides up to 35.5 °C for a wide range of reduced pressures and mass flow rates. These data were compared to the Shah (2009) correlation for pure vapors together with the corrections factors by Bell–Ghaly, Del Col et al., and McNaught. The correction factors of Bell–Ghaly and McNaught gave MAD of 17.7% and 17.6%, respectively. The correction factor by Del Col et al. was evaluated in its heat flux-independent regime and had MAD of 20.1%. The deviations of Bell–Ghaly and McNaught methods are about the same. Also, the Del Col et al. method is applicable only to horizontal tubes. It should also be noted that the Bell–Ghaly method has been evaluated by numerous researchers with mostly good results and is applicable to all types of condensers. There has not been much independent verification of the

McNaught method other than that by Shah et al. (2013) for condensation in tubes mentioned earlier.

2.8.3 Recommendation

Use the Bell–Ghaly method for miscible mixtures for all types of condensers.

2.9 Liquid Metals

2.9.1 Stagnant Vapors

Results of several studies on condensation from stagnant liquid metal vapors were studied by Sukhatme and Rohsenow (1964, 1966). These included those by Misra and Bonilla (1956) with mercury and sodium, Cohn (1960) with cadmium and mercury, and Roth (1962) with rubidium. The heat transfer coefficients measured in those studies were found to be much lower than Nusselt theory as well as modifications of Nusselt theory given in the comprehensive theoretical analyses by Chen (1961a), Sparrow and Gregg (1959a), and Koh et al. (1961). Figure 2.9.1 shows the findings of Sukhatme and Rohsenow. It is seen that all the data are much lower than the theoretical values.

Other studies on condensation of liquid metal vapors include Sukhatme and Rohsenow (1966) with mercury, Bakulin et al. (1967) with sodium, Ivanovskii et al. (1968) with mercury, and Subbotin et al. (1964) with potassium. All of them showed heat transfer coefficients much lower than theory.

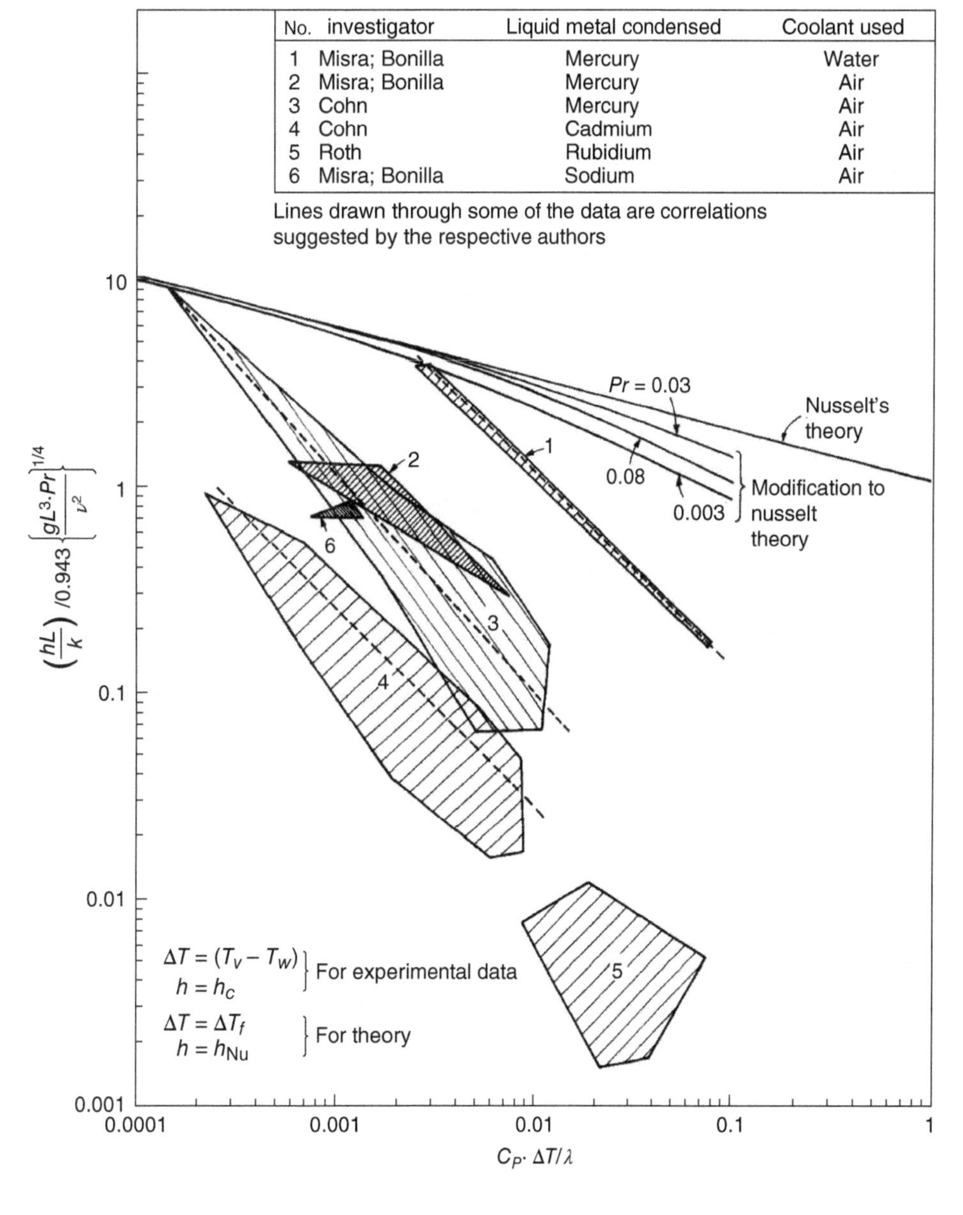

Figure 2.9.1 Data from some early studies on condensation of stagnant liquid metal vapors compared to Nusselt theory and its modifications. λ is the latent heat. Source: From Sukhatme and Rohsenow (1966). © 1966 American Society of Mechanical Engineers.

2.9.2 Interfacial Resistance

The low heat transfer coefficients discussed earlier were attributed to interfacial resistance. For vapor to condense, pressure of vapor has to be higher than that of liquid at the interface. Thus, there is an interfacial resistance. The question is about its magnitude. One theory is that all molecules striking the liquid interface are not condensed. The ratio of molecules condensed to those striking the interface is called the condensing coefficient or the accommodation coefficient. The theory of interfacial resistance was mainly developed by Schrage (1953). Rohsenow (1973a) has given the following simplified theoretical formula for interfacial heat transfer coefficient h_i:

$$h_i = \frac{2\sigma}{2-\sigma}\left(\frac{M}{2\pi R}\right)^{0.5}\frac{i_{fg}^2}{v_{fg}T_g^{3/2}} \tag{2.9.1}$$

This is applicable for

$$\frac{w/A}{\rho_f(2TR/M)^{1/2}} < 0.1 \tag{2.9.2}$$

where σ is the condensation coefficient, A is the area of surface, R is the universal gas constant, M is the molecular mass, and v_{fg} is the specific volume change during condensation. This interfacial resistance is in series to the resistance of liquid film. The actual heat transfer coefficient h is thus

$$\frac{1}{h} = \frac{1}{h_c} + \frac{1}{h_i} \tag{2.9.3}$$

where h_c is the heat transfer coefficient in the absence of interfacial resistance. Many methods to calculate σ have been proposed. Dannon (1962) has given a method to calculate the condensation coefficient. According to it, the condensation coefficient varies from 0.014 for ethyl alcohol to 1.01 for carbon tetrachloride and that for water is 0.051. These were reported to be in good agreement with test data.

From their data on condensation of mercury, Sukhatme and Rohsenow (1966) determined the condensing coefficient to be 0.45. In various studies on condensation, σ was found to be around 1 till a pressure of about 0.05 bar and dropped rapidly to as low as 0.02 (Rohsenow 1973a). However, Wilcox and Rohsenow (1970) and Sakhuja (1970) in their studies obtained condensation coefficients close to 1 at pressures up to 0.2 bar. Huang et al. (1972) performed an analysis that assumes traces of non-condensables that predicts condensing coefficient to be 0.88–0.92 at all pressures up to the atmospheric.

Rohsenow (1973a, 1973b) carried out a careful review of the earlier experimental studies done at M.I.T. as to the extent to which the apparatus had been evacuated and the uncertainty in measurements. He concluded that in all cases in which low condensation coefficient had been found, it could be attributed to inadequate evacuation and/or experimental uncertainties. He pointed out that at pressure above 0.1 bar, $(T_{SAT} - T_i)$ is small compared with $(T_{SAT} - T_w)$ and therefore is difficult to measure with sufficient accuracy. Further, gases may be released from the vessel walls at the higher temperatures. In the later study of Sakhuja (1970), traces of secondary gases were collected in a secondary condenser and bled to the atmosphere. When bleeding was stopped, condensation coefficients fell to the low values in other studies. When bleeding was resumed, the coefficients rose back to 1. He concluded that the condensing coefficient is really always = 1. Berman (1969) also gave the opinion that condensing coefficient is always 1, the lower values reported by some are due to presence of non-condensable gases.

2.9.3 Moving Vapors

Experiments on condensation of potassium containing 1.5% sodium in a 4 mm diameter tube were done by Aladyev et al. (1966). Vapor velocities were 40–290 m s^{-1} and pressures ranged 0.05–1.3 bar. They estimate that the resistance of the condensate film was less than 9% of the total. They correlated their data by the following formula:

$$h_{TP} = 7.94p^{0.22} \tag{2.9.4}$$

where h_{TP} is in W m^{-2} K^{-1} and p is in N m^{-2}. They compared this equation with data for condensation of stagnant vapor from several sources including Misra and Bonilla (1956). All of them agreed with this equation. It would therefore appear that vapor velocity and geometry have no effect. This behavior is quite different from that of non-metallic fluids as was discussed in Section 2.3. It is unlikely that the mechanism of heat transfer of liquid metals be quite different from that of non-metallic fluids. Most probably, the apparent discrepancy is because the resistance of metallic condensate is very small compared with that of non-condensables, which prevents accurate measurement of heat transfer coefficient.

Kroger and Rohsenow (1967) gave a correlation for condensation coefficients that was in good agreement with data for stagnant metal vapors as well as the data of Aladyev et al. for high velocity vapor discussed earlier. This indicated that the data of Aladyev et al. were also affected by non-condensables.

2.9.4 Recommendation

For stagnant vapor, use Eq. (2.9.3) with h_i by Eq. (2.9.1) using $\sigma = 1$, and h_c by the Chen (1961a,1961b) analytical solution represented by Eq. (2.2.16). When using Eq. (2.2.16), calculate h_{Nu} by Nusselt equation appropriate for the condensing surface geometry. Correct for effect of non-condensables as described in Section 2.7.8.

For moving vapors, no verified general method is available. Conservatively, heat transfer coefficient may be calculated assuming stagnant vapor.

2.10 Dropwise Condensation

All discussions in this chapter till now have been on film-type condensation in which condensate forms a continuous layer on the surface. In dropwise condensation, condensation occurs in the form of drops with space between them. Heat transfer coefficients in dropwise condensation are an order of magnitude higher than in film condensation. Hence there is, and has been for a long time, a keen interest in developing heat exchangers with dropwise condensation. As yet durable and economical surfaces with dropwise condensation have not been developed. Ahlers et al. (2019) have reviewed current technologies and discussed future possibilities.

2.10.1 Prediction of Mode of Condensation

It has been observed that dropwise condensation occurs only when the liquid does not wet the surface. Thus pure mercury does not wet metals and always gives dropwise condensation. Steam produces drop condensation on noble metals but on ordinary metals condenses as a film. To achieve dropwise condensation of steam on ordinary metals, promoters have to be applied on the surfaces that render them non-wetting. It is of theoretical and practical interest to be able to predict the mode of condensation. It has been found that dropwise condensation occurs if the liquid–vapor surface tension is greater than the surface free energy or surface tension of the solid surface, Merte (1973). Davies and Ponter (1968) studied data for condensation on Teflon surfaces and found that dropwise condensation occurred when liquid–vapor surface tension was between 0.033 and 0.061 N m^{-1} and film condensation occurred when its value was between 0.019 and 0.02. The surface tension of Teflon is 0.018 N m^{-1} according to Shaffrin and Zisman (1960).

A related topic is transition from dropwise to film condensation. Rose (2002) reviewed the experimental data and theories. He concluded that this phenomenon is still not clearly understood. The important mechanisms appear to be the proximity of active site with increasing temperature difference between vapor and surface and with the speed at which drops coalesce as compared with their growth rate.

2.10.2 Theories of Dropwise Condensation

Theories of dropwise condensation have been reviewed among others by Rose (2002) and Huang et al. (2015).

There are two basic theories of dropwise condensation. According to the first theory, a thin condensate layer is initially formed by the film condensation process. When this film reaches a certain critical thickness, surface tension forces fracture it and rolls it up in the form of drops. This theory was first put forward by Jakob (1936) and has been further developed by others such as Baer and Mckelvey (1958), Welch and Westwater (1961), and Sugawara and Katsuta (1966). Welch and Westwater estimated the critical thickness of film to be 0.5–1 μm. Majumdar and Mezic (1999) actually observed with a microscope that there was a liquid film with a few nanometers thickness on the condenser surface. According to other theories, condensation occurs on random nucleation sites such as pits and scratches. In their experiments, Umur and Griffith (1965) observed that on the surface between drops, the thickness of liquid film was no more than a monomolecular layer, and essentially no condensation occurs on it. This contradicts the critical film thickness theory. Other studies supporting the random nucleation theory include those of Liu et al. (2007) and Mu et al. (2008). As yet there is no consensus on the mechanism of dropwise condensation.

2.10.3 Methods to Get Dropwise Condensations

Most metals have high surface energy and hence are hydrophilic. For dropwise condensation, the surface should be hydrophobic, that is, can repel water. Ordinary metals can be made hydrophobic by treating the surface with substances known as promoters. The method that has been in use from the earliest is to continuously or intermittently inject a promoter into the condenser. The promoters that have been used in steam condensers include stearic acid, montan wax, benzyl mercaptan, oleic acid, dioctadecyle disulfide, and cupric oleate. Another method is to coat the surface with a durable layer of a promoter such as Teflon. However, it has not been possible to achieve a durable layer.

In recent years, a variety of surfaces have been developed in an effort to achieve long-lasting dropwise condensation. This is done by changing the surface characteristics. The techniques used include lithography, plasma treatment, chemical deposition, solution casting, electro-spraying, etc. These have been described and reviewed by Ahlers et al. (2019). However, none of these surfaces has achieved the goal of durability together with economic viability.

2.10.4 Some Experimental Studies

Figure 2.10.1 shows data on dropwise condensation of steam at near atmospheric pressure from 1930 onwards put together by Rose (2002). These were obtained on surfaces with promoters. It is seen that there are very large differences between the data from different sources. According to Rose, the data in the hatched portion are the correct ones. The data with large ΔT are erroneous due to experimental errors and/or presence of non-condensing gases. As was discussed in Section 2.7.8, non-condensable gases greatly reduce heat transfer. Rose points out that careful experimentation shows that heat transfer coefficient increases with increasing temperature difference, the reason being that more nucleation points get activated. The data in the hatched portion of Figure 2.10.1 show this trend, while the data lines with large slope show decreasing heat transfer coefficients with increasing temperature difference.

Koch et al. (1998) studied the effect of orientation of a flat condensing surface. Their results are shown in Figure 2.10.2. It is seen that heat transfer coefficients decrease as inclination changes from vertical to horizontal. Heat transfer coefficient in the horizontal position is about 40% of that in the vertical position. Graham (1969) studied data from three sources and also found similar effect of inclination. This effect is because drainage of drops is retarded as inclination becomes less. Further, a drop sitting on a horizontal surface occupies more area than a drop hanging from a vertical surface and hence has higher thermal resistance.

2.10.5 Prediction of Heat Transfer

There are several resistances involved in the overall heat transfer during dropwise condensation. Griffith (1973) have represented the overall heat transfer coefficient U as follows:

$$\frac{1}{U} = \frac{1}{h_{nc}} + \frac{1}{h_i} + \frac{1}{h_{dc}} + \frac{1}{h_p} + \frac{1}{h_c} \tag{2.10.1}$$

here h_{nc} is the heat transfer coefficient due to presence of non-condensables. Effect of non-condensables on film condensation was discussed in Section 2.7.8. This resistance can be treated in the same way for dropwise condensation. The impact of non-condensable is usually more severe in dropwise condensation as the heat transfer coefficients of pure vapor are much higher than in film condensation.

h_i is the heat transfer coefficient due to interfacial resistance. It can be calculated in the way described in Section 2.9.1. The condensation coefficient is to be taken as 1.

h_{dc} is the heat transfer coefficient for conduction through liquid drops. Its magnitude depends on the size and population of drops. Le Fevre and Rose (1966) developed a method to calculate it. Their method combines an expression for the heat transfer through a drop of given size with an expression for the average distribution of drop sizes. These are integrated overall drop sizes to obtain, for a given vapor–surface temperature difference, the heat flux for the surface. Rose (2002) reports that this method was found to be in good agreement with wide-ranging steam data as well as data for several organics form independent sources. Calculations with this method are quite complex.

Griffith (1973) has given the following correlation for steam on vertical surfaces:

$25 < T_{SAT} < 100\,°C$,

$$h_{dc} = 47\,628 + 2041 T_{SAT} \tag{2.10.2}$$

$T_{SAT} > 100\,°C$,

$$h_{dc} = 255\,150 \tag{2.10.3}$$

The units are T_{SAT} in C and h in W m^{-2}K^{-1}.

Rose (2002) gives the following equation that fits the steam data considered by him to be reliable:

$$q = T_{SAT}^{0.8}[5\Delta T_{SAT} + 0.3(\Delta T_{SAT})^2] \tag{2.10.4}$$

This is applicable to pressures near atmospheric and lower. T is in C and q in kW m^{-2}.

h_p is the promoter heat transfer coefficient. Heat is conducted through the promoter before it reaches the condensing surface. Thus,

$$h_p = \frac{k}{\delta} \tag{2.10.5}$$

where δ is the thickness of promoter layer and k is its thermal conductivity.

Usually this resistance is negligible as the promoter is only a monolayer. For some promoters such as Teflon, the thermal conductivity is low and the layer thickness is significant. The minimum required thickness of Teflon is around 1.5 μm (Griffith 1973), and its thermal conductivity is 0.17 W m K^{-1}. This gives h_p = 113 300 W m^{-2} K^{-1}.

h_c is the constriction heat transfer coefficient attributed to thermal conductivity of condensing surface. Some researchers reported that thermal conductivity has great effect on heat transfer. As the heat of condensation has to pass through the metal surface, heat transfer is "constricted" if the thermal conductivity is low. Rose (2002) examined the data on the effect of thermal conductivity of condensing surface. He concluded that the balance of evidence suggests that this is significant only at very low heat fluxes and for very small condensing surfaces. Most of the data showing large constriction effect are affected by non-condensables and experimental errors. Hence the constriction effect is not pertinent to most practical heat exchangers.

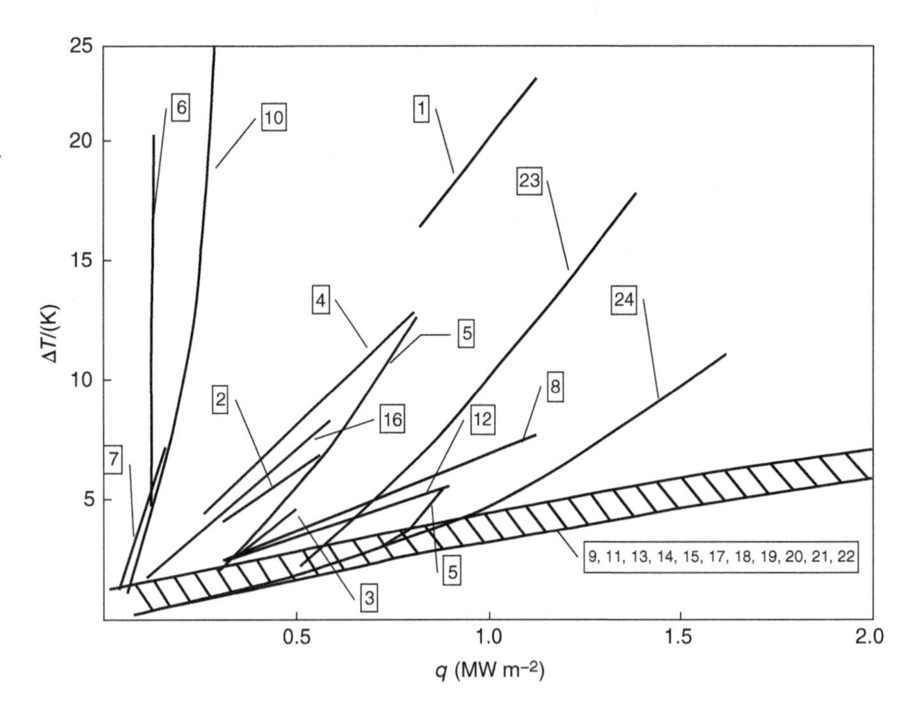

Figure 2.10.1 Heat transfer measurements on dropwise condensation of steam at near atmospheric pressure. Source: From Rose (2002). © 2002 SAGE Publications.

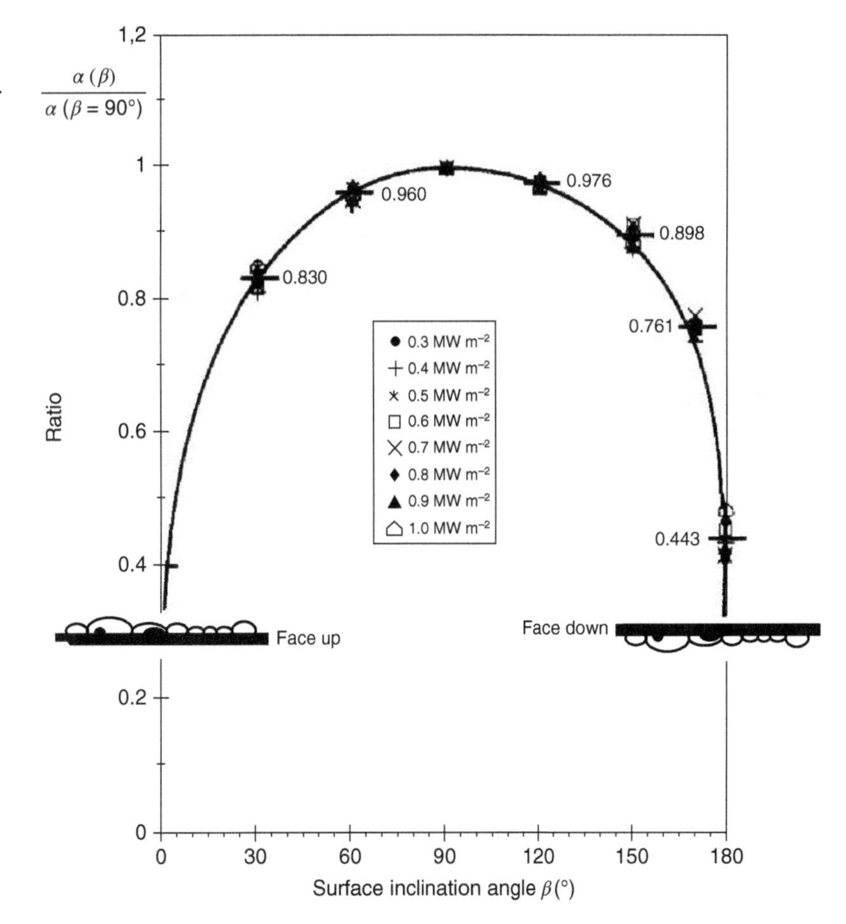

Figure 2.10.2 Effect of surface inclination on heat transfer coefficient α during dropwise condensation. Source: From Koch et al. (1998). © 1998 Elsevier.

2.10.6 Recommendations

For steam near atmospheric pressure and lower pressures, use Eq. (2.10.4) to obtain overall heat transfer coefficient in the absence of non-condensable gases. For other fluids and conditions, use the correlation of Le Fevre and Rose for h_{dc}. If it is a flat surface and it is not vertical, correct for the effect of orientation using Figure 2.10.2. Correct for other resistances as described in Section 2.10.5.

Nomenclature

A	area (m^2)
Bd	Bond number = $g(\rho_L - \rho_G)D^2\sigma^{-1}$ (–)
D	diameter of tube (m)
De_{eq}	equivalent Dean number defined by Eq. (2.7.37) (–)
D_{HP}	equivalent diameter based on perimeter with heat transfer (m)
D_{HYD}	hydraulic equivalent diameter (m)
Fr	Froude number = $\left(\frac{G^2}{\rho^2 gD}\right)$ (–)
Fr_{LT}	Froude number = $\left(\frac{G^2}{\rho_f^2 gD}\right)$ (–)
G	total mass flux (liquid + vapor) (kg m^{-2} s^{-1})
g	acceleration due to gravity (m s^{-2})
h	heat transfer coefficient (W m^{-2} K^{-1})
h_I	heat transfer coefficient in Shah correlation given by Eq. (2.3.16) (W m^{-2} K^{-1})
h_{TP}	two-phase heat transfer coefficient (W m^{-2} K^{-1})
$h_{TP,\theta}$	two-phase heat transfer coefficient at tube inclination θ (W m^{-2} K^{-1})
J_g	dimensionless vapor velocity defined by Eq. (2.3.4) (–)
k	thermal conductivity (W m^{-1} K^{-1})
L	length (m)
MAD	mean absolute deviation
N	number of data points, or number of tube rows (–)
Nu	Nusselt number (–)
p_r	reduced pressure (–)
Pr	Prandtl number, (–)
R	universal gas constant, = 8.134 J m^{-1} K^{-1}
Ra	Raleigh number (–)
Re	Reynolds number = $GD\mu^{-1}$ (–)
Re_{GT}	Reynolds number for all mass flowing as vapor = $GD\mu_G^{-1}$ (–)
Re_{GS}	Reynolds number assuming vapor phase flowing alone = $GxD\mu_G^{-1}$ (–)
Re_{LS}	Reynolds number assuming liquid phase flowing alone, = $G(1-x)D\mu_L^{-1}$ (–)
Re_{LT}	Reynolds number for all mass flowing as liquid = $GD\mu_L^{-1}$ (–)
Su	Sugomel number (–)
T	temperature (K)
ΔT	$(T_w - T_{SAT})$ (K)
ΔT_{SAT}	$(T_w - T_{SAT})$ (K)
u	velocity (m s^{-1})
We_{GT}	Weber number for all mass flowing as vapor, given by Eq. (2.3.54) (–)
w	mass flow rate of condensate (kg s^{-1})
x	vapor quality (–)
X_{tt}	Martinelli parameter defined by Eq. (2.3.6) (–)
Z	Shah's correlating parameter defined by Eq. (2.3.13) (–)

Greek Letters

μ	dynamic viscosity (Pa s)
ν	kinematic viscosity (m^2 s^{-1})
ρ	density (kg m^{-3})
σ	surface tension (N m^{-1})
θ	angle (° or rad)
Γ	condensate mass flow rate per unit width(kg m^{-1} s^{-1})

Subscripts

f	liquid
FC	forced convection
G	vapor
GO	vapor phase flowing alone, same as GS
GS	vapor phase flowing alone, superficial vapor
GT	all mass flowing as vapor
g	vapor
i	interface
L	liquid
LO	liquid phase flowing alone, same as LS
LS	liquid phase flowing alone, superficial liquid
LT	all mass flowing as liquid
m	mean
mix	mixture
Nu	by Nusselt equation
SAT	saturation
w	wall
z	at location z

References

Abdukmanov, K.A. and Mirmov, N.I. (1971). Experimental study of oil-contaminated ammonia vapor on horizontal tubes. *Heat Transfer Sov. Res.* **3** (6): 176–180.

Ackerman, G. (1937). Simultaneous heat and molecular mass transfer by overall temperature and partial pressure differences. *VDI Forschungsh.* **382**: 1–16.

Afroz, H.M.M., Miyarah, A., and Tsubaki, K. (2008). Heat transfer coefficients and pressure drops during in-tube condensation of CO_2/DME mixture refrigerant. *Int. J. Refrig.* **31**: 1458–1466.

Agarwal, R. and Hrnjak, P. (2015). Condensation in two phase and desuperheating zone for R1234ze(E), R134a and R32 in horizontal smooth tubes. *Int. J. Refrig.* **50**: 172–183.

Ahlers, M., Buck-Emden, A., and Bart, H. (2019). Is dropwise condensation feasible? A review on surface modifications for continuous dropwise condensation and a profitability analysis. *J. Adv. Res.* https://doi.org/10.1016/j.jare.2018.11.004.

Akers, W.W., Deans, H.A., and Crosser, O.K. (1959). Condensing heat transfer within horizontal tubes. *Chem. Eng. Prog. Symp. Ser.* **59** (29): 171–176.

Aladyeev, I.T., Kondratyev, N.S., Mukhin, V.A. et al. (1966). Thermal resistance of phase transition with condensation of potassium vapor. In: *Proceedings of the Third International Heat Transfer Conference*, vol. **2**, 312–317. Chicago: Begell House.

Alasadi, A.A.M., Ezzat, A.W., and Muneer, A. (2013). Investigation of steam condensation process on rotating disk condenser at different rotation speed. *Int. J. Comput. Appl.* **84** (14): 10–18.

Alekseev, V.P., Poberezkin, A.E., and Gerasimov, P.V. (1972). Determination of flooding regular packings. *Heat Transfer Sov. Res.* **4**: 159–163.

Al-Hajeri, M., Koluib, A., Mosaad, M., and Al-Kulaib, S. (2007). Heat transfer performance during condensation of R-134a inside helicoidal tubes. *Energy Convers. Manag.* **48**: 2309–2315.

Ali, H.M. (2017). Condensation heat transfer on geometrically enhanced horizontal tube: a review. In: *Heat Exchangers - Advanced Features and Applications* (ed. S.M.S. Murshed). InTech https://doi.org/10.5772/65896.

Altman, M., Staub, F.W., and Norris, R.H. (1959). *Local Heat Transfer and Pressure Drop for Refrigerant 22 Condensing in Horizontal Tubes*. Storrs, CT: ASME-AIChE.

Ananiev, E.P., Boyko, I.D., and Kruzhilin, G.I. (1961). Heat transfer in the presence of steam condensation in horizontal tubes. *Int. Dev. Heat Transfer* **2**: 290–295.

Aroonrat, K. and Wongwises, S. (2018). Condensation heat transfer and pressure drop characteristics of R-134a flowing through dimpled tubes with different helical and dimpled pitches. *Int. J. Heat Mass Transf.* **121**: 620–631.

ASHRAE (2017). *Handbook Fundamentals*, 4.24. Atlanta, GA: ASHRAE.

Awad, M.M., Dalkilic, A.S., and Wongwises, S. (2014). A critical review on condensation heat transfer in microchannels and minichannels. *J. Nanotechnol. Sci. Med.* **5**: 010801-1–010801-25.

Azer, N.Z., Abis, L.V., and Soliman, H.M. (1972). Local heat transfer coefficients during annular flow condensation. *ASHRAE Trans.* **78**: 135–143.

Azzolin, M., Bortolina, S., Nguyen, L.P.L. et al. (2018). Experimental investigation of in-tube condensation in microgravity. *Int. Commun. Heat Mass Transfer* **96**: 69–79.

Bae, S., Maulbetsch, J.L. and Rohsenow, W.M. (1968). Forced Convection Condensation Inside Tubes. M.I.T. Report. *DSR-79760-59*.

Baer, E. and Mckelvey, J.M. (1958). Heat transfer in dropwise condensation. ACS Delaware Science Symposium. Quoted in Huang et al. (2015).

Bakulin, N.V., Ivanovskii, M.N., Sorokin, V.P., and Subbotin, V.I. (1967). Intensity of heat transfer in film condensation of pure sodium vapour. *Teplofiz. Vysok. Temper* **5** (5): 930. Quoted in Necme and Rose (1976).

Beatty, K.O. and Katz, D.L. (1948). Condensation of vapors on outside of finned tubes. *Chem. Eng. Prog.* **44**: 55–70.

Beatty, K.O. & Nandapurkar, S.S. (1959) Condensation on a horizontal rotating disc. Preprint 112, presented at the Third National Heat Transfer Conference, ASME-AIChE, Storrs, Conn., August, 1959. Quoted in Rohsenow (1973a).

Belghazi, M., Bontemp, A., Signe, J.C., and Marvillet, C. (2001). Condensation heat transfer of a pure fluid and binary mixture outside a bundle of smooth tubes, comparison of experimental results and a classical model. *Int. J. Refrig.* **24**: 841–855.

Bell, K.J. and Ghaly, M.A. (1973). An approximate generalized method for multicomponent partial condenser. *AIChe. Symp. Ser.* **69**: 72–79.

Berman, L.D. (1969). Heat transfer during film condensation of vapor on horizontal tubes in transverse flow. In: *Convective Heat Transfer in Two-Phase and One Phase Flow* (eds. V.M. Borishnaskii and I.I. Paleev). Jerusalem: I. P. S. T Press.

Berman, L.D. and Tumanov, Y.A. (1962). Condensation heat transfer in a vapour flow over a horizontal tube. *Teploenergerika* **10**: 77–84. Quoted in Berman (1969).

Berrichon, J.D., Louahlia-Gualous, H., Bandelier, P. et al. (2015). Experimental study of flooding phenomenon in a power plant reflux air-cooled condenser. *Appl. Therm. Eng.* **79**: 214–224.

Bohdal, T., Charun, H., and Sikora, M. (2011). Comparative investigations of the condensation of R134a and R404A refrigerants in pipe minichannels. *Int. J. Heat Mass Transf.* **54** (9–10): 1963–1974.

Briggs, A. and Rose, J.W. (1994). Effect of fin efficiency on a model for condensation heat transfer on a horizontal, integral-fin tube. *Int. J. Heat Mass Transf.* **37** (1): 457–463.

Bromley, L.A. (1952). Effect of heat capacity of condensate. *Ind. Eng. Chem.* **44**: 2966–2968.

Browne, M.W. and Bansal, P.K. (1999). An overview of condensation heat transfer on horizontal tube bundles. *Appl. Therm. Eng.* **19**: 565–594.

Butterworth, D. (1977). Developments in the Design of Shell and Tube Condensers. ASME Paper 77-WA/HT-24.

Carnavos, T.C. (1980). Heat transfer performance of internally finned tubes in turbulent flow. *Heat Transfer Eng.* **1** (4): 32–37.

Carpenter, F.G. and Colburn, A.P. (1951). The effect of vapor velocity in condensation inside tubes. In: *General Discussion on Heat Transfer*, 20–26. Institution of Mechanical Engineers and ASME.

Caruso, G., Vitale Di Maio, D., and Naviglio, A. (2013). Condensation heat transfer coefficient with noncondensable gases inside near horizontal tubes. *Desalination* **309**: 247–253.

Cavallini, A. and Zecchin, R. (1966). High velocity condensation of R-11 vapor inside vertical tubes. *Proceedings of the International Institute of Refrigeration 1966*, Commission 2, Trondheim, Norway.

Cavallini, A. and Zecchin, R. (1974). A dimensionless correlation for heat transfer in forced convection condensation. In: *Proceedings of the Fifth International Heat Transfer Conference*, vol. **3**, 309–313. Tokyo: Begell House.

Cavallini, A., Doretti, L., Klammsteiner, N. et al. (1995). Condensation of new refrigerants inside smooth and enhanced tubes. In: *Proceedings oOf the 19th International Congress of Refrigeration*, vol. **4**, 105–114. The Hague.

Cavallini, A., Del Col, D., Doretti, L. et al. (1999). A new computational procedure for heat transfer and pressure drop during refrigerant condensation inside enhanced tubes. *Enhanced Heat Transfer* **6**: 441–456.

Cavallini, A., Censia, G., Del Col, D. et al. (2003). Condensation inside and outside smooth and enhanced tubes — a review of recent research. *Int. J. Refrig.* **26**: 373–392.

Cavallini, A., Del Col, D., Doretti, L. et al. (2006). Condensation in horizontal smooth tubes: a new heat transfer model for heat exchanger design. *Heat Transfer Eng.* **27** (8): 31–38.

Cavallini, A., Del Col, D., Mancin, S., and Rossetto, L. (2009). Condensation of pure and near-azeotropic refrigerants in microfin tubes: a new computational procedure. *Int. J. Refrig.* **32**: 162–174.

Chamra, L.M. and Mago, P.J. (2006). Modeling of condensation heat transfer of refrigerant mixture in micro-fin tubes. *Int. J. Heat Mass Transf.* **29**: 1915–1921.

Chamra, L.M., Mago, P.J., Tan, M., and Kung, C. (2005). Modeling of condensation heat transfer of pure refrigerant in micro-fin tubes. *Int. J. Heat Mass Transf.* **48**: 1293–1302.

Chandran, R. and Watson, F.A. (1976). Condensation on static and rotating pinned tubes. *Trans. Inst. Chem. Eng.* **54** (2): 65–72.

Chato, J.C., (1960). Laminar condensation inside horizontal and inclined tubes. PhD thesis at the Massachusetts Institute of Technology, Cambridge, MA.

Chen, M.M. (1961a). An analytical study of laminar film condensation part 1- flat plates. *J. Heat Transf.* **83** (1): 48–54.

Chen, M.M. (1961b). An analytical study of laminar film condensation part 1-single and multiple horizontal tubes. *J. Heat Transf.* **83** (1): 55–60.

Chen, S.L., Gerner, F.M., and Tien, C.L. (1987). General film condensation correlations. *Exp. Heat Transfer* **1** (2): 93–107.

Cho, Y., Kim, S., Bae, B. et al. (2013). Assessment of condensation heat transfer model to evaluate performance of the passive auxiliary feedwater system. *Nucl. Eng. Technol.* **45** (6): 759–766.

Cohn, P. D. (1960). MS thesis. Oregon State College. Quoted in Sukhatme and Rohsenow (1964).

Colburn, A.P. and Drew, T.B. (1937). The condensation of mixed vapors. *Trans. Am. Inst. Chem. Eng.* **33**: 197–215.

Colburn, A.P. and Hougen, O.A. (1934). Design of cooler condensers for mixture of vapors with noncondensing gases. *Ind. Eng. Chem.* **26**: 1178–1182.

Collier, J.G. and Thome, J.R. (1994). *Convective Boiling and Condensation*, 3e. Oxford, UK: Oxford University Press.

Da Riva, E., Del Col, D., Garimella, S.V., and Cavallini, A. (2012). The importance of turbulence during condensation in a horizontal circular minichannel. *Int. J. Heat Mass Transf.* **55** (13–14): 3470–3481.

Dalkilic, A.S. and Wongwises, S. (2009). Intensive literature review of condensation inside smooth and enhanced tubes. *Int. J. Heat Mass Transf.* **52**: 3409–3426.

Dannon, F. (1962). Topics Statistical Mechanics. *USAEC Report UCRL 10029.*

Davies, G.A. and Ponter, A.B. (1968). The prediction of the mechanism of on tubes coated with tetrafluoroethylene. *Int. J. Heat Mass Transf.* **11**: 375–377.

Del Col, D., Cavallini, A., and Thome, J.R. (2005). Condensation of zeotropic mixtures in horizontal tubes: new simplified heat transfer model based on flow regimes. *J. Heat Transf.* **127**: 221–230.

Del Col, D., Bortolato, M., Azzolin, M., and Bortolin, S. (2014). Effect of inclination during condensation inside a square cross section minichannel. *Int. J. Heat Mass Transf.* **78**: 760–777.

Del Col, D., Bortolin, S., and Das Riva, E. (2015). Predicting methods and numerical modeling of condensation in microchannels. In: *Encyclopedia of Two-Phase Heat Transfer and Fluid Flow II – Special Topics and Applications, Special Topics in Condensation*, vol. **3** (ed. J.R. Thome), 37–84. New York: World Scientific Publication.

Dobran, F. and Thorsen, R.S. (1980). Forced flow laminar filmwise condensation of a pure saturated vapor in a vertical tube. *Int. J. Heat Mass Transf.* **23**: 161–171.

Dobson, M.K. and Chato, J.C. (1998). Condensation in smooth horizontal tubes. *J. Heat Transf.* **120**: 193–213.

Dorao, C.A. and Fernandino, M. (2018). Simple and general correlation for heat transfer during flow condensation inside plain pipes. *Int. J. Heat Mass Transf.* **122**: 290–305.

Dukler, A.E. (1960). Fluid mechanics and heat transfer in vertical falling film systems. *Chem. Eng. Symp. Ser.* **56** (30): 1–10.

Eckels, S. J. (2005). Effects of Inundation and Miscible Oil Upon Condensation Heat Transfer Performance of R-134a. *ASHRAE report 984.*

Eisenberg, D.M. (1972). An investigation of the variables affecting steam condensation on the outside of a horizontal tube bundle. PhD thesis. University of Tennessee, Knoxville. Quoted in Belghazi et al. (2001).

El Hajal, J., Thome, J.R., and Cavallini, A. (2003). Condensation in horizontal tubes. Part 1: two-phase flow pattern map. *Int. J. Heat Mass Transf.* **46**: 3349–3363.

Eldeeb, R., Aute, V., and Radermacher, R. (2016). A survey of correlations for heat transfer and pressure drop for evaporation and condensation in plate heat exchangers. *Int. J. Refrig.* **65**: 12–26.

Fronk, B.M. and Garimella, S. (2016). Condensation of ammonia and high-temperature-glide ammonia/water zeotropic mixtures in minichannels – part I: measurements. *Int. J. Heat Mass Transf.* **101**: 1343–1356.

Fujii, T. and Oda, K. (1986). Correlation equations of heat transfer for condensate inundation on horizontal tube bundles. *Trans. Jpn. Soc. Mech. Eng. B* **52** (474): 822–826, in Japanese. Quoted by Zeinelabedien et al. (2018).

Fujii, T., Uehera, H., Hirata, K., and Oda, K. (1972). Heat transfer and flow resistance in condensation of low pressure steam flowing through tube banks. *Int. J. Heat Mass Transf.* **15**: 247–260.

Fuks, S.N. and Zernova, E.P. (1970). Heat and mass transfer with condensation pure steam and steam containing air supplied from the side of a bank. *Therm. Eng.* **17** (3): 84–90.

Gacesa, M. (1967). Condensation of steam on a vertically rotating cylinder. Electronic Thesis and Dissertations. https://scholar.uwindsor.ca/etd/6476.

Garimella, S., Killion, J.D., and Coleman, J.W. (2001). An experimentally validated model for two-phase pressure drop in the intermittent flow regime for circular microchannels. In: *Proceedings of NHTC'01*, NHTC2001-20103. ASME.

Garimella, S., Fronk, B.M., Milkie, J.A., and Keinath, B.L. (2014). Versatile models for condensation of fluids with widely varying properties from the micro to macroscale. Proceedings of the 15th International Heat Transfer Conference. IHTC-15, IHTC15-10516, Kyoto, Japan (10–15 August 2014).

Ge, M., Wang, S., Zhao, J. et al. (2016). Condensation of steam with high CO_2 concentration on a vertical plate. *Exp. Thermal Fluid Sci.* **75**: 147–155.

Gerstmann, J. and Griffith, P. (1967). Laminar film condensation on the underside of horizontal and inclined surfaces. *Int. J. Heat Mass Transf.* **10**: 567–580.

Ghiaasiaan, S.M. (2008). *Two Phase Flow, Boiling and Condensation in Conventional and Miniature Systems.* Cambridge: Cambridge University Press.

Graham, C. (1969). The limiting mechanisms of dropwise condensation. Doctoral dissertation. Masschusetts Institute of Technology. Cambridge, MA.

Griffith, P. (1973). Dropwise condensation. In: *Handbook of Heat Transfer* (eds. W.M. Rohsenow and J.P. Hartnett), 12-34–12-47. New York: McGraw-Hill.

Gupta, A., Kumar, R., and Gupta, A. (2014). Condensation of R-134a inside a helically coiled tube-in-shell heat exchanger. *Exp. Thermal Fluid Sci.* **54**: 279–289.

Han, D. and Lee, K.J. (2005). Experimental study on condensation heat transfer enhancement and pressure drop penalty factors in four microfin tubes. *Int. J. Heat Mass Transf.* **48**: 3804–3816.

Han, D., Lee, K., and Kim, Y. (2003). The characteristics of condensation in brazed plate heat exchangers with different chevron angles. *J. Korean Phys. Soc.* **43** (1): 66–73.

Hashimoto, H. and Kaminaga, F. (2002). Heat transfer characteristics in a condenser of closed two-phase thermosyphon: effect of entrainment on heat transfer deterioration. *Heat Transfer Asian Res.* **31**: 212–225.

Hayakawa, T., Kawasaki, J., and Muraki, M. (1975). Condensation of single vapor on a vertical surface. *Heat Transfer Jpn. Res.* **4** (2): 55–62.

Honda, H. and Fujii, T. (1974). Effect of the direction of an oncoming vapor on laminar filmwise condensation on a horizontal tube. In: *Proceedings of the Fifth International Heat Transfer Conference*, vol. **3**, 299–303. Tokyo: Begell House.

Honda, H. and Nozu, B. (1987). A prediction method for heat transfer during condensation on horizontal low integral fin tubes. *J. Heat Transf.* **109**: 218–225.

Honda, H., Nozu, S., and Mitsumori, K. (1983). Augmentation of condensation on finned tubes by attaching a porous drainage plate. In: *Proceedings of the ASME-JSME Thermal Engineering Joint Conference*, vol. **3**, 289–295. ASME.

Honda, H., Nozu, S., Bunken, U., and Fujii, T. (1986). Effect of vapour velocity on film condensation of R-113 on horizontal tubes in a crossflow. *Int. J. Heat Mass Transf.* **29** (3): 429–438.

Honda, H., Nozu, S., and Takeda, Y. (1987). Flow characteristics of condensate on a vertical column of horizontal low finned tubes. In: *ASME-JSME Thermal Engineering Joint Conference*, 517–524. ASME.

Honda, H., Uchima, B., Nozu, S. et al. (1988). Condensation of downward flowing R-113 vapor on bundles of horizontal smooth tubes. *Trans. Jpn. Soc. Mech. Eng.*: 1453–1460. Also in Heat Transfer Japanese Research, 18(6) 1989. Quoted in Collier and Thome (1984).

Howarth, J.A., Poots, G., and Wynne, D. (1978). Laminar film condensation on the underside of an inclined flat plate. *Mech. Res. Commun.* **5** (6): 369–374.

Howe, M. (1972). Heat transfer by steam condensing onto rotating cones. Doctoral Thesis. Durham University. http://etheses.dur.ac.uk/8824.

Huang, Y.S., Lyman, F.A., and Lick, W.J. (1972). Heat transfer by condensation of low pressure metal vapors. *Int. J. Heat Mass Transf.* **15**: 741–754.

Huang, X., Ding, G., Hu, H. et al. (2010). Influence of oil on flow condensation heat transfer of R410A inside 4.18 and 1.6 mm inner diameter horizontal smooth tubes. *Int. J. Refrig.* **33**: 158–169.

Huang, J., Zhang, J., and Wang, L. (2015). Review of vapor condensation heat and mass transfer in the presence of non-condensable gas. *Appl. Therm. Eng.* **89**: 469–484.

Isachenko, V.P., Osipova, V.A., and Sukomel, A.S. (1977). *Heat Transfer*. Moscow: Mir Publishers.

Ivanovskii, M.N., Milovanov, Y.V., and Subbotin, V.I. (1968). Condensation coefficient of mercury vapour. *Atomn. Energ.* **24** (2): 146. Quoted in Necsa and Rose (1976).

Jakob, M. (1936). Heat transfer in evaporation and condensation-II. *Mech. Eng.* **58**: 729–739.

Jakob, M. (1949). *Heat Transfer*, vol. **1**. New York: Wiley.

Jige, D., Inoue, N., and Koyama, S. (2016). Condensation of refrigerants in a multiport tube with rectangular minichannels. *Int. J. Refrig.* **67**: 202–213.

Jouhara, H. and Robinson, A.J. (2010). Experimental investigation of small diameter two-phase closed thermosyphons charged with water, FC-84, FC-77 and FC-3283. *Appl. Therm. Eng.* **30**: 201–211.

Jung, D., Song, K., Cho, Y., and Kim, S. (2003). Flow condensation of heat transfer coefficients of pure refrigerants. *Int. J. Refrig.* **26**: 4–11.

Kedzierski, M.A. and Goncalves, J.M. (1997). Horizontal convective condensation of alternative refrigerants within a micro-fin tube. NISTIR 6095, US Department of Commerce.

Keinath, B.L. and Garimella, S. (2018). High-pressure condensing refrigerant flows through microchannels, part II: heat transfer models. *Heat Transfer Eng.* https://doi.org/10.1080/01457632.2018.1443258.

Kern, D.Q. (1958). Mathematical development of loading in horizontal condensers. *AIChE J* **4**: 157–160.

Kharangate, C.R. and Mudawar, I. (2017). Review of computational studies on boiling and condensation. *Int. J. Heat Mass Transf.* **108**: 1164–1196.

Kim, S.M. and Mudawar, I. (2012). Theoretical model for annular flow condensation in rectangular micro-channels. *Int. J. Heat Mass Transf.* **55**: 958–970.

Kim, S. and Mudawar, I. (2013). Universal approach to predicting heat transfer coefficient for condensing mini/micro-channel flow. *Int. J. Heat Mass Transf.* **56** (1–2): 238–250.

Kirkbride, C.G. (1934). Heat transfer by condensing vapor on vertical tubes. *Int. J. Ind. Eng. Chem.* **26** (4): 425–428.

Knudsen, J.G. and Katz, D.L. (1958). *Fluid Dynamics and Heat Transfer*. New York: McGraw-Hill.

Koch, C., Kraft, K., and Leipertz, A. (1998). Parameter study on the performance of dropwise condensation. *Rev. Gen. Therm.* **37**: 539–548.

Koh, J.C.Y., Sparrow, E.M., and Hartnett, J.M. (1961). The two-phase boundary layer in laminar film condensation. *Int. J. Heat Mass Transf.* **2**: 69–82.

Kondo, C. and Hrnjak, P. (2011). Heat rejection from R744 flow under uniform temperature cooling in a horizontal smooth tube around the critical point. *Int. J. Refrig.* **34** (3): 719–731.

Kondo, C. and Hrnjak, P. (2012a). Heat rejection in condensers close to critical point-Desuperheating, condensation in superheated region and condensation of two phase fluid. International Refrigeration and Air Conditioning Conference, Purdue (16–19 July 2012).

Kondo, C. and Hrnjak, P. (2012b). Condensation from superheated vapor flow of R744 and R410A at sub-critical pressures in a horizontal smooth tube. *Int. J. Heat Mass Transf.* **55**: 2779–2791.

Kondo, C. and Hrnjak, P. (2012c). Effect of microfins on heat rejection in desuperheating, condensation in superheated region and two phase zone. International Refrigeration and Air Conditioning Conference, West Lafayette (Indiana), 16–19 July, 2012. http://docs.lib.purdue.edu/iracc/1345.

Kosky, P.G. and Staub, F.W. (1971). Local condensation heat transfer coefficients in the annular flow regime. *AIChE J* **17** (5): 1037–1043.

Kroger, D.G. and Rohsenow, W.M. (1967). Film condensation of saturated potassium vapor. *Int. J. Heat Mass Transf.* **10**: 1891–1894.

Kuhn, S.Z., Schrock, V.E., and Peterson, P.F. (1997). An investigation of condensation from steam-gas mixtures flowing downward inside a vertical tube. *Nucl. Eng. Des.* **177**: 53–69.

Kumar, R., Varma, H.K., Mohanty, B., and Agrawal, K.N. (2002). Prediction of heat transfer coefficient during condensation of water and R-134a on single horizontal integral-fin tubes. *Int. J. Refrig.* (25): 111–126.

Kuo, W., Lie, Y., Hsieh, Y., and Lin, T. (2005). Condensation heat transfer and pressure drop of refrigerant R-410A flow in a vertical plate heat exchanger. *Int. J. Heat Mass Transf.* **48**: 5205–5220.

Le Fevre, E.J. and Rose, J.W. (1966). A theory of heat transfer by dropwise condensation. In: *Proceedings of 3rd International Heat Transfer Conference*, vol. **2**, 362–375. Chicago.

Lee, K.-Y. and Kim, M.H. (2008). Experimental and empirical study of steam condensation heat transfer with a noncondensable gas in a small-diameter vertical tube. *Nucl. Eng. Des.* **238**: 207–216.

Lee, H., Mudawar, I., and Hasan, M.M. (2013). Experimental and theoretical investigation of annular flow condensation in microgravity. *Int. J. Heat Mass Transf.* **61**: 293–309.

Lips, S. and Meyer, J.P. (2011). Two-phase flow in inclined tubes with specific reference to condensation: a review. *Int. J. Multiphase Flow* **37**: 845–859.

Lips, S. and Meyer, J.P. (2012). Stratified flow model for convective condensation in an inclined tube. *Int. J. Heat Fluid Flow* **36**: 83–91.

Liu, T.Q., Mu, C.F., Sun, X.Y., and Xia, S.B. (2007). Mechanism study on formation of initial condensate droplets. *AIChE J* **53** (4): 1050–1055.

Longo, G.A. (2009). R410A condensation inside a commercial brazed plate heat exchanger. *Exp. Thermal Fluid Sci.* **33** (2): 284–291.

Longo, G.A. (2011). The effect of vapour super-heating on hydrocarbon refrigerant condensation inside a brazed plate heat exchanger. *Exp. Thermal Fluid Sci.* **35** (6): 978–985.

Longo, G.A., Righetti, G., and Zilio, C. (2015). A new computational procedure for refrigerant condensation inside herringbone type brazed plate heat exchangers. *Int. J. Heat Mass Transf.* **82**: 530–536.

Lyulin, Y., Marchuk, I., Chikov, S., and Kabov, O. (2011). Experimental study of laminar convective condensation of pure vapour inside an inclined circular tube. *Microgravity Sci. Technol.*: 439–445.

Majumdar, A. and Mezic, I. (1999). Instability of ultra-thin water films and the mechanism of droplet formation on hydrophilic surfaces. *J. Heat Transf.* **121** (1999): 964–971.

Mancin, S., Del Col, D., and Rossetto, L. (2012). Condensation of superheated vapour of R410A and R407C inside plate heat exchangers: experimental results and simulation procedure. *Int. J. Refrig.* **35** (7): 2003–2013.

Mancin, S., Del Col, D., and Rossetto, L. (2013). R32 partial condensation inside a brazed plate heat exchanger. *Int. J. Refrig.* **36** (2): 601–611.

Marto, P.J. (1988). An evaluation of film condensation on horizontal integral fin tubes. *J. Heat Transf.* **110**: 1287–1305.

Mathews, D. (1962). The transfer of heat to cylinders of various diameters rotating in a steam atmosphere, with varying conditions of temperature, pressure, and rotational speed. PhD Thesis, Imperial College, London, UK.

Mazukewitch, L.V. (1952). Condensation of ammonia on vertical surface. *Cholodilnaja Technika* **29** (2): 50–51. Quoted here from the abstract in Kalteteknikk, 12, 334-335, 1952.

McAdams, W.H. (1954). *Heat Transmission*, 3e. New York: McGraw-Hill.

McNaught, J.M. (1979). Mass-transfer correction term in design methods for multicomponent/partial condensers. In: *Condensation Heat Transfer* (eds. J.P. Marto and P.G. Kroeger), 111–118. New York: ASME.

McNaught, J.M. (1982). Two-phase forced-convection heat transfer during condensation on horizontal tube banks. In: *Proceedings of 7th Int Heat Transfer Conference, 6–10 September, 1982*, Munich, Germany, vol. **5**, 125–131.

McQuillan, K.W. and Whalley, P.B. (1985). A comparison between flooding correlations and experimental flooding data for gas-liquid flow in vertical circular tubes. *Chem. Eng. Sci.* **40** (8): 1425–1440.

Meisenburg, J., Boarts, R.M., and Badger, W.L. (1935). The influence of small concentration of air in steam on the steam film coefficient of heat transfer. *Trans. Am. Inst. Chem. Eng.* **31**: 622–637. Quoted in Ge et al. (2016).

Merte, H. (1973). Condensation heat transfer. In: *Advances in Heat Transfer* (eds. T.F. Irvine and J.P. Hartnett), 180–272. Cambridge, MA: Academic Press.

Meyer, J.P., Dirker, J., and Adelaja, O.K. (2014). Condensation heat transfer in smooth inclined tubes for R134a at different saturation temperatures. *Int. J. Heat Mass Transf.* **70**: 515–525.

Michael, A.G., Rose, J.W., and Daniels, L.C. (1989). Forced convection condensation of steam on a horizontal tube – experience with vertical downflow of steam. *J. Heat Transf.* **111**: 792–797.

Minkowycz, W.J. and Sparrow, E.M. (1966). Condensation heat transfer in the presence of non-condensables, interface resistance, superheating, variable properties, and diffusion. *Int. J. Heat Mass Transf.* **9**: 1125–1144.

Minkowycz, W.J. and Sparrow, E.M. (1969). The effect of superheating on condensation heat transfer in a forced convection boundary layer flow. *Int. J. Heat Mass Transf.* **12**: 147–154.

Mirmov, N.I. and Yemelyanov, Y.V. (1976). Coefficient of heat transfer for ammonia condensers. *Heat Transfer Sov. Res.* **8** (1): 50–51.

Miropolskiy, Z.L., Shneerova, R.I., and Ternakova, L.M. (1974). Heat transfer at superheated steam condensation inside tube. In: *Proceedings of theFifth International. Heat Transfer Conference*, Tokyo, vol. **3**, 246–249.

Misra, B. and Bonilla, C.F. (1956). *Heat transfer in condensation of metal vapors: mercury and sodium up to atmospheric pressure. Chem. Eng. Prog. Symp. Ser.* **52** (18): 7–21.

Miyara, A., Nonaka, K., and Taniguchi, M. (2000). Condensation heat transfer and flow pattern inside a herringbone-type microfin tube. *Int. J. Refrig.* **23**: 141–152.

Mizoshina, T., Ito, R., Yamashita, S., and Kamimua, H. (1977). Film condensation of superheated vapor in a vertical tube. *Heat Transfer Jpn. Res.* **6** (4): 92–101.

Mohseni, S.G., Akhavan-Behabadi, M.A., and Saeedinia, M. (2013). Flow pattern visualization and heat transfer characteristics of R-134a during condensation inside a smooth tube with different tube inclinations. *Int. J. Heat Mass Transf.* **60**: 598–602.

Moser, K.W., Webb, R.L., and Na, B. (1998). A new equivalent Reynolds number model for condensation in smooth tubes. *J. Heat Transf.* **120**: 410–416.

Mozafari, M., Akhavan-Behabadi, M.A., Qobadi-Arfaee, H., and Fakoor-Pakdaman, M. (2015). Condensation and pressure drop characteristics of R600a in a helical tube-in-tube heat exchanger at different inclination angles. *Appl. Therm. Eng.* **90**: 571–578.

Mu, C., Pang, J., Lu, Q., and Liu, T. (2008). Effects of surface topography of material on nucleation site density of dropwise condensation. *Chem. Eng. Sci.* **63** (4): 874–880.

Mudawar, I. (2017). Chapter five - flow boiling and flow condensation in reduced gravity. *Adv. Heat Tran.* **49** (2017): 225–306.

Murata, K. and Hashizume, K. (1992). Prediction of condensation heat transfer coefficients in horizontal integral-fin tube bundles. *Exp. Heat Transfer* **5**: 115–130.

Nicol, A.A. and Gacesa, M. (1970). Condensation of steam on a rotating vertical cylinder. *J. Heat Transf.* **92** (1): 144–151.

Nusselt, W. (1916). Die Oberflachenkondensation des Wasserdampfes. *Zeitscrift Ver. Deutsche Ing.* **60** (541): 569.

Olivier, J.A., Liebenberg, L., and Thome, J.R. (2007). Heat transfer, pressure drop, and flow pattern recognition during condensation inside smooth, helical micro-fin, and herringbone tubes. *Int. J. Refrig.* **30**: 609–623.

Othmer, D.F. (1929). The condensation of steam. *Indian Chem. Eng.* **21**: 57–583.

Owen, R.G., Sardesai, R.G., Smith, R.A., and Lee, W.C. (1983). Gravity controlled condensation on low integral-fin tubes. *AIChe. Symp. Ser.* **75**: 415–428.

Palen, J. and Yang, Z.H. (2001). Reflux condensation flooding prediction: review of current status. *Trans. IChemE* **79** (A): 464–469.

Park, J.E., Vakili-Farahani, F., Consolini, L., and Thome, J.R. (2011). Experimental study on condensation heat transfer in vertical minichannels for new refrigerant R1234ze(E) versus R134a and R236fa. *Exp. Thermal Fluid Sci.* **35**: 442–454.

Rahman, M.M., Kariya, K., and Miyara, A. (2018). An experimental study and development of new correlation for condensation heat transfer coefficient of refrigerant inside a multiport minichannel with and without fins. *Int. J. Heat Mass Transf.* **116**: 50–60.

Razavi, M.D. and Damle, A.S. (1978). Heat transfer coefficients for turbulent filmwise condensation. *Trans. Inst. Chem. Eng.* **56** (2): 81–85.

Rifert, V.G., Barabash, P.A., Solomakha, A.S. et al. (2018). Hydrodynamics and heat transfer in a centrifugal film evaporator. *Bulg. Chem. Commun.* **50** ((Special issue K)): 49–57.

Rohsenow, W.M. (1956). Heat transfer and temperature distribution in laminar film condensation. *Trans. ASME* **78**: 1645–1648.

Rohsenow, W.M. (1973a). Condensation. In: *Handbook of Heat Transfer* (eds. W.M. Rohsenow and J.P. Hartnett), 12-1–12-33. New York: McGraw-Hill.

Rohsenow, W.M. (1973b). Film condensation in liquid metals. In: *Progress in Heat and Mass Transfer*, vol. **7** (ed. O.E. Dwyer), 469–484. London: Pergamon Press.

Rohsenow, W.M., Webber, J.H., and Ling, A.T. (1956). Effect of vapor velocity on laminar and turbulent film condensation. *Trans. ASME*: 1637–1643.

Rose, J.W. (2002). Dropwise condensation theory and experiment: a review. *P. I. Mech. Eng. A-J. Pow.* **216**: 115–128.

Roth, J.A. (1962). Wright-Patterson AFB. Report ASD-TDR-62-738. Quoted in Sukhatme and Rohsenow (1964).

Rudy, T.M. and Webb, R.L. (1985). An analytical model to predict condensate retention on horizontal integral-fin tubes. *J. Heat Transf.* **107** (2): 361–368.

Rufer, C.E. and Kezios, S.P. (1966). Analysis of two-phase one component stratified flow with condensation. *J. Heat Transf.* **88** (3): 265–275.

Saffari, H. and Naziri, V. (2010). Theoretical modeling and numerical solution of stratified condensation in inclined tubes. *J. Mech. Sci. Technol.* **24**: 2587–2596.

Sajjan, S.K., Kumar, R., and Gupta, A. (2015). Experimental investigation during condensation of R-600a vapor over single horizontal integral-fin tubes. *Int. J. Heat Mass Transf.* **88**: 247–255.

Sakhuja, R.K. (1970). Effect of superheat on film condensation of potassium. Sc.D. thesis. Mechanical Engineering Department of MIT Cambridge, MA. Quoted in Rohsenow (1973a).

Salimpour, R.S., Shahmoradi, A., and Khoeini, D. (2017). Experimental study of condensation heat transfer of R-404A in helically coiled tubes. *Int. J. Refrig.* **74**: 584–591.

Sardesai, R.G., Owen, R.G., Smith, R.A., and Lee, W.C. (1983). Gravity controlled condensation on a horizontal low-fin tube. *IChemE Symp. Ser.* **75**: 415–428.

Sarraf, K., Launay, S., and Tadrist, L. (2016). Analysis of enhanced vapor desuperheating during condensation inside a plate heat exchanger. *Int. J. Therm. Sci.* **105**: 96–108.

Schlager, L.M., Pate, M.B., and Bergles, A.E. (1990). Performance predictions of refrigerant-oil mixtures in smooth and internally finned tubes – Part II: design equations. *ASHRAE Trans.* **96** (1): 170–182.

Schrage, R.W. (1953). *A Theoretical Study of Interphase Mass Transfer*. New York: Columbia University Press.

Shaffrin, E.G. and Zisman, W.A. (1960). *J. Phys. Chem.* **64**: 519. Quoted in Merte (1973).

Shah, M.M. (1975). Visual observations in an ammonia evaporator. *ASHRAE Trans.* **81** (1).

Shah, M.M. (1976). A new correlation for heat transfer during boiling flow through pipes. *ASHRAE Trans.* **82** (2): 66–86.

Shah, M.M. (1979). A general correlation for heat transfer during film condensation inside pipes. *Int. J. Heat Mass Transf.* **22**: 547–556.

Shah, M.M. (1981). Heat transfer during film condensation in tubes and annuli, a literature survey. *ASHRAE Trans.* **87** (1): 1086–1105.

Shah, M.M. (2009). An improved and extended general correlation for heat transfer during condensation in plain tubes. *HVAC&R Res.* **15** (5): 889–913.

Shah, M.M. (2013). General correlation for heat transfer during condensation in plain tubes: further development and verification. *ASHRAE Trans.* **119** (2).

Shah, M.M. (2014). A new flow pattern based general correlation for heat transfer during condensation in horizontal tubes. Proceedings of the 15th International Heat Transfer Conference, IHTC-15, Kyoto, Japan, 10–15 August, 2014.

Shah, M.M. (2016a). Prediction of heat transfer during condensation in inclined tubes. *Appl. Therm. Eng.* **94**: 82–89.

Shah, M.M. (2016b). A new correlation for heat transfer during condensation in horizontal mini/micro channels. *Int. J. Refrig.* **64**: 187–202.

Shah, M.M. (2016c). Comprehensive correlations for heat transfer during condensation in conventional and mini/micro channels in all orientations. *Int. J. Refrig.* **67**: 22–41.

Shah, M.M. (2018a). Improved general correlation for heat transfer during gas-liquid flow in horizontal tubes. *J. Therm. Sci. Eng. Appl.* **10**: 051009-1–051009-7.

Shah, M.M. (2018b). General correlation for heat transfer to gas-liquid flow in vertical channels. *J. Therm. Sci. Eng. Appl.* **10**: 061006-1–061006-9.

Shah, M.M. (2019a). Improved correlation for heat transfer during condensation in conventional and mini/micro channels. *Int. J. Refrig.* **98**: 222–237.

Shah, M.M. (2019b). Prediction of heat transfer during condensation in non-circular channels. *Inventions* **2019** (4): 31. https://doi.org/10.3390/inventions4020031.

Shah, M.M., Mahmoud, A.A., and Lee, J. (2013). An assessment of some predictive methods for in-tube condensation heat transfer of refrigerant mixtures, Paper # DE-13-004 presented at ASHRAE meeting in Denver, CO, June 2013, ASHRAE Transactions, **119** (2): 38–51.

Shao, D.W. and Granryd, E. (1995). Heat transfer and pressure drop of HFC134a-oil mixtures in a horizontal condensing tube. *Int. J. Refrig.* **18** (8): 524–533.

Shekriladze, I.G. and Gomelauri, V.I. (1966). Theoretical study of laminar film condensation of flowing vapour. *Int. J. Heat Mass Transf.* **9**: 581–591.

Shen, B. and Groll, E.A. (2005). A critical review of the influence of lubricants on the heat transfer and pressure drop of refrigerants – part II: lubricant influence on condensation and pressure drop. *HVAC&R Res.* **11** (4).

Shiraishi, M., Kikuchi, K., and Yamanishi, T. (1981). Investigation of heat transfer characteristics of a two-phase closed thermosyphon. In: *Advances in Heat Pipe Technology* (ed. D.E. Reay), 95–104.

Shon, B.H., Jung, B.H., Kwon, J.I. et al. (2018). Characteristics on condensation heat transfer and pressure drop for a low GWP refrigerant in brazed plate heat exchanger. *Int. J. Heat Mass Transf.* **122**: 1272–1282.

Short, G.D. (2017). Assessment of Lubricants for Ammonia and Carbon Dioxide Refrigeration Systems. Technical Paper #5, 2017 IIAR Natural Refrigeration Conference and Heavy Equipment Expo, San Antonio, TX.

Siddique, M.S., Golay, M.W., and Kazimi, M.S. (1993). Local heat transfer coefficients for forced-convection condensation of steam in a vertical tube in the presence of a noncondensable gas. *Nucl. Technol.* **102**: 386–402.

Soliman, H.M. (1986). The mist-annular transition during condensation and its influence on the heat transfer mechanism. *Int. J. Multiphase Flow* **12**: 277–288.

Soliman, M., Schuster, J.R., and Berenson, P.J. (1968). A general heat transfer correlation for annular flow condensation. *J. Heat Transf.* **90**: 267–276.

Son, C. and Lee, H. (2009). Condensation heat transfer characteristics of R-22, R-134a and R-410A in small diameter tubes. *Heat Mass Transf.* **45**: 1153–1166.

Sparrow, E.M. and Gregg, J.L. (1959a). Laminar condensation heat transfer on a horizontal tube. *J. Heat Transf.* **81** (4): 291–295.

Sparrow, E.M. and Gregg, J.L. (1959b). A theory of rotating condensation. *J. Heat Transf.* **81**: 113–120.

Sparrow, E.M. and Hartnett, J.P. (1961). Condensation on a rotating cone. *J. Heat Transf.*: 101–102.

Sparrow, E.M., Minkowycz, W.J., and Saddy, M. (1967). Forced convection condensation in the presence of non-condensables and interfacial resistance. *Int. J. Heat Mass Transf.* **10**: 1829–1845.

Spencer, D.L. and Ibele, D.E. (1966). Laminar film condensation of a saturated and superheated vapor on a surface with a controlled temperature distribution. In: *Proceedings Third International Heat Transfer Conference, 7–12 August, Chicago, USA*, 337–347. Begell House.

Sreepathi, L.K., Bapat, S.L., and Sukhatme, S.P. (1996). Heat transfer during film condensation of R-123 vapour on horizontal integral-fin tubes. *J. Enhanc. Heat Transfer* **3**: 147–164.

Stein, R.P., Cho, D.H., and Lambert, G.A. (1985). Condensation on the underside of a horizontal surface in a closed vessel. In: *Multiphase Flow and Heat Transfer*, vol. **47** (eds. V.K. Dhir, J.C. Chen and O.C. Jones). New York: ASME HTD.

Stephan, K. (1992). *Heat Transfer in Condensation and Boiling*. Berlin: Springer Varlag. (English language edition).

Subbotin, V.I., Ivanovskii, M.N., Sorokin, V.P., and Chulkov, B.A. (1964). Heat transfer during the condensation of potassium vapour. *Teplofiz. Vysok. Temper* **2** (4): 616. Quoted in Necme and Rose (1976).

Sugawara, S. and Katsuta, K. (1966). Fundamental study on dropwise condensation. In: *Proceedings of the Third International Heat Transfer Conference*, Chicago, vol. **2**, 354. New York: ASME.

Sukhatme, S.P. and Rohsenow, W.M. (1964). Heat transfer during film condensation of a liquid metal vapor. Massachusetts Institute of Technology. Cambridge, MA. *Report MIT-2995-1*.

Sukhatme, S.P. and Rohsenow, W.M. (1966). Heat transfer during film condensation of a liquid metal vapor. *J. Heat Transf.* **88**: 19–28.

Sur, B. and Azer, N.Z. (1991). Effect of oil on heat transfer and pressure drop during condensation of Refrigerant-113 inside smooth and internally finned tubes. *ASHRAE Trans.* **97** (1): 365–373.

Taylor, R.G. and Krishna, R. (1993). *Multicomponent Mass Transfer*. New York: Wiley.

Tepe, J.B. and Mueller, A.C. (1947). Condensation and subcooling inside an inclined tube. *Chem. Eng. Prog.* **43** (5): 267–278.

Thome, J.R. (1998). Condensation of fluorocarbon and other refrigerants: a state-of-the-art review. Final report prepared for Air-Conditioning and Refrigeration Institute (ARI). Quoted in Shen and Groll (2005).

Thome, J.R., El Hajal, J., and Cavallini, A. (2003). Condensation in horizontal tubes. Part 2: new heat transfer model based on flow regimes. *Int. J. Heat Mass Transf.* **46**: 3365–3387.

Thonon, B. (1995). Design method for plate evaporators and condensers. In: *Proceedings of 1st Int. Conf. Process Intensification for Chemical Industry*, 149–155. *BHR Group Conference Series Publication*.

Tichy, J.A., Macken, N.A., and Duval, W.M.B. (1984). An experimental investigation of heat transfer in forced-convection-condensation of oil-refrigerant mixtures. *ASHRAE Trans.* **90** (1).

Traviss, D.P., Baron, A.B., and Rohsenow, W.M. (1973). Forced convection condensation inside tubes: heat transfer equation for condenser design. *ASHRAE Trans.* **79** (1): 157–165.

Umur, A. and Griffith, P. (1965). Mechanism of dropwise condensation. *J. Heat Transf.* **87** (1965): 275–282.

Varma, V.C. (1977). A study of condensation of steam inside a tube. MSc thesis. Rutgers University, New Jersey.

Vierow, K.M. and Schrock, V.E. (1991). Condensation in a natural circulation loop with noncondensable gases. Part I. Heat transfer. In: *Proceedings of the International Conference on Multiphase Flow*, 183–186. Tsukuba, Japan: International Atomic Energy Agency (IAEA).

Wallis, G.B. (1969). *One-Dimensional Two-Phase Flow*. New York: McGraw-Hill.

Walt, J.V.D. and Kroger, D.G. (1972). Heat transfer during condensation of saturated and superheated Freon 12. In: *Progress in Heat and Mass Transfer*, vol. **6** (eds. G. Hetsroni, S. Sideman and J.P. Hartnett), 75–98. London: Pergamon Press.

Wang, B.X. and Du, X.Z. (2000). Study on laminar film-wise condensation for vapor flow in an inclined small/mini-diameter tube. *Int. J. Heat Mass Transf.* **43**: 1859–1868.

Wang, H.S. and Honda, H. (2003). Condensation of refrigerants in horizontal microfin tubes: comparison of prediction methods for heat transfer. *Int. J. Refrig.* **26**: 452–460.

Wang, H.S. and Rose, J.W. (2005). A theory of film condensation in horizontal noncircular section microchannels. *J. Heat Trans.-T ASME* **127** (10): 1096–1105.

Wang, H.S. and Rose, J.W. (2006). Film condensation in horizontal microchannels: effect of channel shape. *Int. J. Therm. Sci.* **45**: 1205–1212.

Wang, H.S. and Rose, J.W. (2011). Theory of heat transfer during condensation in microchannels. *Int. J. Heat Mass Transf.* **54** (11–12): 2525–2534.

Webb, R.L. (1998). Convective condensation of superheated vapor. *ASME J. Heat Transfer* **120**: 418–421.

Webb, R.L. and Murawski, C.G. (1990). Row effect for R-11 condensation on enhanced tubes. *J. Heat Transf.* **112**: 768–776.

Webb, R.L., Rudy, T.M., and Kedzierski, M.A. (1985). Prediction of the condensation coefficient on horizontal integral-fin tubes. *J. Heat Transf.* **107** (2): 369–376.

Welch, J.F. and Westwater, J.W. (1961). Microscopic study of dropwise condensation. In: *International Developments in Heat Transfer, Part II*, 302–309. New York: ASME.

Wellsandt, S. and Vamling, L. (2005). Prediction method for flow boiling heat transfer in a herringbone microfin tube. *Int. J. Refrig.* **28**: 912–920.

Wilcox, S.J. and Rohsenow, W.M. (1970). Film condensation of potassium using copper condensing block for precise wall-temperature measurement. *J. Heat Transf.* **92C**: 359.

Williams, P.E. and Sauer (1981). Condensation of of refrigerant-oil mixtures on horizontal tubes. *Int. J. Refrig.* **4** (4): 209–222.

Wongwises, S. and Polsongkram, M. (2006). Condensation heat transfer and pressure drop of HFC-134a in a helically coiled concentric tube-in-tube heat exchanger. *Int. J. Heat Mass Transf.* **49**: 4386–4398.

Wurfel, R., Kreutzer, T., and Fratzscher, W. (2003, 2003). Turbulence transfer processes in a diabatic and condensing film flow in inclined tube. *Chem. Eng. Technol.* **26**: 439–448.

Xing, F., Xu, J., Xie, J. et al. (2015). Froude number dominates condensation heat transfer of R245fa in tubes: effect of inclination angles. *Int. J. Multiphase Flow* **71**: 98–115.

Xio, J. and Hrnjak, P. (2018). Flow regime map for condensation from superheated vapor. 17th International Refrigeration and Air Conditioning Conference at Purdue (9–12 July 2018), West Lafayette, Indiana, USA. https://docs.lib.purdue.edu/iracc/1910.

Xu, H., Sun, Z., Gu, H., and Li, H. (2016). Forced convection condensation in the presence of noncondensable gas in a horizontal tube; experimental and theoretical study. *Prog. Nucl. Energy* **88**: 340–351.

Yan, Y.Y., Lio, H.C., and Lin, T.F. (1999). Condensation heat transfer and pressure drop of refrigerant R-134a in a plate heat exchanger. *Int. J. Heat Mass Transf.* **42** (6): 993–1006.

Yanniotis, S. and Kolokotsa, D. (1996). Experimental study of water vapour condensation on a rotating disc. *Int. Commun. Heat Mass Transfer* **23**: 721–729.

Yu, J. and Koyama, S. (1998). Condensation heat transfer of pure refrigerants in microfin tubes. In: *Proceedings of International Refrigeration Conference at Purdue University*, 325–330. Purdue, Indiana, USA, West Lafayette: Purdue University.

Yu, J., Chen, J., Li, F. et al. (2018). Experimental investigation of forced convective condensation heat transfer of hydrocarbon refrigerant in a helical tube. *Appl. Therm. Eng.* **129**: 1634–1644.

Yunxiao, Y. and Li, J. (2015). Experimental investigation on heat transfer coefficient during upward flow condensation of R410A in vertical smooth tubes. *J. Therm. Sci.* **24** (2): 155–163.

Zeinelabdeen, M.I.M., Kamran, M.S., and Briggs, A. (2018). Comparison of empirical models with an experimental database for condensation on banks of tubes. *Int. J. Heat Mass Transf.* **122**: 765–774.

Zhang, J., Martin, R.K., Ommen, T. et al. (2019). Condensation heat transfer and pressure drop characteristics of R134a, R1234ze(E), R245fa and R1233zd(E) in a plate heat exchanger. *Int. J. Heat Mass Transf.* **128**: 136–149.

3

Pool Boiling

3.1 Introduction

Pool boiling is boiling without any forced flow. A typical example is during heating of water in a pan placed over a stove. In pool boiling, any movement of liquid is by natural convection. It is therefore also known as natural convection boiling.

There are several regimes of pool boiling. These were first identified by Nukiyama (1934) by experiments in which he electrically heated a wire in a pool of water. The various regimes of boiling are shown in Figure 3.1.1. Initially at low wall superheat ($T_w - T_{SAT}$), there is no boiling. Heat transfer occurs only by natural convection. When the wall superheat reaches the value needed for initiation or inception of boiling, bubble nucleation starts. As wall superheat increases, heat flux rises. Until point B is the region of partial nucleate boiling in which heat transfer occurs both due to nucleate boiling and natural convection. Beyond point B is the region of fully developed nucleate boiling in which the contribution of natural convection is negligible. This continues until point C where departure from nucleate boiling (DNB) or critical heat flux (CHF) occurs. If heat flux is the independent variable as is the case in electric heating, wall superheat increases sharply with further increase in heat flux until the film boiling line is reached at point E. If the wall temperature is the independent variable (for example, wall heated by a hot liquid), heat flux decreases sharply with increasing wall temperature along the curve CD. Point D is the minimum film boiling (MFB) point. With further increase in wall temperature, heat flux rises along the film boiling curve. The curve CD between DNB and MFB points is known as the transition boiling curve.

There is hysteresis in the heat flux-controlled case. If film boiling is occurring at heat flux higher than at point E and heat flux is reduced, wall superheat decreases along the film boiling curve until the MFB is reached and then suddenly decreases to reach the nucleate boiling curve. On the other hand, if the wall temperature is the one being independently varied, heat flux rises from MFB to DNB along the curve CD with falling wall temperature.

In nucleate boiling, bubbles form at the heated surface and depart into the liquid. In transition boiling, heating surface is in intermittent contact with liquid. In film boiling, the heater surface is covered by vapor, and there is essentially no contact between liquid and heating surface.

In the following, all discussions are for pure non-metallic fluids except where stated otherwise.

3.2 Nucleate Boiling

3.2.1 Mechanisms of Nucleate Boiling

During nucleate boiling, heat transfer coefficients are much higher than in single-phase natural convection. The following three mechanisms are considered to be responsible for the large heat transfer during nucleate boiling.

3.2.1.1 Bubble Agitation

Growing and departing bubbles cause intense agitation of liquid near the heating surface. Growing bubble displaces the superheated liquid around it. On bubble departure, colder liquid moves in its place thus setting up intense convection currents as seen in Figure 3.2.1a. This process involves sensible heat transfer. Hsu and Graham (1976) after studying experimental data concluded that bubble agitation is not a major contributor to heat transfer during nucleate boiling.

3.2.1.2 Vapor–Liquid Exchange

The vapor–liquid exchange model was proposed by Forster and Greif (1959). A slug of hot liquid is pumped away by the growing and departing bubble and a slug of colder liquid takes its place as shown in Figure 3.2.1b. The colder slug gets heated at the wall and then moves back into the bulk liquid. The principal assumption made by Forster and Greif was that the liquid slug moving away was at the film

Two-Phase Heat Transfer, First Edition. Mirza Mohammed Shah.

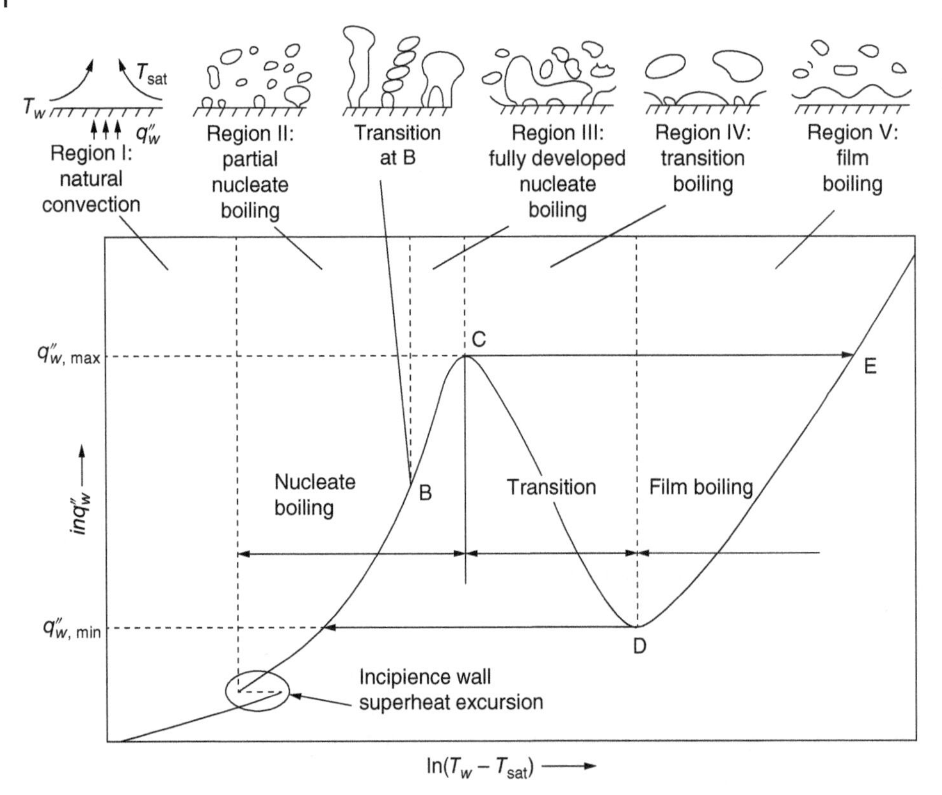

Figure 3.1.1 Regimes of pool boiling and the flow behavior near the heating surface. Source: From Ghiaasiaan (2008).

temperature, which is the average of temperatures of heater surface and bulk liquid. This will require the thickness of the thermal layer at heater surface to be equal the departure radius of bubble. Measurements in fact show that thermal layer is an order of magnitude thinner than the bubble departure diameter. This leads to the conclusion that this mechanism contributes to heat transfer but not as much as Forster and Greif had projected.

3.2.1.3 Evaporative Mechanism

Evaporative mechanism is shown in Figure 3.2.1c. Evaporation occurs in the thin microlayer under the growing bubble. Evaporation also occurs from the superheated layer surrounding the bubble. This mechanism transports latent heat while the other two transport sensible heat. This mechanism is considered to be the major contributor to heat transfer during nucleate boiling.

3.2.2 Bubble Nucleation

Bubble nucleation in principle can be homogeneous or heterogeneous, though it is the latter in most practical applications. Homogeneous nucleation does not require any nucleation sites on the heater surface. The kinetic theory predicts that there are clusters of molecules in gases and clusters of vapor or voids in liquids. These can form nuclei for homogeneous nucleation under certain conditions. Homogeneous nucleation requires very high superheats that have never been seen in practical systems. In practical systems, nucleation is heterogeneous in which nucleation occurs on pre-existing nuclei in the heating surface. These are formed by gases trapped in cavities and crevices on heating surfaces. All metal surfaces contain a wide range of cavities and crevices. In the rest of this chapter, only heterogeneous nucleation is considered.

3.2.2.1 Inception of Boiling

Bubble nucleation starts when the wall temperature exceeds the saturation temperature by a significant amount as discussed in the following.

Most cavities on metal surfaces approximate to conical shapes. Han and Griffith (1965) developed a method to predict boiling incipience considering a hemispherical bubble growing at a conical cavity of radius r_c. For static equilibrium,

$$p_b - p_f = \frac{2\sigma}{r_c} \tag{3.2.1}$$

where p_b is the pressure in the bubble and p_f is pressure of liquid. By using the Clausius–Clapeyron equation, the corresponding temperature difference is

$$(T_b - T_{SAT}) = \frac{T_{SAT}}{i_{fg}}(p_b - p_f)\left(\frac{1}{\rho_g} - \frac{1}{\rho_f}\right) \tag{3.2.2}$$

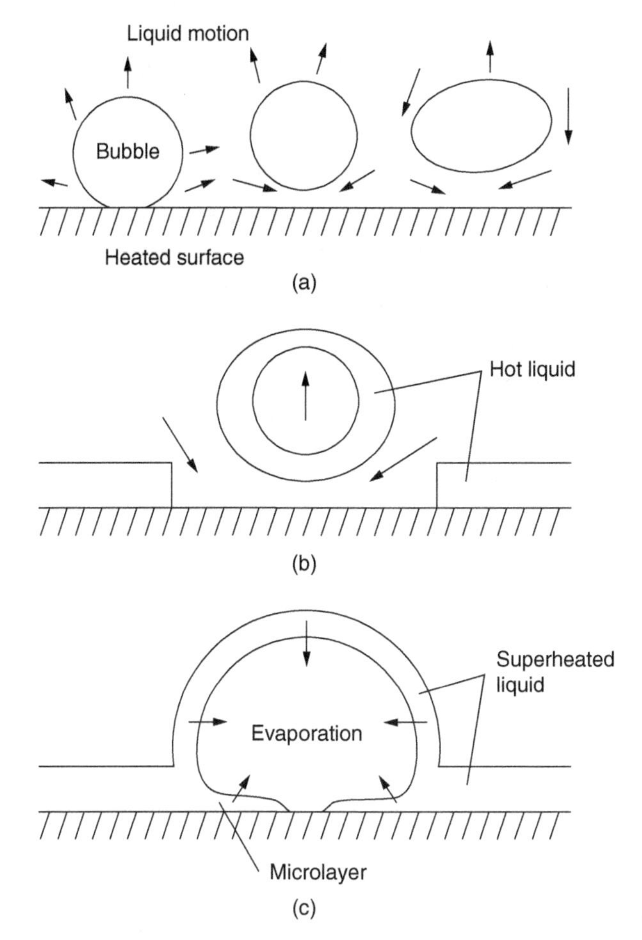

Figure 3.2.1 Mechanisms of nucleate boiling. (a) Bubble agitation, (b) vapor–liquid exchange, and (c) evaporation. Source: From Wieland Heat Transfer Engineering Data Book III. © 2016 Wieland-Werke AG.

T_b is the vapor temperature inside the bubble. By combining Eqs. (3.2.1) and (3.2.2), and further assuming $\rho_f \gg \rho_g$,

$$(T_b - T_{\mathrm{SAT}}) = \frac{2T_{\mathrm{SAT}}\sigma}{i_{\mathrm{fg}}\rho_g r_c} \tag{3.2.3}$$

By assuming that the bubble temperature equals that in the thermal boundary layer at distance 1.5 times the cavity radius, they arrived at the following equation:

$$r_c = \frac{\delta(T_w - T_{\mathrm{SAT}})}{3(T_w - T_\infty)}\left\{1 \pm \left[1 - \frac{12(T_w - T_\infty)T_{\mathrm{SAT}}\sigma}{(T_w - T_{\mathrm{SAT}})^2\delta\rho_g i_{\mathrm{fg}}}\right]^{1/2}\right\} \tag{3.2.4}$$

T_∞ is the temperature outside thermal boundary layer whose thickness is δ to be calculated by natural convection relations. For any superheat, Eq. (3.2.4) gives two possible cavity sizes. Bubble nucleation will occur on any cavity gas embryo of size in between these two extreme values and will not occur outside this range. If the cavity size is larger than the maximum size given by Eq. 3.4, the cavity will be deactivated due to flooding with liquid as the gas pressure in the cavity will not be enough to maintain equilibrium. If the cavity is smaller than the minimum predicted, it will require more superheat than available.

Hsu (1962) and Hsu and Graham (1976) analyzed nucleation of bubbles with the shape of a truncated sphere. Their equations are very similar to Eq. (3.2.4).

Bergles and Rohsenow (1964) performed forced convection experiments on subcooled boiling of water in tubes. They postulated that a bubble of radius r_b will grow only if the liquid at distance r_b from the surface exceeds the temperature of vapor in the cavity. Frost and Dzakowic (1967) introduced the Prandtl number into this criterion. They assumed that the liquid temperature at distance $(r_b \cdot Pr_f)$ should exceed the vapor temperature in the cavity. With further assumptions, they arrived at the following equation:

$$(\Delta T_{\mathrm{SAT}})_{\mathrm{ib}} = \left(\frac{8\sigma T_{\mathrm{SAT}} q_{\mathrm{bi}}}{i_{\mathrm{fg}} k_f \rho_g}\right)^{0.5} Pr_f \tag{3.2.5}$$

The subscript "ib" means boiling inception. For heat flux-controlled systems, the inception superheat is easily calculated. Where heating is by liquid, Eq. (3.2.5) has to be solved along with an equation for single-phase convection with heat transfer coefficient of heating liquid calculated with an appropriate correlation. Frost and Dzakowic compared their correlation with data for many fluids including hydrogen, nitrogen, water, refrigerants, and chemicals. Most of the data were for pool boiling; some forced convection data were also included. The predictions were at the lower limit of data. Actual boiling initiation occurred at higher superheats. Inception superheats can be affected by factors such as aging, surface finish, and the extent to which gases are trapped in the cavities.

3.2.2.2 Bubble Nucleation Cycle

Bubble nucleation cycle includes bubble growth, bubble departure, and a waiting period till a new bubble appears. The reason for the waiting period is that when a bubble departs, liquid moves in to fill the area of its influence, which is an area of a circle with diameter twice that of the departing bubble. This liquid gets heated by transient conduction from the surface until it reaches a temperature high enough to allow growth of a new bubble. The cycle then continues.

Prediction of bubble growth rate, bubble departure diameter, and waiting period are discussed as follows.

Bubble Departure Diameter As the bubble grows, buoyancy, and wake of departing bubbles try to dislodge it from the surface, while surface tension, drag, and inertia try to prevent departure. The earliest formula for the bubble diameter at departure D_d was given by Fritz (1935). It is

$$D_d = 0.02080\theta \left(\frac{\sigma}{g(\rho_f - \rho_g)} \right)^{1/2} \tag{3.2.6}$$

where θ is the contact angle in degrees. This equation is the one most widely used.

Kim and Kim (2006) developed the following correlation. At atmospheric pressure,

$$D_d = \left[45.93 Ja v_f \left(\frac{\rho_f}{\sigma} \right)^{0.5} \right]^2 \tag{3.2.7}$$

At sub-atmospheric pressure,

$$Bd^{0.5} = 0.1649 Ja^{0.7} \tag{3.2.8}$$

where the Bond number Bd is defined as

$$Bd = \frac{g(\rho_f - \rho_g) D_d^2}{\sigma} \tag{3.2.9}$$

Ja is the Jakob number defined as

$$Ja = \frac{\rho_f}{\rho_g} \frac{C_{pf}(T_w - T_{SAT})}{i_{fg}} \tag{3.2.10}$$

They compared this correlation to their own data as well as data from several sources. These included fluids water, refrigerants, and chemicals, subcooled and saturated fluids, and constant heat flux as well as constant temperature heating.

Numerous other formulas, theoretical and empirical, have been proposed. Mohanty and Das (2017) have listed 27 of them and compared many of them to experimental data. Kim and Kim correlation was not included in their evaluation. They found that none of the tested correlations was able to correctly predict data from more than a single source.

Chen et al. (2017) performed experiments on pool boiling of methane at pressures 1–4 bar. They compared the measured departure bubble diameter with six published correlations including that of Kim and Kim. None was found satisfactory. They gave a correlation of their own data. It was not compared with data from other sources.

Bubble Growth Rate Many expressions have been developed for bubble growth rate. Perhaps the earliest correlation is the following by Fritz and Ende (1936):

$$D_b = 2.256 Ja (v_f t)^{0.5} \tag{3.2.11}$$

where D_b is the bubble diameter at time t.

Plesset and Zwick (1954) and Forster and Zuber (1954) have given modifications of Eq. (3.2.11) in which the constant is changed from 2.256 to 3.98 and 3.544, respectively. These modifications were based on the shape of bubbles.

Cooper (1969) developed the following equation considering the bubble growth to be dominated by microlayer evaporation:

$$D_b = 5 \frac{Ja}{Pr_f} (v_f t)^{0.5} \tag{3.2.12}$$

Cooper (1969) recommends it for pressures well-below atmospheric and shows agreement with some such data.

Cole and Shulman (1966) compared their own test data for several chemicals and water at sub-atmospheric pressure to theoretical formulas such as that of Fritz and Ende. They found that the theoretical formulas work well for Jakob numbers up to 100 beyond which they give increasingly large errors. They fitted the following empirical equation to the test data:

$$D = 5 Ja^{3/4} (v_f t)^{0.5} \tag{3.2.13}$$

This agreed with data for Jakob numbers 24–792.

Kim and Kim (2006) presented the following correlation for the time for departure of bubble t_d.

For atmospheric pressure,

$$t_d = 135 Ja v_f \frac{\rho_f D_d}{2\sigma} \tag{3.2.14}$$

For sub-atmospheric pressure,

$$t_d = 5332.4 \frac{\rho_f D_d}{2\sigma} v_f Ja^{-0.194} \tag{3.2.15}$$

The database was the same as for their correlation for bubble departure diameter. This correlation gives the total time for bubble growth, while the others discussed earlier give the time to reach any diameter up to the departure diameter.

Mohanty and Das (2017) have listed and reviewed methods to predict bubble growth rate, including the ones discussed earlier except that of Kim and Kim. None of the predictive methods has been validated with an extensive database.

Waiting Period and Bubble Frequency The total time t_{total} between departure of bubbles from the same site is the sum of waiting period t_{wait} and the bubble departure time t_d. The bubble growth time is the same as the bubble departure time. The bubble frequency f is thus defined as

$$f = \frac{1}{t_{wait} + t_d} \tag{3.2.16}$$

Many prediction methods, theoretical and empirical, for bubble frequency have been proposed. Based on their observations on boiling of water and nitrogen, Jakob and Fritz (1935) gave the following relation:

$$f D_d = 0.078 \tag{3.2.17}$$

Han and Griffith (1965) obtained the minimum waiting period $t_{w,min}$ by noting that it occurs when the two roots of

Eq. (3.2.4) are equal. This occurs when fluid temperature line and bubble equilibrium temperature curve are tangent to each other. The result is

$$t_{w,\min} = \frac{144(T_w - T_\infty)^2 T_{SAT}^2 \sigma^2}{\pi \nu_f \rho_g^2 i_{fg}^2 (T_w - T_{SAT})^4} \tag{3.2.18}$$

Mohanty and Das (2017) have listed 18 formulas including the two above. They compared most of them with data from several sources. None of them was able to predict data other than their own.

Chen et al. (2017) experimentally studied pool boiling of methane at pressures 1–4 bar. They compared the measured bubble frequency with three published correlations. None of them gave satisfactory agreement with their data. They developed a correlation that fitted their own data.

3.2.2.3 Active Nucleation Site Density

While commercial surfaces contain numerous cavities, only some of them become active nucleation sites. Experimental studies show that active site density increases with increasing heat flux. Paul and Abdel-Khalik (1983) found a linear dependence on heat flux during tests on pool boiling of statured water at atmospheric pressure on an electrically heated horizontal platinum wire. They correlated their observations by the following equation:

$$n_s = 1.027 \times 10^{-3} q + 15.74 \tag{3.2.19}$$

where n_s is the active site density in sites m^{-2}, and q is the heat flux in W m^{-2}.

Hibiki and Ishii (2003) developed the following model by mechanistic analysis:

$$n_s = n_{ref} \left\{1 - \exp\left(-\frac{\theta^2}{8\omega^2}\right)\right\} \left[\exp\left\{f(\rho^+)\frac{\lambda}{r_c}\right\} - 1\right] \tag{3.2.20}$$

where $n_{ref} = 4.72 \times 10^5$ sites m^{-2}, θ is the contact angle in radians, $\omega = 0.722$ rad, and $\lambda = 2.5 \times 10^{-6}$ m. The critical cavity radius r_c is given by

$$r_c = \frac{2\sigma\{1 + (\rho_g/\rho_f)\}/p_f}{\exp\{i_{fg}(T_g - T_{SAT})/(R_a T_g T_{SAT})\} - 1} \tag{3.2.21}$$

R_a is the gas constant of fluid $= R/M$, where R is the universal gas constant and M is the molecular weight of the fluid.

$$f(\rho^+) = -0.010\,64 + 0.482\,46\rho^+ - 0.227\,12\rho^{+2} + 0.054\,668\rho^{+3} \tag{3.2.22}$$

$$\rho^+ = \log\left(\frac{\rho_f - \rho_g}{\rho_g}\right) \tag{3.2.23}$$

For $\rho_g \ll \rho_f$ and $[i_{fg}(T_g - T_{SAT})/(R_a T_g T_{SAT})] \ll 1$, Eq. (3.2.24) can be simplified to the following:

$$r_c = \frac{2\sigma T_{SAT}}{\rho_g i_{fg} \Delta T_{SAT}} \tag{3.2.24}$$

This model was verified with data from several sources in the following range. G from 0 to 880 kg m^{-2} s^{-1}, pressure from 0.1 to 19.8 MPa, contact angle from 5° to 90°, and n_s from 1×10^4 to 1.51×10^{10} sites m^{-2}.

Mohanty and Das (2017) have listed and discussed a number of other prediction methods. Some of them required information that was not available. There were six correlations, including the two given earlier, which did not have such limitations. They were compared by them to test data from many sources. The Hibiki and Ishii method gave fair agreement with some data and large deviations from some data. The other correlations gave large deviations with all data.

3.2.2.4 Recommendations

For calculation of inception of boiling, the correlation of Frost and Dzakowic is recommended. It gives the lower limit of required superheat to initiate boiling. Actual superheat needed is likely to be higher.

None of the correlations for bubble departure diameter, bubble frequency, and waiting period can be relied upon beyond the range of data on which they were based.

For active nucleation site density, the model of Hibiki and Ishii is the best available. While it has been shown to agree with data for a wide range of conditions, it has also been found to give large deviations.

3.2.3 Correlations for Heat Transfer

A number of authors gave correlations of the form

$$Nu = C Re^n Pr^m \tag{3.2.25}$$

Re is the Reynolds number. Various parameters are used as characteristic dimension and velocity in *Nu* and *Re*. One of such correlations is by Rohsenow (1952). In that correlation the constant C is specific to each liquid–surface combination and has to be determined experimentally. It is called C_{sf}. As the number of such combinations is limitless, this correlation is of limited use for practical design. However, it is frequently cited in research papers.

A simple correlation is the following by Mostinski (1963):

$$h = 0.00417 q^{0.7} p_c^{0.69} (1.8 p_r^{0.17} + 4 p_r^{1.2} + 10 p_r^{10}) \tag{3.2.26}$$

h is in W m^{-2} K^{-1}, q in W m^{-2}, and the critical pressure p_c in kPa. According to Collier and Thome (1994), it is at least as accurate as any property-based correlation.

Stephan and Abdelsalam (1980) examined a wide range of data from many sources and developed the following correlations by regression analysis in terms of dimensionless parameters.

For water ($10^{-4} \le p_r \le 0.886$, contact angle $\beta = 45°$),

$$Nu = 0.246 \times 10^7 X_1^{0.673} X_4^{-1.58} X_3^{1.26} X_{13}^{5.22} \tag{3.2.27}$$

For hydrocarbons ($5.7 \times 10^{-3} \le p_r \le 0.9$, contact angle $\beta = 35°$),

$$Nu = 0.0546(X_1 X_5^{0.5})^{0.67} X_{13}^{-4.33} X_4^{0.248} \tag{3.2.28}$$

For cryogenic fluids ($4 \times 10^{-3} \le p_r \le 0.97$, contact angle $\beta = 1°$),

$$Nu = 4.82 X_1^{0.624} X_9^{0.117} X_5^{0.257} X_3^{0.374} X_4^{-0.329} \tag{3.2.29}$$

For refrigerants ($3 \times 10^{-3} \le p_r \le 0.78$, contact angle $\beta = 35°$),

$$Nu = 207 X_1^{0.745} X_5^{0.581} X_6^{0.533} \tag{3.2.30}$$

For all fluids ($10^{-4} \le p_r \le 0.97$),

$$Nu = 0.23 X_1^{0.674} X_5^{0.297} X_4^{0.371} X_{13}^{-1.73} X_2^{0.35} \tag{3.2.31}$$

where $Nu = h_{pb} D_d / k_f$ and D_d is the bubble diameter at departure given by the Fritz correlation, Eq. (3.2.6). The other parameters are

$$X_1 = \frac{q D_d}{k_f T_{SAT}} \tag{3.2.32}$$

$$X_2 = \frac{\rho_f \alpha_f^2}{\sigma D_d} \tag{3.2.33}$$

$$X_3 = \frac{C_{pf} T_{SAT} D_d^2}{\alpha_f^2} \tag{3.2.34}$$

$$X_4 = \frac{i_{fg} D_d^2}{\alpha_f^2} \tag{3.2.35}$$

$$X_5 = \frac{\rho_g}{\rho_f} \tag{3.2.36}$$

$$X_6 = \frac{\nu_f}{\alpha_f} \tag{3.2.37}$$

$$X_9 = \frac{(k \rho C_p)_s}{k_f \rho_f C_{pf}} \tag{3.2.38}$$

$$X_{13} = \frac{\rho_f - \rho_g}{\rho_f} \tag{3.2.39}$$

The subscript "s" indicates that the properties are of the heating surface, while α is the thermal diffusivity. The mean absolute deviations of the fluid specific correlations are from 10% to 15%, while that of the equation for all fluids is 23%. The correlations are for roughness $R_p = 1\ \mu m$ according to German standard DIN 4762. Note that roughness according to DIN 4762 is 2.5 times the roughness defined according to ISO 4287, which is more commonly used. For other values of roughness, calculated heat transfer coefficient is to be multiplied by $R_p^{0.136}$. In developing these correlations, only 983 data points from the original 2806 data points were used, the other data showing large deviations from the correlations and data from other sources. Other researchers have generally reported good agreement of their data with this correlation, for example, Fujita and Tsutsui (1994).

Cooper (1984) analyzed a very large and varied database. He noted that there were large differences between data from different sources. By statistical analysis, he arrived at the following simple correlation as the best fit-to-all data:

$$h_{pb} = 55 F p_r^{0.12 - 0.4343 \ln R_p} (-0.4343 \ln p_r)^{-0.55} M^{-0.5} q^{0.67} \tag{3.2.40}$$

This correlation is dimensional. Heat flux q is in W m^{-2} and h_{pb} in W m^{-2} K^{-1}. Surface roughness R_p is according to DIN 4762. If the surface roughness is not known, Cooper recommends using $R_p = 1\ \mu m$. $F = 1.7$ for copper cylinders and $= 1$ for all other heaters of any material or shape. Cooper states that this factor 1.7 is not directly established by test data, is not logical, and may be superseded when more data become available. With substitution of $R_p = 1\ \mu m$ and $F = 1$ in Eq. (3.3.40), the following simplified form of Cooper correlation is obtained:

$$h_{pb} = 55 p_r^{0.12} (-0.4343 \ln p_r)^{-0.55} M^{-0.5} q^{0.67} \tag{3.2.41}$$

This is the form that has been widely used. Shah (2007) compared Eq. (3.3.41) with data for pool boiling on horizontal cylinders from 16 sources. The data included those for copper cylinders from seven sources besides those for other materials. All the data showed fairly good agreement except data from one source for a copper tube, which were high and indicated $F = 2.7$. Recently, Tran et al. (2018) reported good agreement of Eq. (3.2.41) with his data for boiling of refrigerants on a copper block. The data analyzed by Cooper included all types of fluids (water, refrigerants, chemical, and cryogens) and a wide variety of materials. Reduced pressure varied from 0.001 to 0.9 and molecular weight from 2 to 200.

Gorenflo (2013) gave a correlation that is an improved version of his earlier correlations. It is given by the following equation:

$$h_{pb} = h_{ref} \left(\frac{q}{q_{ref}} \right)^n \left(\frac{R_a}{R_{a,ref}} \right)^{2/15} \left(\frac{(k \rho C_p)_s}{(k \rho C_p)_{Cu}} \right)^{0.25} F_p \left(\frac{P_f}{P_{f,ref}} \right)^{0.6} \tag{3.2.42}$$

$h_{ref} = 3580$ W m^{-2} K^{-1} is the reference value of heat transfer coefficient for all fluids. The exponent n is

$$n = 0.95 - 0.3 p_r^{0.3} \tag{3.2.43}$$

P_f is the characteristic boiling parameter of the fluid in $(\mu m\,K)^{-1}$ defined as

$$P_f = \left(\frac{dp_{SAT}/dT}{\sigma}\right)_{p_r=0.1} \tag{3.2.44}$$

$P_{f,ref}$ is the value of $P_f = 1\ (\mu m\,K)^{-1}$. The reference heat flux q_{ref} is 20 000 W m^{-2}.

The reduced pressure-dependent parameter F_p is given by

$$F_p = 0.7p_r^{0.2} + 4p_r + \frac{1.4p_r}{1 - p_r} \tag{3.2.45}$$

R_a is the roughness of heater surface according to ISO standard 4287, and $R_{a,ref}$ is the reference value of roughness at 0.4 μm. The subscripts "*s*" and "Cu" indicate for actual heater material and for copper, respectively. Gorenflo et al. (2014) compared this correlation as well as seven other correlations with a wide range of data for copper cylinders. The Gorenflo correlation was found to give good agreement. They report that the Cooper correlation over-predicted most data. They used the correlation of Cooper with the factor 1.7 for copper cylinders. That factor was considered questionable by Cooper himself and the Cooper correlation is generally used without this factor. They did not compare the data with the Stephan–Abdelsalam correlation.

In their wide-ranging data surveys, Stephan and Abdelsalam and Cooper had found that data of different researchers under apparently identical conditions report widely different heat transfer coefficients. Therefore, the predictions of these correlations and that of Gorenflo must be considered as statistically very probable.

3.2.3.1 Conclusion and Recommendation

The most verified correlations are those of Stephan and Abdelsalam, Cooper, and Gorenflo. Each of them has been validated by these authors with a wide range of data. On the other hand, there are a large amount of data that do not agree with any of them. The present author has seen numerous papers comparing test data of their authors with different correlations. In most cases, these three give reasonable agreement. For some data one of them gives better agreement, while for some data the others give better agreement. To ensure safe design, it is recommended that heat transfer coefficient be calculated with each of these three correlations and the lowest value used for design. The Cooper correlation should be used without the 1.7 factor for copper cylinders.

3.2.4 Multicomponent Mixtures

Boiling of multicomponent mixtures is required in many applications such as chemical, petroleum, and cryogenic processes. Many of the refrigerants being used to replace older ozone-depleting refrigerants are mixtures of two or more components. Hence prediction of heat transfer during boiling of mixtures is of great practical importance. Boiling of miscible mixtures is discussed herein.

3.2.4.1 Physical Phenomena

Figure 3.2.2 shows the process of boiling of a mixture of two liquids. The mole fractions shown are those of the more volatile component. When the mole fraction is 0, the mixture contains 100% of the high boiling temperature (less volatile) component. When the mole fraction is 1, the mixture is 100% low boiling point (more volatile) component. Initially, the liquid is subcooled at point Q. As it is heated at constant pressure, its temperature rises and reaches the point R at the bubble point curve. Boiling starts now. The mole fraction in liquid phase at this condition is seen on the bubble point curve as $\tilde{x}_0$ and the mole fraction in the vapor phase is seen on the dew point curve as $\tilde{y}_0$. As boiling occurs, the more volatile component boils faster than the less volatile component. As a result, the concentration of the less volatile component increases in both the phases. When the temperature of mixture reaches T_2, the mole fractions of liquid and vapor are at points T and V, respectively. When the temperature reaches T_4, the mole fraction of vapor is at point W and is the same as that of the liquid at the start of boiling. All liquid has now been evaporated.

Figure 3.2.3a shows the data of Fujita and Tsutsui (1994) for pool boiling of aqueous mixtures of methanol with various mole fraction of methanol varying from 0 to 1. At zero mole fraction, it is pure water and at mole fraction of one, it is pure methanol. As the concentration of methanol increases, heat transfer coefficient decreases. It reaches a minimum with increasing concentration of methanol and then increases until mole fraction of methanol reaches 1. It is seen that the heat transfer coefficient of mixture is always lower than those of pure water and pure methanol. A liquid with the mean properties of the mixture will have heat transfer coefficients along a straight line joining the heat transfer coefficients at 0 and 1 concentration of methanol.

Figure 3.2.3b shows the data for ethanol–water mixture. Here it is seen that the mixture becomes azeotropic when mole fraction of ethanol is 0.89. In between the azeotropic concentration and zero concentration, the heat transfer coefficients vary as for methanol.

A number of explanations have been put forward to explain why heat transfer coefficients of mixtures are lower than of an equivalent pure fluid with the same properties.

One of the explanations is that the lower boiling point liquid boils out more rapidly and hence the liquid at the heater

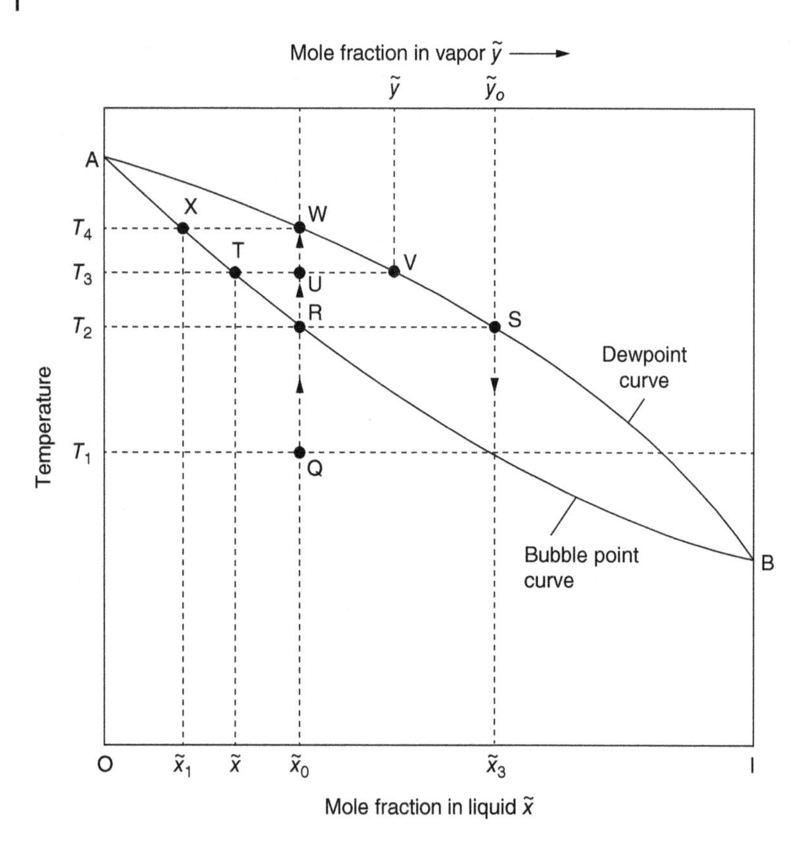

Figure 3.2.2 Phase equilibria during boiling of binary mixture. Source: From Collier and Thome (1994). © 1994 Oxford University Press.

surface is rich in the higher boiling point liquid. The boiling point of mixture, therefore, is higher than of the bulk of the mixture. As a result, the temperature difference between the heater and the liquid close to it is higher than that of a liquid with mixture mean properties and hence the heat transfer of mixture is lower. This explanation was first given by Van Wijk et al. (1956). Another explanation is that the growth of bubbles is impeded due to mass transfer effects. As the more volatile component boils faster, the vapor in the bubble is richer in it, while the liquid surrounding the bubble is richer in the less volatile component. Thus, the bubble point of the liquid surrounding the bubble and in the microlayer is higher than would be in the case of a pure fluid, and therefore the superheat available for growth of bubble is less. Further, diffusion of the more volatile component from the bulk of liquid faces mass transfer resistance in passing through the less volatile component rich liquid at the bubble interface. These factors reduce heat transfer. Some researchers attribute the lower heat transfer coefficients to nonlinear variations of properties of mixtures with evaporation. These and other theories have been discussed in detail by Thome and Shock (1984).

Heat transfer coefficients of mixtures, based on bulk liquid temperature, have been found to decrease with increasing subcooling, Sterman et al. (1966) and Hui (1983).

3.2.4.2 Prediction of Heat Transfer

Based on their experience of boiling of mixtures in kettle reboilers, Palen and Small (1964) gave the following correlation:

$$\frac{h_{\text{mix}}}{h_I} = \exp[-0.027(T_{\text{out}} - T_{\text{in}})] \tag{3.2.46}$$

T_{in} is the bubble point temperature of entering liquid and T_{out} is the temperature of vapor leaving the heat exchanger. If the entire amount of liquid entering is evaporated, it is the dew point of the entering mixture. h_{mix} is the heat transfer coefficient of the mixture and h_I is the heat transfer coefficient of a pure fluid with the properties of the mixture, to be calculated by a method suitable for pure fluids. Note that the heat transfer coefficient of mixtures is calculated based on the bulk mixture temperature in this method as well as in all other methods and experiments.

Stephan and Körner (1969) presented a method that involves a factor different for each mixture and has to be determined experimentally. It can therefore be used only for the mixtures for which this factor is known.

Thome (1983) gave the following simple correlation:

$$\frac{h_{\text{mix}}}{h_I} = \frac{1}{1 + (\Delta T_{\text{BP}}/\Delta T_I)} \tag{3.2.47}$$

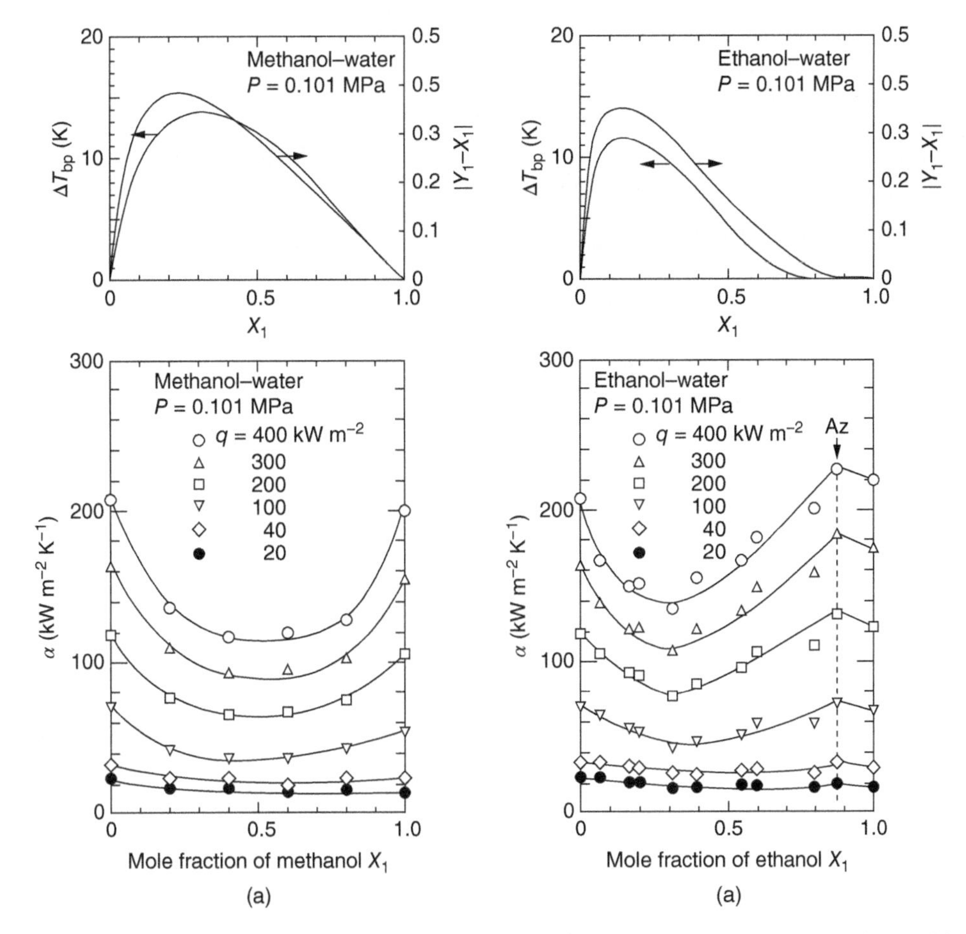

Figure 3.2.3 Heat transfer coefficients of aqueous mixtures of methanol and ethanol during pool boiling. (a) methanol-water (b) ethanol-water. Source: Fujita and Tsutsui (1994). © 1994 Elsevier.

where $\Delta T_{\text{BP}} = (T_{\text{DP}} - T_{\text{BP}})$ of the mixture and $\Delta T_I = q/h_I$ and h_I is calculated by the linear mixing rule as

$$\frac{1}{h_I} = \sum_{i=1}^{n} \frac{x_i}{h_i} \tag{3.2.48}$$

where h_i is the heat transfer coefficient of component "i" whose mole fraction is x_i. This correlation was shown to be in agreement with data for binary mixtures from several sources.

Schlunder (1983) used the film theory diffusion model assuming that the dominant mass transfer resistance is in the liquid layer surrounding the growing bubble. He arrived at the following correlation:

$$\frac{h_{\text{mix}}}{h_I} = \left\{1 + \frac{h_I}{q}\left\{\sum_{i=1}^{n-1}(T_{\text{BP},n} - T_{\text{BP},i})(y_i - x_i)\right.\right.$$
$$\left.\left.\left(1 - \exp\frac{-B_0}{\rho_f i_{\text{fg}} \beta_f}\right)\right\}\right\}^{-1} \tag{3.2.49}$$

$T_{\text{BP},n}$ is the bubble point temperature of the component that has the highest bubble point, and $T_{\text{BP},i}$ is the bubble point of other components. B_0 is a factor related to the fraction of heat flux used for evaporation and was taken = 1. β_f is the mass transfer coefficient of liquid and taken to be 0.0002 m s^{-1}. Schlunder calculated h_I by the linear mixing law, Eq. (3.2.48)

When there are a large number of components in the mixture, as is often the case in chemical processes, the Schlunder method becomes very cumbersome to use. As evaporation proceeds, liquid and vapor compositions change and so do their properties and have to be recalculated. Thome and Shakir (1987) modified the Schlunder correlation to avoid this difficulty. Their correlation is

$$\frac{h}{h_I} = \left\{1 + \frac{h_I}{q}\Delta T_{\text{BP}}\left[1 - \exp\frac{-B_0 q}{\rho_f i_{\text{fg}} \beta_f}\right]\right\}^{-1} \tag{3.2.50}$$

where $\Delta T_{\text{BP}} = (T_{\text{DP}} - T_{\text{BP}})$ for the mixture. $B_0 = 1$ as in the Schlunder correlation but β_f changed to 0.0003 m s^{-1}.

Thome and Shakir calculated h_I directly using mixture properties in a pure fluid correlation as well as by the linear mixing law Eq. (3.2.48). The deviations with the second method were somewhat lower but they recommend the first method as the properties of components of mixture are not needed, thus simplifying the calculations. The pure fluid correlation used by them was Stephan and Abdulsalam (1980), which was presented in Section 3.2.3.1. Thome and Shakir compared their correlation with data for aqueous mixtures of acetone, methanol, and propanol with satisfactory agreement. The boiling range ($T_{DP} - T_{BP}$) of the mixtures was up to 30 K.

Fujita and Tsutsui (1994) performed pool boiling tests on a circular copper plate with five binary mixtures of methanol, ethanol, *n*-butanol, and water at atmospheric pressure. They found the Thome (1983) correlation to deviate from their data at lower heat fluxes. They obtained good fit to their data by modifying the Thome correlation to the following form:

$$\frac{h_{\text{mix}}}{h_I} = \frac{1}{1 + [1 - 0.8\ \exp(-q/10^5)](\Delta T_{\text{BP}}/\Delta T_I)} \tag{3.2.51}$$

The heat flux q is in W m^{-2}. This formula gave good agreement with their own data.

Fujita and Tsutsui (1997) gave this correlation in the following nondimensional form:

$$\frac{h_{\text{mix}}}{h_I} = \frac{1}{1 + \left[1 - \exp\left\{\frac{-60q}{\rho_g i_{fg}}\left(\frac{\rho_g^2}{\sigma g(\rho_f - \rho_g)}\right)\right\}\right](\Delta T_{\text{BP}}/\Delta T_I)} \tag{3.2.52}$$

In the above two correlations, ΔT_I is calculated by the linear mixing rule, Eq. (3.2.48).

Fujita and Tsutsui (2002) measured pool boiling heat transfer of binary and ternary mixtures of R-134a, R-142b, and R-123. They compared these data with correlations of Schlunder, Thome and Shakir, and their own. Good agreement was found with the Thome and Shakir correlation as well as their own, Eqs. (3.2.51) and (3.2.52).

Taboas et al. (2007) compared test data from two sources for ammonia–water mixtures (used in vapor absorption refrigeration machines) with correlations of Thome–Shakir, Schlunder, and Fujita and Tsutsui. The Thome–Shakir correlation worked better at lower ammonia concentrations, while the Schlunder correlation worked better at higher ammonia concentrations. Taboas et al. (2007) gave their own correlation to better fit these data.

3.2.4.3 Recommendation

The correlation of Thome and Shakir is easy to use and it is at least as accurate as any other correlation. It is therefore recommended for use in designs.

3.2.5 Liquid Metals

All discussions till now are applicable only to non-metallic fluids. Liquid metals have very high thermal conductivities and their behavior is very different. Reviews of boiling of liquid metals have been done by Subbotin et al. (1972), Dwyer (1976), Shah (1992), and Wu et al. (2018). The review of Wu et al. is limited to sodium while the others cover all metals.

3.2.5.1 Physical Phenomena

Studies on bubble nucleation during nucleate boiling of liquid metals have shown that the waiting periods are much longer and the bubble growth periods are much shorter than in the boiling of ordinary liquids (Dwyer 1976; Rohsenow 1985). Deane and Rohsenow (1969) estimated that during their tests on sodium, the waiting period was 98% of the total ebullition cycle time, which was on the order of one second. For ordinary fluids, total ebullition cycle time usually falls in the range of 0.01–0.2 seconds; the waiting period and bubble growth period are usually of the same order of magnitude. Very little data on liquid metals are available for parameters such as bubble departure size and frequency; these limited data do not agree with correlations for ordinary fluids. The differences in these basic phenomena suggest that methods for predicting heat transfer to ordinary fluids may not be applicable to liquid metals.

Many studies to measure heat transfer during boiling of single-component liquid metals on a plain surface have been reported. These include studies on sodium, potassium, cesium, lithium, and mercury. In all cases, heater temperatures were found to fluctuate, the fluctuations being greater at lower pressures and heat fluxes. In some cases, the "bumping" phenomenon was observed, in which the heat transfer mechanism alternates between natural convection and nucleate boiling; see, for example, Marto and Rohsenow (1966). This is attributed to quenching of nucleating cavities due to the inrush of subcooled liquid following bubble departure. These wall temperature fluctuations and tendencies toward boiling instability indicate that more data scatter may be expected with liquid metals than with ordinary fluids. The studies with mercury showed that nucleate boiling is usually not obtained initially with pure mercury (Bonilla et al. 1957; Wagner and Lykoudis 1981). However, nucleate boiling is achieved after prolonged operation with some materials that include copper, carbon steel, and stainless steel (Dwyer 1976; Farmer 1952; Bonilla et al. 1957; Wagner and Lykoudis 1981). Nucleate boiling is achieved from the beginning by adding traces of wetting agents such as magnesium (Korneev 1955), magnesium and titanium (Lunardini

1963; Bonilla et al. 1957; Wagner and Lykoudis 1981), and sodium (Bonilla et al. 1965).

Incipient boiling data show large scatter. Most of them show that incipient superheat decreases with increasing pressure and increases with increasing heat flux. The incipient superheats are generally high. In the tests of Holtz and Singer (1968) with sodium boiling on a vertical cylinder, incipient superheats varied from 25 to 83 K.

Fraas et al. (1974) investigated the effect of strong magnetic field on boiling of potassium. Boiling occurred on a 12.5 mm diameter vertical heater. The apparatus was subjected to magnetic field up to 50 000 Gauss. No significant effect on nucleate boiling was found. This research was done in connection with the proposed method of cooling a fusion reactor that involves strong magnetic fields.

3.2.5.2 Prediction of Heat Transfer

Several authors have given dimensional correlations for a single fluid. Among them are Subbotin et al. (1968a) for sodium and Michiyoshi et al. (1986) for potassium.

Subbotin et al. (1970) presented the following correlation:

$$\frac{h_{pb}}{q^{2/3}} = B\left[\frac{k_f i_{fg} \rho_f}{\sigma T_{SAT}}\right]^{2/3} p_r^m \tag{3.2.53}$$

$B = 8$ and $m = 0.45$, for $p_r < 0.001$.

$B = 1$ and $m = 0.15$, for $p_r > 0.001$–0.02.

This correlation was based on their data for sodium, potassium, and cesium boiling on horizontal steel plates, p_r varying from 0.003 to 0.012. This correlation is dimensional. T_{SAT} is in K, h_{pb} in kcal m^{-2} h^{-1} °C^{-1}, q in kcal m^{-2} h^{-1}, i_{fg} in kcal kg^{-1}, ρ_f in kg m^{-3}, and σ in kg m^{-1}.

Based on the analysis of all available data, Shah (1992) presented the following correlation:

$$h_{pb} = Cq^{0.7} p_r^m \tag{3.2.54}$$

For $p_r < 0.001$, $C = 13.7$ and $m = 0.22$.

For $p_r > 0.001$, $C = 6.9$ and $m = 0.12$.

h_{pb} is in W m^{-2} K^{-1} and q is in W m^{-2}. The data used in developing these correlations are listed in Tables 3.2.1 and 3.2.2, for mercury and alkali metals, respectively.

Breitstein and Bonilla (1974) conducted tests on pool boiling of selenium on a horizontal stainless steel plate. They correlated their data by the following dimensional equation:

$$q = (T_w - T_{SAT})^{1.981}(17.1p^{0.5} + 6.348p) \tag{3.2.55}$$

q is in W m^{-2}, p in N cm^{-2}, and T in K. Pressure varied from 0.0133 to 1.01 bar and heat flux from 6×10^4 to 3×10^5 W m^{-2}. They compared their data to correlations for non-metallic fluids and found that the heat transfer coefficients at low pressures were orders of magnitude lower than predictions. They attributed these low heat transfer coefficients to selenium behaving like a multicomponent fluid because it consists of rings of Se_6 and Se_8 rings as well as long chains of atoms that breakdown during boiling. The present author compared their data to the Shah (1992) correlation. The predictions were several times the measurements.

As seen in the discussions earlier, the only correlations that may be considered generally applicable are those of Subbotin et al. and Shah, Eqs. (3.2.53) and (3.2.54), respectively. Shah (1992) compared these two correlations with the data in Tables 3.2.1 and 3.2.2. Figure 3.2.4 shows the comparison with data for pure mercury. It is seen that the Shah correlation is in good agreement, while the Subbotin et al. correlation shows increasing underprediction as the reduced pressure falls below 5×10^{-4}. Most of the data for mercury with promoters were underpredicted by both correlations. Figure 3.2.5 shows comparison with sodium data from five sources. The Shah correlation gives good agreement at all pressures, while the Subbotin et al. correlation shows increasing underprediction as the reduced pressure falls below about 5×10^{-4}. Figure 3.2.6 shows comparison with potassium data. Agreement of both correlations is good except for a few data points at $p_r < 10^{-4}$ that are badly scattered. There was only one data point for lithium from Wadkins (1984). As seen in Figure 3.2.7, it is in good agreement with the Shah correlation, while the Subbotin et al. correlation grossly underpredicts it. Wadkins et al. had also given a value of heat transfer coefficient corrected for interfacial resistance. Data for cesium was available from only one source. It gave excellent agreement with the Subbotin et al. correlation, but agreement with the Shah correlation was also satisfactory. These data were used in developing the Subbotin et al. correlation. The Shah correlation prediction was at the mean of the measured value and that corrected for interfacial resistance. For discussion of interfacial resistance, see Section 2.9.2.

Heat transfer can be affected by surface finish. The data in Tables 3.2.1 and 3.2.2 include normal surfaces as well as roughened or polished surfaces. As Shah's objective was to develop a correlation for normal commercial surfaces, he included data only for those polished and roughened surfaces on which tests were also done with water and shown to be in agreement with correlations for commercial surfaces. Hence the Shah correlation is expected to work well for commercial surfaces.

There is lack of agreement about the temperature of fluid to be used in calculating heat transfer coefficient. Most researchers have used the saturation temperature corresponding to the pressure in the vapor space. At low pressures, the static pressure of liquid above the heater surface can significantly increase the saturation temperature at heater level. Shah (1992) on reviewing the experimental

Table 3.2.1 Some experimental studies on pool boiling of mercury.

Data source	Test heater		Wetting agent	p (bar)	q (kW m^{-2})
	Description	Surface finish			
Bonilla et al. (1965)	70 mm diameter plate of low carbon steel	Mirror smooth	0.1% Na	1	31
					315
Bonilla et al. (1957)	Same as above	Mirror smooth	None or	0.068	38
				2.1	535
Lyon et al. (1955)	19 mm diameter horizontal tube of type 316 stainless steel (SS)	Normal		1	79
					315
Farmer (1952)	Horizontal copper plate	Normal	None	0.0081	126
					441
Korneev (1955)	22 mm diameter carbon steel vertical tubes	Natural	0.01–0.04% Mg	1	60
				10	715
Wagner and Lykoudis (1981)	60 mm diameter horizontal SS	Rough	None or[a)]	0.066	20
	Plate, sandblasted with 0.3 mm sand			1.0	300
Michiyoshi et al. (1975)	Horizontal cylinder	Normal	None	1.0	105
					500
Lunardini (1963)	Horizontal SS plate	Normal		0.68	107
				1.03	315
All				0.0081	20
				10.0	715

a) Wetting agent 0.02% Mg + 0.0001% Ti.
Source: Reprinted from Shah (1992). © 1992, with permission from Elsevier.

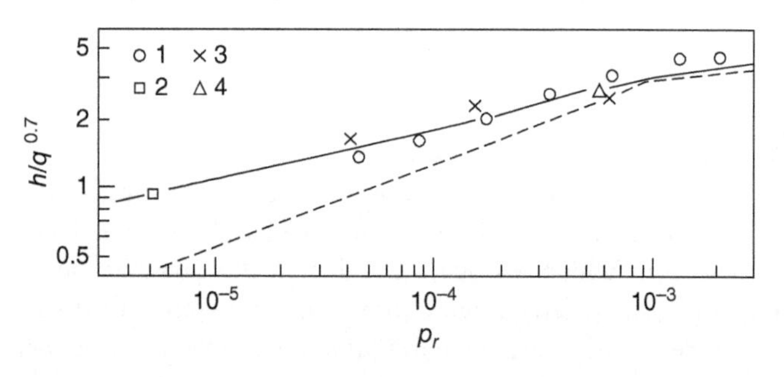

Figure 3.2.4 Comparison of the data for pool boiling of pure mercury from four sources with the correlations of Shah (1992) (continuous line) and Subbotin et al. (1970) (dashed line). Data: (1) Bonilla et al. (1957), (2) Farmer (1952), (3) Wagner and Lykoudis (1981), and (4) Michiyoshi et al. (1975). Source: Reprinted from Shah (1992). © 1992, with permission from Elsevier.

evidence concluded that for pressures exceeding about 700 Pa, the saturation temperature in the vapor space should be used. There is very little data at lower pressures, but these indicate that for system pressures approaching zero, saturation temperature corresponding to the pressure at the heater level should be used.

3.2.5.3 Recommendations

The Shah correlation, Eq. (3.2.54), is recommended for general use for fully developed nucleate boiling of pure metallic liquids on commercial surfaces. In some cases, heat transfer mode fluctuates between nucleate boiling and natural convection. The time-averaged heat transfer in that case will

Table 3.2.2 Some experimental studies on pool boiling of alkali metals.

Data source	Test heater		Fluid	p (bar)	q (kW m^{-2})
	Description	Surface finish			
Lyon et al. (1955)	19 mm diameter horizontal SS tube	Normal	Na	1.03	126
					393
Noyes and Lurie (1966)	9.5 mm diameter horizontal tubes of SS and molybdenum	Fine machined	Na	0.07	346
				0.56	2270
Marto and Rohsenow (1966)	63.5 mm horizontal nickel disks	Mirror smooth or machined	Na	0.08	173
					760
Pethukov et al. (1966)	29.6 mm diameter SS cylinder	Normal	Na	0.011	117
				0.81	1732
Kovalev and Zhukov (1973)	21.5 mm horizontal SS cylinder	Normal	Na	0.01	504
				0.05	1386
Sakurai et al. (1978)	7.6 and 10.7 mm diameter horizontal Inconel cylinders	Normal	Na	0.018	1100
				0.29	2800
Subbotin et al. (1964)	38 mm diameter horizontal disks of Cu, Ni, and SS	Normal	Na	0.14	110
					2200
Fujishiro et al. (1971)	40 mm diameter horizontal nickel plate	5–10 μm roughness	Na	0.02	232
				2.1	698
Borishanskii et al. (1965)	20–40 mm diameter tubes, horizontal and vertical	Normal	Na	0.12	20
				1.25	145
			K	0.04	12
				1.15	175
Bonilla et al. (1964)	75 mm diameter horizontal nickel plate	Normal	K	0.003	38
				2.04	364
Colver and Balzhizer (1964)	9.5 mm diameter horizontal tube of Haynes 25 alloy	Normal	K	0.63	315
				0.98	2200
Michiyoshi et al. (1985, 1986)	40 mm diameter horizontal nickel plate	Normal	K	0.04	120
				1.03	1600
Takenaka (1984)	7.8 mm diameter horizontal cylinder, Inconel	Normal	K	0.04	
				0.13	820
Subbotin et al. (1968b)	38 mm diameter horizontal SS plate	Normal	Cs	0.018	406
				2.96	1624
Wadkins (1984)	9.5 mm diameter SS vertical rod	Normal	Li	0.03	840
All				0.003	12
				2.96	2800

Source: Reprinted from Shah (1992). © 1992, with permission from Elsevier.

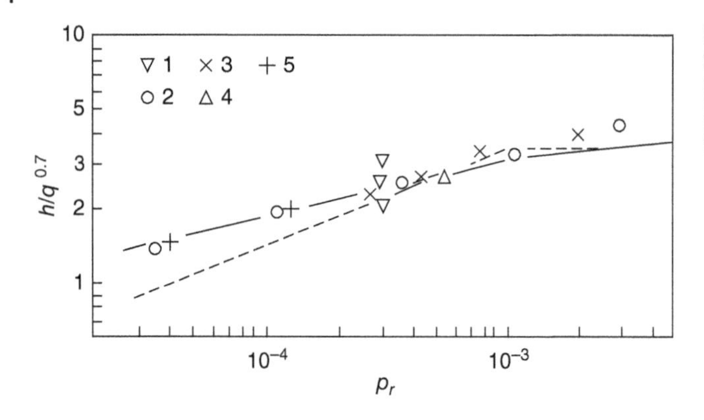

Figure 3.2.5 Comparison of the data for pool boiling of sodium from five sources with the correlations of Shah (1992) (continuous line) and Subbotin et al. (1970) (dashed line). Source: Reprinted from Shah (1992). © 1992, with permission from Elsevier.

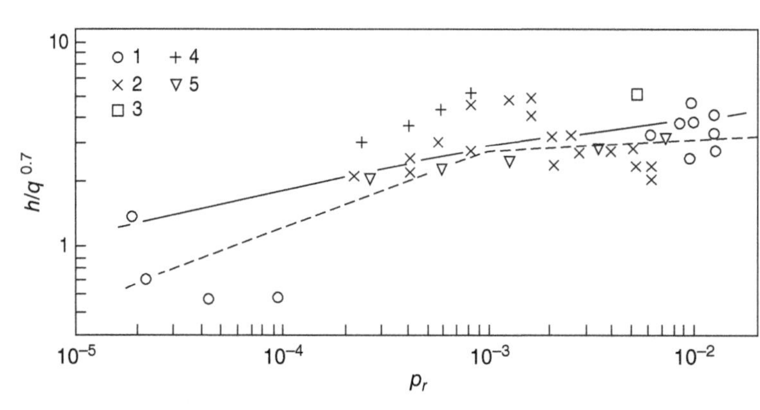

Figure 3.2.6 Comparison of the data for pool boiling of potassium from five sources with the correlations of Shah (1992) (continuous line) and Subbotin et al. (1970) (dashed line). Source: Reprinted from Shah (1992). © 1992, with permission from Elsevier.

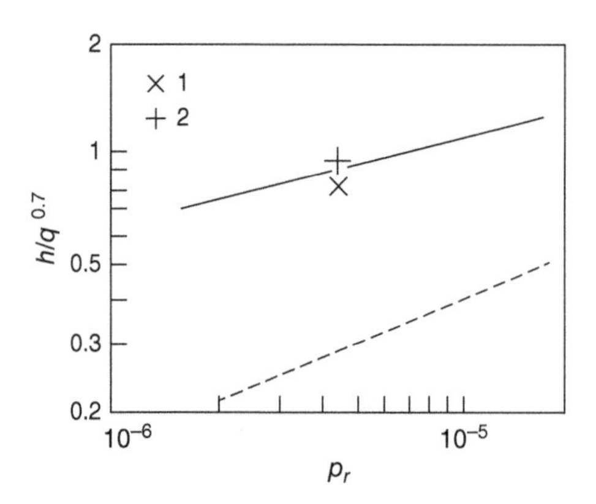

Figure 3.2.7 Data of Wadkins (1984) for lithium compared with the correlations of Shah (1992) (continuous line) and Subbotin et al. (1970) (dashed line). (1) Measured data. (2) Data corrected for interfacial resistance. Source: Reprinted from Shah (1992). © 1992 with permission from Elsevier.

be lower than for steady nucleate boiling. The Shah correlation should not be used for selenium as it behaves like a multicomponent mixture.

3.3 Critical Heat Flux

As was discussed in Section 3.1, heat transfer coefficient initially rises with increasing heat flux but eventually it falls steeply with further increase in heat flux. This phenomenon is known by names such as boiling crisis, DNB, CHF, and dryout. While DNB and dryout refer to particular mechanisms leading to the boiling crisis, all these terms are usually used interchangeably without implying any particular mechanism.

3.3.1 Models of Mechanisms

Five main mechanisms for pool boiling CHF are found in literature. These are bubble interference, hydrodynamic instability, macrolayer dryout, dry spot, and interfacial lift-off models. These are briefly discussed in the following. For a detailed review of these models, see Liang and Mudawar (2018a).

3.3.1.1 Bubble Interference Model

The bubble interference theory was proposed by Rohsenow and Griffith (1955). According to it, CHF occurs when adjacent smaller bubbles join together until they blanket the heater surface. Their analysis yielded the formula below:

$$q_c = 0.012\rho_g i_{fg}\left(\frac{\rho_f - \rho_g}{\rho_g}\right)^{0.6} \tag{3.3.1}$$

3.3.1.2 Hydrodynamic Instability Model

Zuber (1958, 1959) analyzed boiling on a flat horizontal plate. In his hydrodynamic model, *shown in Figure* 3.3.1, vapor leaves the surface in the form of columns on a square grid during intense nucleate boiling. CHF occurs when these jets become unstable and collapse. The distance between the jets equals the most dangerous wavelength due to Taylor instability λ_D given by

$$\lambda_D = 2\pi\sqrt{3}\left(\frac{\sigma}{g(\rho_f - \rho_g)}\right)^{1/2} \quad (3.3.2)$$

Each rising jet of radius R_{jet} carries the entire vapor produced in the square grid whose sides equal λ_D. The heat flux on the grid is therefore the CHF q_c given by

$$q_c = \rho_g u_g i_{\text{fg}} \frac{\pi R_{\text{jet}}^2}{\lambda_D^2} \quad (3.3.3)$$

The jet velocity u_g was considered to be that required by Helmholtz instability as

$$u_g = \frac{2\pi\sigma}{\rho_g \lambda_H} \quad (3.3.4)$$

where the wavelength $\lambda_H = 2\pi R_{\text{jet}}$. It was further assumed that the radius of jet $R_{\text{jet}} = \lambda_D/4$. Combining Eqs. (3.3.2)–(3.3.4) and inserting the value of λ_{H} and R_{jet},

$$q_c = K\rho_g^{0.5} i_{\text{fg}} [\sigma g(\rho_f - \rho_g)]^{0.25} \quad (3.3.5)$$

where $K = 0.131$. Kutateladze (1951) had derived the same equation by dimensional analysis. The constant in his formula was 0.16 ± 0.03. Thus the Zuber constant equals the minimum value of constant given by Kutateladze. Lienhard and Dhir (1973) modified the Zuber theory to arrive at $K = 0.149$ and found better agreement with data on applying this modification.

The Zuber analysis is applicable only to large horizontal plates facing upwards. Lienhard and Dhir (1973) analyzed other shapes and sizes and found that CHF for them could be considerably different as seen in Figure 3.3.2. They developed correction factors in the form

$$\frac{q_c}{q_{\text{c,Zuber}}} = f(L') \quad (3.3.6)$$

where L' is the dimensionless characteristic length of the heater defined as

$$L' = L\sqrt{g(\rho_f - \rho_g)/\sigma} \quad (3.3.7)$$

L is the characteristic dimension of the heater, which may be its length, width, radius, or perimeter. Some of the correction factors given by Lienhard and Dhir (1973) and Lienhard (1981) are given in Table 3.3.1.

3.3.1.3 Macrolayer Dryout Model

This model was proposed by Haramura and Katto (1983). According to it, vapor emanating from stems in the liquid macrolayer forms a large bubble. CHF occurs when the macrolayer dries out just as the bubble lifts off. The interval between the bubbles was considered to be that of the most dangerous wavelength λ_D of Taylor instability given by Eq. (3.3.2). Their analysis lead to the following expression:

$$\frac{q_c}{q_{\text{c,Zuber}}} = 5.5\left(\frac{A_g}{A_w}\right)^{5/8}\left(1 - \frac{A_g}{A_w}\right)^{5/16} \left[\left(1 + \frac{\rho_f}{\rho_g}\right) \Big/ \left(1 + \frac{11\rho_f}{16\rho_g}\right)^{3/5}\right]^{5/16} \quad (3.3.8)$$

where

$$\frac{A_g}{A_w} = 0.0584\left(\frac{\rho_g}{\rho_f}\right)^{0.2} \quad (3.3.9)$$

A_g is the area of vapor stems and A_w is the area of the heater. The above expression is for infinite horizontal plates. They have also given correction factors for finite bodies of different shapes.

3.3.1.4 Dry Spot Model

According to Yagov (1988, 2014), CHF occurs due to irreversible enlargement of dry spots on the heater surface. During normal boiling, there are dry spots under the bubbles. These get quenched by surrounding liquid when the bubble departs. As the heat flux increases, the number of nucleation sites increases and so does the number of dry spots. At CHF, the adjacent dry spots merge, preventing liquid from reaching the heater surface. His analysis led to the following expressions.

For $p_r < 0.001$,

$$q_{\text{c,lp}} = 0.5 \frac{i_{\text{fg}}^{81/55} \sigma^{9/11} \rho_g^{13/110} k_f^{7/110} g^{21/55} f(Pr)}{v_f^{1/2} C_{\text{pf}}^{3/10} R_i^{79/110} T_{\text{SAT}}^{21/22}} \quad (3.3.10)$$

For $p_r > 0.03$,

$$q_{\text{c,hp}} = 0.06 i_{\text{fg}} \rho_g^{3/5} \sigma^{2/5} [g(\rho_f - \rho_g)/\mu_f] \quad (3.3.11)$$

For p_r between 0.001 and 0.03,

$$q_c = (q_{\text{c,lp}}^3 + q_{\text{c,hp}}^3)^{1/3} \quad (3.3.12)$$

$R_i = R/M$ where R is the universal gas constant and M is the molecular weight of fluid. The function $f(Pr)$ for non-metallic fluids is

$$f(Pr) = \left(\frac{Pr_f^{9/8}}{1 + 2Pr_f^{1/4} + 0.6Pr_f^{19/24}}\right) \quad (3.3.13)$$

For liquid metal, $f(Pr) = 0.5$. He shows that for non-metallic fluids, the Zuber–Kutateladze formula gives poor agreement with data at low and intermediate pressures. Yugov states that his high pressure formula gives predictions close to those of the hydrodynamic instability theory.

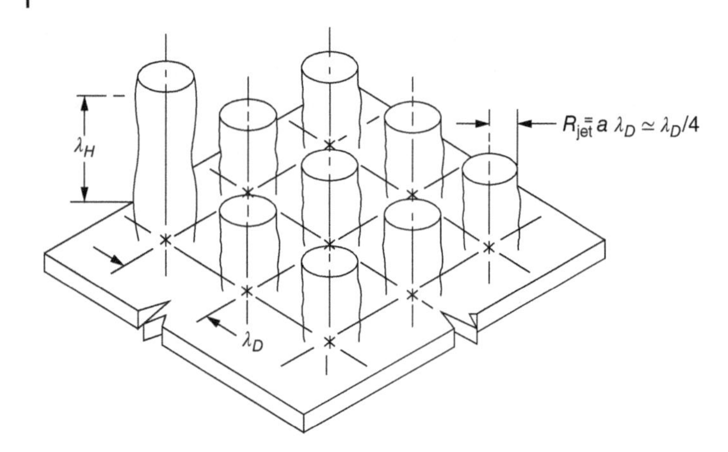

Figure 3.3.1 Zuber's hydrodynamic instability model for CHF in pool boiling. Source: From Lienhard and Dhir (1973). © 1973 American Society of Mechanical Engineers.

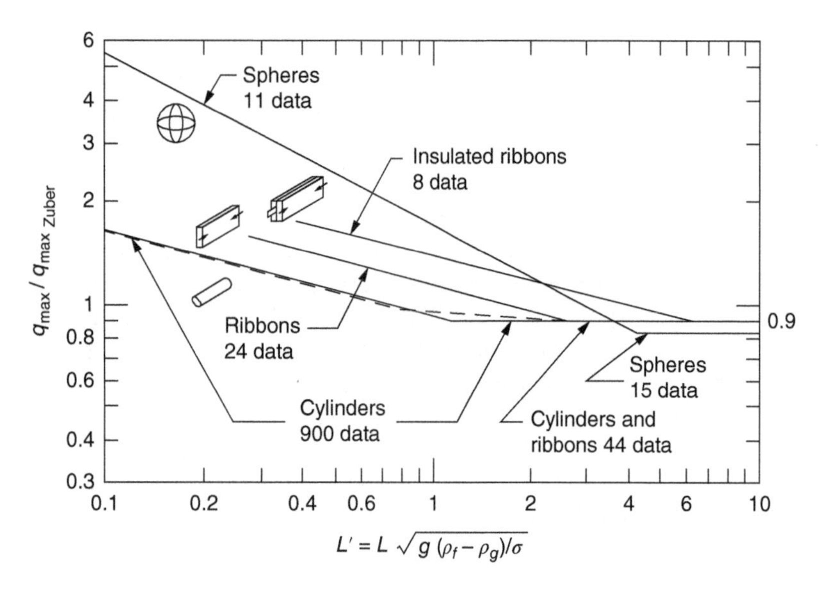

Figure 3.3.2 Effect of shape and size of heaters on CHF; q_{maxz} is the CHF predicted by Zuber formula. Source: Reproduced with permission from Lienhard (1988). © 1988 American Society of Mechanical Engineers.

3.3.1.5 Interfacial Lift-off Model

In this model developed by Mudawar et al. (1997), CHF occurs when vapor momentum becomes strong enough to lift the bulk liquid away from the surface. Their analysis yielded

$$q_c = 0.151\rho_g^{0.5} i_{fg}[\sigma g(\rho_f - \rho_g)]^{0.25} \tag{3.3.14}$$

3.3.2 Correlations for Inclined Surfaces

The various mechanism-based predictive formulas in Section 3.3.1 are for horizontal surfaces. Methods for predicting CHF for other orientations are presented herein.

Vishnev (1973) gave the following formula:

$$q_c = 0.0125(190 - \theta)^{1/2}\rho_g^{0.5} i_{fg}[\sigma g(\rho_f - \rho_g)]^{0.25} \tag{3.3.15}$$

where θ is the inclination to horizontal, in degrees. For horizontal surface facing upwards, the constant in Eq. (3.3.15) becomes 0.167, close to the value given by Kutateladze.

Kandlikar (2001) analyzed a model that takes into account hydrodynamic effects as well as surface–fluid interactions. His result for saturated fluids is

$$q_c = \left(\frac{1 + \cos\beta}{16}\right)\left[\frac{2}{\pi} + \frac{\pi}{4}(1 + \cos\beta)\cos\theta\right]^{0.5} \rho_g^{0.5} i_{fg}[\sigma g(\rho_f - \rho_g)]^{0.25} \tag{3.3.16}$$

where θ is the inclination to horizontal and β is the dynamic receding contact angle. Kandlikar noted that values of dynamic receding contact angles were hard to find. So, he assumed the following values:

- Water/copper system, $\beta = 45°$.
- Water/chromium coated surface, $\beta = 65$.
- Cryogenic liquids/copper, $\beta = 20°$.
- R-113/copper, $\beta = 5°$.

He compared his model with data from eight sources for horizontal and vertical heaters and found better agreement than with the Kutateladze equation. This correlation is applicable to inclinations 0°–90°.

Table 3.3.1 Correction factors to the Zuber formula for CHF.

Heater type	Characteristic length	Range	$q_c/q_{c,\text{Zuber}}$
Large flat plate	Width or diameter	$L' > 2.7$	1.14
Long slender heaters of any cross section	Transverse perimeter, P'	$0.15 \leq P' \leq 5.86$	$1.4/P'^{0.25}$
Cylinder	Radius, R'	$R' \geq 0.15$	$0.89 + 2.27 \exp(-3.44R'^{0.5})$
Large cylinders	Radius, R	$R' \geq 1.2$	0.904
Small cylinders	Radius, R	$0.15 \leq R' \leq 1.2$	$0.94/R'^{0.25}$
Large sphere	Radius, R	$4.26 \leq R'$	0.84
Small sphere	Radius, R	$R' > 4.26$	$1.754/R'^{0.5}$
Large heaters, any type	Length, L	$L' \geq 4$	0.9
Small horizontal ribbon oriented vertically:			
Both sides heated	Height of side, H	$0.15 \leq H' \leq 2.96$	$1.18/H'^{0.25}$
One side insulated	Height of side, H	$0.15 \leq H' \leq 5.86$	$1.4/H'^{0.25}$

Heaters are horizontal.

Chang and You (1996) studied the effect of inclination on boiling of FC-72 on plain and enhanced plates. Inclination angle to horizontal was varied was varied from 0° to 180°. Highest CHF $q_{c,\max}$ was at 0° (horizontal facing up) and lowest at 180° (horizontal facing down). The following formula was fitted to both plain and enhanced surfaces:

$$\frac{q_c}{q_{c,\max}} = 1 - 0.0012\theta \tan(0.414\theta) - 0.122 \sin(0.318\theta) \tag{3.3.17}$$

This correlation was found to also fit data for nitrogen from two sources.

3.3.3 Various Correlations

Wang et al. (2016) studied the data for hydrogen from three sources that covered reduced pressures from 0.1 to almost 1 and gave the following correlation:

$$q_c = [0.18 - 0.14p_r^{5.68}]\rho_g^{0.5} i_{fg}[\sigma g(\rho_f - \rho_g)]^{0.25} \tag{3.3.18}$$

Bewilogua et al. (1975) analyzed the data for several cryogenic fluids (nitrogen, hydrogen, oxygen, argon, neon, helium, and methane) and arrived at the following correlation:

$$\frac{q_c}{q_{c,\max}} = 0.421 + 3.58p_r - 6.19p_r^2 + 2.21p_r^3 \tag{3.3.19}$$

where $q_{c,\max}$ is the maximum CHF on q_c vs. p_r curve. Their study of data showed that $q_{c,\max}$ occurs at p_r close to 0.35 for all fluids. If the CHF at one pressure is known, that at any other pressures can be calculated with this equation. The reduced pressures in the data studied by them included almost the entire range between 0 and 1.

Several other reduced pressure correlations have been proposed. Gorenflo et al. (2010) have given the following correlation:

$$\text{For } p_r < 0.1, \quad \frac{q_c}{q_{c,0.1}} = 1.2(p_r^{0.17} + p_r^{0.8}) \tag{3.3.20}$$

$$\text{For } p_r > 0.1, \quad \frac{q_c}{q_{c,0.1}} = 3.2p_r^{0.45}(1 - p_r)^{1.2} \tag{3.3.21}$$

where $q_{c,0.1}$ is the CHF at $p_r = 0.1$. The data correlated included halogenated refrigerants and some chemicals. Reduced pressures were from 0.015 to 1. Knowing the CHF at any one pressure, that at any other pressures can be calculated by these formulas.

3.3.4 Effect of Subcooling

Subcooling has generally been found to increase CHF. Kutateladze (1952) gave the following correlation:

$$\frac{q_{c,\text{sub}}}{q_{c,\text{sat}}} = 1 + 0.065\left(\frac{\rho_f}{\rho_g}\right)^{0.8}\left(\frac{C_{\text{pf}}\Delta T_{\text{SC}}}{i_{\text{fg}}}\right) \tag{3.3.22}$$

where the subscripts "sub" and "sat" refer to subcooled and saturated conditions, respectively.

Ivey and Morris (1962) gave the following correlation:

$$\frac{q_{c,sub}}{q_{c,sat}} = 1 + 0.1\left(\frac{\rho_f}{\rho_g}\right)^{0.75}\left(\frac{C_{pf}\Delta T_{SC}}{i_{fg}}\right) \tag{3.3.23}$$

This correlation was verified with data for water, ammonia, ethyl alcohol, CCl4, and iso-octane. The data covered pressures from 0.03 to 3.3 MPa.

Elkassabgi and Lienhard (1988) divided subcooled CHF into three regimes. In the low subcooling regime, increasing subcooling increases CHF. In the high subcooling regime, there is no effect of increasing subcooling. In between these two is the intermediate subcooling regime. They developed formulas for each of these regimes that show good agreement with test data. However, no criteria have been provided in identifying these subcooling regimes. Therefore, their formulas cannot be used for design.

Wang et al. (2016) compared the Kutateladze formula, Eq. (3.3.22), to data for hydrogen from two sources at reduced pressures 0.31–0.86. They found better agreement by changing the constant from 0.065 to 0.23. So, their formula is

$$\frac{q_{c,sub}}{q_{c,sat}} = 1 + 0.23\left(\frac{\rho_f}{\rho_g}\right)^{0.8}\left(\frac{C_{pf}\Delta T_{SC}}{i_{fg}}\right) \tag{3.3.24}$$

3.3.5 Various Other Factors Affecting CHF

The effect of surface conditions on CHF is much less than in nucleate boiling heat transfer. Rough and dirty surfaces have slightly higher CHF than smooth and clean surfaces. For example, Berenson (1962) found that the roughest surface tested by him had maximum 15% higher CHF than surface with mirror finish. His tests were done with several chemicals. Lyon (1964) and Nishio and Chandratilleke (1989) reported similar results with cryogenic fluids. Cichelli and Bonilla (1945) found dirty surfaces to have about 15% higher CHF than clean surfaces.

Critical heat flux has been found to increase with heater thickness until a minimum or asymptotic thickness is reached. Grigoriev et al. (1978) reported tests with helium in which CHF increased from 70 to 84 kW m^{-2} as the thickness of copper heaters increased from 0.2 to 0.35 mm. With stainless steel heaters, the CHF changed from 72 to 75 kW m^{-2} for the same change in heater thickness. Thus, the properties of material also have an effect.

Golobič and Bergles (1997) measured CHF on ribbon heaters of varying thicknesses made from 16 different metals in FC-72 pools. They fitted the following equation to their data:

$$\frac{q_c}{q_{c,asy}} = 1 - \exp\left[-\left(\frac{S}{2.44}\right)^{0.8498} - \left(\frac{S}{2.44}\right)^{0.0581}\right] \tag{3.3.25}$$

where $q_{c,asy}$ is the asymptotic CHF and $S = \delta[(k\rho C_p)_{heater}]^{0.5}$ and δ is the half thickness of the heater in meters. Several other correlations have been proposed but none has been verified with data from many sources.

Critical heat flux with short time-pulsed heat flux can be considerably higher than with steady heat flux. See, for example, Johnson (1970) and Kawamura et al. (1970). In their transient tests on quenching of spheres in sodium pools, Farahat et al. (1974) found that their measured CHF were several times higher than given by correlations for CHF of sodium from steady-state tests.

3.3.6 Evaluation of CHF Prediction Methods

Liang and Mudawar (2018b) prepared a very wide-ranging database. It included data from 37 sources and consisted of 800 data points covering 14 working fluids, pressures from 0.0016 to 5.2 MPa, orientation angles from 0° to 180°, and contact angles from 0° to 113°. All data were for saturated fluids. They compared these data to many prediction methods including those given here in this section. For horizontal surfaces facing upwards, best agreement was with the Kutateladze–Zuber equation, Eq. 3.5 with $K = 0.149$ as given by Lienhard and Dhir. Combining this equation with that of Chang and You, Eq. (3.3.15), gave best agreement with data for inclined surfaces. The Kandlikar correlation gave good agreement with data that Liang and Mudawar considered to be affected by contact angle. Liang and Mudawar have not stated what values of dynamic receding contact angles they used in the Kandlikar correlation. This is important to know as there is little published data on receding contact angles.

Liang and Mudawar did not give information on effect of reduced pressure on deviations. This information would have been very useful in view of the fact that Yagov (2004) reports that Kutateladze equation gave large deviations with data from several sources at reduced pressures less than 0.001.

3.3.7 Recommendations

For saturated fluids at $p_r > 0.001$, the model of Lienhard and Dhir is recommended for general use with effect of orientation by the correlation of Chang and You. The resulting equation is

$$q_c = f(L')\{0.131\rho_g^{0.5} i_{fg}[\sigma g(\rho_f - \rho_g)]^{0.25}\}$$
$$(1 - 0.00120\theta\tan(0.414\theta) - 0.122\sin(0.318\theta)) \tag{3.3.26}$$

For $p_r < 0.001$, the correlation of Yagov Eq. (3.3.10) has been shown to be more accurate than hydrodynamic models and is therefore recommended.

For hydrogen, the correlation of Wang et al., Eq. (3.3.18), is recommended.

There is no well-verified general correlation for the effect of subcooling. The three correlations given here should be used in their verified range.

3.3.8 Multicomponent Mixtures

3.3.8.1 Physical Phenomena and Prediction Methods

Experimental studies have shown that CHF of mixtures can be lower or higher than those of pure fluids. This is seen in Figure 3.3.3 that shows the data of Fujita and Bai (1997) for six mixtures boiling on a 0.5 mm platinum wire. In this figure, q_{CHF} is the measured CHF and $q_{CHF,Z}$ is calculated by

$$q_{CHF,Z} = K\rho_g^{0.5} i_{fg}[\sigma g(\rho_f - \rho_g)]^{0.25} \tag{3.3.27}$$

with K given by

$$K = 0.131[f(R')_1 X + f(R')_2(1 - X)] \tag{3.3.28}$$

where X is the molar fraction of the more volatile component. $f(R')_1$ and $f(R')_2$ are the values for more volatile and less volatile component, respectively. R' is calculated by Eq. (3.3.7) with L replaced by R, the heater radius. The value of R' for the various pure fluids was 0.098–0.172. Hence it was mostly outside the range of the formula in Table 3.3.2. Therefore, these were calculated from the experimental data of pure fluids on the same wire heater. Above the figure for each mixture, $\Delta\sigma$ is also plotted where $\Delta\sigma = (\sigma_{DP} - \sigma_{BP})$. It is seen that for all five mixtures, the variations of CHF follow the variations of $\Delta\sigma$. This is the case even when the variation of $\Delta\sigma$ is sinusoidal. Fujita and Bai noted that Hovestreijdt (1963) had proposed that Marangoni flow is induced at the vapor–liquid interface because of surface tension gradient caused by concentration difference and it affects CHF of mixtures. Higher CHF is obtained if surface tension decreases with increasing concentration of the more volatile component; these are called positive mixtures. In the wedge-shaped thin liquid film formed beneath growing bubbles or between rising bubbles, Marangoni flow is expected to be induced. Its direction is from the periphery to the center for positive mixtures. This flow replenishes liquid film being thinned by evaporation and retards both the dryouts beneath growing bubbles and the coalescence between departing bubbles. This brings about increase of CHF. For negative mixtures, the situation is reversed and CHF will decrease. Fujita and Bai therefore felt that a correlation can be obtained in terms of the Marangoni number Ma defined as

$$Ma = \frac{\Delta\sigma}{\rho_f v_f^2}\left[\frac{\sigma}{g(\rho_f - \rho_g)}\right] \tag{3.3.29}$$

On analysis of their data, they obtained the following correlation:

$$q_{CHF} = q_{CHF,Z}\left(1 - 1.83 \times 10^{-3}\frac{|Ma|^{1.43}}{Ma}\right)^{-1} \tag{3.3.30}$$

Figure 3.3.3 shows the predictions of this correlation. It is seen that good quantitative and qualitative agreement is obtained in all cases. Also shown in this figure are the predictions of the correlations of Reddy and Lienhard (1989) and Yang (1987). Both of them show poor agreement with data. No independent evaluation of this correlation came to the author's notice.

Yagov (2004) extended his dry spot model of CHF (discussed in Section 3.3.1.4) to mixtures. He assumed that the liquid flow rate to the intensive evaporation zone in the vicinity of the dry spot boundary is controlled by the capillary pressure gradient. The latter generally comprises the liquid film surface curvature gradient and the surface tension gradient, and therefore the difference in the CHF between pure fluids and mixtures is attributable to change in surface tension. He then gave the following formula:

$$q_{c,m} = q_{c,I}\left(1 + \frac{\Delta\sigma}{\sigma}\right) \tag{3.3.31}$$

where $q_{c,m}$ is the CHF of mixture and $q_{c,I}$ is the CHF of a fluid with the properties of the mixture, to be calculated by a correlation for pure fluids. Yagov compared it to data for mixtures of ethanol and methanol with water from many sources with reasonably good agreement. Yagov did not compare this correlation with any non-aqueous mixtures. He compared it to the data of Fujita and Bai for aqueous mixtures but not their data for non-aqueous mixtures that have different trends. This suggests that his correlation did not correctly predict those data.

McGillis and Carey (1996) also developed a correlation based on Marangoni effect. It fitted their own data for aqueous mixtures. Sakashita et al. (2010) found it to give large deviations with their data for water with 2-propanol. Correlations of Yagov and Reddy and Lienhard gave satisfactory agreement.

3.3.8.2 Recommendation

There is no well-verified general prediction method. The correlation of Fujita and Bai is recommended in its verified range.

3.3.9 Liquid Metals

Critical heat flux in boiling of liquid metals has been reviewed by Subbotin et al. (1972); Dwyer (1976); Shah (1990); and Lee (1999).

Table 3.3.2 Liquid metal CHF data used in developing Shah correlation.

Source	Test heater	Fluid	Pressure (bar)	Heater finish	Liquid depth (mm)	Notes
Sakurai et al. (1978)	7.6 mm diameter horizontal cylinder with Inconel shield	Sodium	0.0058 0.33	?	50 150	Includes subcooled data
Kawamura et al. (1975)	6.5mm diameter vertical cylinder with SS sheath	Sodium	0.10	?	?	Argon gas cover used
Caswell and Balzhiser (1966)	9.5 mm diameter horizontal cylinder of Haynes 25 alloy	Sodium	0.056 0.15	Smooth	90	
Noyes (1963)	9.5 mm diameter horizontal cylinders of SS and molybdenum	Sodium	0.035 0.59	Smooth	150	
Noyes and Lurie (1966)	6.35 mm diameter horizontal cylinders of SS and molybdenum	Sodium	0.041 0.56	Smooth	150	
Carbon (1964)	12.2 mm diameter horizontal molybdenum rod	Sodium	0.031 1.4	Smooth	?	
Colver and Balzhiser (1964)	9.5 mm diameter horizontal cylinder of Haynes 25 alloy	Potassium	0.016 1.54	Smooth	100 150	
Michiyoshi et al. (1986)	40 mm diameter horizontal disk of nickel	Potassium	0.04 1.0	Polished with 6–12 μm paste	30	
Subbotin et al. (1972)	38 mm diameter horizontal disks of SS, nickel, and molybdenum	Sodium, cesium, potassium	0.012 2.45	Smooth	100 203	Argon cover in some tests
	38 mm diameter horizontal copper disk clad with Armco iron	Mercury	0.73 5.8	?	?	Iron cladding found ruptured at end of tests
Subbotin et al. (1974)	38 mm diameter horizontal disks of nickel and SS	Sodium, potassium	0.022 0.4	Smooth	?	
Avksentyuk and Mamontova (1973)	11 mm diameter horizontal SS cylinders	Cesium	0.107 2.45	Smooth	30 60	A few artificial cavities on heater
Takahashi et al. (1980)	20 mm × 100 mm horizontal SS plate	Mercury	4×10^{-6} 0.13	Polished	3 35	
Lee (1968)	19 mm diameter horizontal SS cylinder	Mercury	1.0	?	?	
Lyon et al. (1955)	19 mm diameter horizontal SS tube	Mercury + 0.1% Na	1.0	Normal	?	
Turner and Colver (1971)	9.5 mm diameter horizontal cylinder with silver coating	Mercury	0.003 1.3	Smooth	50 216	

Source: From Shah (1996). © 1996 Taylor & Francis.

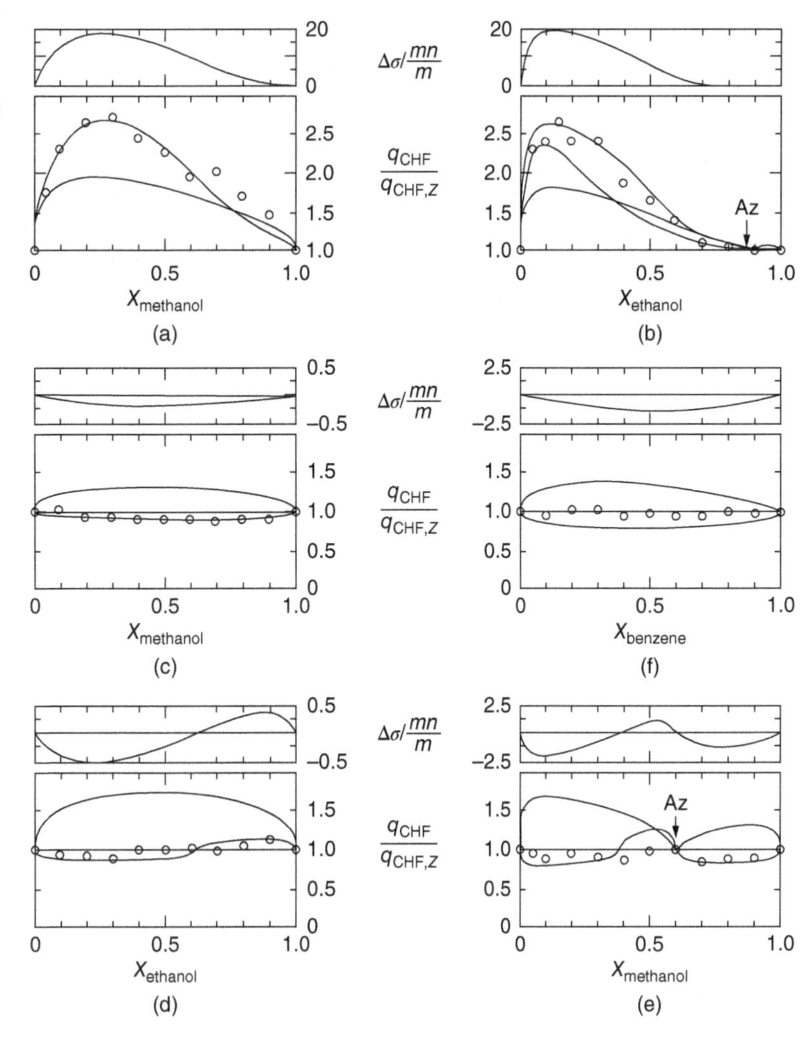

Figure 3.3.3 Effect of surface tension variations on CHF of mixtures. (a) Ethanol/water, (b) methanol/water, (c) methanol/ethanol, (d) ethanol/*n*-butanol, (e) methanol/benzene, and (f) benzene/*n*-heptane. Thick continuous line is the correlation of Fujita and Bai, dashed line is the correlation of Reddy and Lienhard, and thin continuous line is the correlation of Yang. Source: Fujita and Bai (1997). © 1997 Elsevier.

3.3.9.1 Physical Phenomena

Critical heat flux in liquid metals is affected by boiling stability, surface wetting, and cover gases. CHF of alkali metals is always much higher than of ordinary fluids and mechanisms have been proposed to explain it. These are discussed in the following.

Boiling Instability As was discussed in Section 3.2.5.1, nucleate boiling of liquid metals can be steady/stable or unsteady/unstable. Wall temperature fluctuations during stable boiling are small, while the fluctuations are large in unstable boiling. Subbotin et al. (1972) did wide-ranging tests on CHF of liquid sodium, potassium, and cesium on horizontal plates. They found that CHF is higher if boiling prior to CHF was stable. The value of CHF was reduced to half or even lower if boiling had been unstable prior to CHF. Subbotin et al. (1972) also reviewed data from other researchers and found similar effect of wall temperature fluctuations on CHF.

Similar results have been reported by Kutateladze et al. (1973) and Avksentyuk and Mamontova (1973) in boiling of potassium and cesium on horizontal cylinders. CHF was found to vary over a wide range; the lowest values occurred when there was direct transition from single-phase convection to CHF and highest when transition was from stable nucleate boiling.

During unstable boiling, the mode of heat transfer alternates between nucleate boiling and single-phase convection. As liquid metals have long waiting periods, there is always this alternation between convection and nucleate boiling at individual nucleation sites. As long as there are numerous active sites, the instability is small and fluctuations in wall temperature are small. As the number of active sites decreases, instability becomes more pronounced and wall temperature fluctuations increase. During very unstable boiling, vapor generation ceases almost completely for a short time over the entire heater surface. This is followed by very intense vapor generation.

Subbotin et al. (1972) found CHF of alkali metals to decrease as boiling instability increased as indicated by increasing wall temperature fluctuations. Dwyer (1976) noted that low CHF can be caused by inadequate surface wetting or due to too few active nucleation sites. Incomplete wetting often occurs with mercury. Alkali metals generally wet the surface fully. Very smooth surfaces may have only very small cavities that nucleate at high temperature differences; heat transfer then alternates between natural convection and nucleate boiling and is therefore unsteady/unstable.

To explain the increasing instability with prolonged use observed by Subbotin et al. (1972), Dwyer (1976) theorized that with continued boiling, the larger cavities become quenched firstly due to loss of inert gas from them and then reduction of oxides on their walls. This could lead to extremely small cavity sizes, larger superheats, and very unstable boiling. With increasing instability, the rate of vapor generation during the boiling period will be higher. Therefore, the vapor jets could become unstable at smaller heat flux.

Effect of Cover Gases Subbotin et al. (1972) found that presence of inert cover gas during boiling of sodium improved boiling stability and resulted in higher CHF compared with that without cover gas. It may have been due to trapping of gas in cavities and thus increasing nucleation sites. However, they found no effect of inert gas during boiling of sodium–potassium mixtures.

Effect of Surface Wetting Poor wetting of heating surface can cause boiling instability, leading to low CHF as discussed earlier. Mercury often does not wet adequately. The only data obtained under full surface wetting conditions are those of Subbotin et al. (1972) for a horizontal copper disk with iron cladding.

Mechanism for Higher CHF of Metals Many researchers have reported that CHF of alkali metals is always much higher than given by Zuber equation and its modification. Noyes and Lurie (1966) proposed that this discrepancy is because a large amount of heat is removed by conduction and single-phase convection due to the high thermal conductivity of metallic fluids. The hydrodynamic instability model of Zuber assumed that the entire heat flux is used for evaporation, which leaves the surface in the form of jets. As much of the heat flux in boiling of metals is removed by convection and conduction, it will take much higher heat flux to form jets that become unstable. They therefore proposed

$$q_c = q_{c-c} + q_{\text{ev}} \tag{3.3.32}$$

where q_{c-c} is the heat flux removed by conduction and convection and q_{ev} is the heat flux removed by boiling. They proposed that q_{ev} be calculated by the hydrodynamic model such as that of Zuber. For q_{c-c}, they gave a fixed value of 1.3 mW m^{-2} based on their own data for sodium. This obviously cannot be used in all situations. Sakurai et al. (1978) found q_{c-c} to be 1.04 mW m^{-2} for their own sodium data.

Bankoff and Fauske (1974) gave another explanation of why liquid metal CHF is high. They point out that the Zuber analysis does not take into account condensation from the top of the bubbles. Removal of vapor by condensation reduces the amount of vapor flowing in the jets. Hence a higher heat flux will be required to reach the critical jet velocity leading to CHF.

Sakurai et al. (1978) stated that the pressure at the heater level is higher than in the vapor space due to the liquid head. Hence the liquid at the heater level is subcooled and this subcooling increases CHF. Their model is

$$q_c = q_{\text{SC}} + q_{c-c} + q_{\text{ev}} \tag{3.3.33}$$

where q_{sc} is the heat flux due to subcooling; they have not provided any method for calculating it.

3.3.9.2 Prediction of CHF

Various Methods Caswell and Balzhiser (1966) gave the following correlation:

$$\frac{q_c C_{\text{pf}} \sigma}{i_{\text{fg}}^2 \rho_g k_f} = 1.18 \times 10^{-8} \left(\frac{\rho_f - \rho_g}{\rho_g} \right)^{0.71} \tag{3.3.34}$$

The data correlated included their own with sodium and rubidium on 9.5 mm diameter cylinders as well as data for sodium and potassium from other sources. The pressure in the data ranged from 0.104 to 4.1 bar.

Noyes (1963) gave the following correlation based on his sodium data:

$$\frac{q_c}{i_{\text{fg}} \rho_g} = 0.144 \left[\frac{\rho_f - \rho_g}{\rho_g} \right]^{1/2} \left[\frac{g\sigma}{\rho_f} \right]^{1/4} Pr_f^{-0.245} \tag{3.3.35}$$

Krillov (1968) gave the following dimensional equation:

$$q_c = 0.707 k_f^{0.6} p_r^{1/6} \tag{3.3.36}$$

with q_c in mW m^{-2} and k_f in W m^{-1} K^{-1}. This correlation was based on data for sodium, potassium, and cesium covering reduced pressures from 10^{-4} to 3×10^{-3}.

Subbotin et al. (1972) gave the following correlation:

$$\frac{q_c}{q_z} = \left[1 + \left(\frac{C}{p_c} \right) p_r^{-0.4} \right] \tag{3.3.37}$$

The constant C is 45 for stable boiling and 18 for unstable boiling. The critical pressure p_c is in atmospheres. This correlation was developed using data for sodium, potassium, rubidium, and cesium from six sources for horizontal disks as well as horizontal cylinders.

Based on their hypothesis that condensation above bubbles increases CHF, Bankoff and Fauske developed a

Table 3.3.3 The range of data for which the Shah correlation (1996) was verified.

Fluids	Sodium, potassium, cesium, rubidium, mercury
Heater geometry	Horizontal flat plates and cylinders, vertical cylinder
Heater size	Plates 38–40 mm diameter, cylinders 0.35–12.2 mm in diameter
Heater material	Stainless steel, Armco iron, molybdenum, Haynes 25 alloy, nickel
Pressure (bar)	0.0058–5.8
p_r	2.26×10^{-5} to 2.1×10^{-2}
v_f (m^2 s^{-1})	7×10^{-6} to 4.5×10^{-5}

Source: From Shah (1996). © 1996 Taylor & Francis.

calculation method for CHF. Calculations by this method are quite tedious.

Yagov and Sukach (2000) compared Eq. (3.3.10) of the dry spot model of CHF with data for alkali metals from five sources. Predictions were mostly through the middle of data but scatter was considerable.

No prediction method has been published for the effect of subcooling in steady boiling. Farahat et al. (1974) have given the following correlation on the basis of their transient tests on quenching of spheres in subcooled sodium at atmospheric pressure:

$$q_c = 4.1 \times 10^6(1 + 7.8 \times 10^{-3}\Delta T_{SC}) \tag{3.3.38}$$

where q_c is in Btu h^{-1} ft^{-2} and ΔT_{SC} is in °F. The spheres were made of tantalum and had diameters of 25.4, 19.05, and 12.7 mm.

Shah Correlation Shah (1996) developed the following correlation.

For stable boiling on unbounded surfaces,

$$\frac{q_c - q_z}{q_z} = \frac{27\,400 v_f}{p_g^{0.4}} \tag{3.3.39}$$

For unstable boiling,

$$\frac{q_c}{q_z} = \frac{25\,000 v_f}{p_g^{0.156}} \tag{3.3.40}$$

where the thermal diffusivity of liquid v_f is in m^2 s^{-1} and the vapor space pressure p_g is in bar. q_z is calculated with the Zuber–Kutateladze Eq. (3.3.5) with $K = 0.16$.

Shah notes that Eq. (3.3.39) is also valid for non-metallic fluids as their thermal diffusivity is so low that it reduces to $q_c = q_z$.

Shah justified the use of thermal diffusivity as follows. In the theory of Noyes and Lurie, CHF is higher than the hydrodynamic model due to the heat removed by conduction convection. Transient conduction through the liquid layer next to the heater is given by

$$\frac{\partial T}{\partial t} = v_f \frac{\partial T^2}{\partial t^2} \tag{3.3.41}$$

In the Bankoff and Fauske model, the expression for calculation of condensation contribution involves thermal diffusivity. Therefore, the inclusion of thermal diffusivity in the Shah correlation is consistent with the physical phenomena.

The data on which Shah based his correlation are listed in Table 3.3.2. These are from 16 sources, include alkali metals as well as mercury, horizontal and vertical cylinders, and horizontal disks. The summary of the data range is in Table 3.3.3.

Evaluation of Prediction Methods Figure 3.3.4 shows the comparison of various correlations with data for sodium from eight sources. It is seen that the Kutateladze–Zuber formula greatly underpredicts all data. The Shah correlation for stable boiling is at the upper boundary of all data, while the Shah correlation for unstable boiling is at the lower boundary of data. The predictions of the correlation of Subbotin et al. for stable boiling are close to those of Shah for stable boiling. The predictions of Krillov correlation are through the middle of the data.

Figure 3.3.5 compares the data for potassium with various correlations. Results of various correlations are similar to those for sodium shown in Figure 3.3.4.

Figure 3.3.6 shows the comparison with data for cesium. The correlations of Subbotin et al. and Krillov are at the upper boundary of data, while the Shah correlation for stable CHF is a little below these two. Shah's unstable CHF correlation is at the lower boundary of data.

Figure 3.3.7 shows the data of Caswell and Balzhiser for rubidium. No other data for rubidium were found. It is seen that the correlation of Shah for stable boiling shows good agreement with data. The Krillov correlation and that of Subbotin et al. for stable boiling CHF overpredict considerably.

The only data for mercury in which the heater surface was fully wetted are those of Subbotin et al. (1972). These are shown in Figure 3.3.8. The Shah correlation predictions are at the upper boundary of measurements, while Subbotin et al. predicts a little higher than the Shah correlation. The Krillov correlation underpredicts the data. It is interesting that the Zuber formula gives the best agreement with

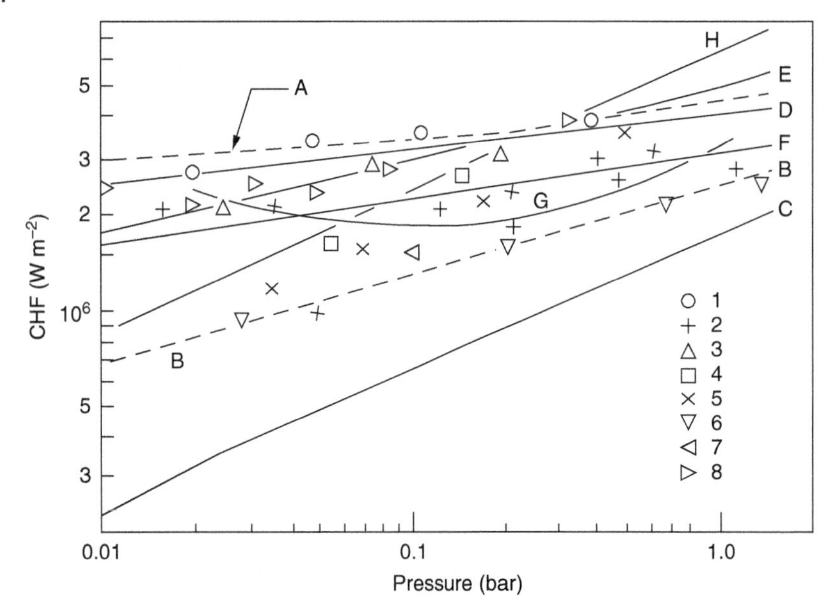

Figure 3.3.4 Comparison of data from various sources with various correlations. Data: (1) Subbotin et al. (1972) with argon cover gas, (2) Subbotin et al. (1972) without cover gas, (3) Subbotin et al. (1974), (4) Caswell and Balzhiser (1966), (5) Noyes and Lurie (1966), (6) Carbon (1964), (7) Kawamura et al. (1975), and (8) Sakurai et al. (1978). Correlations: (a) Shah for stable CHF. (b) Shah for unstable CHF. (c) Kutateladze–Zuber. (d) Subbotin et al. for stable CHF. (e) Caswell and Balzhiser. (f) Krillov. (g) Bankoff Fauske analysis. (h) Noyes. Source: From Shah (1996). © 1996 Taylor & Francis.

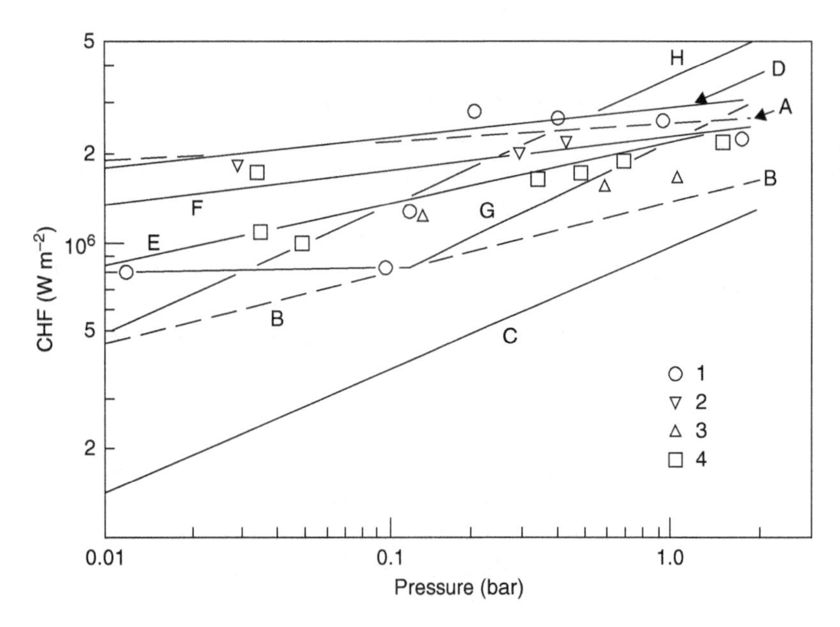

Figure 3.3.5 Critical heat flux data for potassium compared with various correlations. Data: (1) Subbotin et al. (1972), (2) Subbotin et al. (1974), (3) Michiyoshi et al. (1986), and (4) Colver and Balzhizer (1964). Correlations A through H, see caption of Figure 3.3.4. Source: From Shah (1996). © 1996 Taylor & Francis.

this data. It raises the question why the increase in CHF due to conduction–convection contribution postulated by Noyes and Lurie did not occur in this case. Similar question occurs for the Bankoff–Fauske model of heat removal by condensation from bubble boundary.

Figure 3.3.9 shows data of mercury where the heater was not fully wetted and boiling prior to CHF was unstable. Also shown are the correlations of Subbotin et al. and Shah for unstable CHF. The Shah correlation is at the lower boundary of data, while the Subbotin et al. correlation is at the upper limit of most data.

No data for CHF of lithium were found. In the pool boiling tests of Wadkins (1984), heat flux of 1.4 mW m^{-2} was reached without burnout at a pressure of 2.6 Pa. The correlations of Shah and Subbotin et al. predict CHF of 1.6 and 1.7 mW m^{-2}, respectively. Hence these do not contradict nucleate boiling data. The correlation of Krillov predicts an order of magnitude lower CHF and hence contradicts the nucleate boiling data.

The only data found by Shah that was for subcooled condition was that of Sakurai et al. (1978). Shah tried the correlations of Ivey and Morris (1962)) and Kutateladze (1952), Eqs. (3.3.23) and (3.3.22), respectively. The increase in CHF due to subcooling predicted by them far exceeded the measured values. It indicates that correlations for ordinary fluids are not applicable to metallic fluids.

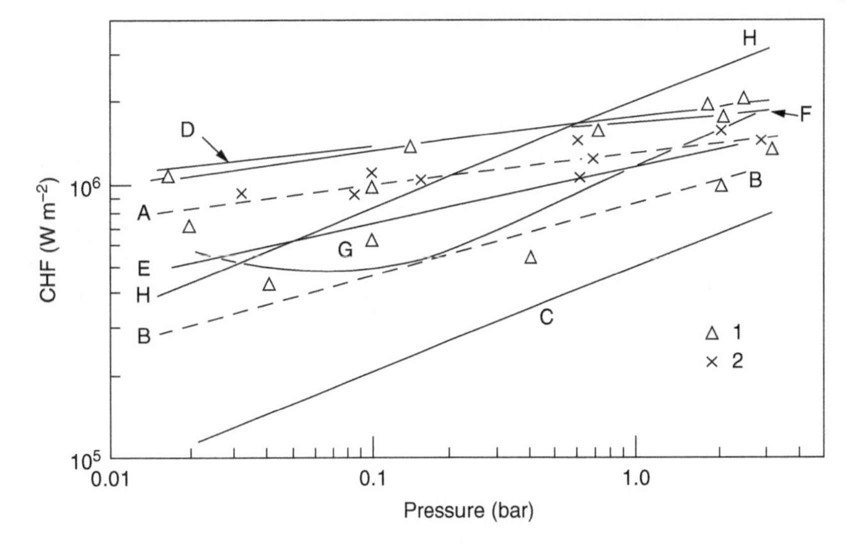

Figure 3.3.6 Cesium CHF data compared with various correlations. Data: (1) Subbotin et al. (1972) and (2) Avksentyuk and Mamontova (1973). Correlations A through H, see caption of Figure 3.3.4. Source: From Shah (1996). © 1996 Taylor & Francis.

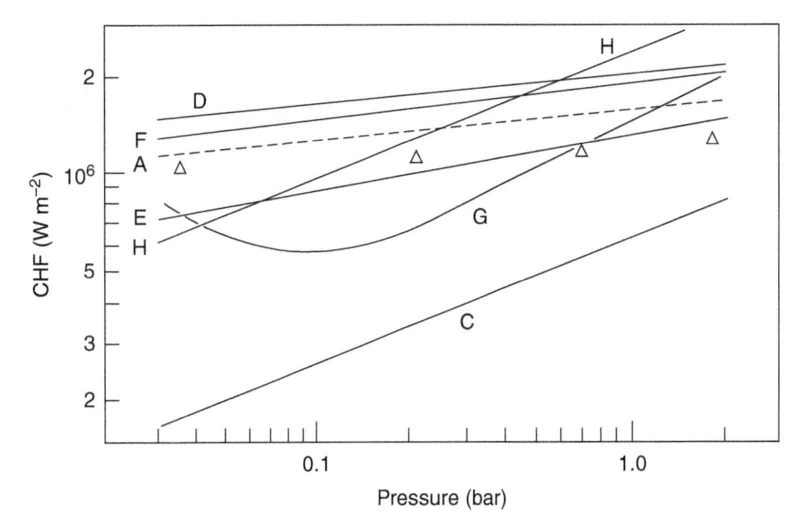

Figure 3.3.7 Data of Caswell and Balzhiser (1966) for rubidium compared to various correlations. See caption of Figure 3.3.4 for A through H. Source: From Shah (1996). © 1996 Taylor & Francis.

Lienhard and Dhir (1973) gave correction factors for various heater shapes and sizes to be applied to the Zuber theory. These are listed in Table 3.3.1. Shah found that these correction factors did not work for liquid metal data. For example, the value of CHF for sodium measured by Subbotin et al. (1972) on 38 mm plates are comparable to those measured by Sakurai et al. (1978) on a 7.8 mm diameter cylinder as seen in Figure 3.3.4. The correction factors of Lienhard and Dhir predicts the CHF of this plate to be about double of that for cylinder. No effect of heater size or shape was found by Shah in his data analysis. Dwyer (1976) also reached the same conclusion. He hypothesized that this is because heater size and shape affect only the part of heat flux removed by evaporation. In liquid metals, a substantial part of heat flux is removed by natural convection.

All the data analyzed by Shah were for horizontal heaters except for one data point from Kawamura et al. (1975) for a vertical cylinder. That data point is about 50% of the CHF predicted by the Shah correlation. The test was done with argon cover gas and there was no indication of instability. The correlation of Vishnev, Eq. (3.3.15), predicts CHF for vertical surfaces to be 72% of that on horizontal surfaces. On the other hand, the formula of Chang and You, Eq. (3.3.17), predicts only 14% lower CHF for vertical orientation. It is therefore uncertain whether orientation caused the low CHF in Kwarmura et al. data point.

From the foregoing, it is seen that the correlation of Shah is the only one that agrees with data for all fluids. The next best is that of Subbotin et al., which works well for sodium, potassium, and cesium, but overpredicts the data for rubidium and mercury. These two are the only correlations that take instability into account and give upper and lower limits of CHF. Shape and size of heaters do not have any effect on liquid metal CHF. Subcooling increases CHF but the formulas for non-metallic fluids overpredict CHF.

Alternative Shah Correlations Shah (1996) also gave two alternative correlations.

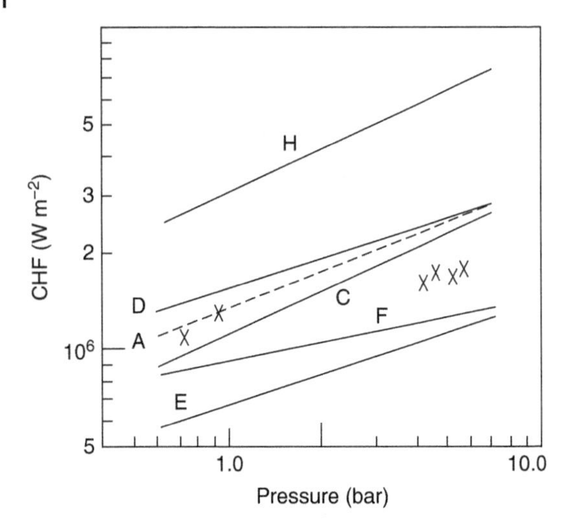

Figure 3.3.8 Data for Subbotin et al. (1972) for mercury with fully wetted heater compared with various correlations. See caption of Figure 3.3.4 for A through H. Source: From Shah (1996). © 1996 Taylor & Francis.

By modifying the correlation of Addoms (1948) for ordinary fluids, Shah gave the following relation for liquid metal CHF:

$$\frac{q_c}{i_{fg}\rho_g} = 0.0021\left(\frac{\rho_f - \rho_g}{\rho_f}\right)^{0.8} (gv_f)^{1/3} Pr_f^{-0.67} \quad (3.3.42)$$

Shah modified the Krillov correlation to improve its accuracy. The resulting equation is

$$q_c = 0.22 k_f^{0.7} p_g^{0.15} \quad (3.3.43)$$

According to Shah, the predictions of these alternative correlation are close to the Shah correlation for stable CHF, Eq. (3.3.39). However, he did not provide a comparison.

3.3.9.3 Recommendations

Calculate stable and unstable boiling CHF for horizontal surfaces by the Shah correlation, Eqs. (3.3.39) and (3.3.40), respectively. These provide the upper and lower limits of CHF. Use a value in between the two based on the expected level of boiling stability. No correction factor is required for heater size or shape. CHF for vertical surfaces may be considerably lower, but there is insufficient evidence as to how much.

3.4 Transition Boiling

In transition boiling, heater surface is intermittently in contact with liquid and vapor. Nucleate boiling occurs on the portion in contact with liquid, while film boiling occurs on the portion in contact with vapor. In Figure 3.3.2, transition boiling curve is that joining the CHF point C with the MFB point D. As seen in this figure, heat flux decreases in this region with increasing wall superheat.

Pioneering experimental work of transition boiling was done by Berenson (1960, 1962). From his observations and analysis, he concluded that transition boiling is a combination of film and nucleate boiling alternately existing at a given location on the heated surface. Any particular spot comes alternately in contact with liquid and vapor. The variation in heat flux with wall superheat is a result of change in the fraction of time each boiling mode is present at a given location.

Increasing surface roughness moves the nucleate boiling and transition boiling curves to the left. Thus, transition boiling heat transfer is higher on rougher surfaces. Decreasing contact angle (more wetting of surface) improves transition heat transfer. This is illustrated in Figure 3.4.1.

It should be noted that there are two different transition curves depending on whether it is obtained by cooling from the film boiling side or by heating from the nucleate boiling side. Heat transfer coefficients are higher if it results from cooling.

Attempts at developing predictive techniques have been made based on estimates of portions of area in contact with liquid and vapor or on the basis of time for which areas are in contact with liquid or vapor. Kalinin et al. (1976) proposed the following relation for q_{tb}, the heat flux during transition boiling:

$$q_{tb} = \delta q_{nb} + (1 - \delta) q_{fb} \quad (3.4.1)$$

δ is the fraction of surface in contact with liquid, q_{nb} is the heat flux due to nucleate boiling, and q_{fb} is the heat flux due to film boiling. For δ, Kalinin et al. (1976) gave the following relation:

$$\delta = (1 - \phi)^7 \quad (3.4.2)$$

$$\phi = \frac{T_w - T_{CHF}}{T_{MFB} - T_{CHF}} \quad (3.4.3)$$

T_{CHF} and T_{MFB} are the wall temperatures at CHF and MFB.

Bjornard and Griffith (1977) proposed

$$q_{tb} = \delta q_{CHF} + (1 - \delta) q_{MFB} \quad (3.4.4)$$

$$\delta = \left(\frac{T_w - T_{MFB}}{T_{CHF} - T_{MFB}}\right)^2 \quad (3.4.5)$$

This correlation was verified with data for forced flow in tubes and annuli. Dhir (1991) states that the previously mentioned two methods give widely different results.

Based on their data from tests with hydrogen in normal and low gravity, Wang et al. (2016) recommend linear interpolation between the CHF point and MFB point. The resulting formula is

$$q_{tb} = q_{CHF} - \frac{\Delta T_{tb} - \Delta T_{CHF}}{\Delta T_{MFB} - \Delta T_{CHF}}(q_{CHF} - q_{MFB}) \quad (3.4.6)$$

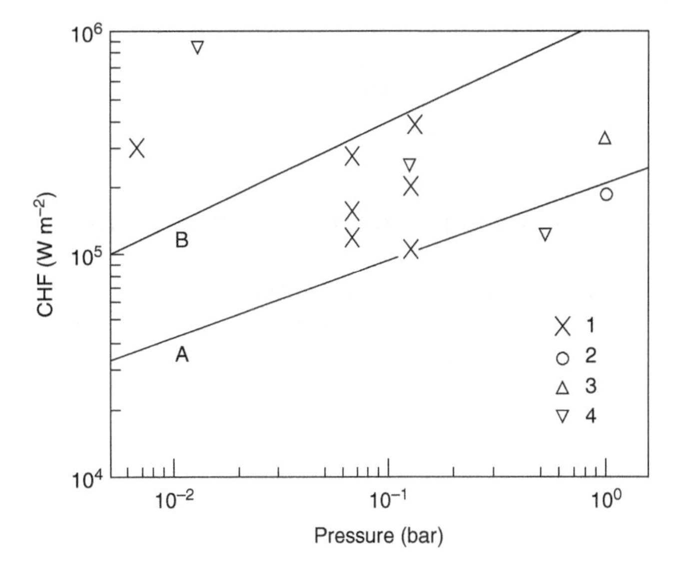

Figure 3.3.9 Critical heat flux data for mercury with partially wetted heater surfaces compared with correlations for unstable boiling. Data: (1) Takahashi et al. (1980), (2) Lyon et al. (1955), (3) Lee (1968), and (4) Turner and Colver (1971). (a) Shah Eq. (3.3.35) and (b) Subbotin et al. Eq. (3.3.33) with $C = 18$. Source: From Shah (1996). © 1996 Taylor & Francis.

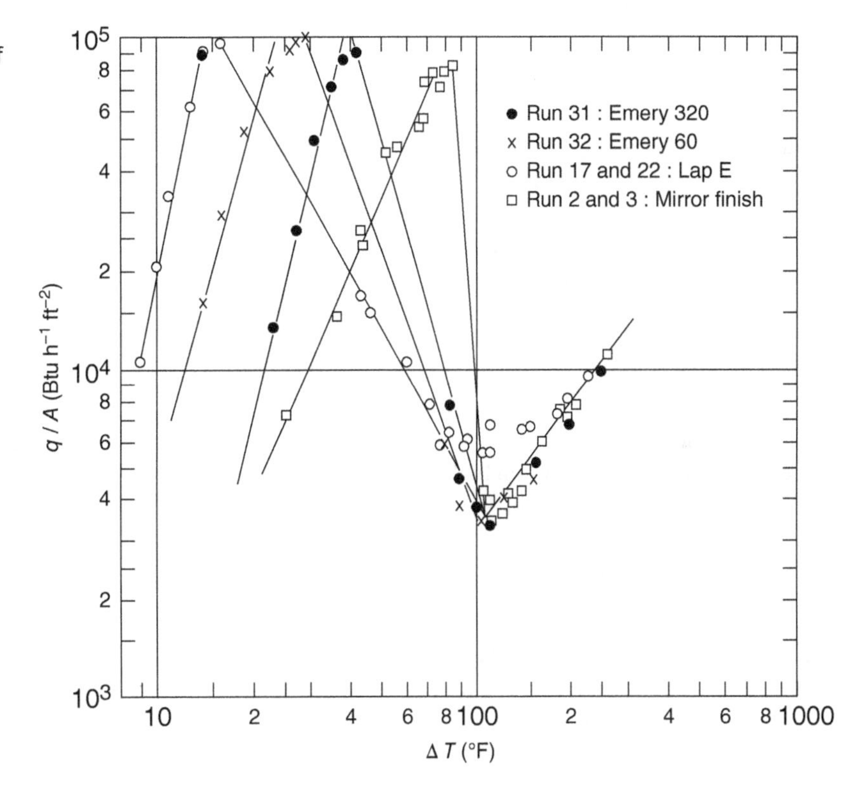

Figure 3.4.1 Pool boiling of pentane on a horizontal copper plate showing the effect of surface roughness. Source: From Berenson (1960). © 1960 Massachusetts Institute of Technology

Berenson (1960, 1962) recommended that for calculating transition boiling, a straight line drawn between CHF and MFB points be plotted on a log–log graph. Figure 3.4.1 shows that this method gives excellent agreement with test data.

Dhir (1991) has reviewed various other approaches for prediction of transition boiling heat transfer but they involve factors that are unknown in most cases.

Ramilison and Lienhard (1987) proposed a model in which the liquid–vapor interface in film boiling takes the form of a cyclically collapsing, two-dimensional, square array of Taylor-unstable waves. With several assumptions, they finally arrived at the following expression:

$$Bi^* = 3.74 \times 10^{-6} K (Ja^*)^2 \tag{3.4.7}$$

Bi^* and Ja^* are modified Biot and Jakob numbers. The latter requires the rewetting temperature T_A, which is given in a graph as a function of advancing contact angle based on their own data for four fluids. Data for the advancing contact angle are hard to get. Hence this correlation has very limited applicability.

In conclusion, no well-verified method is available for predicting transition boiling heat transfer. Linear interpolation between CHF and MFB points by Eq. (3.4.6) appears to be the best approach.

3.5 Minimum Film Boiling Temperature

As seen in Figure 3.3.2, MFB temperature is at point D. It may be considered to be the point at which film boiling starts if it is arrived at by increasing heating. It is the temperature at which film boiling ends if it is reached by cooling from the film boiling side. It is also often called the Leidenfrost temperature though the Leidenfrost effect really is about liquid drops on hot surfaces.

Considering the approach from the transition boiling side, the fraction of heater surface in contact with liquid decreases as more heat is applied and surface temperature rises. Eventually, the temperature rises to the MFB temperature where no part of the heater is in contact with liquid.

During film boiling, the heater surface is surrounded by a film of vapor that supports the liquid. As the heater is cooled, vapor generation decreases until a temperature is reached where vapor layer is unable to support the liquid, which then contacts the heater surface. This is the MFB temperature.

3.5.1 Prediction Methods

3.5.1.1 Analytical Models

If the minimum heat flux for film boiling q_{MFB} and the film boiling heat transfer coefficient h_{fb} are known, the MFB temperature T_{MFB} can be calculated by

$$T_{MFB} = T_{SAT} + \frac{q_{MFB}}{h_{FB}} \tag{3.5.1}$$

Zuber (1959) developed a model to predict MFB heat flux. In his model, vapor leaves in the form of spherical bubbles at the nodes of Taylor instability wave. The nodes are arranged in a square grid. The wavelength is in between the critical wavelength λ_C and the most dangerous wavelength λ_D given by

$$\lambda_C = 2\pi\left(\frac{\sigma}{g(\rho_f - \rho_g)}\right)^{1/2} \tag{3.5.2}$$

$$\lambda_D = 2\pi\sqrt{3}\left(\frac{\sigma}{g(\rho_f - \rho_g)}\right)^{1/2} \tag{3.5.3}$$

Using a wavelength λ within these limits, the radius of the bubbles is $\lambda/4$ and the bubble release occurs when wave peak rises to $\lambda/2$. Two full bubbles rise from each square grid over one cycle. If the heat flux is just enough to form these bubbles, it is the MFB heat flux. With the bubble frequency f, q_{MFB} becomes

$$q_{MFB} = f\rho_g i_{fg}\frac{4\pi}{3}\left(\frac{\lambda}{4}\right)^3\frac{2}{\lambda^2} \tag{3.5.4}$$

The frequency is given by

$$f = 0.4\left[\frac{2}{3}\frac{g(\rho_f - \rho_g)}{(\rho_f + \rho_g)}\right]^{1/2}\left[\frac{g(\rho_f - \rho_g)}{3\sigma}\right]^{1/4} \tag{3.5.5}$$

Combining these equations. The following expression for MFB heat flux is obtained:

$$q_{MFB} = K i_{fg}\rho_g\left[\frac{\sigma g(\rho_f - \rho_g)}{\rho_f^2}\right]^{1/4} \tag{3.5.6}$$

If $\lambda = \lambda_C$, $K = 0.109$. If $\lambda = \lambda_T$, $K = 0.143$. Wang et al. (2016) report that $K = 0.03$ fits their hydrogen data.

Berenson (1960) performed a Taylor instability analysis similar to that of Zuber and arrived at the following relation for MFB heat flux:

$$q_{MFB} = 0.09 i'_{fg}\rho_g\left[\frac{g(\rho_f - \rho_g)}{\rho_f^2}\right]^{1/2}\left[\frac{\sigma}{g(\rho_f - \rho_g)}\right]^{1/4} \tag{3.5.7}$$

where $i_{fg}{}'$ is the latent heat plus sensible heat of vapor at film temperature. The constant 0.09 was based on comparison with test data.

Berenson (1961) developed the following expression for heat transfer coefficient in film boiling:

$$h_{FB} = 0.425\left[\frac{g(\rho_f - \rho_g)i'_{fg}\rho_g k_g^3}{\mu_g(T_w - T_{SAT})[\sigma/g(\rho_f - \rho_g)]^{1/2}}\right]^{1/4} \tag{3.5.8}$$

This equation gave good agreement with data from several sources.

Combining Eqs. (3.5.1), (3.5.7), and (3.5.8), Berenson (1961) obtained the following equation for the MFB temperature:

$$T_{MFB} = T_{SAT} + 0.127\frac{\rho_g i'_{fg}}{k_g}\left[\frac{g(\rho_f - \rho_g)}{\rho_f + \rho_g}\right]^{2/3}\left[\frac{\sigma}{g(\rho_f - \rho_g)}\right]^{1/2}\left[\frac{\mu_f}{(\rho_f - \rho_g)}\right]^{1/3} \tag{3.5.9}$$

The vapor properties are evaluated at the film temperature and the liquid properties are evaluated at the liquid bulk temperature. It was shown to agree with data for *n*-pentane and carbon tetrachloride. However, it was found to give poor agreement with data for liquid metals and cryogens (Baumeister and Simon 1973).

According to Eq. (3.5.9), MFB temperature depends only on the fluid properties. In fact, it has been found to depend on factors such as surface roughness, thermal properties of heater, and contact angle. Several researchers have given prediction methods that take them into account.

3.5.1.2 Empirical Correlations

Ebrahim et al. (2018) performed tests on quenching of vertical rods in saturated and subcooled water at atmospheric pressure. The rods were of stainless steel 316, zirconium, and Inconel-600. They developed the following formula:

$$T_{MFB} = T_{ber} + 1.5(T_{ber} - T_f)\left[1 + \frac{R_a}{R_{ref}}\right]^{0.22} \left[\frac{(k\rho C_p)_f}{(k\rho C_p)_w}\right]^{0.14} \left[\frac{i_{fg}}{C_{p,w}(T_{ber} - T_{SAT})}\right]^{0.44} \tag{3.5.10}$$

T_{ber} is the MFB temperature predicted by Berenson Eq. (3.5.9). R_a is the actual roughness of the surface, while R_{ref} is the reference roughness equal to 1 μm. The roughness of heaters was 0.33–0.86 μm and subcooling was up to 18 °C. The rod diameter was 9.5 mm and $(k\rho C_p)$ of the rods was 4.15×10^7–8.56×10^7 $(\mathrm{J\,m^{-2}\,K^{-2}})^2\,\mathrm{s^{-1}}$. Metal properties are calculated at the MFB temperature.

Henry (1974) gave the following correlation based on his own data, which included water, refrigerants, and liquid metals on several heater materials:

$$T_{MFB} = T_{ber} + 0.42(T_{ber} - T_f) \left[\left[\frac{(k\rho C_p)_f}{(k\rho C_p)_w}\right]^{0.5} \frac{i_{fg}}{C_{p,w}(T_{ber} - T_{SAT})}\right]^{0.6} \tag{3.5.11}$$

Dhir and Purohit (1978) studied MFB during quenching of spheres in water. Subcooling was up to 60 K. Their own data as well as those from two other sources fitted the following equation:

$$T_{MFB} - T_{SAT} = 101 + 8\Delta T_{SUB} \tag{3.5.12}$$

Simon et al. (1968) studied MFB data for water, cryogens, and chemicals from several sources and fitted the following equation:

$$\frac{T_{MFB}}{T_c} = 0.13p_r + 0.84 \tag{3.5.13}$$

Comparison of this equation with data is shown in Figure 3.5.1.

Sakurai et al. (1990a) quenched cylinders of gold and platinum in pools of water, ethanol, iso-propanol, R-11, R-113, and nitrogen to study MFB temperature. Cylinders were horizontal and had diameters 0.7–3 mm. Pressure was up to 20 bar for water and 6 bar for nitrogen. They developed the following correlation of their data:

$$T_I = \frac{(T_{MFB} + \xi T_{SAT})}{(1 + \xi)} \tag{3.5.14}$$

$$T_I = 0.92T_c\left\{1 - 0.26\exp\left[\frac{-20p_r}{1 + 1700/p_c}\right]\right\} \tag{3.5.15}$$

where

$$\xi = \left[\frac{(\rho k C_p)_f}{(\rho k C_p)_w}\right]^{0.5} \tag{3.5.16}$$

T_I is the wall temperature during intermittent wall–liquid contact. In Eq. (3.5.15), p_c is in kPa.

T_{MFB} can be obtained by simultaneous solution of Eqs. (3.5.14) and (3.5.15). In fact, there is little difference between the calculated values of T_I and T_{MFB}.

Baumeister and Simon (1973) have given the following correlation:

$$T_{MFB} = \frac{\frac{27}{32}T_c\left\{1 - \exp\left(-0.52\left[\frac{10^4(\rho_w/A)^{4/3}}{\sigma}\right]^{1/2}\right)\right\} - T_f}{\exp(0.001\,75\Upsilon)\,erfc(0.042\sqrt{\gamma})} + T_f \tag{3.5.17}$$

where $\Upsilon = (k\rho C_p)_w$. A is the atomic number of heater material. Temperatures are in K, density ρ is in $\mathrm{g\,cm^{-3}}$, k in $\mathrm{cal\,m^{-1}\,s^{-1}K^{-1}}$, σ is the liquid–vapor surface tension in $\mathrm{dyne\,cm^{-1}}$, and C_p in $\mathrm{cal\,g^{-1}\,K^{-1}}$. This correlation was shown to be in agreement with data for mercury, potassium, sodium, helium, nitrogen, water, R-11, R-113, CCl_4, ethanol, and pentane. Heater materials included copper, aluminum, platinum, columbium, tantalum, stainless steel, glass, and Teflon. The data included boiling of small drops as well as pool boiling on heaters. Only data for clean and smooth surfaces were used in developing this correlation. Hence effect of surface roughness is not included. For boiling on heaters, the correlation does not include the effect of subcooling; hence T_{SAT} is used in place of T_f.

Farahat et al. (1974) studied quenching of spheres in pools of sodium. From these tests, they obtained the following relations:

$$q_{MFB} = 6.3 \times 10^4 + 1.9\Delta T_{SC} \tag{3.5.18}$$

$$\Delta T_{MFB} = 790 + 12.2\Delta T_{SC} \tag{3.5.19}$$

Temperatures are in °F and q in $\mathrm{Btu\,h^{-1}\,ft^{-2}}$. They note that the Berenson equation underpredicts MFB temperature by an order of magnitude for their sodium data as well as data of others for potassium. On the other hand, the Henry correlation is in good agreement with their data.

Gorenflo et al. (2010) have given the following formula for effect of pressure on MFB heat flux:

$$\frac{q_{MFB}}{q_{MFB,0.1}} = 3.2p_r^{0.45}(1 - p_r)^{1.2} \tag{3.5.20}$$

where $q_{MFB,0.1}$ is the value of q_{MFB} at $p_r = 0.1$. It is shown to agree with a wide range of data covering p_r from 0.04 to 1.

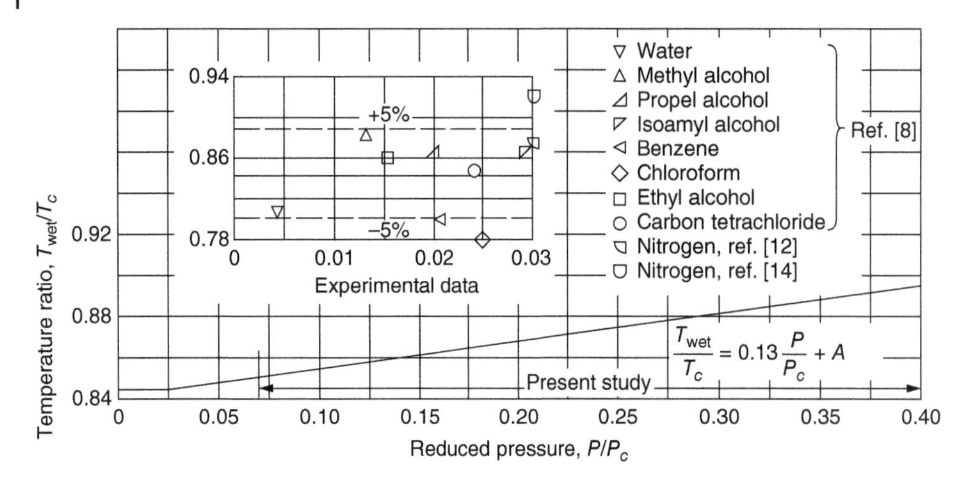

Figure 3.5.1 Minimum film boiling temperature data and correlation. T_{wet} in the figure is T_{MFB}. For references in the figure, see Simon et al. (1968). Source: Simon et al. (1968).

This formula is useful if q_{MFB} is known or can be calculated with confidence at one pressure.

Bernardin and Mudawar (1999) measured the Leidenfrost point for liquid drops on aluminum surfaces. The liquids were water, acetone, benzene, and FC-72. The measurements were compared to a number of prediction methods including those of Simon and Baumeister and Berenson given earlier. The data for polished aluminum surface were in good agreement with the Simon and Baumeister correlation. The Berenson equation also gave fairly good agreement though it was not compared to FC-72 data due to unavailability of needed property data. Roughened surfaces gave higher Leidenfrost temperatures.

Olek et al. (1991) have given a simple correlation, according to which the MFB temperature is the geometric mean of the normal boiling temperature (saturation temperature at atmospheric pressure) T_o and the critical temperature T_c. Thus,

$$T_{MFB} = (T_o \times T_c)^{0.5} \tag{3.5.21}$$

This correlation was compared to data for 17 fluids that included water, halogenated refrigerants, hydrocarbons, alcohols, cryogens, and liquid metals. Data for 16 fluids gave very close agreement. Only the data for potassium gave large deviations; predicted temperature was 1464 K, while measured values from two sources were 1271 and 1588 K. Most of the data were at atmospheric pressure. According to this correlation, MFB temperature is independent of pressure. According to the theoretical and empirical correlations given earlier, it depends on pressure. As most of the data used by Olek et al. were at atmospheric pressure, it is probably applicable only to atmospheric pressure.

3.5.2 Recommendations

Analytical models of Zuber and Berenson are inaccurate. The most verified prediction methods are those of Henry and Baumeister and Simon. These are recommended for clean and smooth surfaces. These do not include effects of roughness and subcooling. For the effect of roughness, the factor in the correlation of Ebrahim et al. may be used. For effect of subcooling, guidance may be taken from results of tests on similar fluids and pressures.

3.6 Film Boiling

In film boiling, the heater surface is completely covered by a vapor layer that is surrounded by liquid. Heat is transmitted to liquid through the vapor layer. Evaporation occurs at the vapor–liquid interface.

3.6.1 Methods for Predicting Heat Transfer

3.6.1.1 Vertical Plates

Bromley (1949, 1950) performed an analysis similar to that by Nusselt for condensation on vertical plates. The resulting equation is

$$h_{FB} = K\left[\frac{\rho_g k_g^3(\rho_f - \rho_g) i'_{fg} g}{\mu_g (T_w - T_{SAT}) L}\right]^{1/4} \tag{3.6.1}$$

L is the height of vertical plate. $K = 0.625$ for stagnant interface and $K = 0.883$ for dynamic interface. The effective latent heat $i_{fg}{}'$ is given by

$$i'_{fg} = i_{fg}\left(1 + 0.34\frac{C_{pg}(T_w - T_{SAT})}{i_{fg}}\right) \tag{3.6.2}$$

This equation has been found to fail when the height exceeds about 20 mm as the vapor film becomes turbulent (Hsu and Graham 1976).

Wang et al. (2016) compared the data for hydrogen boiling on a 25 mm × 556 mm vertical plate and a vertical 76 mm diameter plate with the correlation of Breen and Westwater using height of plate in place of diameter. Predicted heat flux was between +10% and −40% of measurements.

3.6.1.2 Horizontal Cylinders

By an analysis similar to that of Nusselt for condensation on horizontal tubes, Bromley (1949, 1950) gave the following equation for horizontal cylinders:

$$h_{FB} = 0.625\left[\frac{\rho_g k_g^3(\rho_f - \rho_g)i'_{fg}g}{\mu_g(T_w - T_{SAT})D}\right]^{1/4} \tag{3.6.3}$$

As the temperature differences between heater and fluid can be very high, radiation effects should be considerable. Bromley gave the following method to take it into account:

$$h_{total} = h_{FB} + 0.75h_{radiation} \tag{3.6.4}$$

To calculate $h_{radiation}$, the relation for parallel infinite plates is used and liquid is assumed to be a black body. This leads to the following equation:

$$h_{radiation} = \sigma\varepsilon\left[\frac{T_w^4 - T_f^4}{T_w - T_f}\right] \tag{3.6.5}$$

where σ is the Stefan Boltzman constant and ε is the emissivity of surface. Bromley verified this formula with data for water, nitrogen, and a number of chemicals. He noted that this formula is applicable if the thickness of vapor layer is much smaller than the cylinder diameter. He found it to fail on comparison with data for wires 0.1–0.6 mm. Agreement with data for a 1 mm wire was also marginal.

This equation has had extensive verification with a wide range of data for many fluids. For example, Gorenflo et al. (2010) report its good agreement with data for halogenated refrigerants on cylinders of diameters 0.1–25 mm, covering a reduced pressure range from 0.09 to 0.99.

For horizontal plates, Berenson (1960) performed an analysis based on Taylor instability that resulted in Eq. (3.5.8). That equation is very similar to Bromley's Eq. (3.6.3). The main difference is that instead of diameter D, the characteristic length is based on the critical wavelength of Taylor instability.

Breen and Westwater (1962) found that the Bromley equation fails for very small and very large diameter cylinders. They reasoned that if the cylinder diameter is much larger than the critical wavelength of Taylor instability, the cylinder approximates to a flat plate and then Berenson's formula for flat plate should be applicable. They modified the Berenson equation to the following form:

$$h_{FB} = \left[\frac{g(\rho_f - \rho_g)i'_{fg}\rho_g k_g^3}{\mu_g(T_w - T_{SAT})\lambda_c}\right]^{1/4}\left(0.59 - 0.69\frac{\lambda_c}{D}\right) \tag{3.6.6}$$

where λ_C is given by Eq. (3.5.2). This equation was shown to agree with data for λ_C/D from 0.1 to 300, though the data for $\lambda_C/D > 30$ were mostly lower than predictions. Wang et al. (2016) found this equation to be in satisfactory agreement with data for hydrogen boiling on horizontal cylinders and spheres.

Sakurai et al. (1990b) performed an analytical numerical solution for film boiling on horizontal cylinders. They then fitted equations to their numerical results to give the following correlation:

$$\frac{Nu_g}{(1 + 2/Nu_g)} = KM^{*0.25} \tag{3.6.7}$$

where $Nu_g = h_{FB}D/k_g$. K is given by the following equations:

$$\text{For } D' > 6.6, \quad K = 0.415D'^{0.25} \tag{3.6.8}$$

$$\text{For } 1.25 \le D' \le 6.6, \quad K = 2.1D'/(1 + 3D') \tag{3.6.9}$$

$$\text{For } 0.14 \le D' \le 1.25, \quad K = 0.75/(1 + 0.28D') \tag{3.6.10}$$

D' is the non-dimensional diameter given by

$$D' = D[g(\rho_f - \rho_g)/\sigma]^{0.5} \tag{3.6.11}$$

M^* is given by the following relations:

$$M^* = \left[\frac{Gr_g Pr_g i'_{fg}}{C_{pg}\Delta T_{SAT}}\right]\left[\frac{E^3}{1 + E/(SpPr_f)}\right](RPr_f Sp)^{-2} \tag{3.6.12}$$

$$E = (A + C\sqrt{B})^{1/3} + (A - C\sqrt{B})^{1/3} + (1/3)Sc \tag{3.6.13}$$

$$A = (1/27)Sc^3 + (1/3)R^2 SpPr_f Sc + (1/4)R^2 Sp^2 Pr_f^2 \tag{3.6.14}$$

$$B = (-4/27)Sc^2 + (2/3)SpPr_f Sc - \left(\frac{32}{27}\right)SpPr_f R^2 + \left(\frac{1}{4}\right)Sp^2 Pr_f^2 + \left(\frac{2}{27}\right)Sc^3/R^2 \tag{3.6.15}$$

$$C = (1/2)R^2 SpPr_f \tag{3.6.16}$$

$$i'_{fg} = i_{fg} + 0.5C_{pg}\Delta T_{SAT} \tag{3.6.17}$$

$$Sc = 0.93Pr_f^{0.22}C_{pf}\Delta T_{SC}/i'_{fg} \tag{3.6.18}$$

$$Sp = C_{pg}\Delta T_{SAT}/i'_{fg} \tag{3.6.19}$$

$$R = \left[\frac{\rho_g\mu_g}{\rho_f\mu_f}\right]^{0.5} \tag{3.6.20}$$

$$Gr_g = \frac{g(\rho_f - \rho_g)D^3}{\rho_g v_g^2} \tag{3.6.21}$$

This correlation was validated with their own data as well as data from other sources. Their own data included water, argon, nitrogen, R-11, R-113, ethanol, and iso-propanol on cylinders of diameter 0.3–6 mm. Other data included diameters 0.031–48.1 mm. Subcooling was up to 88 K. Comparison of this correlation with data from sources other than those of Sakurai et al. is shown in Figure 3.6.1. Good agreement is seen throughout, with D' varying from 0.02 to 30. They also compared saturated boiling data with the correlations of Bromley and Breen and Westwater. Bromley formula gave fairly good predictions for D' from 0.6 to 10. The Breen and Westwater correlation worked fairly well for $D' > 1$. Neither of these gave good agreement with data of Frederking et al. (1966) for helium even in those ranges, while the Sakurai et al. correlation gave satisfactory agreement with these data.

It should be noted that the Sakurai et al. correlation given earlier does not include the effect of radiation. While they have given their own correlation for the effect of radiation, it can be adequately taken into account with Eq. (3.6.4).

Son and Dhir (2008) did a three-dimensional numerical simulation of film boiling on a horizontal cylinder at Earth gravity and 0.01 times Earth gravity. They compared their results with the correlations of Bromley, Breen and Westwater, and Sakurai et al. The Sakurai et al. predictions agreed with the simulation results for all D' and gravity levels. The Bromley formula predictions agree at Earth gravity with the simulation results at D' of 0.5 and 5 at Earth gravity but are low at $D' = 0.05$. The Breen and Westwater correlation agrees with the simulation only at Earth gravity and then only when $D' = 5$. Predictions of these two correlations differ widely with the results of simulation at 0.01 gravity. These results are in general agreement with the findings of Sakurai et al. on the limits of application of these two correlations.

Michiyoshi et al. (1989) developed a theoretical model applicable to horizontal cylinders, vertical plates, and spheres. It involves constants that depend on velocity and temperature profiles and shear stresses at the vapor–liquid interface.

3.6.1.3 Horizontal Plates

Klimenko (1980) analytically developed the following relations for film boiling on horizontal plates.

For laminar region, $Ga[(\rho_f/\rho_g)-1] < 1\times10^8$,

$$Nu = 0.19\left[Ga\left(\frac{\rho_f}{\rho_g}\right)-1\right]^{\frac{1}{3}} Pr_g^{\frac{1}{3}} f_1(Ja') \qquad (3.6.22)$$

where Ja' is a modified Jakob number defined as

$$Ja' = \left(\frac{i_{fg}}{C_{pg}\Delta T_{SAT}}\right) \qquad (3.6.23)$$

For $Ja' \le 1.4, f_1 = 1$. For $Ja > 1.4$,

$$f_1 = 0.89(Ja')^{1/3} \qquad (3.6.24)$$

For turbulent region, $Ga[(\rho_f/\rho_g)-1] > 1\times10^8$,

$$Nu = 0.0086\left[Ga\left(\frac{\rho_f}{\rho_g}\right)-1\right]^{1/2} Pr_g^{\frac{1}{3}} f_2(Ja') \qquad (3.6.25)$$

For $Ja' < 2.0, f_2 = 1$. For $Ja' > 2.0$,

$$f_2 = 0.71(Ja')^{1/2} \qquad (3.6.26)$$

Ga is the Galileo number defined as

$$Ga = \frac{g\lambda_c^3}{2} \qquad (3.6.27)$$

$$Nu = \frac{h\lambda_c}{k_g} \qquad (3.6.28)$$

λ_c is the critical wavelength in Taylor instability given by Eq. (3.5.2).

Properties of vapor are to be calculated at a film temperature $= T_{SAT} + 0.5\Delta T_{SAT}$.

This correlation was verified with data in the following range:

$Ga[(\rho_f/\rho_g)-1] = 7\times10^4 – 3\times10^8$
$Pr_g = 0.69–3.45$
$Ja' = 0.031–7.3$
$p_r = 0.0045–0.98$
$g/g_{earth} = 1–21.7$

The fluids included in the database were nitrogen, hydrogen, helium, water, R-11, R-113, ethanol, carbon tetrachloride, and pentane. Most of the data were predicted within ±25%. The mean deviations of Berenson correlation were considerably higher.

Hamill and Baumeister (1967) performed a theoretical analysis of film boiling from a horizontal plate with subcooling and radiation. The analysis was based on the postulate that the rate of entropy production is maximized. The model was solved numerically. The results were approximated by the following formula:

$$h_{total} = h_{FB,SAT} + 0.88h_{rad} + 0.12h_{natural\ convection}\left(\frac{T_{SAT}-T_f}{T_w - T_{SAT}}\right) \qquad (3.6.29)$$

3.6.1.4 Inclined Plates

Jung et al. (1987) performed tests on copper plates in R-11. Plate inclination was varied from 0° to 180° to the horizontal. Their results are shown in Figure 3.6.2. It is seen that for a particular heat flux, wall superheat increases with increasing inclination. In other words, heat transfer coefficient decreases with increasing inclination. The decrease is gradual up to 150° inclination but then drops sharply at 180°. Change in inclination from 0° to 90° is seen to decrease heat transfer coefficient by about 12%.

Wang et al. (2016) compared the data for film boiling of hydrogen on a plate 25 mm × 559 mm in horizontal and

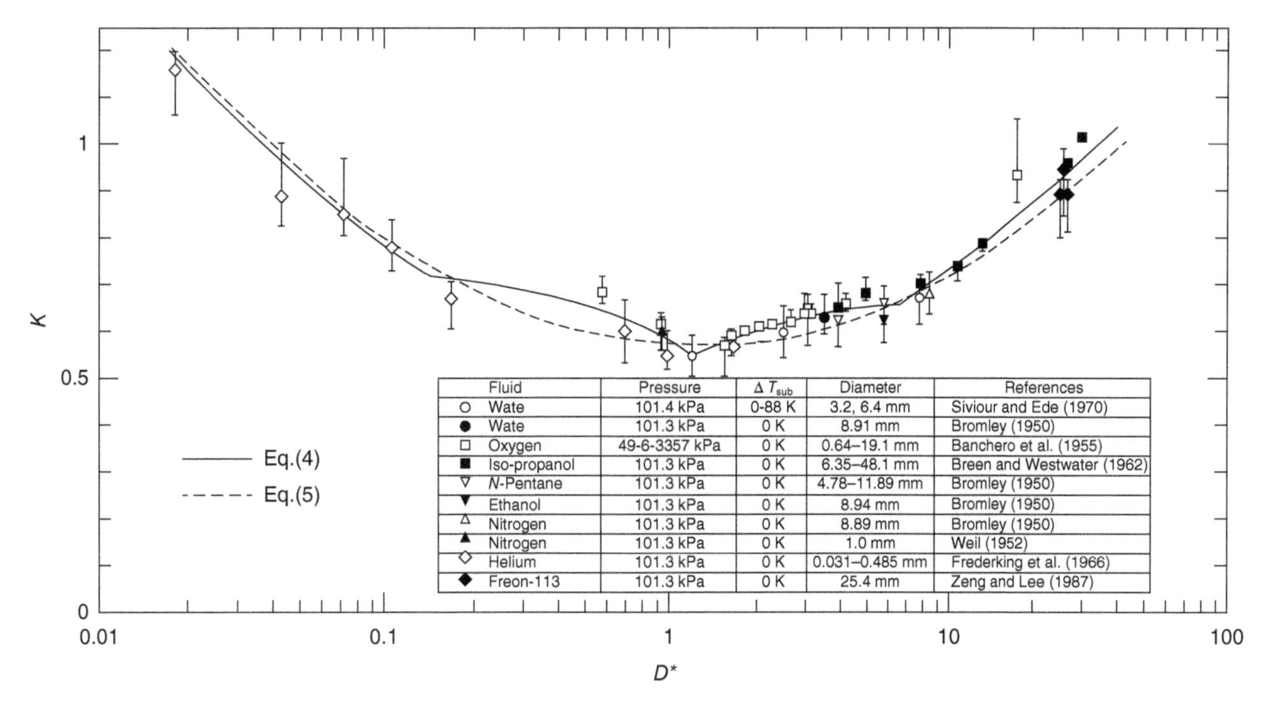

Figure 3.6.1 Comparison of the correlation of Sakurai et al. [shown as Eq. (4)] with data from sources other than their own. Eq. (5) is an approximate form of Sakurai et al. correlation not included in this book. Source: Sakurai et al. (1990b). © 1990 American Society of Mechanical Engineers.

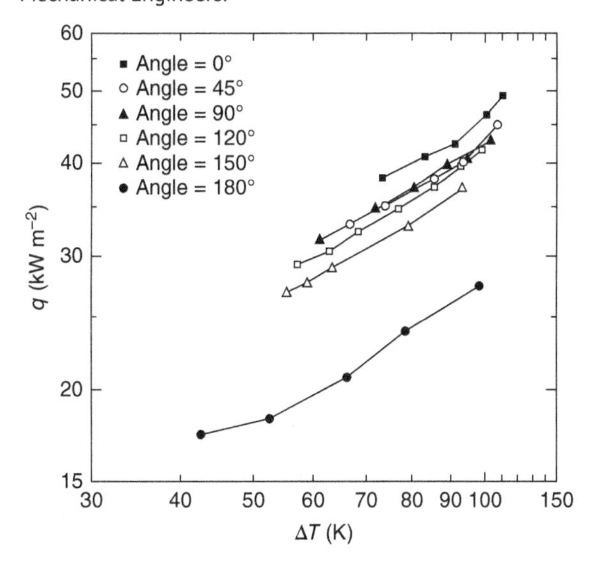

Figure 3.6.2 Effect of inclination of heater on film boiling heat transfer. Source: From Jung et al. (1987). © 1987 Elsevier.

45° inclined position to the Berenson formula. Data for horizontal plates were in good agreement with predictions. Deviations of predictions for inclined plate were within 0% and −40%. Data for a horizontal disk were also in good agreement.

3.6.1.5 Spheres

Yoon and No (2017) quenched spheres in water and have given correlation that fit their data. They point out that heat loss from the rod used to support the sphere during experiments can have considerable effect on the results.

Wang et al. (2016) compared data for film boiling of hydrogen on spheres with the Breen and Westwater correlation. Good agreement was found.

Dhir and Purohit (1978) performed tests on spheres with natural and force convection and have given correlations of their data. These are discussed in Section 7.4.4.

3.6.2 Liquid Metals

Lyon (1953) performed pool boiling tests on a number of liquid metals including mercury and cadmium. The heater was a 19 mm diameter horizontal cylinder. With cadmium and pure mercury, only film boiling was obtained. This was thought to be because of these did not wet the heater. The data are shown in Figure 3.6.3. Also shown is the prediction of Bromley for mercury. The predictions are much lower than data at lower wall superheats but exceed the measurements at the highest wall superheats. As effect of radiation increases with temperature difference and the data have not been corrected for the effect of radiation, this behavior is puzzling.

Padilla (1966) studied film boiling of potassium on a horizontal plate. His data are shown in Figure 3.6.4. Corrections to the measured heat transfer coefficients were made for effects of radiation and chemical reactions. Also shown is the prediction of Berenson theory. It is seen that

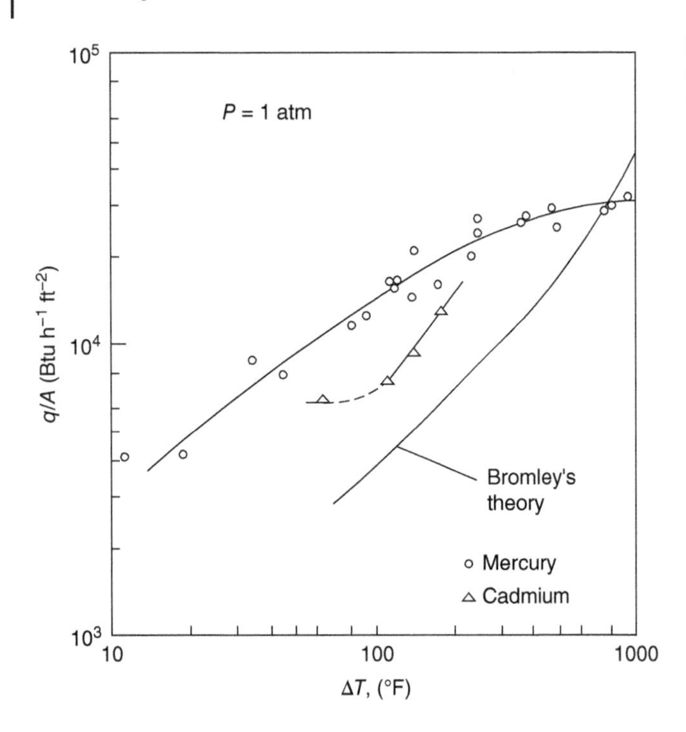

Figure 3.6.3 Data of Lyon (1953) for film boiling of mercury and cadmium on a horizontal cylinder and prediction for mercury. Source: From Padilla (1966). © 1966 University of Michigan.

the corrected heat transfer coefficients are on the average 40% higher than Berenson's predictions.

According to Sakurai (1990b), his correlation presented earlier may be used for liquid metals by replacing Sc given by Eq. (3.6.18) with the following expression:

$$Sc = [0.93Pr_f^{0.22} + 3\exp(-100SpPr_f Sc^{-0.8})]\\ [0.45 \times 10^5 Pr_f Sc/(1 + 0.45 \times 10^5 Pr_f Sc)]\\ (C_{\text{pf}}\Delta T_{\text{SC}}/i'_{\text{fg}}) \tag{3.6.30}$$

No comparison with test data was given. Hence this equation is unverified.

Farahat et al. (1974) have reported data for film boiling on spheres during quenching in a sodium pool.

3.6.3 Recommendations

The following are recommended predictive techniques for non-metallic fluids:

Horizontal cylinders: The Sakurai et al. correlation is the most verified available and is recommended for saturated and subcooled liquids. Alternatives for saturated fluids are Bromley formula for D' from 1 to 10 and Breen and Westwater correlation for $D' > 1$.

Horizontal plates: Use Klimenko correlation.

Spheres: For hydrogen, use the Breen and Westwater correlation with sphere diameter as the characteristic length. For other fluids, use various correlations based on data for spheres.

For metallic fluids, no verified method is available and test data are scarce. Formulas for non-metallic fluids may be used for conservative estimates.

In all cases, contribution due to radiation should be accounted for by Eqs. (3.6.4) and (3.6.5).

3.7 Various Topics

3.7.1 Effect of Gravity

Prediction methods presented in the foregoing were based on Earth-based experiments. Due to activities such as space flights and interest in space colonization, it is important to know whether these are applicable to low and high gravity conditions. If not, methods for application to those conditions are needed. Numerous experimental and theoretical investigations have been done for this purpose. These have been reviewed by Kim (2003), Warrier et al. (2015), and Lei et al. (2016) among others. The phenomena are very complex and the observations of researchers often are in apparent conflict with one another. These reviews may be studied for detailed information. Here the effort is limited to providing some relations useful in predicting the boiling under gravity conditions different from those at Earth gravity.

3.7.1.1 Scaling Method of Raj et al.

Raj et al. (2012a) divided nucleate boiling into two regimes. One is that of "surface tension dominated boiling" (SDB) where heat flux is heater size dependent. The other is the "buoyancy dominated boiling" (BDB) regime in which heat

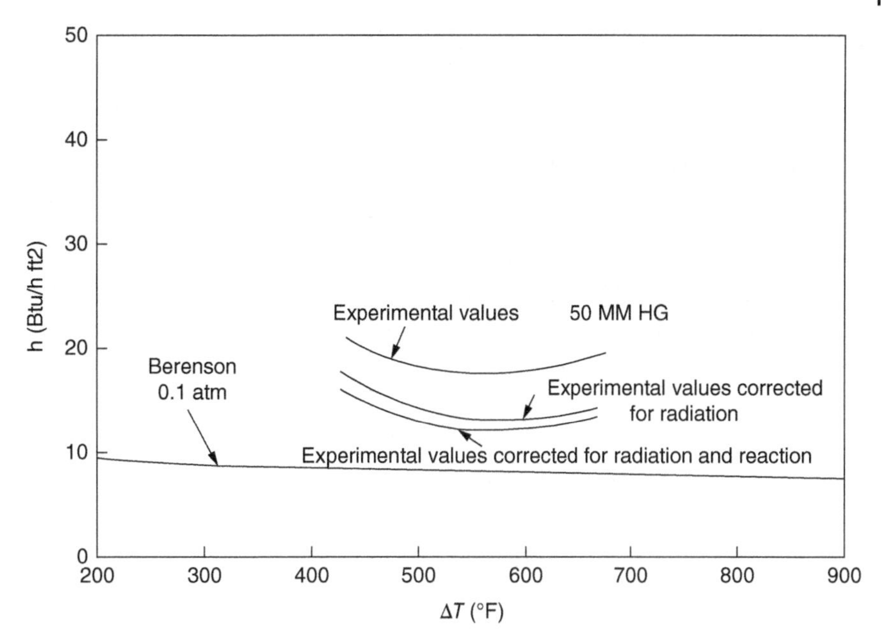

Figure 3.6.4 Film boiling heat transfer of potassium on a horizontal cylinder. Source: From Padilla (1966). © 1966 University of Michigan.

flux is independent of heater size. Transition between these two regimes occurs when $(L_h/L_c = 2.1)$ where L_h is the characteristic length of the heater and L_c is the capillary length. This gives the following expression for the transition acceleration a_{tran}:

$$a_{tran} = \frac{4.41\sigma}{L_h^2(\rho_f - \rho_g)} \tag{3.7.1}$$

This criterion was validated over a wide range of heater sizes, pressures, amounts of dissolved gases, and subcooling.

Raj et al. (2012a) further proposed that effect of gravitational acceleration varies as a^m and m varies continuously from 0 at the onset of nucleate boiling (ONB) to 0.25 at CHF in the BDB regime as follows:

$$m_{BDB} = \frac{0.65T^*}{1 + 1.6T^*} \tag{3.7.2}$$

$$T^* = \frac{T_w - T_{ONB}}{T_{CHF} - T_{ONB}} \tag{3.7.3}$$

The scaling factor in the BDB regime, $a_{tran} \geq 2.1$, is

$$\left(\frac{q_a}{q_{aref}}\right) = \left(\frac{a}{a_{ref}}\right)^{\left(\frac{0.65T^*}{1+1.6T^*}\right)} \tag{3.7.4}$$

This enables calculation of heat flux at any gravity level (a) if heat flux under similar conditions at any other gravity level (a_{ref}) is available. This relation was found to be in excellent agreement with variable gravity data at various dissolved gas concentrations, heaters sizes, and surface morphologies. In the validation calculations, data at Earth gravity were used as reference, while the data at other gravities included 0.3–1.7g.

The scaling of gravity in the SDB regime is now discussed. Under the condition of high subcooling and low dissolved gas concentration, the relation is

$$q_{SDB} = q_{ref}\left[\frac{4.41\sigma}{L_h^2(\rho_f - \rho_g)a_{ref}}\right]^{\left(\frac{0.65T^*}{1+1.6T^*}\right)} \left[\frac{L_h^2(\rho_f - \rho_g)a_{SDB}}{4.41\sigma}\right]^{0.025} \tag{3.7.5}$$

In the case of low subcooling or high concentration of dissolved gases, a jump in heat flux occurs at transition between SDB and BDB. In that case, the following is the relation:

$$q_{SDB} = q_{ref}\left[\frac{4.41\sigma}{L_h^2(\rho_f - \rho_g)a_{ref}}\right]^{\left(\frac{0.65T^*}{1+1.6T^*}\right)} K_{jump}\left[\frac{L_h^2(\rho_f - \rho_g)a_{SDB}}{4.41\sigma}\right]^{0.025} \tag{3.7.6}$$

K_{jump} is given by the following expression:

$$K_{jump} = 1 - e^{-CMa} \tag{3.7.7}$$

where Ma is the Marangoni number defined as

$$Ma = -\frac{d\sigma}{dT}\frac{\Delta T_{SC,app}L_h}{\mu_f \nu_f} \tag{3.7.8}$$

The value of C was determined to be 8.3×10^{-6}. When liquid contains dissolved gas, its partial pressure is lower than the total pressure and hence its saturation temperature is lowered. The subcooling $\Delta T_{SC,app}$ is based on the saturation temperature at the partial pressure. At zero subcooling, Marangoni number becomes zero and hence $K_{jump} = 1$.

Scaling for heater size and subcooling can be done with stepwise calculations using the scaling laws for gravity given earlier. To ease such calculations, scaling laws for gravity have been rearranged by Raj et al. to give the scaling laws for heater size and subcooling. These are given as follows:

If both heaters are in BDB regime (both L_{h1} and $L_{h2} > 2.1\ L_c$),

$$q_{L_{h1}} = q_{L_{h2}} \tag{3.7.9}$$

The subscripts L_{h1} and L_{h2} signify for heaters of characteristic lengths L_{h1} and L_{h2}.

If both heaters are in SDB regime ($L_{h1} < L_{h2} < 2.1L_c$),

$$q_{L_{h1}} = q_{L_{h2}} \left(\frac{1 - e^{-CMa1}}{1 - e^{CMa2}}\right) \left(\frac{L_{h2}}{L_{h1}}\right)^{2\left(\frac{0.65T^*}{1+1.6T^*} - 0.025\right)} \tag{3.7.9}$$

If L_{h1} in the SDB and L_{h2} in the BDB regime ($L_{h1} < 2.1L_c < 2.1L_{h2}$),

$$q_{L_{h1}} = q_{L_{h2}}(1 - e^{-CMa1}) \left(\frac{2.1L_c}{L_{h1}}\right)^{2\left(\frac{0.65T^*}{1+1.6T^*} - 0.025\right)} \tag{3.7.10}$$

The effect of subcooling on CHF is modeled as below.

In the BDB regime,
$$\left(\frac{q_{\text{CHF,SC},a}}{q_{\text{CHF,SC},b}}\right) = \left(\frac{1 + f\Delta T_{\text{SC},a}}{1 + f\Delta T_{\text{SC},b}}\right) \tag{3.7.11}$$

In the SDB regime,
$$\left(\frac{q_{\text{CHF,SC},a}}{q_{\text{CHF,SC},b}}\right) = \left(\frac{1 + f\Delta T_{\text{SC},a}}{1 + f\Delta T_{\text{SC},b}}\right) \times \left(\frac{1 - e^{=CMa_a}}{1 - e^{=CMa_b}}\right) \tag{3.7.12}$$

The factor f is a function of fluid properties and the system.

Raj et al. note that the predictions off these equations explain many observations of various researchers that had seemed puzzling. These equations were validated with data in the range L_h/L_c between 0.3 and 12. The authors expect them to be valid over a wider range.

The foregoing relations are applicable only to flat surfaces facing upwards. In Raj et al. (2012b), the methodology is modified to include other geometries. The modified correlation is

$$q_{\text{SDB}} = q_{\text{BDB}} \left[\frac{a_{\text{tran}}}{a_{\text{BDB}}}\right]^{m_{\text{BDB}}} K_{\text{jump}} \tag{3.7.13}$$

For spheres, transition occurs at $R/L_c = 4.26$, $m_{\text{BDB}} = 0.25$, and $K_{\text{jump}} = 0$. Good agreement was obtained with data for spheres from several sources. They also gave tentative values for other geometries, but there were no data for low gravity to verify them.

3.7.1.2 Scaling for Hydrogen

Lei et al. (2016) applied the method of Raj et al. (2012a) given earlier to CHF and nucleate boiling data for hydrogen in microgravity conditions from two sources. Agreement with plate data was fairly good. The agreement with data of Merte (1970) for spheres was not as good. It will be noted that Raj et al. (2012b) have given a modified correlation for spheres, Eq. (3.7.13).

For film boiling, Lei et al. (2016) gave the following correlation:

$$\frac{q_a}{q_g} = \left(\frac{a}{g}\right)^{0.2} \tag{3.7.14}$$

Subscripts a and g indicate values at accelerations a and g, respectively. It was shown to be in good agreement with the data of Merte (1970).

3.7.1.3 Some Other Studies

Warrier et al. (2015) studied pool boiling data in microgravity conditions from several sources. They found them to be correlated by the following equation:

$$\frac{h}{h_{\text{earth}}} \propto \left(\frac{a}{g}\right)^{1/8} \tag{3.7.15}$$

where h is the heat transfer coefficient at acceleration a and g is gravitational acceleration at Earth. The data covered (a/g) from 10^{-8} to 1.

The Raj et al. gravity scaling method requires the ONB temperature. It may be calculated by simultaneous solution of equations for natural convection and nucleate boiling. This requires prediction of natural convection heat transfer under reduced gravity. Warrier et al. (2015) found the Kobus and Wedekind (2001) correlation to be in good agreement with data for reduced gravity.

Ogata and Nakayama (1977) studied the effect of acceleration on pool boiling of Helium-I on a horizontal plate by revolving it around a displaced axis. Accelerations up to 126g were used. As seen in Figure 3.7.1, acceleration had no effect on the nucleate boiling curve. MFB heat flux and CHF increased with increasing acceleration. The MFB heat flux and CHF were found to increase according to $(a/g)^{0.25}$. Their data for CHF were in agreement with the correlation of Addoms (1948) if it was multiplied by this factor.

The foregoing discussions have been for pure fluids. Behavior of non-azeotropic mixtures is different. It is briefly discussed in the following text.

Abe et al. (1994) performed pool boiling experiments with non-azeotropic water–ethanol mixtures at micro gravity in a drop shaft facility with a 490 m free fall. Nucleate boiling heat transfer coefficients were 20–40% higher than at Earth gravity. For pure fluids, nucleate boiling heat transfer is lower at reduced gravity. CHF was reduced by 20–40% compared to terrestrial conditions. This reduction is smaller than the authors had experienced with pure fluids.

Ahmed and Carey (1998) studied pool boiling of water/2-propanol mixtures under reduced gravity (0.01g),

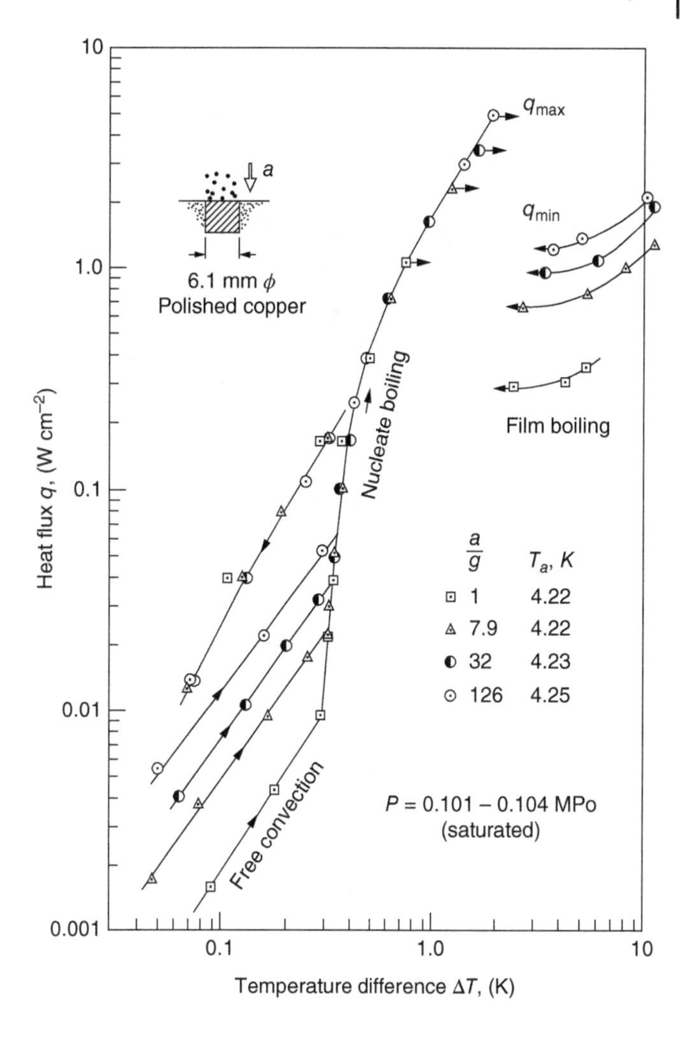

Figure 3.7.1 Effect of acceleration on pool boiling heat transfer of Helium-I. Source: Ogata and Nakayama (1977). © 1977 Elsevier.

normal gravity, and high gravity (2g). The reduced and high gravity experiments were conducted during parabolic flights that provided duration of about 20 seconds for the tests. The CHF values determined under reduced gravity conditions did not change significantly from those measured under 1-g, conditions. The results indicated that the Marangoni mechanism is strong enough in these mixtures to sustain stable nucleate boiling under microgravity. The correlations of Schlunder and Stephan–Korner were found to underpredict CHF of mixtures at micro gravity.

Shatto and Peterson (1999) performed tests on CHF during parabolic flight using a horizontal cylinder with water as the fluid. CHF was found to decrease continuously with decreasing gravity. They have also discussed tests at low gravity done by others. At the present, methods for predicting the effect of gravity on mixtures are not available.

3.7.1.4 Recommendations

The scaling law method of Raj et al. (2012a,b) is the most verified method available for calculation of pool boiling under reduced gravity conditions. It is therefore recommended. For film boiling of hydrogen, the correlation of Lei et al. (2016), Eq. (3.7.14), is recommended.

3.7.2 Effect of Oil in Refrigerants

Most refrigeration systems use compressors in which oil gets mixed with refrigerant and is carried into the system. The larger refrigeration systems use oil separators but they are not 100% effective. The small systems usually do not have an oil separator. Therefore, the effect of oil on heat transfer is of much practical interest. As a result, numerous experimental studies have been performed on the effect of oil on heat transfer during pool boiling and flow boiling. A number of mechanisms have been proposed, and many predictive techniques theoretical and empirical have been also proposed. Reviews of literature on this topic have been done by Wang et al. (2014) and Hung et al. (2016), among others.

Figure 3.7.2 shows typical effect of oil on nucleate pool boiling heat transfer. Tubes number 2–4 had deformed fins, which give the effect of re-entrant cavities. It is seen that oil concentration has greater effect on enhanced tubes than on the plain tube.

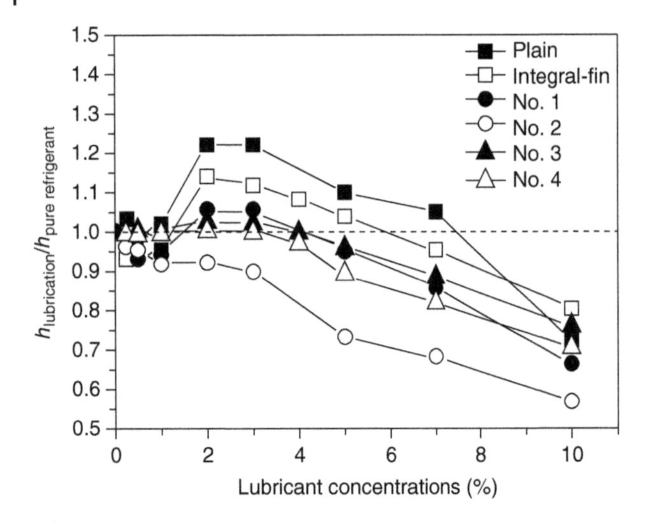

Figure 3.7.2 Effect of oil concentration on heat transfer during nucleate pool boiling of R-134a on plain and enhanced tubes no. 1–4 or different types of enhanced tubes. Source: From Ji et al. (2010). © 2010 American Society of Mechanical Engineers.

3.7.2.1 Mechanisms

Several authors including Jensen and Jackman (1984) and Mitrovic (1998) have suggested the following mechanism. During boiling of refrigerant and lubricant mixture, the refrigerant is more volatile and hence evaporates faster, resulting in an oil-rich layer at the heater surface and around the bubbles. A steep oil concentration gradient forms around the bubble and the surface tension at the liquid–gas surface tension increases. The increase in surface tension requires more work for evaporation and tends to reduce the heat transfer coefficient. However, this also leads to a reduction in bubble size and an increase in bubble frequency. In addition, the high oil viscosity results in a thicker thermal boundary layer at the heated surface, which can increase the number of active nucleation sites and enhance the heat transfer performance. Depending on the relative magnitude of these two effects, heat transfer may increase or decrease.

Kedzierski (2003) developed a model for predicting the effect of oil on heat transfer. According to this model, lubricant accumulates on the heat transfer surface as refrigerant boils out faster. As a result, excess lubricant forms a thin layer on the surface. This reduces the bubble size and increases the bubble frequency. The high oil viscosity induces a thicker thermal boundary layer at the heated surface. The thicker thermal boundary layer can increase the active sites. The overall effect may be an enhancement or degradation in heat transfer. A general predictive model was developed whose inputs include transport and thermodynamic refrigerant properties and the lubricant composition, viscosity, and critical solution temperature with the refrigerant. The model was shown to agree with data for 13 different refrigerant/lubricant mixtures including two different refrigerants and three different lubricants.

Hung et al. (2018) performed pool boiling tests on two different refrigerants (R-134a and R-1234ze) and eight polyolester (POE) lubricants with different miscibility, ISO68–ISO170 viscosity range. Tests were performed on a horizontal copper plate. They found that presence of lubricant decreases the departure bubble diameter. For oil concentrations less than 3%, heat transfer coefficient increased or decreased by a few percent. For higher concentration, heat transfer coefficients decreased. A mechanistic explanation was provided for the observed refrigerant/lubricant boiling phenomena, and a model was presented for the effect of lubricant properties on the heat transfer. According to their model, the presence of lubricant layer on metal surface and surrounding the bubble significantly alters waiting time of boiling, bubble size, bubble departure time, active site density of boiling incipience, and superheat on heating surface.

No information was available for pool boiling heat transfer with immiscible oils. Formation of insulating film of oil on heat transfer surface is likely in such situations with severe reduction in heat transfer. Shah (1975) and Chaddock and Buzzard (1986) had reported such films and reduction in heat transfer during boiling of ammonia containing immiscible oil inside tubes. In their tests with ammonia in a tube, Boyman et al. (2004) found that 0.1% oil reduced heat transfer coefficients up to 50%. Further reductions occurred as oil concentration increased up to 1%. Further increase in oil concentration did not cause further deterioration.

3.7.2.2 Correlations

Jensen and Jackman (1984) studied pool boiling of R-113 with four types of mineral oils on a smooth stainless steel tube. They developed the following correlation:

$$h_{mix} = h_{pure} \exp(-4.095 x_{oil}(1 + 0.0317 \Delta T_{SAT}^{0.753})) - 55.11 x_{oil}(1 + 0.0317 \Delta T_{SAT}^{0.753})^2 \quad (3.7.16)$$

where h_{mix} is the heat transfer coefficient of mixture, h_{pure} is that of pure refrigerant, and x_{oil} is the concentration of oil in the mixture (kg^{-1} kg^{-1} of mixture). These data gave satisfactory agreement with their own data as well as those of Sauer and Dougherty (1974) and Sauer and Chongrungeong (1980) for R-11 and R-113.

Tran et al. (2018) measured pool boiling heat transfer coefficients of R-134a and R-1234ze with and without lubricant oil on a horizontal copper plate. Eight types of lubricants were used. These are POEA68, POEA120, POEA170, POEB68, POEB120, POEB170, POEC170, and POED150. Saturation temperatures were 0 and 10 °C and oil concentrations ranged from 0% to 10% and heat fluxes vary from 10 to 70 kW m^{-2}. With up to 5% oil, the heat transfer coefficients were slightly higher or lower than those of pure refrigerants, depending on the lubricant and operating pressure. However, all the heat transfer coefficients deteriorated appreciably as compared to the pure refrigerants when the oil concentration exceeded 5%. Their data for oil concentrations of up to 5% were in good agreement with the Jensen and Jackman correlation, Eq. (3.7.16). They gave an empirical correlation of their own data.

3.7.2.3 Recommendation

The correlation of Jensen and Jackman agreed with their own data, those of Sauer and Dougherty (1974) and Sauer and Chongrungeong (1980) as well as those of Tran et al. (2018) for halogenated refrigerants with miscible oils. It is therefore recommended for halogenated refrigerants with miscible oils.

When an immiscible oil is used, reduce calculated heat transfer coefficients for pure refrigerant by at least 50%.

3.7.3 Thermosiphons

Closed vertical thermosiphons are considered in this section. In such thermosiphons, evaporation occurs at the bottom. The vapor rises up to the section with cold wall where it condenses. The condensate flows down back into the evaporator section and thus the cycle continues. Jafari et al. (2016) have reviewed the literature on such thermosiphons.

Shiraishi et al. (1981) tested a closed vertical thermosiphon formed by a 37 mm diameter tube with water, ethanol, and R-113. They did not compare their data with any published correlation but fitted an empirical equation to it.

Jouhara and Robinson (2010) studied a vertical closed thermosiphon of 6 mm diameter and 200 mm long. The length of boiling section was 40 mm and that of condensing section 60 mm. Pressure was near atmospheric. They compared the boiling heat transfer data for water as the working fluid with a number of general correlations for pool boiling as well **as** some **correlations** specifically developed for thermosiphons. The general correlation of Stephan and Abdelsalam, presented in Section 3.2.3, showed good agreement with measurements. The correlation of Shiraishi et al. (1981) considerably overpredicted, while that of Chowdhury et al. (1997) gave fairly good agreement.

Kim et al. (2018) studied the effect of filling ratio in a thermosiphon using water as the working fluid. Heat transfer coefficients were in fairly good agreement with the correlation of Rohsenow (1952) at all filling ratios from 0.25 to 1. They did not compare their data with any other correlation.

In the tests of Louahlia-Gualous et al. (2017), the working fluid was water at a pressure of 4–20 kPa. They compared their data with a number of correlations. That of Stephan and Abdelsalam (1980) greatly overpredicted, while Chowdhury et al. (1997) correlation greatly underpredicted. The Cooper correlation came fairly close to the data. Louahlia-Gualous et al. modified it to the following formula that gives excellent agreement with their data:

$$h = 7704M^{-0.5}q^{0.157}\frac{p_r^{0.12}}{[-\log p_r]^{0.55}} \quad (3.7.17)$$

where h is in W m^{-2} K^{-1} and q is in W m^{-2}. The very low exponent of q indicates that nucleate boiling did not occur. Note that this was not a closed thermosiphon. Vapor from the evaporator section was piped to a condenser and the condensate then flowed back to the evaporator by gravity.

From the discussions earlier, it may be concluded that general pool boiling correlations are applicable to vertical thermosiphons. However, the situation at very low pressures is not clear as the data of Louahlia-Gualous et al. at 4–20 kPa disagreed with most general correlations. In fact, it appears that nucleate boiling did not occur.

3.7.4 Effect of Some Organic Additives

Chashchin et al. (1975) studied the effect of small amounts of additives on nucleate boiling of water on a 28 mm diameter brass tube. The additives were propanol, butanol, pentanol, octanol, glycerin, polyvinyl alcohol, and polyacrylamide. A 0.006% addition of the last-mentioned increased heat transfer coefficient by 600% at a heat flux of 40 kW m^{-2}. Addition of 0.182% polyvinyl alcohol increased heat transfer coefficient by 800%. With propanol, increase was 200% at a concentration of 4.3%. The authors attributed the increase in heat transfer to reduction in surface tension due to which there is a reduction in critical diameter of bubbles and in an increase in bubble frequency.

Nomenclature

The following are the most frequently used symbols. Occasionally they have been defined differently as noted in the text at such occasions:

a acceleration (m s^{-2})
C_p specific heat at constant pressure (J kg^{-1} K^{-1})
D diameter (m)
f frequency (s^{-1})
$f(L')$ correction factor for Zuber CHF formula related to geometry (–)

g acceleration due to gravity (m s^{-2})
h heat transfer coefficient (W m^{-2} K^{-1})
h_{pb} pool nucleate boiling heat transfer coefficient (W m^{-2} K^{-1})
h_{fb} film boiling heat transfer coefficient (W m^{-2} K^{-1})
i_{fg} latent heat of vaporization (J kg^{-1})
Ja Jakob number, given by Eq. (3.2.10) (–)
k thermal conductivity (W m^{-1} K^{-1})
L length (m)
L' dimensionless characteristic length (–)
L_c capillary length (m)
M molecular weight
Ma Marangoni number (–)
MFB minimum film boiling
Nu Nusselt number (–)
p pressure (Pa)
p_c critical pressure (Pa)
p_r reduced pressure = p/p_c (–)
Pr Prandtl number (–)
q heat flux (W m^{-2})
r radius (m)
R universal gas constant = 8.314 J mol^{-1} K^{-1}
t time (s)
T temperature (K)
ΔT_{SAT} $(T_w - T_{SAT})$ (K)
ΔT_{SC} $(T_{SAT} - T_f)$ (K)
x mole fraction of component in liquid (–)
y mole fraction of component in vapor (–)

Greek letters

Δ thickness (m)
μ dynamic viscosity (Pa s)
ν kinematic viscosity (m^2 s^{-1})
ρ density (kg m^{-3})
σ surface tension (N m^{-1})
λ wavelength (m)
λ_D most dangerous wavelength in Taylor instability (m)

Subscripts

b bubble
BP bubble point
c cavity or critical
d departure
DP dew point
f liquid
FB Film boiling
g vapor
MFB minimum film boiling
mix mixture
pb pool boiling
SAT saturation
SC subcooled
w wall
z Zuber formula for CHF
Zuber Zuber formula for CHF

References

Abe, Y., Oka, T., Mori, Y.H., and Nagashima, A. (1994). Pool boiling of a non-azeotropic binary mixture under microgravity. *Int. J. Heat Mass Transfer* **37** (16): 2405–2413.

Addoms, J.N. (1948) Heat transfer at high rates to water boiling outside cylinders. ScD thesis. Massachusetts Institute of Technology, Cambridge, MA.

Ahmed, S. and Carey, V.P. (1998). Effects of gravity on the boiling of binary fluid mixtures. *Int. J. Heat Mass Transfer* **41** (16): 2469–2483.

Avksentyuk, B.P. and Mamontova, N.N. (1973). Characteristics of heat transfer crisis during boiling of alkali metals and organic fluids under free convection conditions at reduced pressure. *Prog. Heat Mass Transfer* **7**: 355.

Banchero, J.T., Barker, G.E., and Boll, R.H. (1955). Stable film boiling of liquid oxygen outside single horizontal tubes and wires. *Chem. Eng. Prog. Symp. Ser.* **51**: 21–23. Quoted in Sakurai et al. (1990b).

Bankoff, S.G. and Fauske, H.K. (1974). Improved prediction of critical heat flux in liquid metals. In: *Proceedings of the 5th International Heat Transfer Conference*, vol. **4**, 241–244. Tokyo, Japan: Begell House.

Baumeister, K.J. and Simon, F.F. (1973). Leidenfrost temperature – its correlation for liquid metals, cryogens, hydrocarbons, and water. *J. Heat Transfer* **95**: 166–173.

Berenson, P.J. (1960). Transition boiling heat transfer from a horizontal surface. MIT Technical Report No. 17.

Berenson, P.J. (1961). Film boiling heat transfer from a horizontal surface. *J. Heat Transfer* **83**: 351–358.

Berenson, P.J. (1962). Experiments on pool-boiling heat transfer. *Int. J. Heat Mass Transfer* **5**: 985–999.

Bergles, A.E. and Rohsenow, W.M. (1964). The determination of forced convection surface-boiling heat transfer. *J. Heat Transfer* **86**: 353–372.

Bernardin, J.D. and Mudawar, I. (1999). The Leidenfrost point: experimental study and assessment of existing models. *J. Heat Transfer* **121**: 894–903.

Bewilogua, L., Knoner, R., and Vinzelberg, H. (1975). Heat transfer in cryogenic liquids under pressure. *Cryogenics* **15**: 121–125.

Bjornard, T.A. and Griffith, P. (1977). PWR blowdown heat transfer. In: *Symposium on the Thermal and Hydraulic Aspects of Nuclear Reactor Safety, The ASME Winter Annual Meeting*, Light Water Reactors, vol. **1** (eds. O.C. Jones and S.G. Bankoff), 17–39. New York, NY: ASME.

Bonilla, C.F., Busch, J.S., Staider, A. et al. (1957). Pool boiling heat transfer with mercury. *Chem. Eng. Prog. Symp. Ser.* **53** (20): 51–57.

Bonilla, C.F., Wiener, M.M., and Bilfinger, H. (1964). Pool boiling of potassium. In: *Proceedings of the High Temperature Liquid Metal Heat Transfer Technology Meeting*, vol. **1**, 286–309. Oak Ridge National Laboratory. ORNL-3605.

Bonilla, C.F., Grady, J.J., and Avery, G.W. (1965). Pool boiling heat transfer from scored surfaces. *Chem. Eng. Prog. Symp. Ser.* **61** (57): 280–288.

Borishanskii, V.M., Zhokov, K.A. et al. (1965). Heat transfer in boiling liquid metals. *Atomnaya Energia* **19**: 191.

Boyman, T., Aecherli, P., and Wettstein, A.S.W. (2004). Flow boiling of ammonia in smooth horizontal tubes in the presence of immiscible oil. *International Refrigeration and Air Conditioning Conference. Paper 656, 12-15 July 2004 at West Lafayette, IN.* http://docs.lib.purdue.edu/iracc/656.

Breen, B.P. and Westwater, J.W. (1962). Effect of diameter of horizontal tubes on film boiling heat transfer. *Chem. Eng. Prog.* **58**: 67–72.

Breitstein, L. and Bonilla, C.F. (1974). Heat transfer in pool boiling and condensing selenium. *Proceedings of the Fifth International Heat Transfer Conference, Paper B1.5*, Tokyo, Japan, 3–7 September.

Bromley, L.A. (1949). Heat transfer in stable film boiling. UCRL-122. https://escholarship.org/uc/item/0pj1211q.

Bromley, L.A. (1950). Heat transfer in stable film boiling. *Chem. Eng. Prog.* **46**: 221–227.

Carbon, M.W. (1964). *Boiling Liquid Metal Heat Transfer*. TID-20942. Madison, WI: University of Wisconsin. Quoted by Dwyer (1974).

Caswell, B.F. and Balzhiser, R.E. (1966). The critical heat flux for boiling liquid metal systems. *Chem. Eng. Prog. Symp. Ser.* **62** (64): 41.

Chaddock, J. and Buzzard, G. (1986). Film coefficients for in-tube evaporation of ammonia and R-502 with and without small percentages of mineral oil. *ASHRAE Trans.* **92** (1A): 22–40.

Chang, J.Y. and You, S.M. (1996). Heater orientation effects on pool boiling of microporous-enhanced surfaces in saturated FC-72. *J. Heat Transfer* **118**: 937–943.

Chashchin, P.I., Shigina, L.F., Shavab, N.S., and Sobol, A.D. (1975). An investigation of the effect of certain organic additives on boiling heat transfer. *Therm. Eng.* **22** (8): 73–74.

Chen, H., Chen, G., Zou, X. et al. (2017). Experimental investigations on bubble departure diameter and frequency of methane saturated nucleate pool boiling at four different pressures. *Int. J. Heat Mass Transfer* **112**: 662–675.

Chowdhury, F.M., Kaminaga, F., Goto, K., and Matsumura, K. (1997). Boiling heat transfer in a small diameter tube below atmospheric pressure on a natural circulation condition. *J. Jpn. Assoc. Heat Pipe* **16**: 14–16. Quoted in Jouhara & Robinson (2010).

Cichelli, M.T. and Bonilla, C.F. (1945). Heat transfer to liquids boiling under pressure. *Trans. AIChE* **41**: 755–787.

Cole, R. and Shulman, H.L. (1966). Bubble growth rate at high Jakob number. *Int. J. Heat Mass Transfer* **9**: 1377–1390.

Collier, J.G. and Thome, J.R. (1994). *Convective Boiling and Condensation*, 3e. Oxford, UK: Oxford University Press.

Colver, C. and Balzhizer, R.E. (1964). A study of saturated pool boiling potassium up to burnout heat fluxes. *Chem. Eng. Prog. Symp. Ser.* **61** (59): 253–263.

Cooper, M.G. (1969). The microlayer and bubble growth in nucleate pool boiling. *Int. J. Heat Mass Transfer* **12**: 915–933.

Cooper, M.G. (1984). Heat flow rates in saturated nucleate pool boiling – a wide ranging examination using reduced properties. *Adv. Heat Transfer* **16**: 157–239.

Deane, C.W. and Rohsenow, W.M. (1969). Mechanism and behavior of nucleate boiling heat transfer to the alkali liquid metals. USAEC Report No. DSR 76303-65. MIT. Quoted by Dwyer (1976).

Dhir, V.K. (1991). Nucleate and transition boiling heat transfer under pool and external flow conditions. *Int. J. Heat Fluid Flow* **12** (4): 290–314.

Dhir, V. and Purohit, G. (1978). Subcooled film-boiling heat transfer from spheres. *Nucl. Eng. Des.* **47** (1): 49–66.

Dwyer, O.E. (1976). *Boiling Liquid Metal Heat Transfer*. Hinsdale, IL: American Nuclear Society.

Ebrahim, S.A., Changa, S., Cheunga, F., and Bajorek, S.M. (2018). Parametric investigation of film boiling heat transfer on the quenching of vertical rods in water pool. *Appl. Therm. Eng.* **140**: 139–146.

Elkassabgi, Y. and Lienhard, J.H. (1988). Influence of subcooling on burnout of horizontal cylindrical heaters. *ASME J. Heat Transfer* **110**: 479–486.

Farahat, M.M.K., Eggen, D.T., and Armstrong, D.R. (1974). Pool boiling in subcooled sodium at atmospheric pressure. *Nucl. Sci. Eng.* **53**: 240–253.

Farmer, W.S. (1952). PhD thesis. University of Tennessee. Quoted by Dwyer (1976).

Forster, D.E. and Greif, R. (1959). Heat transfer to a boiling liquid – mechanism and correlation. *J. Heat Transfer* **81** (1): 43–53.

Forster, H.K. and Zuber, N. (1954). Growth of a vapour bubble in a superheated liquid. *J. Appl. Phys.* **25**: 474–478. Quoted in Mohanty & Das (2017).

Fraas, A.P., Lloyd, D.B., and MacPherson, R.E. (1974). Effects of strong magnetic fields on boiling of potassium. Oakridge National Laboratory Report ORNL-TM-4218.

Frederking, T.H.K., Wu, Y.C., and Clement, B.W. (1966). Effects of interfacial instability on film boiling of saturated liquid helium I above horizontal surface. *AIChE J.* **12**: 238–244.

Fritz, W. (1935). Berechnung des maximalvolumes von dampfblasen. *Phys. Z.* **36**: 379–384.

Fritz, W. and Ende, W. (1936). Verdampfungsvorgang nach kinematographischen Aufnahmen an Dampfblasen. *Phys. Z.* **37**: 391.

Frost, W. and Dzakowic, G.S. (1967). An extension of the method for predicting incipient boiling on commercially finished surfaces. ASME Paper 67-HT-61.

Fujishiro, T. et al. (1971). The 8th Japan Heat Transfer Symposium, 153–156. Quoted in Aoki, S. 1973. Current liquid metal heat transfer research in Japan. *Prog. Heat Mass Transfer* **7**: 569–587.

Fujita, Y. and Bai, Q. (1997). Critical heat flux of binary mixtures in pool boiling and its correlation in terms of Marangoni number. *Int. J. Refrig.* **20** (8): 616–622.

Fujita, Y. and Tsutsui, M. (1994). Heat transfer in nucleate pool boiling of binary mixtures. *Int. J. Heat Mass Transfer* **37** (Suppl. 1): 291–302.

Fujita, Y. and Tsutsui, M. (1997). Heat transfer in nucleate boiling of binary mixtures (development of a heat transfer correlation). *JSME Int. J., Ser. B* **40**: 134–141.

Fujita, Y. and Tsutsui, M. (2002). Experimental investigation in pool boiling heat transfer of ternary mixture and heat transfer correlation. *Exp. Therm. Fluid Sci.* **26**: 237–244.

Ghiaasiaan, S.M. (2008). *Two Phase Flow, Boiling, and Condensation in Conventional and Miniature Systems.* Cambridge: Cambridge University Press.

Golobič, I. and Bergles, A.E. (1997). Effects of heater-side factors on the saturated pool boiling critical heat flux. *Exp. Therm. Fluid Sci.* **15**: 43–51.

Gorenflo, D. (2013). *Behaltersieden. Kapitel H2, VDI-Warmeatlas*, 11e. Berlin, Heidelberg: Springer. Quoted in Gorenflo et al. (2014).

Gorenflo, D., Baumhogger, E., Windmann, D., and Herres, G. (2010). Nucleate pool boiling, film boiling and single-phase free convection at pressures up to the critical state. Part I: Integral heat transfer for horizontal copper cylinders. *Int. J. Refrig.* **33**: 1229–1250.

Gorenflo, D., Baumhogger, E., Herres, G., and Kotthoff, S. (2014). Prediction methods for pool boiling heat transfer: a state-of-the-art review. *Int. J. Refrig.* **43**: 203–226.

Grigoriev, V.A., Klimenko, V.V., Pavlov, Y.M., and Amitestov, Y.V. (1978). The influence of some heating surface properties on the CHF in cryogenics liquids boiling. *Proceedings of the 6th International Heat Transfer Conference*, 7–11 August, Toronto, Canada.

Hamill, T.D. and Baumeister, H. (1967). Effect of subcooling and radiation on film-boiling heat transfer from a flat plate. NASA TN D-3925.

Han, C.Y. and Griffith, P. (1965). The mechanism of heat transfer in nucleate pool boiling, part i, bubble initiation, growth and departure. *Int. J. Heat Mass Transf.* **8** (6): 887–904.

Haramura, Y. and Katto, Y. (1983). A new hydrodynamic model of critical heat flux, applicable widely to both pool and forced convection boiling on submerged bodies in saturated liquids. *Int. J. Heat Mass Transfer* **26**: 389–399.

Henry, R.E. (1974). A correlation for the minimum film boiling temperature. *AIChE Symp. Ser.* **70** (138): 81–90.

Hibiki, T. and Ishii, M. (2003). Active nucleation site density in boiling systems. *Int. J. Heat Mass Transfer* **46**: 2587–25601.

Holtz, R.E. and Singer, R.M. (1968). Incipient pool boiling of sodium. *AIChE J.* **14**: 654–656.

Hovestreijdt, J. (1963). The influence of the surface tension difference on the boiling of mixtures. *Chem. Eng. Sci.* **18**: 631–639.

Hsu, Y.Y. (1962). On the size of range of active nucleation cavities on a heating surface. *J. Heat Transfer* **84**: 207–213.

Hsu, Y.Y. and Graham, R.W. (1976). *Transport Processes in Boiling and Two-Phase Flow*. Washington, DC, USA: Hemisphere Publishing corporation.

Hui, T.O. (1983). MS thesis. Michigan State University, East Lansing, MI. Quoted in Thome and Shock (1984).

Hung, J.-T., Chen, Y.-K., Chen, T.-Y., et al. (2016). On the effect of lubricant on pool boiling heat transfer performance. *Proceedings of the International Refrigeration and Air Conditioning Conference*, Purdue, 11–14 July 2016 at West Lafayette, IN. Paper 2131.

Hung, J.T., Chen, Y.K., Shih, H.K., et al. (2018). The effect of refrigeration lubricant properties on nucleate pool boiling heat transfer performance. *International Refrigeration and Air Conditioning Conference*, 9–12 July 2018, Purdue. Paper 1921. https://docs.lib.purdue.edu/iracc/1921.

Ivey, H.J. and Morris, D.J. (1962). On the relevance of the liquid-vapor exchange mechanism for subcooled boiling heat transfer at high pressure. Report AEEW-R137. UKAEA Winfrith.

Jafari, D., Franco, A., Filippeschi, S., and Di Marco, S. (2016). Two-phase closed thermosyphons: a review of studies and solar applications. *Renew. Sust. Energy Rev.* **53**: 575–593.

Jakob, M. and Fritz, W. (1935). Versuche uber den verdampfungsvorgang. *Forsch. auf dem Geb. Des Ing.* **2**: 435–447.

Jensen, M.K. and Jackman, D.L. (1984). Prediction of nucleate pool boiling heat transfer coefficients of refrigerant-oil mixtures. *J. Heat Transfer* **106**: 184–190.

Ji, W.T., Zhang, D.C., Feng, N. et al. (2010). Nucleate pool boiling heat transfer of R134a and R134a-PVE lubricant mixtures on smooth and five enhanced tubes. *J. Heat Transfer* **132**: 111502-1–111502-8.

Johnson, H.A. (1970). Transient boiling heat transfer. *Fourth International Heat Transfer Conference*, Paris-Versailles, France, 31 August–5 September. Vol. 5, paper B3.1.

Jouhara, H. and Robinson, A.J. (2010). Experimental investigation of small diameter two-phase closed thermosyphons charged with water, FC-84, FC-77, and FC-3283. *Appl. Therm. Eng.* **30**: 201–211.

Jung, D.S., Venart, J.E.S., and Sousa, A.C.M. (1987). Effects of enhanced surfaces and surface orientation on nucleate and film boiling heat transfer in R-11. *Int. J. Heat Mass Transfer* **30**: 2627–2639.

Kalinin, E.K., Berlin, I.T., Kostyuk, V.V., and Nosova, E.M. (1976). Heat transfer during transition boiling of cryogenic liquids. *Adv. Cryog. Eng.* **21**: 273–277. Quoted in Dhir (1997).

Kandlikar, S.G. (2001). Theoretical model to predict pool boiling CHF incorporating effects of contact angle and orientation. *J. Heat Transfer* **123**: 1071–1078.

Kawamura, H., Tachibana, F., and Akiyama, M. (1970). Heat transfer and DNB heat flux transient boiling. *Fourth International Heat Transfer Conference*, 31 August–5 September, Paris-Versailles, France. Vol. 5, paper B3.3.

Kawamura, H., Seki, M., Shina, Y., and Sanokawa, K. (1975). Experimental studies on heat transfer by natural convection and pool boiling of sodium in a strong magnetic field. *J. Nucl. Sci. Technol.* **12** (5): 280–286.

Kedzierski, M.A. (2003). A semi-theoretical model for predicting refrigerant and lubricant mixture pool boiling heat transfer. *Int. J. Refrig.* **26** (3): 337–348.

Kim, J. (2003). Review of reduced gravity boiling heat transfer: US research. *J. Jpn. Soc. Microgravity Appl.* **20** (4): 264–271.

Kim, J. and Kim, H.K. (2006). On the departure behaviors of bubble at nucleate pool boiling. *Int. J. Multiphase Flow* **32**: 1269–1286.

Kim, Y., Shin, D.H., Kim, J.S. et al. (2018). Boiling and condensation heat transfer of inclined two-phase closed thermosyphon with various filling ratios. *Appl. Therm. Eng.* **145**: 328–342.

Klimenko, V.V. (1980). Film boiling on a horizontal plate – new correlation. *Int. J. Heat Mass Transfer* **24**: 69–79.

Kobus, C.J. and Wedekind, G.L. (2001). An experimental investigation into natural convection heat transfer from horizontal isothermal circular disks. *Int. J. Heat Mass Transfer* **44**: 3381–3383.

Korneev, M.I. (1955). Pool boiling heat transfer with mercury and magnesium amalgam. *Teploenergetika* **2**: 44. Quoted by Dwyer, O.E. 1976. Boiling Liquid Metal Heat Transfer. American Nuclear Society, Hinsdale, IL.

Kovalev, S.A. and Zhukov, V.M. (1973). Experimental study of heat transfer during sodium boiling under conditions of low pressure and natural convection. *Prog. Heat Mass Transfer* **7**: 347–354.

Krillov, P.L. (1968). A generalized functional relationship between the critical heat flux and pressure in the boiling of metals in large quantities. *Atomnaya Energya* **24**: 143–146.

Kutateladze, S.S. (1951). A hydrodynamic theory of changes in the boiling process under free convection. *Izvest. Akad. Nauk S.S.S.R., Otdel. Tekh. Nauk* **4**: 529.

Kutateladze, S.S. (1952). Heat transfer during boiling and condensation, English translation AEC-TR-3770. Springfield, VA: National Technical Information Service.

Kutateladze, S.S., Moskvicheva, V.N., Bobrovich, G.L. et al. (1973). Some peculiarities of heat transfer crisis in alkali metals boiling under free convection. *Int. J. Heat Mass Transfer* **16**: 705–713.

Lee, Y. (1968). Pool boiling heat transfer with mercury and mercury containing dissolved sodium. *Int. J. Heat Mass Transfer* **11**: 1807–1821.

Lee, Y. (1999). Review of the critical heat flux correlations for liquid metals. Report KAERI/AR-553/99.

Lei, W., Kang, Z., Fushou, X. et al. (2016). Prediction of pool boiling heat transfer for hydrogen in microgravity. *Int. J. Heat Mass Transfer* **94**: 465–473.

Liang, G. and Mudawar, I. (2018a). Pool boiling critical heat flux (CHF) – Part 1: Review of mechanisms, models, and correlations. *Int. J. Heat Mass Transfer* **117**: 1352–1367.

Liang, G. and Mudawar, I. (2018b). Pool boiling critical heat flux (CHF) – Part 2: Assessment of models and correlations. *Int. J. Heat Mass Transfer* **117**: 1368–1383.

Lienhard, J.H. (1981). *A Heat Transfer Textbook*. Englewood Cliffs, NJ: Prentice-Hall.

Lienhard, J.H. (1988). Burnout on cylinders. *J. Heat Transfer* **110**: 1271–1286.

Lienhard, J.H. and Dhir, V.K. (1973). Hydrodynamic prediction of peak pool boiling heat fluxes. NASA-CR-2270.

Louahlia-Gualous, H., Masson, S.L., and Chahed, A. (2017). An experimental study of evaporation and condensation heat transfer coefficients for looped thermosyphon. *Appl. Therm. Eng.* **110**: 931–940.

Lunardini, V. Jr., (1963). An experimental study of the effect of a horizontal magnetic field on the nucleate pool boiling of water and mercury with 0.02% Mg and 0.0001% Ti. PhD thesis. Mechanical Engineering Department, Ohio State University.

Lyon, R.E. (1953). Boiling heat transfer with liquid metals. PhD thesis. University of Michigan. Quoted in Padilla (1966).

Lyon, D.N. (1964). Peak nucleate-boiling heat fluxes and nucleate-boiling heat transfer coefficients for liquid N_2, liquid O_2 and their mixtures in pool boiling at atmospheric pressure. *Int. J. Heat Mass Transfer* **7**: 1097–1116.

Lyon, R.E., Foust, A.S., and Katz, D.L. (1955). Boiling heat transfer with liquid metals. *Chem. Eng. Prog. Symp. Ser.* **51** (17): 41–47.

Marto, P.J. and Rohsenow, W.M. (1966). Effects of surface conditions on nucleate pool boiling of sodium. *J. Heat Transfer* **88**: 196–204.

McGillis, W.R. and Carey, V.P. (1996). On the role of Marangoni effects on the critical heat flux for pool boiling of binary mixtures. *J. Heat Transfer* **118** (1): 103–109.

Merte, H. (1970). Incipient and steady boiling of liquid nitrogen and liquid hydrogen under reduced gravity. ORA Project 07461. Quoted in Lei et al. (2016).

Michiyoshi, I., Takahashi, O., and Serizawa, A. (1975). Effect of magnetic field on pool boiling heat transfer for mercury. Department of Nuclear Engineering Report. Kyoto University, Kyoto, Japan, 1-3. Quoted by Wagner and Lykoudis (1981).

Michiyoshi, I., Takenaka, N., Murata, T. et al. (1985). Effects of liquid level on boiling heat transfer in potassium layers on a horizontal plane heater. *J. Heat Transfer* **107**: 468–472.

Michiyoshi, I., Takenaka, N., and Takahashi, 0. (1986). Dry patch formed boiling and burnout in potassium pool boiling. *Int. J. Heat Mass Transfer* **29**: 689–702.

Michiyoshi, I., Takahashi, O., and Kikuchi, Y. (1989). Heat transfer and the low limit of film boiling. *Exp. Therm. Fluid Sci.* **2** (3): 268–279.

Mitrovic, J. (1998). Nucleate boiling of refrigerant-oil mixtures: bubble equilibrium and oil enrichment at the interface of a growing vapour bubble. *Int. J. Heat Mass Transfer* **41**: 3451–3467.

Mohanty, R.C. and Das, M.K. (2017). A critical review on bubble dynamics parameters influencing boiling heat transfer. *Renew. Sust. Energy Rev.* **78**: 466–494.

Mostinski, I.K. (1963). Application of the rule of corresponding states for calculation of heat transfer and critical heat flux(in Russian). *Teploenergetika* **4**: 66–71. English abstract in British Chemical Engineering, 8(8), 580.

Mudawar, I., Howard, A.H., and Gersey, C.O. (1997). An analytical model for near-saturated pool boiling critical heat flux on vertical surfaces. *Int. J. Heat Mass Transfer* **40** (1997): 2327–2339.

Nishio, S. and Chandratilleke, G.R. (1989). Steady-state pool boiling heat transfer to saturated liquid helium at atmospheric pressure. *JSME Int. J., Ser. II* **32**: 639–645.

Noyes, R.C. (1963). An experimental study of sodium pool boiling heat transfer. *J. Heat Transfer* **85** (2): 125–129.

Noyes, R.C. and Lurie, H. (1966). Boiling sodium heat transfer. In: *Proceedings of the Third International Heat Transfer Conference*, vol. **5**, 92. Chicago, IL: Begell House.

Nukiyama, S. (1934). The maximum and minimum values of heat Q transmitted from metal to boiling water under atmospheric pressure. *J. Jpn. Soc. Mech. Eng.* **37**: 367–374. Available in Int. J. Heat Mass Transfer, (1966), 9, 1419–1433.

Ogata, H. and Nakayama, W. (1977). Heat transfer to subcritical and supercritical helium in centrifugal acceleration fields. 1. Free convection regime and boiling regime. *Cryogenics* **17** (8): 461–470.

Olek, S., Zvirin, Y., and Elias, E. (1991). A simple correlation for the minimum film boiling temperature. *J. Heat Transfer* **113**: 264–265.

Padilla, A. (1966). Film boiling of potassium on a horizontal plate. PhD thesis. University of Michigan.

Palen, J.W. and Small, W. (1964). A new way to design kettle and internal reboilers. *Hydrocarbon Process.* **43** (11): 199–208.

Paul, D.D. and Abdel-Khalik, S.I. (1983). A statistical analysis of saturated nucleate boiling along a heat wire. *Int. J. Heat Mass Transfer* **26**: 509–519.

Pethukov, B.S., Kovalev, S.A., and Zhukov, V.M. (1966). Study of sodium boiling heat transfer. In: *Proceedings of the Third International Heat Transfer Conference*, vol. **5**, 80–91. Chicago, IL: Begell House.

Plesset, M.S. and Zwick, S.A. (1954). The growth of vapor bubbles in superheated liquid. *J. Appl. Phys.* **25**: 493–500. Quoted in Mohanty & Das (2017).

Raj, R., Kim, J., and McQuillen, J. (2012a). On the scaling of pool boiling heat flux with gravity and heater size. *J. Heat Transfer* **134**: 011502.

Raj, R., Kim, J., and McQuillen, J. (2012b). Pool boiling heat transfer on the international space station: experimental results and model verification. *J. Heat Transfer* **134**: 101504.

Ramilison, J.M. and Lienhard, J.H. (1987). Transition boiling heat transfer and the film transition regime. *J. Heat Transfer* **109**: 746–752.

Reddy, R.P. and Lienhard, J.H. (1989). The peak boiling heat flux in saturated ethanol-water mixtures. *J. Heat Transfer* **111**: 480–486.

Rohsenow, W.M. (1952). A method for correlating data for surface boiling of liquids. *Trans. ASME* **74**: 969–975.

Rohsenow, W.M. (1985). Boiling. In: *Handbook of Heat Transfer*, Chapter 12 (eds. W.M. Rohsenow, J.P. Hartnett and E. Ganic). New York, NY: McGraw-Hill.

Rohsenow, W.M. and Griffith, P. (1955). Correlation of maximum heat transfer data for boiling of saturated liquids. *Chem. Eng. Prog. Symp. Ser.* **52**: 47–49.

Sakashita, H., Ono, A., and Nakabayashi, Y. (2010). Measurements of critical heat flux and liquid–vapor structure near the heating surface in pool boiling of 2-propanol/water mixtures. *Int. J. Heat Mass Transfer* **53** (7–8): 1554–1562.

Sakurai, A., Shiotsu, M., and Kataoka, I. (1978). Sodium pool boiling heat transfer. In: *Proceedings of the Sixth International Heat Transfer Confernce*, vol. **1**, 193–198. Toronto, Canada: Begell House.

Sakurai, A., Shiotsu, M., and Hata, K. (1990a). Effect of system pressure on minimum film boiling temperature for various liquids. *Exp. Therm. Fluid Sci.* **3**: 450–457.

Sakurai, A., Shiotsu, M., and Hata, K. (1990b). A general correlation for pool film boiling heat transfer from a horizontal cylinder to subcooled liquid. Part 2: Experimental data for various liquids and its correlation. *J. Heat Transfer* **112**: 441–450.

Sauer, H.J. and Chongrungeong, S. (1980). Nucleate boiling performance of refrigerants and refrigerant-oil mixtures. *ASME J. Heat Transfer* **102**: 701–705.

Sauer, H.J. and Dougherty, R.L. (1974). Nucleate pool boiling of refrigerant-oil mixtures from tubes. *ASHRAE Trans.* **80** (2): 175–193.

Schlunder, E.U. (1983). Heat transfer in nucleate boiling of mixtures. *Int. Chem. Eng.* **23** (4): 589–599.

Shah, M.M. (1975). Visual observations in an ammonia evaporator. *ASHRAE Trans.* **81** (1): 295–306.

Shah, M.M. (1990). Pool boiling. In: *Genium Heat Transfer and Fluid Flow Data Books*, Section 507.2 (ed. F. Krieth), 1–8. Schenectady, NY.

Shah, M.M. (1992). A survey of experimental heat transfer data for nucleate pool boiling of liquid metals and a new correlation. *Int. J. Heat Fluid Flow* **13** (4): 370–379.

Shah, M.M. (1996). Survey of critical heat flux data for pool boiling of liquid metal and new correlations. *Heat Transfer Eng.* **17** (2): 54–66. https://www.tandfonline.com.

Shah, M.M. (2007). A general correlation for heat transfer during saturated boiling with flow across tube bundles. *HVAC&R Res.* **13** (5): 749–768.

Shatto, D.P. and Peterson, G.P. (1999). Pool boiling critical heat flux in reduced gravity. *J. Heat Transfer* **121** (4): 865–873.

Shiraishi, M., Kikuchi, K., and Yamanishi, T. (1981). Investigation of heat transfer characteristics of a two-phase closed thermosyphon. In: *Advances in Heat Pipe Technology* (ed. D.E. Reay), 95–104. Elsevier.

Simon, F.F., Papell, S.S., and Simoneou, R.J. (1968). Minimum film-boiling heat flux in vertical flow of liquid nitrogen. NASA TN D-4307.

Sivior, J.B. and Ede, A.J. (1970). Heat transfer in subcooled pool film boiling. In: *Proceedings of the 4th International Heat Transfer Conference*, vol. **5**, B3.12. Amsterdam: Elsevier Pub. Co.

Son, G. and Dhir, V.K. (2008). Three-dimensional simulation of saturated film boiling on a horizontal cylinder. *Int. J. Heat Mass Transfer* **51**: 1156–1167.

Stephan, K. and Abdelsalam, M. (1980). Heat transfer correlations for natural convection boiling. *Int. J. Heat Mass Transfer* **23** (1): 73–87.

Stephan, K. and Körner, M. (1969). Berechnung des Wärmeübergangs verdampfender binärer Flüssigkeitsgemische. *Chem. Ing. Tech.* **41** (7): 409–417.

Sterman, L.S., Vilemas, J.V., and Abramov, A.I. (1966). On heat transfer and critical heat flux in organic coolants and their mixtures. In: *Proceedings of the International Heat Transfer Conference*, vol. **4**, 258–270. Chicago, IL: Begell House.

Subbotin, V.I., Ushakov, P.A., et al. (1964). Heat transfer from fuel elements of liquid-metal cooled reactors. *Paper No. P/328, 3rd International Conference Peaceful Uses Atomic Energy*, Geneva. Quoted by Dwyer (1976).

Subbotin, V.I., Ovechkin, D.M., Sorokin, D.N., and Kudryaystev, A.P. (1968a). Heat transfer when sodium boils in free convection conditions. *Atomnaya Energiya* **24**: 437–442.

Subbotin, V.I., Ovechkin, D.N., and Kudryavtsev, A.P. (1968b). Heat transfer with cesium boiling in free convection conditions. *Teploenergetika* **15** (6): 63–66.

Subbotin, V.I., Sorokin, D.N., and Kudryavtsev, A.P. (1970). Generalized relations for calculating heat transfer in the developed boiling of alkali metals. *Atomnaya Energiya* **29**: 45–46.

Subbotin, V.I., Sorokin, D.I., Ovechkin, D.M., and Kudryavtsev, A.P. (1972). *Heat Transfer in Boiling Metals by Natural Convection*. AEC-TR-7210. Springfield, VA, USA: National Technical Information Service.

Subbotin, V.L., Sorokin, D.N., Kudryavstev, A.P., and Brigutsa, V.I. (1974). Heat transfer to sodium–potassium alloy in pool boiling. In: *Proceedings of the Fifth International Heat Transfer Confernce*, 325–329. Tokyo, Japan: Begell House.

Taboas, F., Valles, M., Bourouis, M., and Coronas, A. (2007). Pool boiling of ammonia/water and its pure components: comparison of experimental data in the literature with the predictions of standard correlations. *Int. J. Refrig.* **30**: 778–788.

Takahashi, O., Nishida, M., Takenaka, N., and Michiyoshi, I. (1980). Pool boiling heat transfer from horizontal plane heater to mercury under magnetic field. *Int. J. Heat Mass Transfer* **23**: 27–37.

Takenaka, N. (1984). Natural convection and boiling heat transfer in potassium. Doctoral thesis. Department of Nuclear Engineering, Kyoto University, Kyoto, Japan.

Thome, J.R. (1983). Prediction of binary mixture boiling heat transfer coefficients using only phase equilibrium data. *Int. J. Heat Mass Transfer* **26**: 965–974.

Thome, J.R. and Shakir, S. (1987). A new correlation for nucleate pool boiling of aqueous mixtures. *Heat Transf.: Pittsburgh, AIChe. Symp. Ser.* **257** (83): 46–51.

Thome, J.R. and Shock, R.A.W. (1984). Boiling of multicomponent mixtures. In: *Advances in Heat Transfer*, vol. **16** (eds. J.P. Hartnett and T.F. Irvine Jr.,), 59–156. New York, NY: Academic Press.

Tran, N., Sheng, S.-R., and Wang, C.-C. (2018). An experimental study and empirical correlations to describe the effect of lubricant oil on the nucleate boiling heat transfer performance for R-1234ze and R-134a. *Int. Comm. Heat Transf.* **97**: 78–84.

Turner, J.B. and Colver, C.P. (1971). Heat transfer to mercury from horizontal cylinder heater at heat flux up to burnout. *J. Heat Transfer* **93**: 1–10.

Van Wijk, W.R., Vos, A.S., and Van Stralen, S.J. (1956). Heat transfer to boiling binary liquid mixtures. *Chem. Eng. Sci.* **5**: 65–80. Quoted in Thome and Shock (1984).

Vishnev, I.P. (1973). Effect of orienting the hot surface with respect to the gravitational field on the critical nucleate boiling of a liquid. *J. Eng. Phys. Thermophys.* **24**: 43–48. Quoted in Liang & Mudawar (2018b).

Wadkins, R.P. (1984). Low pressure boiling lithium experiments. *AIChE. Symp. Ser.* **80** (236): 50–54.

Wagner, L.R. and Lykoudis, P.S. (1981). Mercury pool boiling under the influence of a horizontal magnetic field. *Int. J. Heat Mass Transfer* **24**: 635–643.

Wang, C.C., Hafner, A., Kuo, C., and Hsieh, W. (2014). Influence of lubricant on the nucleate boiling heat transfer performance of refrigerant – a review. *Heat Transfer Eng.* **35** (6–8): 651–663.

Wang, L., Li, Y., Zhang, F. et al. (2016). Correlations for calculating heat transfer of hydrogen pool boiling. *Int. J. Hydrogen Energy* **41**: 17118–17131.

Warrier, G.R., Dhir, V.K., and Chao, D.F. (2015). Nucleate pool boiling experiment (NPBX) in microgravity: International Space Station. *Int. J. Heat Mass Transfer* **83**: 781–798.

Weil, L. (1952) *Proc. IVe Congr. Intern. Du Chauf. Ind., Paris, No. 210*. Quoted in Sakurai et al. (1990b).

Wu, Y., Luo, S., Wang, L. et al. (2018). Review on heat transfer and flow characteristics of liquid sodium (2): two phase. *Prog. Nucl. Energy* **103**: 151–164.

Yagov, V.V. (1988). A physical model and calculation formula for critical heat fluxes with nucleate pool boiling of liquids. *Therm. Eng.* **35**: 333–339.

Yagov, V.V. (2004). Critical heat flux prediction for pool boiling of binary mixtures. *Trans. IChemE, Part A, Chem. Eng. Res. Des.* **82** (A4): 457–461.

Yagov, V.V. (2014). Is a crisis in pool boiling actually a hydrodynamic phenomenon? *Int. J. Heat Mass Transfer* **73**: 265–273.

Yagov, V.V. and Sukach, A.V. (2000). An approximate model for burnout at low reduced pressures. *Therm. Eng.* **47** (3): 200–204.

Yang, Y.M. (1987). An estimation of pool boiling critical heat flux for binary mixtures. In: *Proceedings of the 2nd ASME/JSME Thermal Engineering Joint Conference*, vol. **5**, 439–446. Honolulu, HI: ASME.

Yoon, S.Y. and No, H.C. (2017). Film boiling heat transfer of a hot sphere in a subcooled liquid pool considering heat loss through its support rod. *Nucl. Eng. Des.* **325**: 97–106.

Zeng, Y. and Lee, Y. (1987). The effect of peripheral wall conduction in pool boiling. In: *Heat Transfer Science & Technology* (ed. B.-X. Wang), 445–452. Hemisphere Publishing. Quoted in Sakurai et al. (1990b).

Zuber, N. (1958). On the stability of boiling heat transfer. *Trans. ASME* **80**: 711–720.

Zuber, N. (1959). Hydrodynamic aspects of boiling heat transfer. PhD thesis. University of California, LA.

4

Forced Convection Subcooled Boiling

4.1 Introduction

Boiling of subcooled liquids can occur if the heating surface temperature exceeds the saturation temperature. Heat transfer coefficients in subcooled boiling are considerably higher than in single-phase convection. Many heat exchangers operate exclusively in the subcooled regime. A notable example is the steam generator of pressurized water nuclear reactor (PWR) for power generation. Newer PWR reactors also operate with subcooled boiling. More commonly, subcooled boiling occurs in part of the heat exchanger near the entrance and saturated boiling in the rest of the heat exchanger.

There are two regimes of subcooled boiling. This is explained by considering the case of flow through channels. Consider subcooled liquid flowing into a uniformly heated tube as shown in Figure 4.1.1. Liquid temperature rises along the length until it reaches the saturation temperature. The saturation temperature drops slightly along the length due to pressure drop. The wall temperature rises until boiling inception/onset of nucleate boiling (ONB) occurs. Then it drops slightly. In the length near the ONB point, there are only a few nucleation sites. Bubbles grow and collapse attached to the wall. The void fraction in this region is very low. This is alternatively called partial boiling region or high subcooling region. Further down the tube, a point is reached where bubbles first leave the surface. This point is often called the point of net vapor generation (NVG) or the point of onset of significant void (OSV). This is where the partial boiling regime ends and fully developed boiling (FDB) regime starts. This regime is also known as the low subcooling regime. The nucleation sites increase rapidly beyond this point and so does the void fraction. The bubbles leave the surface to enter the flow stream where they travel some distance and then condense. The FDB regime ends when the liquid reaches saturation. The wall temperature rises slightly in the partial boiling regime but is essentially constant in the fully developed regime. Once liquid reaches saturation temperature, bulk boiling starts.

All discussions in this chapter pertain to boiling prior to the occurrence of boiling crisis. Furthermore, all discussions pertain to single-component non-metallic fluids except where stated otherwise.

4.2 Inception of Boiling in Channels

The inception of nucleate boiling under natural convection has been discussed in Section 3.2. Numerous measurements of wall superheat during forced convection have been made. All the factors that are significant in pool boiling are also significant in forced convection boiling. Among these are properties of fluids and surface characteristics. Flow velocity also has an important effect. Very high superheats, even exceeding 100 K have been experienced with liquid metals. For ordinary fluids, incipient superheats are generally no more than a few degrees.

Methods for predicting incipient superheats are herein discussed.

4.2.1 Analytical Models and Correlations

Bergles and Rohsenow (1964) developed an analytical model by extending the analysis of Hsu and Graham (1961) for pool boiling to forced convection boiling. The model is simple and has found wide acceptance.

Consider liquid flowing parallel to a heated surface. The temperature profile in the liquid layer is considered linear. Thus, the temperature of liquid T_f at distance y from the surface is given by

$$T_f = T_w - \frac{qy}{k_f} \tag{4.2.1}$$

The wall temperature T_w is obtained by the following equation:

$$q = h_{\mathrm{LT}}(T_w - T_f) \tag{4.2.2}$$

where h_{LT} is the heat transfer coefficient for all mass flowing as liquid calculated by an equation for single-phase forced convection.

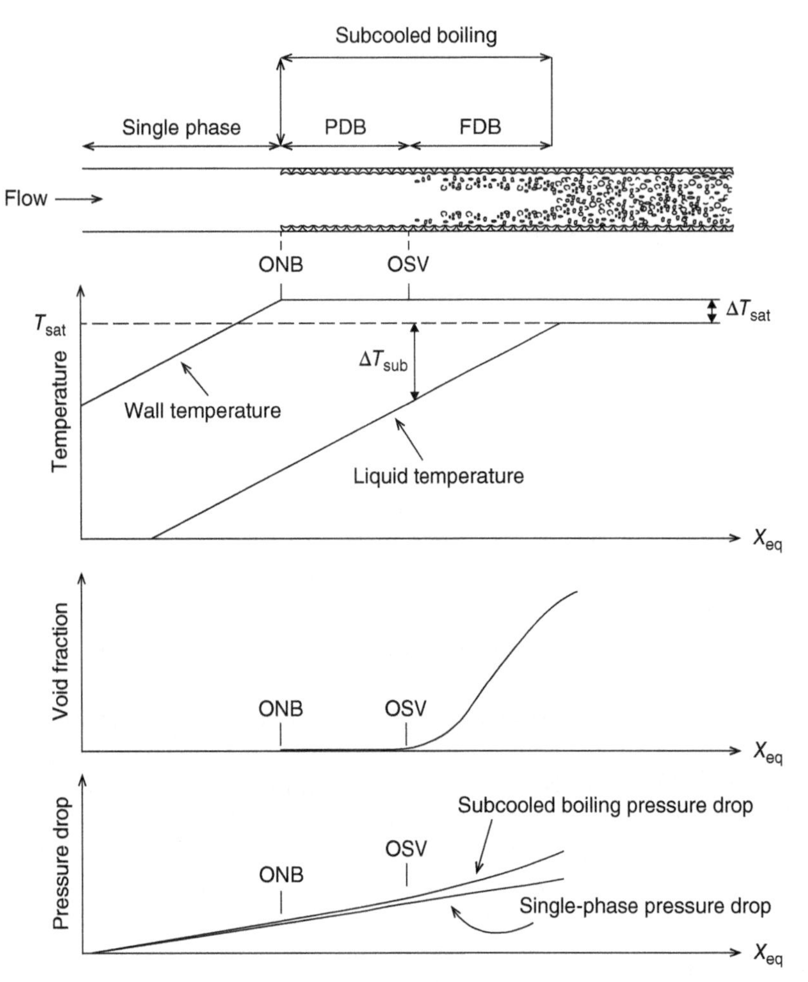

Figure 4.1.1 Regimes of subcooled boiling in a tube. Source: From Delhaye et al. (2004). © 2004 Elsevier.

Equation (3.2.2) was derived considering bubble equilibrium and applying the Clausius–Clapeyron equation. It gives the temperature T_g inside the bubble for a bubble of radius r_c. Bergles and Rohsenow assumed that a bubble of radius r can grow only if the liquid temperature T_f at $y = r$ equals or exceeds the vapor temperature in bubble T_g given by Eq. (3.2.2). Bergles and Rohsenow solved Eqs. (3.2.2) and (4.2.2) graphically for water in the pressure range 1–138 bar to determine the point of tangency. The results were correlated by the following dimensional equation:

$$q_{ib} = 15.6p^{1.156}(\Delta T_{SAT})_{ib}^{2.3/p^{0.0234}} \tag{4.2.3}$$

The subscript "ib" indicates inception of boiling, q is in Btu h^{-1} ft^{-2}, p is in psia, and temperature is in °F. Bergles and Rohsenow found satisfactory agreement of this equation with their own data as well as the data of Rohsenow and Clark (1951a). In SI units, this equation becomes (Collier and Thome 1994)

$$(\Delta T_{SAT})_{ib} = 0.556\left[\frac{q_{ib}}{1082p^{1.156}}\right]^{0.463p^{0.0234}} \tag{4.2.4}$$

where T is in °C, q in W m^{-2}, and p in bar.

Sato and Matsumura (1964) performed an analysis similar to that of Bergles and Rohsenow to obtain the following equation:

$$q_{ib} = \frac{k_f i_{fg} \rho_g}{8\sigma T_{SAT}}(\Delta T_{SAT})_{ib}^2 \tag{4.2.5}$$

In their analysis, Davis and Anderson (1966) introduced the contact angle β to obtain the following equation:

$$q_{ib} = \frac{k_f i_{fg} \rho_g}{8\sigma T_{SAT}(1+\cos\beta)}(\Delta T_{SAT})_{ib}^2 \tag{4.2.6}$$

For a hemispherical bubble, $\beta = 90°$ and then Eq. (4.2.6) becomes the same as Eq. (4.2.5).

As discussed in Section 3.2, Frost and Dzakowic (1967) introduced Prandtl number into Eq. (4.2.5) to give Eq. (3.2.5), which has been verified with a wide range of data for pool boiling as well as forced convection boiling.

Basu et al. (2002) measured inception of boiling superheats during forced convection on a horizontal plate as well as in a tube bundle. They compared their data for plate with Eqs. (4.2.4), (4.2.5), (4.2.6) and found that they underpredict the wall superheat at boiling inception. They pointed out that these equations were based on the criterion that the liquid temperature at the top of the bubble just equals the temperature inside the bubble. However, the cavity corresponding to that bubble diameter may be unavailable on the surface or be flooded. They proposed that the probability of finding an unflooded cavity of the size corresponding to the minimum wall superheat calculated from this criterion diminishes as the wettability of the surface increases. However, the diameter of the available cavity D_c is proportional to that obtained from the minimum superheat criterion, $D_{c,0}$, as expressed by the following relation:

$$D_c = D_{c,0} F \tag{4.2.7}$$

$$D_{c,0} = \left[\frac{8\sigma T_{SAT} k_f}{\rho_g i_{fg} q} \right]^{1/2} \tag{4.2.8}$$

$$F = 1 - \exp\left[-\left(\frac{\pi\beta}{180} \right)^3 - 0.5\left(\frac{\pi\beta}{180} \right) \right] \tag{4.2.9}$$

β is the static contact angle in degrees. F was obtained by analysis of test data.

$$q_{ib} = h_{LT}[(\Delta T_{SAT})_{ib} + \Delta T_{SC}] \tag{4.2.10}$$

$$(\Delta T_{SAT})_{ib} = \frac{4\sigma T_{SAT}}{D_c \rho_g i_{fg}} \tag{4.2.11}$$

With the above equations, the wall superheat at boiling inception can be calculated. Basu et al. compared this correlation with their own data as well as data from seven other sources that included water, refrigerants, and FC-72. Contact angles varied from 1° to 85° and subcooling from 0 to 80 K. Good agreement was found.

Based on their measurements of boiling water in tubes and annuli, Tarasova and Orlov (1969) proposed the following dimensional correlation:

$$(i_f - i_{f,SAT})_{ib} = 135 \frac{q^{1.1} D_{HYD}^{0.2}}{G^{0.9}} \left(\frac{\rho_g}{\rho_f} \right)^{0.3} \tag{4.2.12}$$

where i_f is the actual liquid enthalpy and $i_{f,SAT}$ is the enthalpy of saturated liquid in kcal kg^{-1}. D_{HYD} is in meter, q in kcal m^{-2} h^{-1}, and G in kg m^{-2} h^{-1}.

The previously described models were verified with data for macro channels. Some differences have been reported in minichannels as discussed in Section 4.2.2.

4.2.2 Minichannels

Ghiaasiaan and Chedester (2002) compared data for inception of boiling in tubes of diameter 1–1.45 mm with the predictions of the models of Bergles and Rohsenow (1964) and Sato and Matsumura (1964) given earlier. These were found to underpredict the data. They developed a semi-empirical method to modify the Sato and Matsumura model by incorporating a shape factor using channel turbulence characteristics and experimental data of incipient heat flux. Their correlation is given as follows:

$$q_{ib} = \frac{k_f i_{fg} \rho_g}{C(8\sigma T_{SAT})} (\Delta T_{SAT})_{ib}^2 \tag{4.2.13}$$

$$C = 22 \left[\frac{\sigma_f - \sigma_w}{\rho_f u_L^2 R^*} \right]^{0.765} \tag{4.2.14}$$

$C \geq 1$. σ_f and σ_w are the surface tensions at liquid and wall temperatures, u_L is the average liquid velocity, and R^* is the critical cavity radius given by

$$R^* = \left[\frac{2\sigma T_{SAT} k_f}{q_{ib} \rho_g i_{fg}} \right] \tag{4.2.15}$$

This correlation gave satisfactory agreement with data for tubes of 1–1.45 mm diameter from two sources.

Liu et al. (2005) developed an analytical solution that agreed with their own data for boiling inception in a minichannel. However, others have reported it to be inadequate as seen in the following text.

Qi et al. (2007) studied boiling inception with nitrogen flowing upwards in tubes of diameter 0.53–1.93 mm. Inlet pressure varied from 180 to 920 kPa and mass flux from 440 to 3000 kg m^{-2} s^{-1}. They found that the model of Liu et al. (2005) gave large errors compared to their data. They fitted the following equation to their data:

$$(T_w - T_{SAT})_{ib} = 30.65(q \times 10^{-6})^{0.5} \exp\left(-\frac{p}{8.7 \times 10^6} \right) \tag{4.2.16}$$

where q is in W m^{-2} and p is in Pa. They obtained this equation by changing the constant in the formula given by Thom et al. (1965) for water.

Song et al. (2017) measured boiling inception during downward flow of water in a narrow rectangular channel heated on two sides. The correlations of Bergles and Rohsenow (1964), Sato and Matsumura (1964), and Liu et al. (2005), considerably underpredicted their measurements. They fitted a correlation to their own data.

Once $(\Delta T_{SAT})_{ib}$ is known, the distance from entrance to the point of boiling inception z_{ib} is determined as follows. Energy balance gives

$$q\pi D z_{ib} = C_{pf} G \frac{\pi D^2}{4} (T_{f,ib} - T_{f,e}) \tag{4.2.17}$$

where $T_{f,ib}$ and $T_{f,e}$ are the temperatures of liquid at location of inception of boiling and at entrance to channel, respectively. Further, at the boiling inception location,

$$q = h_{LT}[(\Delta T_{SAT})_{ib} + (\Delta T_{SC})_{ib}] \tag{4.2.18}$$

By definitions,

$$(T_{f,\mathrm{ib}} - T_{f,e}) = (\Delta T_{\mathrm{SC}})_e - (\Delta T_{\mathrm{SC}})_{\mathrm{ib}} \tag{4.2.19}$$

By substituting $(\Delta T_{\mathrm{SC}})_{\mathrm{ib}}$ from Eq. (4.2.17) into Eq. (4.2.18), and with some rearrangement, the following equation is obtained:

$$z_{\mathrm{ib}} = \frac{GDC_{\mathrm{pf}}}{4}\left[\frac{(\Delta T_{\mathrm{SC}})_e - (\Delta T_{\mathrm{SAT}})_{\mathrm{ib}}}{q} - \frac{1}{h_{\mathrm{LT}}}\right] \tag{4.2.20}$$

The subscript "e" indicates "at entrance to channel."

The length of subcooled boiling region z_{scb} can be determined in a similar way by noting that subcooled boiling ends when liquid temperature equals the saturation temperature.

$$z_{\mathrm{scb}} - z_{\mathrm{ib}} = \frac{GDC_{\mathrm{pf}}}{4}\left[\frac{1}{h_{\mathrm{LT}}} - \frac{(\Delta T_{\mathrm{SAT}})_{\mathrm{ib}}}{q}\right] \tag{4.2.21}$$

4.2.3 Effect of Dissolved Gases

Dissolved gases in liquid affect boiling inception. McAdams et al. (1949) carried out experiments with subcooled water in heated annuli. In some of the tests, the water in the reservoir was pressurized with steam, while in some it was pressurized with air. Inception of boiling occurred at much lower wall superheats when pressurization was by air, as shown in Figure 4.2.1. This must have been because pressurizing air got dissolved in water, got trapped in cavities, and thus made more cavities active.

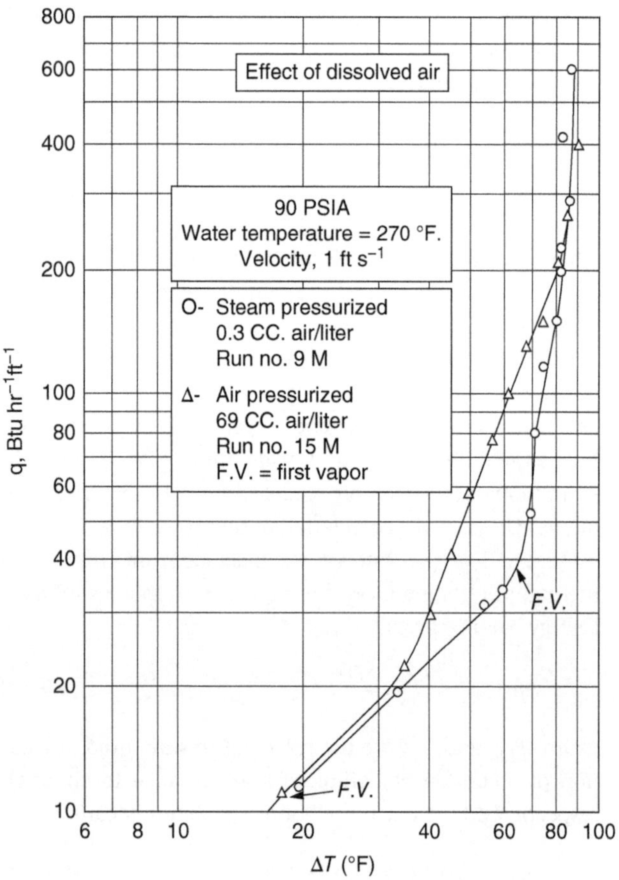

Figure 4.2.1 Effect of dissolved air on inception of boiling in an annulus. Source: From McAdams et al. (1949). © 1949 American Chemical Society.

4.2.4 Recommendations

For conventional channels, the model of Basu et al. (2002) has been verified over a wide range of subcooling and other parameters. It is therefore recommended.

There is no well-verified method for minichannels. Use the various proposed correlations in their verified range.

4.3 Prediction of Subcooled Boiling Regimes in Channels

Physically, the boundary between the high and low subcooling regimes is the location where the bubbles begin to depart from the wall instead of remaining attached to the wall. Saha and Zuber (1974) gave the following criteria to determine the bubble departure point (OSV):

$Pe < 70\,000$,

$$\Delta T_{\mathrm{SC}} = 0.0022(qD/k_f) \tag{4.3.1}$$

$Pe > 70\,000$,

$$\Delta T_{SC} = 153.8q/(GC_{pf}) \tag{4.3.2}$$

If actual ΔT_{SC} is greater than that given by Eqs. (4.3.1) and (4.3.2), the regime is high subcooling, else it is low subcooling. $Pe < 70\,000$ is the thermally controlled regime, and $Pe > 70\,000$ is the hydrodynamically controlled regime. Pe is the Peclet number defined as ($Re_{LT}\ Pr_f$). This correlation is widely used.

Lee and Bankoff (1998) reviewed predictive models for the OSV in forced convection subcooled boiling and compared them with extensive data. Three analytical models and seven empirical correlations were considered by them. They found that the correlation of Saha and Zuber (1974) to be the best. The other prediction methods considered by them included Levy (1967), Dix (1971), Bowring (1962), Staub (1968), Ahmad (1970), and Unal (1975).

Al-Yahia and Jo (2018) studied subcooled boiling of water at atmospheric pressure during upflow in a narrow vertical rectangular channel. Satisfactory agreement was found with the Saha and Zuber correlation.

Kandlikar (1998) developed the following correlation:

$$1058F_{fl}(Gi_{fg})^{-0.7}q_F - q_F^{0.3} - 1058h_{LT}F_{fl}(Gi_{fg})^{-0.7}\Delta T_{SC} = 0 \tag{4.3.3}$$

Iterative solution of this equation with known values of ΔT_{SC} and G yield q_F, which is the heat flux at the intersection of heat flux due to single-phase convection and that due to FDB. The heat flux at OSV q_{OSV} is then obtained by using the Bowring (1962) model as

$$q_{OSV} = 1.4q_F \tag{4.3.4}$$

F_{fl} is a parameter that depends on liquid–surface combination. For water, $F_{fl} = 1$ for all surface materials. It is also $= 1$ for stainless steel surfaces with all materials. For copper surfaces, it varies over a wide range. Single-phase heat transfer coefficient may be calculated with a suitable forced convection equation.

Shah (1977) has given the following correlation for the boundary between low and high subcooling regimes. The regime is high subcooling (partial boiling) if either of the following two conditions is satisfied:

$$\Delta T_{SC}/\Delta T_{SAT} \geq 2 \tag{4.3.5}$$

$$\Delta T_{SC}/\Delta T_{SAT} > 6.3 \times 10^4 Bo^{1.25} \tag{4.3.6}$$

Otherwise, it is low subcooling (FDB) regime. The correlation is shown graphically in Figure 4.3.1. This correlation was developed by analysis of a wide range of subcooled boiling data. The boiling regime was determined according to agreement with the heat transfer correlations for the two regimes. This correlation requires iterative calculations

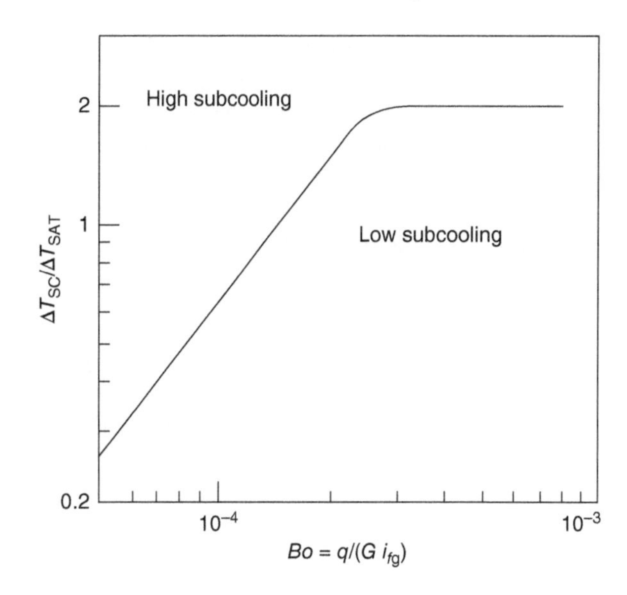

Figure 4.3.1 Boundary between high and low subcooling regimes according to Shah (1977). Source: From Shah (1983). © 1983 Taylor & Francis.

together with a heat transfer correlation to determine ΔT_{SAT}. The Shah heat transfer correlation is presented in Section 4.5.2.

The distance from the channel entrance to the location of bubble departure (OSV), z_{OSV}, can be determined as follows. Heat balance requires that

$$q\pi D z_{OSV} = C_{pf}G\frac{\pi D^2}{4}(T_{f,OSV} - T_{f,e}) \tag{4.3.7}$$

This can be rearranged to the following form:

$$z_{OSV} = \frac{GDC_{pf}}{4q}[(\Delta T_{SC})_e - (\Delta T_{SC})_{OSV}] \tag{4.3.8}$$

Having calculated $(\Delta T_{SC})_{OSV}$ with Saha and Zuber correlation or some other method, its substitution in Eq. (4.3.8) gives z_{OSV}.

4.3.1 Recommendation

The Saha–Zuber correlation is well verified and is easy to use as it does not require any iteration. The Shah correlation is also well verified. It is a good alternative.

4.4 Prediction of Void Fraction in Channels

A number of analytical solutions for prediction of void fraction have been proposed.

Levy (1967) has given the following derivation for void fraction at the bubble departure point, the boundary between high and low subcooling regimes. He assumed that the bubbles are spaced a distance S apart. The number of bubbles around the wetted perimeter P_w is then (P_w/S). The bubbles are assumed to be full spheres of radius r. The distance Y_B to the top of the bubble is then $= 2r$. Their volume in a section of channel S long is $(P_w/S)(4/3)(\pi r^2)$. The void fraction, volume of vapor divided by total volume, at bubble departure point is then

$$\alpha = \left(\frac{P_w}{S}\right)\left(\frac{4}{3}\pi r^2\right)\left(\frac{1}{AS}\right) = \frac{16}{3}\pi\frac{r}{D_{\text{HYD}}}\left(\frac{r}{S}\right)^2 \tag{4.4.1}$$

A is the cross-sectional area of the channel. Levy assumed that the spherical bubbles are packed together in a square array and interfere with one another at $r/s = 0.25$. It follows then

$$\alpha = \frac{\pi}{3}\frac{r}{D_{\text{HYD}}} = \frac{\pi}{6}\frac{Y_B}{D_{\text{HYD}}} \tag{4.4.2}$$

The bubbles are acted upon by buoyancy, drag, and surface tension forces. By analysis and use of experimental data to determine constants, it was shown that

$$Y_B = 0.015\left(\frac{\sigma D_{\text{HYD}}}{\tau_w}\right)^{0.5} \tag{4.4.3}$$

Substituting Eq. (4.4.3) in Eq. (4.4.2), the result is the following expression for the void fraction α_{bd}:

$$\alpha_{\text{bd}} = 0.007\,85\left(\frac{\sigma}{\tau_w D_{\text{HYD}}}\right)^{0.5} \tag{4.4.4}$$

The wall shear stress τ_w is obtained from the following relation:

$$\tau_w = \left[\frac{fG^2}{2\rho_f}\right] \tag{4.4.5}$$

where f is the Fanning friction factor for smooth drawn tube. In deriving this formula, Levy assumed that void fraction in the high subcooling regime is negligible.

Delhaye et al. (2004) have given the following expression for void fraction at the bubble departure point:

$$\alpha_{\text{bd}} = 30\frac{qk_f Pr_f}{h_{\text{LT}}^2(\Delta T_{\text{SC}})_{\text{bd}}}\frac{L_{\text{cap}}}{D^2} \tag{4.4.6}$$

L_{cap} is the capillary length defined as

$$L_{\text{cap}} = \left(\frac{\sigma}{g(\rho_f - \rho_g)}\right)^{0.5} \tag{4.4.7}$$

This was a modification of the correlation of Griffith et al. (1958) for water. The modification was the introduction of the capillary length that resulted in good agreement with data for both water and R-12, while the Griffith et al. formula agreed with water data but had large deviations with R-12 data.

For void fraction in the FDB regime, Zuber and Findlay (1965) developed their drift flux model that may be written as

$$\alpha = \frac{xG\rho_f}{C_o[x\rho_f + (1-x)\rho_g]G + u_{\text{Gj}}\rho_f\rho_g} \tag{4.4.8}$$

where C_o is a distribution parameter and u_{Gj} is the drift velocity that is determined by analysis of experimental data. Zuber and Findlay found C_o to vary between 1 and 1.5. For high pressure steam water system, the average value was 1.2.

Based on the Zuber–Findlay analysis, Rouhani and Axelsson (1970) arrived at the following equation:

$$\alpha = \frac{x}{\rho_g}\left\{C_o\left[\frac{x}{\rho_g} + \frac{1-x}{\rho_f}\right] + \frac{1.18}{G}\left[\frac{\sigma g(\rho_f - \rho_g)}{\rho_f^2}\right]^{1/4}\right\}^{-1} \tag{4.4.9}$$

For subcooled boiling, the equilibrium quality x is replaced by x', the true vapor quality. They developed a methodology to calculate x'. Rouhani and Axelsson (1970) compared Eq. (4.4.9) to a wide range of data. They found that $C_o = 1.12$ except at low mass velocities ($G < 200$ kg m^{-2} s^{-1}), $C_o = 1.54$. Figure 4.4.1 shows comparison of their predictions with some test data. Excellent agreement is seen in subcooled as well as saturated boiling conditions.

Levy (1967) has given the following relation between x' and the equilibrium vapor quality:

$$x' = x - x_{\text{bd}}\exp\left(\frac{x}{x_{\text{bd}}} - 1\right) \tag{4.4.10}$$

where x_{bd} is the equilibrium quality at the start of bubble departure. It can be calculated by taking heat balance as

$$x_{\text{bd}} = -\frac{C_{\text{pf}}(\Delta T_{\text{SC}})_{\text{bd}}}{i_{\text{fg}}} \tag{4.4.11}$$

The equilibrium vapor quality x is defined as

$$x = \frac{i_f - i_{f,\text{SAT}}}{i_{\text{fg}}} \tag{4.4.12}$$

where i_f is the actual enthalpy and $i_{f,\text{SAT}}$ is the enthalpy of saturated liquid. Under subcooled conditions, x is negative.

The subcooling at bubble departure point may be calculated by a suitable correlation such as those presented in Section 4.3.

The Levy model assumes zero quality and void fraction in the partial boiling regime. These assumptions are not quite correct. Manon (2000) has given the following expression to calculate actual quality in both partial and FDB regimes:

$$x' = x + [x'_{\text{bd}} - x_{\text{bd}}]\exp\left[\frac{x}{x_{\text{bd}}} - 1\right] \tag{4.4.13}$$

Figure 4.4.1 Void fraction predictions of the Rouhani–Axelsson model compared with data of Rouhani (1966) for water in a tube. θ_{in} is inlet subcooling, and θ_o is local subcooling. Source: From Rouhani and Axelsson (1970). © 1970 Elsevier.

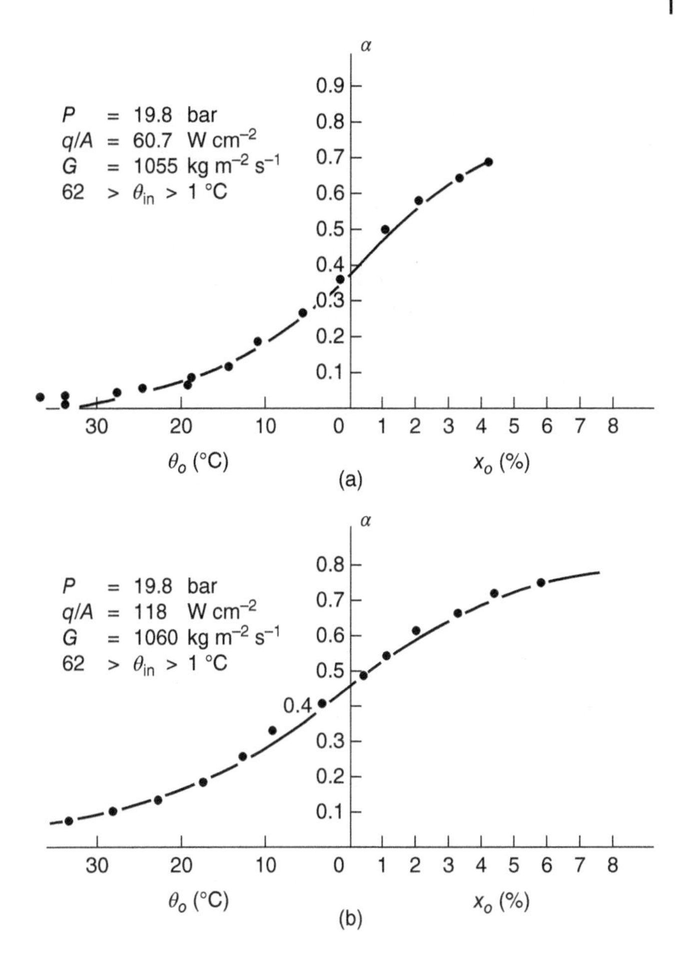

The actual quality at bubble departure x'_{bd} is calculated by

$$x'_{bd} = 0.5\frac{\rho_g}{\rho_f}\alpha_{bd} \tag{4.4.14}$$

4.4.1 Recommendations

There has been no comprehensive evaluation of the equations give earlier, nor of various other formulas that have been proposed. The author suggests the following as these are simple and have had at least as much verification as other methods.

Void fraction at bubble departure point: Levy model, Eq. (4.4.4).

Void fraction FDB regime: Rouhani and Axelsson correlation.

4.5 Heat Transfer in Channels

4.5.1 Visual Observations and Mechanisms

The detailed heat transfer measurements of McAdams et al. (1949) in an annulus were accompanied with visual observations and photographic studies. They observed that an increase in water velocity decreased the number and size of bubbles for a given heat flux. Increase in subcooling had a similar effect. At deep subcooling, cyclic formation and condensation of bubbles at the heater surface was observed. At one instant, numerous bubbles were seen at the surface, and 0.007 seconds later, no bubbles were visible. The presence of dissolved gases made a distinct difference in appearance of surface boiling. The stream of water was cloudy, and many small bubbles were carried out in the water leaving the test section. As seen in Figure 4.2.1, heat transfer with dissolved air was higher than when it was absent.

Rohsenow and Clark (1951b) carefully examined the high-speed motion pictures taken by McAdams et al. (1949) in an attempt to understand the mechanism of heat transport in subcooled boiling. They counted the number of bubbles formed, measured their diameters, and then estimated the amount of heat carried by the bubbles. They found that the amount of heat given up by the bubbles during condensation in the liquid core amounted to less than 2% of the increase in heat transfer during subcooled boiling as compared with single-phase heat transfer. They concluded that latent heat transfer by bubbles has insignificant effect on heat transfer. The enhancement is primarily due to the agitation of the boundary layer caused by the departing bubbles. As the bubble leaves the surface, surrounding cold liquid rushes in. These account for the cyclic bubble formation found in this and other studies.

Gunther (1951) studied boiling in a horizontal rectangular channel made of glass. A horizontal electric heater was placed along the axis of the channel such that it divided the channel into two parts. Water at atmospheric pressure flowed through the channel at velocity of 1.5–12 m s^{-1}. It was seen that at subcooling of 50 K, the bubbles grew and collapsed attached to the heater surface. The bubbles grew while sliding downstream attached to the heater. The sliding velocity was about 80% of the stream velocity. Size of the bubbles was found to decrease with increasing subcooling. At a heat flux of 4.5 MW m^{-2} and a velocity of 3.05 m s^{-1}, the average maximum bubble size decreased from about 0.5 to 0.1 mm as subcooling increased from 33 to 111 K. At the same time, the fraction of surface covered by bubbles decreased from 38% to virtually zero. The bubble life time was also found to decrease with increasing subcooling, from 600 to 110 μs for this run. The bubble population initially remained unchanged with increasing subcooling at about 1.3×10^5 bubbles cm^{-2}. At higher subcooling, the surface became covered with a large number of very small bubbles with a population density of 3×10^5 bubbles cm^{-2}.

Hysteresis often occurs in forced convection subcooled boiling as it does in pool boiling. Heat transfer coefficients with decreasing heat flux are higher than with increasing heat flux. The reason is that with decreasing heat flux, the active nucleation sites continue to be active even at lower heat flux. Hodgson (1968) observed considerable hysteresis during his tests on boiling of water in a tube.

4.5.2 Prediction of Heat Transfer

4.5.2.1 Some Dimensional Correlations

Numerous methods for prediction of heat transfer in channels have been proposed. The early methods were dimensional correlations applicable to a particular fluid in a limited range and did not take into account the physical phenomena. One such correlation is that of Jens and Lottes (1951) for water:

$$\Delta T_{SAT} = 25q^{0.25}e^{-p/62} \tag{4.5.1}$$

where p is in bar, ΔT_{SAT} is in °C, and q in mW m^{-2}. It was based on data from tubes of diameter 3.6–5.7 mm that covered pressures 7–172 bar and mass velocities 11–10 500 kg m^{-2} s^{-1}.

Another well-known correlation for water is that of Thom et al. (1965) who found that the Jens and Lottes correlation overpredicted their measured heat transfer coefficients. They modified it to the following equation:

$$\Delta T_{SAT} = 22.65q^{0.5}e^{-p/87} \tag{4.5.2}$$

This correlation was validated with data for pressures 7–172 bar.

4.5.2.2 The Shah Correlation

Shah (1977) Correlation Shah (1977) gave a general correlation that was shown to agree with a wide range of data for tubes and annuli. It is as follows.

For saturated boiling at zero quality, ψ_0 is the larger of that given by the following two equations:

$$\psi_0 = (h_{TP}/h_{LT})_{x=0} = 230Bo^{0.5} \tag{4.5.3}$$

$$\psi_0 = 1 + 46Bo^{0.5} \tag{4.5.4}$$

Equation (4.5.3) was first given by Shah (1976) as part of his correlation for saturated boiling in tubes. Equation (4.5.4) was added to avoid $\psi_0 < 1$ at low Bo. Bo is the boiling number defined as

$$Bo = \frac{q}{Gi_{fg}} \tag{4.5.5}$$

h_{LT} is the heat transfer coefficient for all mass flowing as liquid. It is calculated by the following equation generally known as Dittus–Boelter equation:

$$h_{LT} = 0.023Re_{LT}^{0.8}Pr_f^{0.4}\frac{k_f}{D} \tag{4.5.6}$$

For low subcooling regime (FDB regime), Eq. (4.5.3) applies. In the high subcooling regime (partial boiling regime), it is postulated that total heat flux is the sum of heat flux removed by single-phase convection q_{spc} and that removed by nucleate boiling q_{nb}. Thus,

$$q = q_{spc} + q_{nb} \tag{4.5.7}$$

The definition of ψ_0 yields the following relation:

$$q = h_{LT}(T_w - T_{SAT}) + h_{LT}(\psi_0 - 1)(T_w - T_{SAT}) \tag{4.5.8}$$

Comparing Eq. (4.5.7) with Eq. (4.5.8), it is seen that

$$q_{nb} = h_{LT}(\psi_0 - 1)(T_w - T_{SAT}) \tag{4.5.9}$$

Shah postulated that Eq. (4.5.9) continues to hold in the high subcooling region and q_{spc} can be calculated by

$$q_{spc} = h_{LT}(T_w - T_f) \tag{4.5.10}$$

Substituting Eqs. (4.5.9) and (4.5.10) into Eq. (4.5.7),

$$q = h_{LT}(T_w - T_f) + h_{LT}(\psi_0 - 1)(T_w - T_{SAT}) \tag{4.5.11}$$

This is rearranged to the following form to give the correlation for the high subcooling regime:

$$\psi = \frac{q}{\Delta T_{SAT} h_{LT}} = \psi_0 + \frac{\Delta T_{SC}}{\Delta T_{SAT}} \tag{4.5.12}$$

Further rearrangement gives the correlation in the following form that gives ΔT_{SAT} explicitly:

In low subcooling regime,

$$\Delta T_{SAT} = q/(h_{LT}\psi_0) \tag{4.5.13}$$

In high subcooling regime,

$$\Delta T_{SAT} = \left(\frac{q}{h_{LT}} - \Delta T_{SC}\right)/\psi_0 \tag{4.5.14}$$

The correlation is shown graphically in Figure 4.5.1. The boundary between high and low subcooling regimes is determined by Eqs. (4.3.5) and (4.3.6), which are shown graphically in Figure 4.3.1.

This correlation was verified with a wide range of data for tubes and annuli. For annuli, D_{HYD} was used as the equivalent diameter if annular gap was greater than 4 mm. For annular gaps ≤4 mm, D_{HP} was used. The hydraulic equivalent diameter is defined as

$$D_{HYD} = (4 \times \text{flow area})/(\text{wetted perimeter}) \tag{4.5.15}$$

D_{HP} is defined as

$$D_{HP} = (4 \times \text{flow area})/(\text{heated perimeter}) \tag{4.5.16}$$

Shah (2017a) Correlation Shah (2017a) reanalyzed the data used in Shah (1977) together with additional data that included minichannels. While the 1977 correlation gave good agreement with all data, he found that better agreement results if a few changes were made as described in the following:

Firstly, the following dimensional equation is used for the high subcooling regime:

$$\Delta T_{SAT} = \frac{0.67q}{\psi_0 h_{LT}} + 1.65(\Delta T_{SC})^{-0.44} \tag{4.5.17}$$

ΔT is in K and q in W m^{-2}.

Secondly, deviations are reduced if D_{HYD} is used as equivalent diameter for annular gaps greater than 3 mm and D_{HP} for gaps ≤3 mm.

Thirdly, the Saha–Zuber correlation, Eqs. (4.3.1) and (4.3.2), was used for determining the boundary between low and high subcooling regimes. The reduction in deviation by its use was small, but it avoids the iterations in using the Shah (1977) method, Eqs. (4.3.5) and (4.3.6).

This correlation was compared to an extensive database whose range is listed in Table 4.5.1. There were 1340 data points from 68 data sets from 37 sources and included many fluids over an extreme range of parameters. The same database was also compared to a number of other correlations. The results are summarized in Table 4.5.2. It is seen that the Shah (2017a) predicted these data with mean absolute deviation (MAD) of 12.2%, with 94.7% of data within ±30%. All other correlations had larger deviations. The deviations are defined as follows:

Mean Absolute Deviation

$$= \frac{1}{N}\sum_{1}^{N} \text{ABS}\left[\frac{h_{TP,predicted} - h_{TP,measured}}{h_{TP,measured}}\right] \tag{4.5.18}$$

Average Deviation

$$= \frac{1}{N}\sum_{1}^{N} \left[\frac{h_{TP,predicted} - h_{TP,measured}}{h_{TP,measured}}\right] \tag{4.5.19}$$

Heat transfer coefficient h_{TP} is defined as

$$h_{TP} = \frac{q}{(T_w - T_f)} = \frac{q}{(\Delta T_{SAT} + \Delta T_{SC})} \tag{4.5.20}$$

Figures 4.5.2–4.5.5 show comparison of data with the Shah correlation as well as several other correlations.

An important point to discuss is the applicability to minichannels whose range is usually considered to be $D_{HYD} \leq 3$ mm. Surface tension forces are said to affect heat transfer in minichannels. The Shah correlation does not have any factor for the effect of surface tension. The data analyzed included diameters as small as 0.176 mm. As seen in Table 4.5.2, data for minichannels were in good agreement with the Shah correlation, and the deviations were no greater than for larger channels. Figures 4.5.3–4.5.5 also show good agreement with data for minichannels. Hence it appears that the boundary where surface tension begins to have effect has not been reached in these data. Shah (2017b) on analysis of data for saturated boiling found that surface tension effects occur when F given by the following equation is greater than 1:

$$F = 2.1 - 0.008We_{GT} - 110Bo \tag{4.5.21}$$

Shah (2017a) has not stated whether he compared the subcooled boiling data with this equation. Looking at the range of Bo and We_{GT} in Table 4.5.1, it seems very unlikely that F would have been greater than one for any of these data. Hence it remains unknown whether Eq. (4.5.21) is applicable to subcooled boiling.

While the Shah correlation was compared to data for 13 diverse fluids, these did not include carbon dioxide. In Shah (2015), data for saturated boiling of CO_2 were compared to Shah (1982) correlation for saturated boiling. It was found that most data gave much better agreement if ψ_0 was calculated by the following equation:

$$\psi_0 = 1820\, Bo^{0.68} \tag{4.5.22}$$

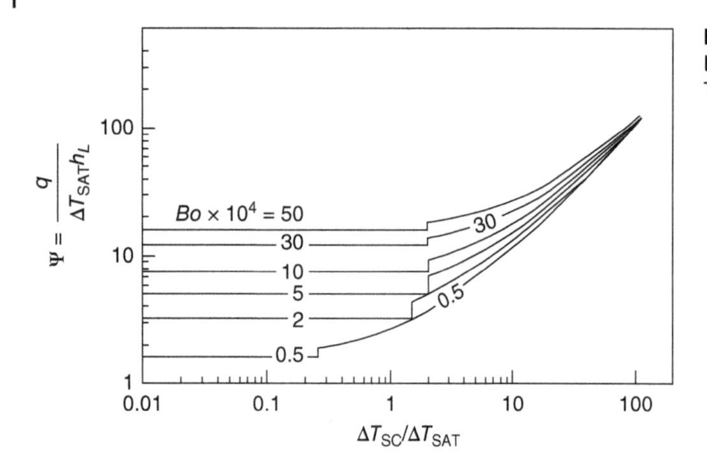

Figure 4.5.1 The correlation of Shah (1977) for subcooled boiling in graphical form. Source: From Shah (1983). © 1983 Taylor & Francis.

Table 4.5.1 Range of data for which Shah (2017a) correlation for subcooled boiling in channels was verified.

	Range
Fluids	Water, ammonia, R-11, R-12, R-113, R-123, R-134a, FC-72, HFE-7100, isopropanol, hexane, cyclohexane, and methanol
Geometry	Single round tubes, single rectangular channel, rectangular multichannels, annuli heated on inner/outer/both tubes, and horizontal and vertical orientations
Tube material	Various stainless steels, copper, brass, zirconium–copper alloy, nickel, Inconel, and glass
D_{HYD} (mm)	0.176–22.8
Annular gap (mm)	0.5–11.4
Aspect ratio of rectangular channels	0.044–0.5
Reduced pressure, p_r	0.0046–0.922
G (kg m^{-2} s^{-1})	59–31 500
Subcooling (°C)	0–165
Re_{LT}	375–1 270 000
$Bo \times 10^4$	0.53–91.2
Bond number	0.025–7100
We_{GT}	158–11 383 366
Peclet number	631–1 112 687
Data points	1340 data points from 68 data sets from 37 sources

Source: Reprinted from Shah (2017a). © 2017, with permission from Elsevier.

Shah (2017a) therefore recommended that this equation replace Eqs. (4.5.3) and (4.5.4) for use with carbon dioxide.

Data for annuli included annular gaps of 0.5–11.4 mm and all possible modes of heating, which are as follows:

1) Only inner tube heated
2) Only outer tube heated
3) Both tubes heated, boiling on inner tube only
4) Both tubes heated, boiling on outer tube only
5) Both tubes heated, boiling on both tubes.

Data over the entire range and all these heating modes were in satisfactory agreement with the Shah correlation. Figure 4.5.5 shows an example of annulus data.

The Dittus–Boelter equation, Eq. (4.5.6), is used at all values of Reynolds number that varied from 375 to 1 270 000 in the data analyzed. Figure 4.5.5 shows good agreement of Shah correlation with data at $Re_{LT} = 375$. Further, all properties are calculated at the bulk liquid temperature, T_f.

Summarizing, the Shah (2017a) correlation is

High subcooling regime: Eq. (4.5.17)
Low subcooling regime: Eq. (4.5.13)

4.5.2.3 Various Correlations

Gungor and Winterton (1986) gave the following correlation:

$$q = h_{LT}(T_w - T_f) + Sh_{pb}(T_w - T_{SAT}) \quad (4.5.23)$$

Table 4.5.2 Results of evaluation of various correlations for subcooled boiling in channels.

Type		No. of data points	Mean absolute deviation (MAD, %) Average deviation (%)					
			Liu and Winterton	Gungor and Winterton (1986)	Haynes and Fletcher	Moles and Shaw	Shah (1977)	Shah (2017a)
Channels (round tubes, rect. single channel, multichannels)	$D_{HYD} \leq 3\,\text{mm}$	265	24.4	62.3	12.8	71.0	15.5	11.5
			−9.2	−37.9	−0.6	68.5	−2.6	−2.7
	$D_{HYD} > 3\,\text{mm}$	518	15.9	43.3	20.0	19.6	13.6	13.3
			−7.4	−42.5	13.0	8.8	−2.5	−1.3
	Multichannels	77	29.5	60.4	23.1	124.7	16.5	14.4
			22.5	−20.0	17.5	124.3	10.6	5.8
	Single channels	706	17.3	48.6	17.8	47.9	14.0	12.5
			−8.9	−43.2	9.4	−42.4	−4.0	−2.6
	All channels	783	18.8	49.7	17.5	37.1	14.2	12.7
			−8.1	−40.9	8.4	29.2	−2.6	−1.8
Annuli	All annuli	557	12.2	53.6	19.7	26.5	13.8	11.4
			−0.8	−51.1	6.4	0.1	−3.6	−3.7
All types	All data	1340	15.9	51.4	18.9	29.6	14.1	12.2
			−3.7	−45.5	12.8	13.7	−3.0	−2.6

Source: Reprinted from Shah (2017a). © 2017, with permission from Elsevier.

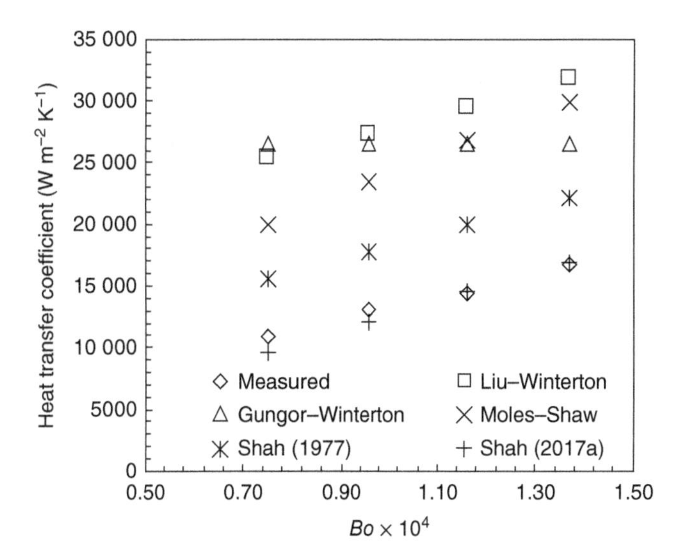

Figure 4.5.2 Data of Peng and Wang (1993) for a multichannel with $D_{HYD} = 0.646$ mm compared to various correlations: $T_{SAT} = 100$ °C, $G = 3237$ kg m^{-2} s^{-1}, subcooling 50 K. Source: Reprinted from Shah (2017a). © 2017, with permission from Elsevier.

S is the suppression factor given by

$$S = (1 + 1.15 \times 10^{-6} Re_{LT}^{1.17})^{-1} \tag{4.5.24}$$

h_{LT} is calculated with Eq. (4.5.6). h_{pb} is the pool boiling heat transfer coefficient calculated by the simplified Cooper correlation, Eq. (3.2.41). For annuli, D_{HP} is used as the equivalent diameter. This correlation was verified with data for water and refrigerants from six sources.

Liu and Winterton (1991) gave the following correlation:

$$q = \sqrt{(h_{LT}(T_w - T_f))^2 + (Sh_{pb}(T_w - T_{SAT}))^2} \tag{4.5.25}$$

$$S = (1 + 0.055 Re_{LT}^{0.16})^{-1} \tag{4.5.26}$$

Calculation of h_{LT} and h_{pb} is done the same way as for Gungor and Winterton correlation. For annuli, D_{HP} is used as the equivalent diameter. This correlation was shown to give

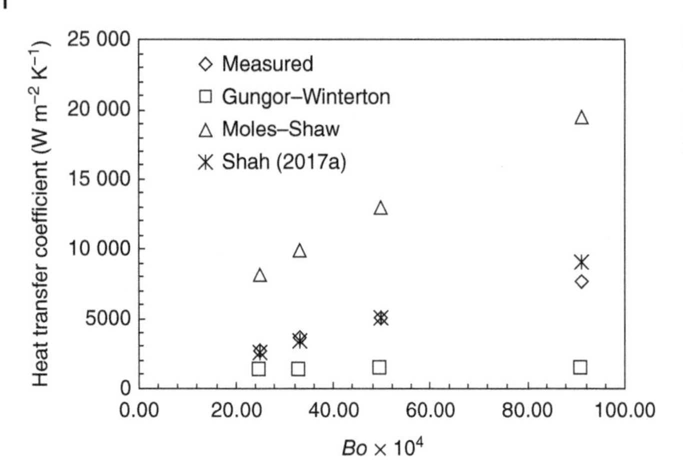

Figure 4.5.3 Data of Lee and Mudawar (2008) for HFE-7100 in a rectangular multichannel with $D_{\text{HYD}} = 0.416$ mm compared to various correlations: $T_{\text{sat}} = 63.6$ °C, $G = 1010$ kg m^{-2} s^{-1}, subcooling 89–93 °C, and $Re_{\text{LT}} = 1300$. Source: Reprinted from Shah (2017a). © 2017, with permission from Elsevier.

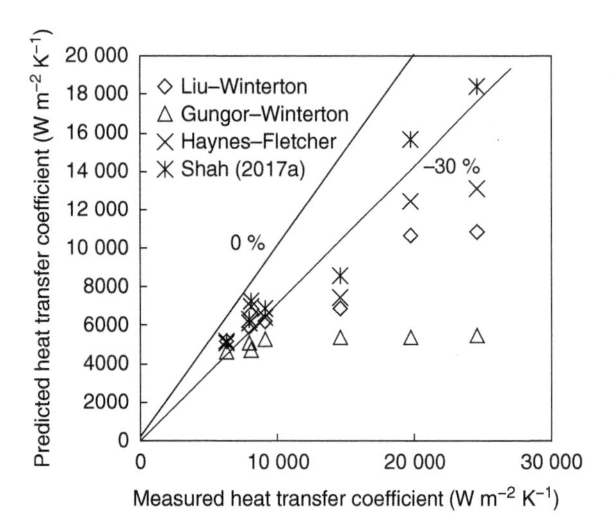

Figure 4.5.4 Data of Qu and Mudawar (2003) for water in a rectangular channel with $D_{\text{HYD}} = 0.35$ mm and $Re_{\text{LT}} = 375$ compared to various correlations. Source: Reprinted from Shah (2017a). © 2017, with permission from Elsevier.

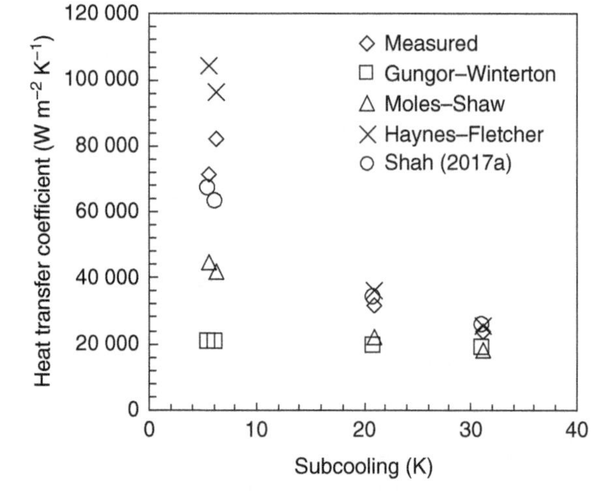

Figure 4.5.5 Data of Alferov and Rybin (1969) for water in an annulus compared to various correlations: inner tube heated, annular gap = 5 mm, $T_{\text{SAT}} = 340.2$ °C, $G = 1870$ kg m^{-2} s^{-1}, and $Bo = 4.25 \times 10^{-4}$. Source: Reprinted from Shah (2017a). © 2017, with permission from Elsevier.

good agreement with data from nine sources that included water, refrigerants, and ethanol.

Moles and Shaw (1972) gave the following correlation based on their own data for water, ammonia, and several chemicals:

$$q = h_{\text{LT}}(T_w - T_f)Pr_f^{0.46}Bo^{0.67}\left[\frac{i_{\text{fg}}}{C_{\text{pf}}(T_w - T_f)}\right]^{0.5}\left(\frac{\rho_g}{\rho_f}\right)^{0.7} \tag{4.5.27}$$

Fluid properties are calculated at saturation temperature, while Prandtl number is calculated at film temperature.

Haynes and Fletcher (2003) gave the following correlation of their own data for R-11 and R-123 in 0.92 and 1.95 mm diameter tubes:

$$q = h_L(T_w - T_f) + h_{\text{pb}}(T_w - T_{\text{SAT}}) \tag{4.5.28}$$

Note that none of the above correlations take into consideration regimes of boiling and apply them to both high and low subcooling regimes. This may be the reason for their large deviations from data seen in Figures 4.5.3–4.5.5, 4.5.6. Among these correlations, that of Liu and Winterton gives the least MAD of 15.9% though it is considerably larger than that of Shah (2017a) correlation at 12.2%.

Kandlikar (1998) has given a correlation that takes into consideration the regimes of subcooled boiling. For FDB, the following equation is used:

$$q = [1058(Gi_{\text{fg}})^{-0.7}F_{\text{fl}}h_{\text{LT}}\Delta T_{\text{SAT}}]^{1/0.3} \tag{4.5.29}$$

In partial boiling regime, the correlation is

$$q = a + b(T_w - T_{\text{SAT}})^m \tag{4.5.30}$$

$$a = q_{\text{ib}} + b(\Delta T_{\text{SAT,ib}})^m \tag{4.5.31}$$

$$b = \frac{q_{FDB} - q_{ib}}{(\Delta T_{SAT,FDB})^m - (\Delta T_{SAT,ib})^m} \tag{4.5.32}$$

$$m = n + \xi q \tag{4.5.33}$$

$$\xi = \left(\frac{1}{0.3} - 1\right) / (q_{FDB} - q_{ib}) \tag{4.5.34}$$

$$n = 1 - \xi q_{ib} \tag{4.5.35}$$

In the above equations, the subscript "ib" means inception of boiling and "FDB" means fully developed boiling. Equation (4.3.4) gives q_{FDB}. F_{fl} is a factor that depends on the type of fluid and tube material. Its values are given in Chapter 5. This correlation was shown to agree with data for water and refrigerants from four sources. For calculation of h_{LT}, Kandlikar specifies the use of the correlations of Pethukov and Gnielinski. These are given in Chapter 1.

4.5.2.4 Recommendations

The Shah (2017a) correlation is the most verified method for prediction of heat transfer, and it takes physical phenomena into consideration. It is recommended for use with all non-metallic fluids over its verified range, which includes reduced pressures from 0.0046 to 0.92 and channel diameters down to 0.176 mm. Note that all fluid properties are calculated at the bulk fluid temperature.

4.6 Single Cylinder with Crossflow

Many heat exchangers, notably kettle reboilers and shell and tube evaporators, involve crossflow over tube bundles. The liquid entering the heat exchangers is usually subcooled. The rational method for designing such heat exchangers is to calculate heat transfer coefficients on a tube-by-tube basis. Computational models for tube-by-tube designs have been developed, for example, the one by Brisbane et al. (1980). To get accurate results with such models, reliable methods for predicting heat transfer to individual tubes under subcooled and saturated flow conditions are needed. With the objective of developing such methods, many experimental studies have been done to study subcooled boiling on single tubes, and a number of methods for prediction of heat transfer have been proposed. These are discussed in the following.

4.6.1 Experimental Studies

McKee (1967) made visual observations and high-speed motion pictures of water at atmospheric pressure flowing across electrically heated horizontal cylinders. Bubbles first appeared on the downstream face (wake region) of the cylinder at a few sites. With increase in heat flux, more sites were activated, and boiling spread throughout the separated flow region and was initiated in the forward section. The distribution of nucleation sites in the forward section was greatest in the vicinity of the stagnation point. Bubbles appeared to be spherical. Measured bubble diameters were 0.63–1.55 mm.

Vliet (1962) also studied boiling of water on horizontal cylinders. He observed that at low subcooling and high velocity, nucleation first occurred on the sides of the cylinder, then spread to the rear side, and then finally on the front side. At high subcooling and high velocities, nucleation first started on the rear side and then gradually spread to the sides and finally to the front. Figure 4.6.1 shows some of his data.

Among other studies are those of Fand et al. (1976), Yilmaz and Westwater (1979), Lemmert and Chawla (1977), Fink et al. (1982), and Bitter (1973).

4.6.2 Prediction of Heat Transfer

Many correlations have been proposed. Among these, the most verified is that by Shah. It is discussed first.

4.6.2.1 Shah Correlation

Shah (1984) developed a correlation for subcooled boiling heat transfer on cylinders that is similar to his correlation for tubes presented in Section 4.5.2.2. Firstly, data for zero vapor quality were analyzed as shown in Figure 4.6.2. This yielded the following equations:

$$\psi_0 = (h_{TP}/h_{LT})_{x=0} = 443 Bo^{0.65} \tag{4.6.1}$$

$$\psi_0 = 19 Bo^{0.27} \tag{4.6.2}$$

The larger of the value of ψ_0 given by these equations is used. The transition between these two occurs at $Bo = 2.5 \times 10^{-4}$

Two regimes of heat transfer were identified, namely, high subcooling and low subcooling regimes. Their boundary is shown in Figure 4.6.3. Mathematically, the regime is high subcooling if either of the following two conditions is satisfied:

$$\Delta T_{SC}/\Delta T_{SAT} \geq 4 \tag{4.6.3}$$

$$\frac{\Delta T_{SC}}{\Delta T_{SAT}} > 7.63 \times 10^4\, Bo^{1.31} \tag{4.6.4}$$

Otherwise, it is low subcooling regime.

In the low subcooling regime, Eqs. (4.6.1) and (4.6.2) apply. In the high subcooling regime, Eqs. (4.5.7)–(4.5.12) apply. Thus, for the high subcooling regime,

$$\Delta T_{SAT} = \left(\frac{q}{h_{LT}} - \Delta T_{SC}\right) / \psi_0 \tag{4.6.5}$$

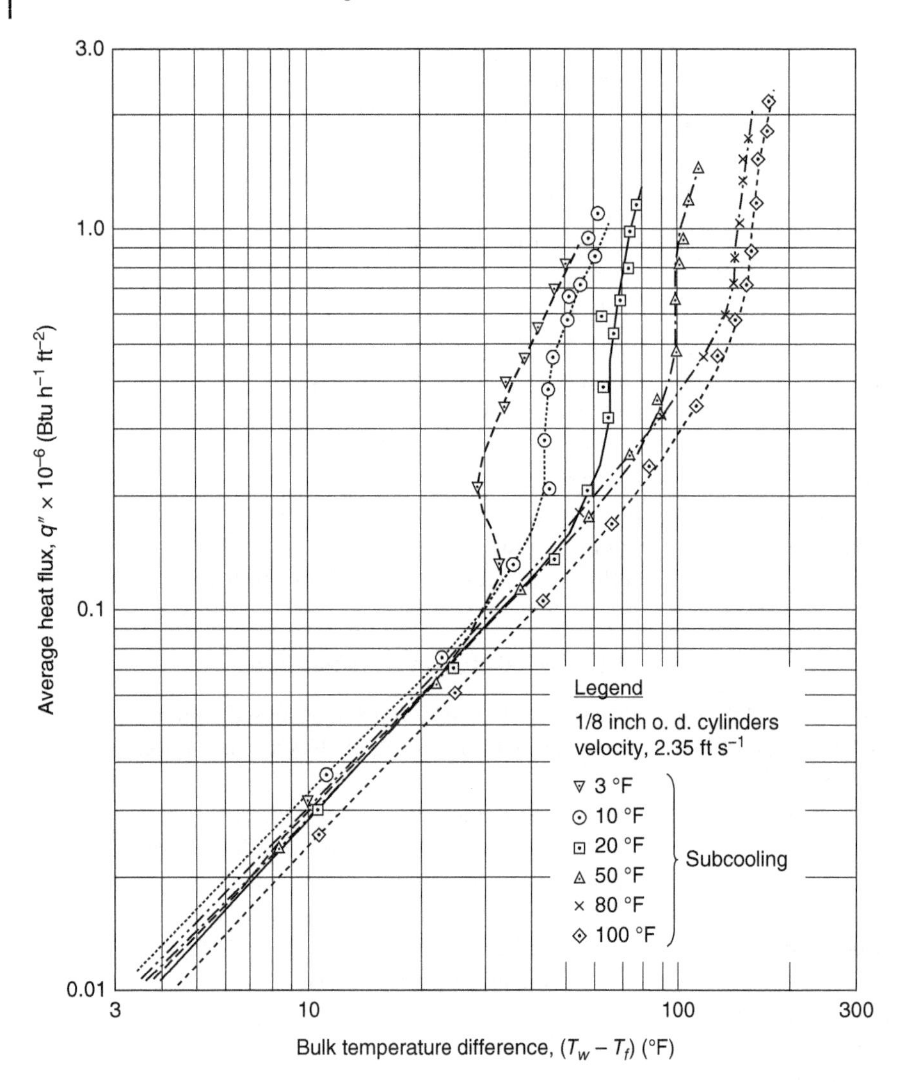

Figure 4.6.1 Effect of subcooling during crossflow boiling on a cylinder. Source: From Vliet (1962).

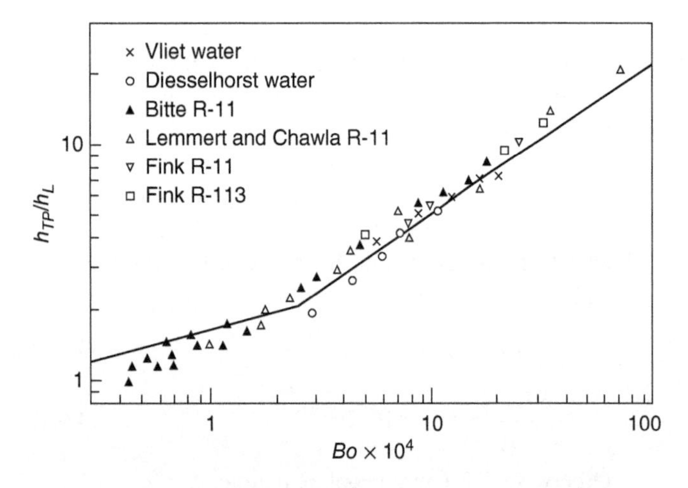

Figure 4.6.2 Correlation of heat transfer during crossflow at zero quality. Source: Reprinted from Shah (1984). © 1984, with permission from Elsevier.

The single-phase heat transfer coefficient h_{LT} is calculated by the following equation:

$$\frac{h_{LT}D}{k_f} = 0.21 Re_{LT}^{0.62} Pr_f^{0.4} \tag{4.6.6}$$

The mass velocity used in Re_{LT} is based on the narrowest gap between the tube and its enclosure or between adjacent tubes. The same velocity is also used in calculating Bo.

This correlation showed good agreement with data for water and refrigerants from 11 sources.

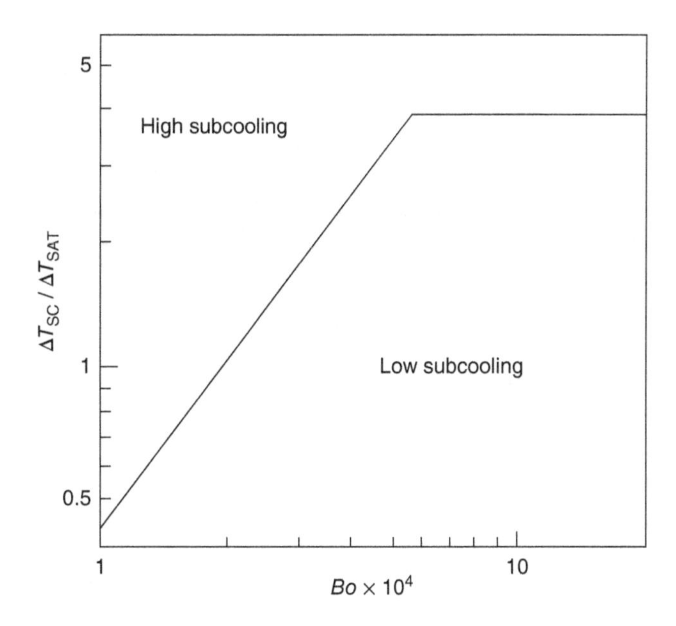

Figure 4.6.3 The boundary between high and low subcooling in crossflow boiling. Source: Reprinted from Shah (1984). © 1984, with permission from Elsevier.

Shah (2005) further developed this correlation. Three changes were made. Firstly, Eqs. (4.6.1) and (4.6.2) were replaced by the following one equation:

$$\psi_0 = 1 + \frac{Bo}{0.000\,216 + 0.041Bo - 1.53Bo^2} \tag{4.6.7}$$

This equation gives results close to those from Eqs. (4.6.1) and (4.6.2). Its use avoids the discontinuity at $Bo = 2.5 \times 10^{-4}$. At very low Bo, which is unlikely to occur in practice, it predicts $\psi_0 < 1$. In such a situation, $\psi_0 = 1$ is used.

The second change was the method for determining subcooling regime. Use of Eqs. (4.6.3) and (4.6.4) to determine it requires iterative calculations. To avoid it, Shah developed the following method similar to that of Saha and Zuber for tubes.

The regime is high subcooling if either of the following two conditions is met; otherwise it is low subcooling regime:

$$\frac{q}{GC_{pf}\Delta T_{SC}} < 38Pe^{-0.38} \tag{4.6.8}$$

$$Bo < 2.5 \times 10^{-4} \tag{4.6.9}$$

Otherwise it is low subcooling. The previously mentioned two changes make the calculations easier but have little effect on predictions. The third change is that Eq. (4.6.6) is used only for $Re_{LT} > 700$. For $Re_{LT} < 700$, the following equation given by Holman (1968) is used:

$$\frac{h_{LT}D}{k_f} = 0.615Re_{LT}^{0.466}Pr_f^{1/3} \tag{4.6.10}$$

Table 4.6.1 The range of data with which the Shah (2005) correlation for subcooled boiling in crossflow was validated.

	Range
Geometry	Single tube alone and tubes in bundles
Fluids	Water, R-11, R-12, and R-113
Tube diameter (mm)	1.2–26.4
(Pitch/diameter) in bundles	1.2–3.0
Reduced pressure	0.005–0.146
Upstream liquid velocity (m s^{-1})	0.001–6.9
Subcooling (K)	0–93
Re_{LT}	56–260 000
$Bo \times 10^4$	0.7–26
Number of data sources	29

Source: Based on Shah (2005).

All properties are calculated at saturation temperature.

This correlation was compared to data for single tubes as well as those in tube bundles. For single tubes, 638 data points from 22 sources were predicted with a mean deviation of 12.1%. For the tube bundles, 77 data points from 7 sources were predicted with a mean deviation of 16.0%. Only 36 of the data points had deviation greater than 30%, 19 of these being from one source. Table 4.6.1 gives the range of data used for validation of this correlation.

4.6.2.2 Other Correlations

Many researchers have attempted correlation of their data by one of the following two equations:

$$q = Fh_{LT}(T_w - T_f) + Sh_{pb}(T_w - T_{SAT}) \tag{4.6.11}$$

$$h_{TP}^m = h_{LT}^m + h_{pb}^m \tag{4.6.12}$$

Equation (4.6.11) is based on Chen's correlation for boiling inside tubes (Chen 1966). F is a two-phase convective enhancement factor and S is a suppression factor for nucleate boiling. Using their own data, these researchers derived values of S and F to fit those data. Among those who have taken this approach are Wege and Jensen (1984), Jensen and Hsu (1988), Hwang and Yao (1986), Huang and Witte (2001), and Polley et al. (1980). The values of S and F found by each of these researchers differ from those found by the others. Wege and Jensen (1984) showed that their correlation also agrees with the data of Polley et al. (1980), while the others have not compared their correlations with any data except their own. Hence, none of these five correlations is well verified.

Equation (4.6.12) was proposed by Kutateladze (1961) with $m = 2$. Rohsenow (1953) proposed $m = 1$. The authors who have attempted to correlate their data using Eq. (4.6.12) include Singh et al. (1983), Hwang and Yao (1984), Gupta et al. (1995), Fink et al. (1982), Lee and Shigechi (1991), Singh et al. (1983), Leppert et al. (1958), and Fand et al. (1976). The values of m found by these authors varied from 0.69 to 5.5. None of them tried to correlate any data except their own. Hence, none of these correlations of the form of Eq. (4.6.12) is well verified.

Lemmert and Chawla (1977) and McKee (1967) have taken another approach. They attempted to distinguish between areas with bubble nucleation and those without bubble nucleation; pool boiling is supposed to occur on the former and single-phase convection on the latter. Thus,

$$qA = q_{pb}A_{nb} + q_{spc}(A - A_{nb}) \tag{4.6.13}$$

A is the total area and A_{nb} is the part of it with nucleate boiling. These authors gave empirical correlations for A_{nb} using only their own data. Their general applicability is therefore unknown.

It is thus clear that none of these correlations has been sufficiently verified to be able to confidently use it for design.

4.6.3 Recommendation

Shah (2005) is the only well-verified correlation available and is therefore recommended for general use. Note that fluid properties are calculated at the saturation temperature.

4.7 Various Geometries

4.7.1 Tube Bundles with Axial Flow

A notable example of vertical tube bundle is the steam generator of PWR. Reactor coolant water flows through vertical tubes enclosed in a shell. Subcooled water enters the shell at the bottom and leaves as steam form the top. Thus, a considerable part of the steam generator has subcooled boiling. Newer PWR also use subcooled boiling. Heat transfer during subcooled boiling can be calculated by correlations for channels given in Section 5.2.2. For single-phase heat transfer also, correlation for tubes can be used with hydraulic equivalent diameter.

4.7.2 Tube Bundles with Crossflow

Shell and tube evaporators used in refrigeration and kettle reboilers used in chemical processing involve crossflow over bundles of horizontal tubes enclosed in shells. Liquid often enters these heat exchangers in a subcooled state at the bottom. Rational design requires tube-by-tube calculations by computational models such as that of Brisbane et al. (1980). Such calculations can be done by the correlations for crossflow across cylinders presented in Section 4.6. As noted therein, the most verified among them is the Shah correlation. In fact, Shah (2005) compared it to data from seven sources for tubes in bundles with satisfactory agreement.

4.7.3 Flow Parallel to a Flat Plate

Lemmert and Chawla (1977) performed tests in which subcooled R-11 flowed parallel to a heated copper plate placed in the middle of a channel 250 mm × 250 mm. They tried to correlate their data with Eq. (4.6.2) with $m = 1$ and $m = 2$ as proposed by Rohsenow (1953) and Kutateladze (1961), respectively. Agreement was poor with both models. They then developed a correlation in the form of Eq. (4.6.13) as described in Section 4.6.2.2. Its general applicability is unknown as they did not compare it to any other data.

4.7.4 Helical Coils

Hardik and Prabhu (2017, 2018) performed tests on vertical helical coils with subcooled and saturated boiling of R-123 and water. Tube inside diameter varied from 5.5 to 10 mm and coil curvature (D_{coil}/D_{tube}) varied from 14.4 to 57.8. Pitch of all coils was 50 mm. Minimum inlet quality was −0.45. The heat transfer coefficient at the inner periphery was found to be the same as for straight tubes, while that at the outer tube was higher than for straight tubes. They attributed it to centrifugal forces pushing liquid toward the outer periphery. They compared their data for circumferentially averaged heat transfer coefficients to a number of correlations for straight tubes. Among them, the best agreement was with the Kandlikar formula for FDB, Eq. (4.5.29), applied to both partial and FDB regimes. For R-123, $F_{fl} = 1.3$ was assumed in calculations. The next best was the Shah (1977) correlation. Both tend to underpredict some of the data for the circumferentially averaged heat transfer coefficients. Hardik and Prabhu did not indicate any effect of coil curvature on accuracy of prediction of heat transfer.

Duchatelle et al. (1976) measured heat transfer to water in coils made of inconoloy heated by liquid sodium. The tube diameter was about 20 mm and the curvature ratio of coils was from 31 to 135. Water pressure varied from 45 to 175 bar. Mass flux ranged from 370 to 3500 kg m^{-2} s^{-1}, and heat flux from 310 to 1500 kW m^{-2}. The data were satisfactorily correlated by the Rohsenow superposition model, Eq. (4.6.2) with $m = 1$. Pool boiling heat transfer coefficient was calculated by the Rohsenow correlation with $C_{sf} = 0.013$. This correlation is discussed in Section 3.2.3.

Owhadi (1966) studied boiling of subcooled and saturated water in two helical coils with vertical axes. Inception of boiling first occurred at the top and bottom of circumference and then on the inner and outer locations. While data for subcooled boiling is shown in a couple of figures, he did not compare it to any prediction method.

There have not been many studies on subcooled boiling in coils. The available information indicates that correlations for straight tubes can be used to calculate circumferentially averaged heat transfer coefficients. Caution should be exercised in using for coil curvature $(D_{coil}/D_{tube}) < 14$ as this was the minimum value in the data of Hardik and Prabhu. Curvature effects may change heat transfer behavior as this ratio decreases.

4.7.5 Bends

Miropoloskiy and Pikus (1969) conducted tests on electrically heated bends of 45°–360° with water flowing through them. They realized that the wall was thinner than average at the outer radius and thicker than average at the inner radius. When voltage is applied across such a bend, heat flux is higher at the inner radius and lower at the outer radius. They therefore used special techniques to determine the local heat flux at the points where wall temperature was measured.

Figure 4.7.1 shows some of their data for a 90° bend. Heat flux at the inner surface was 1.36 times the average heat flux, while the heat flux at the outer surface was 0.7 times the average heat flux. At higher subcooling, outer wall temperature was considerably lower than that at the inner surface. The difference decreased as subcooling decreased. As subcooling approached zero, the heat transfer coefficients at the inner and outer surface came closer.

There is no verified method for calculating heat transfer in bends. From the experience with coils discussed in Section 4.7.4, it appears that the circumferentially averaged heat transfer coefficient in bends can be calculated by correlations for tubes.

4.7.6 Rotating Tube

Marto and Gray (1971) studied boiling on an electrically heated vertical tube rotating on its own axis. The tube inside diameter was 102 mm and it was 50.8 mm long. Water was sprayed on the heated tube surface where it formed an annular layer. Visual observations were made through plexiglass windows at the top of tube. Speeds up to 2660 rpm were used which amounts to $a/g = 400$ at the heater surface. Acceleration a is calculated by

$$a = \left(\frac{\pi r N}{60}\right)^2 / r \tag{4.7.1}$$

where r is the radius of tube and N is the rotational speed in rpm.

Visual observations showed that at low accelerations, there were many active nucleation sites on the boiling surface, giving rise to many large size bubbles in the boiling annulus. The fluid in the boiling annulus had a frothy appearance and the liquid–vapor interface of the annulus was quite irregular. At high accelerations, the number of active sites was sharply reduced and the bubbles were smaller. Also, secondary flow cells disrupted bubble nucleation.

Figure 4.7.2 shows the effect of subcooling and acceleration on heat transfer. It is seen that subcooling has no effect. The heat transfer coefficients at 200g acceleration are higher than that at 25g up to a wall superheat of 15 K. Figure 4.7.3 shows the effect of acceleration up to 400g. It

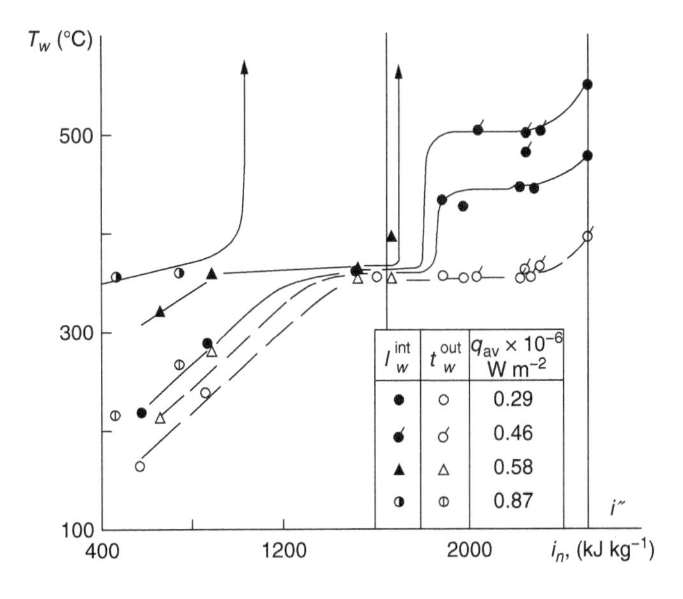

Figure 4.7.1 Wall temperature T_w at inner and outer surface of a 90° bend vs. fluid enthalpy i_n at average heat flux q_{av}: $G = 400$ kg m^{-2} s^{-1}, and $p = 167$ bar. Source: From Miropoloskiy and Pikus (1969).

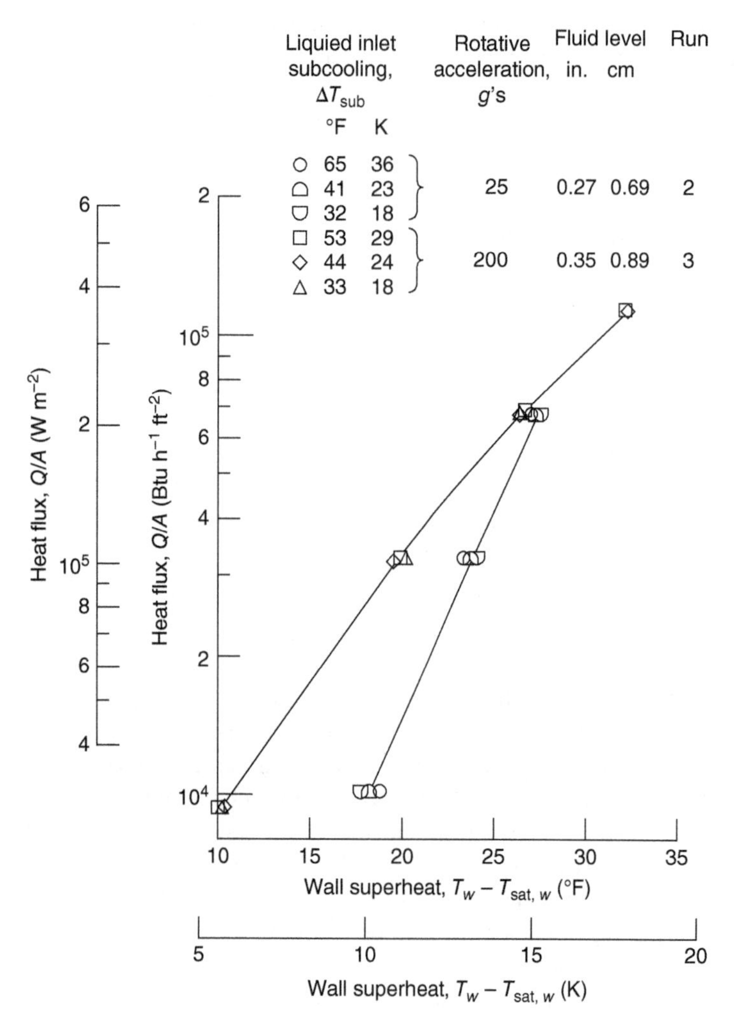

Figure 4.7.2 Effect of subcooling and acceleration on heat transfer to a rotating vertical tube with water at atmospheric pressure. $T_{sat,w}$ is the saturation temperature at boiler wall. Source: From Marto and Gray (1971).

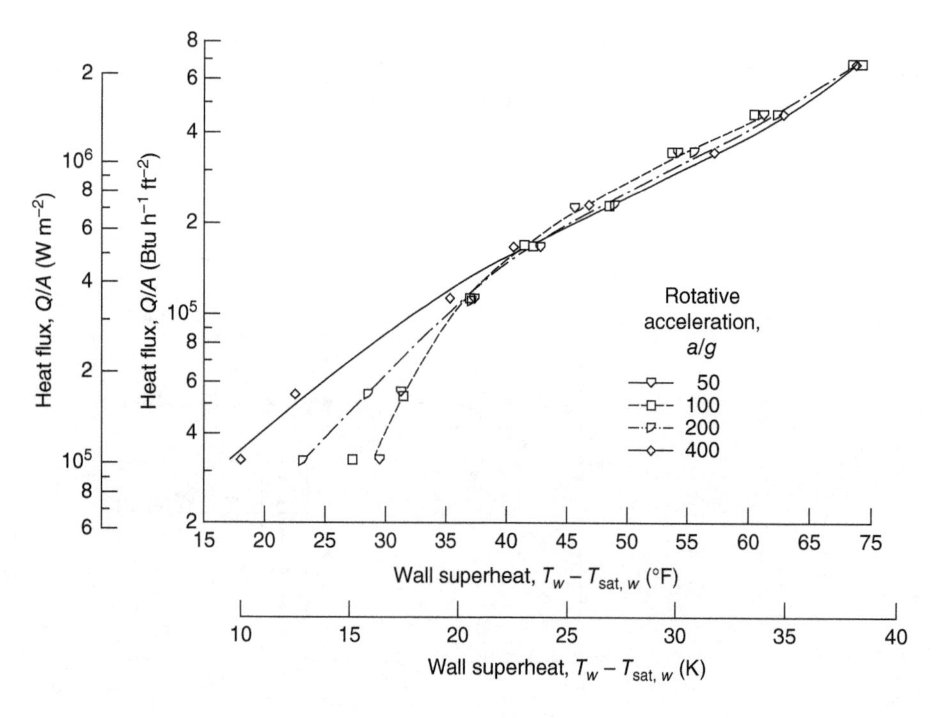

Figure 4.7.3 Effect of acceleration on heat transfer to a rotating vertical tube sprayed with water at atmospheric pressure, inlet subcooling 28 K. Source: Marto and Gray (1971).

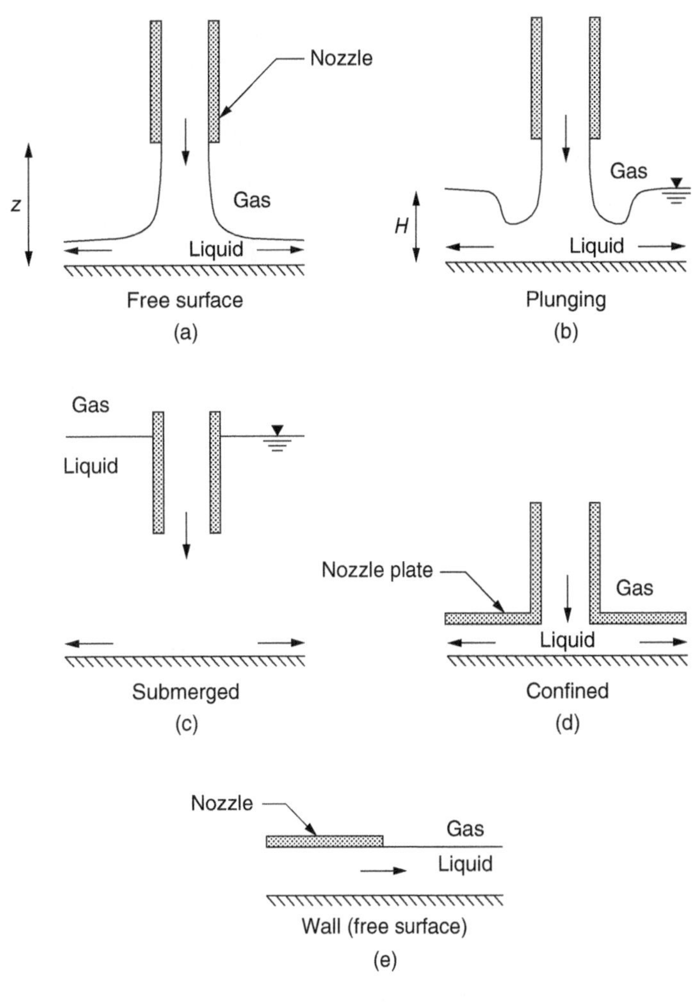

Figure 4.7.4 Various types of jets. Source: From Wolf et al. (1993). © 1993 Elsevier.

is seen that the heat transfer coefficients at 400g are higher than at 200g up to wall superheat of about 23 K.

The authors have not provided any correlation of data, nor have they compared it with any prediction method.

4.7.7 Jets Impinging on Hot Surfaces

Cooling of hot surfaces by impinging jets of liquids has been used for a long time, notably in manufacture of steel sheets. In recent years, there has been a lot of interest in using it for cooling of microelectronic equipment. The common types are free surface, plunging, submerged, confined, and the wall jets. These are schematically shown in Figure 4.7.4. The free surface jet consists of a liquid passing through a gas to impact on a surface. The plunging jet differs from the free surface jet in that the impacted surface is covered by a layer of liquid whose thickness is less than the distance between the nozzle and the surface. In the submerged jet, the nozzle discharging the liquid is located within a liquid pool covering the surface. The confined jet is discharged into the space between the surface and the nozzle plate. The wall jet flows parallel to the surface. It may be submerged within a liquid pool or may be of the free surface type.

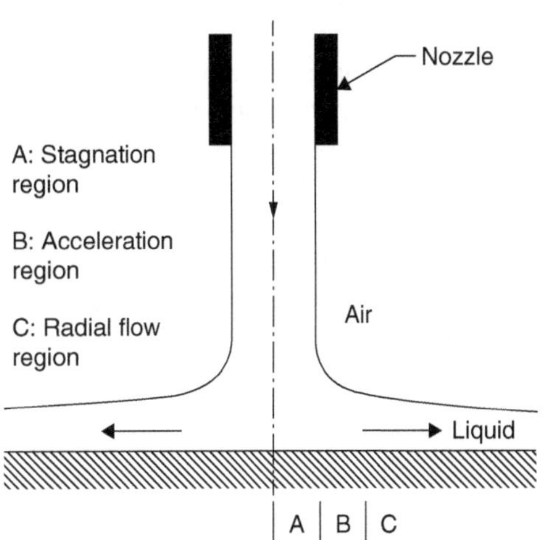

Figure 4.7.5 Various regions in the impingement of a free surface jet on a flat surface. Source: From Karwa et al. (2011). © Elsevier.

Figure 4.7.5 shows a schematic of circular free surface jet impinging normally on a horizontal flat surface. As

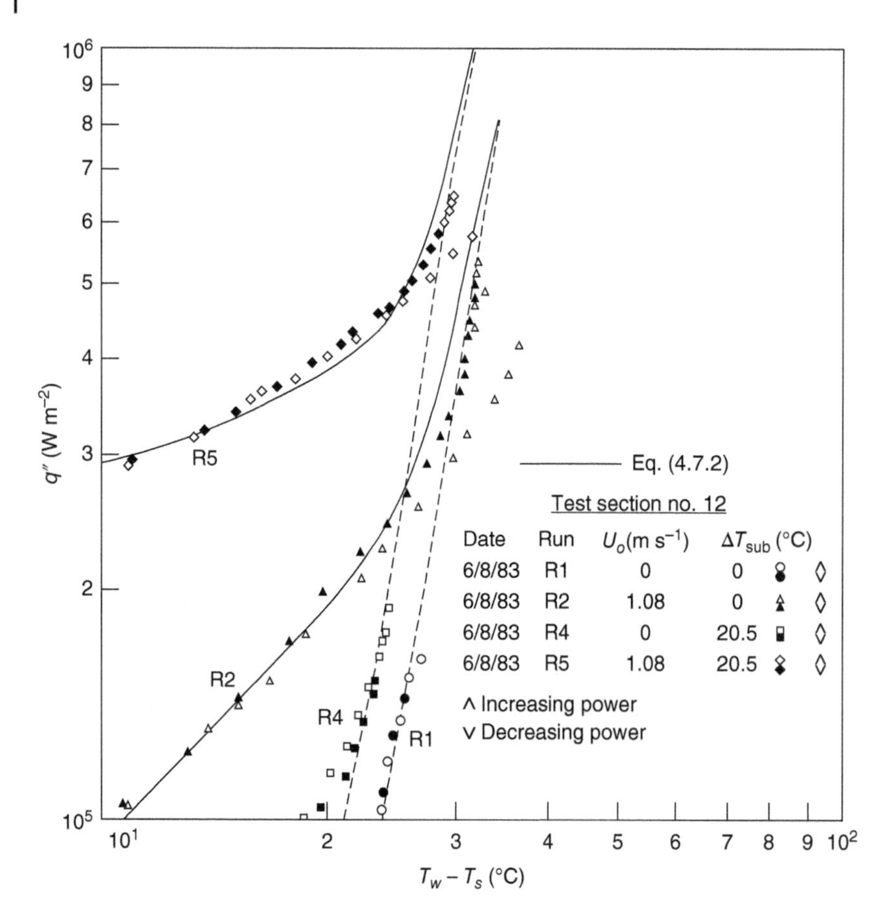

Figure 4.7.6 Effect of subcooling and jet velocity on heat transfer. R-113 submerged jet impinging on a horizontal plate. T_s is the saturation temperature. Source: From Ma and Bergles (1986).

the circular jet impinges onto the surface, the flow is diverted radially outwards along the surface. The flow has three regions, namely, stagnation region, acceleration region, and radial flow region. The stagnation region coincides with the impinging jet. Next to it is the acceleration region that is followed by the radial flow region. The pressure is maximum at the stagnation point and decreases with distance from it. The fluid continues to accelerate in the acceleration region until its free surface velocity approaches the jet velocity. The entire stagnation and acceleration region is called the impingement region. In the radial flow region, the velocity decreases due to effects of viscosity.

4.7.7.1 Experimental Studies and Correlations

Wolf et al. (1993) reviewed the earlier literature on this subject, while Qiu et al. (2015) have reviewed the more recent literature. They have listed the range of parameters in various experimental studies, discussed their findings, and proposed prediction methods. The results of various studies often differ. None of the proposed correlations was verified with data from many sources. A few studies and correlations are discussed in the following.

Single Jets Ma and Bergles (1986) studied submerged jets of R-113 at atmospheric pressure impinging normally on vertical heated strips. Jet diameters used were 1.07 and 1.81 mm. The heated test strips were 3 mm × 3 mm or 5 mm × 5 mm. Distance of jet outlet to heated surface was two diameters. The jet axis was aligned to the middle of the heater surface. Jet velocity was up to 10 m s^{-1}, while the subcooling was up to 20 K. Pool boiling tests (those without jet flow) were also done. Figure 4.7.6 shows the boiling curves; wall temperature was measured at the middle of the heater surface. It is seen that increase in subcooling shifts the boiling curves to the left. Thus, subcooling increases heat transfer coefficients in both pool boiling and boiling with jet impingement. Extrapolation of pool boiling curves at 0 and 20 K subcooling coincide with the corresponding curves for FDB with jet impingement. To predict the heat transfer over the entire range including partial boiling, they used the following superposition equation:

$$q = [q_{\text{spc}}^2 + (q_{\text{FDB}} - q_{\text{ib}})^2]^{1/2} \tag{4.7.2}$$

where

$$q_{\text{spc}} = h_{\text{spc}}(T_w - T_f) \tag{4.7.3}$$

h_{spc} is the heat transfer coefficient in single-phase convection without boiling. Good agreement is seen between measurements and Eq. (4.7.2).

Ma and Bergles found their inception of boiling data to be in agreement with the Bergles–Rohsenow model, described in Section 4.2.1, when the actual cavity sizes on the surface were taken into consideration.

Ma and Bergles also did some tests in which the jet impingement point was moved horizontally two diameters away from the center of the heater. It resulted in liquid after impact flowing parallel to the surface at the location of the thermocouple measuring the wall temperature. At low velocity, the FDB curves for normal and parallel flow merged. At higher velocities the normal flow boiling curve was to the right of the curve for parallel flow.

Liu et al. (2004) studied free surface vertical jets of water impacting on horizontal surfaces of the same diameter. A wide range of subcooling, jet diameters, jet velocities, and distance between jet and surface, were tried. Figure 4.7.7 shows the effect of subcooling at a fixed velocity and jet diameter. Also shown is the pool boiling line according to the Stephan and Abdelsalam correlation that was described in Section 3.2.3. It is seen that at FDB, the curves at all subcooling levels approximately merge together into the extension of Stephan and Abdelsalam pool boiling correlation. This indicates that FDB is entirely a nucleate boiling process. In partial boiling, heat flux at constant ΔT_{SAT} increases with increasing subcooling. Comparison of Figures 4.7.6 and 4.7.7 shows a marked difference in the effect of subcooling. In Figure 4.7.6, the FDB curves for different subcooling do not merge, while they merge in Figure 4.7.7. It is to be noted that there are differences in the type of jet and the relative sizes of jet and heated surface. Ma and Bergles used submerged jets, while those of Liu et al. were free surface jets. The size of surface in the tests of Ma and Bergles was much larger than the jet diameter, while the sizes of jet and heater surface were the same in the tests of Liu et al. Another possible explanation is that the extrapolation of the pool boiling data by Ma and Bergles to much higher heat fluxes was not accurate.

Wolf et al. (1996) performed test with a free surface jet of water at atmospheric pressure impinging vertically on a horizontal plate. The nozzle was 10.2 mm × 102 mm at a distance of 102 mm. The plate was made of Haynes Alloy 230 and measured 35.7 mm × 260 mm × 0.297 mm. Wall temperature was measured at several points along the plate. Water had constant subcooling of 50 K. Jet velocities of 2 and 5 m s^{-1} were used. Single-phase and partial boiling heat transfer coefficients increased with jet velocity. The distance from the stagnation line had no effect on FDB. However, heat transfer coefficients decreased with distance from stagnation line in single-phase convection and partial

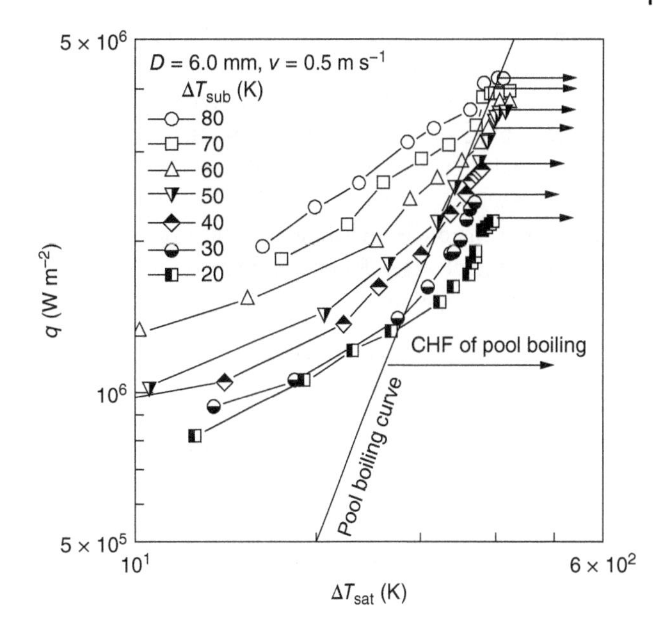

Figure 4.7.7 Effect of subcooling with on heat transfer with a vertical water jet impinging on a horizontal disk. Source: From Liu et al. (2004). © 2004 American Society of Mechanical Engineers.

boiling. The FDB curves for both velocities merged and were represented by the following equation:

$$q = 63.7(\Delta T_{SAT})^{2.95} \tag{4.7.4}$$

where q is in W m^{-2} and ΔT_{SAT} is in K. This equation predicts about 30% lower heat flux than the Stephan and Abdelsalam correlation for pool boiling.

Li et al. (2014) studied free surface vertical jets of water impinging on a copper disk as well as two other specially prepared surfaces. The diameters of the jet and surfaces were 3 mm each. Water was at atmospheric pressure. Subcooling up to 30 K and velocities up to 40 m s^{-1} were used. Wall superheat needed for boiling inception increased with increasing velocity and subcooling. The data for all levels of subcooling and velocities merged into Eq. (4.7.4). Their data for heat transfer before inception of boiling was correlated by the following equation given by Gabour and Lienhard (1994):

$$\frac{h_{spc}D_{jet}}{k_f} = 0.27Re_{jet}^{0.633}Pr_f^{1/3} \tag{4.7.5}$$

Re_{jet} is defined as

$$Re_{jet} = \frac{\rho_f u_{jet} D_{jet}}{\mu_f} \tag{4.7.6}$$

D_{jet} is the diameter of jet nozzle and u_{jet} is the velocity of jet. Fluid properties are calculated at the film temperature. Reynolds numbers in their tests ranged from 25 000 to 85 000.

The previously described tests of Li et al. were done on a normal copper surface with static contact angle of 60°. They also did tests with a hydrophilic surface (contact angle 5°) and a hydrophobic surface (contact angle 105°). Compared to the normal copper surface, the FDB heat flux was lower for the hydrophilic surface and higher for the hydrophobic surface. Thus the nucleate boiling heat flux decreases with increasing contact angle. Based on their data, they arrived at the following relation:

$$h_{FDB} = \frac{q}{\Delta T_{SAT}} \propto (1 - \cos\beta)^{0.5} \quad (4.7.7)$$

where β is the static contact angle. For $\beta < 15°$, $\beta = 15°$ is to be used. For inception of boiling, they have given the following correlations of their data:

$$\Delta T_{SAT,ib} = 15\ \ln(2 + u_{jet})(1 + 0.006\,06\Delta T_{SC})(1 - 0.002\,81\beta) \quad (4.7.8)$$

$$q_{ib} = h_{spc}[15\ \ln(2 + u_{jet})(1 + 0.006\,06\Delta T_{SC})(1 - 0.002\,81\beta) + \Delta T_{SC}] \quad (4.7.9)$$

where h_{spc} is calculated with Eq. (4.7.5).

Qiu et al. (2018) studied the effect of jets on a staggered bundle of horizontal rods. The 3 mm diameter horizontal rods were confined in a shell of 42 mm diameter. One end of the rods was attached to a heated copper plate, while the other end was insulated. A jet discharged water at atmospheric pressure into the shell from the bottom and steam–water mixture flowed out at the top. Inlet jet velocity was 0.17–0.74 m s^{-1}, which gave the inlet Reynolds number from 3470 to 15 100 based on inlet diameter of 6 mm. Inlet subcooling of 2–8 K was studied. Heat flux increased at the same ΔT_{SAT} with increasing subcooling. Heat flux also increased with entrance velocity at the same ΔT_{SAT} and ΔT_{SC}. They also performed a numerical solution that agreed with their data.

Clark et al. (2019) studied vertical confined jet of HFE-7100 on a horizontal copper surface 20 mm × 20 mm. HFE-7100 was at atmospheric pressure and with a constant subcooling of 8 K. The jet issued through a single 2 mm diameter orifice, at jet exit velocities of 1–9 m s^{-1}, into a confinement gap with a spacing of three jet diameters between the orifice and the heated surface. Additional orifice-to-target spacings of 0.5, 1, and 10 jet diameters are tested at the lowest highest jet velocities. Pool boiling tests were also done on the same surface. Measured wall temperature was the average for the surface. Increasing jet velocity increased the heat transfer coefficient before inception of boiling. For FDB, the boiling curves at all jet velocities combined together and coincided with the pool boiling curve or its extension. This has been the experience of most researchers. The orifice-to-surface distance had no effect with the exception that at distance of 0.5 times diameter with jet velocity of 9 m s^{-1}, heat transfer deteriorated beyond a heat flux of 460 kW m^{-2} and the critical heat flux (CHF) was much lower than with other spacings. CHF increased linearly with velocity, rising from 21 W cm^{-2} for pool boiling to 85 W cm^{-2} at 9 m s^{-1} velocity.

Cardenas and Narayanan (2012) studied submerged jet of FC-72 impinging vertically on a flat horizontal surface 27.6 mm diameter. Jet nozzles of diameters 1.16, 2.29, and 3.96 mm were used. All experiments are performed at saturation conditions at atmospheric pressure. For all experiments, the surface-to-nozzle height remained fixed at six jet diameters. For a particular jet diameter, boiling curves at all jet velocities collapsed into a single curve though there was some divergence near CHF. The boiling curves for different jet velocities did not merge, heat flux being higher for smaller diameter jets. Strong hysteresis was observed.

Mani et al. (2012) studied submerged water jet impinging vertically down on a horizontal surface. Local surface temperatures were measured. The boiling curve for the area in the stagnation zone was found to be much higher than the curve based on area average temperature. The curves for other locations on the surface were fairly close to the surface average curve.

Karwa et al. (2011) studied quenching of a horizontal stainless steel type 314 plate by a vertical jet of water. The plate was heated to 900 °C before quenching. Temperatures of the surface during quenching were measured at various points. Boiling curves were drawn for various locations. It was found that the curves near the center of jet impact were the highest and the curves got lower with distance from the center. Post-CHF curves also had the same trend.

Lamvik and Iden (1982) studied free surface water jets impinging on a heated aluminum disk 150 mm diameter, 10 mm thick. Single jets as well as array of jets were used. Jets were vertical up and down as well as horizontal, with surface orientation changed such that jets were always at 90° to it. Most tests were done with 0.7 mm diameter jets though a few tests with 1.4 and 2 mm diameter jets were also done. They found heat transfer coefficient to increase with jet diameter as shown in Figure 4.7.8. For a vertical surface impacted by a horizontal jet, the relation is

$$h = 3700 D_{jet}^{0.35} \quad (4.7.10)$$

D_{jet} is in mm and h is in W m^{-2} K^{-1}. The effect of velocity on heat transfer coefficient in various orientations was given by the following relations.

$$\text{Jet vertical downward}: h = 340 + 99u_{jet} \quad (4.7.11)$$

$$\text{Jet vertical upward}: h = 31u_{jet}^{1.4} \quad (4.7.12)$$

$$\text{Jet horizontal}: h = 23u_{jet}^{1.5} \quad (4.7.13)$$

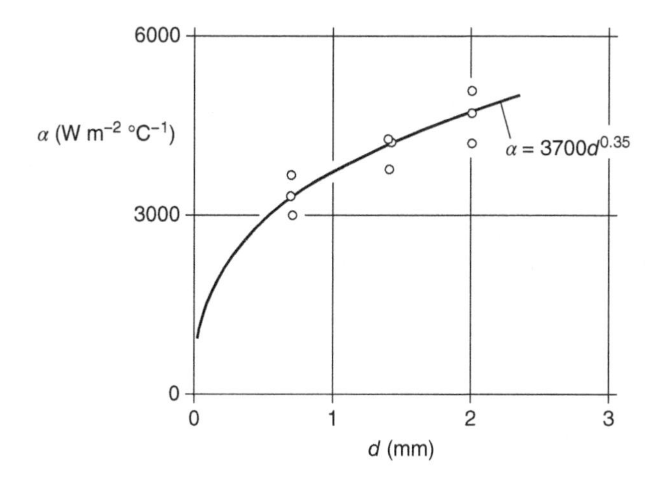

Figure 4.7.8 Effect of nozzle diameter d on mean heat transfer coefficient with horizontal jet on a vertical plate: plate temperature 500 °C, jet temperature 12 °C, and jet velocity 27.4 m s^{-1}. Source: Reprinted from Lamvik and Iden (1982). © 1982, with permission from Begell House, Inc.

h is in W m^{-2} K^{-1} and u_{jet} is in m s^{-1}. Jet velocity varied from 12 to 35 m s^{-1}.

Wolf et al. (1993) have discussed several studies in which no effect of jet diameter or surface orientation on heat transfer was found.

Array of Jets Besides the single jet tests described earlier, Lamvik and Iden (1982) also studied array of jets in all three orientations. Jets were 10 mm apart in most tests but tests were also done with spacing up to 20 mm. While heat transfer coefficients for single jets consistently increased with increasing jet velocity, with multiple jets, heat transfer coefficient initially increased with jet velocity and then decreased with further increase in velocity. The optimum velocity was in the 15–25 m s^{-1} range. The heat transfer coefficients at high velocities were much lower than for single jets. At a velocity of 30 m s^{-1}, heat transfer coefficient of jet arrays was less than one-third of that for single jets. The explanation given by the authors is that as velocity increases, droplets from the impact of jets interfere with flow of adjacent jets, cause a reduction in their impulse, thus making film boiling more likely. Heat transfer coefficient was found to decrease with distance from the stagnation point.

Monde et al. (1980) studied up to four free surface jets of saturated R-113 or water impacting vertically on a horizontal surface. Several arrangements of jet location were tried. Jet velocity varied from 9 to 15 m s^{-1}. The authors concluded that the scatter in data was typical of nucleate boiling and hence heat transfer coefficient is not affected by number or arrangement of jets.

Nonn et al. (1988) used one to nine free stream jets of FC-72 in the velocity range 2.8–11.3 m s^{-1} impinging on a plate. They found that number of jets did not affect heat transfer.

It should be noted that jet velocity in the tests of Nonn et al. and Monde et al. did not exceed 15 m s^{-1} while most of the data of Lamvik and Iden were for velocities greater than 15 m s^{-1}. Hence the results of the last mentioned do not necessarily contradict those of the other two.

4.7.7.2 Recommendations

As is seen in the previously described studies as well as in the review paper by Qiu et al. (2015), experimental studies often show conflicting trends, and there is no prediction method that has been validated over a wide range of data from many studies. It is therefore suggested that designs be based on the results of the experimental studies whose conditions resemble the application. If no such experimental study is available, the following method is suggested for a conservative estimate. Calculate q_{FDB} by a pool boiling correlation, and for the partial boiling regime, use Eq. (4.7.2).

4.7.8 Spray Cooling

Cooling of hot surfaces by spray of liquid droplets has been widely used for rapid cooling of hot surfaces and has been considered for applications such as cooling of computer chips and refrigeration evaporators. Sprays of fine droplets are obtained by passing pressurized liquids through nozzles. Many designs of nozzles have been developed for this purpose.

Nucleate boiling heat transfer has been found to depend on factors that include subcooling, nature of fluid, droplet diameter, volumetric flow rate, and angle of spray. Many analytical and empirical methods have been proposed for predicting nucleate boiling heat transfer, but none has been validated with more than a couple of data sets covering a very limited range of data. A couple of them are given herein.

Mudawar and Valentine (1989) performed tests with water sprays on a 0.5 cm^2 metal surface. The test conditions included volumetric spray flux 0.6×10^{-3}–9.96×10^{-3} m^3 s^{-1} m^{-2}, mass mean drop diameter 0.434–2.005 mm, and mean drop velocity 10.6–26.5 m s^{-1}. Correlations were given for water temperatures from 23 to 80 °C. For water temperature 22.5–23.5 °C, their correlation is

$$q = 1.87\times10^{-5}(T_w - T_f)^{5.55} \tag{4.7.14}$$

with q in W m^{-2} and temperatures in K.

Pereira et al. (2013) gave the following correlation of their data for water with subcooling 30–75 °C obtained with arrays of four and five nozzles:

$$q = 2067(T_w - T_{SAT})^{5.55} \tag{4.7.15}$$

For more information, the review paper of Liang and Mudawar (2017) may be consulted.

Nomenclature

a acceleration (m s^{-2})
Bo boiling number (–)
C_p specific heat at constant pressure (J kg^{-1} K^{-1})
D diameter of tube or jet (m)
D_{HP} equivalent diameter based on heated perimeter (m)
D_{HYD} hydraulic equivalent diameter (m)
g acceleration due to gravity at Earth (m s^{-2})
h heat transfer coefficient (W m^{-2} K^{-1})
i_{fg} latent heat of vaporization (J kg^{-1})
k thermal conductivity (W m^{-1} K^{-1})
L length (m)
L_c capillary length (m)
M molecular weight (–)
N number of data points (–)
Nu Nusselt number (–)
Pe Peclet number = (Re Pr) (–)
Pr Prandtl number (–)
p pressure (Pa)
p_c critical pressure (Pa)
p_r reduced pressure = p/p_c (–)
q heat flux (W m^{-2})
Re Reynolds number = (GD/μ) (–)
r radius (m)
T temperature (K)
ΔT_{SAT} $(T_w - T_{SAT})$ (K)
ΔT_{SC} $(T_{SAT} - T_f)$ (K)
t Time (s)
u velocity (m s^{-1})
x equilibrium vapor quality (–)
x' actual vapor quality (–)
z length or distance (m)

Greek letters

α void fraction (–)
β contact angle (°)
μ dynamic viscosity (Pa s)
ν kinematic viscosity (m^2 s^{-1})
ρ density (kg m^{-3})
σ surface tension (N m^{-1})
ω angular velocity (rad s^{-1})

Subscripts

bd Bubble departure
f liquid
e entrance to channel
FDB fully developed boiling
g vapor
ib inception of boiling
jet jet
LT all mass flowing as liquid
OSV onset of significant void
pb pool boiling
SAT saturation
SC subcooled
spc single-phase convection
w wall

References

Ahmad, S.Y. (1970). Axial distribution of bulk temperature and void fraction in a heated channel with inlet subcooling. *J. Heat Transfer* **92**: 595–609.

Alferov, N.S. and Rybin, R.A. (1969, 1969). Heat transfer in annular channel. In: *Convective Heat Transfer in Two-Phase and One-Phase Flows* (eds. V.M. Borishanskii and I.I. Paleev), 115–134. Jerusalem: Israel Program For Scientific Translations.

Al-Yahia, O.S. and Jo, D. (2018). ONB, OSV, and OFI for subcooled flow boiling through a narrow rectangular channel heated on one-side. *Int. J. Heat Mass Transfer* **116**: 136–151.

Basu, N., Warrier, G.R., and Dhir, V.K. (2002). Onset of nucleate boiling and active nucleation site density during subcooled flow boiling. *J. Heat Transfer* **124**: 717–728.

Bergles, A.E. and Rohsenow, W.M. (1964). The determination of forced convection surface-boiling heat transfer. *J. Heat Transfer* **86**: 353–372.

Bitter, R.C. (1973). Zum Warmeubergang von einem querangestromten Rohr an siedendes R-11 bei Ein-und Zweiphasenstromung. Dr. Ing. thesis. Technical University Clausthal, Germany.

Bowring, R.W. (1962). Physical model based on bubble detachment and calculation of steam voidage in the subcooled region of a heated channel. Institutt for Atomenergi, Halden, Norway, Report No. HPR-10.

Brisbane, T.W.C., Grant, I.D.R., and Whalley, P.B.A. (1980). Prediction method for kettle reboiler performance. ASME Paper 80-HT-42, ASME, New York, NY.

Cardenas, R. and Narayanan, V. (2012). Heat transfer characteristics of submerged jet impingement boiling of saturated FC-72. *Int. J. Heat Mass Transfer* **55**: 4217–4231.

Chen, J.C. (1966). Correlation for boiling heat transfer to saturated fluids in convective flow. *Ind. Eng. Chem. Proc. Des. Dev.* **5** (3): 322–329.

Clark, M.D., Weibel, J.A., and Garimella, S.V. (2019). Identification of nucleate boiling as the dominant heat transfer mechanism during confined two-phase jet impingement. *Int. J. Heat Mass Transfer* **128**: 1095–1101.

Collier, J.G. and Thome, J.R. (1994). *Convective Boiling and Condensation*, 3e. Oxford, UK: Oxford University Press.

Davis, E.J. and Anderson, G.H. (1966). The incipience of nucleate boiling in forced convection flow. *AlChE J.* **12** (4): 774–780.

Delhaye, J.M., Maugin, F., and Ochterbeck, J.M. (2004). Void fraction predictions in forced convective subcooled boiling of water between 10 and 18 MPa. *Int. J. Heat Mass Transfer* **47** (19–20): 4415–4425.

Dix, G. E. (1971). Vapor void fraction for forced convection with subcooled boiling at low flow rates. PhD thesis. University of California, Berkeley, CA. Quoted by Lee and Bankoff (1998).

Duchatelle, L., Nucheze, L., and Robin, M.G. (1976). Heat transfer in helical tube sodium heated steam generator. In: *Future Energy Production Systems*, vol. **1** (eds. I.C. Denton and N.H. Afghan), 269–285. New York, NY: Academic Press.

Fand, R.M., Keshwani, K.K., Jotwani, M.M., and He, B.C.C. (1976). Simultaneous boiling and forced convection heat transfer from a horizontal cylinder to water. *J. Heat Transfer* **98**: 395–400.

Fink, J., Gaddis, E.S., and Vogelpohl, A. (1982). Forced convection of a mixture of Freon 11 and Freon 113 flowing normal to a cylinder. In: *Proceedings of the 7th International Heat Transfer Conference*, paper FB5, 207–212. Munich: Begell House.

Frost, W. & Dzakowic, G.S. (1967). An extension of the method for predicting incipient boiling on commercially finished surfaces. ASME Paper 67-HT-61.

Gabour, L.A. and Lienhard, J.H. (1994). Wall roughness effects on stagnation point heat transfer beneath an impinging liquid jet. *ASME J. Heat Transfer* **116**: 81–87.

Ghiaasiaan, S.M. and Chedester, R.C. (2002). Boiling incipience in microchannels. *Int. J. Heat Mass Transfer* **45**: 4599–4606.

Griffith, P., Clark, J.A., Rohsenow, W.M. (1958) Void volumes in subcooled boiling systems. ASME Paper 58-HT-19, ASME, New York, NY.

Gungor, K.E. and Winterton, R.H.S. (1986). A general correlation for flow boiling in tubes and annuli. *Int. J. Heat Mass Transfer* **29**: 351–358.

Gunther, F.C. (1951). Photographic study of surface boiling heat transfer to water with forced convection. *Trans. ASME* **73** (2): 115–123.

Gupta, A., Saini, J.S., and Varma, H.K. (1995). Boiling heat transfer in small horizontal tube bundles at low cross-flow velocities. *Int. J. Heat Mass Transfer* **38** (4): 599–605.

Hardik, B.K. and Prabhu, S.V. (2017). Boiling pressure drop and local heat transfer distribution of helical coils with water at low pressure. *Int. J. Therm. Sci.* **114**: 44–63.

Hardik, B.K. and Prabhu, S.V. (2018). Heat transfer distribution in helical coil flow boiling system. *Int. J. Heat Mass Transfer* **117**: 710–728.

Haynes, B.S. and Fletcher, D.F. (2003). Subcooled flow boiling heat transfer in narrow passages. *Int. J. Heat Mass Transfer* **46**: 3673–3682.

Hodgson, A.S. (1968). Forced convection subcooled boiling heat transfer with water in an electrically heated tube at 100–550 lb/in^2. *Trans. Inst. Chem. Eng.* **46**: T25–T-31.

Holman, J.P. (1968). *Heat Transfer*, 2e. New York, NY: McGraw-Hill.

Hsu, Y.Y. and Graham, R.W. (1961). An analytical and experimental study of the thermal boundary layer and ebullition cycle in nucleate boiling. NASA TN-D-594.

Huang, L. and Witte, L.C. (2001). Highly subcooled boiling in crossflow. *J. Heat Transfer* **123**: 1080–1085.

Hwang, T.H. and Yao, S.C. (1984). *Boiling Heat Transfer of a Horizontal Cylinder at Low Quality Crossflow*, vol. **HTD-38**, 9–17. New York: ASME.

Hwang, T.H. and Yao, S.C. (1986). Forced convection boiling in tube bundles. *Int. J. Heat Mass Transfer* **29** (5): 785–795.

Jens, W.H. and Lottes, P.A. (1951). Analysis of heat transfer burnout, pressure drop and density data for high pressure water. Argonne National Laboratory Report ANL-4627.

Jensen, M.K. and Hsu, J.T. (1988). A parametric study of boiling heat transfer in a horizontal tube bundle. *J. Heat Transfer* **110**: 976–981.

Kandlikar, S.G. (1998). Heat transfer characteristics in partial boiling, fully developed boiling, and significant void flow regions of subcooled flow boiling. *J. Heat Transfer* **120**: 395–401.

Karwa, N., Gambaryan-Roisman, T., Stephan, P., and Tropea, C. (2011). Experimental investigation of circular free-surface jet impingement quenching: transient hydrodynamics and heat transfer. *Exp. Therm Fluid Sci.* **35**: 1435–1443.

Kutateladze, S.C. (1961). Boiling heat transfer. *Int. J. Heat Mass Transfer* **4**: 31–45.

Lamvik, M. and Iden, B.A. (1982). Heat transfer coefficient by water jet impinging on a hot surface. In: *Proceedings of the 7th International Heat Transfer Conference*, 369–375. Munich, Germany: Begell House.

Lee, S.C. and Bankoff, S.G. (1998). A comparison of predictive models for the onset of significant void at low pressures in forced-convection subcooled boiling. *KSME Int. J.* **12** (3): 504–513.

Lee, J. and Mudawar, I. (2008). Fluid flow and heat transfer characteristics of low temperature two-phase micro-channel heat sinks – Part 2. Subcooled boiling pressure drop and heat transfer. *Int. J. Heat Mass Transfer* **51**: 4327–4341.

Lee, Y. and Shigechi, T. (1991). Conjugated cross flow boiling on a horizontal cylinder. In: *ASME/JSME Thermal Engineering Conference Proceedings*, vol. **2**, 153–158. New York, NY: ASME.

Lemmert, M. and Chawla, J.M. (1977). Influence of flow velocity on surface boiling heat transfer coefficients. In: *Heat Transfer in Boiling* (eds. E. Hahne and U. Grigull), 237–247. Washington, DC: Hemisphere Publishing Corporation.

Leppert, G., Costello, C.P., and Hoglund, B.M. (1958). Boiling heat transfer to water containing a volatile additive. *Trans. ASME* **80**: 1395–1403.

Levy, S. (1967). Forced convection subcooled boiling prediction of vapour volumetric fraction. *Int. J. Heat Mass Transfer* **10**: 951–965.

Li, Y., Chen, Y., and Liu, Z. (2014). Correlations for boiling heat transfer characteristics of high-velocity circular jet impingement on the nano-characteristic stagnation zone. *Int. J. Heat Mass Transfer* **72**: 177–185.

Liang, G. and Mudawar, I. (2017). Review of spray cooling – Part 1: Single-phase and nucleate boiling regimes, and critical heat flux. *Int. J. Heat Mass Transfer* **115**: 1174–1205.

Liu, Z. and Winterton, R.H.S. (1991). A general correlation for saturated and subcooled flow boiling in tubes and annuli, based on a nucleate pool boiling equation. *Int. J. Heat Mass Transfer* **34**: 2759–2827.

Liu, Z., Tong, T., and Qiu, Y. (2004). Critical heat flux of steady boiling for subcooled water jet impingement on the flat stagnation zone. *J. Heat Transfer* **126**: 179–183.

Liu, D., Lee, P.-S., and Garimella, S.V. (2005). Prediction of the onset of nucleate boiling in microchannel flow. *Int. J. Heat Mass Transfer* **48**: 5134–5149.

Ma, C.F. and Bergles, A.E. (1986). Jet impingement nucleate boiling. *Int. J. Heat Mass Transfer* **29** (8): 1095–1101.

Mani, P., Cardenas, R., and Narayanan, V. (2012). Comparison of area-averaged and local boiling curves in pool and jet impingement boiling. *Int. J. Multiphase Flow* **42**: 115–127.

Manon, E. (2000). Contribution a l'analyse et a la modelisation locale des ecoulements bouillants sous-satures dans les conditions des reacteurs a eau sous pression. These de doctorat. Ecole Centrale de Paris, 2000. Quoted in Delhaye et al. (2004).

Marto, P.J. and Gray, V.H. (1971). Effects of high accelerations and heat fluxes on nucleate boiling of water in an axisymmetric rotating boiler. NASA TN-D-6307.

McAdams, W.H., Kennel, W.E., Minden, C.S. et al. (1949). Heat transfer at high rates to water with surface boiling. *Ind. Eng. Chem.* **41**: 1945–1953.

McKee, H.R. (1967). Forced convection boiling from a cylinder normal to the flow. PhD thesis. Chemical Engineering Department, Oklahoma State University.

Miropoloskiy, Z.L. and Pikus, V.Y. (1969). Critical boiling heat flux in channels. *Heat Transfer Sov. Res.* **1**: 74–79.

Moles, F.D. and Shaw, J.F.G. (1972). Boiling heat transfer to subcooled liquids under condition of forced convection. *Trans. Inst. Chem. Eng.* **50**: 76–84.

Monde, M., Kusuda, H., and Uehara, H. (1980). Burnout heat flux in saturated forced convection boiling with two or more impinging jets. *Heat Transfer Jpn. Res.* **9** (3): 1834–1843.

Mudawar, I. and Valentine, W.S. (1989). Determination of the local quench curve for spray-cooled metallic surfaces. *J. Heat. Treat.* **7**: 107–121.

Nonn, T., Dagan, Z., and Jiji, L.M. (1988). Boiling jet impingement cooling of simulated microelectronic sources. ASME Paper 88-WA/EEP-3.

Owhadi, A. (1966). Boiling in self-induced radial acceleration fields. PhD thesis. Oklahoma State University, Stillwater.

Peng, X.F. and Wang, B.X. (1993). Forced convection and flow coining heat transfer for liquid flowing through microchannels. *Int. J. Heat Mass Transfer* **36** (14): 3421–3427.

Pereira, R.H., Filho, E.P.B., Braga, S.L., and Parise, J.A.R. (2013). Nucleate boiling in large arrays of impinging water sprays. *Heat Transfer Eng.* **34**: 479–491.

Polley, G.T., Ralston, T., and Grant, I.D.R. (1980). *Forced Crossflow Boiling in An Ideal In-Line Tube Bundle*. ASME Paper 8-HT-46. New York, NY: ASME.

Qi, S.L., Zhang, P., Wang, R.Z. et al. (2007). Flow boiling of liquid nitrogen in micro-tubes: Part I – The onset of nucleate boiling, two-phase flow instability and two-phase flow pressure drop. *Int. J. Heat Mass Transfer* **50**: 4999–5016.

Qiu, L., Dubey, S., Choo, F.H., and Duan, F. (2015). Recent developments of jet impingement nucleate boiling. *Int. J. Heat Mass Transfer* **89**: 42–58.

Qiu, L., Dubey, S., Choo, F.K., and Duan, F. (2018). Confined jet impingement boiling in a chamber with staggered pillars. *Appl. Therm. Eng.* **131**: 724–733.

Qu, W. and Mudawar, I. (2003). Flow boiling heat transfer in two-phase micro-channel heat sinks – I. Experimental investigation and assessment of correlation methods. *Int. J. Heat Mass Transfer* **46**: 2755–2771.

Rohsenow, W.M. (1953). Heat transfer associated with nucleate boiling. In: *Proceedings of the Heat Transfer and Fluid Flow Institute*, 123. Stanford University Press.

Rohsenow, W.M. and Clark, J.A. (1951a). Heat transfer and pressure drop for high heat flux densities to water at high subcritical pressures. Heat Transfer and Fluid Flow Preprints, Stanford University Press. Quoted by Bergles and Rohsenow (1964).

Rohsenow, W.M. and Clark, J.A. (1951b). A study on the mechanism of flow boiling. *Trans. ASME* **73**: 609–620.

Rouhani, S.Z. (1966). Void measurements in the region of subcooled and low quality boiling, Part 2. Report AE-239, Aktiebolaget Atomenergi, Stockholm, Sweden. Quoted in Rouhani and Axelsson (1970).

Rouhani, S.Z. and Axelsson, E. (1970). Calculation of void volume fraction in the subcooled and quality boiling regions. *Int. J. Heat Mass Transfer* **13**: 383–393.

Saha, P. and Zuber, N. (1974). Point of net vapor generation and vapor void. In: *Proc. 5th Int. Heat Transfer Conf.*, 3-7 September, Tokyo, Japan, Paper B4.7. Begell House Publishers.

Sato, T. and Matsumara, H. (1964). On the conditions of incipient subcooled boiling with forced convection. *Bull. JSME* **7** (26): 392–398.

Shah, M.M. (1976). A new correlation for heat transfer during boiling flow through pipes. *ASHRAE Transactions* **82** (2): 66–86.

Shah, M.M. (1977). A general correlation for heat transfer during subcooled boiling in pipes and annuli. *ASHRAE Trans.* **83** (1): 205–215.

Shah, M.M. (1982). CHART correlation for saturated boiling heat transfer: equations and further study. *ASHRAE Trans.* **88** (1): 165–196.

Shah, M.M. (1983). Generalized prediction of heat transfer during subcooled boiling in annuli. *Heat Transfer Eng.* **4** (1): 24–31. https://www.tandfonline.com.

Shah, M.M. (1984). A correlation for heat transfer during subcooled boiling on a single tube with forced crossflow. *Int. J. Heat Fluid Flow* **5** (1): 13–20.

Shah, M.M. (2005). Improved general correlation for subcooled boiling heat transfer during flow across tubes and tube bundles. *HVAC&R Res.* **11** (2): 285–303.

Shah, M.M. (2015). Evaluation of correlations for predicting heat transfer during boiling of carbon dioxide inside channels. Paper No. 8435, IHTC 15, Kyoto, Japan.

Shah, M.M. (2017a). New correlation for heat transfer during subcooled boiling in plain channels and annuli. *Int. J. Therm. Sci.* **112**: 358–370.

Shah, M.M. (2017b). Unified correlation for heat transfer during boiling in plain mini/micro and conventional channels. *Int. J. Refrig* **74**: 604–624.

Singh, R.L., Saini, J.S., and Varma, H.K. (1983). Effect of crossflow in boiling heat transfer. *Int. J. Heat Mass Transfer* **26** (12): 1882–1885.

Song, H., Lee, J., Chang, S.H. et al. (2017). Onset of nucleate boiling in narrow, rectangular channel for downward flow under low pressure. *Ann. Nucl. Energy* **109**: 498–506.

Staub, F.W. (1968). The void fraction in subcooled boiling: prediction of the initial point of net vapor generation. *J. Heat Transfer* **90**: 151–157.

Tarasova, N.V. and Orlov, V.M. (1969). Heat transfer and hydraulic resistance during surface boiling of water in annular channels. In: *Convective Heat Transfer in Two-Phase and One-Phase Flows* (eds. V.M. Borishanskii and I.I. Paleev). Israel Program for Scientific Translations, Inc.

Thom, J.R.S., Walker, W.M., Fallon, T.A., and Reising, G.F.S. (1965). Boiling in subcooled water during flow up heated tubes or annuli. *Symposium on Boiling Heat Transfer in Steam Generating Units and Heat Exchangers*, Manchester (September 1965). London: IMechE. Quoted in Qi et al (2007).

Unal, H.C. (1975). Determination of the initial point of net vapor generation in flow boiling system. *Int. J. Heat Mass Transfer* **18**: 1095–1099.

Vliet, C.C. (1962). Local boiling peak heat flux for water flowing normal to cylinder. PhD thesis. Mechanical Engineering Department, Stanford University, San Francisco, CA, USA.

Wege, M.E. and Jensen, M.K. (1984). Boiling heat transfer from a horizontal tube in an upward flowing two-phase crossflow. *J. Heat Transfer* **106**: 849–855.

Wolf, D.H., Incropera, F.P., and Viskanta, R. (1993). Jet impingement boiling. In: *Advances in Heat Transfer* (eds. P.H. James and F.I. Thomas), 1–132. Elsevier.

Wolf, D.H., Incropera, F.P., and Viskanta, R. (1996). Local jet impingement boiling heat transfer. *Int. J. Heat Mass Transfer* **39**: 1395–1406.

Yilmaz S. B. and Westwater J. W. (1979) Effect of velocity on heat transfer to boiling Freon-113. ASME Paper 79-WA/HT-35.

Zuber, N. and Findlay, J. (1965). Average volumetric concentration in two-phase flow systems. *J. Heat Transfer* **87**: 453–462.

5

Saturated Boiling with Forced Flow

5.1 Introduction

This chapter is about boiling of saturated liquids prior to critical heat flux (CHF). As vapor quality increases at a constant heat flux, heat transfer coefficient generally increases until a limit is reached when heat transfer coefficient begins to decrease. This is the CHF point. Beyond this point, the surface is partially in contact with liquid and partly with vapor. As the thermal conductivity of vapor is much less than that of liquid, heat transfer deteriorates. When quality reaches 100%, the entire surface is covered by vapor and heat transfer coefficient is that for forced convection vapor flow.

A variety of heat exchangers are considered in this chapter. This includes circular and non-circular channels, tube bundles, coils, etc. As was discussed in Chapters 2–4, the behavior of multicomponent mixtures differs greatly from that of pure fluids. Further, heat transfer of liquid metals is very different from that of non-metallic fluids. Therefore, these topics have been addressed separately in Sections 5.7 and 5.8, respectively. Section 5.9 deals with various problems such as effect of gravity. The other sections deal only with single-component pure non-metallic fluids unless otherwise noted.

5.2 Boiling in Channels

5.2.1 Effect of Various Parameters

The effect of various parameters is illustrated by the data of Chawla (1967) shown in Figure 5.2.1. It shows the effect of quality and mass flux at fixed heat flux and saturation temperature. It is seen that at higher flow rates, heat transfer coefficient increases with quality till about $x = 0.8$ where CHF is reached. The increase in heat transfer is because of the acceleration caused by increasing amount of vapor that has much higher specific volume than liquid. After the CHF point, heat transfer decreases sharply. This is the region of transitional boiling in which tube surface is partly in contact with liquid and partly with vapor. As thermal conductivity of vapor is much lower than that of liquid, heat transfer decreases. At 100% quality, no liquid is left and heat transfer is entirely by vapor forced convection.

At lower mass flux, heat transfer coefficient initially remains constant or decreases with increasing quality. This is because with slowly moving liquid, bubble nucleation is strong and heat transfer is mainly by this mechanism. When more vapor is generated, there is large increase in velocity, which causes suppression of bubble nucleation. At the lower qualities, the increase in heat transfer due to convection is less than the decrease in heat transfer due to bubble suppression. The net effect is a decrease in heat transfer coefficient. With further increase in quality, convection becomes the dominant mode and heat transfer coefficient increases with further increase in quality. At the lowest mass flux, heat transfer coefficient decreases slowly with increasing quality till CHF is reached and then drops sharply.

Figure 5.2.2 shows two distinct regimes of heat transfer. At lower heat flux, heat transfer coefficient is essentially independent of heat flux. As heat flux increases at a fixed flow rate, a point is reached where heat transfer increases sharply with increasing heat flux. This is the regime in which nucleate boiling is the dominant mode.

Another interesting point to note is that heat transfer coefficient is seen to decrease with increasing tube diameter. In single-phase heat transfer, h varies as $D^{-0.2}$. This will suggest that h in 6 mm tube will be 1.3 times that for the 25 mm diameter tube. Actually, the ratio is seen to be about 2.4 in the convective regime. The explanation of this phenomenon is found in the observations of flow patterns reported by Chawla. In the 6 mm tube, flow patterns were such that the entire circumference was wetted by liquid. In the 25 mm diameter tube, flow patterns were such that part of the upper circumference was in contact with vapor. The low thermal conductivity of vapor causes low heat transfer on the part of tube in contact with it. In a vertical

Two-Phase Heat Transfer, First Edition. Mirza Mohammed Shah.

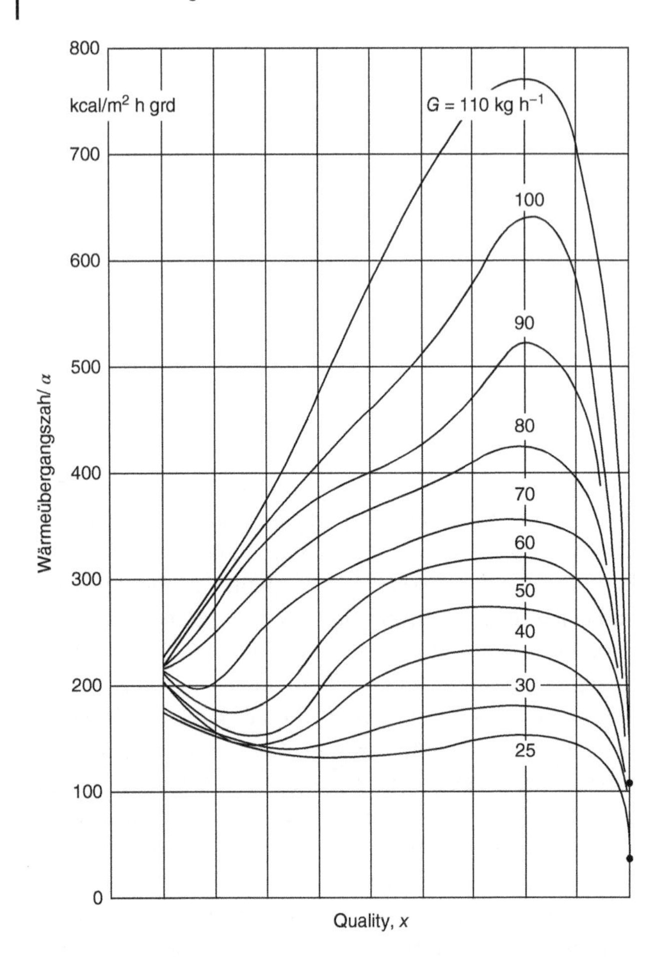

Figure 5.2.1 Effect of quality and mass flow rate on heat transfer coefficient: R-11 in 25 mm tube, $q = 10\,000\,\text{kcal m}^{-2}\,\text{h}^{-1}$, and $T_{SAT} = 10\,°\text{C}$. Source: From Chawla (1967). Reproduced with permission of the Verein Deutscher Ingenieure e. V.

tube, liquid is uniformly distributed around the circumference. Hence the heat transfer coefficients in vertical tubes will be higher under conditions causing stratification in horizontal tubes.

All of Chawla's data are at a constant saturation temperature and hence do not shed light on the effect of pressure. Figure 5.2.3 shows the data of Mumm (1954) for water flowing in a horizontal tube. It is seen that at lower vapor qualities, h is higher at higher pressure. At high qualities the reverse is the case. This behavior becomes comprehensible when it is noted that at low vapor qualities, the dominant mode is nucleate boiling, while at higher qualities, heat transfer is dominated by forced convection. As was seen in Chapter 3, pool boiling heat transfer increases with pressure. Hence at low qualities, heat transfer at higher pressure are higher. On the other hand, specific volume of vapor decreases with increasing pressure. Hence the acceleration due to vapor generation is lower at higher pressures. As a consequence, heat transfer in convection dominated regime is higher at lower pressure.

5.2.2 Prediction of Heat Transfer

A very large number of correlations have been proposed. Most of them are based on the authors' own data and fail outside the range of those data. Here only those correlations are discussed, which were verified with a wide range of data. The earlier correlations were based on data with tube diameters larger than 3 mm that are considered macro or conventional channels. In recent years, the use of minichannels ($D \leq 3$ mm) has greatly increased and many correlations specific to minichannels have been proposed.

5.2.2.1 Correlations for Macro Channels

Among the well-known correlations for macro channels are those of Chen (1966), Shah (1976, 1982), Winterton and coworkers, Kandlikar (1990), Steiner and Taborek (1992), and flow pattern-based correlation from researchers at EPFL.

Chen Correlation The first correlation that was verified with data from many sources is that of Chen (1966). It is expressed by the following equation:

$$h_P = Eh_{LS} + Sh_{pb} \tag{5.2.1}$$

The factor F represents the enhancement of single-phase heat transfer due to two-phase convection. It was determined by data analysis and shown graphically. The following equations fit the curve, Collier (1981):

$$\text{For } X_{tt} \geq 10, \quad E = 1 \tag{5.2.2}$$

$$\text{For } X_{tt} < 10, \quad E = 2.35\left(0.213 + \frac{1}{X_{tt}}\right)^{0.736} \tag{5.2.3}$$

X_{tt} is the Martinelli parameter for turbulent–turbulent flow given by the following equation:

$$X_{tt} = \left(\frac{\rho_g}{\rho_f}\right)^{0.5}\left(\frac{\mu_f}{\mu_g}\right)^{0.1}\left(\frac{1-x}{x}\right)^{0.9} \tag{5.2.4}$$

S is a suppression factor representing the effect of suppression of nucleate boiling due to two-phase convection. It was determined by data analysis and shown graphically. The curve is represented by the following equation, Collier (1981):

$$S = [1 + (2.56 \times 10^{-6})(Re_{LS}E^{1.25})^{1.17}]^{-1} \tag{5.2.5}$$

The pool boiling heat transfer coefficient is calculated by the Forster and Zuber (1955) correlation. That correlation is in the form of Eq. (3.2.25). Chen expressed it as below:

$$h_{pb} = 0.00\,122\left[\frac{k_f^{0.79}C_{pf}^{0.45}\rho_f^{0.49}}{\sigma^{0.5}\mu_f^{0.29}i_{fg}^{0.24}\rho_g^{0.24}}\right]\Delta T_{SAT}^{0.24}\Delta p_{SAT}^{0.75} \tag{5.2.6}$$

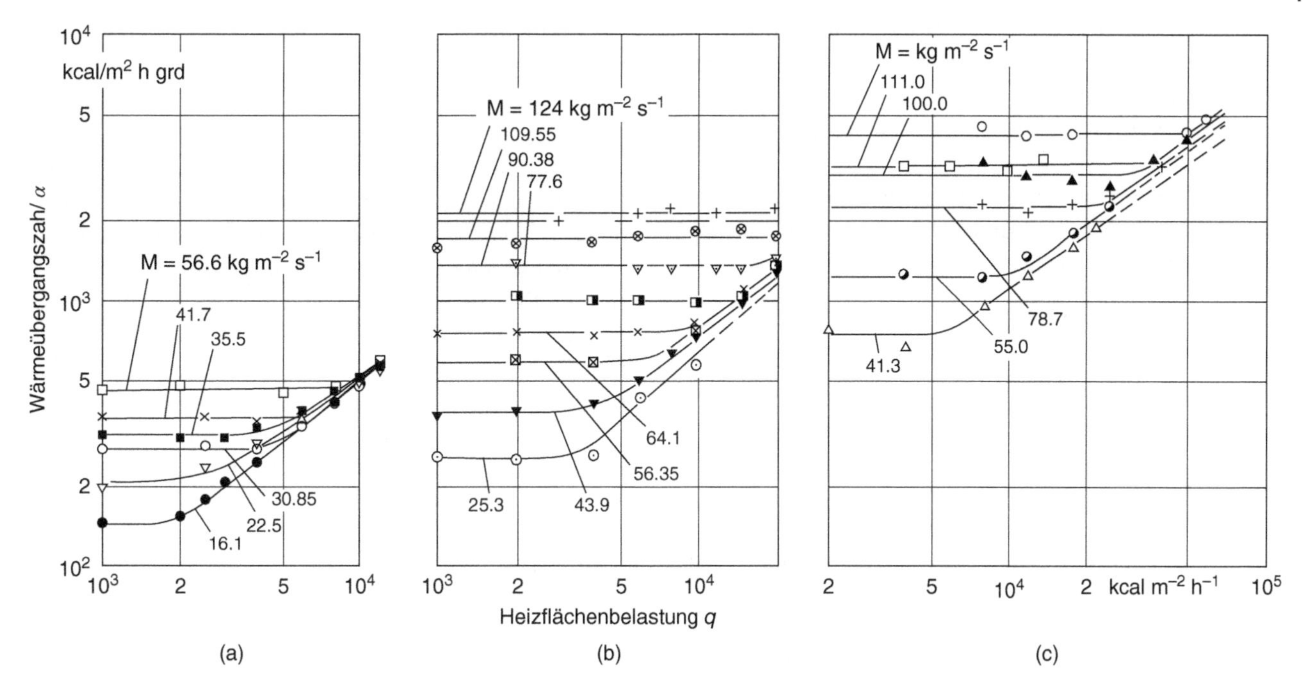

Figure 5.2.2 Effect of heat flux q and mass flux M on heat transfer coefficients during evaporation of R-11 in horizontal tubes at $x = 0.5$ and $T_{SAT} = 10\,°C$. Tube diameters are (a) 25 mm, (b) 14 mm, (c) 6 mm. Source: From Chawla (1967). Reproduced with permission of the Verein Deutscher Ingenieure e. V.

Figure 5.2.3 Effect of pressure on heat transfer for water flowing in a horizontal tube: $D = 11.8$ mm, $G = 518\ \text{kg m}^{-2}\ \text{s}^{-1}$, and $q = 79.4\ \text{kW m}^{-2}$. Source: Based on Mumm (1954).

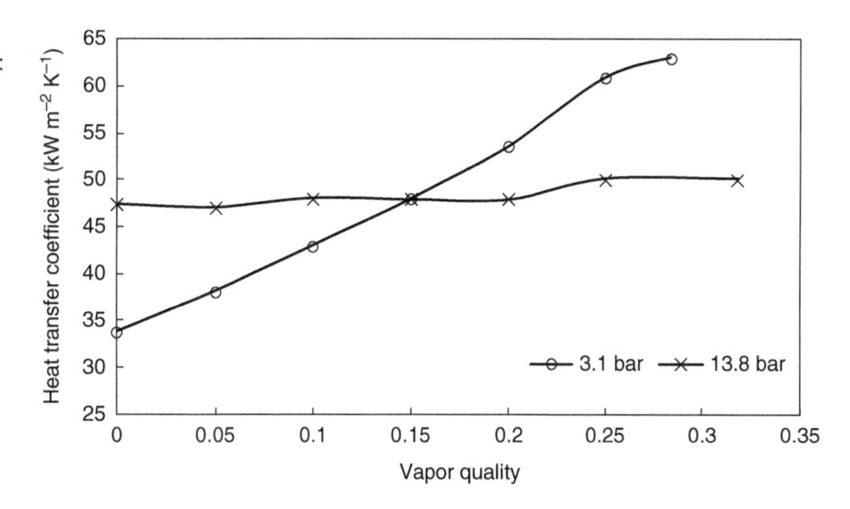

In the above equation,

$$\Delta T_{SAT} = T_w - T_{SAT} \tag{5.2.7}$$

$$\Delta p_{SAT} = p_{SAT}(T_w) - p \tag{5.2.8}$$

Thus, Δp_{SAT} is the saturation pressure difference corresponding to the wall superheat.

The superficial heat transfer coefficient of liquid phase h_{LS} is calculated by the Dittus and Boelter (1930) correlation.

$$h_{LS} = 0.023 Re_{LS}^{0.8} Pr_f^{0.4} k_f / D \tag{5.2.9}$$

Re_{LS} is the superficial Reynolds number of liquid phase given by

$$Re_{LS} = \frac{G(1-x)D}{\mu_f} \tag{5.2.10}$$

Chen verified this correlation with data for vertical tubes and annuli. The data were for water from four sources and chemicals from one source. The water data were for pressure of 1–34 bar. These data were reported to be in excellent agreement with the correlation. For annulus data, the heated equivalent diameter D_{HP} was used, which is defined below:

$$D_{HP} = \frac{4 \times (\text{flow area})}{\text{heated perimeter}} \tag{5.2.11}$$

It is seen that ΔT_{SAT} appears on the right-hand side of Eq. (5.2.6), while it is what is to be determined. Therefore, iterative calculations with assumed values of ΔT_{SAT} have to be done till convergence is achieved, noting that $h_{TP} = q/\Delta T_{SAT}$.

Bennet and Chen (1980) modified the Chen correlation as follows. They used a modified Chilton–Colburn analogy to give the following expression for E:

$$E = \left(\frac{Pr_f + 1}{2}\right)^{0.444}\left[\frac{(\Delta p)_{TP}}{(\Delta p)_{LS}}\right]^{0.444} \quad (5.2.12)$$

Note that Δp is the frictional pressure drop. The two-phase pressure drop $(\Delta p)_{TP}$ can be calculated by the correlations given in Chapter 1.

The correlations of Chen and Bennet and Chen have been reported by other researchers to give poor agreement with data. Gungor and Winterton (1986) compared these correlations to a wide-ranging database. The Chen correlation had mean absolute deviation (MAD) of 57.7%, while that of Bennet and Chen had MAD of 71.5%. Liu and Winterton (1990) reported a MAD of about 39% of the Chen correlation with their database. Kandlikar (1990) compared the Chen correlation to his wide-ranging database. The MAD for refrigerant data was 48.6%, while MAD for water data was 29.6%. MAD is defined as

$$\text{Mean absolute deviation} = \frac{1}{N}\sum_{1}^{N} \text{ABS}\left[\frac{h_{TP,predicted} - h_{TP,measured}}{h_{TP,measured}}\right] \quad (5.2.13)$$

N is the number of data points. The poor performance of these correlations is not too surprising as these were based on a very limited database.

From the foregoing discussion, it is clear that the Chen and Bennet–Chen correlations are quite erratic and hence not suitable for use in design. However, the Chen correlation has had a great influence on research on boiling heat transfer as numerous authors have developed correlations for many types of heat exchangers in the form of Eq. (5.2.1). It is therefore an important part of literature.

Shah Correlation The first correlation applicable to both horizontal and vertical channels was presented by Shah (1976). For this correlation, Shah introduced the convection number Co that is defined as

$$\text{Co} = \left(\frac{1}{x} - 1\right)^{0.8}\left(\frac{\rho_g}{\rho_f}\right)^{0.5} \quad (5.2.14)$$

The other factors in this correlation are

$$\psi = \frac{h_{TP}}{h_{LS}} \quad (5.2.15)$$

$$\text{Bo} = \frac{q}{G i_{fg}} \quad (5.2.16)$$

$$Fr_{LT} = \frac{G^2}{\rho_f^2 g D} \quad (5.2.17)$$

h_{LS} is calculated by the Dittus–Boelter equation, Eq. (5.2.9). Shah first analyzed data for vertical tubes in terms of ψ, Co, and Bo and developed a graphical correlation. When it was compared to data for horizontal tubes, it was found that this tentative correlation agreed with data at higher mass flux and smaller tube diameters, but overpredicted data at lower flow rates and larger tube diameters. Study of the flow pattern data of Chawla (1967) revealed that his data, which were being overpredicted, were for flow patterns in which the upper part of the tube did not have contact with liquid. As Froude number is the ratio of inertia forces and gravity forces, it was expected that it may be able to take into account the effects of stratification. Data analysis confirmed this expectation. A graphical correlation was developed that is shown in Figure 5.2.4. For horizontal tubes with $Fr_{LT} > 0.04$ and for vertical tubes at all Froude numbers, Froude number lines are ignored. For horizontal tubes with $Fr_{LT} < 0.04$, a vertical line is drawn from the Co line to the applicable Froude number line. From the intersection a horizontal line is drawn to intersect line AB. From this intersection, a vertical upwards line is drawn to the applicable Bo line. If the actual Bo is below the intersection with the line AB, ψ is that at the intersection with line AB.

Shah (1982) gave equations for the graphical correlation to make it easier to use it in computerized calculations. These are given below in a rearranged form as given in Shah (1984a):

$$\psi_0 = 230\text{Bo}^{0.5} \geq 1 \quad (5.2.18)$$

If Eq. (5.2.18) predicts $\psi_0 < 1$, use $\psi_0 = 1$.

$$\psi_{cb} = \frac{1.8}{J^{0.8}} \geq 1 \quad (5.2.19)$$

If Eq. (5.2.19) predicts $\psi_{cb} < 1$, use $\psi_{cb} = 1$.

$$\text{For } J > 1, \quad \psi_{nb} = \psi_0 \quad (5.2.20)$$

Ψ equals the larger of ψ_{nb} and ψ_{cb} given by the above equations. Thus if $\psi_{nb} > \psi_{cb}$, $\psi = \psi_{nb}$. Otherwise $\psi = \psi_{cb}$.

$$\text{For } 0.1 < J \leq 1.0, \quad \psi_{bs} = F\psi_0 \exp(2.74J^{-0.1}) \quad (5.2.21)$$

Ψ is the larger of ψ_{bs} and ψ_{cb}.

$$\text{For } J \leq 0.1, \quad \psi_{bs} = F\psi_0 \exp(2.47J^{-0.15}) \quad (5.2.22)$$

$F = 0.064$ for $\text{Bo} \geq 11 \times 10^{-4}$ and $F = 0.067$ for $\text{Bo} < 11 \times 10^{-4}$.

J is given by the following equation:

$$J = (0.38 Fr_{LT}^{-0.3})^n \text{Co} \quad (5.2.23)$$

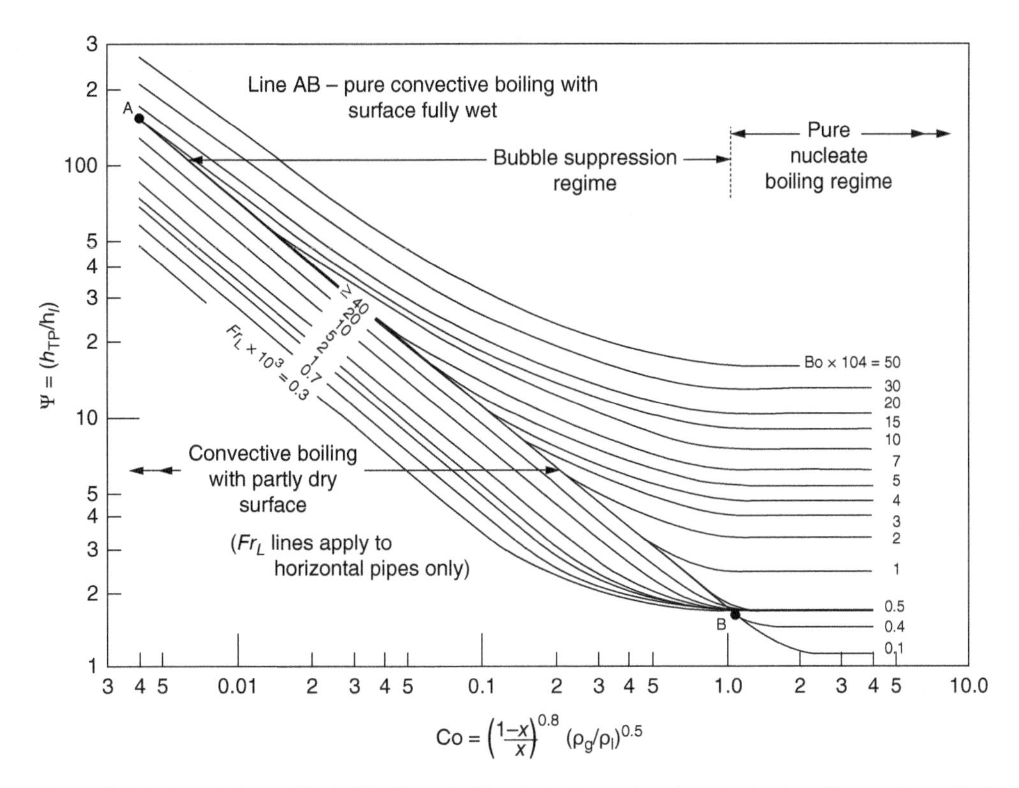

Figure 5.2.4 Correlation of Shah (1976) for boiling in horizontal and vertical tubes. Source: From Shah (1982). © ASHRAE.

$n = 0$ for vertical tubes at all Fr_{LT} and for horizontal tubes at $Fr_{LT} \geq 0.04$. $n = 1$ for horizontal tubes at $Fr_{LT} < 0.04$.

For annuli, D is replaced by hydraulic equivalent diameter D_{HYD} when annular gap >4 mm and D_{HP} is used when annular gap <4 mm. This was based on analysis of data from several sources for internally, externally, and bilaterally heated annuli. All properties are calculated at the saturation temperature.

It is interesting and useful to study the trends predicted by the Shah correlation. Figure 5.2.5a shows the predictions for $(\rho_g/\rho_f) = 0.001$. It is seen that heat transfer coefficient increases with quality up to critical quality at all values of boiling number. This indicates that convective effects are dominant. Figure 5.2.5b shows the predictions for $(\rho_g/\rho_f) = 0.1$. It is seen that heat transfer coefficient remains constant with increasing quality or decreases. This shows that at higher pressures, nucleate boiling is the dominant mode. The trends shown in these figures are similar to those in the data of Chawla shown in Figures 5.2.1 and 5.2.2.

Shah (2015a) studied data for boiling of carbon dioxide from many sources. It was found that for most data sets, good agreement was obtained if Eq. (5.2.19) is replaced by the following equation:

$$\psi_0 = 1820\text{Bo}^{0.68} \geq 1 \tag{5.2.24}$$

It is therefore recommended for use with carbon dioxide.

Shah (2006) compared a varied database with several correlations that included Shah (1982), Kandlikar (1990), Gungor and Winterton (1987), Liu and Winterton (1991), and Steiner and Taborek (1992). The data included 30 fluids (water, halogenated refrigerants, ammonia, cryogens, and chemicals), reduced pressures 0.0053–0.78, and diameters 1.1–27.1 mm. The correlations of Shah and Gungor and Winterton (1987) gave good agreement with data with both having a MAD of about 17.5%. Giving equal weight to each data set, the MAD of Liu and Winterton, Steiner and Taborek, and Kandlikar were 37.5%, 26.5%, and 55.0%, respectively. These correlations are described later in this section.

Shah (2016) has given a modification to his correlation that extends it to very small channel sizes. That is presented in Section 5.2.2.3.

Kandlikar Correlation The Kandlikar correlation uses the same correlating parameters as the Shah correlation except that it introduces a parameter F_{fl}. The original correlation was given in Kandlikar (1990). The equations were rearranged in Kandlikar (1991). The correlation is

$$h_{TP,NB} = 0.6683\text{Co}^{-0.2}(1-x)^{0.8} f_{Fr} h_{LT} + 1058\text{Bo}^{0.7}(1-x)^{0.8} F_{fl} h_{LT} \tag{5.2.25}$$

$$h_{TP,CB} = 1.136\text{Co}^{-0.9}(1-x)^{0.8} f_{Fr} h_{LT} + 667.2\text{Bo}^{0.7}(1-x)^{0.8} F_{fl} h_{LT} \tag{5.2.26}$$

h_{TP} is larger than the $h_{TP,NB}$ and $h_{TP,CB}$ given by the above equations.

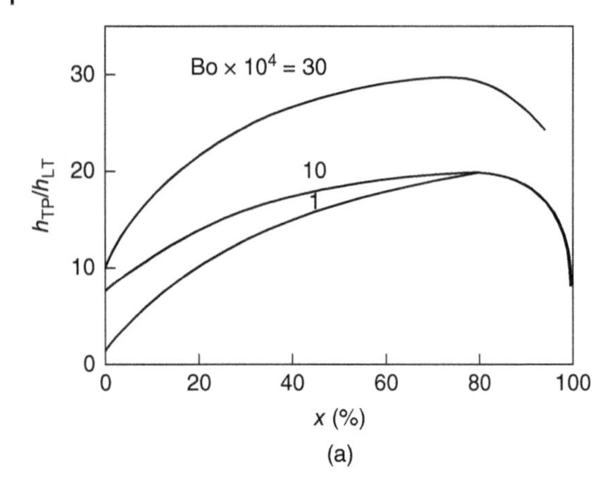

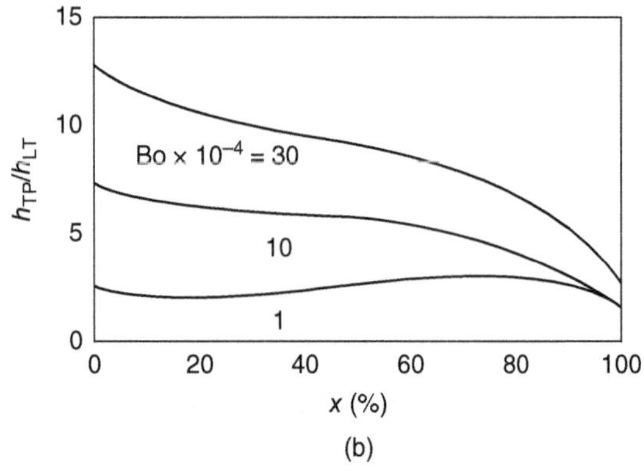

Figure 5.2.5 (a) Predictions of Shah correlation for (a) $(\rho_g/\rho_f) = 0.001$ (b) $(\rho_g/\rho_f) = 0.1$. Source: From Shah (1982). © ASHRAE.

For horizontal tubes with $Fr_{LT} < 0.04$, f_{Fr} is given by

$$f_{Fr} = (25Fr_{LT})^{0.3} \tag{5.2.27}$$

For horizontal tubes at $Fr_{LT} \geq 0.04$ and for vertical channels, $f_{Fr} = 1$.

The all-liquid single-phase heat transfer coefficient h_{LT} is calculated by the Dittus–Boelter correlation, Eq. (5.2.9) with $x = 0$. Values of F_{fl} are listed in Table 5.2.1. These apply to all tube materials.

Kandlikar and Steinke (2003) and Kandlikar and Subramanian (2004) made two changes. Firstly, $F_{fl} = 1$ for stainless steel tubes for all fluids. Secondly, h_{LT} for macro channels is to be calculated as below.

For $D > 3$ mm:

For $10^4 \leq Re_{LT} \leq 5 \times 10^6$, use the correlation of Pethukov.
For $3000 \leq Re_{LT} \leq 10^4$, use the correlation of Gnielinski.

These correlations are given in Chapter 1.

Kandlikar (1990) compared his correlation with a database of about 10 000 data points from many sources and reported good agreement.

Table 5.2.1 Values of F_{fl} in the Kandlikar et al. correlation for surfaces other than stainless steel for which $F_{fl} = 1$ for all fluids.

Fluid	F_{fl}
Water	1.00
R-11	1.30
R-12	1.50
R-13B1	1.31
R-22	2.20
R-113	1.30
R-114	1.24
R-134a	1.63
R-152a	1.10
R-32/R-132	3.30
R-141b	1.80
R-124	1.00
Kerosene	0.488
Nitrogen	4.7
Neon	3.5

Source: From Kandlikar and Steinke (2003). © 2003 ASHRAE.

Many other researchers have found the Kandlikar (1990) correlation to be erratic. These include Shah (2006), Gungor and Winterton (1986, 1987), and Liu and Winterton (1991).

Correlations of Winterton et al. Winterton and coworkers have published three well-known correlations that are described in the following text.

Gungor and Winterton (1986) gave a correlation in the form of the Chen correlation, Eq. (5.2.1). Their expressions for E and S are

$$E = 1 + 24\,000\text{Bo}^{1.16} + 1.37X_{tt}^{-0.86} \tag{5.2.28}$$

$$S = (1 + 1.15 \times 10^{-6} E^2 Re_{LS}^{1.17})^{-1} \tag{5.2.29}$$

Pool boiling heat transfer is calculated with the simplified Cooper correlation, Eq. (3.2.41).

For horizontal tubes with $Fr_{LT} < 0.05$, E should be multiplied by E_2, which is

$$E_2 = Fr_{LT}^{(0.1 - 2Fr_{LT})} \tag{5.2.30}$$

And S should be multiplied by S_2 given by

$$S_2 = \sqrt{Fr_{LT}} \tag{5.2.31}$$

For annuli, equivalent diameter is determined the same way as in the Shah correlation.

Gungor and Winterton (1986) compared this correlation with data from many sources, which included water, halogenated refrigerants, and ethylene glycol. Agreement was good. They reported that Shah (1982) correlation was also in good agreement with these data, while several other correlations were found erratic.

Gungor and Winterton (1987) gave a correlation using boiling number, which is as below:

$$h_{TP} = EE_2 h_{LS} \tag{5.2.32}$$

$$E = 1 + 3000\text{Bo}^{0.86} + 1.12\left(\frac{x}{1-x}\right)^{0.75}\left(\frac{\rho_f}{\rho_g}\right)^{0.41} \tag{5.2.33}$$

If the tube is horizontal and $Fr_{LT} < 0.05$, E_2 is given by Eq. (5.2.30). For horizontal tubes with $Fr_{LT} > 0.05$ and for vertical tubes, $E_2 = 1$. For annuli, D_{HP} is used as the equivalent diameter. h_{LS} is by Eq. (5.2.9). All properties are at saturation temperature.

This correlation was compared to a varied database, which included water, halogenated refrigerants, and organics. All data were predicted with a MAD of 20.8%. The Shah correlation also gave good agreement, while the Chen correlation had large deviations for both horizontal and vertical channels.

Liu and Winterton (1991) gave the following correlation:

$$h_{TP} = [(Eh_{LT})^2 + (Sh_{pb})^2]^{1/2} \tag{5.2.34}$$

$$E = \left[1 + xPr_f\left(\frac{\rho_f}{\rho_g} - 1\right)\right]^{0.35} \tag{5.2.35}$$

$$S = (1 + 0.055E^{0.1}Re_{LT}^{0.16})^{-1} \tag{5.2.36}$$

For horizontal tubes with $Fr_{LT} < 0.05$, E is multiplied by E_2 from Eq. (5.2.30) and S is multiplied by S_2 from Eq. (5.2.31). Pool boiling heat transfer is calculated with the simplified Cooper correlation, Eq. (3.2.41), and h_{LT} is by Eq. (5.2.9) with $x = 0$. All properties are at saturation temperature. They evaluated this correlation with the database of Gungor and Winterton (1987) together with some data for cryogenic fluids. MAD was a little less than that of Gungor and Winterton (1987) correlation. MAD of Shah correlation was a little higher.

Steiner–Taborek Correlation Steiner and Taborek (1992) presented the following correlation:

$$h_{TP} = [(F_{fc}h_{LT})^3 + (h_{nb,0}F_{nb})^3]^{1/3} \tag{5.2.37}$$

h_{LT} is calculated by a single-phase correlation such as that of Pethukov. $h_{nb,0}$ is a reference heat transfer coefficient for nucleate boiling at reduced pressure of 0.1, tube diameter $D_0 = 0.01$ m, mean surface roughness $R_{p,0} = 1\ \mu$m, and a standard heat flux $q_{nb,0}$. Values of $h_{nb,0}$ and $q_{nb,0}$ are unique for each fluid and a table of such values for many fluids is provided. F_{nb} is calculated by the following equation:

$$F_{nb} = F_{pr}\left(\frac{q}{q_{nb,0}}\right)^n\left(\frac{D}{D_0}\right)^{-0.4}\left(\frac{R_p}{R_{p,0}}\right)^{0.133}F_{MW} \tag{5.2.38}$$

F_{pr} is a factor that depends on reduced pressure and F_{MW} is a function of molecular weight. F_{fc} is the enhancement factor due to forced convection. Formulas are given for calculation of these factors.

This correlation was developed exclusively with data for vertical channels. Shah (2006) compared several correlations with a wide range of data for 30 fluids. Steiner–Taborek correlation worked well for vertical tubes and annuli with MAD of 20.4%, while those for Shah (1982) and Gungor and Winterton (1987) correlation were 18.0% and 15.9%, respectively. For horizontal channels, Steiner–Taborek correlation had MAD of 36.8%. Hence this correlation should not be used for horizontal channels.

The major difficulty with this correlation is that it requires values of $h_{nb,0}$ and $q_{nb,0}$ for each fluid. This difficulty is even greater for non-azeotropic mixtures as the composition of mixture changes along the tube as evaporation occurs. Therefore, different $h_{nb,0}$ and $q_{nb,0}$ will be needed as the evaporation proceeds. As well-verified correlations applicable to all fluids without needing such information are available, this correlation is of limited interest.

EPFL Flow Pattern-Based Correlations Researchers at EPFL in Lausanne, Switzerland, have presented flow pattern-based correlations for horizontal tubes that are briefly described herein.

Kattan et al. (1998) presented a flow pattern-based correlation for boiling heat transfer in horizontal channels that was developed from a varied database. It was later modified by Wojtan et al. (2005). This correlation was verified with data for eight halogenated refrigerants from several sources. Cheng et al. (2008) found the Wojtan et al. (2005) correlation unsatisfactory when compared to CO_2 data. They therefore modified it to suit the carbon dioxide data. A useful feature of these correlations is that they also cover the post-CHF region where transition boiling occurs. Other correlations, such as those of Shah and Gurgor–Winterton are applicable to pre-CHF conditions only. Shah (2015a) compared pre-CHF data for carbon dioxide from many sources to several correlations including Cheng et al. (2008) correlation. It gave good agreement with data.

As there is lack of reliable correlations for the transition boiling region, these correlations could be useful there. This is especially the case for carbon dioxide as with it, CHF often occurs at low qualities and a large part of the tube is under post-CHF conditions.

5.2.2.2 Correlations for Minichannels

Numerous correlations for minichannels have been published in recent years. Some of the more verified ones are discussed in the following. Among these, the most verified is that of Kim and Mudawar (2013), which is described first.

Kim–Mudawar Correlation Kim and Mudawar (2013) developed the following correlation based on a large and varied database:

$$h_{\text{TP}} = (h_{\text{nb}}^2 + h_{\text{cb}}^2)^2 \tag{5.2.39}$$

$$h_{\text{nb}} = \left[2345\left(\text{Bo}\frac{P_H}{P_{\text{wet}}}\right)^{0.7} p_r^{0.38}(1-x)^{-0.51}\right] h_{\text{LS}} \tag{5.2.40}$$

$$h_{\text{cb}} = \left[5.2\left(\text{Bo}\frac{P_H}{P_{\text{wet}}}\right)^{0.08} We_{\text{LT}}^{-0.54} + 3.5\left(\frac{1}{X_{\text{tt}}}\right)^{0.94}\left(\frac{\rho_g}{\rho_f}\right)^{0.25}\right] h_{\text{LS}} \tag{5.2.41}$$

$$We_{\text{LT}} = \frac{G^2 D}{\rho_f \sigma} \tag{5.2.42}$$

P_{H} is the heated perimeter and P_{wet} is the wetted perimeter. For non-circular channels, hydraulic diameter is used throughout. h_{LS} is calculated by the Dittus–Boelter equation.

The heat transfer coefficients predicted are for circular channel with all perimeter heated. If a rectangular channel is heated on three sides only, a correction factor is applied by using Eq. (2.3.47)–(2.3.49) as in the Kim and Mudawar correlation for condensation.

This correlation was compared to a pre-dryout database consists of 18 fluids, hydraulic diameters of 0.19–6.5 mm, mass velocities of 19–1608 kg m^{-2} s^{-1}, Re_{LT} of 57–49 820, qualities of 0–1, and reduced pressures of 0.005–0.69. It had a MAD of 20.3%. The same database was also compared to several other correlations for macro channels and minichannels. All of them had greater deviations. Very large deviations were found with correlations for minichannels, some of which are presented in the following.

Various Correlations for Minichannels Only those correlations that were based on wide-ranging databases are discussed herein.

According to Li and Wu (2010b), minichannels are those for which ($Re_{\text{LO}}{}^{0.5}$ Bd) ≤ 200. They gave the following correlation:

$$\text{For } (Re_{\text{LS}}{}^{0.5}\, Bd) \le 200, \quad \frac{h_{\text{TP}}D}{k_f} = 22.9(Re_{\text{LS}}^{0.5} Bd)^{0.355} \tag{5.2.43}$$

$$\text{For } (Re_{\text{LS}}{}^{0.5}\, Bd) > 200, \quad \frac{h_{\text{TP}}D}{k_f} = 30 Re_{\text{LS}}^{0.857}\text{Bo}^{0.714} \tag{5.2.44}$$

Equation (5.2.44) was given by Lazarek and Black (1982). This correlation was stated to agree well with a very wide-ranging database for minichannels while all other correlations they tried performed poorly. However, comparison with extensive database by Shah (2017a) found this correlation to give large deviations with most data sets.

Li and Wu (2010a) gave the following correlation:

$$\frac{h_{\text{TP}}D}{k_f} = 334\text{Bo}^{0.3}(Re_{\text{LS}}^{0.36} Bd)^{0.4} \tag{5.2.45}$$

They did not give any criterion for the boundary between minichannels and macro channels. They analyzed a large database with D_{HYD} from 0.19 to 3.0 mm and reported good agreement with this correlation, while other correlations they tried had much larger deviations. Kim and Mudawar (2013) found this correlation to be very erratic on comparison with their database.

Sun and Mishima (2009) developed the following correlation:

$$h_{\text{TP}} = \frac{6 Re_{\text{LT}}^{1.05}\text{Bo}^{0.54}}{We_{\text{LT}}^{0.191}\left(\frac{\rho_f}{\rho_g}\right)^{0.142}} \frac{k_f}{D_{\text{HYD}}} \tag{5.2.46}$$

This correlation was compared to data for 11 fluids including halogenated refrigerants, carbon dioxide, and water in channels with hydraulic diameter from 0.21 to 6.5 mm. They reported that it had a MAD of 30.8%, which is really not good enough.

They report that all other correlations tested by them had larger deviations. Shah (2017a,b) found it unsatisfactory on comparison with a large database as described in Section 5.2.2.3.

Bertsch et al. (2009) have given the following correlation:

$$h_{\text{TP}} = h_{\text{pb}}(1-x) + h_{C,\text{TP}}[1 + 80(x^2 - x^6)e^{-0.6/Bd}] \tag{5.2.47}$$

h_{pb} is calculated by the Cooper correlation, Eq. (3.2.40). They have not stated whether 1.7 factor for copper is to be used or not. $h_{C,\text{TP}}$ is calculated by the following equation:

$$h_{C,\text{TP}} = h_{\text{LT}}(1-x) + h_{\text{GT}}x \tag{5.2.48}$$

h_{LT} and h_{GT} are the single-phase heat transfer coefficients for all mass flowing as liquid and vapor, respectively. These are calculated by the following laminar flow correlation of Hausen (1943):

$$h = \left(3.66 + \frac{0.0668\frac{D_{\text{HYD}}}{L} RePr}{1 + 0.04\left[\frac{D_{\text{HYD}}}{L} RePr\right]^{2/3}}\right)\frac{k}{D_{\text{HYD}}} \tag{5.2.49}$$

They compared this correlation to database of 3899 data points from 14 studies in the literature covering 12 different wetting and non-wetting fluids, hydraulic diameters ranging from 0.16 to 2.92 mm, and Bond numbers from 0.025 to 3.3. MAD was 28%. Kim and Mudawar (2013) report MAD of this correlation with their database of 30.8%.

Thome et al. (2004) developed a three-zone model for heat transfer in the elongated bubble regime in minichannels. The heat transfer model predicts the transient variation in local heat transfer coefficient during the cyclic passage of a liquid slug, an evaporating elongated bubble, and a vapor slug from which the time-averaged heat transfer coefficient is calculated. Bertsch et al. (2009) compared this model to a wide-ranging database and found it very erratic, its MAD being 51%. Ong and Thome (2011) compared this model with their data for refrigerants boiling in channels of diameter 1.03–3.04 mm. They found that the model does not give good agreement with the assumed minimum liquid film thickness of 300 nm but gave good agreement when the measured surface roughness was used as minimum liquid film thickness. The measured surface roughness of channels varied from 596 to 897 nm. Independent verification of this revised model has not come to the present author's attention.

From the foregoing, it may be concluded that among the many correlations specifically for minichannels, only that of Kim and Mudawar has been shown to be reliable.

5.2.2.3 Correlations for Both Minichannels and Macrochannels

Kandlikar et al. Correlation Kandlikar and Steinke (2003) and Kandlikar and Subramanian (2004) extended the Kandlikar (1990) correlation to minichannels by making three changes. Firstly, single-phase heat transfer coefficient h_{LT} is calculated as below:

For $Re_{LT} > 3000$, use the Gnielinski equation.

For $Re_{LT} < 1600$, $Nu = K$ (a constant)
For Re_{LT} between 1600 and 3000, linearly interpolate between the values at Re_{LT} of 1600 and 3000.

The Nusselt numbers at $Re_{LT} \leq 1600$ are as follows:
For circular tubes, constant heat flux,

$$Nu_H = 4.36 \tag{5.2.50}$$

For circular tubes, constant temperature boundary,

$$Nu_T = 3.66 \tag{5.2.51}$$

For square cross section, $Nu_H = 3.61$ and $Nu_T = 2.98$. A detailed listing for other geometries is given by Kakac et al. (1987). Nu_H is the Nusselt number under constant heat flux condition and Nu_T is the Nusselt number with constant wall temperature.

The second change is that Froude number has no effect for minichannels. Thus, $f_{Fr} = 1$.

The third change is that for $Re_{LT} < 100$, heat transfer due to convection is considered negligible.

$$\text{For } Re_{LT} < 100, \quad h_{TP} = h_{TP,NB} \tag{5.2.52}$$

With these modifications, data for channels of diameters down to 0.19 mm were satisfactorily correlated. Bertsch et al. (2009) compared this correlation to data for minichannels from 14 sources. Its MAD was 66%.

Shah Correlation Shah (2017a) compared his 1982 correlation for macro channels to a large amount of data for minichannels. The results showed fairly good agreement with most data. However, some data at low flow rates were considerably underpredicted. It seemed likely that this may be due to the effect of surface tension. At high mass flow rates, inertia force is dominant, while at low flow rates the surface tension force dominates. Weber number is the ratio between surface tension force and inertia force. It seemed likely that this behavior may be quantified by its use. For all mass flowing as vapor, it is defined as

$$We_{GT} = \frac{G^2 D}{\rho_g \sigma} \tag{5.2.53}$$

Analysis showed that We_{GT} was indeed having an effect. In analyzing the data for condensation in minichannels, Shah (2016) had found that Weber number had effect when it was less than 100. The boiling data showed that boiling number was also having an effect. The transition Weber number decreased as boiling number increased. Data analysis yielded the following relation:

$$E_s = \frac{h_{TP}}{h_{Shah}} = 2.1 - 0.008 We_{GT} - 110\text{Bo} \tag{5.2.54}$$

If $E_s < 1$, use $E_s = 1$. h_{Shah} is the prediction of Shah (1982) correlation. Figure 5.2.6 shows the transition Weber number, when $E_s = 1$, predicted by Eq. (5.2.54). Figure 5.2.7 illustrates the effect of this factor on heat transfer.

Shah recommended that for partially heated channels, D_{HYD} be used in Weber number and Froude number. D_{HP} is to be used in all other parameters. This recommendation was based on his research on condensation heat transfer, Shah (2016), and physical reasoning. He noted that in the data analyzed for boiling, there was not sufficient difference between D_{HYD} and D_{HP} to have much effect. Hence he recommended that this choice should be reviewed when data for widely differing D_{HYD} and D_{HP} become available.

This correlation was verified with data for 30 fluids from 81 sources, which included diameters from 0.38 to 26.2 mm. Figures 5.2.8 and 5.2.9 show the comparison of various correlations with data from two sources. The

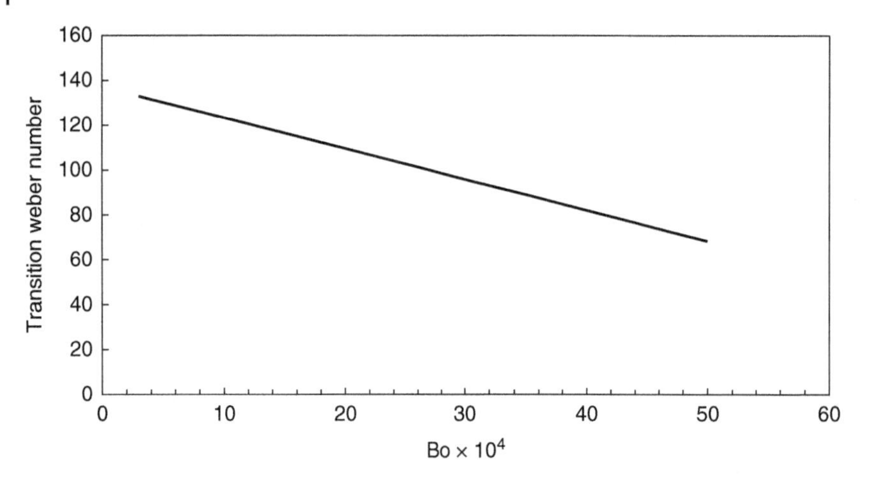

Figure 5.2.6 Transition Weber number during boiling in channels. Source: From Shah (2017a). © 2017 Elsevier.

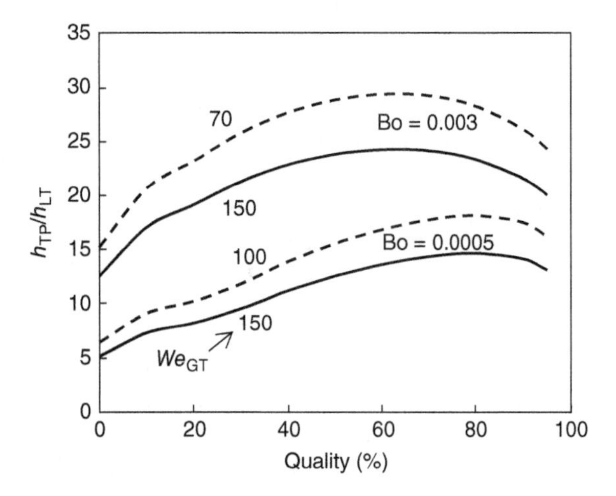

Figure 5.2.7 Effect of Weber number on heat transfer at $p_r = 0.0512$, corresponds to R-134a at −40 °C. Source: From Shah (2017a). © 2017 Elsevier.

complete range of data analyzed is shown in Table 5.2.2. The channels included round tubes as well as channels of other shapes such as rectangular and triangular, fully or partially heated. All data were predicted with MAD of 18.5%.

Table 5.2.3 shows the deviations of this correlation as well as those of several other correlations in various ranges. It is seen that for minichannels ($D \leq 3$ mm), several correlations give good agreement when $E_s \leq 1$ but their deviations increase greatly for $E_s > 1$. For example, the MAD of Gungor and Winterton (1987) correlation increases from 20.1% to 30.6% and that of Shah (1982) from 19.8% to 33.7%. Even for the correlation of Kim and Mudawar that was developed for minichannels, the MAD increases from 22.5% to 25%. A similar trend is seen in the data for $D > 3$ mm. The MAD of the Liu–Winterton correlation increases from 23.4% to 33.1% when $E_s \leq 1$ and $E_s > 1$, respectively. The deviations of the Gungor and Winterton (1987) and Shah (1982) correlations also increase significantly.

Table 5.2.2 The range of data analyzed by Shah (2017a).

Parameter	Range
Fluids	Water, CO_2, R-11, R-12, R-22, R-32, R-113, R-123, R-114, R-123, R-134a, R-152a, R-1234yf, R-236fa, R-245fa, ammonia, propane, isobutane, carbon tetrachloride, isopropyl alcohol, ethanol, methanol, *n*-butanol, cyclohexane, benzene, heptane, pentane, argon, hydrogen, nitrogen, and helium (30 fluids)
Channels	Single and multiport channels of various shapes (round, rectangular, and triangular), fully or partly heated, horizontal and vertical upflow
D_{HYD} (mm)	0.38–27.1
Aspect ratio	0.25–1
G (kg m^{-2} s^{-1})	15–2437
p_r	0.0046–0.787
We_{GT}	14–1.6×10^6
Fr_{LT}	0.0018–821
Bond number	0.15–527
Number of data points	4852
Number of data sets	137
Number of data sources	81

Source: From Shah (2017a). © 2017 Elsevier.

It is therefore clear that the effects of surface tension start when $E_s > 1$ irrespective of the channel diameter. Correlations for macro channels become inapplicable when $E_s > 1$ and even the correlations specifically developed for minichannels begin to be erratic in this range.

From the results in Table 5.2.3, it may be concluded that macro channel correlation can be used only when $E_s \leq 1$ and the most accurate in this range is Shah (1982)

Table 5.2.3 Deviations of various correlations for boiling heat transfer in different ranges of parameters.

D_{HYD} (mm)	Limit	Number of data points	Deviation (%) Mean absolute verage							
			GW 87	GW 86	Liu Winterton	Sun and Mishima	Li and Wu (2010b)	Kim Mudawar	Shah (1982)	Shah (2017a)
All	Li and Wu criterion[a]	1621	23.0	38.6	24.8	32.9	40.1	24.5	22.4	19.1
			−7.9	30.4	−12.2	−16.9	14.1	−13.0	−12.3	−6.3
≤3	$E_s \le 1$	2155	20.1	38.6	23.0	30.3	33.7	22.5	19.8	19.8
			−2.0	34.5	−9.0	−6.7	−2.0	−7.1	−8.1	−8.1
	$E_s > 1$	348	30.6	33.9	27.6	34.7	44.7	25.0	33.7	17.5
			−29.8	14.8	−24.6	−19.8	37.2	−21.2	−33.0	−5.4
	All E_s	2503	21.6	37.9	23.6	30.9	35.2	22.9	21.8	19.5
			−5.9	31.8	−11.1	−8.5	3.3	−9.0	−11.5	−7.7
>3	$E_s \le 1$	2263	20.7	22.4	23.4	41.9	38.9	39.7	17.6	17.6
			−8.8	13.2	6.9	6.6	−16.5	−24.5	−5.5	−5.5
	$E_s > 1$	86	24.8	20.2	33.2	39.3	59.7	29.8	20.6	12.5
			−24.8	−16.5	−33.2	−31.1	18.8	−28.5	−20.2	2.3
	All E_s	2349	20.9	22.3	23.8	41.8	39.6	39.3	17.7	17.4
			−9.4	12.1	5.4	5.2	−15.2	−24.7	−6.0	−5.2
All	$E_s > 1$	453	29.0	30.7	28.0	37.1	47.2	26.5	30.9	17.9
			−27.3	8.8	−25.4	−18.1	31.8	−20.1	−28.4	−1.6
All	All E_s	4852	21.2	30.4	23.7	36.2	37.4	30.8	19.8	18.6
			−7.6	22.3	−3.1	−1.9	−5.7	−6.6	−8.9	−6.4

a) Li and Wu criterion for minichannels, $(Re_{LO}\,Bd^{0.5}) \le 200$.

Source: From Shah (2017a). © 2017 Elsevier.

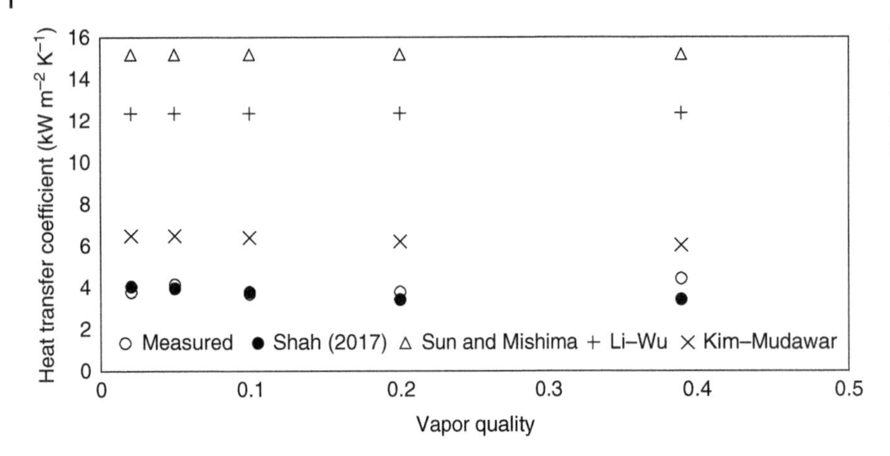

Figure 5.2.8 Comparison of various correlations with the data of Ogata and Sato (1974) for helium in a vertical tube: $D = 1.09$ mm, $G = 86$ kg m^{-2} s^{-1}, $p_r = 0.478$, and $Bo = 3.6 \times 10^{-4}$. Source: From Shah (2017a). © 2017 Elsevier.

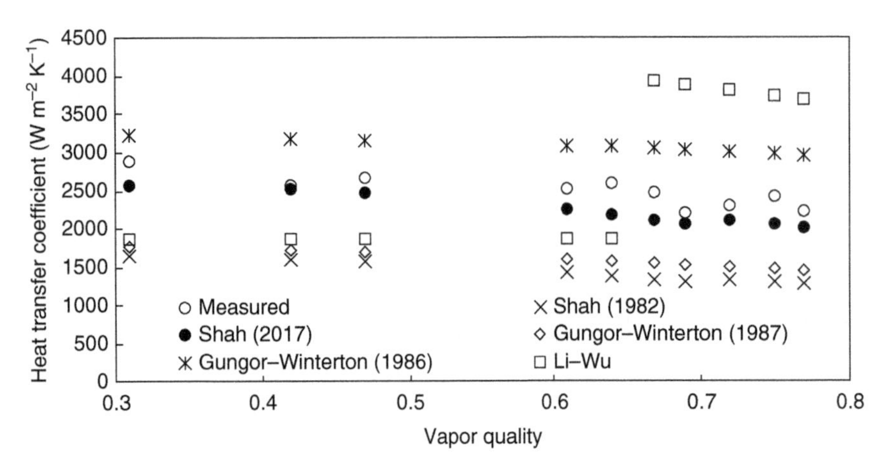

Figure 5.2.9 Comparison of various correlations with the data of Tran et al. (1996) for R-12 in a horizontal tube: $D = 2.46$ mm, $G = 89$ kg m^{-2} s^{-1}, $p_r = 0.1977$, $We_{GT} = 56$, and $Bo = 6.3 \times 10^{-4}$. Source: From Shah (2017a). © 2017 Elsevier.

and Gungor and Winterton (1987). For $E_s > 1$, only Shah (2017a) is reliable.

5.2.2.4 Recommendations

Shah (2017a) correlation is the most accurate available for both minichannels and macro channels over the entire range of parameters. It is therefore the first choice for application. For $E_s \leq 1$, the Gungor and Winterton (1987) is a good alternative except for carbon dioxide. Among the correlations for minichannels, the only one that is fairly accurate is that of Kim and Mudawar. Its accuracy begins to deteriorate at $E_s > 1$ and hence should be used only when $E_s \leq 1$.

5.3 Plate-Type Heat Exchangers

Literature on boiling in plate-type heat exchangers has been reviewed among others by Eldeeb et al. (2016), Almalfi et al. (2016a), and Ayub et al. (2019).

There are many types of plate heat exchangers. The most widely used type are those with herringbone or chevron plates. Other types include plain plate, serrated fin, and plate fin types. These are discussed in the following.

5.3.1 Herringbone Plate Type

The geometry of herringbone plates is shown in Figure 2.7.1. However, the chevron angle β used in this chapter is measured from the vertical axis while in Figure 2.7.1, it is measured from the horizontal axis. For use in boiling and condensation, the plates are brazed together to ensure leak-tightness. These are therefore often known as brazed plate heat exchangers (BPHEs).

An important point is the definition of equivalent diameter D_{eq} and hydraulic diameter D_{HYD}. Almalfi et al. (2016a) use the following definitions:

$$D_{eq} = 2b \tag{5.3.1}$$

$$D_{HYD} = \frac{2b}{\varphi} \tag{5.3.2}$$

φ is the enlargement factor.

Many other authors also use the same definitions, while some have these definitions interchanged. For example, Longo et al. (2015) have $D_{HYD} = 2b$. In the equations given below, definitions are according to Eqs. (5.3.1) and (5.3.2).

Kumar (1992) tested several PHEs, using R-22 and ammonia as test fluids, and found that the heat transfer coefficient kept on rising with increasing outlet vapor quality until it exceeded 0.7. Then it decreased considerably. This behavior is similar to that in round tubes. However most available correlations do not take into accounts the modes of heat transfer in the heat exchanger and just give an average heat transfer. A few of the most verified correlations are discussed in the following.

5.3.1.1 Longo et al. Correlation

Longo et al. (2015) have given the following correlation. Firstly, heat transfer coefficient is the larger of convective and nucleate boiling ones:

$$h_{\mathrm{TP}} = \text{maximum}\,(h_{\mathrm{cb}}, h_{\mathrm{nb}}) \tag{5.3.3}$$

The convective heat transfer coefficient is calculated as

$$h_{\mathrm{cb}} = 0.122\varphi\left(\frac{k_f}{D_{\mathrm{eq}}}\right) Re_{\mathrm{eq}}^{0.8} Pr_f^{1/3} \tag{5.3.4}$$

$$Re_{\mathrm{eq}} = G[(1-x) + x(\rho_f/\rho_g)^{1/2}]D_{\mathrm{eq}}/\mu_f \tag{5.3.5}$$

h_{nb} is calculated by a modified form of Gorenflo (1993) correlation as below:

$$h_{\mathrm{nb}} = 0.58\varphi h_{\mathrm{ref}} F_p \left(\frac{q}{q_{\mathrm{ref}}}\right)^{0.467} \left(\frac{R_a}{R_{a,\mathrm{ref}}}\right)^{2/15} \tag{5.3.6}$$

$$F_p = 1.2p_r^{0.27} + [2.5 + 1/(1-p_r)]p_r \tag{5.3.7}$$

The reference surface roughness $R_{a,\mathrm{ref}}$ is 1.4 μm as defined in ISO4287/1. The reference heat flux is 20 000 W m^{-2}. The reference heat transfer coefficient h_{ref} at the reference heat flux and reference reduced pressure 0.1 is listed for many fluids in Gorenflo (1993). As described in Section 3.2.3, Gorenflo has given an improved version of his correlation that does not require reference heat transfer coefficients. It may be used to calculate the reference heat transfer coefficients if they are not available for some fluids. Note that the heat transfer coefficients are based on the projected area of the plates.

If superheating occurs in the heat exchanger, the heat transfer coefficient of superheated vapor h_{sh} is calculated by the following equation:

$$h_{\mathrm{sh}} = 0.277(k_g/D_{\mathrm{eq}})Re_{\mathrm{GT}}^{0.766} Pr_g^{0.33} \tag{5.3.8}$$

This correlation was compared with authors' own data as well as data from five other studies. The fluids were R-600a, R-290, R-1270, and several halogenated refrigerants. Chevron angles ranged from 28° to 65° and corrugation height from 2 to 3.65 mm. Their own data were predicted with MAD of 13.6%. Data from other sources had larger deviations but deviations appear to be within ±30% for most data.

5.3.1.2 Almalfi et al. Correlation

By analyzing data from many sources, Almalfi et al. (2016b) have given the following correlation:

$$Bd < 4, \quad \frac{h_{\mathrm{TP}} D_{\mathrm{HYD}}}{k_f} = 982\left(\frac{\beta}{\beta_{\max}}\right)^{1.101} \times We_m^{0.315} \mathrm{Bo}^{0.315} \left(\frac{\rho_f}{\rho_g}\right)^{-0.224} \tag{5.3.9}$$

$$Bd > 4, \quad \frac{h_{\mathrm{TP}} D_{\mathrm{HYD}}}{k_f} = 18.495\left(\frac{\beta}{\beta_{\max}}\right)^{0.248} \times Re_{\mathrm{GS}}^{0.135} Re_{\mathrm{LT}}^{0.315} Bd^{0.235} \mathrm{Bo}^{0.198} \left(\frac{\rho_f}{\rho_g}\right)^{-0.223} \tag{5.3.10}$$

$$We_m = \frac{G^2 D_{\mathrm{HYD}}}{\rho_m \sigma} \tag{5.3.11}$$

where ρ_m is the mixture density for homogeneous flow. See Chapter 1 regarding homogeneous model of flow. Channels with $Bd < 4$ are considered minichannels and those with $Bd > 4$ are considered macro channels; $\beta_{\max} = 70°$.

They compared this correlation with data from 10 sources, which included halogenated refrigerants, water, ammonia, and hydrocarbons. Hydraulic diameters ranged from 1.7 to 8.0 mm and chevron angles from 27° to 70°. The verified range of this correlation is listed in Table 5.3.1. The 1903 data points were predicted with MAD of 22.1%. Many other correlations were also compared to the same database; all had much larger deviations.

Longo et al. (2019) measured heat transfer in a BPHE with chevron angle 65° using R1234ze(Z) and R1233zd(E). Measurements were in excellent agreement with the Almalfi et al. correlation as well as the Longo et al. (2015) correlation.

Table 5.3.1 Verified range of the Almalfi et al. (2016b) correlation for boiling heat transfer in herringbone-type plate heat exchangers.

Parameter	Range
Fluids	Water, ammonia, water–ammonia mixture, halogenated refrigerants
Chevron angle (°)	27–70
D_{HYD} (mm)	1.7–8
Re_{LT}	41–5360
Re_{LS}	23–5320
ρ_f/ρ_g	19–1350
Bond number	1.9–79
We_{m}	0.027–162

Source: Based on Almalf et al. (2016b).

5.3.1.3 Ayub et al. Correlation

Ayub et al. (2019) developed the following correlation:

$$\frac{h_{TP}D_{HYD}}{k_f} = \left(1.8 + 0.7\frac{\beta}{\beta_{max}}\right) Re_{eq}^{\left(0.49-0.3\frac{\sigma}{\sigma_{ammonia}}\right)} Bo_{eq}^{-0.2} \quad (5.3.12)$$

β_{max} is 65°. $\sigma_{ammonia}$ is the surface tension of ammonia. Equivalent mass flux G_{eq} is calculated as

$$G_{eq} = G(1 - x_{mean}) + x_{mean}\left(\frac{\rho_f}{\rho_g}\right)^{0.5} \quad (5.3.13)$$

where x_{mean} is the mean of the inlet and outlet qualities.

$$Re_{eq} = \frac{G_{eq}D_{HYD}}{\mu_f} \quad (5.3.14)$$

$$Bo_{eq} = \frac{q}{G_{eq}i_{fg}} \quad (5.3.15)$$

This correlation was verified with data for ammonia, R-134a, R-1234yf, R-236fa, and R-410A from 10 sources. Chevron angles for herringbone plates were from 27° to 65° though data from two of the sources for ammonia was for plain plates, $\beta = 0°$. Their figures showing comparison of data with their correlation show a significant amount of data outside the ±30% band. They found the Almalfi et al. (2016b) correlation to be less accurate than their own in comparison with the same database.

5.3.1.4 Recommendation

The three discussed earlier are the most verified correlations. Among these, that of Almalfi et al. has been verified with more fluids and a wider data range than others. It is therefore recommended as the first choice.

5.3.2 Plane Plate Heat Exchangers

Arima et al. (2010) boiled ammonia in a vertical plain plate heat exchanger. It consisted of a plate heated by water on one side. The flow passage was formed by an unheated plate. Mass flux was 7.5–15 kg m^{-2} s^{-1}, pressure was 0.7–0.9 MPa, and quality at inlet was 0.1–0.4. They found that the Shah (1982) and Kandlikar (1990) correlations grossly underpredicted their data. Further, their measured heat transfer coefficients were close to the pool boiling heat transfer coefficients for ammonia measured by Arima et al. (2003), while the predictions of Stephan and Abdelsalam correlation for pool boiling (see Chapter 3) were much lower. It therefore appears that the reason for the failure of the Shah and Kandlikar correlations is that nucleate boiling was unusually strong.

They fitted the following correlation to their local heat transfer data:

$$\frac{h_{TP}}{h_{LS}} = 16.4\left(\frac{1}{X_{vv}}\right)^{1.08} \quad (5.3.16)$$

X_{vv} is the Martinelli parameter for laminar–laminar flow given by the following equation:

$$X_{vv} = \left(\frac{1-x}{x}\right)^{0.5}\left(\frac{\rho_g}{\rho_f}\right)^{0.5}\left(\frac{\mu_f}{\mu_g}\right)^{0.5} \quad (5.3.17)$$

h_{LS} is calculated by the Dittus–Boelter equation.

Koyama et al. (2014) boiled ammonia in a plate heat exchanger formed by an electrically heated titanium plate and an insulated plate on the other side. The flow gap was alternately, 1, 2, and 5 mm. Mass flux was 5–7.5 kg m^{-2} s^{-1}, and saturation pressure was 7–9 bar. They fitted the following correlation to their data:

$$\frac{h_{TP}}{h_{LS}} = B\left(\frac{1}{X_{vv}}\right)^{n} \quad (5.3.18)$$

For flow gap of 1 mm, $B = 52.2$ and $n = 0.90$. For gaps of 2 and 5 mm, $B = 48.6$ and $n = 0.79$.

The data of Arima et al. and Koyama et al. were satisfactorily predicted by the correlation of Ayub et al. (2019) as discussed in Section 5.3.1.

There is no verified general predictive technique for plane plate heat exchangers. The correlations of Arima et al. and Koyama et al. should not be used outside the range of data on which they are based. It is possible that the correlation of Ayub et al. may be applicable beyond the range of these data but it needs verification.

5.3.3 Serrated Fin Plate Heat Exchangers

Figure 5.3.1 shows a typical rectangular serrated or offset-fin plate-fin heat exchanger. The perforated fin type has perforations on the fins.

Robertson and Wadekar (1988) performed tests on boiling of cyclohexane in a perforated fin-type heat exchanger. They found the heat transfer behavior very similar to that during their tests in a round tube. Heat transfer coefficients at higher flow rates were in agreement with the Chen (1966) correlation for tubes but did not agree with it at low flow rates. The authors attributed this difference to the much smaller hydraulic diameter of their plate heat exchanger.

Feldman et al. (2000) performed tests with R-114 flowing in electrically heated serrated-fin and perforated fin test sections to measure local boiling heat transfer coefficients with mass flux up to 45 kg m^{-2} s^{-1}, heat flux up to 3500 W m^{-2}, and pressure up to 3 bar. Several fin heights and fin lengths were used. Hydraulic diameter ranged from 1.67 to 2.06 mm. Figure 5.3.2 shows the data for the

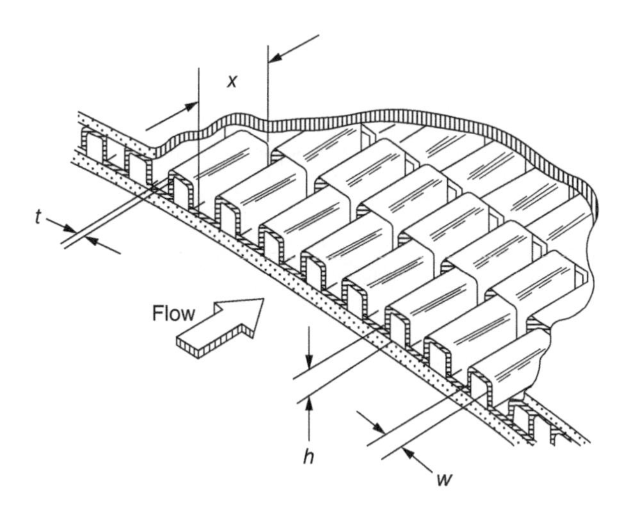

Figure 5.3.1 Typical rectangular offset-fin plate-fin heat exchanger. Source: From Wieting (1975). © 1975 American Society of Mechanical Engineers.

perforated fin test section. It is seen that the behavior is very similar to that in tubes shown in Figure 5.2.2. There are two regimes of boiling. In the nucleate boiling regime, heat transfer coefficients increase with increasing heat flux and quality has no effect. At each heat flux, a quality is reached when heat transfer coefficient increases with increasing quality and heat flux has no effect. This is the convective regime.

They developed a correlation as follows. The convective heat transfer coefficient h_{cb} is given by

$$h_{\text{cb}} = Fh_{\text{LS}} \tag{5.3.19}$$

The single-phase heat transfer coefficient of liquid phase is calculated by the correlation of Wieting (1975) for the offset fin and by the correlation of Shah and London (1978) for the perforated fin type. The Wieting correlation is as follows:

For $Re_{\text{LS}} < 1000$,

$$\frac{h_{\text{LS}}}{G(1-x)C_{\text{pf}}} Pr_f^{2/3} = 0.483\left(\frac{X}{D_{\text{HYD}}}\right)^{-0.162} A_R^{-0.184} Re_{\text{LS}}^{-0.536} \tag{5.3.20}$$

For $Re_{\text{LS}} > 1000$,

$$\frac{h_{\text{LS}}}{G(1-x)C_{\text{pf}}} Pr_f^{2/3} = 0.243\left(\frac{X}{D_{\text{HYD}}}\right)^{-0.0.322} \left(\frac{t}{D_{\text{HYD}}}\right)^{0.089} Re_{\text{LS}}^{-0.0.368} \tag{5.3.21}$$

Referring to the definitions shown in Figure 5.3.1, X is the length of the fin, t is the thickness of fin, A_R is the aspect ratio of fin = w/h, and $D_{\text{HYD}} = [2(w \times h)/(w + h)]$. $Re_{\text{LS}} = G(1-x)\, D_{\text{HYD}}/\mu_f$.

The factor F was found by data analysis to be

$$F = 1 + X_{\text{tt}}^{-0.79} \tag{5.3.22}$$

The nucleate boiling heat transfer coefficient h_{nb} is calculated by the simplified Cooper correlation, Eq. (3.2.41). Heat transfer coefficient is the higher of h_{cb} and h_{nb}. This correlation gave excellent agreement with their own data.

There is no well-verified method for calculation of heat transfer in serrated fin plate heat exchangers. The correlation of Feldman et al. (2000) may be used in its verified range.

5.3.4 Plate Fin Heat Exchangers

In this type, fins are formed on the plate surface itself. Typically, the plate has round or square microfins, in-line or staggered.

McNeil et al. (2010) used a plain surface as well as a pin-fin surface comprised of 1 mm square pin fins that were 1 mm high and located on a 2 mm² pitch array covering the base. The channel was 1 mm high and had a glass top plate. The boiling fluid was R-113 at atmospheric pressure. For

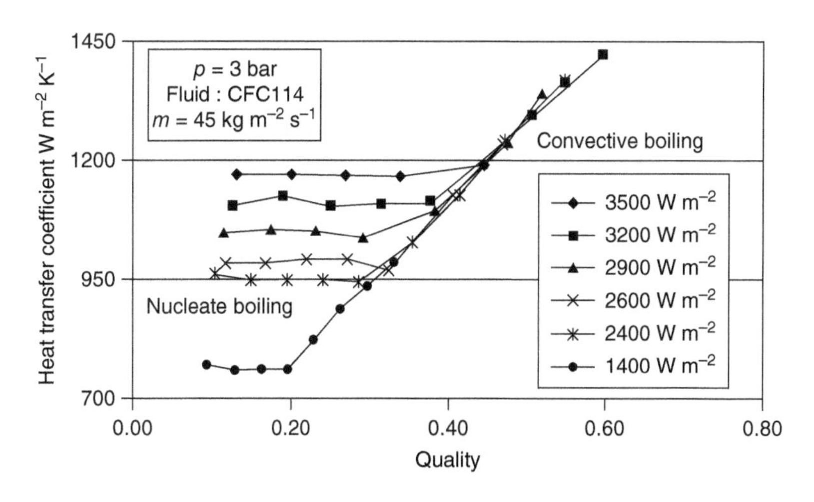

Figure 5.3.2 Heat transfer in a perforated fin plate heat exchanger. Source: Data of Feldman et al. (2000). © Elsevier

both surfaces, the mass flux range was 50–250 kg m^{-2} s^{-1} and the heat flux range was 5–140 kW m^{-2}. Heat transfer coefficient for the surface with pins was based on the total area including fins, corrected for the fin efficiency. The heat transfer coefficient of the surface with fins were about the same as that of the plain surface. However, much more heat could be transferred with the finned plate because of its greater surface area.

Many other experimental studies on pin-fin heat exchangers have been done, and many correlations have been proposed. For example, Krishnamurthy and Peles (2008) performed tests with boiling water and gave a correlation of their data in the form of the Chen correlation for tubes. Qu and Siu-Ho (2009) also did tests with water in a similar heat exchanger. Their data showed poor agreement with the correlation of Krishnamurthy and Peles. Therefore, they developed a new correlation to fit their data.

Despite numerous experimental studies and correlating attempts, there is no general method for predicting heat transfer in plate-fin heat exchangers.

5.4 Boiling in Various Geometries

5.4.1 Helical Coils

Helical coils with boiling are used in many applications including refrigeration and air conditioning, nuclear and conventional power plants, and chemical processes. Hence ability to predict heat transfer in them is of much practical interest. Many experimental studies have been done to measure heat transfer during boiling in helical coils and many correlations for predicting heat transfer coefficients have been proposed. These have been reviewed by Fsadini and Whitty (2016) and Shah (2019).

Experimental studies have shown that heat transfer coefficient is highest at the outer periphery of coil (furthest from the coil axis) and smallest at the inner periphery. An example from the study of Owhadi (1966) is seen in Figure 5.4.1. Similar behavior has been reported by Hardik and Prabhu (2017, 2018). This appears to be related to the fact that centrifugal force moves the liquid, which is heavier, to the outer side of the coil. However, there is no generally agreed upon explanation of the mechanisms involved. According to Owhadi et al. (1968), flow along the curved tube results in a secondary flow caused by the centrifugal force. This force drives the fluid from the center to the outer wall of the tube. This in turn induces a motion from the outer wall around the tube and back to the center. The net result is a pair of symmetrical recirculation patterns superposed on the main flow, which bring about the observed heat transfer behavior.

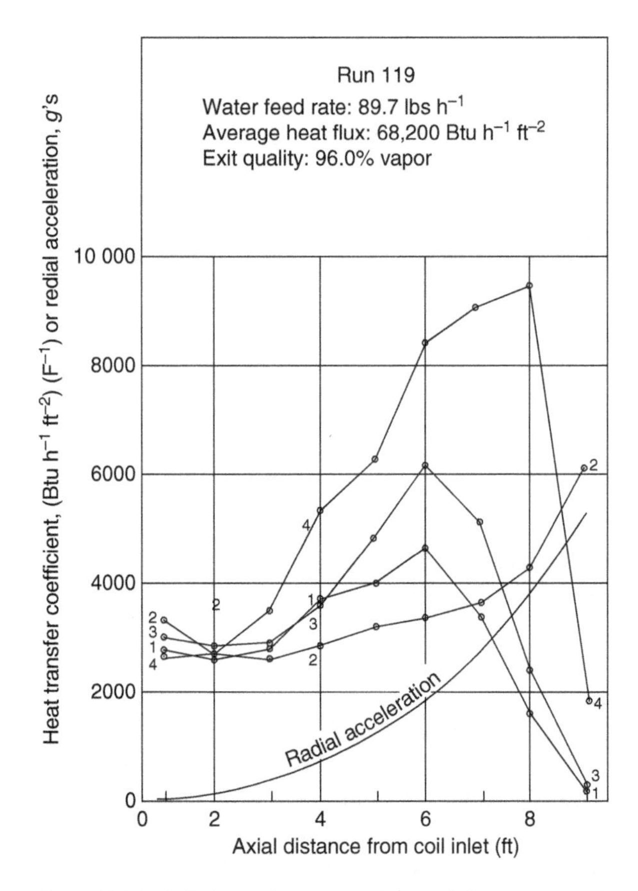

Figure 5.4.1 Variations of heat transfer coefficients around the circumference of a coil. "4" is outer periphery of coil, "2" is inner periphery, "1" and "4" are top and bottom. Source: From Owhadi (1966).

A number of researchers have reported that circumferentially averaged heat transfer coefficients are satisfactorily predicted by correlations for straight tubes. Among such authors are Owhadi et al. (1968), Hardik and Prabhu (2017, 2018), Xiao et al. (2018), Santini et al. (2016), and Hwang et al. (2014). On the other hand, several researchers have given correlations for coils. Some of these are given in the following text.

5.4.1.1 Correlations for Heat Transfer

Based on their own data for helium in a coil, La Harpe et al. (1969) proposed the correlation below:

$$Nu_{TP} = 0.023[GD_t/\mu_{mix}]^{0.85} Pr_L^{0.4} [D_t/D_c]^{0.1} \tag{5.4.1}$$

Zhao et al. (2003) gave the following correlation based on their data for water:

$$h_{TP} = h_{LT} \left[\frac{1.11}{X_{tt}^{0.66}} + 7400\text{Bo} \right] \tag{5.4.2}$$

where h_{LT} is calculated by Eq. (5.2.9) with $x = 0$.

Kaji et al. (1998) have given following correlation based on their data for R-113:

$$\frac{h_{\mathrm{TP}}}{h_{\mathrm{LT}}} = [(40\,000\mathrm{Bo}Re_{\mathrm{LT}}^{-0.12})^3 + (2.6X_{\mathrm{tt}}^{-0.95})^3]^{1/3} \quad (5.4.3)$$

where

$$h_{\mathrm{LT}} = \frac{Pr_L^{0.4}}{41} Re_{\mathrm{LT}}^{5/6} \left(\frac{D_t}{D_c}\right)^{1/12} \left\{1 + \frac{0.061}{\{Re_{\mathrm{LT}}(D_t/D_c)^{2.5}\}^{1/6}}\right\} \frac{k_L}{D_t} \quad (5.4.4)$$

D_c is the coil diameter and D_t is the tube diameter.

The following correlations are based on authors' own data for R-134a:

Wongwises and Polsongkram (2006):

$$Nu_{\mathrm{TP}} = 6896Dn^{0.432}Pr_L^{-5.055}(\mathrm{Bo} \times 10^4)^{0.132}X_{\mathrm{tt}}^{-0.0238} \quad (5.4.5)$$

where Dn is the modified Dean number given by

$$Dn = \left[Re_{\mathrm{LS}} + Re_{\mathrm{GS}} \left(\frac{\mu_G}{\mu_L}\right)\left(\frac{\rho_G}{\rho_L}\right)^{0.5}\right] \quad (5.4.6)$$

Aria et al. (2012):

$$Nu_{\mathrm{TP}} = 7850Dn^{0.432}Pr_L^{-5.055}(\mathrm{Bo} \times 10^4)^{0.125}X_{\mathrm{tt}}^{-0.036} \quad (5.4.7)$$

Chen et al. (2011):

$$\frac{h_{\mathrm{TP}}}{h_{\mathrm{LT}}} = 2.84X_{\mathrm{tt}}^{-0.27} + (46\,162\mathrm{Bo}^{1.15} - 0.88) \quad (5.4.8)$$

where

$$h_{\mathrm{LT}} = 0.023Re_{\mathrm{LT}}^{0.85}Pr_L^{0.4}(D_t/D_c)^{0.1}k_L/D_t \quad (5.4.9)$$

5.4.1.2 Evaluation of Correlations

Shah (2019) collected a database that included all available data for coils whose range is listed in Table 5.4.1. He compared these data from 12 sources with eight correlations for straight tubes and all six correlations for coils listed earlier. The correlations for straight tubes were Shah (2017a), Gungor and Winterton (1986, 1987), Liu and Winterton (1991), Kandlikar (1990), Steiner and Taborek (1992), Fang et al. (2017), and a modified version of the Chen correlation in which nucleate pool boiling heat transfer coefficient was calculated at full wall heat flux using the simplified Cooper correlation, Eq. (3.2.41). The Zhao et al. correlation came out best among the coil correlation with MAD of 28.7%; the other four had very large deviations. Among the straight tube correlations, two were found unsatisfactory. The Fang et al. correlation gave very large deviations. The Steiner–Taborek correlation had MAD of 32.1%. The other straight tube correlations had MAD of 19.8% to 24.8%. Figure 5.4.2 shows a typical comparison of data with correlations.

Table 5.4.1 The range of data for boiling in helical coils analyzed by Shah.

Parameter	Range
Fluids	Water, R-134a, R-123, helium
Tube diameter, mm	2.8–14.5
D_c/D_t	12.4–107
Coil axis orientation	Vertical, horizontal
p_r	0.0046–0.7857
G (kg m^{-2} s^{-1})	80–1250
x (%)	0–99
$Bo \times 10^4$	0.16–13.6
Fr_{LT}	0.03–37.4
We_{GT}	173–70 705
Number of data sources	12

Source: From Shah (2019). © 2019 American Society of Mechanical Engineers.

5.4.1.3 Discussion

The data analyzed included (D_c/D_t) from 12.4 to 107. The larger this ratio, the more the coil resembles a straight tube. Hence applicability of straight tube correlations to larger (D_c/D_t) cannot be in doubt. The smaller the (D_c/D_t), the more the behavior is likely to be different from that of straight tubes. Hence applicability of straight tube correlations for smaller (D_c/D_t) remains to be investigated.

In the vertical coils, tube is near horizontal and hence stratification could occur at low flow rates. In the correlations of Shah and Kandlikar, stratification starts at $Fr_{\mathrm{LT}} < 0.04$. In the data analyzed, there were only a few data point $Fr_{\mathrm{LT}} = 0.03$ and all other data were $Fr_{\mathrm{LT}} > 0.04$. Hence it remains unknown whether heat transfer is affected by stratification at low Froude numbers.

As seen in Section 5.2.2.3.2, Shah (2017a) found that general correlations other than his own fail when surface tension effects become significant, and this occurs when the factor $E_s > 1$. This factor depends on We_{GT} and Bo. As is evident from Figure 5.2.6, $E_s \leq 1$ when $We_{\mathrm{GT}} > 140$. In the data analyzed here, the minimum value of We_{GT} was 173. While the Shah correlation (2017) has been verified for straight tubes at $E_s > 1$, it remains to be seen whether it also works for coils at that condition. This is likely to occur in coils made from small diameter tubes and with fluids of high surface tension.

5.4.1.4 Recommendation

Use the Shah (2017a) correlation for straight tubes to calculate heat transfer coefficient in coils. Caution should be exercised for $D_c/D_t < 12$, $E_s > 1$, and $Fr_{\mathrm{LT}} < 0.04$ as these are outside the range of data analyzed by Shah. Other

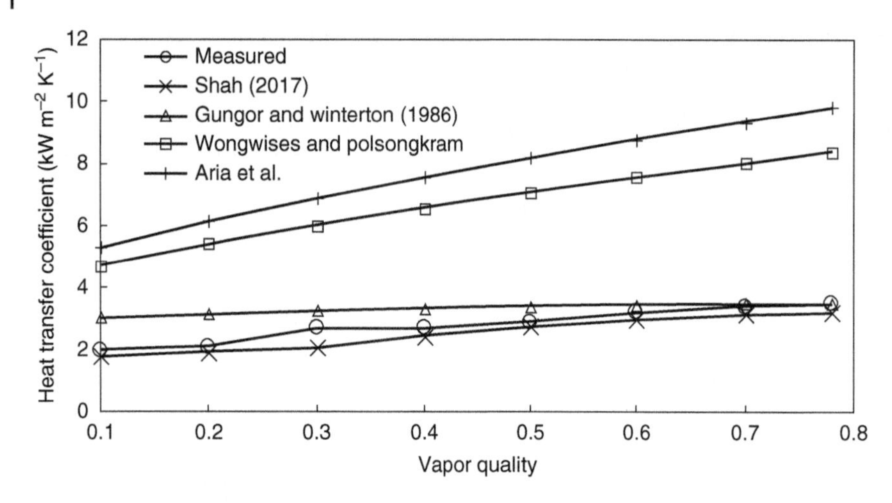

Figure 5.4.2 Data for R-134a boiling in a helical coil compared to various correlations. $T_{SAT} = 20\,°C$, $G = 600\,kg\,m^{-2}\,s^{-1}$, $q = 7.1\,kW\,m^{-2}$. Data of Chen et al. (2011). From Shah (2019). © 2019 American Society of Mechanical Engineers.

correlations found satisfactory in Shah (2019) may be used in the range of data in Table 5.4.1.

5.4.2 Rotating Disk

Evaporators with a rotating disk are used for the concentration of liquids in the food, pharmaceutical industries, and bioindustry. They are also of interest for water recovery from liquid waste in life support systems for spacecraft and space stations.

Kolokosta and Yanniotis (1996) studied evaporation of water and corn syrup on a horizontal rotating disk of 30 cm diameter. Liquid flowed out of the shaft on which the disk was mounted on to the disk surface. Experiments were at speeds from 0 to 1000 rpm and liquid feed rate of 1–5 l min^{-1}. The boiling temperature varied between 40 and 50 °C. The heat transfer coefficient (mean for the disk area) for water at heat flux of 10–30 kW m^{-2} increased from about 2 to 9 kW m^{-2} K^{-1} when the speed of rotation increased from 0 to 1000 rpm. When the range of heat flux was 60–100 kW m^{-2}, heat transfer coefficient increased from 5 to 16 kW m^{-2} K^{-1} for the same increase in the speed of rotation. The rate of liquid feed had no effect on heat transfer coefficient in the speed range of 200–600 rpm. With the 60° Brix corn syrup, the heat transfer coefficient at 10 °C temperature difference increased from about 0.3 to about 2.3 kW m^{-2} K^{-1} when the speed of rotation increased from 0 to 1000 rpm. Liquid feed rate did not have any significant effect except that with stationary disk, increasing feed rate slightly increased heat transfer coefficient at low heat flux.

They derived the following expression for mean heat transfer coefficient h_m, using the film thickness according to the Nusselt theory for condensation on vertical plates, assuming no bubble nucleation, and heat transfer by conduction through liquid film:

$$h_m = 1.28\left(\frac{\rho_f^2 k_f^3 \omega^2}{\mu_f}\right)^{1/3}\left(\frac{m_{in} - m_{out}}{m_{in}^{4/3} - m_{out}^{4/3}}\right)\left(\frac{R_o^{8/3} - R_i^{8/3}}{R_o^2 - R_i^2}\right) \quad (5.4.10)$$

where m_{in} and m_{out} are the mass flow rates of liquid at entrance and outlet from the disk periphery, R_i and R_o are the disk radii at liquid inlet and at disk periphery, and ω is the angular velocity of disk. They converted it to dimensionless form and then simplified it considering the usual conditions in industry, to the following form:

$$h_m\left(\frac{\mu_f^2}{\rho_f^2 k_f^3 \omega^2 R_o}\right)^{1/3} = 0.87\left(\frac{R_o}{R_i}\right)^{1/3} Re_i^{-1/3} \quad (5.4.11)$$

Re_i is the inlet Reynolds number defined as

$$Re_i = \frac{4\Gamma_i}{\mu_f} \quad (5.4.12)$$

where Γ_i is the flow rate per unit perimeter at inlet to the disk. This equation gave reasonable agreement with their data for water and syrup for ΔT_{SAT} less than 5 K. They did not consider data for higher wall superheats as they felt that nucleate boiling may have occurred. Prandl number of corn syrup was from 60 to 350 and inlet Reynolds number from 10 to 90. For water, inlet Reynolds number was from about 700 to 2000.

For local heat transfer coefficient, they gave the following equation:

$$h\left(\frac{\mu_f^2}{\rho_f^2 k_f^3 \omega^2 R}\right)^{1/3} = 1.1 Re_R^{-1/3} \quad (5.4.13)$$

where Re_R is the liquid Reynolds number at radius R.

Kolokosta and Yanniotis (2010) studied boiling of water film on a similar apparatus under similar conditions. Visual observations were also made They observed bubble nucleation at wall superheats of 7 and 9 K. The number of bubbles and mean bubble diameter were found to decrease with increasing speed.

Rifert et al. (2018) have reviewed the various experimental and theoretical studies on this subject. They have not identified any generally applicable correlation for heat transfer.

In conclusion, there is no generally applicable method for calculation of heat transfer during boiling/evaporation on rotating disks. The correlations from various studies may be used in the range of data on which they are based.

5.4.3 Cylinder Rotating in a Liquid Pool

A number of experimental studies have been made on heat transfer during boiling on heated cylinders/tubes rotating in liquid pools. Some of them are discussed herein.

Nicol and Mclean (1968) used an electrically heated horizontal copper cylinder 25.2 mm diameter submerged in a pool of saturated water. They found that heat transfer coefficient initially increased with speed of rotation, reached a peak around 200 rpm, and then fell with increasing speed whose maximum was 800 rpm. Figure 5.4.3 shows their data. They theorized that the initial increase in heat transfer was due to activation of more nucleation sites. The decrease at higher speeds was believed to be due to formation of a thick vapor layer around the cylinder causing resistance to heat transfer.

Tang and McDonald (1971) used horizontal cylinders of brass and copper of 28.6 mm diameter rotating in pools of saturated R-113 and water at atmospheric pressure. Speeds up to 1800 rpm were used. Some of their test data is shown in Figure 5.4.4. Heat transfer coefficients are seen to increase with heat flux. There is little effect of speed except at high Reynolds numbers where there is an upwards trend to merge into the lines for heat transfer without boiling as predicted by two published correlations. They did not find any peak in heat transfer as in tests of Nicol and Mclean. Figure 5.4.4 also shows the data of Nicol and Mclean that are much lower than those of Tang and McDonald who suggest that the measurements of Nicol and Mclean are erroneous. Further, they suggest that nucleate boiling heat transfer on rotating tube can be calculated with correlations for stationary surfaces. They correlated their stationary tube data by the Rohsenow (1952) correlation for pool boiling. However, the constants required by the Rohsenow correlation were derived from their own data. So, it really does not prove applicability of this correlation or that of other pool boiling correlations.

Garg et al. (1980) used horizontal copper cylinders of 30, 42, and 57 mm diameter in a pool of water at atmospheric pressure. Heat flux was up to 45 kW m^{-2}. Heat transfer coefficients increased sharply up to a speed of 180 rpm and then dropped sharply. No effect of cylinder diameter was found.

Garg and Tripathy (1981) employed a copper cylinder of 42 mm diameter in pools of water and aqueous solutions of carboxymethyl cellulose (CMC) and polyvinyl acetate (PVA). Results were similar to those of Garg et al. (1980).

5.4.3.1 Recommendation

There is no verified method for calculation of heat transfer. Rotation increases convective heat transfer while it also suppresses nucleation. The resulting heat transfer coefficient should not be lower than that due to either of these mechanisms. Therefore, for design purposes, the following approach is suggested. Calculate heat transfer coefficients

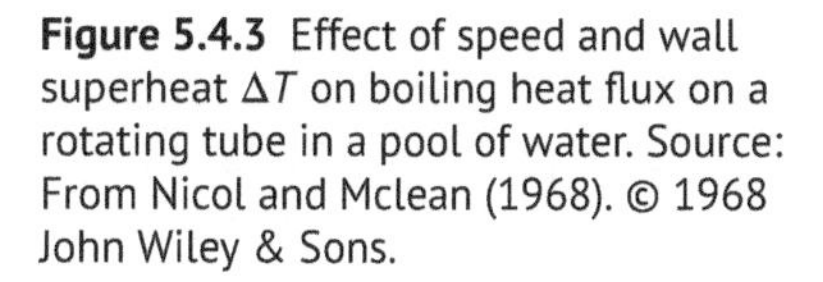
Figure 5.4.3 Effect of speed and wall superheat ΔT on boiling heat flux on a rotating tube in a pool of water. Source: From Nicol and Mclean (1968). © 1968 John Wiley & Sons.

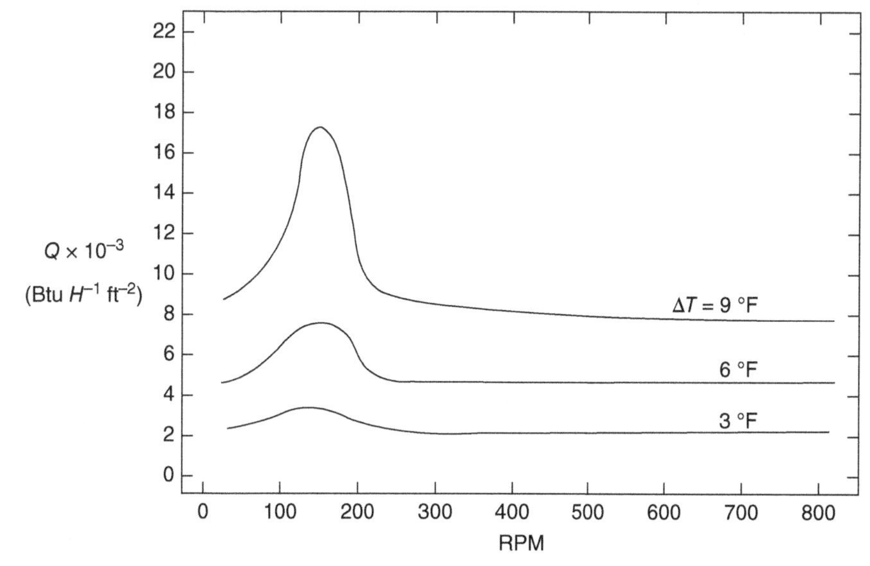

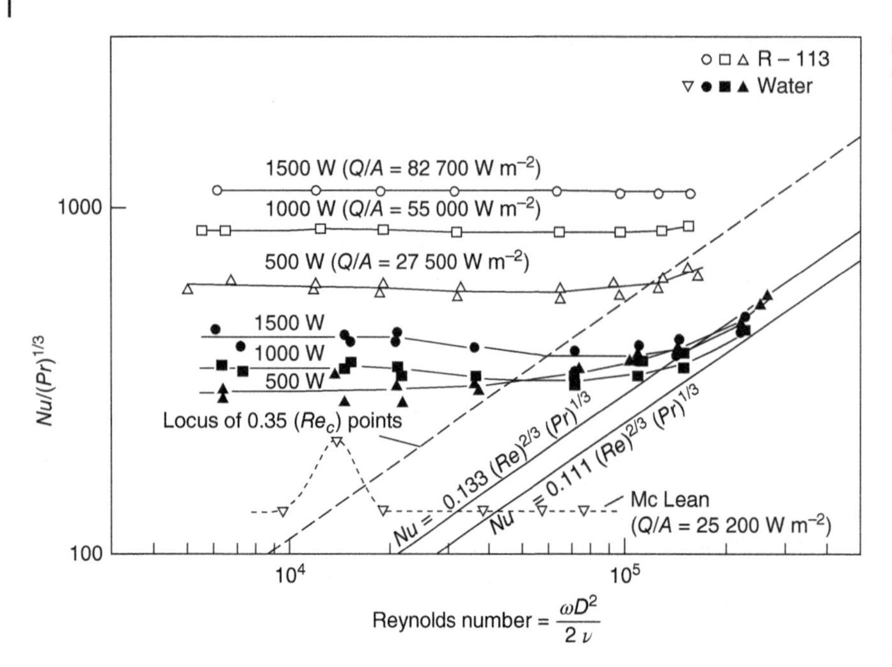

Figure 5.4.4 Heat transfer during boiling on a rotating cylinder in water. Source: From Tang and McDonald (1971). © 1971 Elsevier.

separately by equations for non-boiling convection and by a pool boiling correlation. Use the larger of the two.

5.4.4 Bends

The subcooled boiling data of Miropolskiy and Pikus (1969) for bends were discussed in Section 4.7.5. Those tests also included saturated boiling. Heat transfer coefficients during saturated boiling were also higher at the outer periphery than at the inner periphery but the difference between the two was smaller than in subcooled boiling.

Hasan and Rhodes (1982) measured heat transfer in a U-bend with water boiling inside it. Heat transfer coefficients at the outer periphery of bend were twice as high as those on the inner periphery of the bends. Circumferentially averaged heat transfer coefficients were 10–15% lower than in straight tubes.

As was seen in Section 5.4.1 on helical coils, heat transfer in coils is also higher on the outer surface than on the inner surface but the circumferentially averaged heat transfer coefficient is about the same as in a straight tube. The data of Hasan and Rhodes on bends also show the same behavior.

From the foregoing, the conclusion is that the circumferentially averaged heat transfer coefficients in bends can be calculated in the same way as for straight tubes.

5.4.5 Spiral Wound Heat Exchangers (SWHE)

Spiral wound heat exchangers (SWHEs) are widely used in liquefied natural gas (LNG) plants for condensation of natural gas. The gas to be condensed flows upwards in spiral wound coils enclosed in a shell. An evaporating liquid flows downwards in the shell. For the design of such heat exchangers, methods for predicting heat transfer of evaporating fluid on the shell side are needed. The available information is discussed in the following.

Barbe et al. (1971) studied boiling on the shell side of a model of wound coil evaporator with three layers of coiled electrical cable. Coil diameter was 250 mm and it was 2.5 m high. Boiling fluids used were propane and mixtures of propane with ethane and methane. Data for boiling of propane were correlated by the following equation:

$$\frac{h_{\mathrm{TP}}}{h_{\mathrm{LS}}} = \left(1 + \frac{1}{X_{\mathrm{tt}}}\right)^{0.8} \tag{5.4.14}$$

The data included in this correlation were for pressure 2–5 bar, mass flux 44–114 kg m^{-2} s^{-1}, and heat flux 2.1–4.9 kW m^{-2}.

Ding et al. (2017) studied the evaporation of propane in a SWHE with three layers of spiral coils. Tube outside diameter was 12 mm. Pressure was 2.5 bar, propane flow rate was 40–80 kg m^{-2} s^{-1}, and heat flux was 4–10 kW m^{-2}. Heat transfer coefficient was seen to increase with quality up to about 85% and then drop sharply. Heat transfer coefficient increased with heat flux but its effect diminished with increasing mass flux. They compared their data with the correlation of Gupte and Webb (1995) for upward flow boiling in tube bundles and that of Chien and Tsai (2011) for falling film evaporation on tube bundles. Large deviations were found. They therefore developed a correlation for their own data.

Ding et al. (2018) did further tests in which they tested longitudinal and radial tube pitches 2–6 and 1–4 mm, respectively. They found that as the longitudinal tube pitch and radial tube pitch increase, heat transfer coefficient decreases. They modified their 2017 correlation to take this into account.

Hu et al. (2019a,b) evaporated mixtures of ethane and propane in a SWHE similar to that used by Ding et al. (2017, 2018). Heat transfer coefficients increased with vapor qualities up to 0.7–0.8 and then dropped sharply. They compared their data to the Ding et al. (2017, 2018) correlations but agreement was not good. So, they modified those correlations to a new correlation that gave satisfactory agreement.

In conclusion, the correlations mentioned earlier are limited to the authors' own data. These may be used only in the range of these data. The present author is not aware of other data or correlations in the open literature.

5.4.6 Falling Thin Film on Vertical Surfaces

Evaporators using thin films of liquid flowing down under gravity over vertical surfaces such as tubes and plates are widely used in industrial applications such as desalination and processing of chemicals and foods. At low heat fluxes, heat transfer coefficients for evaporating films are comparable to those without evaporation. At higher heat fluxes, nucleate boiling occurs and higher heat transfer coefficients are obtained.

5.4.6.1 Various Studies and Correlations

Chun and Seban (1971) studied heat transfer to water films falling down the external surface of vertical tubes. They correlated their test data by the following equations:

Laminar flow:

$$\frac{h_{\mathrm{TP}}}{k_f}\left(\frac{v_f^2}{g}\right)^{1/3} = 0.606\left(\frac{Re}{4}\right)^{-0.22} \tag{5.4.15}$$

Turbulent film:

$$\frac{h_{\mathrm{TP}}}{k_f}\left(\frac{v_f^2}{g}\right)^{1/3} = 0.0038 Re^{0.4} Pr_f^{0.65} \tag{5.4.16}$$

Reynolds number $Re = 4\Gamma/\mu_f$ where Γ is the film mass flow rate per unit perimeter of tube. The transition between laminar and turbulent flow was considered to occur when heat transfer coefficient given by Eqs. (5.4.15) and (5.4.16) is equal. This gives

$$Re_{\mathrm{transition}} = \frac{5800}{Pr_f^{0.6}} \tag{5.4.17}$$

Fujita and Ueda (1978) studied boiling of water films flowing down the outer surfaces of tubes 16 mm OD, 600 or 1000 mm long. This tube was enclosed in a glass tube whose diameter was varied from 25 to 46 mm. Steam generated from the water film was removed from the top in some tests and from the bottom in some tests. For heat flux up to 70 kW m^{-2}, there were only a few bubbles and their data were in good agreement with the Chun and Seban correlation, Eqs. (5.4.15)–(5.4.17). At heat flux of 160–280 kW m^{-2}, flow rate had no effect and the heat transfer coefficients were correlated by the following equation:

$$h_{\mathrm{TP}} = 1.24 q^{0.741} \tag{5.4.18}$$

q is in W m^{-2} and h_{TP} is in W m^{-2} K^{-1}. This equation predicts about 30% lower than the Stephan and Abdelsalam correlation described in Chapter 3. These researchers found that dry spots formed near the bottom of the tube when flow of liquid at outlet dropped to 0.01–0.02 kg ms^{-1}. Dryout by flooding was also observed due to steam flowing out at the top.

Cerza and Sernas (1983) studied heat transfer to falling liquid films on a heated vertical tube. Reynolds number ranged from 600 to 4900. Their data at Reynolds numbers higher than 3000 were in good agreement with the Chun and Seban correlation.

Shinkawa (1980) experimented with liquid films flowing down on the inside surface of a vertical tube. Steam and unevaporated liquid were removed from the bottom. Water and water mixed with methanol were used. Heat transfer coefficient was unaffected by liquid flow rate but increased with increasing vapor flow. They correlated their data by the following equation:

$$\frac{h_{\mathrm{TP}} D}{k_f} = 0.82 Re_{\mathrm{GS}}^{0.46} Pr_f^{1/3} \tag{5.4.19}$$

Shmerler and Mudawar (1988) measured heat transfer coefficients with water film flowing down a vertical tube with heated length of 781 mm and outside diameter of 25.4 mm. They found that the developing boundary layer length was long. Heat transfer coefficients for the lower part of tube were correlated by the following equation:

$$\frac{h_{\mathrm{TP}} v_f^{2/3}}{k_f g^{1/3}} = 0.0038 Re^{0.35} Pr_f^{0.95} \tag{5.4.20}$$

The range of data was Re 5000–37 500 and Prandtl number 1.75–5.4. This equation predicts close to the Chun and Seban correlation at higher Prandtl numbers but predicts considerably lower at Prandtl number of 1.75.

Volodin et al. (2017) studied heat transfer to a mixture of R114 and R21 refrigerants on three vertical cylinders with differently micro-structured outer surfaces. Heat transfer coefficients were up to three times those for a smooth tube. CHF of enhanced tubes was also higher than of plain tube.

5.4.6.2 Recommendation

There is no verified generally applicable method for falling films outside tubes. The following method is suggested.

Calculate heat transfer coefficients by the Chun and Seban correlation and also by a pool boiling correlation (see Chapter 3). Use the higher of the two.

5.4.7 Vertical Tube/Rod Bundles with Axial Flow

Such bundles consist of vertical tubes or rods enclosed in a shell. Boiling liquid flows upwards on the shell side and vapor leaves from the top. Examples of such heat exchangers include boiling water nuclear reactors and steam generators of pressurized water nuclear plants. Fuel rods provide the heating in the reactors, while in the steam generators, hot water from the nuclear reactor flowing through the tubes provides the heating.

Shah (1976) compared his correlation for boiling in tubes with the data of Matzner (1963) for water boiling in a bundle of electrically heated rods 11.2 mm diameter and 432 mm long. The rod arrangement was triangular and the pitch to diameter ratio was 1.05. Single-phase heat transfer coefficient was calculated by the Dittus–Boelter equation using the hydraulic equivalent diameter. The predictions were about 30% higher than measurements. According to Kays and Perkins (1973), single-phase heat transfer coefficients for rod bundles with this P/D ratio are about 20% lower than for round tubes. With this correction to the single-phase heat transfer coefficient, the predictions of the Shah correlation come close to data of Matzner. This suggests that while applying boiling heat transfer correlation for tubes to rod bundles, single-phase heat transfer coefficient should be based on a correlation suitable for rod bundles.

Bjornard and Griffith (1977) recommend for nuclear reactor analysis the use of the Chen (1966) correlation for tubes with hydraulic equivalent diameter. Their recommendation is based entirely on the results for round tubes and annuli in the Chen (1966) paper; no comparison with data for bundles is mentioned.

It is recommended to use a reliable boiling heat transfer correlation for tubes (for example, Shah correlation) with single-phase heat transfer coefficient calculated by a method suitable for tube bundles. See Chapter 1 for single-phase correlations.

5.4.8 Spiral Plate Heat Exchangers

Spiral plate heat exchangers are used for single-phase heat transfer as well as for boiling and condensation. The most common design consists of two plates that are wound around a central core to form two spiral channels. Various designs of such heat exchangers have been described by Minton (1970), and he has given formulas for calculation of heat transfer. No information on the bases of those formulas has been provided.

Yilmaz et al. (1983) studied vertical thermosiphon boiling in two spiral plate heat exchangers. The pure fluid tested was *p*-xylene. They compared their data with results for shell and tube thermosiphons obtained earlier. It was concluded that spiral plate heat exchangers have much higher heat transfer coefficients.

5.5 Horizontal Tube Bundles with Upward Crossflow

Heat exchangers consisting of bundles of horizontal tubes with boiling on the outer surface of tubes are widely used in the industry. Examples are flooded refrigerant evaporators, boilers, and kettle reboilers. Because of the practical importance of such heat exchangers, many experimental studies have been done to observe and measure heat transfer on tube bundles and single tubes with crossflow. Further, many correlations for predicting heat transfer have been published. These experimental studies and prediction methods have been reviewed among others by Browne and Bansal (1999), Casciaro and Thome (2001), Ribatski and Thome (2007), and Swain and Das (2014).

5.5.1 Physical Phenomena

Liquid enters the bottom in saturated or subcooled state. If it is subcooled, heat transfer first occurs by single-phase convection and subcooled boiling. Saturated boiling starts once saturation is reached and a mixture of vapor and liquid moves upwards. With increasing quality, flow patterns change from bubbly to cap bubbly, then to churn, and then to annular. Dryout may occur if high vapor quality is reached. If all liquid is not evaporated, liquid leaving the bundle may recirculate if there are gaps between the shell and tube bundle as is the case in many heat exchangers.

Cornwell et al. (1980) performed tests on a model of a reboiler tube bundle. It consisted of 241 electrically heated tubes enclosed in a shell with one side of glass that allowed visual observations. R-113 at atmospheric pressure was the fluid that entered the bundle from bottom. Figure 5.5.1 shows the velocity vectors and streamlines observed by them at two values of heat flux. Figure 5.5.2 shows the variations of heat transfer coefficients in the tube bundle.

Many attempts have been made to develop flow pattern maps, most recently by Mao and Hibiki (2017). Only two of the data sets correlated by them were for boiling, and the rest being for air–water flow. Several other flow pattern maps have been proposed as discussed in Section 1.5.6. No well-verified comprehensive flow pattern map is as yet available.

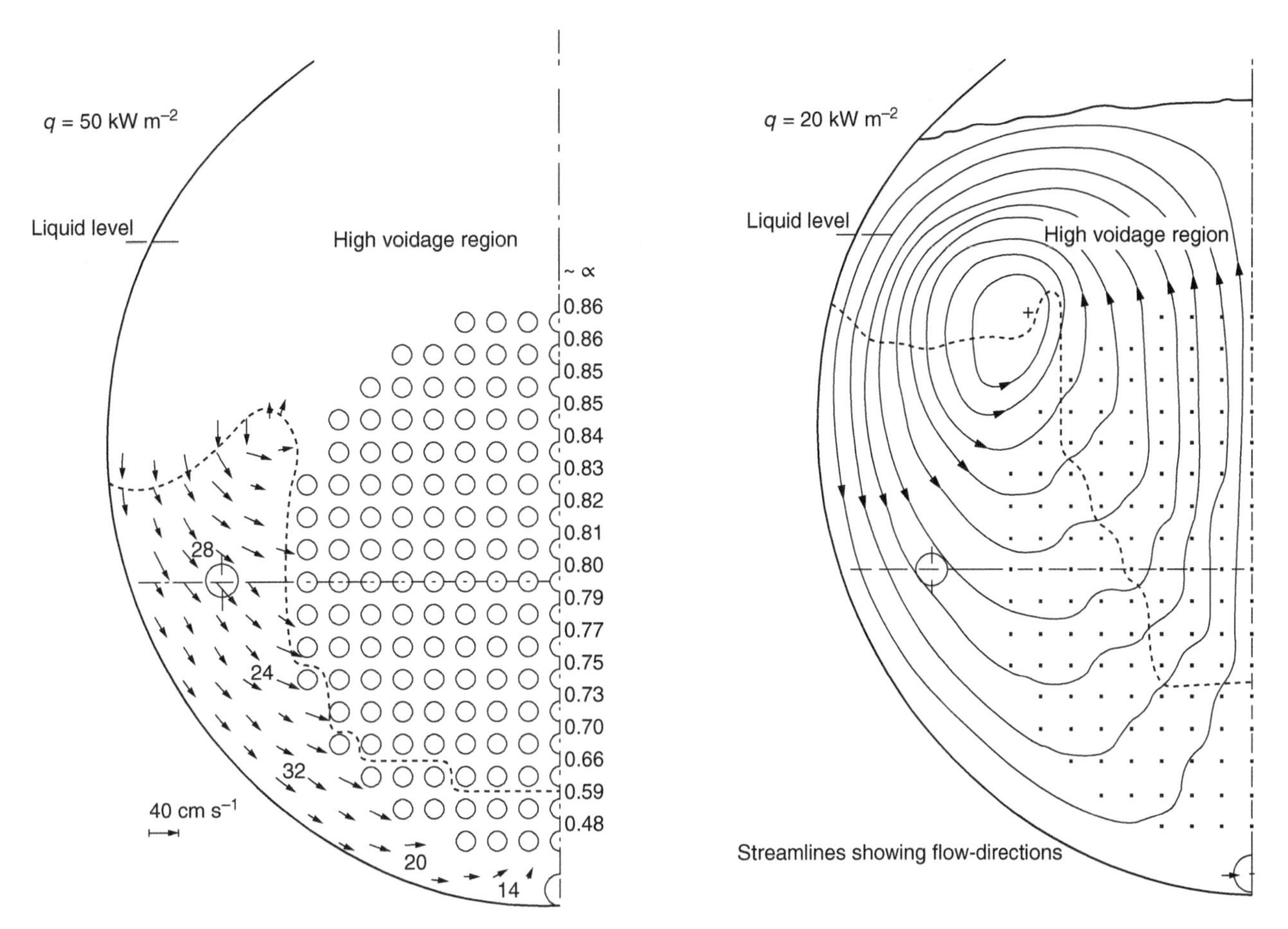

Figure 5.5.1 Velocity vectors and stream lines during boiling of R-113 in a tube bundle. Source: From Cornwell et al. (1980). © 1980 American Society of Mechanical Engineers.

5.5.2 Prediction Methods for Heat Transfer

Early researchers gave bundle average heat transfer correlations. Gilmour (1959) gave a correlation according to which bundle average heat transfer coefficient was always lower than the pool boiling heat transfer coefficient for a single tube. Palen and Taborek (1962) found the Gilmour method to generally underpredict their own data for refinery kettle reboilers. They gave a new correlation. Palen et al. (1972) found that correlation to predict low. These correlations are clearly unsatisfactory.

Abbas and Ayub (2017) measured heat transfer during boiling of ammonia in a flooded evaporator that had a bundle of steel tubes 19.1 mm OD on triangular pitch. Saturated liquid entered the bundle at the bottom. They correlated their bundle average data by the following correlation:

$$h_{TP} = 70q^{0.9-0.4p_r^{0.1}}p_r^{0.55}(-\log p_r)^{-0.6} \tag{5.5.1}$$

Their data covered the range $T_{SAT} = -20$ to $-1.7\,°C$, $q = 5\text{–}45\ kW\,m^{-2}$. h_{TP} is in $W\,m^{-2}\,K^{-1}$.

A bundle may have subcooled boiling, saturated boiling prior to dryout, and post-dryout heat transfer. Heat transfer coefficients usually vary from point to point as seen in Figure 5.5.2. Hence correlations for bundle mean heat transfer cannot be generally applicable. For reliable design, calculations have to be done for individual tubes taking into consideration local conditions such as velocity and vapor quality. For such calculations, models have been proposed such as those by Brisbane et al. (1980), Burnside et al. (2001), and Kumar et al. (2003).

Figure 5.5.3 shows the circulation model of Brisbane et al. (1980). The liquid is assumed to circulate around a rectangular bundle of the same total cross-sectional area as the actual bundle. The liquid level in the model is the same as in the actual bundle. This is important as it is the difference in hydrostatic head that causes flow. The liquid enters the bundle at the bottom and together with recirculating liquid, flows upwards. The liquid vaporizes partially as it moves upwards and emerges at the top as a mixture of liquid and vapor. Recirculation flow rate is calculated by considering the pressure drop of two-phase mixture flowing through the bundle from A to C to equal the hydrostatic head of liquid between A and C. Calculations based on this model showed fairly good agreement with the data of Leong and

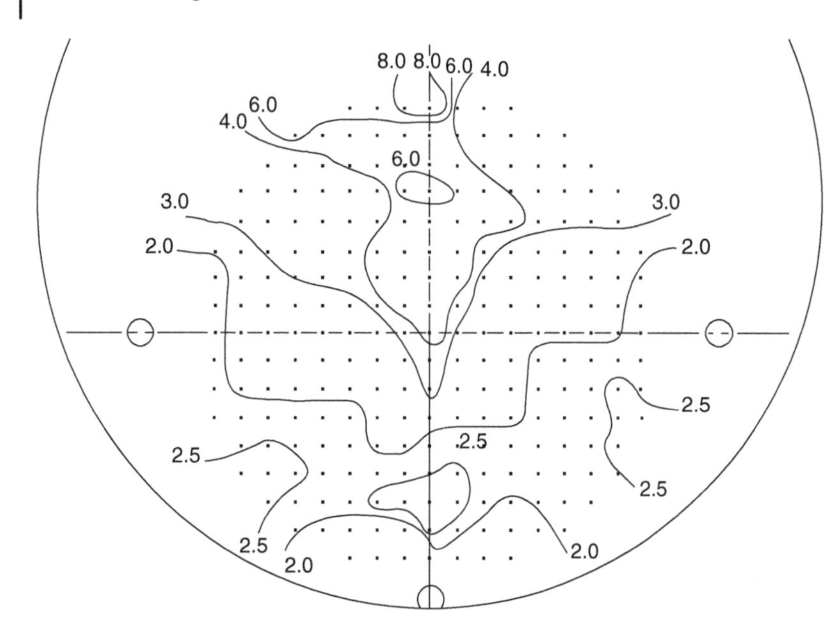

Figure 5.5.2 Lines of constant heat transfer coefficients in a tube bundle with R-113 at atmospheric pressure, $q = 20\ \text{kW m}^{-2}$. Heat transfer coefficients in kW m^{-2}. Source: From Cornwell et al. (1980). © 1980 American Society of Mechanical Engineers.

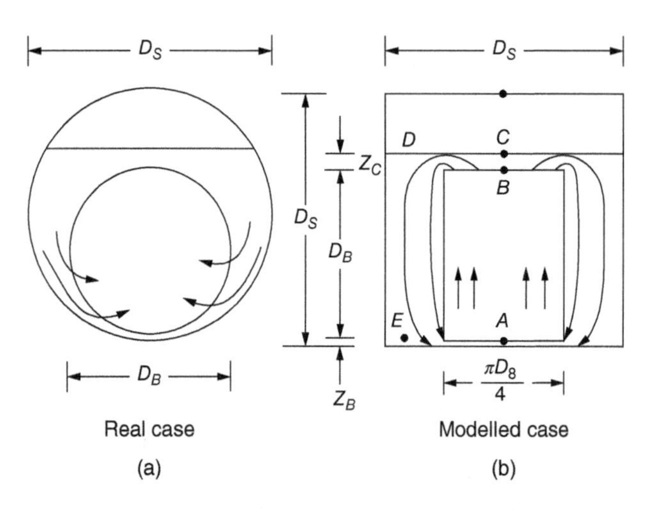

Figure 5.5.3 The recirculation model of Brisbane et al. (1980). Source: From Brisbane et al. (1980). © 1980 American Society of Mechanical Engineers.

Cornwell (1979). Two-phase heat transfer coefficients were calculated by the method of Polley et al. (1980), Eq. (5.5.2).

The model of Brisbane et al. is one-dimensional while it is apparent from Figure 5.5.1 that flow is two-dimensional. However, its reported agreement with data is satisfactory. More complex models have been proposed, for example, by Burnside et al. (2001) and Kumar et al. (2003). In any of these models, correlations for local heat transfer on individual tubes are needed. Some of those are discussed in the following text.

Polley et al. (1980) gave the following correlation:

$$h_{\text{TP}}^n = h_{\text{pb}}^n + \left[h_{\text{LS}}\left(\frac{1}{1-\alpha}\right)^{0.744}\right]^n \tag{5.5.2}$$

where h_{LS} is the heat transfer coefficient of liquid phase flowing alone, which was calculated by the ESDU method (1977) for tube bundles. α is the void fraction that is calculated by the following equation:

$$\alpha = \frac{0.833x}{x + (1-x)\rho_f/\rho_g} \tag{5.5.3}$$

With $n = 1$ in Eq. (5.5.2), their own data for an in-line bundle with R-113 were satisfactorily correlated by this method. Brisbane et al. (1980) used this method in calculations with their circulation model using $n = 2$ to obtain good agreement with the data of Leong and Cornwell (1979) and Cornwell et al. (1980).

Many researchers have proposed correlations by combining pool boiling and convective boiling in various ways. The correlations have the form

$$h_{\text{TP}} = ((Sh_{\text{pb}})^n + (Fh_{\text{LS}})^n)^{1/n} \tag{5.5.4}$$

where S is a suppression factor and F an enhancement factor similar to those in the Chen (1966) correlation for boiling inside tubes. The exponent n has been given various values from 1 to 3. If $n = 1$, it is called the superposition model. If $n > 1$, it is called the asymptotic model. A variety of formulas have been proposed for S and F. Gupte and Webb (1994) presented a correlation of this form with $S = 1$ that agreed with their own data for GEWA and Turbo enhanced tubes. Among correlations of this type are Chien and Wu (2004) and Hwang and Yao (1986).

Thome and Robinson (2006) presented separate correlations for plain, finned, and Turbo tube bundles based on their own data. The correlation for finned tubes uses Eq. (5.5.4) with $S = 1$ and $n = 2$. Their correlation for Turbo tube bundle neglected convective effects and had factors for modifying pool boiling heat transfer coefficient. Their correlation for plain tubes is also based on Eq. (5.5.4). These correlations are applicable only to staggered bundles. The authors compared these correlations only to their own data and showed good agreement.

Van Rooyen et al. (2012) gave a correlation for plain tubes that showed agreement with their own data. It is similar to that of Thome and Robinson (2006). Van Rooyen and Thome (2014) have given a correlation that agrees with their own data for R-134a and R-236fa in Gewa and Turbo tube bundles.

All the correlations mentioned earlier were validated with the authors' own data only. A correlation verified with data from many sources was presented by Shah (2007, 2017b). It is discussed in the following text.

5.5.2.1 Shah Correlation

Shah (2007) presented a general correlation that was validated with data for bundles of plain tubes from many sources. Shah (2017b) further developed that correlation to include enhanced tubes; it is described as follows.

The boiling intensity parameter Y_{IB} is defined as

$$Y_{IB} = BoFr_{LT}^{0.3} \tag{5.5.5}$$

Heat transfer coefficient is calculated by the following three equations:

$$h_{TP} = F_{pb}\, h_{cooper} \tag{5.5.6}$$

$$\psi = \psi_0 \tag{5.5.7}$$

$$\psi = \frac{2.3}{Z^{0.08} Fr_{LT}^{0.22}} \tag{5.5.8}$$

When $Y_{IB} > 0.008$, h_{TP} is given by Eq. (5.5.6).

When $Y_{IB} \leq 0.008$, h_{TP} is the largest of those given by Eqs. (5.5.6), (5.5.7), and (5.5.8).

ψ_0 is the value of ψ when $x = 0$. It is the highest of that calculated by the following relations:

$$\psi_0 = 443 Bo^{0.65} F_{pb} \tag{5.5.9}$$

$$\psi_0 = 31 Bo^{0.33} F_{pb} \tag{5.5.10}$$

$$\psi_0 = 1 \tag{5.5.11}$$

The various parameters in the above equations are as below:

$$Z = \left(\frac{1-x}{x}\right)^{0.8} p_r^{\;0.4} \tag{5.5.12}$$

$$\psi = h_{TP}/h_{LT} \tag{5.5.13}$$

$$Fr_{LT} = \frac{G^2}{gD\rho_f^2} \tag{5.5.14}$$

$$F_{pb} = h_{pb,actual}/h_{Cooper} \tag{5.5.15}$$

h_{Cooper} is the pool boiling heat transfer coefficient calculated by the simplified Cooper correlation, Eq. (3.2.41).

$h_{pb,actual}$ is the same as h_{Cooper} unless pool boiling test data is available for the tubes to be used in the heat exchanger or there is reason to believe that another correlation may be more appropriate than the Cooper correlation. In that case $h_{pb,actual}$ is calculated from the test data or the preferred correlation. Thus $F_{pb} = 1$ unless test data or an alternative correlation is used.

With $F_{pb} = 1$, Eqs. (5.5.9) and (5.5.10) are the same as developed by Shah (2005) through analysis of varied data from many sources for flow across single plain tubes and plain tube bundles at zero vapor quality.

h_{LT} is the heat transfer coefficient for all mass flowing as liquid. It is calculated by the following equation (Shah 1984b):

$$\frac{h_{LT} D}{k_f} = 0.21 Re_{LT}^{0.62} Pr_f^{0.4} \tag{5.5.16}$$

$$Re_{LT} = \frac{GD}{\mu_f} \tag{5.5.17}$$

All fluid properties are calculated at the saturation temperature. Mass flux G is based on the narrowest gap between adjacent tubes. For enhanced tubes, D is the diameter at the outer periphery of tube such as tip of fin or other types of enhancement, and the gap used in G is based on this diameter. Heat transfer coefficient of enhanced tube is also based on the nominal area (πDL) using this D. Note that x is the quality approaching the tube for which heat transfer coefficient is being calculated.

This correlation was compared to data for plain and enhanced tubes whose range is given in Table 5.5.1. Summary of the results of comparison is given in Table 5.5.2. It is seen that that data for 12 fluids of widely varying properties and all types of tubes (plain, finned, and enhanced) are predicted with MAD of 15.2%. Abbas and Ayub (2017) have also reported good agreement with their data for ammonia with the Shah (2007) correlation.

The data analyzed by Shah included in-line and staggered tube bundles. No significant difference in deviations from the correlation was found. This is in agreement with experimental studies in which the effect of tube arrangement was tested. An example is the study by Andrews and Cornwell (1987) in which an in-line bundle with square pitch was rotated 90° to form a staggered bundle. The difference in heat transfer coefficients in the two cases was found negligible. Palen et al. (1972) also did not find any difference between the performance of in-line and staggered bundles.

Table 5.5.1 Range of data with which the Shah (2017b) correlation for horizontal tube bundle was verified.

Fluids	Water, ammonia, R-11, R-12, R-22, R-113, R-134a, R-236fa, R-410A, R-507A, isobutane, *n*-pentane
Tube diameter (mm)	3–25.4
Tube type	Plain, enhanced (integral fin, Gewa-SE, Tubo-B, deformed integral fin with pores and connecting gaps)
Tube material	Copper, brass, copper-nickel alloy, steel, stainless steel, nickel-coated porcelain, nickel-coated ceramic
Bundle arrangement	Single tube and multi-tube. Inline, staggered, P/D 1.17–2.0, electric, or liquid heating
Reduced pressure	0.005–0.2866
Mass flux (kg m^{-2} s^{-1})	0.15–1391
Heat flux (kW m^{-2})	1–1000
Vapor quality	0–0.98%
Re_{LT}	7–52 117
$Bo \times 10^4$	0.04–2632
Y_{IB}	0.07–106
Number of data points	2173 from 51 data sets from 28 sources
Mean absolute deviation (MAD)	15.2%

Source: From Shah (2017b). © 2017 Elsevier.

The tube bundle data analyzed by Shah included pitch to diameter ratios from 1.17 to 2.0. Application to smaller (P/D) should be made with much caution as tests by Liu and Qiu (2004) show that behavior changes drastically as the gap between adjacent tubes approaches zero. They found that as the gap between tubes approaches zero, heat transfer coefficients at lower heat fluxes increase and dryout heat flux is lowered.

In analyzing the data for bundles of enhanced tubes, Shah used pool boiling heat transfer coefficient calculated

Table 5.5.2 Summary of comparison of data for horizontal tube bundles with the Shah (2017b) correlation.

	Number of data points	Mean absolute deviation (%) Average deviation (%)	Percent of data predicted within ±30%
Single tubes	143	14.0 −4.2	89.5
Plain tube bundles	1239	17.1 −8.1	85.2
Enhanced tube bundles	791	12.6 −3.1	91.1
All types	2173	15.2 −6.0	87.7

Source: Adapted from Shah (2017b). © 2017 Elsevier.

using the correlations for pool boiling data for the tubes used in those bundles. No general correlation for pool boiling on enhanced tubes is available. However, there have been numerous experimental studies on pool boiling of a wide variety of enhanced tubes. Hence suitable data or correlations for tubes to be used can usually be found in literature.

Figures 5.5.4 and 5.5.5 show comparison of the Shah correlation with some test data for plain and enhanced tube bundles.

5.5.3 Conclusion and Recommendation

The Shah (2017b) correlation is by far the most verified prediction method for local heat transfer coefficients in tube bundles. It is recommended for bundles of plain and enhanced tubes for all fluids within its verified range of parameters given in Table 5.5.1. It can be used with any circulation model of tube bundles such as that of Brisbane et al. (1980)

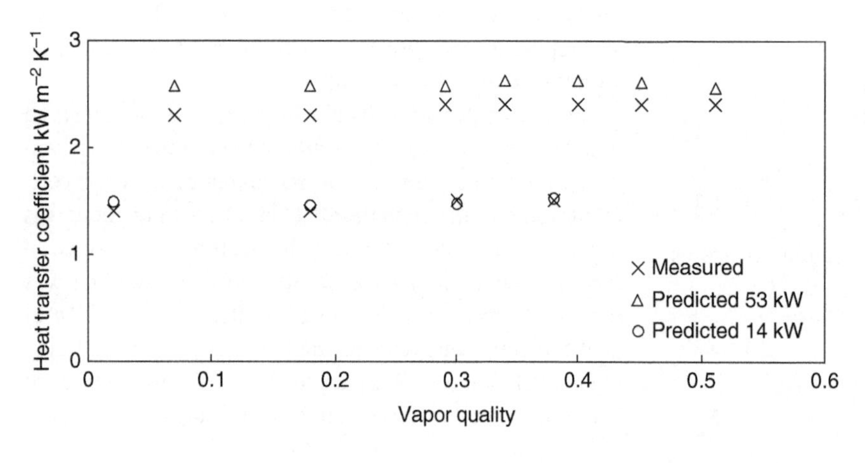

Figure 5.5.4 Comparison of the data of Webb and Chien (1994) for R-113 in a plain tube bundle at two levels of heat flux with the Shah (2017b) correlation. $T_{SAT} = 19\,°C$, $G = 15.6$ kg m^{-2} s^{-1}. Source: From Shah (2017b). © 2017 Elsevier.

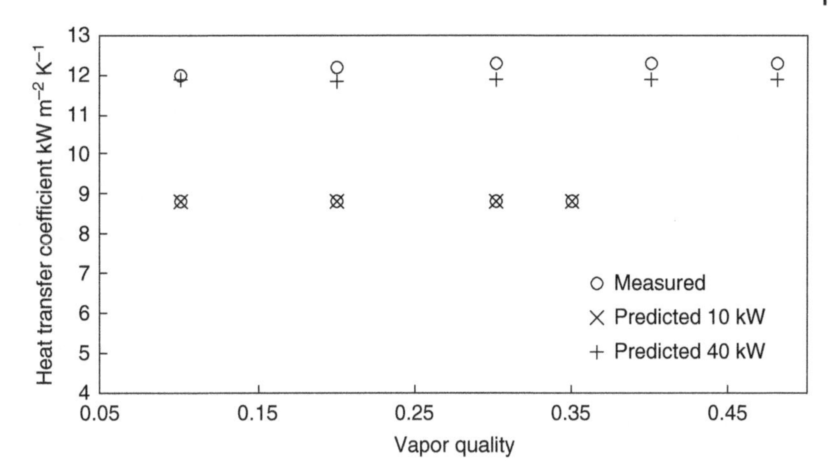

Figure 5.5.5 Comparison of the Shah (2017b) correlation with the data of Kim et al. (2002) for a bundle of tubes with pores and gaps at two values of heat flux. Pore diameter 0.23 mm. R-123, $T_{SAT} = 26.7\,°C$, $G = 26\,kg\,m^{-2}\,s^{-1}$. Source: From Shah (2017b). © 2017 Elsevier.

5.6 Horizontal Tube Bundles with Falling Film Evaporation

Falling film evaporators are also known as spray evaporators. In these evaporators, liquid is sprayed on the tubes from a distributor located above the tube bundles and flows down as film around the tubes. Figure 5.6.1 shows a schematic representation of such evaporators. Falling film evaporators have been used in desalination, food processing, petrochemical industries, and (Ocean Thermal Energy Conversion (OTEC) systems. In the refrigeration industry, these are increasingly replacing the flooded evaporators that have upwards crossflow, described in Section 5.5. The advantages offered by falling film evaporators include higher heat transfer coefficients and lower charge of refrigerant. This brings about size and cost reductions as well as less environmental impact in case of leakage. Numerous experimental studies have been done and correlations proposed. These have been reviewed among others by Ribatski and Jacobi (2005), Fernandez-Seara and Pardinas (2014), and Thome (2017).

5.6.1 Flow Patterns/Modes

Mitrovic (1986) identified three falling flow modes or patterns. These are droplet, jet, and sheet. The jet pattern is also known as column pattern. Hu and Jacobi (1995) identified two intermediate regimes, namely, jet–sheet and droplet–jet. In the droplet mode, liquid leaves the bottom of tube in the form of drops. In the jet mode, there are continuous columns of liquid falling from the bottom of tubes. In the sheet mode, liquid leaves the tube in the form of a continuous sheet. These modes are illustrated in Figure 5.6.2. Based on their own experimental data, Hu and Jacobi developed the following expressions for the transition between these modes:

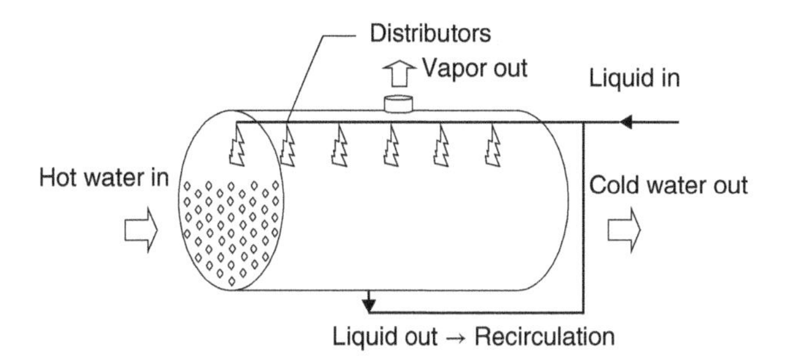

Figure 5.6.1 Schematic of a falling film evaporator. Source: From Wieland Data Book III. © 2016 Wieland-Werke AG.

Between sheet and sheet–jet,

$$Re = 1.488Ga^{*0.236} \tag{5.6.1}$$

Between sheet–jet and jet,

$$Re = 1.414Ga^{*0.233} \tag{5.6.2}$$

Between jet and jet–droplet,

$$Re = 0.096Ga^{*0.301} \tag{5.6.3}$$

Between jet–droplet and droplet,

$$Re = 0.074Ga^{*0.302} \tag{5.6.4}$$

Ga^* is a modified Galileo number defined as

$$Ga^* = \frac{\sigma^3 \rho_f}{\mu_f^4 g} \tag{5.6.5}$$

Re is the film Reynolds number defined as

$$Re = \frac{4\Gamma}{\mu_f} \tag{5.6.6}$$

Γ is the liquid film flow rate kg ms^{-1} on one side of the tube. The total flow rate on both sides of tube is therefore 2Γ. Hu and Jacobi compared these equations to data from several other sources. Agreement was generally good. On the other hand, formulas proposed by some other researchers were found unsatisfactory when compared to their data.

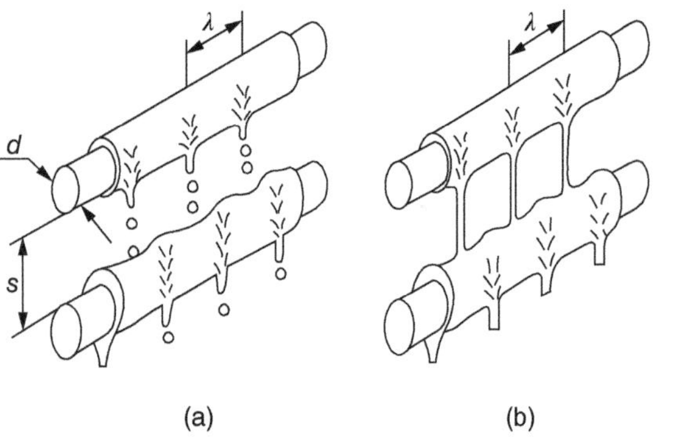

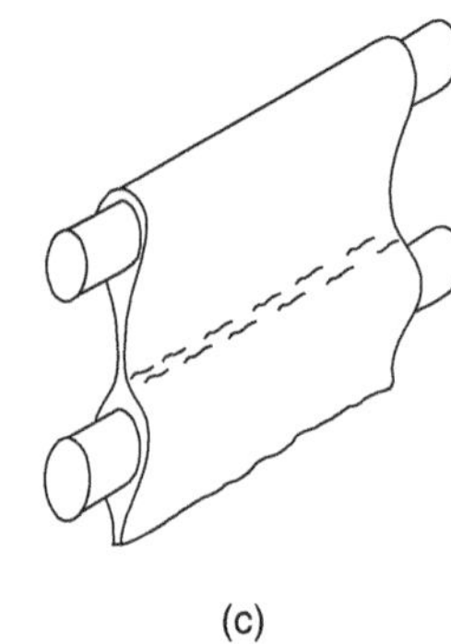

Figure 5.6.2 Flow modes in falling film evaporators according to Mitrovic (1986). (a) drop mode (b) jet/column mode (c) sheet mode From Hu and Jacobi (1995). Source: From Mitrovic (1986).

5.6.2 Heat Transfer

Numerous studies have been done on heat transfer to single tubes and many correlations have been given. The most verified among them is that of Zhao et al. (2016). Their correlation has two regimes, a fully wetted regime and a partially dryout regime. The correlation for the partial dryout regime is

$$\frac{h_{\mathrm{TP}}D}{k_f} = 0.00464Re^{1.51}Bo^{0.43}Pr_f^{0.15}We^{-0.45} \tag{5.6.7}$$

$$Bo = \frac{qD}{\Gamma i_{\mathrm{fg}}} \tag{5.6.8}$$

$$We = \frac{\Gamma^2}{\pi^2(\rho_f - \rho_g)D\sigma} \tag{5.6.9}$$

For the fully wetted regime, the formula is

$$\frac{h_{\mathrm{TP}}D}{k_f} = 3.58 \times 10^{-9}Re^{2.89}Bo^{0.37}Pr_f^{0.2}We^{-1.13} \tag{5.6.10}$$

The boundary between the two regimes is defined by the transition of Reynolds number Re_{tran} given by

$$Re_{\mathrm{tran}} = 5.36 \times 10^4 Bo^{0.0045}Pr_f^{-0.52}We^{0.5} \tag{5.6.11}$$

This correlation gave reasonable agreement with their own data for R-134a on plain copper tubes of 16–25.3 mm diameter as well as data from eight other sources for other halogenated refrigerants.

Jin et al. (2018) performed tests on a single plain tube with R-32 and R-410A. They report that their data for R-410A were in reasonable agreement with the correlation of Zhao et al. (2016) described earlier but their data for R-32 had large deviations with that correlation.

Results for single tubes are generally not directly applicable to tube bundles. Many experimental studies have been done on tube bundles to understand the effects of various parameters.

Zhao et al. (2018) tested a bundle formed by enhanced tubes with three-dimensional fins. Seven horizontal tubes were arranged in a vertical column with two dummy tubes above them. R-134a was sprayed from a distributor at the top. Figure 5.6.3 shows some of the test data. It is seen that there are two regions of heat transfer. At higher Reynolds numbers, heat transfer coefficients decrease slowly with decreasing Reynolds number or are constant. With further decrease in Reynolds number, a point is reached when heat transfer coefficients decrease sharply. The transition Reynolds number increases with increasing heat flux. Heat transfer coefficients decrease from top to bottom. This shows that correlations for single tube cannot be used to predict bundle heat transfer.

Jin et al. (2018) performed tests on a bundle of enhanced tubes with R-32 and R-410A. Their results are similar to those of Zhao et al. (2018).

Chyu et al. (1995) performed tests with ammonia in bundles of plain and low-finned tubes. Tests were done on a single tube, a 3×3 square-pitch tube bundle, and a 3–2–3 triangular-pitch tube bundle. Both pool boiling and spray evaporation tests were performed. Commercial standard-angle and wide-angle spray nozzles were used to distribute liquid ammonia on the tube or tube bundle during spray evaporation tests. Saturation temperature test range was from −23 to 10 °C and heat flux ranged from 3.15 to 31.5 kW m^{-2}. Figure 5.6.4 shows their data for an in-line bundle of plain tubes. It is seen that heat transfer coefficients at the top row are significantly higher than at the bottom row. Further, heat transfer coefficients with spray evaporation are considerably higher than with pool boiling. Similar results were obtained with the triangular arrangement bundle. Heat transfer coefficients for the bundles of low-finned tube (GEWA K40) also showed higher heat transfer coefficients for the upper tubes. Spray rate and type of nozzle did not have much effect. They have given correlations of their data for single tubes and mean heat transfer coefficients for each bundle. For plain single tube, their correlation is

$$Nu = 0.0518Re^{0.039}Pr_f^{0.278}p_r^{0.385}\varphi^{0.753} \tag{5.6.12}$$

where

$$Nu = \frac{h_{\mathrm{TP}}}{k_f}\left(\frac{v_f^2}{g}\right)^{1/3} \tag{5.6.13}$$

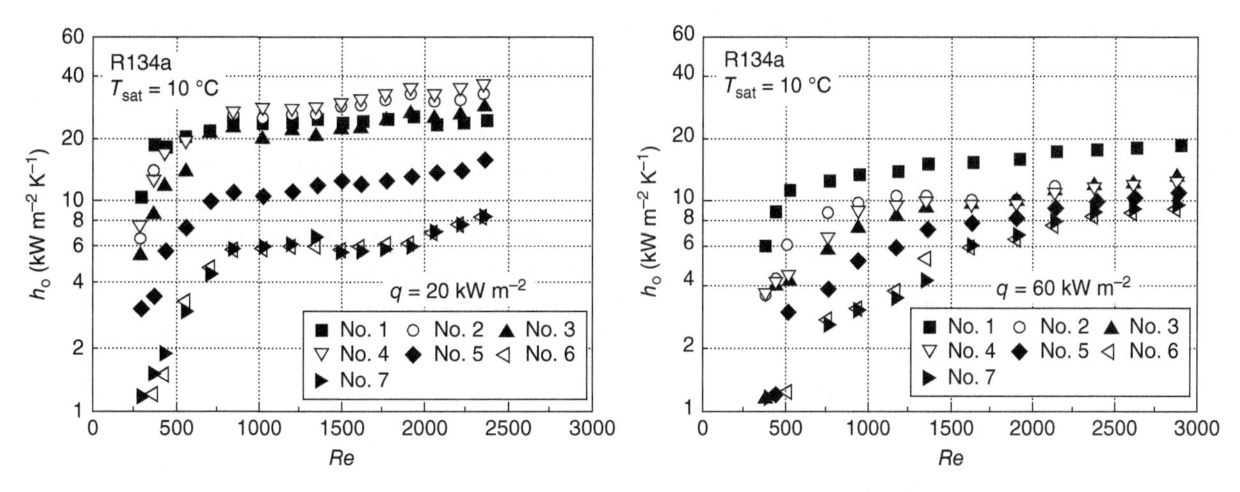

Figure 5.6.3 Heat transfer in a vertical row of horizontal enhanced tubes. Tube 1 is on top, and tube 7 at bottom. Source: From Zhao et al. (2018). © 2018 Elsevier.

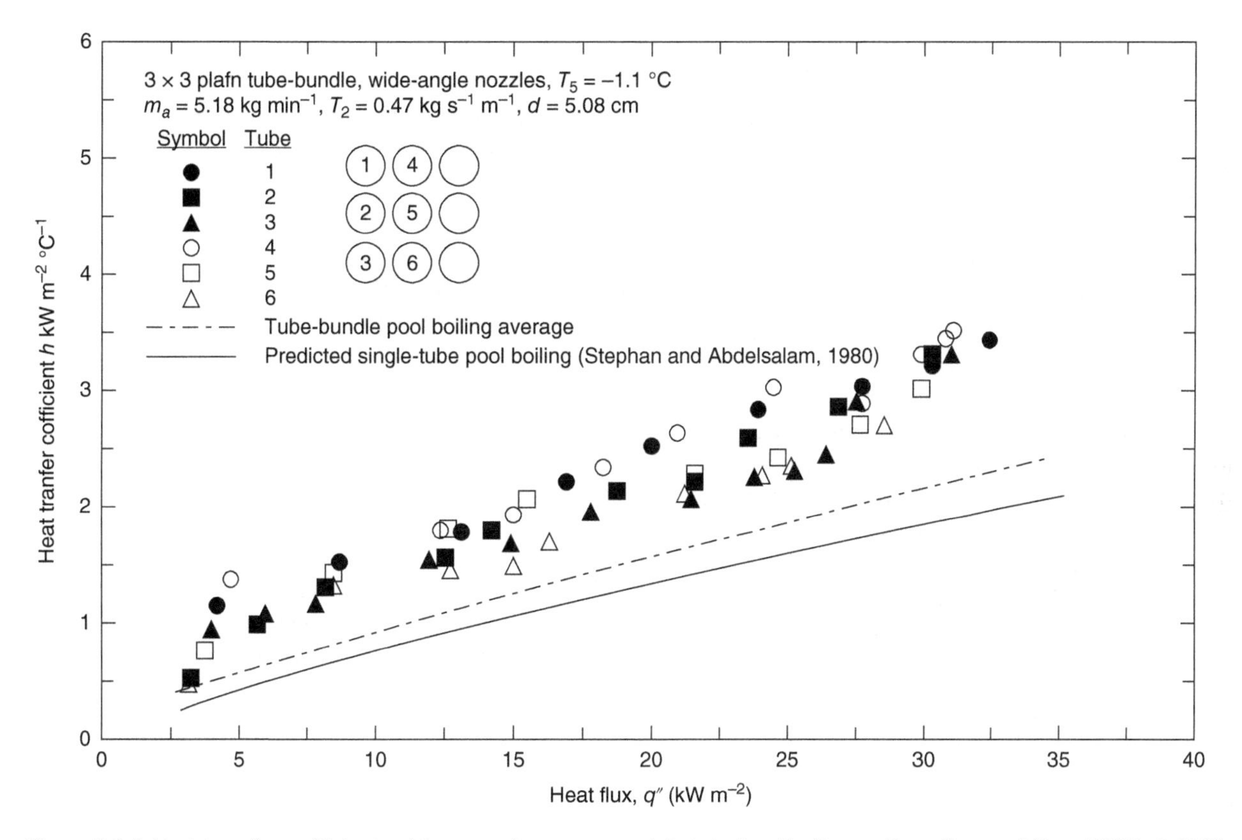

Figure 5.6.4 Heat transfer coefficients with ammonia spray on a plain tube bundle. Source: From Zeng and Chyu (1995). © 1995.

And φ is the dimensionless heat flux defined as

$$\varphi = \frac{qD}{(T_c - T_{SAT})k_f} \tag{5.6.14}$$

T_c is the critical temperature of fluid.

For square-pitch plain tube bundle, their correlation is

$$Nu = 0.00495 Re^{-0.00399} Pr_f^{0.209} p_r^{0.261} \varphi^{0.722} \tag{5.6.15}$$

For triangular-pitch tube bundle, their correlation is

$$Nu = 0.0678 Re^{0.049} Pr_f^{0.296} p_r^{0.456} \varphi^{0.704} \tag{5.6.16}$$

For GEWA K40 single tube, the formula is

$$Nu = 0.0568 Re^{-0.0058} Pr_f^{0.193} p_r^{0.323} \varphi^{1.034} \tag{5.6.17}$$

Note that the diameter used for finned tube is that at the tip of the fin. Heat transfer coefficient and heat flux for finned tube are also based on a tube of this diameter.

For square-pitch finned tube bundle, the correlation is

$$Nu = 0.0622Re^{-0.00035}Pr_f^{0.108}p_r^{0.127}\varphi^{0.773} \tag{5.6.18}$$

For triangular-pitch finned tube bundle, the following equation was fitted to data:

$$Nu = 0.0556Re^{0.034}Pr_f^{0.147}p_r^{0.179}\varphi^{0.758} \tag{5.6.19}$$

5.6.3 Conclusion and Recommendation

While numerous prediction methods for heat transfer coefficients on single tubes and tube bundles have been published, none of them can be considered generally applicable. The recommendation is to base design on results of the experimental study done under conditions close to those of the heat exchanger being designed.

5.7 Boiling of Multicomponent Mixtures

Heat transfer during forced convection boiling of mixtures needs to be calculated for many applications that include evaporators for refrigeration and air conditioning, heat exchangers for cryocoolers, LNG vaporizers, reboilers for chemical processes, etc. As discussed in Section 5.2, heat transfer during forced convection boiling involves both nucleate boiling and forced convection effects. As described in Section 3.2.4.1, nucleate pool boiling heat transfer of mixtures deteriorates due to mass transfer resistance. Forced convection heat transfer also deteriorates due to mass transfer resistance and also due to sensible heat transfer. Heat transfer coefficients of mixtures are lower than of a fluid with mixture mean properties. Many methods for predicting heat transfer of mixtures during forced convection boiling have been proposed. Some of them are discussed in the following.

5.7.1 Boiling in Tubes

As the mixture gets heated during its passage through the tube, the low boiling point component (called the light component) evaporates faster than the higher boiling point component (called the heavy component). This changes the composition of the liquid and vapor phases and raises the bubble point. This affects nucleate boiling as well as convective boiling, i.e. evaporation without bubble nucleation. In the nucleate boiling region, the liquid at the interface with a growing bubble is richer in the heavier component and hence its bubble point is higher than that of the bulk liquid. For bubble growth to continue, the lighter component from the bulk liquid has now to diffuse through this heavy component-rich layer. This mass transfer resistance causes a decrease in the heat transfer coefficient in the bubble nucleation region.

The convective region where there is no nucleation is now considered. As the mixture flows along the heated tube, the bubble point temperature increases due to faster evaporation of the lighter component. Hence heat has to be supplied for sensible heating of the liquid and vapor phases in addition to the heat for evaporation. Further, the concentrations of the components at the interface are different from those in the bulk vapor and liquid; mass transfer resistance has to be overcome in both liquid and vapor. These two effects lower the heat transfer coefficient compared with an equivalent pure fluid.

Most of the proposed predictive methods modify the nucleate boiling contribution to heat transfer in correlations for single-component fluids. Examples are Palen and Small (1964) and Kandlikar (1998). Thome (1996) had also suggested the same approach, using the Thome and Shakir correction factor for pool boiling to correlations for single-component fluids. Thus Eq. (5.2.1) that is used in many single component correlations becomes

$$h_{\mathrm{TP}} = Eh_{\mathrm{LS}} + F_{\mathrm{TS}}Sh_{\mathrm{pb}} \tag{5.7.1}$$

where F_{TS} is the correction factor for pool boiling by the Thome–Shakir method given by

$$F_{\mathrm{TS}} = \frac{h}{h_I} = \left\{1 + \frac{h_I}{q}\Delta T_{\mathrm{BP}}\left[1 - exp\left(\frac{-B_0 q}{\rho_f i_{\mathrm{fg}} \beta_f}\right)\right]\right\}^{-1} \tag{5.7.2}$$

For details about Eq. (5.7.2), see Section 3.2.4.2.

There have been a few comparisons of this method with data, some showing good agreement and some showing poor agreement. For example, Grauso et al. (2011) found this method unsatisfactory, while Zurcher et al. (1998) showed that this method worked well for their R-407C data. Ardhpurkar et al. (2014) compared their data for mixtures of nitrogen and hydrocarbons with the Gungor and Winterton (1987) correlation modified by this method. Results were unsatisfactory.

Shah (2015b) noted that the heat transfer in convection without nucleation also deteriorates due to mass transfer resistance and sensible heating of the mixture as bubble point rises with evaporation. Noting that the heat transfer process during condensation and evaporation without nucleation are similar, he proposed using the correction factor of Bell and Ghaly that is widely used for condensation heat transfer. The Bell and Ghaly method is described in Section 2.8.2.

The method proposed by Shah (2015b) is expressed by the following equation.

$$h_{\mathrm{TP,mix}} = F_{\mathrm{TS}} h_{\mathrm{nb}} + \left(\frac{1}{h_{\mathrm{cb}}} + \frac{Y}{h_{\mathrm{GS}}}\right)^{-1} \quad (5.7.3)$$

where Y is given by

$$\mathrm{Y} = \frac{q_{\mathrm{sens}}}{q_{\mathrm{tot}}} = xC_{\mathrm{pg}}\frac{dT_{\mathrm{BP}}}{dH} \quad (5.7.4)$$

q_{sens} is the sensible heat flux and q_{tot} is the total heat flux. T_{BP} is the bubble point and H is the enthalpy of mixture. h_{nb} is the heat transfer coefficient due to nucleate boiling and h_{cb} is the heat transfer coefficient due to forced convection; these are calculated by correlations for pure fluid using the mixture mean properties. The superficial heat transfer coefficient of gas phase h_{GS} is calculated as

$$h_{\mathrm{GS}} = 0.023\left(\frac{GxD}{\mu_g}\right)^{0.8}\frac{\mathrm{Pr}_g^{0.4}k_g}{D} \quad (5.7.5)$$

Properties used are mixture mean.

In the correlations of Chen (1966) and Gungor and Winterton, h_{nb} and h_{cb} are clearly identified and hence they are easily adapted to use Eq. (5.7.3). The Liu and Winterton correlation becomes

$$(h_{\mathrm{TP,mix}})^2 = (F_{\mathrm{TS}}h_{\mathrm{nb}})^2 + \left(\frac{1}{h_{\mathrm{cb}}} + \frac{Y}{h_{\mathrm{GS}}}\right)^{-2} \quad (5.7.6)$$

The Shah correlation for pure fluids is described in Sections 5.2.2.1.2 and 5.2.2.3.2. It has three regimes: nucleate boiling, convective boiling, and bubble suppression in which both nucleate and convective boiling have influence. The equations for mixture using the Shah correlation are

$$h_{\mathrm{nb,mix}} = h_{\mathrm{LS}}F_{\mathrm{TS}}\Psi_0 \quad (5.7.7)$$

$$h_{\mathrm{cb,mix}} = \left(\frac{J^{0.8}}{1.8h_{\mathrm{LS}}} + \frac{Y}{h_{\mathrm{GS}}}\right)^{-1} \quad (5.7.8)$$

$$h_{\mathrm{bs,mix}} = \left[(h_{\mathrm{bs}} - h_{\mathrm{nb,mix}})^{-1} + \frac{Y}{h_{\mathrm{GS}}}\right]^{-1} \quad (5.7.9)$$

Heat transfer coefficient of the mixture h_{mix} is the largest of those given by Eqs. (5.7.7), (5.7.8), and (5.7.9).

This method was evaluated by comparison with an extensive database consisting of 878 data points for 45 mixtures of 19 fluids from 21 independent studies. The fluids included halogenated refrigerants, hydrocarbons, nitrogen, and carbon dioxide. The data include tube diameters 0.19–14 mm, horizontal and vertical orientations, flow rates 50–930 kg m^{-2} s^{-1}, reduced pressures from 0.05 to 0.63, saturation temperatures down to −180 °C, and temperature glides up to 156 °C. This database was compared to several correlations for pure fluids in two ways:

1. Applying the Thome–Shakir correction factor to nucleate boiling component only.
2. Applying the Thome–Shakir correction factor to nucleate boiling component and the Bell–Ghaly correction factor to the convective boiling component. That is, apply Eq. (5.7.3).

The pure fluid correlations tested in this way were Shah (1982), Gungor and Winterton (1986, 1987), Liu and Winterton (1991), and Chen (1966) with Cooper correlation to calculate the pool boiling heat transfer coefficient. For all mixtures, except LNG, the second method gave better agreement with all correlations. The mean deviations of the correlations of Shah, Liu and Winterton, and Gungor and Winterton (1987) with this method were 19.5%, 20.4%, and 20.7%, respectively. The other two correlations gave significantly larger deviations (28.4% and 31.2%). The LNG data were from only one source. They were in good agreement with the Shah and Gungor–Winterton (1987) correlations without any correction factor. Use of either of the two correction methods greatly increased deviations.

Figures 5.7.1 and 5.7.2 show the comparison of the Shah correlations for experimental data of mixture with corrections for mixture effects. It is seen that the second method gives much better predictions.

It may be concluded that the second method, applying correction factors to both nucleate boiling and convective boiling, has been well verified. The case of LNG is an exception that needs further investigation.

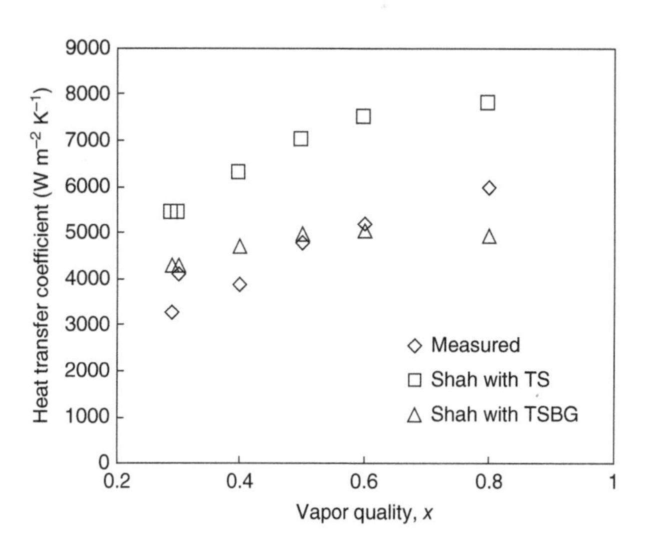

Figure 5.7.1 Comparison of the data of Park et al. (2011) for a mixture of R123–R134a–R22 with the predictions of the Shah correlation with correction by Thome–Shakir factor only (TS) and by both Thome–Shakir and Bell–Ghaly factors (TSBG): $D = 0.19$ mm, $G = 392$ kg m^{-2} s^{-1}, $q = 15$ kW m^{-2}, $T_{\mathrm{SAT}} = 33$ °C, and glide = 16.8 K. Source: Reprinted from Shah (2015b). © 2015, with permission from Elsevier.

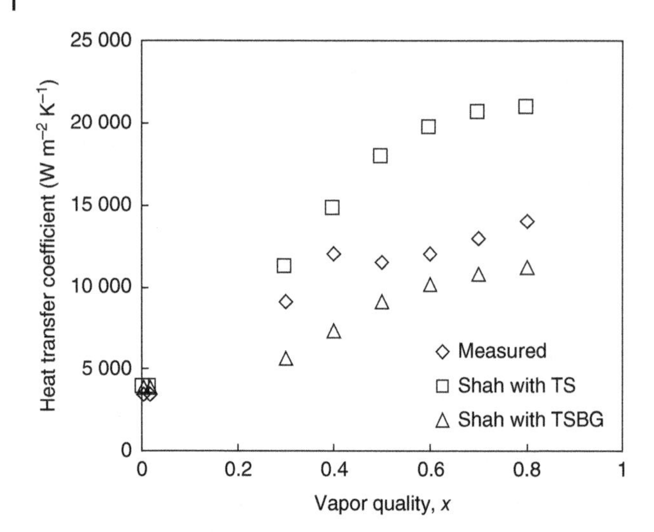

Figure 5.7.2 Comparison of data of Nellis et al. (2005) for mixture for cryocoolers with predictions of Shah correlation with correction by Thome–Shakir factor only (TS) and by both Thome–Shakir and Bell–Ghaly factors (TSBG). Glide = 156 °C, T_{SAT} = −160 °C, G = 840 kg m^{-2} s^{-1}, q = 79.5 kW m^{-2}, D = 8.4 mm. Source: From Shah (2015b). Copyright Elsevier, reproduced with their permission.

5.7.2 Boiling in Various Geometries

Besides boiling inside tubes, boiling of mixture also occurs in a variety of heat exchangers such as plate heat exchangers, flooded evaporators, spiral wound evaporators, etc. These are discussed in the following text.

Li et al. (2018) measured heat transfer of hydrocarbon mixtures in a plate fin heat exchanger. They compared their data for several correlations for pure fluid in tube. Data were severely overpredicted. They also tried Gungor and Winterton (1986) correlation modified with Thome–Shakir factor but deviations were large. They correlated their data by the following equation:

$$h_{TP} = 0.42q^{0.96} \tag{5.7.10}$$

q is in W m^{-2} and h_{TP} is in W m^{-2} K^{-1}.

Taboas et al. (2010) studied boiling of ammonia–water mixtures in a herringbone plate heat exchanger with chevron angle 60°, pitch 9.6 mm, and corrugation height 2 mm. Heat transfer coefficients were measured. In Taboas et al. (2012), these data were compared to many correlations including those for pool boiling and for plate heat exchangers. None was found satisfactory. They therefore developed the following correlation to fit their own data:

$$h_{nb} = 5Bo^{0.15}h_{LT} \tag{5.7.11}$$

$$h_{cb} = \left(1 + \frac{3}{X_{tt}} + \frac{1}{X_{tt}^2}\right)^{0.2} h_{LT} \tag{5.7.12}$$

If $u_{GS} < -111.88\, u_{LS} + 11.848$ m s^{-1}, $h_{TP} = h_{nb}$, else h_{TP} is the larger of h_{cb} and h_{nb}.

$$h_{LT} = 0.2092\left(\frac{k_f}{D_{HYD}}\right)Re_{LT}^{0.78}Pr_f^{1/3}\left(\frac{\mu_f}{\mu_{f,wall}}\right)^{0.14} \tag{5.7.13}$$

The range of data covered is heat flux between 20 and 50 kW m^{-2}, mass flux between 70 and 140 kg m^{-2} s^{-1}, mean vapor quality from 0.0 to 0.22, pressure between 7 and 15 bar, and ammonia concentration between 0.42 and 0.62 by mass.

Hu et al. (2019a) studied boiling of ethane–propane mixtures on the shell side of a spiral wound evaporator. Heat transfer coefficients were found to decrease with increasing fraction of ethane. Dryout occurred at a vapor quality of 0.7. No comparison with any prediction method was made.

5.7.3 Conclusions and Recommendations

For boiling of mixtures inside tubes, the method proposed by Shah (2015a,b) was verified with a very wide range of data. This method consists of applying the Thome–Shakir correction to the nucleate boiling component and the Bell–Ghaly correction to the convective boiling component. It is expressed by Eq. (5.7.3). This method is therefore recommended for boiling inside tubes of all mixtures. Exception is made for LNG as the only data set analyzed did not agree with it.

For other heat exchangers, there is no well-verified general predictive technique. Where data or correlations are available in the range of interest, those may be used. Otherwise, use Eq. (5.7.3) with h_{nb}, h_{cb}, and h_{GS} based on correlations for pure fluids for those geometries.

5.8 Liquid Metals

Both subcooled and saturated boiling of liquid metals are addressed in this section.

As was discussed in Chapter 3 on pool boiling, boiling of liquid metals is different from that of ordinary (non-metallic) liquids. The correlations for ordinary fluids are generally inapplicable to liquid metals.

5.8.1 Inception of Boiling

Aladyev (1973) studied several studies on inception of boiling of alkali metals during forced convection in tubes. He reached the following conclusions:

1. There is no incipient superheat if two-phase mixture enters the tube.

2. Incipient superheat decreases with increasing pressure, as it does in pool boiling.
3. Incipient superheat does not change with increasing heat flux though the location of inception point moves closer to the inlet. With decreasing heat flux, location of boiling inception moves toward the exit.

Studies on potassium show that if liquid enters tube in subcooled state, incipient superheats are high and fluctuations in wall temperature occur. When wall temperature reaches some value greater than the saturation temperature, rapid boiling occurs, and the wall temperature rapidly drops to the saturation temperature. The cycle then keeps on repeating. Inception superheat decreases with increasing pressure but frequency of cycles increases.

Hoffman and Krakoviak (1964) studied boiling of potassium in a stainless steel tube and found inception superheats exceeding 160 K and wall temperature fluctuation at 0.025 cycles per second. During their experiments with boiling of potassium and sodium, Morozov et al. (1988) also found high incipient superheats and temperature fluctuations of wall.

Schleisiek (1970) experimented with boiling of sodium in a 9 mm diameter tube. Pressure was 0.3–1 bar and inlet subcooling was 85–170 K. The incipient boiling wall superheat was between 40 and 125 °C. It increased with increasing heat flux and decreased with increasing sodium velocity. In the whole range of parameters, there was violently pulsating boiling with partial discharge and subsequent refilling of the test section and strong oscillations of wall temperature.

Chen (1968) studied boiling of potassium in tubes at pressure 0.55–0.96 bar, deactivation subcooling of 6–211 K, and boiling temperatures of 975–1144 K. Measured incipient superheats were 9–65 K. He found the superheats to be related to conditions prior to boiling. He developed an analytical model that agreed with his own data.

Based on their experiments with potassium in straight tubes and coils, Dodonov et al. (1979) gave the following correlation for incipient superheat:

$$(\Delta T_{SAT})_{ib} = 100\exp(-50p_r) \tag{5.8.1}$$

ΔT_{SAT} is in K. There was very large scatter of data around this correlation at pressures below 1.7 bar. Hence this formula gives a very rough estimate of incipient superheat. Their conclusion was that conditions prior to start of boiling do not affect the incipient superheat. This is in contrast to the finding of Chen (1968) noted earlier. Their range of data included pressures from 0.3 to 3.6 bar. Mass flux varied from 0 to 270 kg m^{-2} s^{-1}, heat flux from 32 to 300 kW m^{-2}, and subcooling from 43 to 448 K but variations of these parameters had no effect on incipient superheat.

Peppler (1977) measured incipient superheats with sodium boiling in single channels and pin bundles. He found that new test sections had zero incipient superheats but after many cycles incipient superheats reached up to 25 K. Henry et al. (1974) measured inception superheats of up to 75 K during their tests with sodium in an annulus.

Qiu et al. (2013) studied inception of boiling of sodium during upflow in a vertical annulus consisting of a 6 mm diameter heater surrounded by a tube of 10 mm diameter. The heat flux was from 128 to 846 kW m^{-2}, with inlet subcooling from 63 to 288 °C, mass flow rate from 7 to 122 kg h^{-1} and system pressure from 0.85 to 28.79 kPa. They fitted the following equation to their data:

$$(\Delta T_{SAT})_{ib} = 3.2765\times10^{-5}q^{1.188}p^{-0.098}\Delta T_{SC}^{0.296}Re_{LT}^{-0.257} \tag{5.8.2}$$

ΔT_{SAT} and ΔT_{SC} (inlet subcooling) are in K, q in W m^{-2}, p in Pascal. This formula predicted their data within ±15%. They did not compare this correlation to data from other sources.

Xiao et al. (2006) performed experiments on boiling of sodium during upflow in a vertical annulus with a 12.3 mm diameter heater and 19 mm diameter outside tube. The main experimental parameters are temperature, 250–270 °C; *Pe* number, 125–860; and *Re* number, 0.20×10^5–1.35×10^5. They fitted the following equation to their test data:

$$(\Delta T_{SAT})_{ib} = 0.921[q - 8348.1 - 31(\varepsilon Pe)^{0.779}]^{0.32}p^{-0.27}Pe^{-0.544} \tag{5.8.3}$$

ε is the ratio of eddy diffusivity of momentum and heat and *Pe* is the Peclet number. All their data were within 19% of this equation.

On the basis of their study of data on inception of boiling from several sources as well as their own data for boiling of sodium, Dwyer et al. (1973a,b) concluded that boiling inception superheat increases with rapidity of increase in heat flux. The faster the increase in heat flux, the higher the inception superheat. They further concluded that inception superheat decreases with increasing flow rate and increases with increasing heat flux.

Tang et al. (1964) performed tests on mixtures of potassium and mercury in tubes. Mori et al. (1970) tested mixtures of mercury, bismuth, and lead. Both tests involved transition from single-phase flow to subcooled boiling. Neither of them found any significant boiling inception superheat or instabilities.

From the foregoing discussion, the following conclusions may be reached:

1. Inception of boiling of subcooled pure liquids is often unsteady involving large wall temperature fluctuations. However, in some studies little or no inception superheat was found.
2. When two-phase mixture enters the tube, boiling inception superheat is zero.
3. There is no inception superheat in boiling of mixtures.
4. There is no generally applicable method for predicting boiling inception superheats. The formulas given earlier should be used only for the conditions on which they are based.

5.8.2 Heat Transfer

Many experiments on heat transfer during forced convection boiling of liquid metals have been reported, and correlations for prediction have been proposed.

5.8.2.1 Sodium

The literature on boiling of sodium has been reviewed by Wu et al. (2018).

Qiu et al. (1993) performed tests on boiling of sodium in an annular channel and gave the following correlation based on their own data:

$$h_{TP} = 0.832q^{0.768}p^{0.253} \tag{5.8.4}$$

where h_{TP} is in W m^{-2} K^{-1}, q in W m^{-2}, and p in Pa. The range of data was q = 157–4450 kW m^{-2}; p = 0.85–50 kPa; mass flow rate 40–320 kg m^{-2} s^{-1}; and inlet subcooling 63–214 K. These data were predicted with MAD of 14.7%.

Based on their tests in an annulus, Qiu et al. (2015) gave the following correlation:

$$h_{TP} = 5q^{0.7}p^{0.15} \tag{5.8.5}$$

with q in W m^{-2}, h_{TP} in W m^{-2} K^{-1}, and p in Pascal. This formula agrees with their test data within ±25%. The annulus was 1000 mm long, 8 mm inner diameter, and 12 mm outer diameter. The heat flux varied from 80 to 500 kW m^{-2}, inlet subcooling from 63 to 285 °C, inlet flow velocity from 0.02 to 0.5 m s^{-1}, and system pressure from 3.67 to 103 kPa.

A correlation for boiling in tubes has been given by Kirillov et al. (1984) and a correlation for boiling in reactor core was given by Martsiniouk and Sorokin (2000).

Noyes and Lurie (1966) performed subcooled and saturated boiling tests during upflow in a vertical annulus 6.35 mm ID, 12.7 mm OD. Subcooled boiling heat transfer coefficients were around 60 kW m^{-2} K^{-1} and were only a little higher than the predictions of Dwyer (1962) correlation for single-phase flow in annulus. Figure 5.8.1 shows a comparison of their subcooled boiling data with data for pool boiling on stainless steel heaters.

No and Kazimi (1982) developed an analytical correlation that was shown to be in reasonable agreement with the

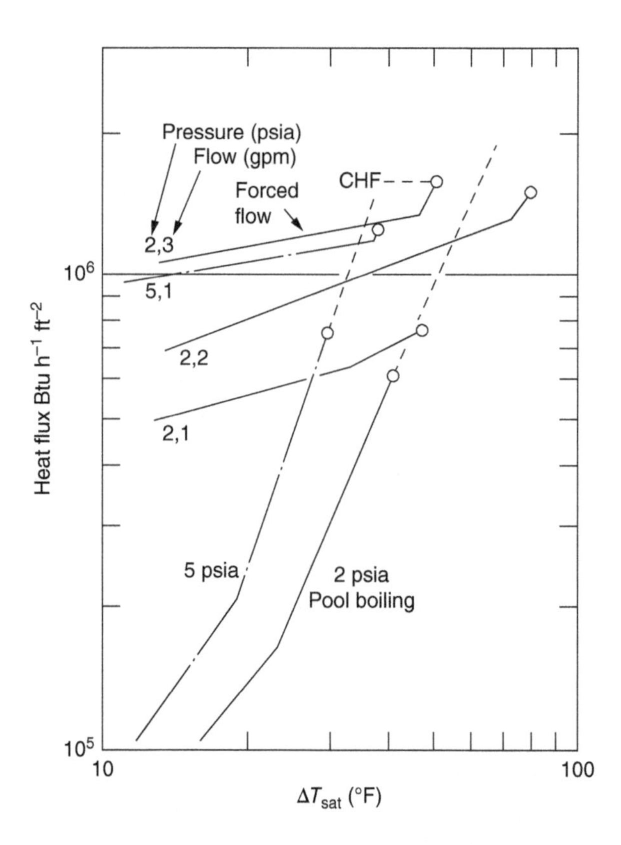

Figure 5.8.1 Subcooled boiling heat transfer of sodium compared to pool boiling. Source: From Noyes and Lurie (1966). © 1966 Begell House Inc.

data of Zeigarnick and Litinov (1980) for boiling of sodium in a vertical tube. This correlation assumes no nucleate boiling, while Eqs. (5.8.4) and (5.8.5) that are based on experimental data show strong nucleate boiling. Tests with potassium also show strong nucleate boiling as discussed in the following text.

5.8.2.2 Potassium

Hoffman and Krakoviak (1964) studied heat transfer to boiling potassium in vertical tubes. They report that boiling heat transfer coefficients were 5–10 times those with single-phase flow.

Peterson (1967) performed tests on annuli in which potassium flowed up in the inner tube, while hot sodium flowed up or down in the annular space. The inner tube diameters were 23.4 and 17 mm. Tests were done with or without vortex generating inserts inside the tubes. Boiler exit temperatures were 815–954 °C. Figure 5.8.2 shows some of their data for the 17 mm diameter tube without insert. It is seen that heat transfer coefficient increases with exit quality until it reaches a maximum and then falls. This is similar to the behavior of ordinary fluids. Further heat transfer coefficient increases with increasing mass flow rate. The inserts brought about significant increase in heat

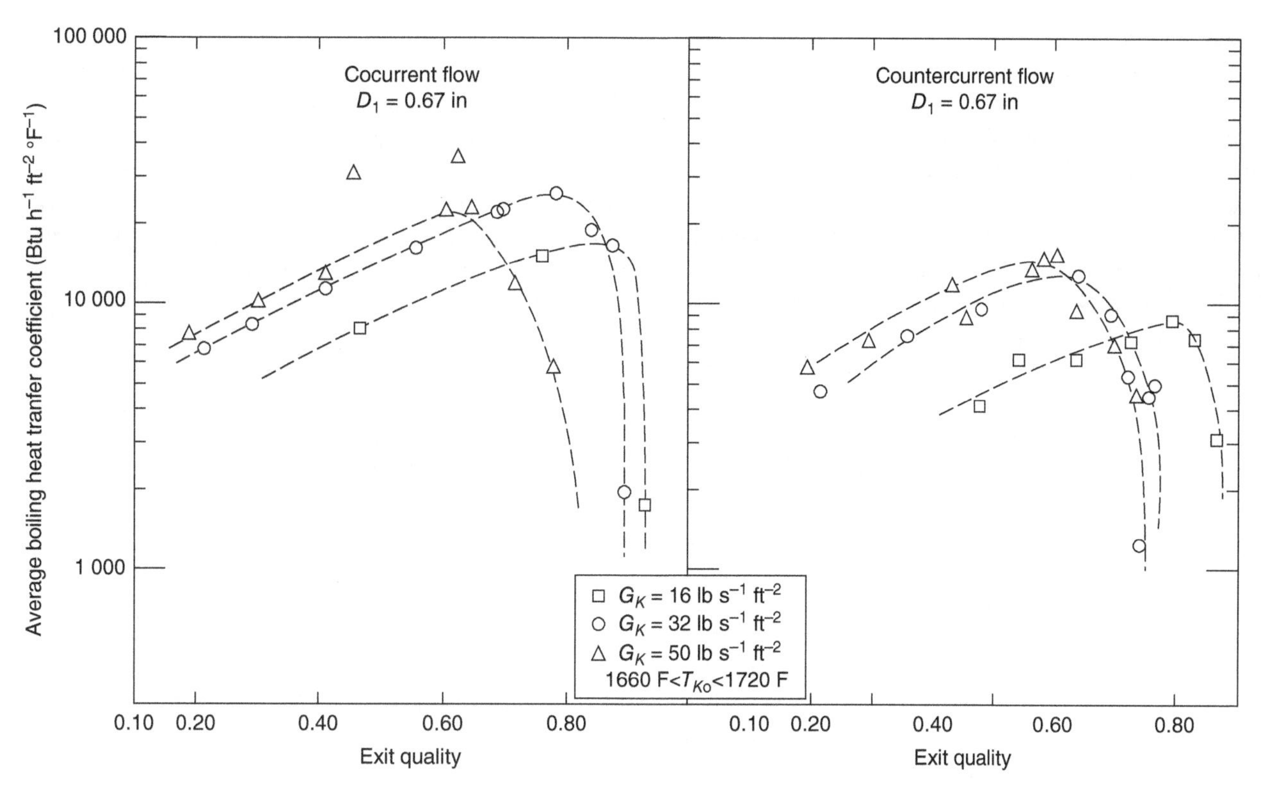

Figure 5.8.2 Effect of flow rate and direction of heating fluid on boiling of potassium in an annulus. G_K is mass flux of potassium, and T_{Ko} is the saturation temperature of potassium. Source: From Peterson (1967).

transfer coefficients. As noted by Peterson, the increase in heat transfer coefficient with increasing exit quality may be due to increasing heat flux.

Aladyev et al. (1969) performed tests on potassium boiling during upflow in electrically heated vertical tubes of 4 mm diameter. Conditions included p = 0.15–4.1 bar, q = 100–1400 kW m^{-2}, G = 22–70 kg m^{-2} s^{-1}, and x = 4–80%. No effect of quality and mass flux was found. The following formula predicted their data within ±30%:

$$h_{TP} = 0.57q^{0.7}p^{0.15} \tag{5.8.6}$$

where h_{TP} is in W m^{-2} K^{-1}, q in W m^{-2}, and p in Pa. The authors state that this equation is in reasonable agreement also with test data from other sources for pool boiling and flow boiling in tubes with diameters 8.2–23.6 mm.

Aladyev et al. (1974) gave the following correlation based on their tests on electrically heated vertical coils:

$$h_{TP} = 0.13q^{0.85}p^{0.15} \tag{5.8.7}$$

where h_{TP} is in W m^{-2} K^{-1}, q in W m^{-2}, and p in Pa. The two coils had tube diameters of 4 and 10 mm with coil diameters of 64 and 162 mm, respectively. Pressure was 0.9–8 bar, heat flux 30–510 kW m^{-2}, and quality 0.03–0.84. Equation (5.8.7) predicts a little higher than Eq. (5.8.6), which is for straight tubes. In Section 5.4.1, it was seen that heat transfer coefficients of ordinary fluids in coil are close to those in straight tubes. In this respect, the behavior of liquid metals seems to be similar to that of ordinary liquids.

Various researchers have found different effects of vapor quality on heat transfer coefficients. As noted earlier, Aladyev et al. (1969, 1974) found no effect of quality. Borishanskii et al. (1966) also reported no effect of quality. Grachev et al. (1968) found heat transfer coefficients to increase with increasing quality in some cases and decrease in some cases. Berenson and Killakly (1964) report that heat transfer coefficients decreased as quality increased from 0.2 to 0.65.

Dodonov and Koroleva (1973) examined data for boiling of alkali metals from several sources and compared them with data for boiling of ordinary liquids. They concluded that heat transfer in boiling of liquid metals differs from that in ordinary liquids in that the interfacial resistance is the controlling factor in boiling of liquid metals, while liquid film resistance is the controlling factor in boiling of ordinary liquids. As was discussed in Section 2.5.2, similar views were prevalent about condensation of liquid metals. However, careful examination of experimental data showed that the resistance attributed to interfacial resistance was in fact due to the presence of non-condensable gases.

5.8.2.3 Mercury

As mercury normally does not wet metal surfaces, attaining nucleate boiling with it is difficult. Flow oscillations usually occur. In the tests of Schmucker and Grigull (1973) with a natural circulation loop, nucleate boiling could not be achieved. There were severe oscillations.

Romie et al. (1960) studied boiling in an electrically heated vertical carbon steel tube with natural circulation by thermosiphon action. They were unable to attain nucleate boiling until they coated the tube with a thin copper layer. They also added traces of titanium hydride and magnesium. Their measured heat transfer coefficients were in the range of 23–53 kW m^{-2}. Heat flux varied from 880 to 1890 kW m^{-2} and pressure was 0.7–2.3 bar.

Gelman and Kopp (1968) boiled mercury in a 16 mm diameter electrically heated carbon steel tube with pressures in the range of 1.5–20 bar. It took some 30 hours of operation at wall temperatures exceeding 800 °C before nucleate boiling occurred and brought down the wall temperature to nucleate boiling levels. Their measurements covered heat flux from 70 to 3000 kW m^{-2}. They have shown their data in the form of a h_{TP} vs. q graph. The mean through those data is represented by the following equation fitted by the present author:

$$h_{TP} = 0.05q^{0.95} \tag{5.8.8}$$

where h_{TP} is in kW m^{-2} K^{-1} and q in kW m^{-2}.

Hsia (1970) performed tests in a horizontal tantalum tube 17 mm diameter heated by a mixture of sodium and potassium in an annular passage around it. The tube had a helical coil insert with $P/D = 2$ in the first 3.66 m length and a helical insert with $P/D = 0.97$ in the last 1.22 m length. Saturation temperatures were 800–879 K. The tube wall was fully wetted at these temperatures. They correlated their data by the following formula within ±20%:

$$h_{TP} = 3.09 \times 10^{-4} G^{0.65} p^{0.54} q^{0.85} \tag{5.8.9}$$

The units are h_{TP} in Btu h^{-1} ft^{-2} °F^{-1}, G in lb ft^{-2} s^{-1}, p in psi, and q in Btu h^{-1} ft^{-2}. The range of parameters covered by this correlation included G from 397 to 936 kg m^{-2} s^{-1} and q from 78 to 650 kW m^{-2}. Some of their test data is shown in Figure 5.8.3.

5.8.2.4 Cesium and Rubidium

Fisher (1964) and Fisher et al. (1964) performed tests on boiling of cesium and rubidium in a 7.1 mm diameter tube made of columbium–zirconium alloy. The following equation was fitted to the data for rubidium:

$$h_{TP} = Bq^n \tag{5.8.10}$$

The units are h_{TP} in Btu h^{-1} ft^{-2} F^{-1} and q in Btu h^{-1} ft^{-2}. The values of B and n are listed in Table 5.8.1. The range of parameters covered by the tests was as follows: mass flux 158–1192 kg m^{-2} s^{-1}, saturation temperature 977–1255 K,

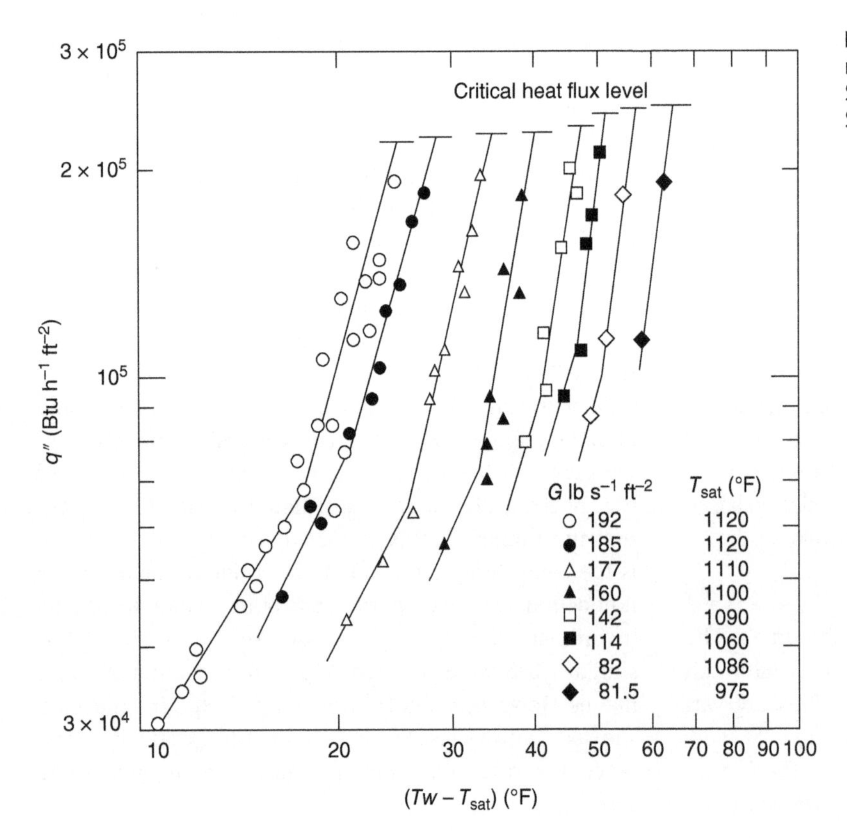

Figure 5.8.3 Heat transfer to boiling of mercury in a tube with swirl insert ($P/D = 2$). Source: From Hsia (1970). © 1970 American Society of Mechanical Engineers.

Table 5.8.1 Values of B and n in Eq. (5.8.10), the correlation of Fisher (1964) for boiling of rubidium.

Temperature range (°F)	n	$\text{Log}_{10}\ B$
1350–1400	1.126	−2.046
1400–1450	0.917	−0.918
1450–1500	1.181	−2.294
1500–1550	1.325	−3.044
1550–1600	1.405	−3.414
1600–1650	1.333	−3.128
1650–1700	0.958	−1.268
1700–1750	1.582	−4.427

subcooling 111–222 K, heat flux up to 103 kW m^{-2}, and outlet quality 0–0.8.

Fisher et al. have given tabulations of their data for cesium but no correlation was attempted.

5.8.2.5 Mixtures of Liquid Metals

Tang et al. (1964) performed tests on potassium amalgams (mixtures of potassium and mercury) in a vertical electrically heated stainless steel tube of 7.75 mm diameter. Two amalgams with 14.7% and 44.5% potassium were used. Deeply subcooled liquid entered the tube and was partially evaporated at the exit. Pressure range was 0.1–0.9 bar. The authors do not mention any incipient superheat or temperature/flow fluctuations. Heat transfer coefficients during saturated boiling were 30–60% below those for non-boiling flow. Heat transfer coefficients during subcooled boiling showed a lot of scatter, but they were not higher than single-phase flow heat transfer coefficients. Heat transfer rates for the 14.7% K amalgams were consistently lower than those for 44.5% K amalgams. This is similar to the behavior of mixtures of ordinary fluids. Figure 5.8.4 shows some of their data together with data from other sources. They noted that the slope of their data is lower than for nucleate boiling data from other sources; that led the authors to conclude that film boiling prevailed during their tests.

Mori et al. (1970) used a mixture of mercury–bismuth–lead (15–48–37% by weight, respectively) flowing through an alloy steel vertical tube 19.6 mm ID. The tube was inserted in a furnace and thus received heat by radiation. Pressure was 1.1–1.6 bar, and flow rate was 30–750 kg h^{-1}. Liquid entered the test section in a subcooled state and left it as a two-phase mixture. No flow or temperature fluctuations were observed. Their data is shown along with data for pure fluids and mixtures from other sources in Figure 5.8.4. It is seen that the slope of their data is close

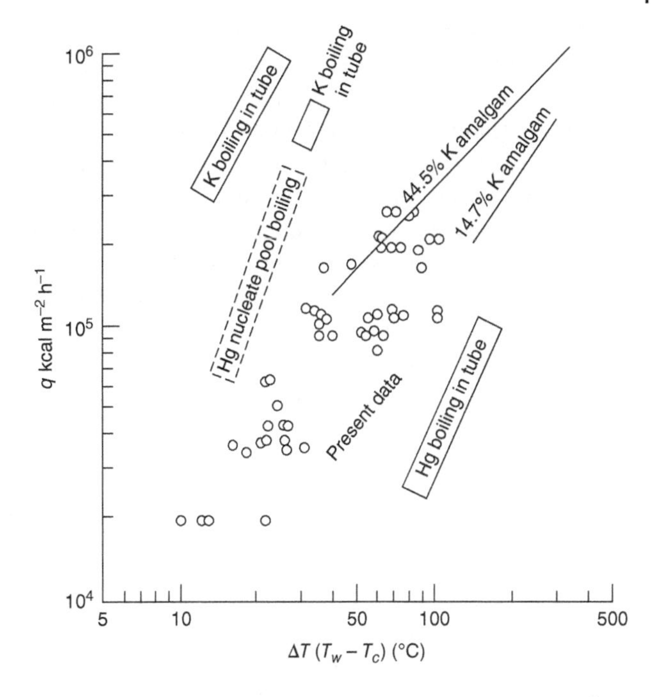

Figure 5.8.4 Data for boiling of mixtures and pure liquid metals from several sources. T_c is saturation temperature. "Present Data" are those of Mori et al. (1970) for a mixture. Amalgam data are those of Hsia (1970). Source: From Mori et al. (1970). © 1970 American Society of Mechanical Engineers.

to that of potassium amalgams' data of Hsia. It is noted that for both mixture data, the required wall superheat is higher than for pure potassium or mercury. This behavior is similar to that of mixtures of ordinary fluids as discussed in Section 5.7.

5.8.3 Conclusions and Recommendations

1. Inception of boiling superheats are high and wall temperature fluctuations occur during boiling of pure metals. With liquid metal mixtures, there is no boiling inception superheat, and there are no flow or temperature fluctuations.
2. Heat transfer coefficients of liquid metal mixtures are lower than those of their pure components. This is similar to the behavior of mixtures of ordinary liquids.
3. There are no general predictive techniques for inception of boiling or heat transfer coefficients. The correlations listed herein should be used only in their verified range.

5.9 Effect of Gravity

To support the efforts for space exploration and colonization, methods for calculating heat transfer under conditions of low and high gravity are needed. Most of the experimental work till now has been on pool boiling. Those efforts were reviewed in Section 3.7.1. As noted therein, the difficulty with pool boiling is that under low gravity conditions buoyancy decreases linearly with decreasing gravity. This hinders removal of bubbles that grow to large sizes and cause severe deterioration in heat transfer. Hence there is increasing interest in forced convection system where the inertia of moving liquid carries the bubbles away and thus prevents formation of large bubbles that cause early CHF. As noted in the review paper by Konishi and Mudawar (2015), the amount of experimental work done till now on forced convection systems is very limited. Much of the work done is on the study of flow pattern transitions. The work on the effect on heat transfer is discussed in the following.

5.9.1 Experimental Studies

Saito et al. (1994) studied heat transfer to water in a horizontal flow channel of 25 mm × 25 mm cross section with an 8 mm diameter rod-type electrical heater located at its center. Gravity levels of $0.01g_e$ to $2.5g_e$ were attained for 20 seconds during parabolic flights. The various parameters during these tests were as follows: inlet fluid velocity, 3.7–22.9 cm s^{-1}; inlet fluid temperature, 86.1–112.8 °C; heat flux, 5.3–18.6 W cm^{-2}; and system pressure, 0.9–2.04 bar. Under Earth gravity, there were many small bubbles that detached frequently and flowed on the upper part of the channel. Under low gravity, bubbles rarely detached. They flowed downstream attached to the heater and growing larger on the way. Despite these large differences in the flow behavior, heat transfer coefficients in the low gravity condition were about the same as at Earth gravity.

Ohta (1997) studied boiling of R-113 in an 8 mm diameter vertical glass tube coated with a gold film on its inner surface through which current was passed to provide heating. Parabolic flights provided 0.01 g_e for 20 seconds, which was considered sufficient to achieve steady state. There was no effect of low gravity on heat transfer coefficient except at low heat flux and moderate quality where it deteriorated. Under this condition, the annular film became thicker and smooth. As shown in a figure in the paper, heat transfer dropped from 1200 W m^{-2} K^{-1} at $1g_e$ to 1000 W m^{-2} K^{-1} at $0.01g_e$ under the following conditions: $p = 1$ bar, $G = 150$ kg m^{-2} s^{-1}, $q = 10$ kW m^{-2}, and $x = 0.29$. At the same conditions, heat transfer coefficient rose to 1400 W m^{-2} s^{-1} at $2g_e$.

Ma and Chung (2001) studied subcooled flow boiling of FC-72 across a heated 0.254 mm platinum wire using a 2.1 second drop tower. They found that the CHF significantly declines in microgravity. However, the differences in both the heat transfer rate and CHF between microgravity and normal gravity decrease with increasing flow rate.

Luciani et al. (2008) studied subcooled boiling of HFE-7100 in three channels (6.0 × 0.254mm^2, 6.0 × 0.454mm^2, and 6.0 × 0.654 mm^2) with a heated length of 50 mm. Tests were done in parabolic flights in which temperatures and various other parameters were recorded. They employed inverse methods to estimate the local heat transfer coefficient from the collected data. Their estimated heat transfer coefficients are shown in Figure 5.9.1. As seen in it, heat transfer coefficients under microgravity condition were higher near the inlet compared with $1g_e$ and $1.8g_e$ but dropped sharply to be about the same as at $1g_e$ from around the middle of the channel to the exit. Heat transfer coefficients at hypergravity were about the same as at $1g_e$. The authors state that the heat transfer coefficients at all gravity levels in the second half of the channel were comparable to those for vapor only flow.

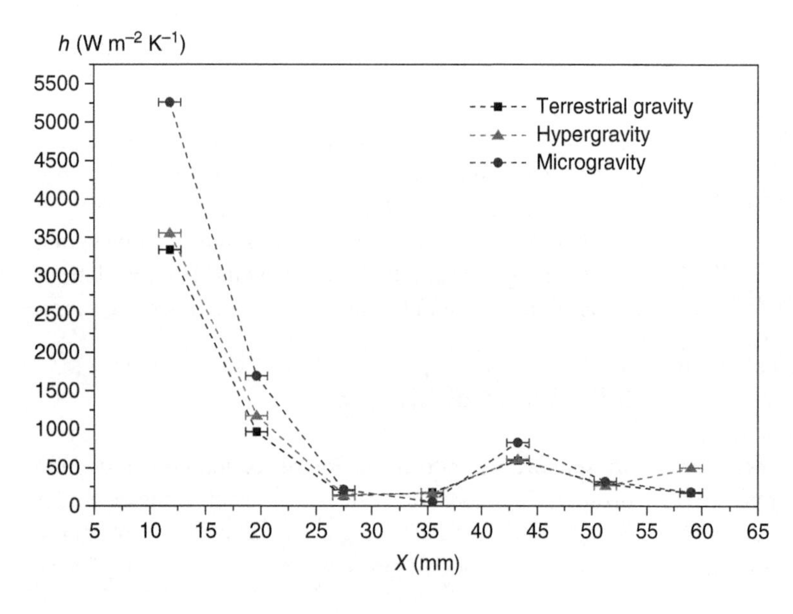

Figure 5.9.1 Effect of gravity on heat transfer during boiling of HFE-7100 in a minichannel 6 mm × 0.254 mm. $q = 32$ kW m^{-2}, flow rate 0.26 g s^{-1}, and exit quality 0.26. X is the distance from entrance. Source: From Luciani et al. (2008). © 2008 American Society of Mechanical Engineers.

Celata and Zummo (2009) describe tests with FC-72 flowing up in heated vertical tubes of 2.4 and 6 mm diameter. Tests were done during parabolic flights, which provided $0.01g_e$ for 22 seconds. They found that there was no effect of gravity at vapor qualities greater than 0.3. They found no effect of gravity when liquid velocity was higher than 25 cm s^{-1}; the velocities in the tests were 3–30 cm s^{-1}. Heat transfer coefficients decreased under microgravity at lower flow rates and qualities.

Konishi et al. (2015) studied boiling of FC-72 in a channel 5 mm high and 2.5 mm wide during parabolic flights. They found that during a flight parabola, heated wall temperatures decrease slightly as the aircraft enters the hypergravity phase of the parabola. The heated wall temperatures then increase slightly as the aircraft enters the microgravity phase and decrease once again during the hypergravity phase. They concluded that these temperature variations point to enhancement in flow boiling heat transfer with increasing gravity, and degradation in microgravity.

Zhang et al. (2018) studied flow boiling heat transfer of subcooled FC-72 at atmospheric pressure on micro-pin-finned surfaces in earth gravity as well as microgravity by utilizing a drop tower that provided 3.6 seconds of low gravity. Velocity was 0.5 and 1.0 m s^{-1}. It was found that there was no effect of gravity on bubble behavior at low and moderate heat flux. At high heat flux, a large bubble was formed near the outlet, which caused CHF. Heat transfer coefficients under microgravity were the same as at Earth gravity.

Lebon et al. (2019) performed subcooled flow boiling measurements using HFE-7000 in a vertical 6 mm ID sapphire tube during upward and downward flow at various gravity levels including hypergravity and microgravity performed in parabolic flights and on ground. The flights provided 22 seconds of microgravity (gravity levels less than $0.02g_e$). In microgravity, there was a decrease in heat transfer coefficient compared with downward and upward conditions at lower mass and heat flux. The effect of gravity decreased with increasing mass flux and heat flux. At the highest mass flux, there was no effect of gravity. Similarly, at the highest heat flux, there was no effect of gravity. In hypergravity, heat transfer coefficients increased with gravity level at lower mass and heat flux, but there was no effect of gravity at higher mass or heat flux.

5.9.2 Conclusions and Recommendation

The experimental studies are limited and the results are in some cases conflicting. There is no method to predict the limit beyond which there is no effect of gravity on heat transfer nor any method to predict the effect of gravity below that limit. The only thing that is clear is that effects of gravity disappear at higher velocities, which may be attributed to sweeping away of bubbles due to liquid inertia. Hence the velocities should be kept high to avoid deterioration in heat transfer.

5.9.3 Effect of Oil in Refrigerants

Most of the compressors used in vapor compression refrigeration system allow some oil to get mixed with refrigerants that are carried into the system components including the evaporator. Effect of oil on pool boiling was discussed in Section 3.7.2. Herein, the effect of oil in forced convection systems is discussed. Literature on this subject has been reviewed among others by Bandarra Filho et al. (2009), Momenifar et al. (2015), and Gao et al. (2019).

Most of the oils are fully or partially miscible with refrigerants but some oils are immiscible. The use of immiscible oils is limited to some ammonia and carbon dioxide systems.

5.9.3.1 Heat Transfer with Immiscible Oils

Shah (1974, 1975) performed tests on a horizontal ammonia evaporator that was 140 m long made of 26.2 mm steel tube. Flow patterns were observed through sight glasses at 13 locations along the length. Flow rate was 60–3000 kg h^{-1} and evaporation temperature was −40–5 °C. The compressor used mineral oil (Mobil Arctic 300), which is insoluble in ammonia. Single-phase ammonia liquid entered the evaporator and evaporated during flow along the tube, exiting at vapor qualities up to 100%. Thick apparently solid oil films were seen in some tests; these were accompanied by high pressure drop and very low heat transfer coefficients. In most tests, liquid oil films were seen. In some tests, no oil film was observed. Single-phase heat transfer coefficients were well below the Dittus–Boelter equation, Eq. (5.2.9), and were instead given by the following equation:

$$\frac{h_{\mathrm{LT}} D}{k_f} = 0.1825 Re_{\mathrm{LT}}^{0.509} Pr_f^{0.4} \tag{5.9.1}$$

Based on his visual observations, Shah assumed that the difference between Eq. (5.9.1) and the Dittus–Boelter equation is due to the resistance of an oil film around the tube circumference. Thus the heat transfer coefficient of the mixture h_{mix} is given by the following equation:

$$\frac{1}{h_{\mathrm{mix}}} = \frac{1}{h_{\mathrm{pure}}} + \frac{1}{k_{\mathrm{oil}}/\delta} \tag{5.9.2}$$

where h_{pure} is the heat transfer coefficient of pure refrigerant and δ is the thickness of oil film.

The calculated thickness of this oil film is given by the following equation:

$$\frac{\delta}{D} = \frac{0.028}{Re_{\mathrm{LT}}^{0.23}} \tag{5.9.3}$$

where δ is the oil film thickness. The range of Reynolds number in the data was 3×10^3–11.5×10^5. The calculated thickness of the film from this equation in this range was 0.04–0.1 mm. Shah (1976) used the resistance of the oil film calculated by Eq. (5.9.3) to correct the predictions of his general correlation and found reasonable agreement with the data for ammonia from Shah (1974, 1975).

Chaddock and Buzzard (1986) performed tests on a horizontal ammonia evaporator with diameter 13.4 mm. Tests were done with pure ammonia as well as with up to 4.3% mineral oil. Thick oil films similar to those reported by Shah were observed in some tests. Heat transfer coefficients with oil were as much as seven times lower than of pure ammonia.

Boyman et al. (2004) studied heat transfer of ammonia flowing in a 14 mm diameter horizontal tube. Mass velocities ranged from 40 to 170 kg m^{-2} s^{-1}, saturation temperatures from −10 to +10 °C, and heat fluxes from 10 to 50 kW m^{-2}. Test were done with 0–3% immiscible mineral oil. They found that 0.1% oil reduced average heat transfer coefficients up to 50% compared with oil-free ammonia. Further reductions occurred as oil concentration increased up to 1%. Further increase in oil concentration did not cause further deterioration.

5.9.3.2 Heat Transfer with Miscible Oils

Presence of miscible oils has been found to increase or decrease heat transfer or to have no effect. Figure 5.9.2 shows some data of Momenifar et al. (2015) for R-600a with miscible oil in a horizontal tube 8.7 mm diameter. It is seen that presence of oil increases heat transfer at qualities up to about 0.5 and cause decrease at higher qualities. Similar results have been reported in several other studies with halogenated refrigerants, for example, Hu et al. (2008) for R-410A in a microfin tube and Zürcher et al. (1997) for R-134a in a plain tube.

Effect of miscible oils on refrigerant heat transfer has been analyzed in two different ways. One way is to consider oil as a contaminant. All calculations are done based on the properties of pure fluid, and correction factors are developed for the effect of oil. In another approach, refrigerant–oil mixture is treated in the same way as mixtures of refrigerants as described in Section 5.7. Heat transfer of mixtures in this approach is defined as

$$h_{\mathrm{TP}} = \frac{q}{T_w - T_{\mathrm{BP}}} \tag{5.9.4}$$

Properties used are those of refrigerant–oil mixture. Thome (1995) has given a generalized method for calculation of mixture properties including bubble point. Oil concentration ω is defined as

$$\omega = \frac{m_{\mathrm{oil}}}{m_{\mathrm{oil}} + m_{\mathrm{refrigerant}}} \tag{5.9.5}$$

where m is the mass flow rate. As vapor quality increases along the tube, liquid flow rate decreases and hence oil concentration in liquid increases. At 100% quality, $\omega = 1$.

There are numerous correlations for hydrocarbons and halogenated refrigerants mixed with oil. Examples are Momenifar et al. (2015) and Hu et al. (2008), but they are based only on the author's own data and cannot be used under other conditions. Indeed, most of the papers offering correlations of their own data show that other correlations perform poorly with their data.

Li et al. (2014) performed tests with carbon dioxide with PAG 100 oil that is partially miscible. Tests were done in a horizontal 2 mm diameter stainless steel tube. Mass flux was 360–1440 kg m^{-2} s^{-1}, heat flux was from 4.5 to 36 kW m^{-2}, evaporating temperature was 15 °C, and vapor quality was 0% to near 100%. Oil concentration was 0.5–5%. They based the measured heat transfer coefficients on the saturation temperature of pure carbon dioxide. Heat transfer coefficients were found to decrease with increasing inlet oil concentration. They have given a correlation that agrees with their own data. Agreement with data from some other studies was also reported to be fairly good. Other correlations for effect of oil on heat transfer of carbon dioxide were found to give large deviations with data.

Gao et al. (2019) studied boiling of ammonia mixed with a miscible oil. Tests were done in an 8 mm diameter horizontal tube at oil concentration from 0% to 5.78%, mass flux from 51 to 99.5 kg m^{-2} s^{-1}, heat fluxes of 9 and 21 kW m^{-2}, and temperature from −5.5 to −5 °C. Heat transfer coefficients were based on the bubble point temperature and were found to decrease with increasing oil content at all qualities. The decrease was generally in the range of 5–20%. They compared their data with several general correlations using the properties of ammonia–oil mixture. The MAD with the correlations of Shah (1982), Gungor and Winterton (1986), and Gungor and Winterton (1987) were 23.2%, 12.9%, and 26.7%, respectively. These results indicate that this approach is satisfactory for ammonia mixed with miscible oils.

5.9.3.3 Conclusions and Recommendations

For refrigerants with immiscible oil, heat transfer is severely deteriorated by the presence of oil. The recommendation is to assume an oil film thickness of 0.04–1 mm and correct the pure refrigerant heat transfer coefficient with the resistance of this oil film using Eq. (5.9.2).

For carbon dioxide with miscible oil, the correlation of Li et al. (2014) has had more verification than others and is therefore preferable.

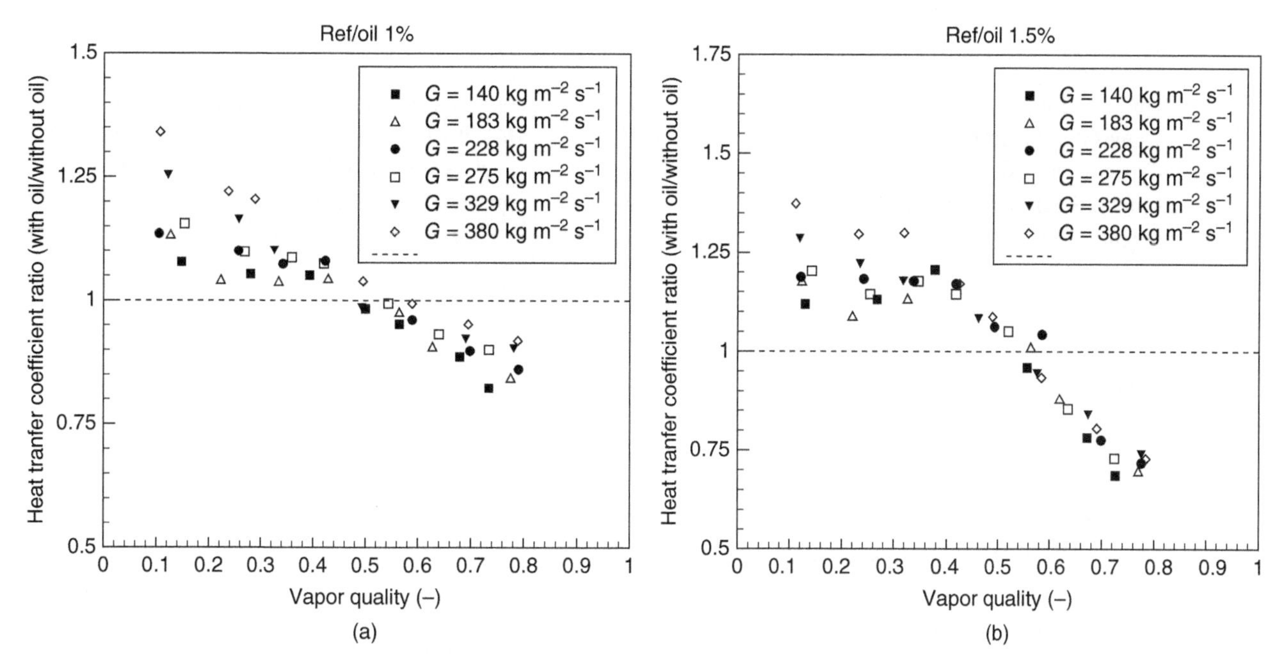

Figure 5.9.2 Effect of 1% and 1.5% oil concentration on heat transfer to R-600a in a horizontal tube. Source: From Momenifar et al. (2015). © 2015 Elsevier.

For ammonia with miscible oil, use correlations for pure fluids but use oil–ammonia mixture properties for the liquid phase.

For other refrigerants with miscible oils, heat transfer coefficient increases with oil concentration at lower qualities and decreases at higher qualities, similar to the trends shown in Figure 5.9.2. There is no generally applicable prediction method. Guidance should be taken from experimental studies that are close to the design conditions.

Nomenclature

The following are the most frequently used symbols. Occasionally they have been defined differently as noted in the text at such occasions:

Bd	Bond number = $g(\rho_f - \rho_g)D^2/\sigma$ (–)
Bo	boiling number (–)
Co	convection number, given by Eq. (5.2.14) (–)
C_p	specific heat at constant pressure (J kg^{-1} K^{-1})
D	diameter of tube (m)
D_{HP}	equivalent diameter based on heated perimeter (m)
D_{HYD}	hydraulic equivalent diameter (m)
E_s	factor in Shah correlation given by Eq. (5.2.54) (–)
F_{TS}	Thome–Shakir correction factor (–)
Fr	Froude number (–)
G	mass flux (kg m^{-2} s^{-1})
g	acceleration due to gravity (ms^{-2})
h	heat transfer coefficient (W m^{-2} K^{-1})
i_{fg}	latent heat of vaporization (J kg^{-1})
k	thermal conductivity (W m^{-1} K^{-1})
L	length (m)
M	molecular weight (–)
m	mass flow rate (kg s^{-1})
N	number of data points (–)
Nu	Nusselt number (–)
P	Perimeter (m)
p	Pressure (Pa)
p_c	critical pressure (Pa)
p_r	reduced pressure, = p/p_c (–)
Pe	Peclet number, = $(Re{\cdot}Pr)$ (–)
Pr	Prandtl number (–)
q	heat flux (W m^{-2})
Ra	Rayleigh number (–)
Re	Reynolds number = (GD/μ) (–)
t	time (s)
T	temperature (K)
ΔT_{SAT}	$(T_w - T_{SAT})$ (K)
ΔT_{SC}	$(T_{SAT} - T_f)$ (K)
u	velocity (m s^{-1})
We	Weber number (–)
X_{tt}	Martinelli parameter for turbulent–turbulent flow, given by Eq. (5.2.4) (–)
x	equilibrium vapor quality (–)
z	length or distance (m)

Greek

α	void fraction, (–)
β	Chevron angle from vertical (°)
μ	dynamic viscosity (Pa s)
ν	kinematic viscosity ($m^2\ s^{-1}$)
ρ	density ($kg\ m^{-3}$)
φ	enlargement factor of herringbone plate (–)
ψ	h_{TP}/h_{LS} (–)
σ	surface tension ($N\ m^{-1}$)
ω	angular velocity (radian s^{-1})

Subscripts

bd	bubble departure
BP	bubble point
cb	convective boiling
f	liquid
e	on Earth
GS	vapor phase flowing alone, also called superficial vapor
GT	all mass flowing as vapor
g	vapor
ib	inception of boiling
mix	mixture
nb	nucleate boiling
LS	liquid phase flowing alone, also called superficial liquid
LT	all mass flowing as liquid
pb	pool boiling
SAT	saturation
SC	subcooled
w	wall, at wall temperature
0	at zero vapor quality

References

Abbas, A. and Ayub, Z.H. (2017). Experimental study of ammonia flooded boiling on a triangular pitch plain tube bundle. *Applied Thermal Engineering* **121**: 484–491.

Aladyev, I.T. (1973). On the nature of liquid superheat. In: *Advances in Heat Transfer* (ed. O.E. Dwyer). London: Pergamon Press.

Aladyev, I.T., Gorlov, I.G., Dodonov, L. et al. (1969). Heat transfer to boiling potassium in uniformly heated tubes. *Heat Transf. Sov. Res.* **1** (4): 14–26.

Aladyev, I.T., Petrov, V.I. et al. (1974). Investigation of heat transfer to potassium in coil tubes and sodium-potassium coil tube generator. In: *Proceedings of the Fifth International Heat Transfer Conference*3-7 September 1974, Tokyo, vol. **4**, 330–334.

Almalfi, R.L., Vakili-Farahani, F., and Thome, J.R. (2016a). Flow boiling and frictional pressure gradients in plate heat exchangers. Part 1: review and experimental database. *Int. J. Refrig.* **61**: 166–184.

Almalfi, R.L., Vakili-Farahani, F., and Thome, J.R. (2016b). Flow boiling and frictional pressure gradients in plate heat exchangers: part 2, comparison of literature methods to database and new prediction methods. *Int. J. Refrig.* **61**: 185–203.

Andrews, P.R. and Cornwell, K. (1987). Cross-sectional and longitudinal heat transfer variations in a reboiler. *Chem. Eng. Res. Des.* **65**: 127–130.

Ardhpurkar, P.M., Sridharan, A., and Atrey, M.D. (2014). Flow boiling heat transfer coefficients at cryogenic temperatures for multi-component refrigerant mixtures of nitrogen–hydrocarbons. *Cryogenics* **59**: 84–92.

Aria, H., Akhavan-Behabadi, M.A., and Shemirani, F.M. (2012). Experimental investigation on flow boiling heat transfer and pressure drop of hfc-134a inside a vertical helically coiled tube. *Heat Transf. Eng.* **33** (2): 79–87.

Arima, H., Monde, M., and Mitsutake, Y. (2003). Heat transfer in pool boiling of ammonia/water mixture. *Heat Mass Transf.* **39**: 535–543.

Arima, H., Kim, J.H., Okamoto, A., and Ikegami, Y. (2010). Local boiling heat transfer characteristics of ammonia in a vertical plate evaporator. *Int. J. Refrig.* **33**: 359–370.

Ayub, A., Khan, T.S., Salam, S. et al. (2019). Literature survey and a universal evaporation correlation for plate type heat exchangers. *Int. J. Refrig.* **99**: 408–418.

Bandarra Filho, E.P., Cheng, L., and Thome, J.R. (2009). Flow boiling characteristics and flow pattern visualization of refrigerant/lubricant oil mixtures. *Int. J. Refrig.* **32**: 185–202.

Barbe, C., Grange, A., and Roger, D. (1971). Echanges de chaleur et pertes de charges en ecoulement diphasique dans la calandre de echangeurs bobines. *Proc. XIII Int. Cong. Refrig.* **2**: 223–234.

Bennet, D.L. and Chen, J.C. (1980). Forced convection boiling in vertical tubes for saturated pure components and binary mixtures. *AICHE J.* **26**: 454–461.

Berenson, P.I. and Killakly, I.I. (1964). Experimental investigation of forced convection vaporization of potassium. ORNL-3605, Vol. 2.

Bertsch, S.S., Groll, E.A., and Garimella, S.V. (2009). A composite heat transfer correlation for saturated flow boiling in small channels. *Int. J. Heat Mass Transf.* **52**: 2110–2118.

Bjornard, T.A. and Griffith, P. (1977). PWR blowdown heat transfer. In: *Symposium on the Thermal and Hydraulic Aspects of Nuclear Safety*, Vol.1: Light Water Reactors (eds. O.C. Jones and S.G. Bankoff). New York: ASME.

Borishanskii, V.M., Andreevsky, A.A., Zhokhov, K.A. et al. (1966). Heat transfer with boiling potassium in tubes in the region of moderate vapor content. *Atomnaya Energia* **21** (1). Quoted by Dodonov and Koroleva (1973).

Boyman, T., Aecherli, P., and Wettstein, A.S.W. (2004). Flow boiling of ammonia in smooth horizontal tubes in the presence of immiscible oil. International Refrigeration and Air Conditioning Conference. http://docs.lib.purdue.edu/iracc/656.

Brisbane, T.W.C., Grant, I.D.R. and Whalley, P.B. (1980). A Prediction Method for Kettle Reboiler Performance. ASME Paper 80-HT-42, ASME, New York.

Browne, M.W. and Bansal, P.K. (1999). Heat transfer characteristics of boiling phenomenon in flooded refrigerant evaporators. *Applied Thermal Engineering* **19** (6): 595–624.

Burnside, B.M., Miller, K.M., McNeil, D.A., and Bruce, T. (2001). Heat transfer coefficient distributions in an experimental kettle reboiler thin slice. *Chemical Engineering Research and Design* **79** (4): 445–452.

Casciaro, S. and Thome, J.R. (2001). Thermal performance of flooded evaporators, part 1: Review of boiling heat transfer studies. *ASHRAE Trans.* **107** (1): 903–918.

Celata, G.P. and Zummo, G. (2009). Flow boiling heat transfer in microgravity: recent progress. *Multiph. Sci. Technol.* **21** (3): 187–212.

Cerza, M. and Sernas, V. (1983). Nucleate boiling heat transfer in developing laminar falling water films. In: *Proceedings of ASME-JSME Thermal Engineering Joint Conference*20-24 March 1983, Honululu, Hawaii, vol. 1, pp. 111. ASME.

Chaddock, J. and Buzzard, G. (1986). Film coefficients for in-tube evaporation of ammonia and R502 with and without small percentages of mineral oil. *ASHRAE Trans.* **92** (1A): 22–40.

Chawla, J.M. (1967). Warmeubergang und druckabfall in wagrechten rohren bei der stromung von verdampfenden kaltmitteln. *VDI Forschungsheft* **523**.

Chen, J.C. (1966). A correlation for boiling heat transfer to saturated fluids in convective flow. *Ind. Eng. Chem. Proc. Des. Dev.* **5** (3): 322–329.

Chen, J.C. (1968). Incipient boiling superheats in liquid metals. *J. Heat Transf.* **90** (3): 303–312.

Chen, C.N., Han, J.T., Jen, T.C., and Shao, L. (2011). Chemical characteristics of r134a flow boiling in helically coiled tubes at low mass flux and low pressure. *Thermochim. Acta* **512** (1): 163–169.

Cheng, L., Ribatski, G., and Thome, J.R. (2008). New prediction methods for CO_2 evaporation inside tubes: part II – an updated general flow boiling heat transfer model based on flow patterns. *Int. J. Heat Mass Transf.* **51**: 125–135.

Chien, L.H. and Tsai, Y.L. (2011). An experimental study of pool boiling and falling film vaporization on horizontal tubes in R-245fa. *Appl. Therm. Eng.* **31**: 4044–4054.

Chien, L.H. and Wu, J.S. (2004). Convective evaporation on plain tube and low-fin tube banks using R-123 and R-134a. *ASHRAE Trans.* **110**: 101–108.

Chun, K.R. and Seban, R.A. (1971). Heat transfer to evaporating liquid films. *J. Heat Transf.* **93**: 391–395.

Chyu, M.C., Zeng, X., and Ayub, Z.H. (1995). Nozzle-sprayed flow rate distribution on a horizontal tube bundle. *ASHRAE transactions* **1995** (2).

Collier, J.G. (1981). *Convective Boiling and Condensation*, 2e. New York: McGraw Hill.

Cornwell, K., Duffin, N.W., and Schuller, R.B. (1980). An experimental study of the effects of fluid flow on boiling within a kettle reboiler tube bundle. ASME Paper 80-HT-45. Presented at the ASME/AIChE National Heat Transfer Conference, Orlando, Florida, (27–30 July 1980).

Ding, C., Hu, H., Ding, G. et al. (2017). Experimental investigation on downward flow boiling heat transfer characteristics of propane in shell side of LNG spiral wound heat exchanger. *Int. J. Refrig.* **84**: 13–25.

Ding, C., Hu, H., Ding, G. et al. (2018). Influences of tube pitches on heat transfer and pressure drop characteristics of two-phase propane flow boiling in shell side of LNG spiral wound heat exchanger. *Appl. Therm. Eng.* **131**: 270–283.

Dittus, P.W. and Boelter, L.M.K. (1930). Heat transfer in the radiators of the tubular type. *Univ. Calif. Eng. Pub.* **2** (13): 443–461.

Dodonov, L.D. and Koroleva, V.S. (1973). Some features of heat transfer to alkali metals in vapor generating tubes. *Progr. Heat Mass Transf.* **7**: 385–395.

Dodonov, L.D., Koroleva, V.S., and Razayev, A.I. (1979). Experimental study of incipient boiling superheats for potassium in channel flow. *Heat Transfer Sov. Res.* **11** (4).

Dwyer, O.E. (1962). Eddy Transport of Liquid Metal Heat Transfer. Brookhaven National Laboratory Report BNL-6149. Quoted in Noyes and Lurie (1966).

Dwyer, O.E., Stickland, G., Kalish, S. et al. (1973a). Incipient boiling superheats for sodium in turbulent channel flow:

effect rate of temperature rise. *J. Heat Transf.* **95** (2): 159–165.

Dwyer, O.E., Stickland, G., Kalish, S. et al. (1973b). Incipient boiling superheats for sodium in turbulent, channel flow: effects of heat flux and flow rate. *Int. J. Heat Mass Transf.* **16**: 971–984.

Eldeeb, R., Aute, V., and Radermacher, R. (2016). A survey of correlations for heat transfer and pressure drop for evaporation and condensation in plate heat exchangers. *Int. J. Refrig.* **65**: 12–26.

Fang, X., Wu, Q., and Yuan, Y. (2017). A general correlation for saturated flow boiling heat transfer in channels of various sizes and flow directions. *Int. J. Heat Mass Transf.* **107**: 972–981.

Feldman, A., Marvillet, C., and Lebouche, M. (2000). Nucleate and convective boiling in plate and fin heat exchangers. *Int. J. Heat Mass Transf.* **43**: 3433–3442.

Fernandez-Seara, J. and Pardinas, A.A. (2014). Refrigerant falling film evaporation review: description, fluid dynamics and heat transfer. *Appl. Therm. Eng.* **64**: 155–171.

Fisher, C.R. (1964). Heat Transfer and Pressure Drop Characteristics for Boiling Rubidium in Fforced Convection. *Oakridge National Laboratory Report ORNL–3605*, vol. 2, pp. 64–75.

Fisher, C.R., Clark, L.T., and Moskowitz, J.H. (1964). Forced Convection Boiling Rubidium and Cesium Heat Transfer and Pressure Drop. *Argonne National Laboratory Report AGN-8099*.

Forster, H.K. and Zuber, N. (1955). Dynamics of vapor bubbles and boiling heat transfer. *AIChE J.* **1** (4): 531–535.

Fsadini, M.S. and Whitty, J.P.M. (2016). A review on the two-phase heat transfer characteristics in helically coiled tube heat exchangers. *Int. J. Heat Mass Transf.* **95**: 551–565.

Fujita, T. and Ueda, T. (1978). Heat transfer to falling liquid films and film breakdown – II, saturated liquid film with nucleate boiling. *Int. J. Heat Mass Transf.* **21**: 109–118.

Gao, Y., Shao, S., Feng, Y., and Tian, C. (2019). Heat transfer and pressure drop characteristics of ammonia/miscible oil mixture during flow boiling in an 8 mm horizontal smooth tube. *Int. J. Therm. Sci.* **138**: 341–350.

Garg, N.S. and Tripathy, G. (1981). Boiling heat transfer from rotating cylinder to non-Newtonian fluid. *Ind. J. Technol.* **18** (2): 131–134.

Garg, N.S., Shankar, U., and Tripathy, G. (1980). Pool boiling heat transfer from rotating cylinders. *Ind. J. Technol.* **19** (4): 53–56.

Gelman, L.I. and Kopp, I.Z. (1968). Heat transfer in nucleate boiling of mercury at thermal loads upto 2.10^6 W/m^2. *High Temp.* **6** (3): 532–533.

Gilmour, C.H. (1959). Performance of vaporizer: heat transfer analysis of plant data. *Chem. Eng. Prog. Symp. Ser.* **55** (29): 67–78.

Gorenflo, D. (1993). Pool boiling, VDI Heat Atlas, Dusseldorf, Germany, Ha1-25.

Grachev, N.S., Zelensky, V.N., Kirillov, P.L. et al. (1968). Heat transfer and hydrodynamics in boiling potassium in tubes. *Teplofizika Vyasokikn Temperature* **8** (2). Quoted in Dodonov and Koroleva (1973).

Grauso, S., Mastrulio, R., Mauro, A.W., and Vanoli, G.P. (2011). CO_2 and propane blends: experiments and assessment of predictive methods for flow boiling in horizontal tubes. *Int. J. Refrig.* **34** (4): 1028–1039.

Gungor, K.E. and Winterton, R.H.S. (1986). A general correlation for flow boiling in tubes and annuli. *Int. J. Heat Mass Transf.* **29**: 351–358.

Gungor, K.E. and Winterton, R.H.S. (1987). Simplified general correlation for saturated flow boiling and comparisons of correlations with data. *Chem. Eng. Res. Des.* **65**: 148–156.

Gupte, N.S. and Webb, R.L. (1994). Convective evaporation of pure refrigerants in enhanced and integral-fin tube banks. *Enhanc. Heat Transf.* **4** (1): 351–364.

Gupte, N.S. and Webb, R.L. (1995). Shell-side boiling in flooded refrigerant evaporators, part I: integral finned tubes. *HVAC&R Res.* **1**: 35–47.

Hardik, B.K. and Prabhu, S.V. (2017). Boiling pressure drop and local heat transfer distribution of helical coils with water at low pressure. *Int. J. Therm. Sci.* **114**: 44–63.

Hardik, B.K. and Prabhu, S.V. (2018). Heat transfer distribution in helical coil flow boiling system. *Int. J. Heat Mass Transf.* **117**: 710–728.

Hasan, R. and Rhodes, E. (1982). Boiling two-phase flow in a horizontal bend. *Chem. Eng. Commun.* **18**: 191–209.

Hausen, H. (1943). Darstellung des Wärmeüberganges in Rohren durch verallgemeinerte Potenzbeziehungen. *Z. VDI Beiheft Verfahrenstechnik* **4**: 91–102.

Henry, R.E., Singer, R.M., Quin, D.J. et al. (1974). Incipient superheat in in a convective sodium system. *Proc. Fifth Int. Heat Transfer Conf.* **4**: 305–309.

Hoffman, W.H. and Krakoviak, A. (1964). Convective boiling with liquid potassium. In: *Proceedings Heat Transfer and Fluid Flow Institute*, 19–37. Stanford University Press.

Hsia, E.S. (1970). Forced convective annular flow boiling with liquid mercury under wetted and swirl flow conditions. In: *Liquid Metal Heat Transfer and Fluid Dynamics* (eds. J.C. Chen and A.A. Bishop), 76–83. New York: ASME.

Hu, X. and Jacobi, A.M. (1995). The Intertube Falling-film Modes: Transition, Hysteresis, and Effects on Heat Transfer. *Report ACRC-CR-5*, Air Conditioning and Refrigeration Center, University of Illinois.

Hu, H., Ding, G., and Wang, K. (2008). Heat transfer characteristics of R410A–oil mixture flow boiling inside a 7 mm straight microfin tube. *Int. J. Refrig.* **31**: 1081–1093.

Hu, H., Ding, C., Ding, G. et al. (2019a). Experimental study of heat transfer and pressure drop characteristics of two-phase mixed hydrocarbon refrigerants flow boiling in shell side of spiral wound heat exchanger. *IOP Conf. Ser.: Mater. Sci. Eng.* **502** https://doi.org/10.1016/j.ijheatmasstransfer.2020.119733.

Hu, H., Ding, C., Ding, G. et al. (2019b). Heat transfer characteristics of two-phase mixed hydrocarbon refrigerants flow boiling in shell side of LNG spiral wound heat exchanger. *Int. J. Heat Mass Transf.* **131**: 611–622.

Hwang, T.H. and Yao, S.C. (1986). Forced convection boiling in tube bundles. *Int. J. Heat Mass Transf.* **29** (5): 785–795.

Hwang, K.W., Kim, D.E., Yang, K.H. et al. (2014). Experimental study of flow boiling heat transfer and dryout characteristics at low mass flux in helically-coiled tubes. *Nucl. Eng. Des.* **273**: 529–541.

Jin, P., Zhao, C., Ji, W., and Tao, Q. (2018). Experimental investigation of R410A and R32 falling film evaporation on horizontal enhanced tubes. *Appl. Therm. Eng.* **129**: 502–511.

Kaji, M., Mori, K., Matsumoto, T. et al. (1998). Forced convective boiling heat transfer characteristics and critical heat flux in helically coiled tubes. *Nippon Kikai Gakkai Ronbunshu* **64**: 3343–3349.

Kakac, S., Shah, R.K., and Aung, W. (1987). Laminar convective heat transfer in ducts. In: *Handbook of Single-Phase Convective Heat Transfer*, Chapter 3, 3-1–3-137. New York: Wiley.

Kandlikar, S.G. (1990). A general correlation for saturated two-phase flow boiling heat transfer inside horizontal and vertical tubes. J. *Heat Transf.* **112**: 219–228.

Kandlikar, S.G. (1991). Development of a flow boiling map for subcooled and saturated flow boiling of different fluids inside circular tubes. *J. Heat Transf.* **113** (1): 190–200.

Kandlikar, S.G. (1998). Boiling heat transfer with phase change: part II – flow boiling in plain tubes. *J. Heat Transf.* **120**: 388–394.

Kandlikar, S.G. & Steinke, M.E. (2003) Predicting heat transfer during flow boiling in minichannels and microchannels. Paper CH-03-13-1. *ASHRAE Annual Meeting*, Chicago, January 24–29, 2003. Also in ASHRAE Transactions 109(1).

Kandlikar, S.G. and Subramanian, P. (2004). An extension of the flow boiling correlation to transition, laminar, and deep laminar flows in minichannels and microchannels. *Heat Transf. Eng.* **25** (3): 86–93.

Kattan, R., Thome, J.R., and Favrat, D. (1998). Flow boiling in horizontal tubes: Part-3: development of a new heat transfer model based on flow patterns. J. *Heat Transf.* **120**: 156–165.

Kays, W.M. and Perkins, H.C. (1973). Forced convection, internal flow in ducts. In: *Handbook of Heat Transfer* (eds. W.M. Rohsenow and J.P. Hartnett), 7–136. New York: McGraw-Hill.

Kim, S. and Mudawar, I. (2013). Universal approach to predicting saturated flow boiling heat transfer in mini/micro-channels –Part II. Two-phase heat transfer coefficient. *Int. J. Heat Mass Transf.* **64**: 1239–1256.

Kim, N.H., Cho, J.P., and Youn, B. (2002). Forced convective boiling of pure refrigerants in a bundle of enhanced tubes having pores and connecting gaps. *Int. J. Heat Mass Transf.* **45**: 2449–2463.

Kirillov, P., Yur'ev, Y.S., and Bobkov, V. (1984). *Handbook on Thermohydraulic Calculations (Nuclear Reactors, Heat Exchangers, Steam Generators)* [in Russian]. Moscow: Energoatomizdat. Quoted in Wu et al. (2018).

Kolokosta, D. and Yanniotis, S. (1996). Boiling on the surface of a rotating disc. *J. Food Eng.* **30**: 313–325.

Kolokosta, D. and Yanniotis, S. (2010). Experimental study of the boiling mechanism of a liquid film flowing on the surface of a rotating disc. *Exp. Thermal Fluid Sci.* **34**: 1346–1352.

Konishi, C., Lee, H., Mudawar, I. et al. (2015). Flow boiling in microgravity: part 1 – interfacial behavior and experimental heat transfer results. *Int. J. Heat Mass Transf.* **81**: 705–720.

Konishi, C. & Mudawar, I. (2015) Review of flow boiling and critical heat flux in microgravity. International Journal of Heat and Mass Transfer **80**, 469–493.

Koyama, K., Chiyoda, H., Arima, H., and Ikegami, Y. (2014). Experimental study on thermal characteristics of ammonia flow boiling in a plate evaporator at low mass flux. *Int. J. Refrig.* **38**: 227–235.

Krishnamurthy, S. and Peles, Y. (2008). Flow boiling of water in a circular staggered micro-pin fin heat sink. *Int. J. Heat Mass Transf.* **51**: 1349–1364.

Kumar, H. 1992. The design of plate heat exchangers for refrigerants. Proceedings of the Conference in Institute of Refrigeration, 5.1–5.2. Quoted by Almalfi et al. (2016a).

Kumar, S., Jain, A., Mohanty, B., and Gupta, S.C. (2003). Recirculation model of kettle reboiler. *Int. J. Heat Mass Transf.* **46**: 2899–2909.

La Harpe, A., Le Hongre, S., Mollard, J., and Johannes, C. (1969). Boiling heat transfer and pressure drop of liquid helium-I under forced circulation in a helically coiled tube. In: *Advances in Cryogenic Engineering*, vol. **14** (ed. K.D. Timmerhaus). Boston, MA: Springer.

Lazarek, G.M. and Black, S.H. (1982). Evaporative heat transfer, pressure drop and critical heat flux in a small vertical tube with R-113. *Int. J. Heat Mass Transf.* **25** (1982): 945–960.

Lebon, M.T., Hammer, C.F., and Kim, J. (2019). Gravity effects on subcooled flow boiling heat transfer. *Int. J. Heat Mass Transf.* **128**: 700–714.

Leong, L.S. and Cornwell, K. (1979). Heat transfer in a reboiler tube bundle. *Chem. Eng.* **343**: 219–221.

Li, W. and Wu, Z. (2010a). A general correlation for evaporative heat transfer in micro/minichannels. *Int. J. Heat Mass Transf.* **53**: 1778–1787.

Li, W. and Wu, Z. (2010b). A general criterion for evaporative heat transfer in micro/mini-channels. *Int. J. Heat Mass Transf.* **53**: 1967–1976.

Li, M., Dang, C., and Hihara, E. (2014). Flow boiling heat transfer of carbon dioxide with PAG-type lubricating oil in pre-dryout region inside horizontal tube. *Int. J. Refrig.* **41**: 45–59.

Li, R., Liu, J., Liu, J., and Xu, X. (2018). Measured and predicted upward flow boiling heat transfer coefficients for hydrocarbon mixtures inside a cryogenic plate fin heat exchanger. *Int. J. Heat Mass Transf.* **123**: 75–88.

Liu, Z. and Qiu, Y. (2004). Boiling characteristics of R-11 in compact tube bundles with smooth and enhanced tubes. *Experimental Heat Transfer* **17**: 91–102.

Liu, Z. and Winterton, R.H.S. (1991). A general correlation for saturated and subcooled flow boiling in tubes and annuli based on a nucleate pool boiling equation. *Int. J. Heat Mass Transf.* **34** (11): 2759–2766.

Longo, G.A., Mancin, S., Righetti, G., and Zilio, C. (2015). A new model for refrigerant boiling inside brazed plate heat exchangers (BPHEs). *Int. J. Heat Mass Transf.* **91**: 144–149.

Longo, G.A., Mancin, S., Righetti, G., and Zilio, C. (2019). Boiling of the new low-GWP refrigerants R1234ze(*Z*) and R1233zd(*E*) inside a small commercial brazed plate heat exchanger. *Int. J. Refrig.* **104**: 376–385.

Luciani, S., Brutin, D., Le Niliot, C. et al. (2008). Flow boiling in minichannels under normal, hyper-, and microgravity: local heat transfer analysis using inverse methods. *J. Heat Transf.* **130**: 1–13.

Ma, Y. and Chung, J.N. (2001). An experimental study of critical heat flux (CHF) in microgravity forced-convection boiling. *Int. J. Multiphase Flow* **27** (10): 1753–1767.

Mao, K. and Hibiki, T. (2017). Flow regime transition criteria for upward two-phase cross-flow in horizontal tube bundles. *Appl. Therm. Eng.* **112**: 1533–1546.

Martsiniouk, D. and Sorokin, A. (2000). The questions of liquid metal two-phase flow modelling in the FBR core channels. LMFR Core Thermohydraulics Status Prospects 327. Quoted in Wu et al. (2018).

Matzner, B. (1963) Heat Transfer and Hydraulic Studies for SNAP-Fuel Element Geometries. *Topical Report No. 2*, Task 15 of Contract AT(30-3)-187. Columbia University, New York.

McNeil, D.A., Raeisi, A.H., Kew, P.A., and Bobbili, P.R. (2010). A comparison of flow boiling heat-transfer in in-line mini pin fin and plane channel flows. *Appl. Therm. Eng.* **30**: 2412–2425.

Minton, P.E. (1970). Designing spiral-plate heat exchangers. *Chem. Eng.* **4**: 103–112.

Miropolskiy, Z.L. and Pikus, V.Y. (1969). Critical boiling heat flux in curved channels. *Heat Transfer Sov. Res.* **1** (1): 74–79.

Mitrovic, J. (1986). Influence of tube spacing and flow rate on heat transfer from a horizontal tube to a falling liquid film. In: *Proceedings of the 8th International Heat Transfer Conference*, 17-22 August,, 1949–1956. San Francisco: Begell House Publisher.

Momenifar, M.R., Akhavan-Behabadi, M.A., Nasr, M., and Hanafizadeh, P. (2015). Effect of lubricating oil on flow boiling characteristics of R-600a/oil inside a horizontal smooth tube. *Appl. Therm. Eng.* **91**: 62–72.

Mori, Y., Harada, T., Uchida, M., and Hara, T. (1970). Convective boiling of a binary liquid metal. In: *Liquid Metal Heat Transfer and Fluid Dynamics* (eds. J.C. Chen and A.A. Bishop). New York: ASME.

Morozov, Y.D., Privalov, A.N., Prisnyakov, V.F., and Belogurov, S.A. (1988). Mechanism of the transition to two-phase flow and flow regimes in the boiling of liquid metals in a once-through steam boiler. *High Temp.* **26** (6): 905–912.

Mumm, J.F. (1954). Heat Transfer to Boiling Water Forced Through a Uniformly Heated Tube. *Argonne National Laboratory Report ANL-5276.*

Nellis, G., Hughes, C., and Pfotenhauer, J. (2005). Heat transfer coefficient measurements for mixed gas working fluids at cryogenic temperatures. *Cryogenics* **45**: 546–556.

Nicol, A.A. and McLean, J.T. (1968). Boiling heat transfer from a rotating horizontal cylinder. *Can. J. Chem. Eng.* **46**: 304–308.

No, H.C. and Kazimi, M.S. (1982). Wall heat transfer coefficients for condensation and boiling in forced convection of sodium. *Nucl. Sci. Eng.* **81**: 319–324.

Noyes, R.C. and Lurie, H. (1966). Boiling sodium heat transfer. In: *Proceedings of Third International Heat Transfer Conference*, 7-12 August 1966, Chicago, vol. **5**, 92–100. Begell House.

Ogata, H. and Sato, S. (1974). Forced convection heat transfer to boiling helium in a tube. *Cryogenics* **14**: 375–380.

Ohta, H. (1997). Experiments on microgravity boiling heat transfer by using transparent heaters. *Nucl. Eng. Des.* **175**: 167–180.

Ong, C.L. and Thome, J.R. (2011). Macro-to-microchannel transition in two-phase flow: part 2 – flow boiling heat transfer and critical heat flux. *Exp. Therm. Fluid Sci.* **35** (2011): 873–886.

Owhadi, A. (1966). *Boiling in self-induced radial acceleration fields*. Doctorate thesis: Oklahoma State University.

Owhadi, A., Crain, B., Bell, K.J., and Crain, B. (1968). Forced convection boiling inside helically coiled tubes. *Int. J. Heat Mass Transfer* **11** (12): 1779–1793.

Palen, J.W. and Small, W.M. (1964). A new way to design kettle and internal reboilers. *Hydrocarb. Reprocess.* **43** (11): 199–208.

Palen, J.W. and Taborek, J. (1962). Refinery kettle reboilers – proposed method for design and optimization. *Chem. Eng. Prog.* **58** (7): 37–46.

Palen, J.W., Yarden, A., and Taborek, J. (1972). Characteristics of boiling outside large scale horizontal multi-tube bundles. *AIChE Symp. Symp. Ser.* **68** (118): 50–61.

Park, I., Sehwan, I., and Jeong, S. (2011). Flow boiling heat transfer of ternary mixture in a micro-channel. Proceedings of the ASME/JSME 2011 8th Thermal Engineering Joint Conference, Honolulu, Hawaii (13–17 March 2011). Paper AJTEC2011-44079.

Peppler, W. (1977). Sodium boiling in fast reactors: a state of the art review. In: *Thermal and Hydraulic Aspects of Nuclear Reactor Safety*, Vol. 2: Liquid Metal Fast Breeder Reactor (eds. O.C. Jones and S.G. Bankoff). New York: ASME.

Peterson, J.R. (1967). High-performance "once-through" boiling of potassium in single tubes at saturation temperatures of 1500° to 1750° F. NASA CR-842.

Polley, G. T., Ralston, T., and Grant, I.D.R. (1980). Forced Crossflow Boiling in an Ideal In-line Tube Bundle, *ASME Paper 8-HT-46*. ASME, New York.

Qiu, S.Z., Jia, D.N., Yu, Z.W. et al. (1993). Experimental study on the boiling mechanism of liquid sodium in an annular channel. *Chin. J. Nucl. Sci. Eng.* **13**: 298–303. Quoted in Wu et al. (2018).

Qiu, Z.C., Ma, Z.Y., Qiu, S.Z. et al. (2013). Experimental research on the incipient boiling wall superheat of sodium. *Prog. Nucl. Energy* **68**: 121–129.

Qiu, Z.C., Ma, Z.Y., Qiu, S.Z. et al. (2015). Experimental research on the thermal hydraulic characteristics of sodium boiling in an annulus. *Exp. Thermal Fluid Sci.* **60**: 263–274.

Qu, W. and Siu-Ho, A. (2009). Experimental study of saturated flow boiling heat transfer in an array of staggered micro-pin-fins. *Int. J. Heat Mass Transf.* **52**: 1853–1863.

Ribatski, G. and Jacobi, A.M. (2005). Falling-film evaporation on horizontal tubes – a critical review. *Int. J. Refrig.* **28**: 635–653.

Ribatski, G. and Thome, J.R. (2007). Two-Phase flow and heat transfer across horizontal tube bundles-a review. *Heat Transfer Engineering* **28** (6): 508–524.

Rifert, V.G., Barabash, P.A., Solomakha, A.S. et al. (2018). Hydrodynamics and heat transfer in a centrifugal film evaporator. *Bulg. Chem. Commun.* **50** ((Special Issue K)): 49–57.

Robertson, J.M. and Wadekar, V.V. (1988). Boilinh characteristics of cyclohexane in vertical upflow in perforated plate-fin passages. *AIChE Symp. Ser* **84** (263): 120.

Rohsenow, W.M. (1952). A method for correlating data for surface boiling of liquids. *Trans. ASME* **74**: 969–975.

Romie, F.E., Brovarney, S.W., and Geidt, W.H. (1960). Heat transfer to boiling mercury. *J. Heat Transfer* **82** (4): 387–388. also AEC R& D Report No. ATL-A-102, Adv. Tech. Labs., October, 1959.

Saito, M., Yamaoka, N., Miyazaki, K. et al. (1994). Boiling two-phase flow under microgravity. *Nucl. Eng. Des.* **146**: 451–461.

Santini, L., Cioncilini, A., Butel, M.T., and Ricotti, M.E. (2016). Flow boiling heat transfer in a helically coiled steam generator for nuclear power applications. *Int. J. Heat Mass Transf.* **92**: 91–99.

Schleisiek, K. (1970). Heat transfer and boiling during forced convection of sodium in an induction-heated tube. *Nucl. Eng. Des.* **14**: 60–68.

Schmucker, H. and Grigull, U. (1973). Boiling of mercury in a vertical tube under forced flow conditions. In: *Progress in Heat and Mass Transfer*, vol. **7** (ed. O.E. Dwyer), 363–376. Oxford: Pergamon Press.

Shah, M.M. (1974). Heat transfer and pressure drop in ammonia evaporators. *ASHRAE Trans.* **80** (2): 238–254.

Shah, M.M. (1975). Visual observations in an ammonia evaporator. *ASHRAE Trans.* **81** (1): 295–306.

Shah, M.M. (1976). A new correlation for heat transfer during boiling flow through pipes. *ASHRAE Trans.* **82** (2): 66, 66–86, 86.

Shah, M.M. (1982). Chart correlation for saturated boiling heat transfer: equations and further study. *ASHRAE Trans.* **88** (1): 185–196.

Shah, M.M. (1984a). Prediction of heat transfer during boiling of cryogenic fluids flowing in tubes. *Cryogenics* **24** (5): 231–236.

Shah, M.M. (1984b). A correlation for heat transfer during subcooled boiling on a single tube with forced crossflow. *Int. J. Heat Fluid Flow* **5** (1): 13–20.

Shah, M.M. (2005). Improved general correlation for subcooled boiling heat transfer during flow across tubes and tube bundles. *Int. J. HVAC&R Res.* **11** (2): 285–304.

Shah, M.M. (2006). Evaluation of general correlations for heat transfer during boiling of saturated liquids in tubes and annuli. *HVAC&R Res.* **12** (4): 1047–1063.

Shah, M.M. (2007). A general correlation for heat transfer during saturated boiling with flow across tube bundles. *HVACR Res.* **13** (5): 749–768.

Shah, M.M. (2015a). Evaluation of correlations for predicting heat transfer during boiling of carbon dioxide inside channels. Paper # 8435, IHTC 15, Kyoto, Japan.

Shah, M.M. (2015b). A method for predicting heat transfer during boiling of mixtures in tubes. *Appl. Therm. Eng.* **89**: 812–821.

Shah, M.M. (2016). A new correlation for heat transfer during condensation in horizontal mini/micro channels. *Int. J. Refrig.* **64**: 187–202.

Shah, M.M. (2017a). Unified correlation for heat transfer during boiling in plain mini/micro and conventional channels. *Int. J. Refrig.* **74**: 604–624.

Shah, M.M. (2017b). A correlation for heat transfer during boiling on bundles of horizontal plain and enhanced tubes. *Int. J. Refrig.* **78**: 47–59.

Shah, M.M. (2019). Prediction of heat transfer during saturated boiling in coils. *J. Therm. Sci. Eng. Appl.* **11**: 031013-1–031013-7.

Shah, R.K. and London, A.L. (1978). *Laminar Flow Forced Convection in Ducts*. New York: Academic Press. Quoted in Feldman et al. (2000).

Shinkawa, T. (1980). Heat transfer coefficient in a downward flow vertical liquid film type evaporator. *Heat Transf. Jap. Res.* **9** (3): 1–17.

Shmerler, J.A. and Mudawar, I. (1988). Local heat transfer coefficient in wavy free-falling turbulent liquid films undergoing uniform sensible heating, Int. J. *Heat Mass Transfer* **31**: 67–77.

Steiner, D. and Taborek, J. (1992). Flow boiling heat transfer in vertical tubes correlated by an asymptotic model. *Heat Transfer Eng.* **13** (2): 43–68.

Sun, L. and Mishima, K. (2009). An evaluation of prediction methods for saturated flow boiling heat transfer in mini-channels. *Int. J. Heat Mass Transf.* **52** (23–24): 5323–5329.

Swain, A. and Das, M.K. (2014). A review on saturated boiling of liquids on tube bundles. *Heat Mass Transfer* **50**: 617–637.

Taboas, F., Valles, M., Bourouis, M., and Coronas, A. (2010). Flow boiling heat transfer of ammonia/water mixture in a plate heat exchanger. *Int. J. Refrig.* **33**: 695–705.

Taboas, F., Valles, M., Bourouis, M., and Coronas, A. (2012). Assessment of boiling heat transfer and pressure drop correlations of ammonia/water mixture in a plate heat exchanger. *Int. J. Refrig.* **35**: 633–644.

Tang, S. and McDonald, T.W. (1971). A study of boiling heat transfer from a rotating horizontal cylinder. *Int. J. Heat Mass Transf.* **14**: 1643–1657.

Tang, Y.S., Ross, P.T., Nicholson, R.C., and Smith, C.R. (1964). Forced convection boiling of potassium-mercury systems. *AICHE J.* **10** (5): 617–620.

Thome, J.R. (1995). Comprehensive thermodynamic approach to modelling refrigerant-oil mixtures. *HVAC&R Res.* **1** (2): 110–126.

Thome, J.R. (1996). Boiling of new refrigerants: a state of the art review. *Int. J. Refrig.* **19** (7): 435–457.

Thome, J.R. (2017). A review on falling film evaporation. *J. Enhanc. Heat Transf.* **24**: 483–498.

Thome, J.R. and Robinson, D.M. (2006). Prediction of local bundle boiling heat transfer coefficients: pure refrigerant boiling on plain, low fin, and turbo-bii HP tube bundles. *Heat Transfer Eng.* **27** (10): 20–29.

Thome, J.R., Dupont, V., and Jacobi, A.M. (2004). Heat transfer model for evaporation in microchannels, part I: presentation of the model. *Int. J. Heat Mass Transf.* **47**: 3375–3385.

Tran, T.N., Wambsganss, M.W., and France, D.M. (1996). Small circular and rectangular-channel boiling with two refrigerants. *Int. J. Multiphase Flow* **22**: 485–498.

Van Rooyen, E. and Thome, J.R. (2014). Flow boiling data and prediction method for enhanced boiling tubes and tube bundles with R-134a and R-236fa including a comparison with falling film evaporation. *Int. J. Refrig.* **41**: 60–71.

Van Rooyen, E., Thome, J.R., Agostini, F., and Borhani, N. (2012). Boiling on a tube bundle: part II – heat transfer and pressure drop. *Heat Transfer Eng.* **33** (11): 930–946.

Volodin, O., Pecherkin, N., Pavlenko, A., and Zubkov, N. (2017). Heat transfer and crisis phenomena at boiling of refrigerant films falling down the surfaces obtained by deformational cutting. *Interfacial Phenom. Heat Transf.* **5** (3): 215–222.

Webb, R.L. and Chien, L.H. (1994). Correlation of convective boiling on plain tubes using refrigerants. *Heat Transfer Engng.* **15** (3): 57–69.

Wieting, A.R. (1975). Empirical correlations for heat transfer and flow friction characteristics of rectangular offset fin plate heat exchangers. *J. Heat Transf.* **97** (3): 488–490.

Wojtan, L., Ursenbacker, T., and Thome, J.R. (2005). Investigation of flow boiling in horizontal tubes: part II – development of a new heat transfer model for stratified-wavy, dryout and mist flow regimes. *Int. J. Heat Mass Transf.* **48**: 2970–2985.

Wongwises, S. and Polsongkram, M. (2006). Evaporation heat transfer and pressure drop of hfc-134a in a helically coiled concentric tube-in-tube heat exchanger. *Int. J. Heat Mass Transf.* **49** (3): 658–670.

Wu, Y., Luo, S., Wang, L. et al. (2018). Review on heat transfer and flow characteristics of liquid sodium (2): two phase. *Prog. Nucl. Energy* **103**: 151–164.

Xiao, Z.J., Zhang, G.Q., Shan, J.Q. et al. (2006). Experimental research on heat transfer to liquid sodium and its incipient boiling wall superheat in an annulus. *Nucl. Sci. Tech.* **17**: 177–184.

Xiao, Y., Hu, Z., Chen, S., and Gu, H. (2018). Experimental investigation of boiling heat transfer in helically coiled tubes at high pressure. *Ann. Nucl. Energy* **113**: 409–419.

Yilmaz, S., Moliterno, A., and Samuelson, B. (1983). Vertical thermosiphon boiling in spiral plate heat exchanger. *AIChe. Symp. Ser.* **79** (225): 47–53.

Zeigarnick, Y.A. and Litinov, V.D. (1980). Heat transfer and pressure drop in sodium boiling in tubes. *Nucl. Sci. Eng.* **73**: 19–28.

Zeng, X. and Chyu, M. (1995). Heat Transfer and Fluid Flow Study of Ammonia Spray Evaporators. Report ASHRAE *RP-725*.

Zhang, Y., Liu, B., Zhao, J. et al. (2018). Experimental study of subcooled flow boiling heat transfer on micro-pin finned surfaces in short-term microgravity. *Exp. Thermal Fluid Sci.* **97**: 417–430.

Zhao, L., Guo, L., Bai, B. et al. (2003). Convective boiling heat transfer and two-phase flow characteristics inside a small horizontal helically coiled tubing once-through steam generator. *Int. J. Heat Mass Transf.* **46** (25): 4779–4788.

Zhao, C., Ji, W., Jin, P., and Tao, W. (2016). Heat transfer correlation of the falling film evaporation on a single horizontal smooth tube. *Appl. Therm. Eng.* **129**: 502–511.

Zhao, C., Ji, W., Jin, P. et al. (2018). Experimental study of the local and average falling film evaporation coefficients in a horizontal enhanced tube bundle using R134a. *Appl. Therm. Eng.* **129**: 502–511.

Zürcher, O., Thome, J.R., and Favrat, D. (1997). Flow boiling and pressure drop measurements for r-134a/oil mixtures part 2: evaporation in a plain tube. *HVAC&R Res.* **3** (1): 54–64.

Zurcher, O., Thome, J.R., and Favrat, D. (1998). In-tube flow boiling of R-407C and R-407C/oil mixtures part II: plain tube results and predictions. *HVAC&R Res.* **4** (4): 373–379.

6

Critical Heat Flux in Flow Boiling

6.1 Introduction

Critical heat flux (CHF) during pool boiling was discussed in Chapter 3. Similar phenomena also occur during forced convection boiling. A sharp deterioration in heat transfer occurs at some point with increasing heat flux or quality. The heat flux at which the deterioration in heat transfer occurs is known as the CHF and the corresponding quality is known as the critical quality. The mechanisms and prediction of CHF for various conditions and types of heat exchangers are discussed in this chapter. Included are flow inside channels such as tubes and annuli, coils, single tubes, tube bundles, plates, coils, falling films, jets, sprays, etc.

Except where stated otherwise, all discussions are for single-component non-metallic fluids.

6.2 CHF in Tubes

6.2.1 Types of Boiling Crisis and Mechanisms

As shown in Figure 5.2.1, heat transfer during boiling in tubes keeps rising with increasing quality until it drops sharply with further increase in quality. The quality at which it occurs is called the critical quality and the corresponding heat flux is called the CHF. The decrease in heat transfer at higher qualities occurs due to the drying out of the liquid film in contact with the tube surface, and hence the CHF under this condition is usually called dryout. When heat flux is raised under subcooled or low quality conditions, a sudden rise in wall temperature occurs. This is caused by the inability of liquid to reach the tube surface due to intense vapor generation. This type of CHF is usually called departure from nucleate boiling (DNB). The temperature rise in this type of CHF is high and can cause actual destruction of the tube wall. For this reason, it is often called burnout. A general term applicable to all types of CHF is boiling crisis. The terms CHF, boiling crisis, DNB, and dryout are usually used interchangeably without implying any particular mechanism, and this is also the case in this book.

Attempts have been made on developing analytical models for prediction of CHF with limited success. On the other hand, some empirical methods have been developed, which have been validated with a wide range of data.

6.2.2 Prediction Methods

6.2.2.1 Analytical Models

Three main models of CHF have been widely used. These are boundary layer separation, bubble crowding, and liquid film dryout.

According to the boundary layer separation model, CHF is essentially a hydrodynamic phenomenon. It occurs when boundary layer separates from the wall. The stagnant liquid at the wall evaporates, leaving a vapor blanket at the wall. It results in rise of wall temperature. When the local wall temperature exceeds the Leidenfrost point, the flow boiling crisis occurs. Tong (1968) used this model to develop the formula below for predicting CHF that agreed with data for water from several sources:

$$q_c = 1.76 - 7.433x + 12.222x^2 \frac{i_{fg}\mu_f^{0.6}}{D_{HYD}^{0.6}}(\rho_f U_o)^{0.4} \tag{6.2.1}$$

U_o is the mainstream velocity ft s^{-1}. The other units are q_c in Btu h^{-1} ft^{-2}, i_{fg} in Btu lb^{-1}, μ_f in lb h^{-1} ft^{-1}, ρ in lb ft^{-3}, and D_{HYD} in ft. μ_f is calculated at the saturation temperature.

The bubble crowding model was proposed by Weisman and Pei (1983), among others. This model applies to high velocity subcooled and low-quality flow. According to it, a layer of oblong bubbles forms at the wall. CHF occurs when so much bubble crowding occurs at the heated wall that turbulent fluctuations in the core liquid are unable to provide adequate liquid supply to the wall. The model equations

Two-Phase Heat Transfer, First Edition. Mirza Mohammed Shah.

developed by them were compared to data for water, R-11, R-113, ammonia, and nitrogen, and good agreement was found.

In the liquid film dryout model, annular flow is assumed. Mass balance is taken on liquid gain and loss from the liquid film. Liquid film loses liquid by evaporation and entrainment and gains liquid by deposition of drops from the entrained liquid in vapor core. This is expressed by the following equation:

$$\frac{dG_{LF}}{dz} = \frac{4}{D}\left(D_d - E - \frac{q}{i_{fg}}\right) \tag{6.2.2}$$

G_{LF} is the mass flow of liquid in the liquid film per unit cross-sectional area of channels; z is the distance along the channel; D_d is the droplet deposition rate per unit area of the tube wall; and E is the rate of entrainment per unit area of the tube wall. Whalley et al. (1974) assumed that annular flow starts when quality reaches 1%. E and D were calculated using empirical correlations. Equation (6.2.2) was integrated from the point of onset of annular flow along the tube until G_{LF} became zero. This indicated the dryout point. Good agreement with data for water and R-12 from several sources was reported. A more recent analysis based on this model is by Ahmad et al. (2013), which uses more recent correlations for entrainment and deposition.

Galloway and Mudawar (1993) and Sturgis and Mudawar (1999) proposed the interfacial lift-off model. According to this model, vapor bubbles coalesce to form a wavy layer along the wall. Wetting of wall occurs only at the troughs of this wavy layer. CHF occurs when intense evaporation of liquid at the wetting fronts lifts the interface off the wall and liquid can no longer reach the wall.

Kandlikar (2010) has presented a theoretical force balance model for prediction of CHF during saturated boiling. It was compared to data from 10 sources for water and halocarbon refrigerants in channels of hydraulic equivalent diameters of 0.127–3.36 mm with a mean absolute deviation (MAD) of 19.7%. Calculation with this model require the receding contact angle. This angle depends on liquid–surface combination and has to be experimentally determined. Very limited information is presently available for these contact angles. This makes it difficult to use this model.

Ravellin and Thome (2008) and Kosar (2009) have developed a theoretical model of dryout specifically for minichannels.

Various CHF models have been reviewed, among others, by Habib et al. (2014).

While all these models were shown to be in agreement with some data, none of them has been validated with a wide-ranging database. Besides, they are difficult to use for design.

6.2.2.2 Lookup Tables of CHF

A number of authors have presented tables that can be used to predict CHF for water flowing up in uniformly heated tubes. These list values of CHF for an 8 mm diameter tube at intervals of pressure, mass flux, and quality. CHF at in-between values of these parameters is obtained by linear interpolations. Formulas are provided to calculate CHF for other diameters. Such tables include those by Doroshchuk et al. (1975), Kirillov et al. (1991), and Groeneveld et al. (2007). The last mentioned is known as the 2006 CHF lookup table and covers the widest range. It is discussed in the following text.

The range of parameters included in the 2006 lookup table is as follows: pressure 1–210 bar, mass flux 0–8000 kg m^{-2} s^{-1}, and critical quality −0.5 to 1. The CHF for tube diameters other than 8 mm is obtained by multiplying the listed CHF values by $(8/D)^{0.5}$. Calculation with this table can be done in two ways. One is called the direct substitution method (DSM). This method is based on the local condition hypothesis, according to which CHF depends only on the local conditions and is independent of the upstream conditions. With this method, CHF is read from the table at the required critical quality, interpolating as needed. The other method is based on the hypothesis that CHF depends on the inlet quality and L/D. This method is also known as the heat balance method (HBM). In this method, calculations are done as follows. Assume a value of CHF, q_c. The critical quality x_c can then be calculated by the heat balance equation

$$x_c = x_{in} + \frac{\pi L_c D q_c}{\left(\frac{\pi}{4}D^2\right) G i_{fg}} \tag{6.2.3}$$

where L_c is the length of tube between inlet and location of CHF. Rearrangement gives

$$x_c = x_{in} + \frac{4L_c}{D} Bo \tag{6.2.4}$$

where Bo is the boiling number. CHF is now obtained from the lookup table for this value of x_c. If it agrees with the assumed value, it is the correct CHF. Otherwise, x_c is calculated using this value of CHF, and calculations are repeated until adequate convergence is obtained.

Kalimullah et al. (2012) have given the following method to predict CHF for mass flux up to 30 000 kg m^{-2} s^{-1}. Calculate CHF at $G = 8000$ from the lookup table. Then CHF at higher G is obtained by multiplying it by $(G/8000)^{0.376}$. They recommend this method for all pressures covered by the 2006 lookup table.

It should be noted that the heat balance equation, Eq. (6.2.4), has to be satisfied irrespective of whether calculations are based on local condition method (DSM) or upstream condition method (HBM). This equation gives the relation between x_{in}, x_c, Bo, and L_c/D. If three of them are fixed, the fourth one is also fixed. If Bo is determined using the critical quality, it should be checked that the corresponding x_{in} and L_c/D satisfy Eq. (6.2.2). This point is applicable not only to calculations with CHF table but also to all other methods.

6.2.2.3 Dimensional Correlations for Water

Many dimensional correlations applicable to water only have been published. Notable among them are Thompson and Macbeth (1964), Bertoletti et al. (1965), Biasi et al. (1967), and Bowring (1972). The Biasi et al. correlation is represented by the following equations:

For low qualities,

$$q_c = \frac{1883}{D^n G^{1/6}} \left[\frac{f(p)}{G^{1/6}} - x_c \right] \tag{6.2.5}$$

For high qualities,

$$q_c = \frac{3780F(p)}{D^n G^{0.6}} (1 - x_c) \tag{6.2.6}$$

where $n = 0.4$ for $D \geq 1$ cm, and $n = 0.6$ for $D < 1$ cm.

$$f(p) = 0.7249 + 0.099p \exp(-0.032p) \tag{6.2.7}$$

$$F(p) = -1.159 + 0.149p \exp(-0.019p) + \frac{8.99p}{10 + p^2} \tag{6.2.8}$$

Critical heat flux is the higher of those given by the intersection of Eqs. (6.2.5) and (6.2.6) with the heat balance equation, Eq. (6.2.4). For $G < 30$ g cm^{-2} s^{-1}, Eq. (6.2.6) is always used.

Critical heat flux is in W cm^{-2}. The other units and the validated range of this correlation is

$D = 0.3$–3.75 cm
$L_c = 20$–600 cm
$p = 2.7$–140 bar
$G = 10$–600 g cm^{-2} s^{-1}

$$1/(1 + \rho_f/\rho_g) < x_c < 1$$

The correlation had RMS deviation of 7.3% with 4500 data points.

6.2.2.4 General Correlations

Vertical Tubes Despite intense research around the world, there was no generally applicable correlation for CHF for a long time. Then Katto (1978) published his general correlation and that was followed a few months later by the Shah (1979) correlation. Katto and Ohno (1984) and Shah (1987) are improved versions of these correlations. These are the most verified correlations and are described herein.

Katto–Ohno Correlation The original Katto correlation had many regimes based on some physical considerations, and CHF formulas were given for each regime. Katto and Ohno discarded the regime concept and gave a purely empirical correlation. The Katto–Ohno correlation is represented by the following equation:

$$q_c = q_{co}(1 - Kx_{in}) \tag{6.2.9}$$

q_{co} is the CHF when $x_{in} = 0$.

For $(\rho_g/\rho_f) < 0.15$, q_{co} is calculated with the following equations:

$$\frac{q_{co}}{Gi_{fg}} = C\left(\frac{\sigma\rho_f}{G^2 L_c}\right)^{0.043} \frac{D}{L_c} \tag{6.2.10}$$

$$\frac{q_{co}}{Gi_{fg}} = 0.1\left(\frac{\rho_g}{\rho_f}\right)^{0.133} \left(\frac{\sigma\rho_f}{G^2 L_c}\right)^{1/3} \frac{1}{1 + 0.0031L_c/D} \tag{6.2.11}$$

$$\frac{q_{co}}{Gi_{fg}} = 0.098\left(\frac{\rho_g}{\rho_f}\right)^{0.133} \left(\frac{\sigma\rho_f}{G^2 L_c}\right)^{0.433} \frac{(L_c/D)}{1 + 0.0031L_c/D} \tag{6.2.12}$$

Designating q_{co} given by Eqs. (6.2.10), (6.2.11), and (6.2.12) as $q_{co}(10)$, $q_{co}(11)$, and $q_{co}(12)$, respectively, the choice among these equations is made as follows:

If $q_{co}(10) < q_{co}(11)$, $q_{co} = q_{co}(10)$. Otherwise q_{co} is the smaller of $q_{co}(11)$ and $q_{co}(12)$.

C in Eq. 6.10 is calculated as follows:

$$\text{For } L_c/D < 50, \quad C = 0.25 \tag{6.2.13}$$

For $L_c/D = 50$ to 150,

$$C = 0.25 + 0.0009[(L_c/D) - 50] \tag{6.2.14}$$

$$\text{For } L_c/D > 150, \quad C = 0.34 \tag{6.2.15}$$

K in Eq. (6.2.9) is calculated by the following equations:

$$K = \frac{1.043}{4C\left(\frac{\sigma\rho_f}{G^2 L_c}\right)^{0.043}} \tag{6.2.16}$$

$$K = \frac{5}{6} \frac{0.0124 + D/L_c}{\left(\frac{\rho_g}{\rho_f}\right)^{0.133} \left(\frac{\sigma\rho_f}{G^2 L_c}\right)^{1/3}} \tag{6.2.17}$$

The larger of the values of K given by Eqs. (6.2.16) and (6.2.17) is used in Eq. (6.2.9).

For $(\rho_g/\rho_f) > 0.15$, the following two equations are used in addition to Eq. (6.2.10):

$$\frac{q_{co}}{Gi_{fg}} = 0.234\left(\frac{\rho_g}{\rho_f}\right)^{0.513}\left(\frac{\sigma\rho_f}{G^2L_c}\right)^{0.433}\frac{(L_c/D)^{0.27}}{1+0.0031L_c/D} \tag{6.2.18}$$

$$\frac{q_{co}}{Gi_{fg}} = 0.0384\left(\frac{\rho_g}{\rho_f}\right)^{0.6}\left(\frac{\sigma\rho_f}{G^2L_c}\right)^{0.173}\frac{1}{1+0.28\left(\frac{\sigma\rho_f}{G^2L_c}\right)^{0.233}L_c/D} \tag{6.2.19}$$

Designating the q_{co} given by Eqs. (6.2.18) and (6.2.19) as $q_{co}(18)$ and $q_{co}(19)$, respectively, q_{co} at $(\rho_g/\rho_f) > 0.15$ is determined as follows:

If $q_{co}(10) < q_{co}(18), q_{co} = q_{co}(10)$. Otherwise,

q_{co} is the larger of $q_{co}(18)$ and $q_{co}(19)$.

For the calculation of K at $(\rho_g/\rho_f) > 0.15$, the following equation is also needed:

$$K = 1.12\frac{1.52\left(\frac{\sigma\rho_f}{G^2L_c}\right)^{0.233} + D/L_c}{\left(\frac{\rho_g}{\rho_f}\right)^{0.6}\left(\frac{\sigma\rho_f}{G^2L_c}\right)^{0.173}} \tag{6.2.20}$$

Designating K from Eqs. (6.2.16), (6.2.17), and (6.2.20), as $K(16)$, $K(17)$, and $K(20)$, respectively, K at $(\rho_g/\rho_f) > 0.15$ is calculated as follows:

If $K(16) > K(17)$, then $K = K(16)$. Otherwise,

K is the smaller of $K(17)$ and $K(20)$.

This correlation was evaluated by Shah (1987, 2017) against a very wide range of data. It was found erratic for subcooled burnout, helium, and very short tubes. However, satisfactory agreement was found with subcooled data if L_c was replaced with boiling length. Other than those, its predictions were satisfactory. These results are discussed along with the Shah correlation that is described next.

Shah Correlation This correlation was originally given in Shah (1979) in graphical form. It was presented in equation form in Shah (1987). This correlation consists of two correlations that are used in different ranges. These are the upstream condition correlation (UCC) and the local condition correlation (LCC). In the UCC, the CHF at a location depends on the upstream conditions, namely, the inlet quality and the distance from the tube inlet. In the LCC, CHF depends only on the local quality except for very short tubes.

The UCC is as follows:

$$Bo = 0.124\left(\frac{D}{L_c}\right)^{0.89}\left(\frac{10^4}{Y}\right)^n(1-x_{IE}) \tag{6.2.21}$$

Y is a correlating parameter defined as

$$Y = \left(\frac{GDC_{pf}}{k_f}\right)\left(\frac{G^2}{\rho_f^2 gD}\right)^{0.4}\left(\frac{\mu_f}{\mu_g}\right)^{0.6} \tag{6.2.22}$$

When $Y \le 10^4$, $n = 0$ for all fluids.

For $Y > 10^4$, n is given by the following relations:

For helium at all values of $Y > 10^4$,

$$n = \left(\frac{D}{L_E}\right)^{0.33} \tag{6.2.23}$$

For all fluids other than helium,

$$\text{For } Y \le 106, \quad n = \left(\frac{D}{L_E}\right)^{0.54} \tag{6.2.24}$$

$$\text{For } Y > 106, \quad n = \frac{0.12}{(1-x_{IE})^{0.5}} \tag{6.2.25}$$

The effective length L_E and effective inlet quality x_{IE} are defined as follows:

When $x_{in} \le 0$, $L_E = L_c$ and $x_{IE} = x_{in}$
When $x_{in} > 0$, $L_E = L_B$ and $x_{IE} = 0$

L_B is the boiling length defined as the length of tube between $x = 0$ and $x = x_c$. For uniformly heated tubes, heat balance yields the following expression for it:

$$\frac{L_B}{D} = \frac{x_c}{4Bo} = \frac{L_C}{D} + \frac{x_{in}}{4Bo} \tag{6.2.26}$$

The UCC for $Y < 10^4$ is shown graphically in Figure 6.2.1.

The LCC is given by the expression

$$Bo = F_E \cdot F_x \cdot Bo_0 \tag{6.2.27}$$

F_E is the entrance effect factor given by the equation as follows:

$$F_E = 1.54 - 0.032\left(\frac{L_E}{D}\right) \tag{6.2.28}$$

If Eq. (6.2.28) gives $F_E < 1$, use $F_E = 1$.

Bo_0 is the boiling number at $x_c = 0$ and is the highest of those given by the following three equations:

$$Bo_0 = 15Y^{-0.612} \tag{6.2.29}$$

$$Bo_0 = 0.082Y^{-0.3}(1+1.45p_r^{4.03}) \tag{6.2.30}$$

$$Bo_0 = 0.000\,24Y^{-0.105}(1+1.15p_r^{3.39}) \tag{6.2.31}$$

For $x_c > 0$, F_x is given by the following expression:

$$F_x = F_3\left[1 + \frac{(F_3^{-0.29}-1)(p_r-0.6)}{0.35}\right]^c \tag{6.2.32}$$

When $p_r \le 0.6$, $c = 0$. When $p_r > 0.6$, $c = 1$.

$$F_3 = \left(\frac{1.25\times 10^5}{Y}\right)^{0.833x_c} \tag{6.2.33}$$

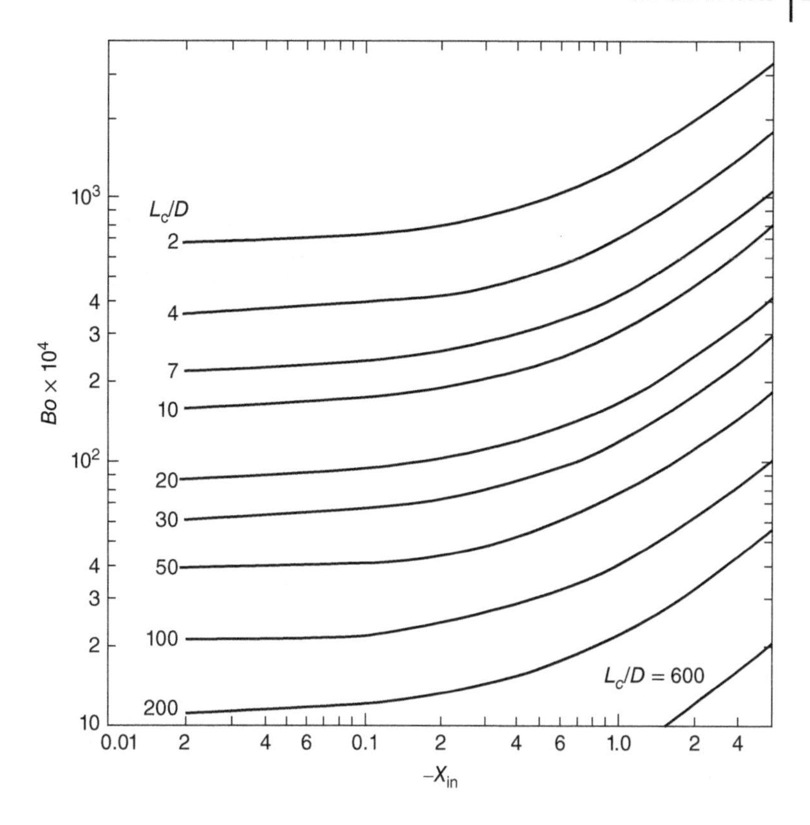

Figure 6.2.1 Upstream condition correlation (UCC) of Shah for $Y < 10^4$. Source: Reprinted from Shah (1987). © 1987, with permission from Elsevier.

When $x_c < 0$, F_x is calculated by the following equation:

$$F_x = F_1\left[1 - \frac{(1 - F_2)(p_r - 0.6)}{0.35}\right]^b \tag{6.2.34}$$

When $p_r \le 0.6$, $b = 0$, otherwise $b = 1$.

$$\text{When } Y \le 1.4 \times 10^7, \quad F_1 = 1 + 0.0052(-x_c^{0.88})Y^{0.41} \tag{6.2.35}$$

When $Y > 1.4 \times 10^7$, use Eq. (6.2.35) with $Y = 1.4 \times 10^7$.

$$\text{When } F_1 \le 4, \quad F_2 = F_1^{-0.42} \tag{6.2.36}$$

$$\text{When } F_1 > 4, \quad F_2 = 0.55 \tag{6.2.37}$$

The choice between UCC and LCC is made as follows. For helium, always use UCC. For all other fluids, select as follows.

For $Y \le 10^6$, use the UCC.

For $Y > 10^6$, use the correlation that gives the lower value of *Bo*. The only exception to this rule is that UCC is used if $L_E > (160/p_r^{1.14})$.

The LCC is shown graphically in Figures 6.2.2 and 6.2.3. These can be used for manual calculations as follows. For a given value of Y and x_c, read Bo_o from Figure 6.2.2. From Figure 6.2.3, read F_x at the given x_c. This value of F_x is for $p_r \le 0.6$. For higher p_r, use the p_r lines on the right side of the figure. Thus if F_x at $p_r \le 0.6$ is 5, it is 3 at $p_r = 0.9$.

Shah compared this correlation with a database whose range is shown in Table 6.2.1. It included 23 fluids (water, refrigerants, cryogens, chemicals, and liquid metals), tube diameters 0.315–37.5 mm, tube length 1.3–940 times diameter, mass flux 4–29 051 kg m^{-2} s^{-1}, reduced pressures 0.0014–0.96, inlet quality −4 to +0.85, and critical quality −2.6 to +1. The same data were also compared to the Bowring (1972) correlation for water and that of Subbotin et al. (1985) for helium as those were the best available for those fluids. The results are shown in Table 6.2.2. It is seen that the correlation of Bowring et al. was the most accurate in its verified range but the Shah correlation was better outside that range. The Subbotin et al. correlation was satisfactory in its verified range but showed large deviations beyond it. The Katto–Ohno correlation had large deviations with subcooled data and with helium data. Considering all data, the MAD of the Shah and Katto–Ohno correlations is 16.0% and 22.3%, respectively. Note that the deviations of Katto–Ohno correlation shown in Table 6.2.2 are using the boiling length instead of the critical length. If critical length is used, as in the published Katto–Ohno correlation, deviations are much larger.

It is well known that N_2O_4 decomposes when heated. In comparing his correlation with data for N_2O_4, Shah assumed that it decomposes completely at the burnout point to form NO_2. Therefore, properties used were of NO_2 with an effective latent heat, which was the sum of latent heat of NO_2 and the heat of decomposition of N_2O_4. This resulted in good agreement with CHF data. Whether

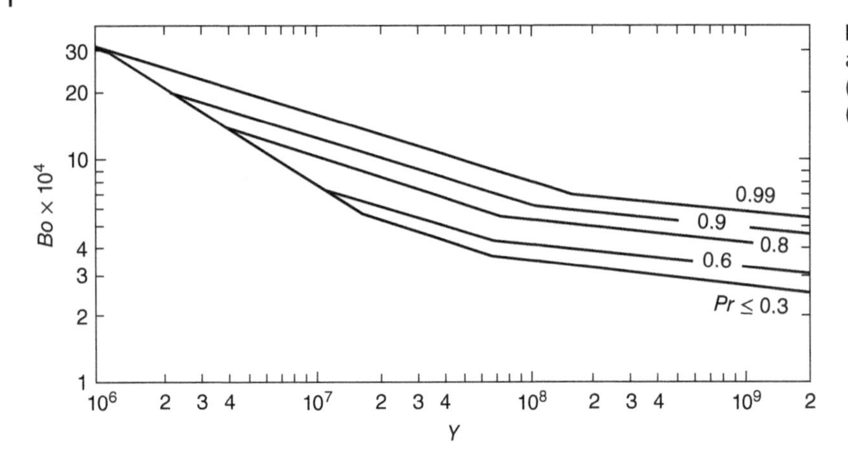

Figure 6.2.2 Boiling number at $x_c = 0$ according to the local condition correlation (LCC) of Shah. Source: Reprinted from Shah (1987). © 1987, with permission from Elsevier.

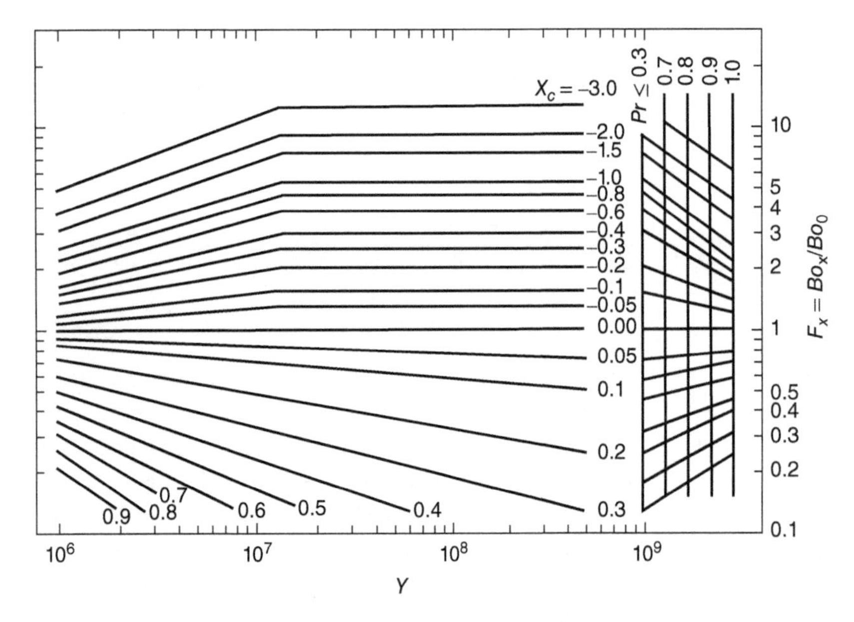

Figure 6.2.3 Ratio of boiling numbers at $x_c = x$ and $x_c = 0$ according to the local condition correlation (LCC) of Shah. Source: Reprinted from Shah (1987). © 1987, with permission from Elsevier.

this method can be used for other decomposing fluids is unknown.

Shah (2017) compared the Shah (1987) correlation with a wide-ranging database for CHF in minichannels ($D \leq 3$ mm). Good agreement was found with tube diameters down to 0.13 mm. This is further discussed in Section 6.2.2.4.5.

Some authors, such as Doroshchuk et al. (1970), have defined a limiting quality at which dryout occurs independent of the heat flux and is thus a purely hydrodynamic phenomenon. This is called CHF of the second kind. Figure 6.2.4 shows some test data in which CHF drops along a vertical line at constant quality and thus represents the CHF of the second kind. It is seen that the Shah correlation is in satisfactory agreement with these data. This is because as CHF falls at constant critical quality in a tube of fixed length, the inlet quality is increasing, which reduces the CHF predicted by the Shah correlation. In this region, CHF is by dryout. CHF drops sharply at constant quality because the very strong evaporation from the liquid film blocks the deposition of entrained droplets onto the liquid film. While the limiting quality is an interesting phenomenon, such data are satisfactorily predicted by the usual methods for calculating CHF. Katto (1982) has shown that data for the limiting quality are satisfactorily predicted by his general correlation Katto (1978).

Horizontal Tubes Many heat exchangers in use have horizontal channels. Most of the evaporators used in the air conditioning and refrigeration industry have horizontal tubes. The devices to remove heat from miniature electronic circuits and computer chips have horizontal channels. Many boilers, kettle reboilers, and some nuclear reactors use horizontal tubes. Hence the ability to predict CHF in horizontal channels is of great practical importance.

While flow distribution in vertical channels is always symmetrical about the axis of the tube, it can be asymmetrical in horizontal tubes due to the effect of gravity.

Table 6.2.1 Range of CHF data analyzed in Shah (1987).

Fluids	Water, R-11, R-12, R-21, R-22, R-113, R-114, ammonia, hydrazine, N_2O_4, MIPD, CO_2, helium, nitrogen, hydrogen, acetone, benzene, diphenyl, ethanol, ethylene glycol, *o*-terphenyl, potassium, and rubidium
D (mm)	0.32–37.8
p_r	0.0014–0.962
G (kg m^{-2} s^{-1})	4–29 051
q_c (kW m^{-2})	0.11–45 000
L_c/D	1.3–940
x_{in}	−4.0 to +0.81
x_c	−2.6 to +1.0
Y	6–720 000 000

Source: Reprinted from Shah (1987). © 1987, with permission from Elsevier.

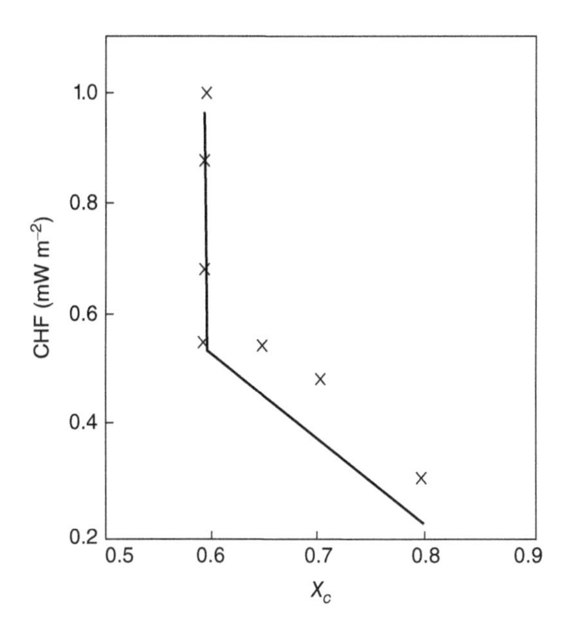

Figure 6.2.4 Comparison of the Shah correlation with the data of Hewitt and Kearsey (1966) for water in a vertical tube: $D = 12.62$ mm, $L_C/D = 290$, $G = 1355$ kg m^{-2} s^{-1}, and $p_r = 0.31$.

Table 6.2.2 Results of comparison of CHF data with various correlations.

Fluid	Data range	No. of data	Correlation of	Deviation (%) Mean	Deviation (%) Average	No. of data with deviation >30%
Water	All data	427	Shah	14.4	−0.9	40
			Katto	16.1	+3.1	51
			Bowring	18.6	−9.6	86
	Verified range of Bowring correlation	251	Shah	14.4	−0.3	20
			Katto	14.1	−3.2	17
			Bowring	11.7	−3.9	19
Helium	All data	167	Shah	17.8	−2.6	29
			Katto	32.7	+29.3	73
			Subbotin	74.0	+54.2	74
	Verified range of Subbotin correlation	67	Shah	20.3	+7.9	18
			Katto	25.3	+22.1	25
			Subbotin	22.3	−3.5	21
All fluids	$x_c < 0$	397	Shah	16.9	−2.1	56
			Katto	30.7	+21.8	141
	$x_c \geq 0$	1046	Shah	15.4	−3.2	129
			Katto	18.8	+11.8	187
	All data	1443	Shah	16.0	−2.9	185
			Katto	22.3	+14.6	328

Source: Reprinted from Shah (1987). © 1987, with permission from Elsevier.

Gravitational forces pull both phases in a direction perpendicular to the flow direction. The liquid being heavier is affected more. This causes asymmetric phase distribution, liquid becoming thinner at the top portion of channel. During low quality and subcooled conditions, bubbles coalesce and form a continuous vapor layer at the top, preventing liquid from reaching the channel surface, thus causing CHF. At low qualities, slug flow can occur similar to vertical channels but with liquid film surrounding the bubble thinner at the top. Hence CHF in horizontal tubes during slug flow will be lower than in vertical channels at lower flow rates. At moderate vapor qualities, wavy flow with alternating splashing waves can occur. During the surge, top of the channel is coated with a thin liquid film that begins to dry due to drainage and evaporation. At sufficiently high heat flux, the top part of the channel may completely dry out before the next splashing wave comes along, resulting in CHF.

At higher qualities, annular flow occurs. The phenomena are similar to those in vertical channels except that the liquid film is thinner at the top and thicker at the bottom. The thin liquid film at the top gets completely depleted at sufficiently high heat flux, and CHF occurs at the upper surface. Due to the asymmetry of film distribution around the circumference, the CHF at the upper surface of channel is lower in a horizontal channel than in a vertical channel with identical conditions.

When flow is asymmetric due to the effect of gravity, CHF occurs first at the top part of the tube. Further downstream, it spreads to the lower parts of the circumference until it reaches the bottom and the entire tube is in CHF condition. This is illustrated by Figure 6.2.5. While this figure is for tube inclined upwards by 15°, behavior in horizontal tube is very similar.

While numerous correlations have been published for CHF in vertical tubes, there are very few for horizontal tubes. These are discussed in the following text.

Merilo (1979) developed the following correlation for CHF in horizontal tubes by modifying the fluid-to-fluid modeling technique of Ahmad (1973):

$$Bo = 575\, Re_{LT}{}^{-0.34}(Z^3 Bd)^{0.358} (\mu_f/\mu_g)^{-0.218}(L/D)^{-0.511}\left(\frac{\rho_f}{\rho_g}-1\right)^{1.27}(1-x_{in})^{1.64} \tag{6.2.38}$$

where Z is the Ohnesorge number $= \mu_f \cdot (\sigma D \rho_f)^{-0.5}$ and Bd is the Bond number.

Merilo compared this formula with data for water and R-12 from four sources with good agreement. However, Wong et al. (1990) and Pioro et al. (2002) found it to be inaccurate when compared with more data.

Groeneveld et al. (1986) developed a method in which the CHF in vertical tubes was multiplied by a factor K_{hor}, which was 0 for fully stratified flow and increased linearly up to 1 with increasing mass flux. Wong et al. (1990) found this method to be inadequate. They developed an analytically based expression for K_{hor} that involves calculation of void fraction and vapor quality taking into consideration non-equilibrium effects using various correlations. With vertical tube CHF calculated from the CHF lookup table of Groeneveld et al. (1986), good agreement was found with data for R-12 and water from six sources.

Kefer et al. (1989) performed CHF measurements with water in tubes of diameters 12.5 and 24.3 mm in horizontal and vertical upwards flow as well as with various inclinations in between the two. They developed a correlation

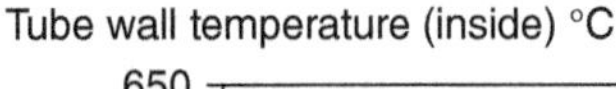

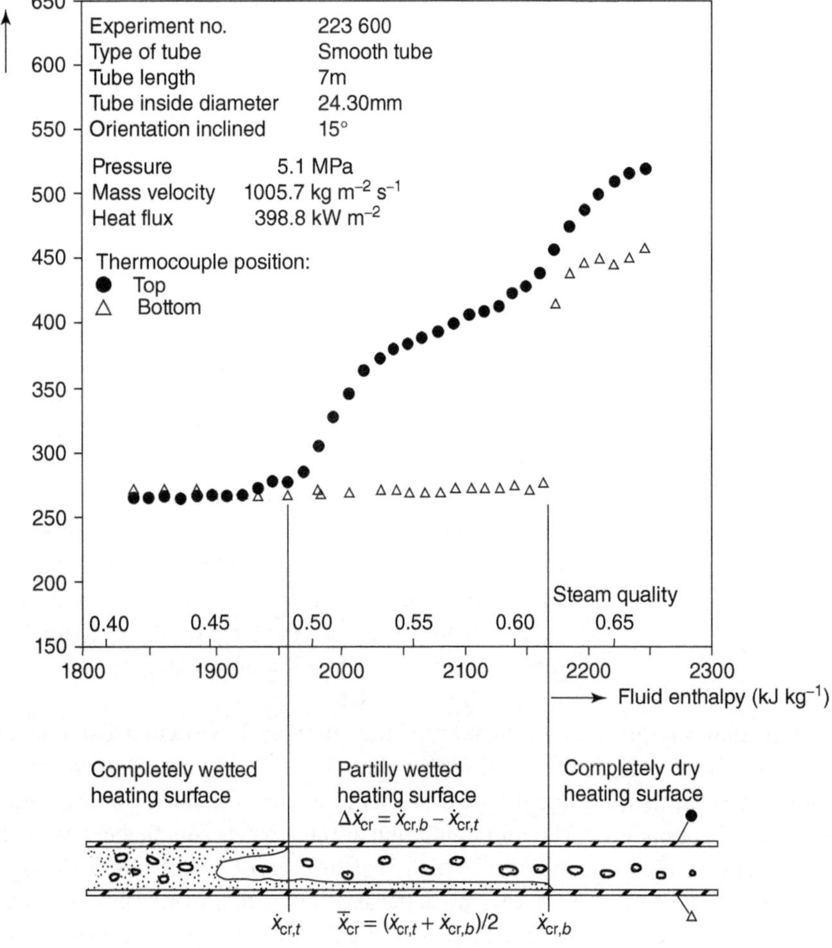

Figure 6.2.5 Wall temperatures during boiling of water in a tube inclined 15° to horizontal. Source: Kefer et al. (1989). © 1989 Elsevier.

using a two-phase Froude number defined as

$$Fr_{TP} = x_c G/\{gD\cos\theta\rho_G(\rho_f - \rho_g)\}^{0.5} \tag{6.2.39}$$

θ is the inclination to horizontal. At $Fr_{TP} \geq 10$, according to this correlation, the CHF in all orientations including horizontal and vertical is the same. For lower values of Fr_{TP}, their correlation is

$$(x_{c,\text{bottom}} - x_{c,\text{top}})/2 = x_{c,\text{vertical}} \tag{6.2.40}$$

$$(x_{c,\text{bottom}} - x_{c,\text{top}}) = 16/(2 + Fr_{TP})^2 \tag{6.2.41}$$

where $x_{c,\text{bottom}}$ and $x_{c,\text{ top}}$ are, respectively, the qualities for CHF at bottom and top of a horizontal tube and $x_{c,\text{vertical}}$ is the critical quality for the vertical tube. Fr_{TP} is calculated with Eq. (6.2.39) using $x_{c,\text{vertical}}$. They compared this correlation only with their own data mentioned earlier. This correlation cannot be used for subcooled burnout as Fr_{TP} becomes negative.

Shah (2015a) presented a general correlation that provides a correction factor to his vertical tube correlation to make it applicable to horizontal tubes. It is described in the following text.

Shah Correlation Shah noted that that CHF at lower flow rates in horizontal tubes is lower than in vertical tubes because of the stratification caused by gravity. The inertia force tends to keep the distribution of phases symmetrical, while the gravitational force tends to pull the heavier phase (liquid) down. Froude number gives the ratio of inertia and gravitational forces. It was therefore felt that it may be possible to quantify the effect of stratification using Froude number. In his correlation for heat transfer, Shah (1976), he had used the all liquid Froude number Fr_{LT} defined as

$$Fr_{LT} = \frac{G^2}{\rho_f^2 gD} \tag{6.2.42}$$

Shah used Fr_{LT} and Fr_{TP} to analyze the data for horizontal tubes to determine the factor K_{hor} in the equation

$$q_{c,\text{hor}} = K_{\text{hor}} q_{c,\text{vert}} \tag{6.2.43}$$

where $q_{c,\text{hor}}$ is the CHF in a horizontal tube and $q_{c,\text{vert}}$ is the CHF in the same tube with vertical upflow with exactly the same conditions. The vertical CHF was calculated with the Shah (1987) correlation. This analysis led to the development of the following correlating equations:

$$\text{For } x_{in} < 0 \text{ when } L/D < 10, \quad K_{\text{hor}} = 1 \tag{6.2.44}$$

$$\text{For } x_c \leq 0.05, \quad K_{\text{hor}} = 0.725 Fr_{LT}^{0.082} \leq 1 \tag{6.2.45}$$

$$\text{For } x_c > 0.05, \quad K_{\text{hor}} = 0.64 Fr_{TP}^{0.15} \leq 1 \tag{6.2.46}$$

If K_{hor} calculated with Eq. (6.2.43) or (6.2.44) that is greater than 1, use $K_{hor} = 1$. According to Eq. (6.2.45), $K_{hor} = 1$ at $Fr_{LT} \geq 50$. According to Eq. (6.2.46), $K_{hor} = 1$ at $Fr_{TP} \geq 20$.

The transition from Eq. (6.2.43) to Eq. (6.2.44) at $x_c = 0.05$ was arrived at by trial and error but was considered to be approximately the value at which CHF mechanism changes. Shah notes that the flow patterns and mechanisms of CHF change as vapor quality goes above a low value. At high flow rates, DNB occurs under subcooled conditions and at low qualities in both vertical and horizontal channels. At low flow rates with subcooling or low qualities in horizontal channels, flow pattern is bubbly and CHF occurs due to vapor blanketing caused by coalescence of bubbles in the upper portion of the tube. In their experimental study with R-134a in a horizontal 0.5 mm diameter channel, Revellin et al. (2006) report that bubble and bubble–slug flows occurred at vapor qualities of a few percent and changed to slug and slug–annular at higher qualities. Hence $x_c = 0.05$ may be approximating to the region of change in CHF mechanisms.

Shah compared this correlation with an extensive database whose range is given in Table 6.2.3. It included single round tubes as well as multichannels with round and rectangular channels. Some of the channels were heated only on three sides. In such cases D_{HP} was used as the equivalent diameter.

The same data were also compared to a number of other correlations including Merilo, the critical quality correlation of Kim and Mudawar (2013), and minichannel

Table 6.2.3 Range of data used to validate the Shah correlation for CHF in horizontal channels.

Fluids	Water, FC-72, HFE-7100, R-12, R-32, R-113, R-134a, R-123, R-236fa, and R-245fa
Geometry	Single tubes and multichannels, round and rectangular
Heating flux type	Uniform, axially non-uniform, circumferentially non-uniform. Liquid heated and electric heated
D_{hp}	0.13–24.3 mm
L/D_{hp}	1.97–488
p_r	0.0053–0.900
G	20–11 390 kg m^{-2} s^{-1}
Fr_L	0.01–1229
Fr_{TP}	0.19–79.3
x_{in}	−1.05 to 0.72
x_c	−0.2 to 0.99
$Bo \times 10^4$	0.9–1320
Data points	878 data points from 39 data sets from 18 sources

Source: Shah (2015a). © 2015 Elsevier.

correlations of Zhang et al. (2006), Wu et al. (2011), and Wojtan et al. (2006). The summary of results is given in Table 6.2.4. The Shah correlation has a MAD of 15.4% with all 878 data points. All other correlations performed poorly except that the correlation of Zhang et al. (2006) had MAD of only 14.7% with minichannel data. Calculations were also done with the Shah horizontal tube correlations, while using the Katto–Ohno correlation to calculate the vertical tube CHF. The MAD by this method was 24.8%, considerably higher than 15.4% while using the Shah correlation for vertical CHF.

Inclined Tubes Inclined orientations include all orientations other than horizontal and vertical up that have already been discussed in the foregoing. The other orientations are discussed in the following text.

Bertoni et al. (1976) measured CHF with R-12 at 17.5 bar flowing vertically up and down in a 17.8 mm diameter tube. CHF for downflow was about 10% lower than for upflow at mass flux 150–1000 kg m^{-2} s^{-1}.

Ohno et al. (1999) studied CHF of R-22 flowing in a 9 mm diameter tube at various inclinations from horizontal to vertically upwards. Reduced pressures were from 0.6 to 0.84 and mass velocity from 400 to 2600 kg m^{-2} s^{-1}. They found four characteristic regimes. In the first regime, CHF generally increases with increasing inclination to horizontal. In the second regime, inclination has no effect on CHF. In the third and fourth regimes, CHF is equal to that for horizontal or vertical orientation, respectively. They have given maps to distinguish these regimes.

Ami et al. (2011, 2014) measured CHF during upward flow of water in inclined stainless steel tubes of 20 mm diameter. Tube inclinations varied from 15° to 90° to the horizontal. Tube lengths were from 450 to 1800 mm. Figure 6.2.6 shows some of their test data. It is seen that CHF falls increasingly below the vertical flow CHF as the inclination angle decreases. The authors developed the following correlation that fits their own data, those of Cumo et al. (1978) as well as those of Ohno et al. (1999) for R-22 described earlier:

$$K_{\text{inc}} = \frac{q_{c,\text{inc}}}{q_{c,\text{vert}}} = 1 - \exp(-1.2\sqrt{Fr_{\text{TP}}}) \tag{6.2.47}$$

where $q_{c,\text{inc}}$ is the CHF of inclined tube.

At $Fr_{\text{TP}} = 2$, K_{inc} is about 0.8 and drops sharply as Fr_{TP} becomes lower. K_{inc} rises gradually as Fr_{TP} rises above 2 and asymptotically reaches 1 at $Fr_{\text{TP}} = 10$.

The correlation of Kefer et al. (1989) is given by Eqs. (6.2.38) and (6.2.39). It is based on their data for water in tubes inclined 0°–90° (upflow).

Zhang et al. (2002) studied CHF with FC-72 flowing in a rectangular channel with one side heated. Orientations varied from 0° (horizontal with heater at the bottom) with anti-clockwise increments of 45°–315°. They found that CHF was very sensitive to orientation at velocities less than 0.2 m s^{-1}. At higher velocities, there was reasonably good agreement with correlations for flow boiling. At velocities 0.2 m s^{-1} and lower, CHF for orientations with upward facing heater (0°, 45°, and 315°) was near the predictions of Zuber pool boiling CHF model (described in Chapter 3), while those for orientations with downward facing heater (135°, 180°, and 225°) were fairly close to the predictions of Nejat (1981) for flooding CHF. Figure 6.2.7 shows the CHF at various orientations observed by them.

Nejat (1981) measured CHF in vertical tubes closed at the bottom. Tube diameters were 8, 10, and 14 mm. L/D were 18.4,15, and 10.7. The experiments were conducted with water, carbon-tetrachloride, normal-hexane, and ethyl alcohol. He first compared his results with the flooding correlation of Wallis, Eq. (2.7.28), but the agreement was not

Table 6.2.4 Results of comparison of data for CHF in horizontal channels with various correlations by Shah (2015a).

Channel type	N	Mean absolute deviation (%) Average deviation (%)						
		Shah (2015a) correlation with vertical tube CHF by correlation of		Correlations of				
		Shah	Katto and Ohno	Merilo	Zhang et al.	Wojtan et al.	Kim and Mudawar	Wu et al.
Single channels	771	14.7	23.6	134.3	46.6	43.9	45.3	88.5
		0.4	15.8	111.7	43.1	−42.6	11.6	74.5
Multichannels	107	20.3	33.4	43.4	14.7	33.4	43.6	1835.7
		−18.4	30.2	−10.3	−3.0	30.2	−43.4	1835.7
All channels	878	15.4	24.8	123.1	42.7	42.6	45.1	510.5
		−1.6	17.5	96.7	37.4	−19.7	4.8	499.9

Source: Shah (2015a).

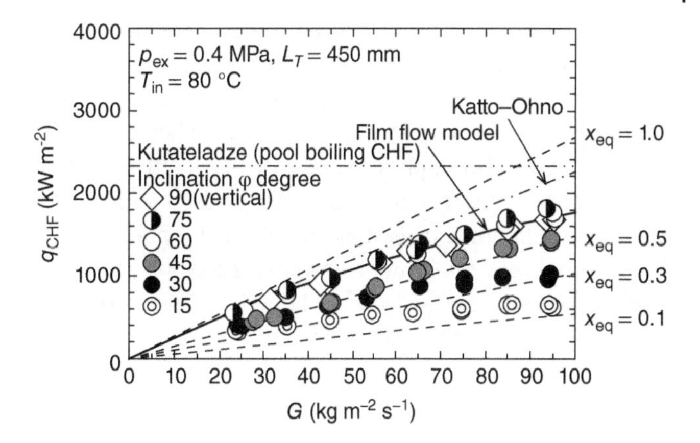

Figure 6.2.6 Effect of inclination angle on CHF of water flowing in a 20 mm ID tube. Source: Ami et al. (2014). © 2014 Taylor & Francis.

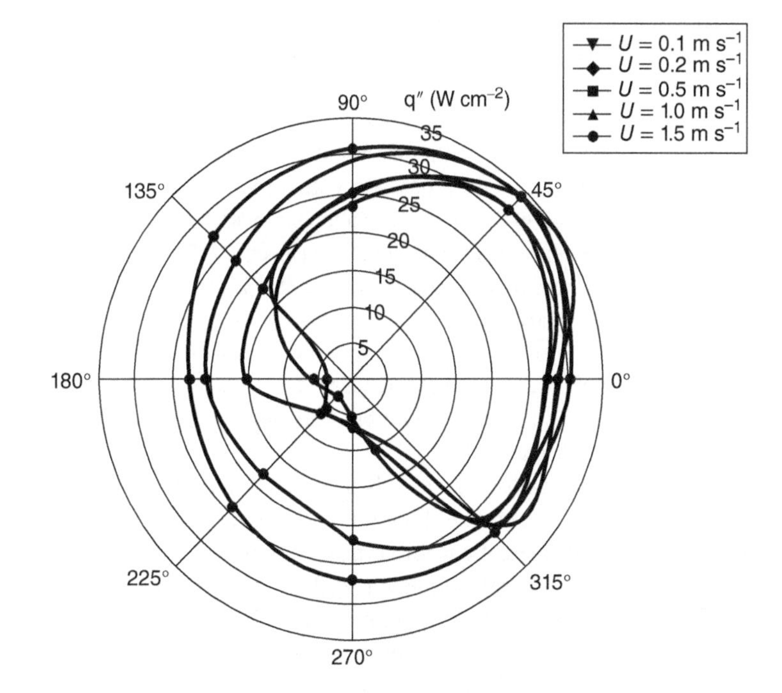

Figure 6.2.7 Effect of velocity and orientation on CHF in a rectangular channel with one side heated. Exit subcooling 3 K. Source: Zhang et al. (2002). ©2002 Elsevier.

satisfactory. He therefore modified the Wallis correlation to obtain

$$q_c = 0.6\left(\frac{L}{D}\right)^{0.1} \rho_g i_{fg} \left[\frac{(\rho_f - \rho_g)gD}{\rho_g}\right]^{1/2} \left[1 + \left(\frac{\rho_g}{\rho_f}\right)^{1/4}\right]^{-2} \tag{6.2.48}$$

Nejat also analyzed the data of Griffith et al. (1962) on a vertical heated tube and Kusuna and Imura (1974) on a vertical thermosyphon and found good agreement. Those data included L/D from 11 to 35.

Sudo et al. (1985) performed tests with water near atmospheric pressure flowing up and down vertical 2 mm wide channels formed by two heated plates. Based on these data as well as data from several other studies of low-pressure water flowing upward and downward in annuli, tubes, and rectangular channels, they give the following equations:

$$q_c^* = 0.7\frac{A}{A_h}\frac{\sqrt{D/\lambda}}{[1 + (\rho_g/\rho_f)^{1/4}]^2} \tag{6.2.49}$$

$$q_c^* = 0.005(G^*)^{0.611} \tag{6.2.50}$$

$$q_c^* = -x_{in}\frac{A}{A_h}G^* \tag{6.2.51}$$

where

$$q_c^* = \frac{q_c}{i_{fg}\sqrt{\lambda\rho_g(\rho_f - \rho_g)g}} \tag{6.2.52}$$

$$G^* = \frac{G}{\sqrt{\lambda\rho_g(\rho_f - \rho_g)g}} \tag{6.2.53}$$

$$\lambda = \sqrt{\frac{\sigma}{(\rho_f - \rho_g)g}} \tag{6.2.54}$$

A is the total flow area and A_h is the heated area. For rectangular channels, D is the gap between the sides. All test data were for subcooled liquid at the entrance to channel. The limits of applicability of these equations are shown graphically in their paper and are as follows. Equation (6.2.49) is applicable to both upflow and downflow when $G^* < 20$. This equation also applies to tubes with closed bottom. In downflow, this equation is considered to represent CHF due to flooding caused by vapor velocity pulling the liquid film from the bottom upwards. Equation (6.2.51) applies to downflow when G^* is between 20 and 10 000. Equation (6.2.50) applies to upflow for $G^* > 20$ and to both upflow and downflow at $G^* > 10\,000$. The authors note that these equations are not the best fit to data but are intentionally kept close to the lower limit of data so as to provide conservative estimates of CHF. The scatter in data is large. Note that Eqs. (6.2.49) and (6.2.51) were developed by Mishima (1984).

Lu et al. (2014) performed tests on vertical rectangular channels formed with two heated plates with gap in between as the flow passage with cross section 60 mm × 2 mm and 80 mm × 2 mm. Subcooled water near atmospheric pressure was used. Their measured CHF for downflow are in reasonable agreement with the Sudo et al. correlation given earlier. G^* in their data was 4–25. The data at $G^* < 20$ were overpredicted but were within −33% of the prediction. Lu et al. have discussed several other correlations for rectangular channels and also given a correlation of their own data.

From the foregoing discussions, it is clear that orientation does not affect CHF at higher flow rates and hence it can be predicted with correlations for vertical channels. There is no well-verified general method for predicting this limiting flow rate where orientation begins to have effect. There is also no well-verified method to predict CHF when orientation begins to have effect. The various correlations given here may be used in their verified range.

Critical Quality Correlations Many correlations have been published for determining the critical quality. Once the critical quality x_c is known, CHF can be calculated by the heat balance equation, Eq. (6.2.4).

Wojtan et al. (2005) have given the following correlation for critical quality at the completion of CHF:

$$x_c = 0.61\,\exp[0.57 - 0.0058\,We_{GT}^{0.38}\,Fr_{GT}^{0.15}(\rho_f/\rho_g)^{0.09} \times (q_c/q_K)^{0.27}] \tag{6.2.55}$$

where

$$q_K = 0.131\rho_g^{0.5} i_{fg}[g(\rho_f - \rho_g)\sigma]^{0.25} \tag{6.2.56}$$

$$We_{GT} = G^2 D/(\rho_g \sigma) \tag{6.2.57}$$

$$Fr_{GT} = G^2/(\rho_g^2 g D) \tag{6.2.58}$$

This correlation was developed with data for R-22 and R-410A in horizontal tubes of 8.0 and 13.8 mm diameter.

Kim and Mudawar (2013) have given the following correlation for the calculation of incipient critical quality at CHF:

$$x_c = 1.4\,We_{LT}^{0.03} p_r^{0.08} - 15.0\left(Bo\frac{D_{HYD}}{D_{HP}}\right)^{0.15} Ca^{0.35}\left(\frac{\rho_g}{\rho_f}\right)^{0.06} \tag{6.2.59}$$

where

$$We_{LT} = \frac{G^2 D_{HYD}}{\rho_f \sigma} \tag{6.2.60}$$

$$Ca = \frac{G\mu_f}{\sigma \rho_f} \tag{6.2.61}$$

Ca is called the capillary number. This correlation was developed using data from 26 sources for 13 fluids in channels of hydraulic diameters from 0.51 to 6 mm. They compared nine other correlations for critical quality with the same database; none of them gave good agreement.

Minichannels Many correlations have been developed specifically for minichannels, which are generally considered to be those with hydraulic diameter ≤ 3 mm though some regard the limit be ≤6 mm as discussed in Chapter 1. Such correlations are discussed herein together with applicability of macro channel correlations to minichannels.

The Kim and Mudawar correlation for critical quality, Eq. (8.2.59), was intended for minichannels and was based on data for channels up to 6 mm.

Wu et al. (2011) analyzed data for several fluids (nitrogen, water, and refrigerants) in minichannels from many sources and gave the following correlation:

$$Bo = 0.6\left(\frac{L}{D_{HP}}\right)^{-1.19} x_c^{0.817} \tag{6.2.62}$$

The limits of its applicability were given as follows: $x_c > 0.05$, $Bo \cdot Re_{LS}^{0.5} \leq 200$, $G \leq 4000\ \text{kg m}^{-2}\,\text{s}^{-1}$, $L/D_{HP} \leq 250$, and $p_r \leq 0.25$. Equivalent diameters ranged from 0.29 to 2.98 mm.

Wojtan et al. (2006) gave a modified form of Katto–Ohno correlation that agrees with their own data for refrigerants in 0.5 and 0.8 mm tubes. Their correlation is

$$Bo = 0.437\left(\frac{\rho_g}{\rho_f}\right)^{0.073}\left(\frac{G^2 L}{\sigma \rho_f}\right)^{-0.24}\left(\frac{L}{D}\right) \tag{6.2.63}$$

Zhang et al. (2006) analyzed data from 13 sources for water in channels of diameters 0.33–6.22 mm. These included both horizontal and vertical channels. They gave the following correlation:

$$Bo = 0.0352\left[\frac{G^2D}{\sigma\rho_f} + 0.0119\left(\frac{L}{D}\right)^{2.31}\left(\frac{\rho_f}{\rho_f}\right)^{0.361}\right]^{-0.295}$$
$$\times\left(\frac{L}{D}\right)^{-0.311}\left[2.05\left(\frac{\rho_g}{\rho_f}\right)^{0.17} - x_{in}\right] \quad (6.2.64)$$

Bowers and Mudawar (1974) gave a correlation of their own data in channels of diameter 0.5 and 2.54 mm with R-113. Qu and Mudawar (2004) presented a modified form of Katto–Ohno correlation that agreed with their data for R-113 and water in minichannels as well as the data of Bowers and Mudawar. Kosar and Peles (2007) found these two correlations to give large deviations with their own data for R-123 in a multichannel of 0.29 mm diameter. They gave a new correlation that agreed with their own data. Qi et al. (2007) presented a correlation that agreed with their own data for nitrogen in 0.531, 0.834, and 1.042 mm diameter channels.

Revellin et al. (2009) compared a database for minichannels that had data from 13 sources for 9 diverse fluids with several predictive techniques. They found large deviations with the correlations of Qu and Mudawar (2004) and Qi et al. (2007). The correlations of Wojtan et al. Eq. (6.2.63) and Zhang et al., Eq. (6.2.64); and the theoretical model of Revellin and Thome (2008) gave fairly good agreement.

Shah (2017) compared a database for minichannels with the Shah (1987) correlation as well as several other correlations. Included in this database are data for 13 fluids from 30 sources for single channels and multichannels, round and rectangular, of diameters 0.13–3.0 mm. Data included uniform as well as axially and circumferentially non-uniform heat flux. The complete range of parameters in this database is in Table 6.2.5. The MAD of the correlations of Shah (1987), Katto and Ohno (1984), Zhang et al. (2006), and Wojtan et al. (2006) were 18.9%, 33.2%, 23.4%, and 59.7%, respectively.

The horizontal channel data analyzed by Shah (2015a) included a considerable number of data for minichannels. The results are listed in Table 6.2.4. It is seen that for multichannels, Shah (2015a) and Zhang et al. (2006) gave good agreement, while the deviations of Katto and Ohno (1984), Wojtan et al. (2006), Kim and Mudawar (2013), and Wu et al. (2011) correlations were 33.4%, 33.4%, 43.6%, and 1836%, respectively.

From the preceding text, it may be concluded that the general correlations of Shah for vertical and horizontal channels, Shah (1987) and Shah (2015a), respectively, are the most reliable. The correlation of Zhang et al. (2006) is the next best. It is also seen that as far as prediction of CHF is concerned, there is no difference between minichannels and macro channels as the Shah correlation satisfactorily predicts the data for channels of diameters for 0.13–37.8 mm.

Table 6.2.5 Range of data for minichannels analyzed by Shah (2017).

Fluids	Water, nitrogen, helium, propane, ethanol, FC-72, HFE-7100, R-12, R-32, R-113, R-123, R-134a, R-152a, R-236fa, and R-245fa
Geometry	Single tubes and multichannels, round and rectangular
Orientation	Horizontal and vertically upwards
D_{hp}	0.13–3.0 mm
L/D_{hp}	1.66–484
p_r	0.009–0.887
G	10–41 810 kg m^{-2} s^{-1}
x_{in}	−2.34 to 0.32
x_c	−0.96 to 1.04
$Bo \times 10^4$	3.9–203
Data points	577 data points from 55 data sets from 30 sources

Source: Shah (2017).

6.2.2.5 Fluid-to-Fluid Modeling

Testing of models of boiling water reactors (BWRs) is very expensive and uses very large amounts of power. Costs can be greatly reduced if tests could be done with low latent heat fluids. To enable such tests, fluid-to-fluid modeling techniques were developed.

Ahmad (1973) developed a compensated distortion model using dimensional analysis. He arrived at the following relation:

$$\Psi_{CHF} = \left(\frac{GD}{\mu_f}\right)\left(\frac{\mu_f^2}{\sigma D\rho_f}\right)^{2/3}\left(\frac{\mu_g}{\mu_f}\right)^{1/5} \quad (6.2.65)$$

For the same inlet quality, L/D, and ρ_g/ρ_f, critical boiling number will be the same if ψ_{CHF} is the same. Ahmad showed the agreement of this technique by comparison with a wide range of data. The following alternative expression was also given:

$$\Psi_{CHF} = \left(\frac{GD}{\mu_f}\right)\left(\frac{\gamma^{1/2}\mu_f}{D\rho_f^{1/2}}\right)^{2/3}\left(\frac{\mu_g}{\mu_f}\right)^{1/8} \quad (6.2.66)$$

where

$$\gamma = \frac{\partial(\rho_g/\rho_f)}{\partial p} \quad (6.2.67)$$

The two expressions give almost the same result. The choice between the two depends on which properties are available. Note that this method has been verified only for vertical tubes.

Among other fluid-to-fluid modeling methods, a well-known one is the graphical correlation of Stevens and Kirby (1964).

A possible application of the fluid-to-fluid modeling method is to use it to predict the CHF of a fluid from the CHF data of another fluid. The Groeneveld et al. (2007) lookup table provides CHF values for water over a vast range, and CHF of other fluids could in principle be calculated from them or from other compilations of water data. However, the present author has not seen any comprehensive verification of this method.

6.2.2.6 Non-uniform Heat Flux

The vast majority of experimental data for CHF has been obtained in electrically heated channels with uniform heat flux and correlations developed mostly using such data. In many heat exchangers, heat flux is non-uniform. For example, heat flux in most evaporators used in refrigeration and air conditioning varies along the length. For finned tube evaporators for cooling air, it may also vary around the circumference. Heat flux in nuclear reactors varies along the length in the form of a sinusoidal wave. Hence it is important that applicability of correlations to non-uniform heat flux be discussed.

Circumferentially Non-uniform Heat Flux Shah (2017) analyzed the data of Leontiev et al. (1981) for water in a 6 mm tube that had circumferentially non-uniform heat flux. The heat flux at the top of tube was higher, with $q_{max}/q_{avg} = 1.5$. Heat flux was axially uniform. The CHF reported was the average heat flux for the tube length and the circumference. These data had MAD of 16.2% and 14.5%, respectively, with the correlations of Shah and Katto–Ohno.

Shah (2017) analyzed data for multichannels from five sources. In all cases, heat was directly applied to one side only. Four of these multichannels had rectangular channels. Heat was conducted into the sides of the channels that act like fins. Heat flow in a fin decreases with distance from its base. Thus the heat flux was non-uniform around the perimeter though it was uniform along channel length. The fifth multichannel had round channels. Heat was conducted from the top face into the channels and thus decreased with distance from the top, causing non-uniform heat flux around the circumference. The data from all five sources had MAD of 20.3% with the Shah correlation and only 14.5% with the Zhang et al. (2006) correlation. Note that for partially heated channels, CHF is based on the area of channel, which transfers heat. Area of sides made of non-conducting materials such as plexiglass is excluded.

After analyzing data for circumferentially non-uniform heat flux in tubes from many sources, Hewitt (1982) made the following recommendations:

1) For flux tilt less than 1.3, average CHF for non-uniform heat flux may be assumed to be the same as for uniform heat flux.
2) For flux tilt > 1.3, reduce the uniform heat flux CHF by 10–20%
3) For side heated tubes, reduce uniform heat flux CHF by up to 30%.

Flux tilt is the ratio of maximum to minimum heat flux on the surface. These recommendations are for channels made entirely of conducting material.

The data analysis by Shah indicates that general correlations for CHF can correctly predict the circumferentially averaged CHF for channels with circumferentially non-uniform heat flux. However, it will be prudent to follow the recommendations by Hewitt to ensure safe design.

Axially Non-uniform Heat Flux Four approaches have been most commonly considered for predicting CHF in channels with axially varying heat flux. These are the total or overall power hypothesis, local condition hypothesis, the F-factor method, and the boiling length average (BLA) method.

According to the overall power hypothesis proposed by Lee and Obertelli (1963), the total power for a particular tube will be the same for the same inlet quality irrespective of whether the heat flux is uniform or non-uniform though they recommend using 10% lower power for the non-uniform heat flux case. The location of CHF cannot be predicted with this method.

The local condition hypothesis was originally proposed by Barnett (1964). It is assumed that there is a unique relation between local quality and CHF. CHF occurs at a location along the channel where the quality equals that given by the relation between quality and CHF. Barnett assumed a linear relation between quality and CHF. Kirby (1966) improved the prediction accuracy by using the actual quality-CHF relation instead of the linear relation assumed by Barnett.

The F-factor approach was proposed by Tong et al. (1966). It takes into consideration the effect of the upstream heat flux profile on the local CHF. The F-factor is defined as

$$F = \frac{q_{c,u}}{q_{c,nu}} \tag{6.2.68}$$

where $q_{c,u}$ is the CHF for uniform heating and $q_{c,nu}$ is the CHF for axially non-uniform heating at the same local quality. F is expressed by the following relation:

$$F = \frac{K}{1 - \exp(-Kz_c)} \int_0^{z_c} \frac{q(z)}{q(z_c)} \exp[-K(z_c - z)]dz \tag{6.2.69}$$

where z is the axial location and z_c is the axial location of CHF. Various expressions for K have been given by Tong as well as by others. Tong and Tang (1997) have given the following expression:

$$K = 0.15\frac{(1-x_c)^{4.31}}{(G/10^6)^{0.478}} \tag{6.2.70}$$

G is in lb h^{-1} ft^{-2}, z is in inches.

Among the previously mentioned three methods, Collier and Thome (1994) recommend the F-factor method as the most accurate. The F-factor method is more suitable for DNB-type CHF, which occurs at negative or low qualities.

The BLA method has been described by several authors including Hewitt (1982). As given by Groeneveld et al. (1999), it is as follows.

Boiling length average heat flux q_{BLA} is defined as

$$q_{\mathrm{BLA}}(z_c) = \frac{1}{L_B}\int_{z=0}^{z_c} q(z)dz \tag{6.2.71}$$

A factor K_5 is defined as

$$K_5 = \frac{q(z_c)}{q_{\mathrm{BLA}}(z_c)} \tag{6.2.72}$$

Critical heat flux is postulated to occur where q_{BLA} equals the equivalent uniform critical flux $q_{c,u}$. This results in the following relation:

$$q_{c,\mathrm{nu}} = K_5 q_{c,u} \tag{6.2.73}$$

The four methods for calculating CHF with non-uniform heat flux are shown schematically in Figure 6.2.8.

Yang et al. (2006) performed test with HFC-134a during upflow in tubes of 5.46 mm diameter with one uniform and four non-uniform axial profiles. The ratio of maximum heat flux to average heat flux was 1.267. Pressure varied from 17 to 24 bar and mass flux from 2827 to 4687 kg m^{-2} s^{-1}. They found that with deeply subcooled conditions, variations of axial heat flux profile do not have much effect on CHF. They compared their measurements with the four methods mentioned earlier. All of them predicted the data well. The overall power method underpredicted by up to 15% at the highest subcooling at inlet while at near zero inlet qualities, deviations were about +10% to −15%. The local condition method overpredicted slightly at lower inlet qualities and had scatter up to +24% at higher qualities. The results with the F-factor method were very similar to those for the overall power method. With the BLA method, Yang et al. tried definitions of boiling length other than the standard definition, length between $x = 0$ and $x = x_c$. They found that if the boiling length is defined as starting at the location of the start of annular flow, RMS of data becomes 2.4% compared with 3.4% with the conventional definition. The RMS errors with overall power, local quality, and F-factor methods were 7.4%, 5.4%, and 4.3%.

Del Col and Bortolin (2012) measured CHF for three refrigerants in a 0.96 mm diameter water-heated tube in which heat flux increased axially from inlet to exit. The ratio of maximum to minimum heat flux along the length reached up to 4. Shah (2017) compared these data to his general correlation. The CHF analyzed was an averaged value $q_{c,\mathrm{avg}}$ obtained by the relation

$$q_{c,\mathrm{avg}} = Q_c/(\pi DL) \tag{6.2.74}$$

where Q_c is the total heat input to the tube up to the location of CHF. The MAD of the Shah correlation for the data

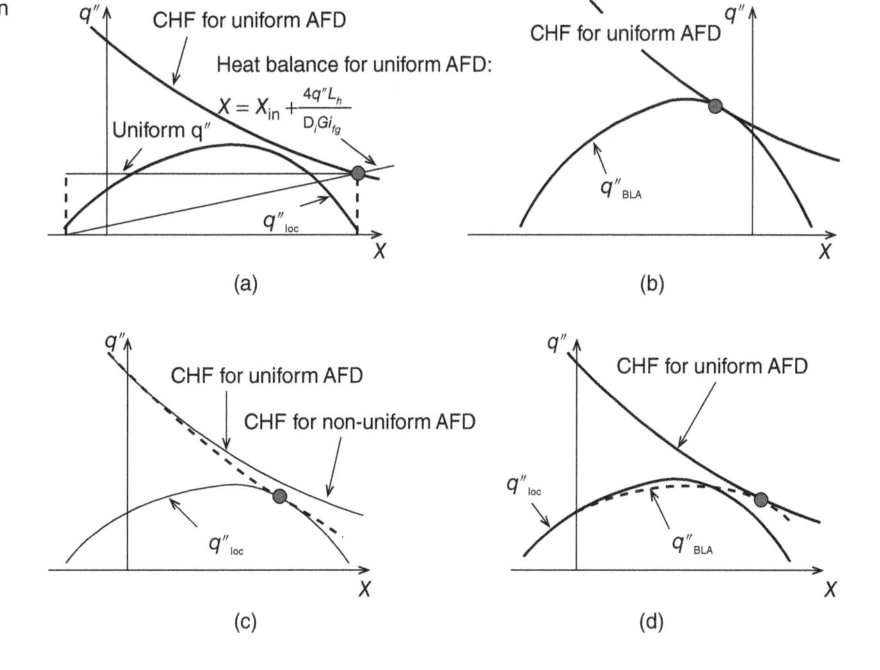

Figure 6.2.8 Various methods for calculation of CHF in channels with non-uniform axial flux distribution (AFD): (a) total power hypothesis, (b) local condition hypothesis, (c) F factor method, and (d) BLA method. Source: Yang et al. (2006). ©2006 Elsevier.

for the three refrigerants was 9.6%, 10.6%, and 22.2%. This result suggests that the Shah correlation may be applicable to axially varying heat flux using the total power hypothesis.

From the foregoing discussions, the following conclusions may be reached. For circumferentially varying heat flux, the circumferential average CHF can be calculated with general CHF correlations such as that of Shah, but it will be prudent to apply the deductions recommended by Hewitt (1982) given earlier. For axially non-uniform heat flux, none of the prediction methods has been demonstrated to be generally applicable but the F-factor and BLA methods appear to be more accurate.

6.2.3 Recommendations

1) For CHF in vertical upflow, use the Shah (2015a) correlation. For CHF in horizontal channels, use the Shah (2017) correlation. These are recommended for all types of pure fluids including liquid metals for channels with $D_{HYD} \geq 0.13$ mm.
2) For circumferentially non-uniform heat flux, reduce calculated CHF by 15% if heat flux tilt exceeds 1.3, and by 30% for tubes heated on one side only.
3) For axially non-uniform heat flux, use the F-factor method if expression for F verified for design conditions is available. Otherwise, use the BLA method for higher qualities.
4) For orientations other than vertical up and horizontal, there is no well-verified general correlation. Use the correlations given here in their verified range.

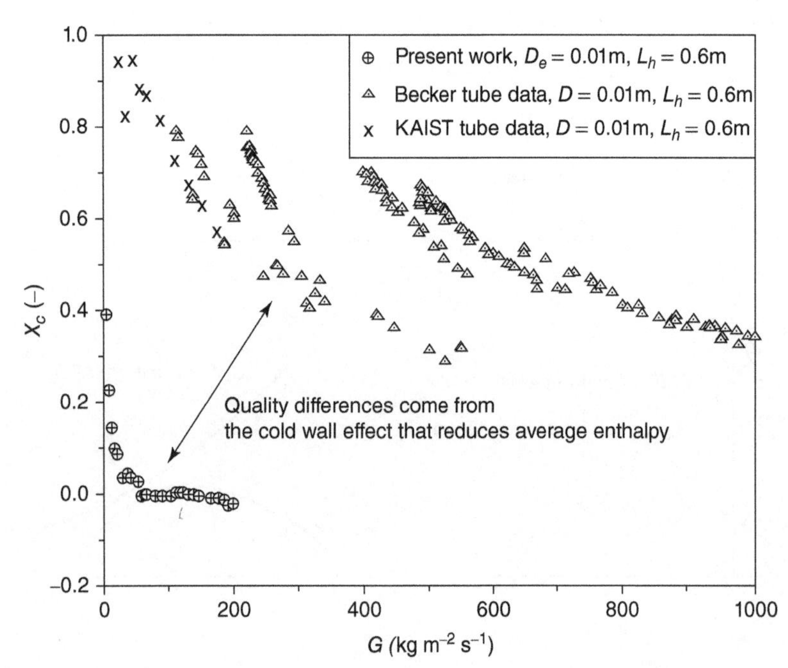

Figure 6.3.1 Comparison of quality at CHF in an annulus with D_{HYD} = 10 mm and round tubes of the same diameter. Source: Park et al. (1997). ©1997 Elsevier.

6.3 CHF in Annuli

There can be several types of heating in annuli. Only the inner tube may be heated, or the outer tube alone may be heated, or both tubes may be heated (bilateral heating). In addition, heat flux may be uniform or non-uniform. Heat transfer coefficients in annuli can be predicted with correlations for tubes, using an equivalent diameter. However, this is generally not the case for CHF as the behavior in annuli is different from that in tubes. Figure 6.3.1 shows the comparison of critical quality during tests with water in an internally heated annulus and tubes of the same tubes of the same diameter as the hydraulic diameter of the annulus. It is seen that the critical quality is much lower in the annulus. This is because the liquid flowing along the cold wall does not remove any heat from the surface where CHF occurs. Barnett (1964) compared data for annuli and tubes of equal hydraulic diameter (D_{HYD}) and equivalent diameter based on heated perimeter (D_{HP}) and found that CHF in annuli with either equivalent diameter are quite different from those in tubes. Shah (1980) tried to apply to annuli his correlation for CHF in tubes (Shah 1979) alternately with D_{HYD} and D_{HP} without success. Hence correlations specifically for annuli have been developed.

6.3.1 Vertical Annuli with Upflow

6.3.1.1 Dimensional Correlations for Water

Barnett (1966) developed the following correlation for water flowing up in uniformly heated annuli (internal tube heated) at 69 bar:

$$\frac{q_c}{10^6} = \frac{A + B(\Delta i_{SC})_{inlet}}{C + L} \tag{6.3.1}$$

where Δi_{SC} is the enthalpy difference between subcooled liquid and saturated liquid and

$$A = 67.45 D_{HP}^{0.68} (G \times 10^{-6})^{0.192} \left[1 - 0.774 \exp(-6.512 D_{HYD}) \left(\frac{G}{10^6}\right)\right] \quad (6.3.2)$$

$$B = 0.2587 D_{HP}^{1.261} \left(\frac{G}{10^6}\right)^{0.817} \quad (6.3.3)$$

$$C = 185 D_{HP}^{1.415} \left(\frac{G}{10^6}\right)^{0.212} \quad (6.3.4)$$

The correlation is in British units. The range of parameters is as follows:

$L = 24–108$ in.
$D_i = 0.375–3.798$ in.
$D_o = 0.551–4.006$ in.
$D_{HYD} = 0.127–0.875$ in.
$D_{HP} = 0.258–3.791$ in.
$G \times 10^{-6} = 0.14–6.2$ lb h^{-1} ft^{-2}
$(\Delta i_{SC})_{inlet} = 0–412$ Btu lb^{-1}

Doerffer et al. (1994) compared his wide-ranging database with several correlations and found that the Barnett correlation predicts negative CHF at high critical qualities.

Bertoletti et al. (1965) have given the following correlation (known as the CISE correlation) based on the boiling length concept. It is given as follows:

$$x_c = \frac{W_B}{G A_f i_{fg}} = \frac{P_h}{P_t} \frac{1 - p_r}{(G/100)^{1/3}} \frac{L_B}{L_B + 0.315 (p_r^{-1} - 1)^{0.4} D_{HYD}^{1.4} G} \quad (6.3.5)$$

where W_B is the total power input to the boiling length in kW, G is the mass velocity g cm^{-2} s^{-1}, i_{fg} is in cal g^{-1}, L_B is in cm, A_f is the flow area in cm^2, D_{HYD} is in cm, P_h is the heated perimeter, and P_t is the wetted perimeter. This correlation was tested with considerable data for water from many sources for tubes, annuli, and rod bundles. Both uniform and non-uniform heat flux were included. Annuli data included internal, external, and bilateral heating. The range of applicability of this correlation is

$$P > 45 \text{ kg cm}^{-2}$$

$$G > 100(1 - p_r)^3 \left(\frac{P_h}{P_t}\right)^3$$

$$x_c > 0$$

$$x_{in} > 0.5 \left(\frac{P_h}{P_t}\right)^{1/3} \frac{1 - p_r}{(G/100)^{1/3}}$$

Doerffer et al. (1994) have given two methods to use the lookup table for CHF in tubes to internally heated annuli. The simpler of the two is given herein. In this correlation, CHF in annuli occurs at quality x_c when the critical quality in tube is x_c'. These are related by the following equation:

$$x_c' = x_c + 0.658 - 0.33 p^{0.428} G^{0.108} \delta^{-0.453} D_{HP}^{0.37} + 0.208 \exp[-18.5 (x_c - 0.35)^2] \quad (6.3.6)$$

D_{HP} is in cm, annular gap δ is in mm, p is in MPa, and G is in Mg m^{-2} s^{-1}.

To determine the CHF at quality x_c from the lookup table, the following relation applies:

$$\text{CHF}_{\text{annulus}}(p, G, x_c) = \text{CHF}_{D=8}(p, G, x_c') \quad (6.3.7)$$

Doerffer et al. recommend this correlation for x_c from −0.1 to +0.5. Comparison was made with data for water from three sources and data for R-12 from one source converted to water equivalent by Ahmad's fluid-to-fluid modeling method. Annular gap in all data varied from 1.61 to 11.1 mm.

6.3.1.2 General Correlations

The first attempt at developing a general correlation applicable to all fluids appears to have been by Bernath (1960). It was intended only for subcooled burnout. Gambill (1963) found it to perform poorly. Gambill (1963) presented his own correlation for subcooled CHF, according to which CHF was the sum of pool boiling CHF $q_{c,pool}$ and heat flux removed by forced convection.

$$q_c = q_{c,pool} + h_{LT}(T_{w,chf} - T_f) \quad (6.3.8)$$

The wall temperature at CHF, $T_{w,chf}$, was determined from a graph given in Bernath (1960). While Gambill showed it to agree with several data sets for tubes and annuli, it was found to fail for annular gaps smaller than 2 mm. No independent evaluation of this correlation has come to the present author's notice.

General correlations verified with data for many fluids have been presented by Katto (1979) and Shah (1980, 2015b). These are discussed in the following text.

Katto Correlation Katto (1979) presented a correlation for annuli that has a form similar to his correlation for tubes. This correlation involves several regimes whose boundaries are determined as follows. The boundary between L and H regime is given by the following relation. It is L regime if

$$\frac{L_c}{D_{HP}} > \frac{1}{0.48 \left(\frac{\rho_g}{\rho_f}\right)^{0.133} \left(\frac{\sigma \rho_f}{G^2 L_c}\right)^{0.0.29} - 0.0081} \quad (6.3.9)$$

The boundary between H and N regimes is defined as follows. The regime is H if

$$\frac{L_c}{D_{HP}} > 0.0288 \frac{1}{\left(\frac{\sigma \rho_f}{G^2 L_c}\right)^{0.584}} \quad (6.3.10)$$

As for the tubes, CHF is given by the following relation:

$$q_c = q_{co}(1 - Kx_{in}) \quad (6.3.11)$$

In the H regime q_{co} is given by the following equation:

$$\frac{q_{co}}{Gi_{fg}} = 0.12\left(\frac{\rho_g}{\rho_f}\right)^{0.133}\left(\frac{\sigma \rho_f}{G^2 L_c}\right)^{1/3}\frac{1}{1 + 0.0081 L_c/D_{HP}} \quad (6.3.12)$$

In the L regime,

$$\frac{q_{co}}{Gi_{fg}} = 0.25\left(\frac{\sigma \rho_f}{G^2 L_c}\right)^{0.043}\frac{D_{HP}}{L_c} \quad (6.3.13)$$

For the N regime,

$$\frac{q_{co}}{Gi_{fg}} = 0.22\left(\frac{\rho_g}{\rho_f}\right)^{0.133}\left(\frac{\sigma \rho_f}{G^2 L_c}\right)^{0.433}\frac{(L_c/D_{HP})^{0.171}}{1 + 0.0081 L_c/D_{HP}} \quad (6.3.14)$$

For internally heated annuli, the following relations are given for K:

$$\text{In the L regime,} \quad K = 1. \quad (6.3.15)$$

In the H regime, K is obtained by the following relation:

$$K = 0.057\left(\frac{69.2}{L_c/D_{HP}}\right)^{11(\rho_g/\rho_f)}\left(\frac{\sigma \rho_f}{G^2 L_c}\right)^{-1/3} \quad (6.3.16)$$

K for N and HP regimes is not given. Nor is the boundary of HP regime given.

Katto correlation for externally heated annuli is as follows. Determination of regimes and calculation of CHF is done using the Katto (1978) correlation for tubes with D replaced by D_{HP} based on the diameter of external tube. Agreement was shown with a limited amount of data in H, N, and HP regimes.

For bilateral heating, an approximate method is suggested for use when heating on one of the tubes is very small compared with that on the other tube. No comparison with test data is given.

Doerffer et al. (1994) compared the Katto correlation with their database for water. They found its predictions to be in poor agreement with data and that it predicts incorrect trends for the effect of L/D and annular gap width. Shah (2015b) compared the Katto correlation to a wide range of data for many fluids. Agreement was not good as is discussed in the following text.

Shah Correlation Shah (1980) presented a correlation that was verified with data for four fluids from many sources. The data included annuli with internal, external, and bilateral heating. The correlation was in graphical form. To facilitate its use in computerized calculations, Shah (2015b) converted it to equation form and presented comparison with a much wider range of data. It is described in the following.

The correlating parameters are L/D_{hp}, inlet quality x_{in}, Y, and Bo. If the quality at CHF, x_c, is negative, it also becomes a correlating parameter. When $x_{in} \le 0$, L is the heated length of tube up to the CHF location. If $x_{in} > 0$, L equals the boiling length, which is the tube length between $x_{in} = 0$ and the location of CHF. L_B is given by

$$L_B = L_c + x_{in} D_{hp}/(4Bo) \quad (6.3.17)$$

The use of L_B implies that $x_{in} = 0$ has been assumed and this value has to be used when $x_{in} > 0$ in all equations that follow.

Y is given by Eq. (6.2.20) as for tubes, with D the diameter of the tube on which CHF occurs.

D_{HP} is the equivalent diameter based on heated perimeter and is defined as flow area divided by heated perimeter. Thus, for cylindrical annuli,

$$D_{HP} = \frac{\text{Abs}(D^2 - {D_2}^2)}{D} \quad (6.3.18)$$

D is the diameter of the tube on which CHF occurs, and D_2 is the diameter of the other tube of the annulus.

The boiling number at CHF is given by

$$Bo = F_1 F_2 Bo_1 \quad (6.3.19)$$

Bo_1 is the boiling number for $Y < 10^4$ at actual inlet quality. It is calculated as follows:

$$Bo_1 = E_1 Bo_0 \quad (6.3.20)$$

Bo_0 is the boiling number at $x_{in} = 0$ with $Y < 10^4$, given by

$$Bo_0 = 0.045[K(L/D_{HP})^{0.56}]^{-1} \quad (6.3.21)$$

The factor $K = 1$ for $(L/D_{hp}) \le 120$.

$$\text{For } (L/D_{hp}) > 120, \quad K = 0.336 + 0.0055(L/D_{hp}) \quad (6.3.22)$$

If K from Eq. (6.3.22) > 2.2, use $K = 2.2$.

For $(L/D_{HP}) \leq 30$ and $-x_{in} \geq 0.6$,

$$E_1 = 0.61\left(\frac{L}{D_{HP}}\right)^{0.404}(1 - x_{in})^{0.66} \quad (6.3.23)$$

For $(L/D_{HP}) \leq 30$ and $-x_{in} < 0.6$,

$$E_1 = \left[0.832\left(\frac{L}{D_{HP}}\right)^{0.404} - 1\right]\left(-\frac{x_{in}}{0.6}\right) + 1 \quad (6.3.24)$$

For $(L/D_{HP}) \geq 85$ and $-x_{in} \geq 1.14$,

$$E_1 = 2.9 - 2.53x_{in} \quad (6.3.25)$$

For $(L/D_{HP}) \geq 85$ and $-x_{in} < 1.14$,

$$E_1 = 1 - 4.2x_{in} \quad (6.3.26)$$

For L/D_{HP} between 30 and 85, calculate E_1 using equations for $L/D_{HP} < 30$ and $L/D_{HP} > 85$ and use the larger of the two.

F_1 is calculated as follows:

$F_1 = 1$ for $Y \leq 10^4$. For $Y > 10^4$,

$$F_1 = (10^{-4}Y)^{(-0.73-0.073\ \ln(Bo_1))} \quad (6.3.27)$$

F_2 is calculated as follows:

When $x_c \geq 0$, $F_2 = 1$.

For $x_c < 0$, F_2 is calculated as follows:

$F_2 = 1$ for $Y < 3.5 \times 10^5$ for all values of x_c.

For $Y > 3.5 \times 10^5$,

$$\text{For } (-x_c) > 0.2, \quad E_2 = 2.54(-x_c)^{0.355} \quad (6.3.28)$$

$$\text{For } (-x_c) \leq 0.2, \quad E_2 = (1 + 4.12x_c)^{(-0.194)} \geq 1 \quad (6.3.29)$$

$$\text{For } 5.5 \times 10^5 < Y < 10^7, \quad F_2 = E_2 \quad (6.3.30)$$

$$\text{For } Y > 10^7, \quad F_2 = E_2(10^{-7}Y)^n \quad (6.3.31)$$

where

$$n = -1.143x_c \leq 0.24 \quad (6.3.32)$$

For Y between 3.5×10^5 and 5.5×10^5, interpolate linearly between $F_2 = 1$ at $Y = 3.5 \times 10^5$ and F_2 at $Y = 5.5 \times 10^5$ calculated as given earlier. As there were no test data in this range, this recommendation to interpolate is tentative.

The above equations are applicable to annuli with one heated tube. The same are also applicable to bilaterally heated annuli by adjusting the vapor inlet quality to take into account the vapor generated at the other tube. To do so, x_{in} in the above equations is replaced by an effective inlet quality $x_{in,ef}$ defined as

$$x_{in,ef} = x_{in} + 4\left[\frac{q_2}{Gi_{fg}}\right] L/D_{HP,2} \quad (6.3.33)$$

where $D_{HP,2}$ is the heated equivalent diameter of the tube other than that on which CHF occurs and q_2 is the heat flux on it. If $x_{in,ef}$ from Eq. (6.3.33) is positive, $x_{in,ef}$ replaces x_{in} in Eq. (6.3.17) for calculating the boiling length, and all other calculations are the same as for annuli with one tube heated. Using this effective inlet quality, calculations proceed as if only one tube is heated. $D_{HP,2}$ is calculated as

$$D_{HP,2} = \frac{\text{Abs}(D^2 - D_2^{\,2})}{D_2} \quad (6.3.34)$$

If heat fluxes on both tubes are high, calculations should be done for each tube in this way to determine on which one the CHF occurs.

Figure 6.3.2a–c show the correlation in graphical form. These may be used for hand calculations.

Shah recommends that for water, this correlation be used only when

$$G \geq \frac{50\,000}{p^{1.35}} \quad (6.3.35)$$

where G is in kg m^{-2} s^{-1} and p is the pressure in bar.

The reason for this exclusion is that predicted CHF were found to be much higher than measured CHF for water at low pressure and low flow. Similar experience with other correlations has been reported by researchers. For example, Park et al. (1997) compared data for water from many sources with a number of published correlations for water. While most data gave good agreement with several correlations, data for $G \leq 500$ kg m^{-2} s^{-1} at pressures of 1–10 bar showed poor agreement with all correlations. Shah therefore concluded that a separate correlation is needed for the low pressure–low flow region.

This correlation was compared to a wide range of data from many sources whose range is shown in Table 6.3.1. Figure 6.3.3 shows the range of water data satisfactorily predicted. This includes $G \geq 100$ kg m^{-2} s^{-1}, and pressure between 15 and 200 bars, subject to the restriction of Eq. (6.3.35). The same data for internally and externally heated annuli were also compared to the Katto correlation. Katto's correlation for bilateral heating has very limited range and hence could not be evaluated. Barnett's correlation is applicable only to water and is not applicable to bilaterally heated annuli. The results of data analysis are summarized in Table 6.3.2. It is seen that the Shah correlation gives good agreement with all types of annuli. The Katto correlation performs very poorly. As was noted earlier, Doerffer et al. (1994) had found the Katto correlation to give large deviations as well as to predict wrong trends. The Barnett correlation was based only on data for water at 69 bars. However, its predictions at other pressures were not too bad. The data analyzed by Shah included those of Larsen et al. (1966) in which the external channel was

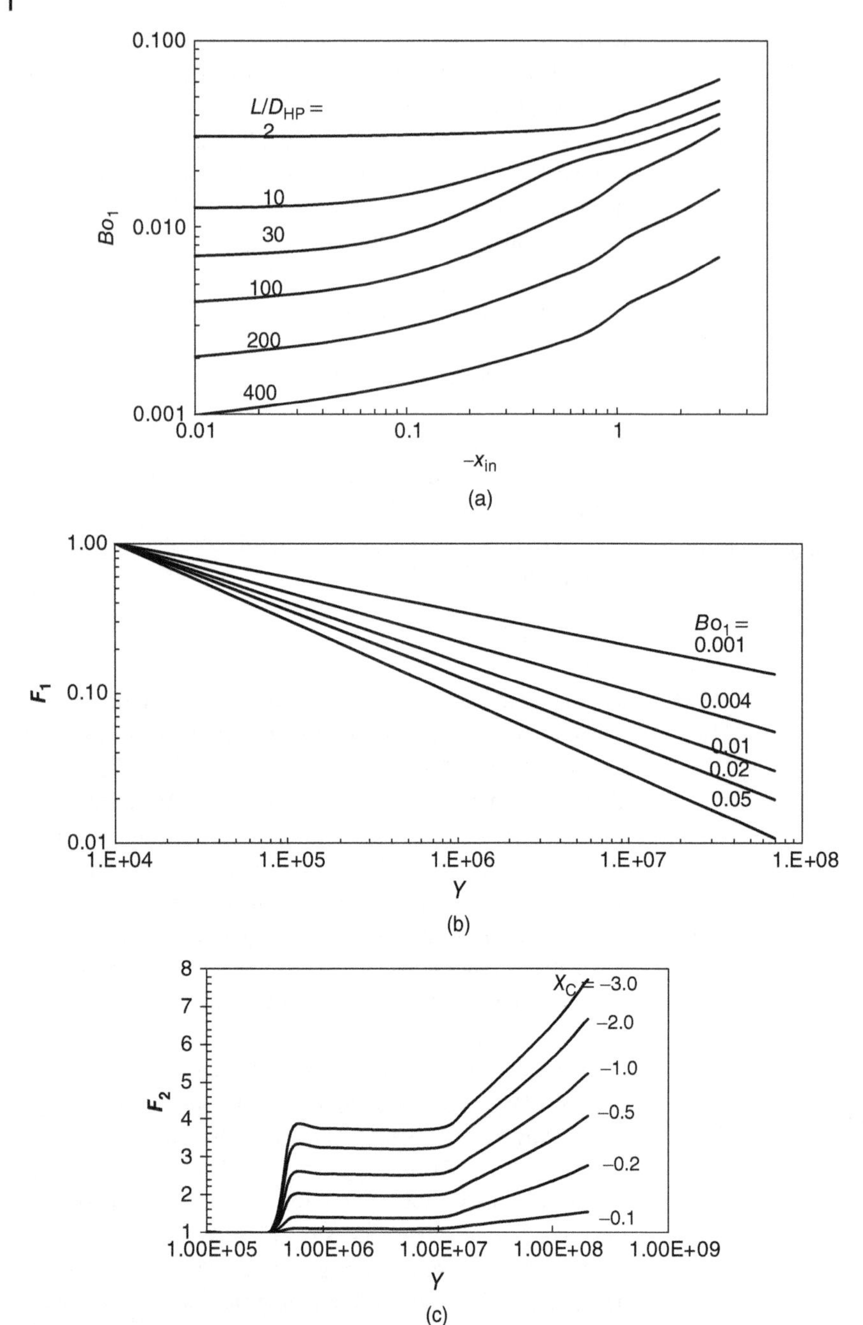

Figure 6.3.2 (a) Part 1 of the Shah correlation for annulus, critical boiling number at $Y \leq 10^4$. Source: Shah (2015b). © 2015 Taylor & Francis. (b) Part 2 of the Shah correlation for annuli. Source: Shah (2015b). © 2015 Taylor & Francis. (c): Part 3 of the Shah correlation for CHF in annuli. Source: Shah (2015b). © 2015 Taylor & Francis.

square, while the inner tube was round. These data were in good agreement with the Shah correlation.

The data listed in Table 6.3.2 include 10 fluids of very diverse properties. However, no data for liquid metals are included. Shah states that the only data for metallic fluids that he could find were those of Noyes and Lurie (1966) for sodium. These data were much lower than the Shah correlation. Rohsenow (1973) reviewed these data and the test setup on which they were obtained. He concluded that severe instabilities occurred during these tests and that caused these data to be low. The Shah correlation for annuli is similar to the Shah correlation for CHF in tubes described in Section 6.2.2.4.1. That correlation showed good agreement with data for several liquid metals from several sources. Hence it seems likely that the Shah annulus correlation is applicable to liquid metals. This needs verification with test data under stable conditions.

6.3.1.3 Recommendations

The dimensional correlations for water may be used in their verified ranges. The correlations of Barnett and Doerffer et al. were verified only for internally heated annuli. The Bertoletti et al. correlation was verified with data for internal, external, and bilateral heating.

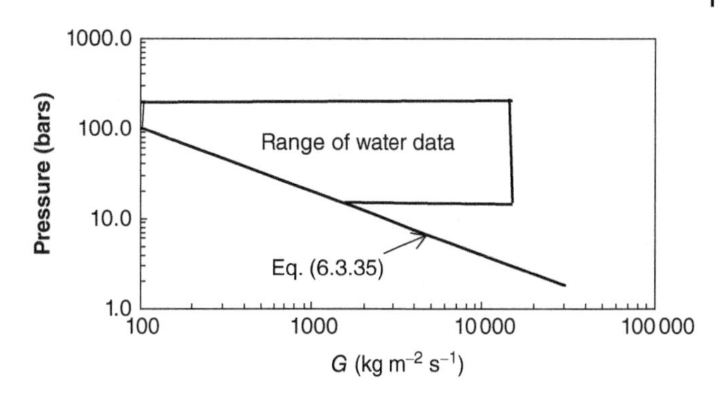

Figure 6.3.3 Applicability limit of the Shah correlation for CHF in annuli to water and the range of water data in satisfactory agreement with the Shah correlation. Source: Shah (2015b). © 2015 Taylor & Francis.

Table 6.3.1 The range of data satisfactorily predicted by the Shah correlation for annuli.

Fluids	Water, deuterium oxide, R-12, R-113, R-114, R-134a, diphenyl, acetone, benzene, and toluene
Heated tube	Inner, outer, both
Tube with CHF	Inner, outer
Annulus ID (mm)	1.5–96.5
Annulus OD, (mm)	2.5–99.7
Annular gap (mm)	0.3–16.2
L/D_{hp}	1.3–394
p_r	0.016–0.906
G (kg m^{-2} s^{-1})	100–15 759
q_c (kW m^{-2})	33–20 372
Inlet quality	−3.3 to +0.91
Critical quality	−2.7 to +0.95
q_2/q_c	0.1–1.0 (for bilateral heating)
$Bo \times 10^{-4}$	1.6–240
Y	5100–240 000 000
No. of data points	1146
No. of data sources	25

Source: Shah (2015b). © 2016 Taylor & Francis.

The Shah correlation is recommended for all fluids over its verified range that includes internal, external, and bilateral heating. Caution should be exercised in using it for liquid metals as it has not been verified with such data.

6.3.2 Horizontal Annuli

Stoddart et al. (2002) measured CHF in a horizontal annulus with water. The test section was an annulus with 6.45 and 7.77 mm inner and outer diameters, respectively (0.66 mm gap width), and an 18.5 cm long heated section. Both tubes of annulus were heated. The experimental parameters included the following ranges: test section exit pressure, 0.344–1.034 MPa; mass flux, 100–380 kg m^{-2} s^{-1}; wall heat flux, 0.231–1.068 MW m^{-2}; and water inlet temperature, 30–65 °C. The measured CHF values were considerably lower than what was expected for vertical orientation. In all the tests, CHF occurred at relatively high equilibrium qualities and was preceded by flow stratification that caused dryout of the upper surface of the flow channel. They have given a correlation in which CHF for horizontal annulus is obtained by Eq. (6.2.43), i.e. multiplying CHF in vertical annulus by the factor K_{hor}, which is given by the relation as follows:

$$K_{hor} = 0.52 + 4.4 \times 10^{-5} \frac{GD_{HYD}}{\mu_f} \tag{6.3.36}$$

The vertical annulus CHF was calculated with the CISE correlation, Eq. (6.3.5).

6.3.3 Eccentric Annuli

Eccentricity has the effect of decreasing CHF. Doerffer et al. (1994) examined data for eccentric annuli with vertical upflow from three sources. The inner tubes were heated. They developed the following correlation. It uses a void migration parameter Υ defined as

$$\gamma = \frac{\delta_{norm}}{D_o + D_i} \tag{6.3.37}$$

where δ_{norm} is the annular gap of a concentric annulus. Eccentricity ε is defined as

$$\varepsilon = 1 - \frac{\delta_{min}}{\delta_{norm}} \tag{6.3.38}$$

δ_{min} is the minimum gap between the inner and outer tubes. The correlation is expressed by the following relation:

$$q_{c,eccentric} = q_{c,concentric}(1 - k_\varepsilon)k_\gamma k_x \tag{6.3.39}$$

$$k_\varepsilon = 0.991 - 2.771\varepsilon^{2.5} + 2.3\varepsilon^3 \tag{6.3.40}$$

For $\Upsilon < 0.0133$, $k_\Upsilon = 2.07$. For $\Upsilon > 0.1183$, $k_\Upsilon = 0.26$. For Υ between 0.0133 and 0.1183,

$$k_\gamma = -3.403 + 5.7588 \exp\left(-\frac{\gamma}{0.2615}\right) \tag{6.3.41}$$

For $x_c \leq 0.05$, $k_x = 1.05$. For $x_c \geq 0.5$, $k_x = 0.6$.

Table 6.3.2 Summary of results of data analysis by Shah (2015b) for CHF in annuli.

Heating mode	Fluid	No. of data sources	No. of data points	Deviation (%) Mean absolute Average		
				Shah	Katto	Barnett
Internally heated	Water	14	675	15.9	257.5	31.2
				−7.3	236.0	−2.3
	All other	9	265	12.7	171.9	
				−2.7	162.7	
Externally heated	All	3	58	20.4	69.4	42.3
				−8.0	40.1	25.2
Bilaterally heated	Water	4	148	24.0		
				−21.8		
All type	All	25	1146	16.5		
				−8.0		

Soure: Data from Shah (2015b).

For $0.05 < x_c < 0.5$,

$$k_x = 1.2516 - 0.9111x_c^{0.5} \tag{6.3.42}$$

This correlation was in good agreement with data from three sources in which eccentricity was from 0.126 to 0.974.

Balino and Converti (1994) measured CHF in a horizontal annulus with R-12 as the fluid. The outer tube diameter was 54 mm and the inner tube was 11.2 mm. Tests were done with the inner tube concentric or moved up and down by up to 15 mm. Moving the tube down had very little effect but CHF decreased sharply when the tube was moved upwards, becoming about 1/3 when the inner tube was moved up by 15 mm. The CHF occurred at the upper surface. A predictive model was developed based on the assumption of intermittent flow pattern.

6.4 CHF in Various Geometries

6.4.1 Single Cylinder with Crossflow

The following discussions are for liquid flowing vertically upwards across a horizontal cylinder except where stated otherwise.

Cochran and Andracchio (1974) studied peak heat flux on cylindrical heaters. The test fluids were water and R-113. Heaters of 0.049–0.181 cm diameter were tested over a fluid velocity range of 10.1–81.1 cm s^{-1}. The experimental results were observed to fall within two regions based on the vapor removal geometry: jets or sheets. Figure 6.4.1 shows some of their test data. They gave dimensional formulas to fit their own data.

Lienhard and Eichhorn (1976) performed an analysis of CHF during crossflow. They concluded that for low velocities, mechanisms remain the same as in pool boiling and CHF may be calculated by the hydrodynamic model as given in Section 3.3.1.2. With increasing velocity, a transition in mechanism occurs when the velocity reaches the limit given by the following equation:

$$(We_{\mathrm{GT}})_{\mathrm{transition}} = 4\left(\frac{\rho_f}{\rho_g}\right)R' \tag{6.4.1}$$

where

$$We_{\mathrm{GT}} = \frac{G^2 D}{\rho_g \sigma} \tag{6.4.2}$$

R′ is dimensionless radius of the cylinder given by

$$R' = R\left[\frac{\sigma}{g(\rho_f - \rho_g)}\right]^{-0.5} \tag{6.4.3}$$

The flow pattern then changes from individual three-dimensional jets to a continuous two-dimensional vapor sheet. Sausage like bubbles break off the top of the vapor sheet as seen in Figure 6.4.2. CHF occurs when this configuration becomes unstable and is given by the following relation:

$$\frac{\pi q_c}{\rho_g i_{fg} u_\infty} = 1 + \left(\frac{4}{We_{\mathrm{GT}}}\right)^{1/3} \tag{6.4.4}$$

where u_∞ is the velocity of liquid approaching the cylinder. With further increase in approaching liquid velocity, a second transition occurs. The flow in the sheet and its terminal velocity can then be described by a potential flow model. This second transition velocity is given by the following relation:

$$\frac{\pi q_c}{\rho_g i_{fg} u_\infty} = 0.275\left(\frac{\rho_f}{\rho_g}\right)^{0.5} + 1 \tag{6.4.5}$$

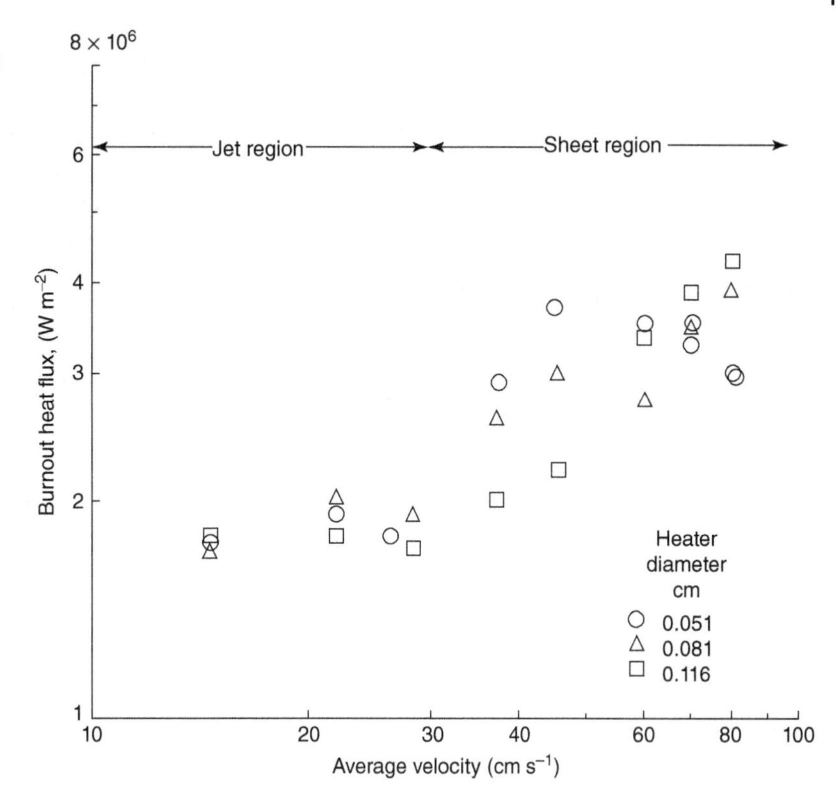

Figure 6.4.1 Effect of velocity of water flowing across single cylinders on CHF. Source: Cochran and Andracchio (1974).

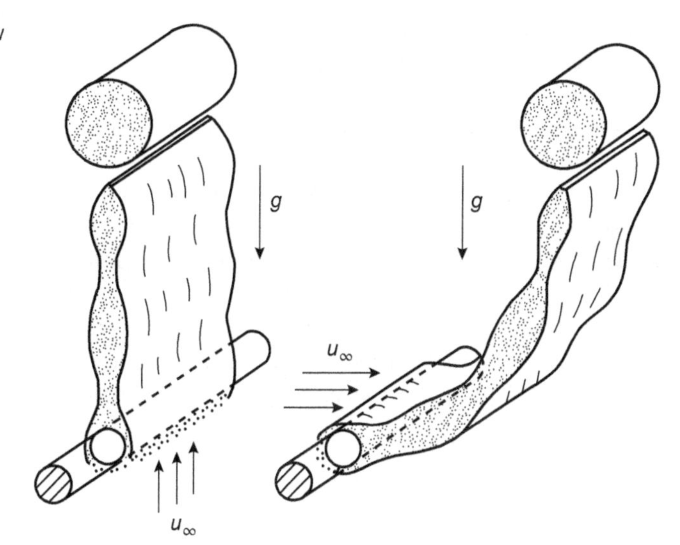

Figure 6.4.2 Vapor removal pattern during upflow and side flow on a heated cylinder. Source: Lienhard (1988). © 1988 ASME.

The CHF after the second transition is represented by the following equation:

$$\frac{\pi q_c}{\rho_g i_{fg} u_\infty} - \frac{1}{169}\left(\frac{\rho_f}{\rho_g}\right)^{3/4} = \frac{1}{19.2}\left(\frac{\rho_f}{\rho_g}\right)^{0.5} \frac{1}{We_{GT}^{1/3}} \tag{6.4.6}$$

These correlations were developed from data for water, methanol, isopropanol, and R-113, from four sources. R′ was from 0.174 to 0.888.

Hasan et al. (1981) measured CHF with vertical upflow and downflow to investigate the influence of gravity. The cylindrical heaters were of 14, 20, and 24 gage (0.5–1.6 mm diameter) nichrome wires, 13 cm long. Fluids were methanol and isopropanol. They concluded that the influence of gravity is negligible when $G^* \geq 10$, where

$$G^* = \left[\frac{u_\infty^4 \rho_f^2}{g(\rho_f - \rho_g)\sigma}\right]^{1/4} \tag{6.4.7}$$

For $G^* \geq 10$, they give the following equation for CHF:

$$\frac{\pi q_c}{\rho_g i_{fg} u_\infty} = 0.000\,919\left(1 + \frac{16.3}{We_{LT}^{1/3}}\right) \tag{6.4.8}$$

We_{LT} is defined as

$$We_{LT} = \frac{G^2 D}{\rho_f \sigma} \tag{6.4.9}$$

Equation (6.4.8) was shown to be in satisfactory agreement with data for water, methanol, isopropanol, and R-113 from several sources.

Harmura and Katto (1983) derived the following equation for CHF at zero quality assuming sheet like flow of vapor:

$$\frac{q_{co}}{G i_{fg}} = 0.151\left(\frac{\rho_g}{\rho_f}\right)^{0.463}\left(\frac{\sigma \rho_f}{G^2 D}\right)^{1/3} \tag{6.4.10}$$

This equation gave satisfactory agreement with the data of Vliet and Leppert (1964) for water on a 3.18 mm cylinder as well as the data of Yilmaz and Westwater (1980) for R-113 on a 6 mm cylinder. Both data were for atmospheric pressure.

Sadasivan and Lienhard (1986) measured CHF with liquid upflow as well as side flow on a cylinder. They found that there was negligible difference between the CHF in the two cases. The reason given for it was that gravity affects burnout by thinning the vapor sheet and burnout occurs at the same sheet thickness in either case. Figure 6.4.2 shows the vapor removal pattern in the two cases.

Yao and Hwang (1989) measured CHF with R-113 flowing across a cylinder of 19.1 mm OD, 61.6 mm long. They also performed tests with this tube in an unheated in-line tube bundle. Their data for single cylinder near zero quality was about 30% lower than the Katto and Harmura correlation, Eq. (6.4.10). They attributed this to the difference in the width of the total flow area. They gave the following correlation:

$$\frac{q_c}{q_{c,KH}} = \left[1 - \left(\frac{D}{W}\right)^{0.86}\right] \tag{6.4.11}$$

W is the width of the passage in which the cylinder is located and $q_{c,KH}$ is the CHF predicted by the Katto–Harmura correlation, Eq. (6.4.10). This agreed with the single cylinder data of McKee and Bell (1969), Sun and Lienhard (1970), and Katto and Harmura (1983). It also agreed with their own data for tube in a bundle. For the bundle, W is $(2 \times P)$, where P is the tube pitch. The data included D/W from near 0 to 0.33.

6.4.2 Horizontal Tube Bundles

Bundles of horizontal tubes with upward crossflow are discussed herein.

Cumo et al. (1980) performed tests on a bundle of nine tubes in a triangular arrangement. Tube diameter was 13.6 mm and pitch was 17 mm. Only the central tube of the bundle was heated. Fluid was R-12 at a pressure of 7.85 bar. They found that the CHF increased with two-phase velocity by up to 20% above pool boiling CHF. They correlated their data by the following equation:

$$\frac{q_c}{q_{c,pb}} = \left(1 + 0.002\frac{i_{fg}\rho_g G^2}{q_{c,pb}^2 \rho_{TP}}\right)^{1/2} \tag{6.4.12}$$

The pool boiling CHF, $q_{c,pb}$, is calculated by the hydrodynamic model as given in Chapter 3. The two- phase density ρ_{TP} is calculated by the homogeneous model. The mass velocity G is based on the narrowest gap between tubes. Figure 6.4.3 shows the comparison of this correlation with their data.

Dykas and Jensen (1992) and Leroux and Jensen (1992) measured CHF in tube bundles with R-113. The bundles were in-line square arrays (5×27) with tube pitch-to-diameter ratios of $P/D = 1.30$ and 1.70 and an equilateral triangular array (5×27) with $P/D = 1.30$. CHF data were obtained from a single tube in the center column of the 25th row from the bottom of the bundles. Pressure was 1.5 and 5 bars quality from 0% to 70% and mass fluxes (based on the minimum flow area) of 50–500 kg m^{-2} s^{-1}. Figure 6.4.4 shows some of the data of Leroux and Jensen (1992). CHF is seen to increase with flow rate.

Jensen and Tang (1994) have given a correlation for CHF of tubes in horizontal tube bundles. They have defined three regions of CHF. For in-line tube bundles, the relations are as follows:

The boundary between Regions 1 and 2:

$$x_{1-2} = 0.137 B^{0.369} \tag{6.4.13}$$

The boundary between Regions 2 and 3:

$$x_{2-3} = 0.354 B^{0.075} \tag{6.4.14}$$

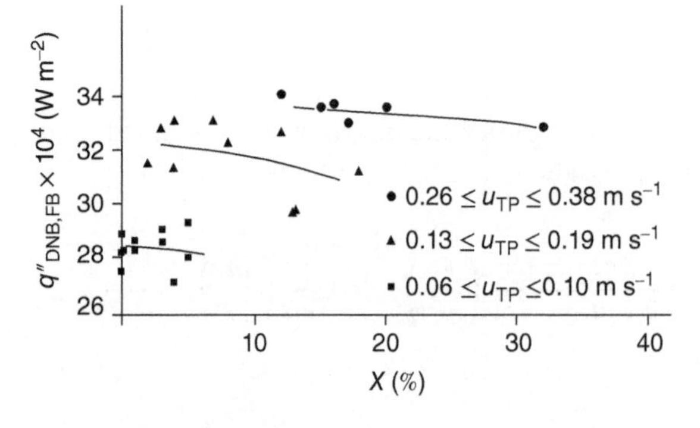

Figure 6.4.3 Comparison of the correlation of Cumo et al. for CHF in tube bundles with their test data. Source: Cumo et al. (1980). ©1980 Taylor & Francis.

For the staggered tube bundle, the following are the relations:

The boundary between Regions 1 and 2:

$$x_{1-2} = 0.242B^{0.369} \tag{6.4.15}$$

The boundary between Regions 2 and 3:

$$x_{2-3} = 0.432B^{0.0.098} \tag{6.4.16}$$

B is defined as

$$B = \frac{g\rho_{TP}(\rho_f - \rho_g)D}{G^2} \tag{6.4.17}$$

where ρ_{TP} is the homogeneous two-phase density, x_{1-2} is the quality at the boundary between regimes 1 and 2, and x_{2-3} is that at the boundary between regimes 2 and 3. G is based on the minimum flow area.

Their correlation for the in-line bundle is given by the following equations:

Region 1:

$$\frac{q_c}{q_{c,pb}} = \exp\left[\left(0.126 - \frac{1}{\Psi^{0.193}}\right)\left(\frac{P}{D}\right)^{-0.651}\right] \tag{6.4.18}$$

Region 3:

$$Bo = 2.04 \times 10^{-5} B^{0.429} Re_{LT}^{0.0136}\left(\frac{P}{D}\right)^{-0.651} \tag{6.4.19}$$

The correlation for staggered tube bundles is given by the following equations:

For in-line bundles:

Region 1:

$$\frac{q_c}{q_{c,pb}} = \exp\left(-0.0322 - \frac{10.1}{\Psi^{0.585}}\right) \tag{6.4.20}$$

Region 2:

$$Bo = 1.97 \times 10^{-5} B^{0.165} Re_{LT}^{0.0858} \tag{6.4.21}$$

q_{pb} is the CHF during pool boiling by the hydrodynamic theory (see Chapter 3) and G is the mass flux based on the narrowest gap between tubes. The parameter ψ is defined as

$$\Psi = \left(\frac{D\rho_{TP}}{\mu_f}\right)\left[\frac{\sigma g(\rho_f - \rho_g)}{\rho_f^2}\right]^{1/4} \tag{6.4.22}$$

The CHF in Region 2 is obtained by linear interpolation between the values of CHF at the transitions between Regions 1 and 2 and Regions 2 and 3. Thus $q_{c,1}$ and $q_{c,3}$ and the CHF at x_{1-2} and x_{2-3}, respectively, are calculated by the above equations. Then q_c, the CHF in Region 2 at quality x, is calculated by the following relation:

$$q_c = q_{c,1} + \frac{(q_{c,3} - q_{c,1})(x - x_{1-2})}{(x_{2-3} - x_{1-2})} \tag{6.4.23}$$

This correlation was in excellent agreement with the data of Dykas and Jensen (1992) and Leroux and Jensen (1992). However, it did not agree with the data of Cumo et al. (1980) in a 3 × 3 bundle. They were unable to explain this lack of agreement though noted that the method of determining the occurrence of CHF used by Cumo et al. was different from their own. Figure 6.4.4 shows some of the data of Leroux and Jensen.

Khalil (1982) measured CHF of electrically heated tube bundles placed in a pool of helium. The CHF of a single tube was in satisfactory agreement with the predictions of the hydrodynamic model of pool boiling CHF (Chapter 3). The CHF of bundles was the same as that of a single tube as long as the gap between tubes was greater than the bubble diameter given by the Fritz correlation, Eq. (3.2.6). It decreased when gap became smaller as shown in Figure 6.4.5.

Hasan et al. (1982) studied the effect of unheated cylinders near a heated cylinder. They found that there was no

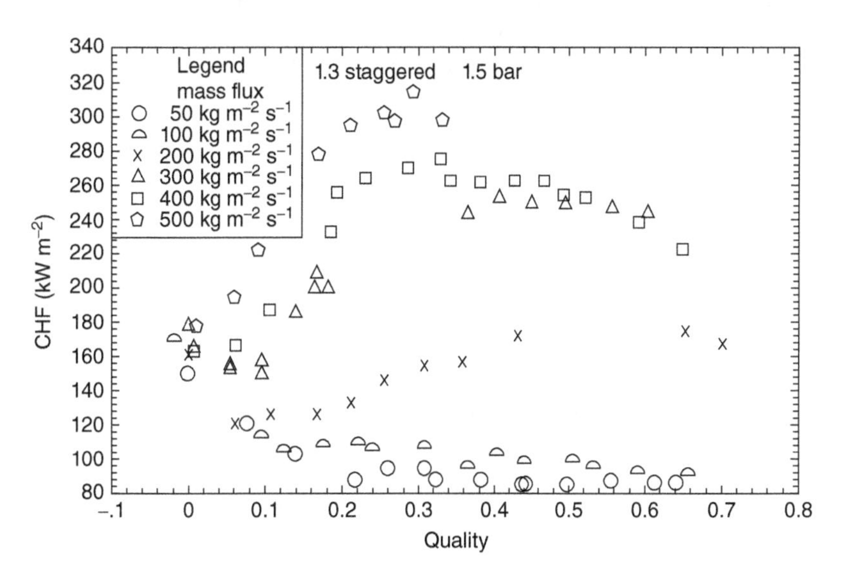

Figure 6.4.4 Effect of mass flux on CHF in a horizontal tube bundle with $P/D = 1.3$. Source: Leroux and Jensen (1992). ©1992 ASME.

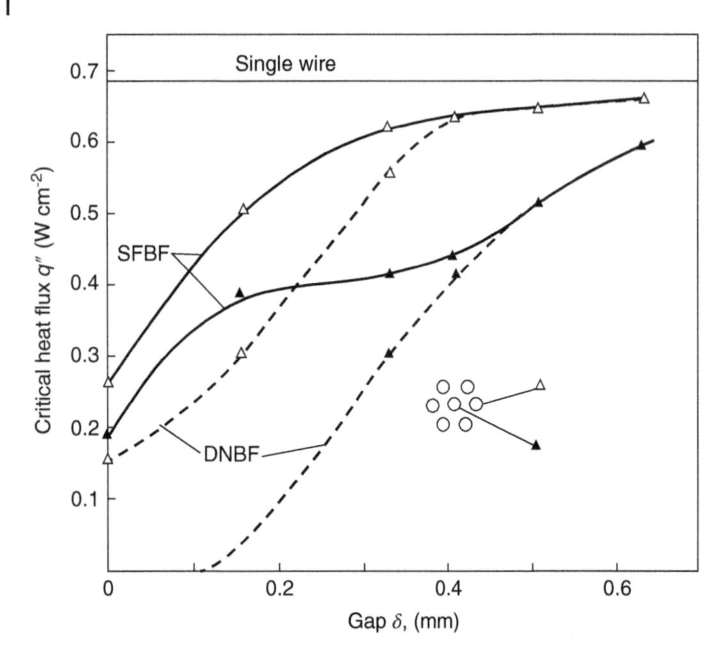

Figure 6.4.5 Effect of gap on the departure from nucleate boiling (DNBF) and transition to film boiling (SFBF) for horizontal bundles ($D = 1.25$ mm) cooled by helium. Source: Khalil (1982). © 1982 Elsevier.

effect on CHF if $(P/D) > 4$. At $(P/D) < 4$, the unheated cylinder had effect only if placed directly upstream of the heated cylinder. In that position, CHF was reduced severely, even as low as 10% of single tube CHF. They have given a graphical correlation for the effect of P/D. Their correlating curve can be represented by the following equation fitted to it by the present author:

$$\frac{q_c}{\rho_f i_{fg} U_\infty} \times 10^4 = 0.264 \left(\frac{P}{D} \right)^{2.36} \tag{6.4.24}$$

U_∞ is the velocity of the liquid approaching the bundle. This equation is based on data for P/D from 2.14 to 3.8. It should not be extrapolated to lower or higher P/D.

Palen and Small (1964) studied data from many large kettle reboilers with in-line and staggered tube arrangements. Based on these data, they have given correlations for bundle average CHF. The correlation of Palen and Small is shown in Figure 6.4.6. For $\phi \le 0.06$, it is expressed by the following equation:

$$q_c = 176 \psi \phi \tag{6.4.25}$$

ϕ is the factor for tube density in the bundle and ψ is a fluid property factor.

$$\phi = \frac{D_b L}{\pi D_o N L} = 0.359 \left(\frac{P}{D_o} \right) \sqrt{\frac{\text{Sin}\alpha}{N}} \tag{6.4.26}$$

D_b is the bundle diameter, L the tube length, D_o the outside diameter of tubes, P is the tube pitch, N the number of tubes, and α the tube layout angle.

$$\psi = \rho_g i_{fg} \left[\frac{g\sigma(\rho_f - \rho_g)}{\rho_g^2} \right]^{0.25} \tag{6.4.27}$$

Density ρ is in lb ft^{-3}, g is in ft s^{-2}, i_{fg} is in Btu lb^{-1}, q_c is in Btu h^{-1} ft^{-2}, and σ is in lbf ft^{-1}. They recommend as safety factor multiplying the calculated CHF by 0.7. This correlation was based on data for large kettle reboilers using hydrocarbons and their mixtures. Equation (6.4.24) holds for $\phi \le 0.06$. For a single tube, $\phi = 1/\pi$. For ϕ between 0.06 and $1/\pi$, the curves in Figure 6.4.6 apply.

To gain insight into the Palen and Small correlation, Eq. (6.4.25) is put in the following dimensionless form:

$$\frac{q_c}{q_{c,\text{Zuber}}} = 6.8 \left(\frac{D_b L}{\pi D_o N L} \right) \tag{6.4.28}$$

where $q_{c,\text{Zuber}}$ is the CHF given by the Zuber formula, Eq. (3.3.5). This relation shows that the bundle average bundle CHF is always lower than the CHF on a single tube. For a bundle of fixed diameter, the bundle average CHF decreases with increasing number of tubes. Increasing the number of tubes decreases the gap between tubes. As discussed earlier, other researchers have also reported decrease in CHF with decreasing gap between tubes. See, for example, the data of Khalil (1982) in Figure 6.4.5.

Palen et al. (1972) studied data from many more kettle reboilers. The fluids included straight chain hydrocarbons, aromatics, oxygenated hydrocarbons, OH group hydrocarbons, and R-113. While they have not reported a comparison with the Palen and Small correlation, the results shown in their paper appear to be in qualitative and quantitative agreement with it.

6.4.2.1 Recommendation

There is no well-verified general method for predicting CHF on individual tubes, and the results of various studies are often conflicting. The correlations for individual tubes

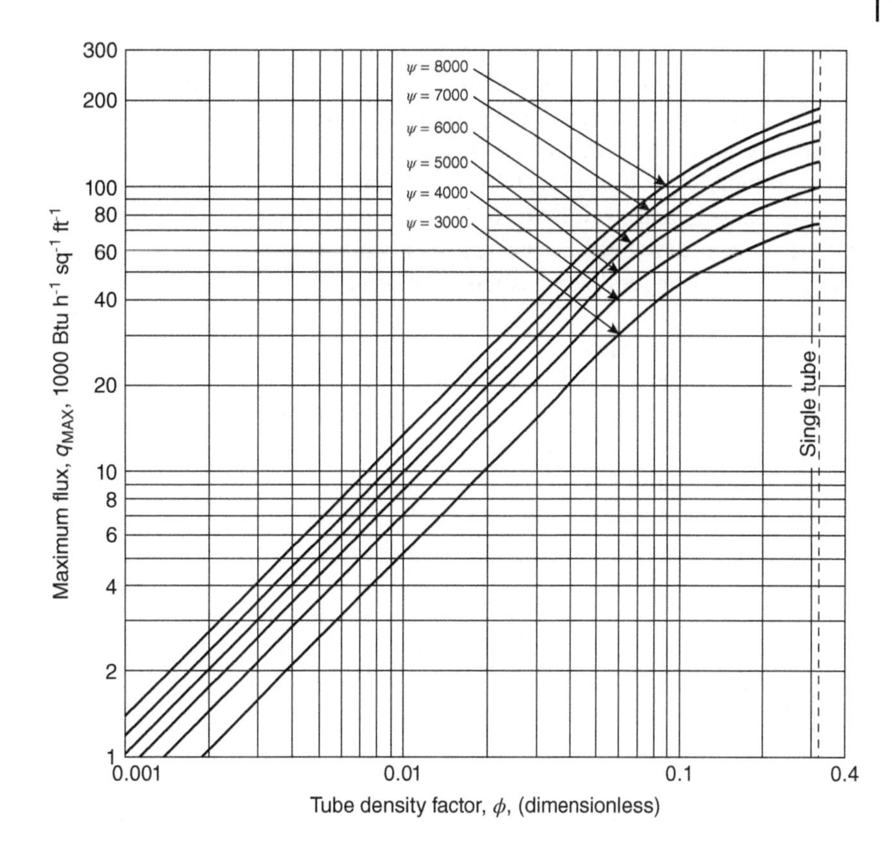

Figure 6.4.6 Correlation of Palen and Small (1964) for bundle average CHF. Source: Palen and Small (1964).

in bundles given in the foregoing should be used only in their verified range. The bundle average CHF correlation of Palen and Small may be used for kettle reboilers with various types of hydrocarbons, refrigerants, and their mixtures with glide less than 3 °C.

6.4.3 Vertical Tube/Rod Bundles

In this section, vertical tube/rod bundles with upflow are discussed. The most notable examples of such bundles are the light water-cooled nuclear reactors of the pressurized water reactors (PWRs) and BWRs types, which consist of nuclear fuel rods spaced from each other by grid type spacers or spirally wrapped wires. As failure of nuclear reactors can have disastrous consequences, a tremendous amount of research has been done to measure and predict CHF in them. The various methods of prediction include mixed flow analyses, subchannel analyses, and phenomenological analyses. These are discussed in the following.

6.4.3.1 Mixed Flow Analyses

In this approach, the variations of flow across the channel are ignored, and the conditions along each horizontal section are considered to be uniform. An example of this approach is the CISE correlation of Bertoletti et al. (1965), Eq. (6.3.5). These authors compared this correlation to data for rod bundles from many sources. These included bundles with uniform and non-uniform heating, 90% of the data were predicted within ±25%. They note that the spacers may increase (discrete spacers) or decrease (wrapped wire spacers) CHF but their effect is comparatively small unless they are very numerous.

Barnett (1966, 1968) showed that his correlation for annuli, Eqs. (6.3.1)–(6.3.4), can be applied to rod bundles by defining the diameters as follows. The rod diameter D_{rod} is used as the diameter of the inner tube of annulus. For the diameter D_o of this equivalent annulus, the following relation is used:

$$D_o = [D_{rod}(D_{rod} + D_{HP,*})]^{0.5} \tag{6.4.29}$$

where

$$D_{HP,*} = \frac{4 \times (\text{flow area})}{S \times (\text{heated rod perimeter})} \tag{6.4.30}$$

$$S = \sum \frac{q}{q_{max}} \tag{6.4.31}$$

where q is the heat flux on a rod and q_{max} is the maximum heat flux in the bundle. If all the rods have the same heat flux, $S = 1$ and $D_{HP*} = D_{HP}$. This method was shown to be in excellent agreement with a wide variety of data.

Other mixed flow correlations that have been shown to agree with a wide range of data include those by Macbeth (1964), Becker (1966), and Bowring (1977). Bowring showed that his correlation was significantly more accurate than the Macbeth and Barnett correlations.

Reactor manufacturers have developed correlations that are applicable to their own reactors. Among such correlations are those of Westinghouse (Tong 1967) and General Electric (Janssen and Levy 1962; Roy 1966). Correlations have also been developed by the reactor manufacturers Babcock and Wilcox, and Combustion Engineering.

6.4.3.2 Subchannel Analysis

In this approach, the rod bundle is divided into a number of parallel subchannels. Interchange of mass, heat, and momentum between the subchannels is allowed. Using the mass, momentum, and energy conservation equations, local quality and mass flow rate at each location is calculated. Many computer codes have been developed to calculate the fluid distribution and conditions in subchannels. The best known among them is the COBRA code that has many versions. A recent code is VIPRE-01. Knowing the local quality and mass flux, empirical correlations can be used to determine CHF at any location in a subchannels. A variety of subchannel configurations have been used in different analyses. For the square lattice in a nuclear reactor, the subchannels are formed by straight lines joining the centers of four adjacent rods. Gaspari et al. (1970, 1974) used rod-centered subchannels formed by lines equidistant from the rod and the nearest surface that may be another rod or the wall of the reactor vessel as seen in Figure 6.4.7. They showed good agreement with the CISE correlation using these subchannel configuration.

Evangelisi et al. (1972) analyzed data for 9 and 16 rod bundles from several sources. The bundles were divided into subchannels similar to those shown in Figure 6.4.7. Correlations of Macbeth (1964), Becker (1966), Barnett (1966, 1968), and Bertoletti et al. (1965) were applied to the subchannels. The Becker correlation gave the best agreement. Bertolleti et al. was the next best, followed by the other two.

Critical heat flux correlations have also been developed for subchannel analysis. These include the WSC-2 correlation given by Bowring (1979), KfK correlation of Dalle Donne (1991), and the EPRI correlation by Fighetti and Reddy (1983). The KfK correlation is an adaptation of the Bowring (1979) WSC-2 correlation.

Cheng (2005) performed tests with R-12 in a 37-rod bundle with $P/D = 1.178$. The data were converted to water equivalent using Ahmad's fluid-to-fluid modeling technique. Subchannel flow conditions were determined using the subchannel analysis code COBRA-IV-TUBS. The KfK and EPRI correlations underpredicted the data by about 25%. On the other hand, the data for the bundle were in good agreement with data for R-12 in a tube whose diameter was close to the hydraulic diameter of the rod bundle.

Recent works on subchannel analysis, CFD analysis, and other topics related to CHF in nuclear reactors have been reviewed by Yang et al. (2019).

6.4.3.3 Phenomenological Analyses

In this approach, models based on the physical phenomena are analyzed. The analysis of film dryout CHF during annular flow in tubes by Whalley et al. (1974) was discussed in Section 6.2.2.1. Whalley (1976, 1978) extended this analysis

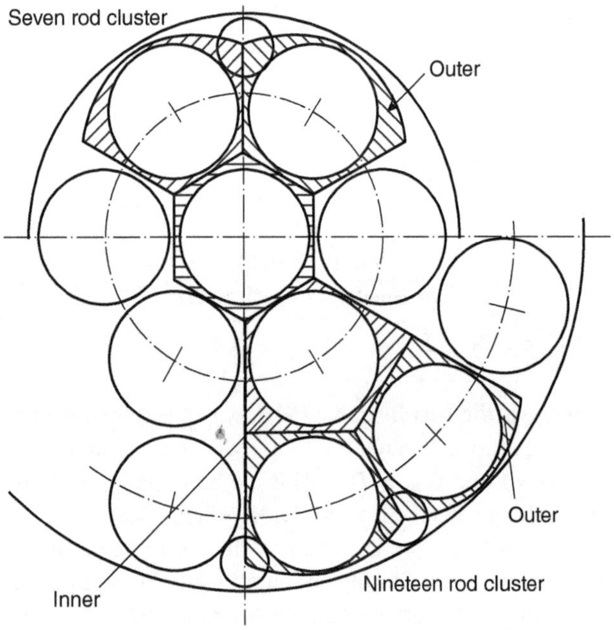

Figure 6.4.7 Rod-centered subchannels according to Gaspari et al. Source: Gaspari et al. (1970).

to annuli and tube bundles. Good agreement with data was reported.

Weisman and Ying (1985) extended the methodology of Weisman and Pei (1983) for DNB in tubes to rod bundles. Subchannel conditions were calculated by the COBRA code. Good agreement was found with data for uniform and non-uniform heat flux.

6.4.4 Falling Films on Vertical Surfaces

Ueda et al. (1981) studied liquid films falling down on outer surface of an electrically heated vertical stainless steel tube 8 mm diameter, 180 mm long. It was enclosed in a glass tube 70 mm diameter. The liquids used were water, R-11, and R-113 at atmospheric pressure, slightly subcooled. Their test data is shown in Figure 6.4.8 together with the data of Fujita and Ueda (1978) on a 16 mm diameter tube. Based on these data and visual observations, they identify three regimes of CHF. In Regime I, film flow rate decreases with increasing heat flux. This type of CHF occurs when film flow rate is less than the minimum needed to wet the entire surface. It is somewhat similar to the dryout type CHF in annular flow. In Regime II, CHF increases with increasing film flow rate. With increasing heat flux, flow near the bottom gets distorted. A large dry patch forms near the bottom of the tube in a thinned film area that causes a large increase in wall temperature. In the region of type III, there was a considerable amount of liquid flowing at the exit end. The film flowed down covering the tube periphery almost entirely. When the heat flux was increased near the critical value, the main part of the film on the exit end of the heating section appeared to be separated from the heating surface.

For the upper limit of velocity u_c at the boundary of Regime I, they gave the following formula:

$$\rho_f u_c^2 y_c / 2 = \sigma(1 - \cos\theta) \tag{6.4.32}$$

where u_c is the mean film velocity at the CHF location, y_c is the mean film thickness at the CHF location, and θ is the contact angle assumed to be 15°.

The boundary between Regimes II and III is defined by the following equation:

$$\left(\frac{\rho_g}{\rho_f}\right)^{0.78} \left(\frac{u_c^2 \rho_f}{\sigma}\right)^{-0.43} = 4.44 \times 10^{-4} (y_c^{0.1} L^{0.33}) \tag{6.4.33}$$

L is the length of the tube up to the CHF location.

Critical heat flux in Regime II is given by the following equation:

$$\frac{q_c}{i_{fg} \rho_f u_c} = 6 \times 10^{-6} \left(\frac{\rho_g}{\rho_f}\right)^{-0.7} \left(\frac{\rho_f y_c u_c^2}{\sigma}\right)^{0.1} \tag{6.4.34}$$

For CHF in Regime III, the following equation was given:

$$\frac{q_c}{i_{fg} \rho_f u_c} = 0.0135 \left(\frac{\rho_g}{\rho_f}\right)^{0.08} \left(\frac{\rho_f L u_c^2}{\sigma}\right)^{-0.33} \tag{6.4.35}$$

Mudawwar et al. (1987) studied CHF of falling films of FC-72 on vertical copper plates 25.4 mm wide, with lengths of 12.7, 25.4, 63.5, and 127 mm. Tests were done at atmospheric pressure and subcooling did not exceed 6 °C. They concluded that CHF in flowing liquid films is triggered by the Helmholtz instability that collapses the vapor jets. This instability separates the main part of the liquid film, leaving a thin liquid layer in contact with the wall. Burnout occurs when the latent heat of the liquid entering this layer

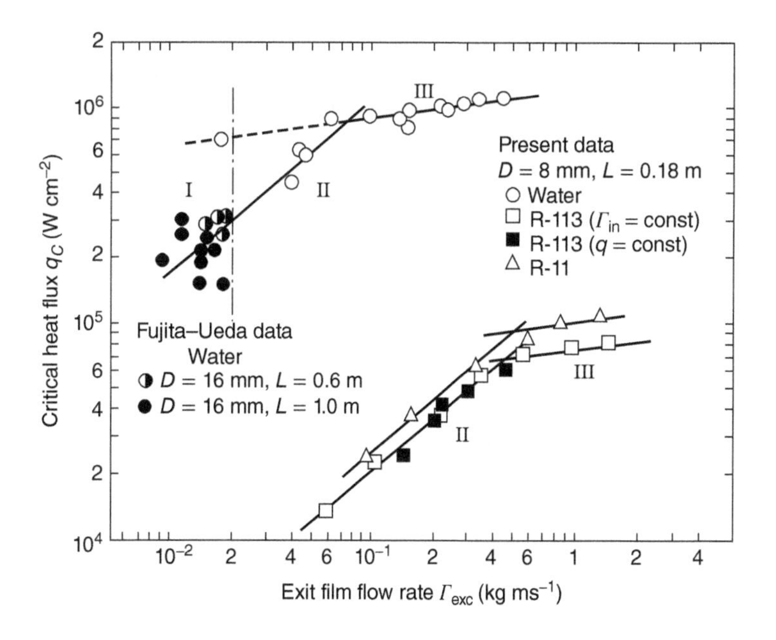

Figure 6.4.8 Critical heat flux vs. exit film flowrate for falling films on vertical tubes. Source: Ueda et al. (1981). ©1981 Elsevier.

is less than the total heat supplied at the wall. Based on these mechanisms and some empirical adjustments, they gave the following correlation of their data:

$$\frac{q_c}{i_{fg}\rho_g U} = 0.121\left(\frac{\rho_g}{\rho_f}\right)^{-2/3}\left(\frac{\sigma}{\rho_f U^2 L}\right)^{0.42}\left[1+\frac{C_{pf}\Delta T_{SC}}{i_{fg}}\right]^{1/3}\left[1+0.16\frac{C_{pf}\rho_f\Delta T_{SC}}{i_{fg}}\right]^{2/3} \tag{6.4.36}$$

U is the mean velocity of liquid film and L is the height of plate. This correlation agreed well with their own data.

6.4.5 Flow Parallel to a Flat Plate

Haramura and Katto (1983) analyzed the case of a saturated liquid flowing upward parallel to a vertical flat plate. Vapor stems from the surface maintain a continuous vapor blanket over the liquid while the liquid flows under the vapor blanket. The thickness of the liquid film decreases along the length due to evaporation. It is assumed that CHF occurs when the heat being transferred from the heated surface just balances with latent heat of the total evaporation of the liquid flowing into the liquid film. This leads to the following equation for CHF:

$$\frac{q_c}{Gi_{fg}} = 0.175\left(\frac{\rho_g}{\rho_f}\right)^{0.467}\left(1+\frac{\rho_g}{\rho_f}\right)^{1/3}\left(\frac{\sigma\rho_f}{G^2 L}\right)^{1/3} \tag{6.4.37}$$

L is the length of plate and $G = U\rho_f$, where U is the upstream velocity. This equation shows satisfactory agreement with the data of Katto and Kurata (1980) obtained with saturated water and R-113 at atmospheric pressure on uniformly heated plates of 10, 15, and 20 mm lengths.

Katto and Kurata (1980) gave the following empirical correlation of their data:

$$\frac{q_c}{Gi_{fg}} = 0.186\left(\frac{\rho_g}{\rho_f}\right)^{0.559}\left(\frac{\sigma\rho_f}{G^2 L}\right)^{0.264} \tag{6.4.38}$$

The predictions of Eqs. (6.4.37) and (6.4.38) are quite close at the low pressures that the data were obtained.

The velocity U in the data of Katto and Kurata varied from 1.3 to 9.1 m s^{-1}, and the plates were vertical with flow upwards. Equations. (6.4.37) and (6.4.38) are likely to be applicable to other orientations at higher velocities. The data of Zhang et al. (2002) discussed in Section 6.2.2.4.3 shows that at low velocities, CHF tends to be close to pool boiling CHF.

6.4.6 Helical Coils

Cumo et al. (1972) measured CHF in a straight vertical tube and a coil with vertical axis. Both were made from tubes 4.75 mm ID and were 2 m long. The diameter of coil was 180 mm and the pitch 70 mm. The test fluid was R-12, which entered from the bottom of both test sections. Tests were done over a wide range of flow rates and pressures up to and beyond critical. Figure 6.4.9 shows some of their data. It is seen that for identical flow rate, pressure, and inlet quality, CHF in the coil occurred at much higher quality. This may be attributed to the mixing action of centrifugal forces in the coil. Because of the higher critical quality, much more power can be fed to a coil than a straight tube. It is also seen that CHF occurred at a lower quality on the inner surface (that closer to the coil axis) of the coil than at the outer surface. This is because centrifugal force pushes the liquid to the outer periphery and hence the thinner liquid film at the inner periphery dries out earlier. The difference between the inner and outer surface CHF disappears at higher flow rates though the wall temperature of the inner surface remains higher in the post-CHF region.

Similar behavior was also reported by other researchers. Jensen (1980) and Jensen and Bergles (1981, 1982) tested a straight horizontal tube as well as a number of vertical coils with the curvature ratio (D_c/D_t) ranging from 15 to 55. The fluid was R-113. Some of the tests were done in which heat flux was higher on one side of the coil. They found that subcooled CHF in coils was lower than that in straight tube. For the saturated region, they found that CHF in coils was higher than in straight tubes. They presented correlations of their own data.

Hardik and Prabhu (2017) measured CHF on six vertical electrically heated coils. Tube diameters varied from 6 to 10 mm and (D_c/D_t) varied from 14 to 58. In their tests, CHF started on the inner surface of coil and then spread to the entire circumference until the tube burst. They compared their measured CHF in coils to those predicted by the lookup table for straight tubes, Groeneveld et al. (2007), using DSM. Based on this comparison, they concluded that in the low quality region, CHF in a helical coil is lower than that in the straight vertical tube. However, for high quality region, CHF in a helical coil is higher than the straight vertical tube CHF. They compared their test data to a number of correlations for CHF in coils but none of them gave satisfactory agreement. Hardik and Prabhu (2018) measured CHF in vertical coils with R-123. They compared their data with many correlations for coils and for straight tubes. None of the correlations for coils was found satisfactory. However straight tube correlations of Shah (1987, 2015a) and Zhang et al. (2006) gave reasonable agreement with their data.

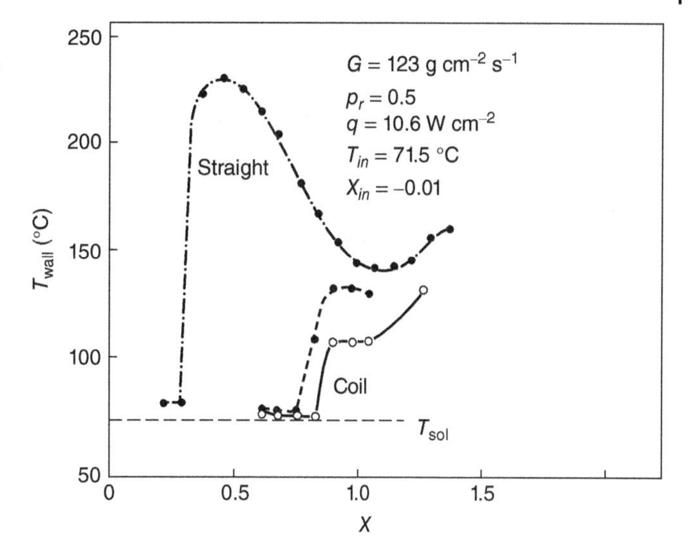

Figure 6.4.9 Variation of wall temperatures with quality X in a straight tube and a coil with the same tube diameter, flow rate, and inlet quality. Source: Cumo et al. (1972). © 1972 Elsevier.

A number of correlations have been proposed for CHF in coils.

Jensen and Bergles (1981) gave the following correlation based on their own data.

For subcooled CHF:

$$\text{When } \frac{Fr_{\text{LT}}}{(D_t/D_c)} > 10, \quad \frac{q_{c,\text{coil}}}{q_{c,\text{straight}}} = 0.769 Fr_{\text{LT}}^{-0.26} \tag{6.4.39}$$

$$\text{When } \frac{Fr_{\text{LT}}}{(D_t/D_c)} \le 10, \quad \frac{q_{c,\text{coil}}}{q_{c,\text{straight}}} = 1 \tag{6.4.40}$$

They gave a correlation to fit their data for straight tubes.

For quality CHF in coils:

For $G > 950$ kg m^{-2} s^{-1},

$$Bo_c = 17\,126 Re_{\text{GT}}^{-1.143} x_c^{-0.436} (D_t/D_c)^{0.31} \tag{6.4.41}$$

For $G \le 950$ kg m^{-2} s^{-1},

$$Bo_c = 0.000\,004\,09 Re_{\text{GT}}^{0.5} x_c^{-0.46} (D_t/D_c)^{0.17} \tag{6.4.42}$$

Ma et al. (1995) gave the following correlation based on their own data:

$$q_c = A - Bx_c \tag{6.4.43}$$

where:

For $G \le 440$ kg m^{-2} s^{-1},

$$A = 42.3(D_t/D_c)^{0.11}(G - 179.2) \tag{6.4.44}$$

$$B = 38.8(D_t/D_c)^{0.06}(G - 184.7) \tag{6.4.45}$$

For $G > 440$ kg m^{-2} s^{-1},

$$A = 25.7(D_t/D_c)^{0.11}(G - 16.2) \tag{6.4.46}$$

$$B = 21.6(D_t/D_c)^{0.06}(G + 11.3) \tag{6.4.47}$$

Kaji et al. (1998) gave the following correlation of their own data:

$$Bo_c = 0.34\left(\frac{\sigma \rho_f}{G^2 L_c}\right)^{0.043} \frac{L_{\text{cr}}}{D_t}\left\{1 + 3.5\left(\frac{D_t}{D_c}\right)^{1.2}\right\} \tag{6.4.48}$$

Hardik and Prabhu (2017) gave the following correlation of their own data for water:

$$\frac{q_{c,\text{coil}}}{q_{c,\text{straight tube}}} = 1.637x_c + 0.568 \tag{6.4.49}$$

The range of data on which these correlations were based is given in Table 6.4.1.

The CHF of straight tube was calculated by the lookup table of Groeneveld et al. (2007) using the DSM method.

Berthoud and Jayanti (1990) gave a correlation for critical quality in coils. Hardik and Prabhu (2017) found its predictions to be inaccurate when compared to their data.

The present author compared a number of correlations for coils as well as straight tubes with data for coils from many sources. The results are summarized in Table 6.4.2. It is seen that for coils with $D_c/D_t \ge 23$, the straight tube correlations of Shah (1987, 2015a), Zhang et al. (2006), and Katto and Ohno (1984) give fairly good agreement but give large deviations for D_c/D_t 13–19. It may be concluded that straight tube correlations can be used when $D_c/D_t \ge 23$. This conclusion is supported by the data of Chen et al. (2011) shown in Figure 6.4.10. It is seen that when D_c/D_t is greater than about 25, it has little or no effect on CHF.

It is seen in Table 6.4.2 that all the correlations for coils give large deviations considering all data. However, they were found to give good agreement with the data on which they were based.

The Hardik and Prabhu correlation, Eq. (6.4.49), was developed using the Groeneveld et al. (2007) CHF lookup

Table 6.4.1 The range of data on which some correlations for CHF in coils were based.

Correlation of	Orientation of coil axis	Fluid	D_t (mm)	D_c/D_t	L_c/D_t	p_r	G	x_{in}	x_c
Hardik and Prabhu (2017)	Vertical	Water	6.0	14	43	0.0055	92	−0.28	0.03
			10.0	58	287	0.0258	1241	0.53	0.91
Ma et al. (1995)	Horizontal	Water	10.0	13	221	0.0906	300	−0.46	0.78
			11.0	23	756	0.1813	730	0.61	0.96
Jensen and Bergles (1982)	Vertical	R-113	7.4	15	83.5	0.2801	566	−1.33	−0.499
			7.6	55	175		5474	−0.12	0.928
Kaji et al. (1998)	Vertical	R-113	10.0	16		0.11	305	−06.5	
				32			1560	−0.1	

Table 6.4.2 Summary of results of comparison of correlations with data for CHF in coils.

Coil axis orientation	No. of sources	D_c/D_t	No. of data sets	N	Deviation (%) Mean absolute Average							
					Shah Horizontal	Shah Vertical.	Zhang et al.	Katto and Ohno	Merilo	Jensen and Bergles	Ma et al.	Kaji et al.
V	4	14	6	140	54.2	54.1	41.2	63.1	372.9	183.4	7224	106.8
		19			11.1	24.9	23.7	31.8	331.2	160.6	7178	78.8
V	7	23	16	240	26.2	25.5	18.9	32.2	179.5	47.2	6267	38.8
		100			−4.3	5.6	7.9	15.7	147.1	19.7	6187	24.2
H	2	13	6	126	18.7	18.5	25.3	26.3	38.5	132.4	2048	48.3
		36			−9.1	8.5	16.4	15.2	−14.1	131.7	2013	−7.5
V and H	6	13	8	159	52.0	54.6	46.6	65.7	336.3	183.5	7097	107.2
		19			12.9	28.9	30.4	37.3	298.2	163.2	7056	80.0
V and H	8	23	16	347	23.0	21.2	17.6	27.2	134.4	70.7	4741	38.3
		100			−7.7	3.7	7.0	18.5	93.7	51.4	4673	9.2

table to calculate CHF in straight tubes by the DSM method. The present author used this correlation with straight tube CHF by the Shah (1987) correlation. It was found that all Hardik and Prabhu data were overpredicted, even those for larger curvature ratios that agree with the Shah (1987) correlation. All of the data of Hardik and Prabhu was for $Y < 10^6$ and hence in the UCC of Shah correlation which calculates CHF based on the inlet quality. Hardik and Prabhu applied the lookup table with DSM, which gives the CHF based on the critical quality. It therefore appears that the lookup table using DSM is not applicable to helical coils.

6.4.6.1 Recommendation

For coils with $D_c/D_t \geq 23$, use correlations for straight tubes: Zhang et al. (2006) for water and Shah (1987) for other fluids.

The correlations for coils listed earlier should be used only in the range of data on which they were based.

6.4.7 Spiral Wound Heat Exchangers (SWHE)

The available experimental data on CHF in spiral wound heat exchangers (SWHE) has been discussed in Section 5.4.5. No method for predicting CHF is available.

6.4.8 Rotating Liquid Film

Rotating liquid films are important in applications such as turbine blade cooling. Mudawwar et al. (1985) carried out an experimental study on heat transfer to water films flowing through a narrow passage over a heater mounted on the face of a vertical rotating disk. The variables investigated included static pressure (1–5.41 atm),

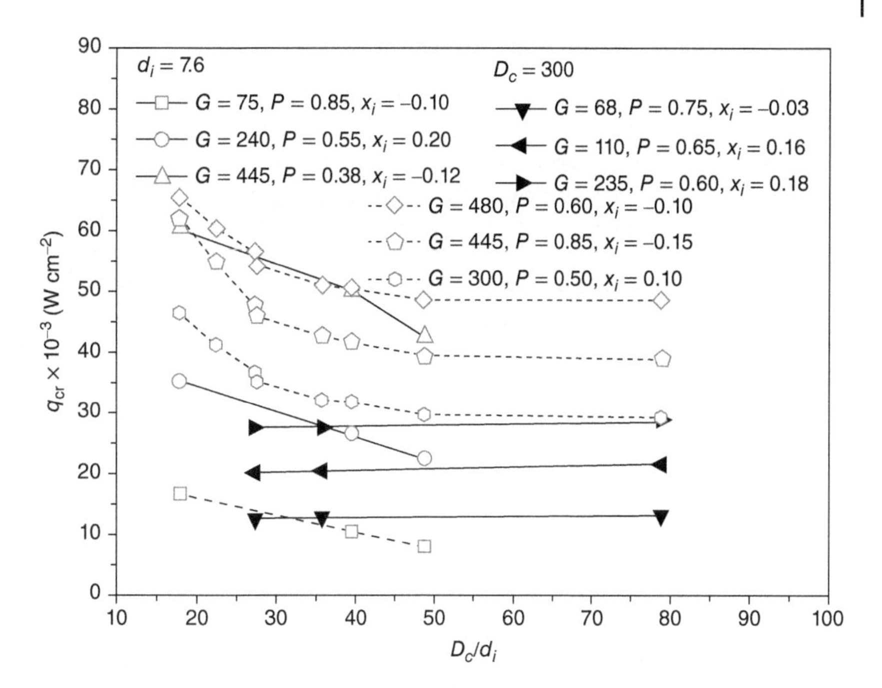

Figure 6.4.10 Effect of coil to tube diameter ratio on CHF (d_i is tube inside diameter). Fluid is R-134a. Source: Chen et al. (2011) © Elsevier.

film Reynolds number (7000–72 000), centrifugal acceleration (a/g = 36.5–460), and subcooling (8–24 °C). The test channel was mounted on a 34.3 cm OD rotating disk. They found that CHF increased with pressure as well as rotational speed. They developed a CHF model based upon the balance between the Coriolis forces and the vapor drag acting upon drops formed from the shattered liquid film. It is represented by the following equation:

$$q_c = 0.69\rho_g i_{fg} \left[\frac{(\rho_f - \rho_g)\sigma\omega(\omega^2 R v_f)^{1/3}}{\rho_g^2} \right]^{1/4} \tag{6.4.50}$$

R is the radius of rotation. This model gave excellent agreement with their data for saturated liquid. Subcooling of 24 °C increased CHF by 22%.

They fitted the following equation to their data for fully developed nucleate boiling:

$$q = 1.04 \times 10^4 (T_w - T_{SAT})^{1.75} \tag{6.4.51}$$

with q in W m^{-2} and T in °C.

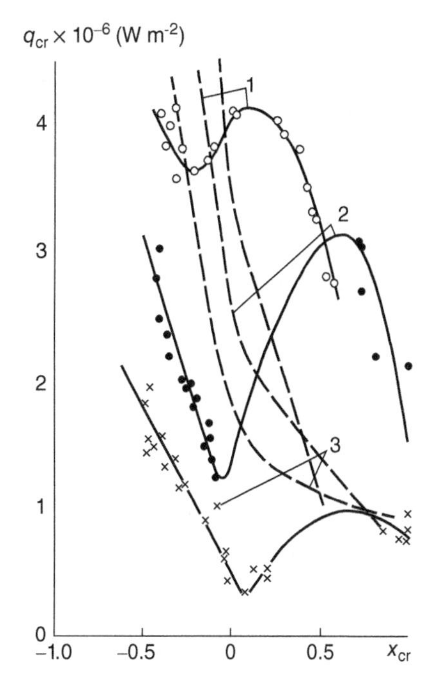

Figure 6.4.11 Critical heat flux with water in a 180° bend at various flow rates compared to CHF in straight tubes. G (kg m^{-2} s^{-1}) = 1–2000, 2–800, 3–400. Dashed lines are for straight tubes. Pressure 98 bar. Source: Miropolskiy and Pikus (1969).

6.4.9 Bends

Heat transfer during subcooled boiling in bends was discussed in Section 4.7.5. Data of Miropolskiy and Pikus (1969) on a 90° bend were shown in Figure 4.7.2. This figure also shows the sharp rise of temperature indicating occurrence of CHF. Figure 6.4.11 shows their data for CHF in a 180° bend. It is seen that for subcooled flow, CHF decreases with increasing quality. At positive qualities, CHF first increases with increasing quality, reaches a maximum, and then falls with further increase in quality. This is contrary to the trend in straight tubes in which CHF always decreases with increasing quality. They found the same trend in bends with other angles.

6.4.10 Jets Impinging on Hot Surfaces

The different types of jets and heat transfer with them have been discussed in Section 4.7.7. The CHF with jets is discussed herein. Various experimental studies and predictive techniques have been reviewed by Wolf et al. (1993) and Qiu et al. (2015), among others.

6.4.10.1 Correlations for CHF in Free Stream Jets

Starting with Monde (1980), Monde and coworkers developed a number of correlations for circular jets. Their final form was given in Monde (1987). Four regimes were identified, namely, I, V, L, and HP. The formula for the V regime is

$$\frac{q_{co}}{u\rho_{fg}i_{fg}} = 0.221\left(\frac{\rho_f}{\rho_g}\right)^{0.645} \left\{\frac{2\sigma}{\rho_f u^2(D_s - D_{jet})}\right\}^{0.343}\left(1 + \frac{D_s}{D_{jet}}\right)^{-0.364} \tag{6.4.52}$$

q_{co} is the CHF for saturated liquid jets. D_s is the diameter of disk, D_{jet} is the diameter of nozzle, and u is the velocity of jet at nozzle outlet. The V regime covers higher velocities and moderate pressures. According to Mitsutake and Monde (2003), this equation is applicable in the following range: ρ_g/ρ_f = 0.000 62–0.19, D_s/D_{jet} = 5–30. It can be applied to surface of any shape by regarding D_s as twice the distance between the stagnation point and the furthest point from it on the heated surface affected by the jet.

Katto and Yokoya (1988) analyzed the case of a vertical jet of saturated liquid impacting a heated disk normally and developed the following correlation:

$$\frac{q_{co}}{Gi_{fg}} = \left[0.0166 + 7\left(\frac{\rho_g}{\rho_f}\right)^{1.12}\right] \left[\frac{\sigma\rho_f}{G^2(D_s - D_{jet})}\frac{1}{1 + D_s/D_{jet}}\right]^m \tag{6.4.53}$$

G is the mass velocity of jet at nozzle exit. The exponent m is given by the following equations:

$$\text{For } \rho_g/\rho_f \leq 0.004\,03, \quad m = 0.374\left(\frac{\rho_g}{\rho_f}\right)^{0.0155} \tag{6.4.54}$$

$$\text{For } \rho_g/\rho_f \geq 0.004\,03, \quad m = 0.532\left(\frac{\rho_g}{\rho_f}\right)^{0.0794} \tag{6.4.55}$$

This correlation was verified with data for water, R-12, and R-113, in the following range: ρ_g/ρ_f = 0.000 624–0.189, jet velocity 0.3–60.0 m s^{-1}, disk diameter D_s = 10–60 mm, nozzle diameter D_{jet} = 0.7–4.1 mm, and D_s/D_{jet} = 3.9–53.9. The data included up-facing and down-facing disks, i.e. upward and downward jets. The same data were also compared to the correlations of Sharan and Lienhard (1985) and Monde (1985, 1987). Their RMS errors were 20.2% and 16.2%, respectively, compared to 15.7% for the Katto–Yokoya correlation. The Monde correlation equations evaluated were for his V and I regimes. The data analyzed and satisfactorily predicted by Katto and Yokoya correlation included those considered by Monde (1987) to be in V, I, L, and HP regimes. Monde and his coworkers, for example, Mitsutake and Monde (2003), have not mentioned any regimes and referred to Eq. (6.4.52) as the general correlation. Hence formulas given by Monde for regimes other than V are not given here.

Katto and Yokoya point out that their correlation will fail if the jet velocity is below a transition velocity for which they have given the following tentative relation:

$$\left(\frac{q_c}{Gi_{fg}}\right)_{\text{transition}} = K\left(\frac{D_{jet}}{D_s}\right)^2 \tag{6.4.56}$$

The factor K was given in a graphical correlation. The correlating curve can be expressed by the following equation:

$$K = \frac{0.58}{D_{jet}^{1.5}} \tag{6.4.57}$$

D_{jet} is in mm. It is applicable for D_s = 40–60 mm and $D_{jet} > 0.8$ mm.

Mahmoudi et al. (2012) performed tests with very low velocity free surface jets falling normally on horizontal surfaces. They found that at low velocities, distance between the nozzle outlet and surface has effect on CHF. This was attributed to the effects of gravity becoming important compared with jet inertia. The jet diameter decreases as it falls due to acceleration caused by gravity and hydraulic jump occurs with jet diameters smaller than the surface diameter. At the lowest flow rate of 20 cc min^{-1} from a 15 gauge needle (0.056 m s^{-1} nozzle outlet velocity), they found that at 50 mm distance between nozzle and surface, CHF approached the pool boiling CHF. At a flow rate of 60 cc min^{-1} (velocity 0.17 m s^{-1}), there was no effect of nozzle height on CHF. The minimum velocity in the data correlated by Katto and Yokoya was 0.3 m s^{-1}.

Li et al. (2014) used a rectangular free stream jet of water 12 mm × 1 mm wide impacting on a 10 mm × 1 mm wide electrically heated nickel foil. The CHF was correlated by the following equation:

$$\frac{q_{co}}{Gi_{fg}} = 0.16\left(\frac{\sigma\rho_f}{G^2W}\right)^{1/3}\left(\frac{\rho_g}{\rho_f}\right)^{1.4/3} \tag{6.4.58}$$

W is the width of jet. Jet velocity was from 4 to 43 m s^{-1}. Subcooling up to 90 K was used. The effect of subcooling was expressed by the following relation:

$$\frac{q_c}{q_{co}} = 1 + 0.26\left(\frac{\rho_g}{\rho_f}\right)^{-0.55}\left(\frac{C_{pf}\Delta T_{SC}}{i_{fg}}\right)^{0.64} \tag{6.4.59}$$

The authors state that CHF varies with contact angle and the static contact angle for the nickel surface they used was 90°. Hence the constant in Eq. (6.4.58) will be different for surfaces with different contact angles.

Several correlations have been proposed for the effect of subcooling. Monde et al. (1994) gave the following correlation:

$$\frac{q_c}{q_{co}} = \frac{1 + \sqrt{1 + 4CJa}}{2} \tag{6.4.60}$$

where

$$Ja = \left(\frac{\rho_f}{\rho_g}\right)\left(\frac{C_{pf}\Delta T_{SC}}{i_{fg}}\right) \tag{6.4.61}$$

$$C = \frac{0.95(D_{jet}/L)^2(1 + L/D_{jet})^{0.364}}{(\rho_f/\rho_g)^{0.43}(2\sigma/\rho_f u^2(L - D_{jet}))^{0.343}} \tag{6.4.62}$$

L is the length of the heater. This equation was verified with data from tests on heaters 40 and 60 mm long and 7 mm wide. Fluids were water, R-22, and R-113 at pressures of 1–26 bar, subcooling from 0 to 113 K, and jet velocity u from 5 to 34 m s^{-1}. CHF at zero subcooling was in agreement with Eq. (6.4.52) with D_s replaced by L, the length of heaters.

Nonn et al. (1988) gave the following correlation:

$$\frac{q_c}{q_{co}} = 1 + 0.952\left(\frac{C_{pf}\Delta T_{SC}}{i_{fg}}\right)^{1.414}\left(\frac{\rho_g}{\rho_f}\right)^{-0.118} \tag{6.4.63}$$

Hall et al. (2001) performed tests in which a heated copper disk of 107 mm diameter was quenched by a circular free stream jet of water at 25 °C from a nozzle of 5.1 mm diameter. Figure 6.4.12 shows their CHF data as a function of the radial distance from the stagnation point. In steady-state tests, CHF occurs at the point furthest away from the stagnation point. During these quenching tests, CHF first occurred at the stagnation point and then moved outwards as the surface cools down. Hall et al. compared their data to the Monde correlation, Eq. (6.4.52), with the effect of subcooling alternately with Eqs. (6.4.60) and (6.4.63). Good agreement was found using either of the subcooling effect formulas.

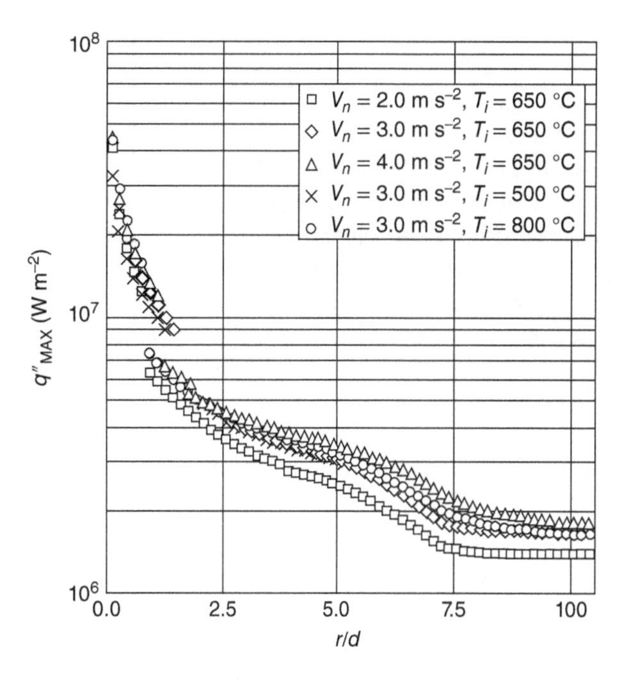

Figure 6.4.12 Critical heat flux of a disk during quenching by a free surface round jet. r is the radial distance from the stagnation point, d is the jet diameter, V_n is the jet velocity, and T_i is the initial surface temperature. Source: Hall et al. (2001). ©2001 ASME.

6.4.10.2 Effect of Contact Angle

By theoretical analysis of a circular jet, Qiu and Liu (2010) developed the following correlation for the effect of contact angle:

$$\frac{q_{c,\theta}}{q_{c,55}} = 1.49 - 0.0089\theta \tag{6.4.64}$$

where θ is the static contact angle between liquid and a solid surface, $q_{c,55}$ is the CHF when contact angle is 55°, which is the contact angle for water on smooth copper surfaces. This equation was shown to be in fairly good agreement with data for water jet from one source in which contact angle was 25° and 55°. More verification is needed to gain confidence in its generality.

Li et al. (2013) measured CHF with circular jets of water on nano surfaces with contact angles of 5°–105°. They gave the following correlation that shows the effect of contact angle θ (degree):

$$q_{co} = (0.191 - 0.0556\theta)i_{fg}(\sigma\rho_f^2)^{1/3}\left(\frac{\rho_g}{\rho_f}\right)^{1.4/3}\left(\frac{u}{D_{jet}}\right)^{1/3} \tag{6.4.65}$$

The correlations of Monde and Katto–Yokoya have been found to be in agreement with data for many fluids on many types of surfaces. The contact angle would have varied considerably in those data. Yet these correlations do not include contact angle. This suggests that the effect of contact angle predicted by Eq. (6.4.64) should be checked with data from many sources.

6.4.10.3 Multiple Jets

Monde and Mitsutake (1996) measure CHF with two or four circular water jets impacting on stainless steel rectangular plates 45–46 mm long and 15 mm wide. Subcooling was up to 80 K. They found that CHF could be satisfactorily predicted by the Monde et al. equations for single jets by using as characteristic length that of the zone dominated by each jet. Figure 6.4.13 shows the characteristic length for their two and four jet arrangements. L from this figure is used in Eq. (6.4.62) and replaces D_s in Eq. (6.4.52). Thus L is twice the distance from the stagnation point to the furthest point from it on the heated surface affected by the jet. For other jet configurations, zones of influence of individual jets should be identified in a similar way.

6.4.10.4 Effect of Heater Thickness

Mitsutake and Monde (2003) used nickel heaters of thickness 0.03, 0.05, 0.1, and 0.3 mm, 5–10 mm long, and 4 mm wide heaters with subcooled jets of water from a 2 mm diameter nozzle. Experiments were carried out at jet velocities of 5–60 m s^{-1}, a jet temperature of 20 °C, and system pressures of 0.1–1.3 MPa. The degree of subcooling was varied from 80 to 170 K with increasing system pressure. They compared their data with the Monde correlation, Eqs. (6.4.52) and (6.4.60). Heaters of thickness less than 0.3 mm were found to have lower CHF than predictions. They concluded that the Monde correlation is applicable for $\rho C_p \delta > 1$ kJ m^{-2} K^{-1}, where δ is the thickness of heater.

6.4.10.5 Confined Jets

The tests of Clark et al. (2019) on a confined jet of HFE-7100 were discussed in Section 4.7.7. While heat transfer coefficient was little affected by velocity, CHF increased linearly with velocity, rising from 21 W cm^{-2} for pool boiling to 85 W cm^{-2} at 9 m s^{-1} velocity.

6.4.10.6 Submerged Jets

Cardenas and Narayan (2012) studied CHF with submerged jets of FC-72. They compared their data to a number of correlations for free stream jets. None of them gave satisfactory agreement. They did not identify any correlation for submerged jets.

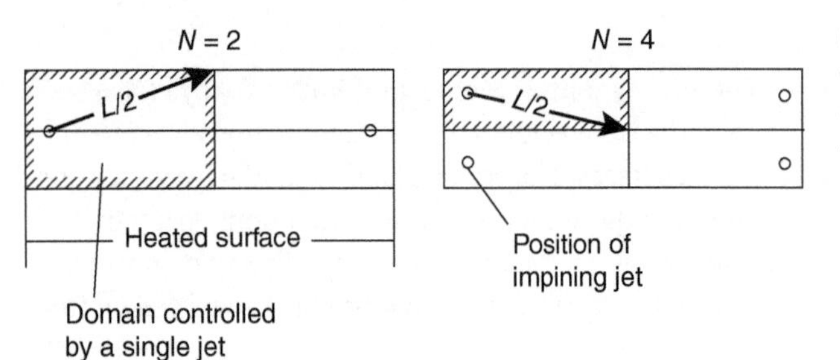

Figure 6.4.13 Location of jets and definition of characteristic length in the tests of Monde and Mitsutake (1996). *N is the number of jets.* Source: Monde and Mitsutake (1996). ©1996 ASME.

6.4.10.7 Recommendations

For free surface circular jets, calculate q_{co} by the Monde correlation, Eq. (6.4.52). For the effect of subcooling, use the Monde et al. (1994) correlation Eq. (6.4.60). For multiple jets, use the same method with characteristic length determined by the zone of influence of each jet.

6.4.11 Spray Cooling

Sprays of fine droplets are widely used for cooling hot surfaces. These are produced using air-assisted nozzles or by pressure nozzles. The former use air to assist in breaking up liquid into droplets, while the latter use only the momentum of liquid to form the droplets. Only sprays from pressure nozzles are discussed herein. Nozzles can be full-cone spray type that wet the entire impact area or of other types that wet only part of the impact area. Only the full-cone spray type are discussed herein. The impact area of such sprays is a circle.

An important parameter affecting CHF is the mean droplet diameter. The most commonly used mean diameter is D_{32}, commonly called the Sauter mean diameter of spray droplets defined as

$$D_{32} = \frac{\sum_i n_i D_i^3}{\sum_i n_i D_i^2} \tag{6.4.66}$$

where n_i is the number of droplets with diameter D_i. Many correlations for determining it have been proposed. A simple one is the following by Estes and Mudawar (1995):

$$\frac{D_{32}}{D_o} = 3.67(We_o^{1/2} Re_o)^{-0.259} \tag{6.4.67}$$

D_o is the diameter of the nozzle orifice. We_o is orifice Weber number defined as

$$We_o = \frac{\rho_g (2\Delta p/\rho_f) D_o}{\sigma} \tag{6.4.68}$$

Re_o is the orifice Reynolds number defined as

$$Re_o = \frac{\rho_g (2\Delta p/\rho_f)^{1/2} D_o}{\mu_f} \tag{6.4.69}$$

Δp is the pressure drop through the nozzle. This correlation was based on data for FC-72 and water in nozzles of diameter 0.772–1.7 mm.

Numerous correlation for prediction of CHF have been proposed. One of the more verified for sprays normal to plain flat surfaces is the following by Thiagarajan et al. (2014):

$$\frac{q_c}{\rho_g i_{fg} Q''} = 1.449\left(\frac{\rho_f}{\rho_g}\right)^{0.3}\left(\frac{\rho_f Q''^2 D_{32}}{\sigma}\right)^{-0.3371}\left(1 + 0.0058\frac{\rho_f C_{pf}\Delta T_{SC}}{\rho_g i_{fg}}\right) \tag{6.4.70}$$

Q'' is liquid flow rate divided by the area of a square enclosing the impact area of spray (which is a circle) whose sides are just touching it. q_c is also based on the area of this square. Liquid and vapor properties are evaluated at the average liquid inlet and saturation temperatures. This correlation was based on their own data for HFE-7100. It was also shown to give satisfactory agreement with data from five other sources that included dielectric fluids FC-72, FC-87, and PF-5060. It was found to fail with data for water and methanol.

Thiagarajan et al. also performed tests on a microporous copper surface. Heat transfer coefficients were much higher than on plain surface and were correlated by the following equation:

$$\frac{q_c}{\rho_g i_{fg} Q''} = 2.3\left(\frac{\rho_f}{\rho_g}\right)^{0.3}\left(\frac{\rho_f Q''^2 D_{32}}{\sigma}\right)^{-0.35}\left(1 + 0.005\frac{\rho_f C_{pf}\Delta T_{SC}}{\rho_g i_{fg}}\right) \tag{6.4.71}$$

This correlation was also in good agreement with data from two other sources for PF-5060 on enhanced surfaces.

Visaria and Mudawar (2008) have given the following correlation that was verified with data for up and down vertical sprays as well as inclined sprays on plain surfaces:

$$\frac{q_c}{\rho_g i_{fg} Q''} = 2.3\left(\frac{\rho_f}{\rho_g}\right)^{0.3}\left(\frac{\rho_f Q''^2 D_{32}}{\sigma}\right)^{-0.35}\left(1 + 0.005\frac{\rho_f C_{pf}\Delta T_{SC}}{\rho_g i_{fg}}\right) \tag{6.4.72}$$

The data included six different nozzle types and four fluids (FC-72, FC-77, water, and PF-5052). Subcooling was 15–77 K. Q'' in this correlation is based on the actual impact area of spray and q_c is for the outer periphery of the impact area.

For large surfaces, a single spray nozzle is not sufficient and therefore arrays of nozzles have to be used. Many experimental studies on arrays of nozzles have been done and many correlations have been proposed. However, none of them has had much verification beyond the data of their authors.

More information can be found in the review paper by Liang and Mudawar (2017).

6.4.12 Effect of Gravity

Effect of gravity on CHF in pool boiling was discussed in Section 3.7.1. It was noted that during pool boiling at microgravity, large bubbles form, which cause CHF to be low. This is because buoyancy becomes weaker as gravity decreases, causing increasingly greater difficulty in removing the bubbles which keep on growing until flow of liquid to surface becomes so low that CHF occurs. To overcome this difficulty, some means to remove the vapor bubbles is needed. Flowing liquid can be the means. If flow boiling is used, flowing liquid can wash away the vapors before they grow into large bubbles. Most of the experimental research till now has been devoted to pool boiling and only a few studies have been done on flow boiling at low gravity. The work in this field has been reviewed by Konishi and Mudawar (2015), Mudawar et al. (2015), and Mudawar (2017). Effect of gravity on flow boiling heat transfer was discussed in Section 5.9.1 of the present book.

6.4.12.1 Terrestrial Studies

As experiments in microgravity are expensive and time consuming, efforts have been made to gain insight through terrestrial experiments. One such study was done by Zhang et al. (2002), which was discussed in Section 6.2.2.4.3. Their test data is shown in Figure 6.2.7. It is seen that at velocities up to 0.5 m s^{-1}, CHF for heated surface facing downward is much lower than for heated surface facing upward. This is because buoyancy cannot remove vapor bubbles when the heated surface is facing downwards. At higher velocities, the CHF in all orientations becomes the same. This shows that if the velocity is sufficiently high, vapor can be removed without the help of buoyancy. This indicates that the formation of large vapor bubbles at low gravity can be avoided if the velocity is sufficiently high. Konishi et al. (2013a) extended these studies to positive inlet qualities. It was found that increasing inlet quality resulted in CHF for downward orientations became equal to that in upward orientations at lower mass flux. For example, as inlet quality increases from 0.03 to 0.19, CHF at inlet velocity of 0.712 m s^{-1} in all orientations became equal. At 0.03 inlet quality, CHF in downward orientations was lower. This behavior may be attributed to the accelerating effect due to the presence of vapor as void fraction increases with increasing quality.

Zhang et al. (2004) performed an analysis to determine the criteria for avoiding the influence of gravity on CHF during subcooled flow boiling. They proposed three criteria, all of which should be satisfied simultaneously to avoid the effect of gravity. These are (i) overcoming the effect of the component of gravity perpendicular to the heated wall, (ii) overcoming the effect of the component of gravity opposite to the direction of fluid flow, and (iii) ensuring that the wavelength associated with instability of the liquid–vapor interface is smaller than the heated length so that liquid can contact wall. The criterion for the first requirement is

$$\frac{Bd}{We^2} = \frac{(\rho_f - \rho_g)(\rho_f + \rho_g)^2 \sigma g}{\rho_f^2 \rho_g^2 U^4} \leq 0.09 \tag{6.4.73}$$

U is the mean velocity of liquid entering the channel, We is the Weber number, and Bn is the Bond number.

The criterion for meeting the second requirement amounts to avoiding flooding and is given by

$$\frac{1}{Fr} = \frac{(\rho_f - \rho_g) g D_{\text{HYD}}}{\rho_f U^2} \leq 0.13 \tag{6.4.74}$$

The third criterion was developed based on the interfacial lift-off model and is expressed by the following equation:

$$We = \frac{\rho_f \rho_g (U_g - U_f)^2 L}{(\rho_f + \rho_g)\sigma} \geq 2\pi \tag{6.4.75}$$

U_g is the mean velocity of vapor layer, U_f is the mean velocity of liquid layer, and L is the heated length of channel. These criteria were developed using Zhang et al.'s own data for FC-72. No other verification is provided.

Konishi et al. (2013b) extended the preceding analysis of Zhang et al. (2004) to positive inlet qualities. Their modified criteria are given as follows:

$$\frac{Bd}{We^2} = \frac{(\rho_f - \rho_g)(\rho_f + \rho_g)^2 \sigma g}{\rho_f^2 \rho_g^2 (G/\rho_f)^4} \leq 0.23 e^{(-67.03 x_{\text{in}})} + 0.002\,49 \tag{6.4.76}$$

$$\frac{1}{Fr} = \frac{(\rho_f - \rho_g) g D_{\text{HYD}}}{\rho_f (G/\rho_f)^2} \leq 0.0184 e^{(-37.07 x_{\text{in}})} + 0.0016 \tag{6.4.77}$$

$$We = \frac{\rho_f \rho_g (G/\rho_f)^2 (G/\rho_f)^2 L}{(\rho_f + \rho_g)\sigma} \geq 1015 e^{(8.682 x_{\text{in}})} - 595.2 \tag{6.4.78}$$

These criteria were developed using their own data for FC-72, Konishi et al. (2013a). No other verification is provided. Figure 6.4.14 shows the predictions of these criteria for various gravity levels. It is seen that the flooding criterion is dominant only at high gravity levels. The heated length criterion is dominant at low gravities. It is also seen that required minimum velocity becomes lower with increasing inlet quality for all three criteria.

6.4.12.2 Experimental Studies at Low Gravities

Most of the experimental studies have been done in parabolic flights, which provide 20–30 seconds of low gravity (about 0.01 g_e) and a short period of high gravity. A few studies have been done in drop towers that give about 2–10 seconds of low gravity ($<1 \times 10^{-4}$ g_e). Most of these studied flow patterns and heat transfer. Those directly for CHF are discussed in the following text.

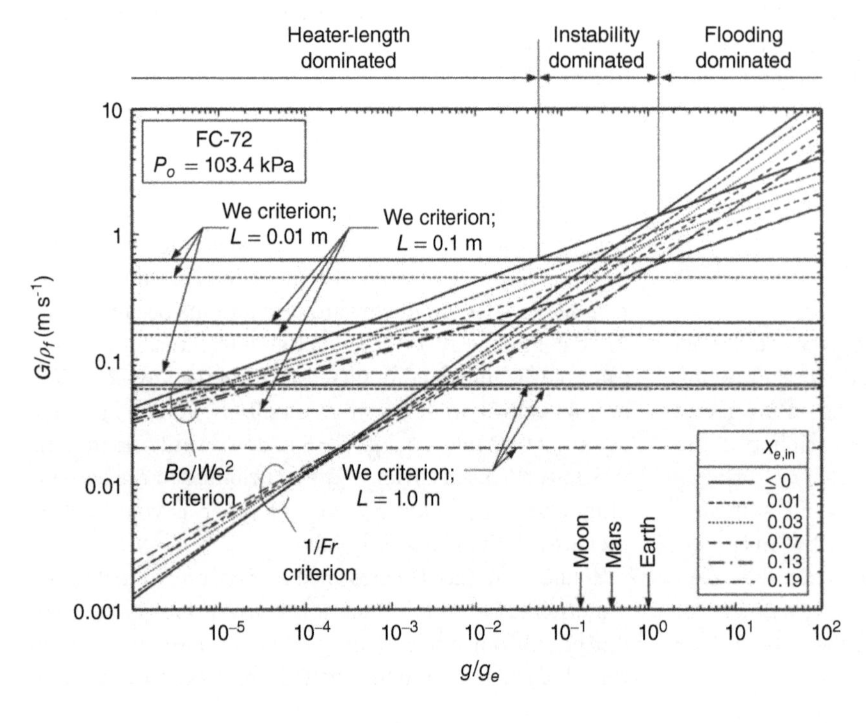

Figure 6.4.14 Minimum flow rate required to overcome effect of gravity at various gravity levels $X_{e,in}$ is the inlet quality. . Source: Konishi et al. (2013b). © 2013 Elsevier.

Ma and Chung (2001) studied boiling on a heated wire with crossflow of FC-72 at Earth gravity as well as in a drop tower. At Reynolds number = 50, CHF at low gravity was about 50% lower than in Earth gravity. The difference narrowed considerably as Reynolds number rose to 200.

Zhang et al. (2005) studied boiling of slightly subcooled FC-72 flowing in a rectangular channel during parabolic flights. At the lowest velocity, CHF at microgravity was only 50% of that at Earth gravity. The difference narrowed as flow velocity increased. At 1.5 m s^{-1} velocity, there was no difference between CHF at low gravity and Earth gravity. The visual observations supported the interfacial lift-off theory of CHF.

6.4.12.3 CHF Prediction Methods

Kharangate et al. (2016) state that CHF in channels at low gravity can be calculated by the interfacial lift-off model, which was briefly described in Section 6.2.2.1. However, the only verification is with a few data points for FC-72 during parabolic flights. Hence further verification is needed to confirm general applicability.

6.4.12.4 Recommendation

There is no well-verified method to calculate CHF under low gravity. It is advisable to keep the inlet velocity above the minimum velocity needed to avoid the effect of low gravity, calculated by the method of Konishi et al., Eqs. (6.4.76)–(6.4.78). As these equations have had very little verification, a large safety factor is advisable.

Nomenclature

Bd	Bond number = $g(\rho_f - \rho_g)D^2/\sigma$ (–)
Bo	boiling number at CHF = $q_c/(Gi_{fg})$ (–)
C_p	specific heat at constant pressure (J kg^{-1} K^{-1})
D, D_t	diameter of tube (m)
D_c	coil diameter (m)
D_{HP}	equivalent diameter based on heated perimeter (m)
D_{HYD}	hydraulic equivalent diameter (m)
Fr	Froude number (–)
G	mass flux (kg m^{-2} s^{-1})
g	acceleration due to gravity (m s^{-2})
g_e	acceleration due to gravity on Earth (m s^{-2})
i	enthalpy (J kg^{-1})
i_{fg}	latent heat of vaporization (J kg^{-1})
k	thermal conductivity (W m^{-1} K^{-1})
L	length, (m)
L_B	boiling length (length between $x = 0$ and $x = x_c$) (m)
P	perimeter (m)
p	pressure (Pa)
p_c	critical pressure (Pa)
p_r	reduced pressure = p/p_c (–)
Pr	Prandtl number (–)
q	heat flux (W m^{-2})
q_{co}	CHF at zero inlet quality (W m^{-2})
Re	Reynolds number = (GD/μ) (–)
T	temperature (K)
ΔT_{SAT}	$(T_w - T_{SAT})$ (K)
ΔT_{SC}	$(T_{SAT} - T_f)$ (K)
u	velocity (m s^{-1})
We	Weber number (–)
x	equilibrium vapor quality (–)
z	length or distance (m)

Greek letters

α	void fraction, (–)
μ	dynamic viscosity (Pa s)
ν	kinematic viscosity (m^2 s^{-1})
ρ	density (kg m^{-3})
σ	surface tension (N m^{-1})
ω	Angular velocity (rad s^{-1})

Subscripts

c	critical, coil
e	effective
f	liquid
g	vapor
GT	all mass flowing as vapor
h	heated
i	inner
in	at inlet
jet	at jet outlet from nozzle
LS	liquid phase flowing alone, superficial liquid
LT	all mass flowing as liquid
loc	local
o, 0	zero vapor quality
SAT	saturation
SC	subcooled
t	total
TP	two-phase
∞	free stream

References

Ahmad, S.Y. (1973). Fluid to fluid modeling of critical heat flux: a compensated distortion model. *Int. J. Heat Mass Transfer* **16**: 641–662.

Ahmad, M., Chandraker, D.K., Hewitt, G.F. et al. (2013). Phenomenological modeling of critical heat flux: the GRAMP code and its validation. *Nucl. Eng. Des.* **254**: 280–290.

Ami, T., Umekawa, H., Ozawa, M. et al. (2011). CHF in a circumferentially non-uniformly heated tube under low-pressure and low-mass-flux condition (inclined upward flow). *Heat Transfer Asian Res.* **40** (2): 125–139.

Ami, T., Hirose, T., Nakamura, N. et al. (2014). CHF in a circumferentially nonuniformly heated tube under low-pressure and low-mass-flux condition (influence of inclined angles under high-heat-flux condition). *Heat Transfer Eng.* **35** (5): 430–439. https://www.tandfonline.com.

Balino, J.L. and Converti, J. (1994). Saturated CHF in horizontal eccentric annuli for low mass fluxes: measurement and modelling. *Int. J. Multiphase Flow* **20** (5): 901–913.

Barnett, P.G. (1964). The prediction of heat transfer in non-uniformly heated rod clusters from data for uniformly heated tubes. Report AEEW-R 362. Quoted in Collier and Thome (1994).

Barnett, P.G. (1966). A correlation of burnout data for uniformly heated annuli and its use for predicting burnout in uniformly heated rod bundles. Report AEEW-R 463. Quoted in Collier and Thome (1994).

Barnett, P.G. (1968). A comparison of the accuracy of some correlations for burnout in annuli and rod bundles. Report AEEW-R 558. Quoted in Collier and Thome (1994).

Becker, K.M. (1966). A correlation for burnout predictions in vertical rod bundles. AB Atomenergi Report S-349.

Bernath, L. (1960). A theory of local boiling burnout and its application to existing data. *Chem. Eng. Prog. Symp. Ser.* **56** (3): 95–116.

Berthoud, G. and Jayanti, S. (1990). Characterization of dryout in helical coils. *Int. J. Heat Mass Transfer* **33** (7): 1451–1463.

Bertoletti, S., Gaspari, G.P., Lombardi, C. et al. (1965). Heat transfer crisis with steam-water mixtures. *Energia Nucleare* **12** (1): 121–172.

Bertoni, R., Cipriani, R., Cumo, M. et al. (1976). Upflow and downflow burnout. *Energia Nucleare.* CNEN, Report No. RT/ING **76** (24).

Biasi, L., Clerici, G.C., Garibba, S., and Sara, R. (1967). Studies on burnout, part 3. *Energia Nucleare* **14** (9): 530–536.

Bowers, M.B. and Mudawar, I. (1974). High flux boiling in low flow rate, low pressure drop in mini-channel and micro-channel heat sinks. *Int. J. Heat Mass Transfer* **37**: 321–332.

Bowring, R.W. (1972). A simple but accurate round tube, uniform heat flux, dryout correlation over the pressure range 10–2500 psia. Report AEEW-R789, UKAEA, Winfrith, UK.

Bowring, R.W. (1977). A new mixed flow cluster dryout correlation for pressures in the range 0.6–15.5 MN/m^2 (90–2250 psia) for use in transient blowdown code. *Paper C217/77 Presented at Conference on Heat and Fluid Flow in Reactor Safety*, IMechE, Manchester (13–15 September).

Bowring, R.W. (1979). WSC-2: a subchannel dryout correlation for water-cooled clusters over the pressure range 3.4–15.9 MPa. Winfrith: Atomic Energy Establishment Report AEEW-R-983.

Cardenas, R. and Narayan, V. (2012). Heat transfer characteristics of submerged jet impingement boiling of saturated FC-72. *Int. J. Heat Mass Transfer* **55**: 4217–4231.

Chen, C.N., Han, J.T., Jen, T.C., and Shao, L. (2011). Dry-out CHF correlation for R134a flow boiling in a horizontal helically coiled tube. *Int. J. Heat Mass Transfer* **54**: 739–745.

Cheng, X. (2005). Experimental studies on critical heat flux in vertical tight 37-rod bundles using Freon-12. *Int. J. Multiph. Flow* **31**: 1198–1219.

Clark, M.D., Weibel, J.A., and Garimella, S.V. (2019). Identification of nucleate boiling as the dominant heat transfer mechanism during confined two-phase jet impingement. *Int. J. Heat Mass Transfer* **128**: 1095–1101.

Cochran, T.H. and Andracchio, C.R. (1974). Forced-convection peak heat flux on cylindrical heaters in water and refrigerant 113. NASA TN D-7553.

Collier, J.G. and Thome, J.R. (1994). *Convective Boiling and Condensation*, 3e. Oxford, UK: Oxford University Press.

Cumo, M., Farello, G.E., and Ferrari, G. (1972). The influence of curvature in post-dryout heat transfer. *Int. J. Heat Mass Transfer* **15**: 2045–2062.

Cumo, M., Fabrizi, F., and Palazzi, G. (1978). The influence of inclination on CHF in steam generators channels. RT/ING 78(11). Quoted in Ami et al. (2015).

Cumo, M., Farello, G.E., Gasiorowski, J. et al. (1980). Quality influence on the departure from nucleate boiling in cross flows through bundles. *Nucl. Technol.* **49**: 337–346.

Dalle Donne, M. (1991). CHF-KfK-3: a critical heat flux correlation for triangular arrays of rods with tight lattices. Research Center Karlsruhe Report KfK-4826.

Del Col, D. and Bortolin, S. (2012). Investigation of dryout during flow boiling in a single microchannel under non-uniform axial heat flux. *Int. J. Therm. Sci.* **57**: 25–36.

Doerffer, S., Groeneveld, D.C., Cheng, S.C., and Rudzinski, K.F. (1994). A comparison of critical heat flux in tubes and annuli. *Nucl. Eng. Des.* **149**: 167–175.

Doroshchuk, V.E., Lantsman, F.P., and Levitan, L.L. (1970). A peculiar type of burnout in evaporator tubes. In: *Heat Transfer 1970*, Paper B6.1. Amsterdam, Netherlands: Elsevier Pub. Co.

Doroshchuk, V.E., Levitan, L.L., and Lantzman, F.P. (1975). Investigation into burnout in uniformly heated tubes. *ASME Paper 75-WA/HT-22, ASME Winter annual Meeting*, Houston, Texas (30 November–4 December).

Dykas, S. and Jensen, M.K. (1992). Critical heat flux on a tube in a horizontal tube bundle. *Exp. Therm Fluid Sci.* **5**: 34–39.

Estes, K.A. and Mudawar, I. (1995). Correlation of Sauter mean diameter and critical heat flux for spray cooling of small surfaces. *Int. J. Heat Mass Transfer* **38**: 2985–2996.

Evangelisti, R., Gaspari, G.P., and Vanoli, G. (1972). Heat transfer crisis data with steam-water mixture in a sixteen rod bundle. *Int. J. Heat Mass Transfer* **15**: 387–402.

Fighetti, C.F. and Reddy, D.G. (1983). *A Generalized Subchannel CHF Correlation for PWR and BWR Fuel Assemblies*, EPRI-NP-2609-Vol. 2. Electric Power Research Institute.

Fujita, T. and Ueda, T. (1978). Heat transfer to falling liquid films and film breakdown – I. Subcooled liquid films. *Int. J. Heat Mass Transfer* **21**: 97–108.

Galloway, J.E. and Mudawar, I. (1993). CHF mechanism in flow boiling from a short heated wall – Part 2. Theoretical CHF model. *Int. J. Heat Mass Transfer* **36** (1993): 2527–2540.

Gambill, W.R. (1963). Generalized prediction of burnout heat flux for subcooled flowing liquids. *Chem. Eng. Prog. Symp. Ser.* **41** (59): 71–87.

Gaspari, G.P., Hassid, A., and Lucchini, F. (1970). Some considerations on the on the critical heat flux in rod clusters in annular dispersed vertical dispersed upwards two-phase flow. In: *Proceedings of the Fourth International Heat Transfer Conference*, vol. **6**. Begell House.

Gaspari, G.P., Hassid, A., and Lucchini, F. (1974). A rod-centered sub-channel analysis with turbulent enthalpy mixing for critical heat flux prediction in rod clusters cooled by boiling water. In: *Proceedings of the Fifth International Heat Transfer Conference*, vol. **4**, 295. Begell House.

Griffith, P., Schumann, W.A., and Neustal, A.D. (1962). Flooding and burn-out in closed-end vertical tubes. In: *Symposium on Two-Phase Flow, Proceedings of the Institution of the Mechanical Engineers*, 35–39. Quoted in Nejat (1981). SAGE Publisher.

Groeneveld, D.C., Cheng, S.C., and Doan, T. (1986). AECL-UO critical heat flux look-up table. *Heat Transfer Eng.* **7**: 46–62.

Groeneveld, D.C., Leung, L.K.H., Aksan, N. et al., (1999) A general method of predicting critical heat flux in advanced water cooled reactors. *Ninth International Topical Meeting on Nuclear Reactor T/H (NURETH-9)*, 3-8 October, San Francisco, CA, USA. Quoted in Yang et al. (2006).

Groeneveld, D.C., Shah, J.Q., Vasic, A.Z. et al. (2007). The 2006 CHF look-up table. *Nucl. Eng. Des.* **237**: 1909–1922.

Habib, M.A., Nemitallah, M.A., and El-Nakla, M. (2014). Current status of CHF predictions using CFD modeling technique and review of other techniques especially for non-uniform axial and circumferential heating profiles. *Ann. Nucl. Energy* **70**: 188–207.

Hall, D.E., Incropera, F.P., and Viskanta, R. (2001). Jet impingement boiling from a circular free-surface jet during quenching: Part 1: Single-phase jet. *J. Heat Transfer* **123**: 901–910.

Haramura, Y. and Katto, Y. (1983). A new hydrodynamic model of critical heat flux, applicable widely to both pool and forced convection boiling submerged bodies in saturated liquids. *Int. J. Heat Mass Transfer* **26**: 389–399.

Hardik, B.K. and Prabhu, S.V. (2017). Critical heat flux in helical coils at low pressure. *Appl. Therm. Eng.* **112**: 1223–1239.

Hardik, B.K. and Prabhu, S.V. (2018). Boiling pressure drop, local heat transfer distribution and critical heat flux in helical coils with R123. *Int. J. Therm. Sci.* **125**: 149–165.

Hasan, M.Z., Hasan, M.M., Eichhorn, R. et al. (1981). Boiling burnout during crossflow over cylinders beyond the influence of gravity. *J. Heat Transfer* **103**: 479–484.

Hasan, M.M., Eichhorn, R., and Lienhard, J.H. (1982). Boiling burnout during crossflow across a small cylinder influenced by parallel cylinders. In: *Proceedings of the 7th International Heat Transfer Conference*, 285–290. Begell House.

Hewitt, G.F. (1982). Burnout. In: *Handbook of Multiphase Systems* (ed. G. Hetsroni), 6-66–6-141. Washington, DC, USA: Hemisphere Publishing Corporation.

Hewitt, G.F. and Kearsey, H.A. (1966). Heat transfer to steam-water mixtures at high pressures – studies of burnout in round tubes. In: *Proceedings of the Third International Heat Transfer Conference*, vol. **3**, 160–174. Begell House.

Janssen, E. and Levy, S. (1962). Burnout limit curves for boiling water reactors. General Electric Co. Report APED-3892.

Jensen, M.K. (1980). Boiling heat transfer and critical heat flux in helical coils. PhD thesis. Iowa State University, Ames, Iowa.

Jensen, M.K. and Bergles, A.E. (1981). Critical heat flux in helically coiled tubes. *J. Heat Transfer* **103** (4): 660–666.

Jensen, M.K. and Bergles, A.E. (1982). Critical heat flux in helical coils with a circumferential heat flux tilt toward the outside surface. *Int. J. Heat Mass Transfer* **25** (9): 1383–1395.

Jensen, M.K. and Tang, H. (1994). Correlations for CHF condition in two-phase crossflow through multitube bundles. *J. Heat Transfer* **116** (3): 780–783.

Kaji, M., Mori, K., Matsumoto, T. et al. (1998). Forced convective boiling heat transfer characteristics and critical heat flux in helically coiled tubes. *Nippon Kikai Gakkai Ronbunshu* **64**: 3343–3349.

Kalimullah, M., Feldman, E.E., Olson, A.P., et al. (2012). An evaluation of subcooled CHF correlations and databases for research reactors operating at 1 to 50 bar pressure. *RERTR 2012 – 34th International Meeting on Reduced Enrichment For Research And Test Reactors*, 14-17 October 2012, Warsaw, Poland.

Kandlikar, S. (2010). A scale analysis based theoretical force balance model for critical heat flux chf during saturated flow boiling in microchannels and minichannels. *J. Heat Transfer* **132**: 081501-1–081501-13.

Katto, Y. (1978). A generalized correlation of critical heat flux for the forced convection boiling in uniformly heated vertical tubes. *Int. J. Heat Mass Transfer* **21**: 1527–1542.

Katto, Y. (1979). Generalized correlation of critical heat flux for the forced convection boiling in vertical uniformly heated annuli. *Int. J. Heat Mass Transfer* **22**: 575–584.

Katto, K. and Kurata, C. (1980). Critical heat flux of saturated convective boiling on uniformly heated plates in a parallel flow. *Int. J. Multip. Flow* **6** (6): 575–582.

Katto, Y. (1982). A study on limiting exit quality of CHF of forced convection boiling in uniformly heated vertical channels. *J. Heat Transfer* **104**: 40–47.

Katto, K. and Haramura, Y. (1983). Critical heat flux on a uniformly heated horizontal cylinder in an upward cross flow of saturated liquid. *Int. J. Heat Mass Transfer* **26**: 1199–1205.

Katto, Y. and Ohno, H. (1984). An improved version of the generalized correlation of critical heat flux for the forced convection boiling in uniformly heated vertical tubes. *Int. J. Heat Mass Transfer* **27** (9): 1641–1648.

Katto, Y. and Yokoya, S. (1988). Critical heat flux on a disk heater cooled by a circular jet of saturated liquid impinging at the center. *Int. J. Heat Mass Transfer* **31**: 219–227.

Kefer, V., Kohler, W., and Kastner, W. (1989). Critical heat flux (CHF) and post-CHF heat transfer in horizontal and inclined evaporator tubes. *Int. J. Multiphase Flow* **15** (3): 385–392.

Khalil, A. (1982). Steady state heat transfer of helium cooled cable bundles. *Cryogenics* **22** (6): 277–281.

Kharangate, C.R., Konishi, C., and Mudawar, I. (2016). Consolidated methodology to predicting flow boiling critical heat flux for inclined channels in earth gravity and for microgravity. *Int. J. Heat Mass Transfer* **92**: 467–482.

Kim, S. and Mudawar, I. (2013). Universal approach to predicting saturated flow boiling heat transfer in mini/micro-channels – Part 1. Dryout incipience quality. *Int. J. Heat Mass Transfer* **64**: 1226–1238.

Kirby, G.J. (1966). A new correlation for non-uniformly heated round tube burnout data. Report AEEW-R 500.

Kirillov, P.L., Bobkov, P.E., Katan, E.A. et al. (1991). New CHF tables for water in round tubes. *At. Energ.* **70**: 18–28. (in Russian). Quoted in Collier and Thome (1994).

Konishi, C. and Mudawar, I. (2015). Review of flow boiling and critical heat flux in microgravity. *Int. J. Heat Mass Transfer* **80**: 469–493.

Konishi, C., Mudawar, I., and Hasan, M.M. (2013a). Investigation of the influence of orientation on critical heat flux for flow boiling with two-phase inlet. *Int. J. Heat Mass Transfer* **61**: 176–190.

Konishi, C., Mudawar, I., and Hasan, M.M. (2013b). Criteria for negating the influence of gravity on flow boiling critical heat flux with two-phase inlet conditions. *Int. J. Heat Mass Transfer* **65**: 203–218.

Konishi, C., Lee, H., Mudawar, I. et al. (2015). Flow boiling in microgravity: Part 1 – Interfacial behavior and experimental heat transfer results. *Int. J. Heat Mass Transfer* **81**: 705–720.

Kosar, A. (2009). A model to predict saturated critical heat flux in minichannels and microchannels. *Int. J. Therm. Sci.* **48**: 261–270.

Koşar, A. and Peles, Y. (2007). Critical heat flux of r-123 in silicon-based microchannels. *J. Heat Transfer* **129**: 844–851.

Kusuna, H. and Imura, H. (1974). Stability of a liquid film in a countercurrent annular two-phase flow. *Bull. JSME* **17**: 1613–1618.

Larsen, P.S., Kitzes, A.S., Stormer, T.D. et al. (1966). DNB measurements for upward flow of water in an unheated square channel with a single uniformly heated rod at 1600–2300 PSIA. In: *Proceedings of the Third Internaitonal Heat Transfer Conference*, vol. **5**, 143–148. Begell House.

Lee, D.H. and Obertelli, J.D. (1963). An experimental investigation of forced convection burnout in high pressure water Part II. Report AEEW-R 309. Quoted in Collier and Thome (1994).

Leontiev, A.I., Mostinsky, I.L., Polonsky, V.S. et al. (1981). Experimental investigation of the critical heat flux in horizontal channels with circumferentially variable heating. *Int. J. Heat Mass Transfer* **24** (5): 821–828.

Leroux, K.M. and Jensen, M.K. (1992). Critical heat flux in horizontal tube bundles in vertical crossflow of R113. *J. Heat Transfer* **114**: 179–184.

Li, Y.Z., Liu, Z., Wang, G., and Pang, L. (2013). Experimental study on critical heat flux of high velocity circular jet impingement boiling on the nano-characteristic stagnation zone. *Int. J. Heat Mass Transfer* **67**: 560–568.

Liang, G. and Mudawar, I. (2017). Review of spray cooling – Part 1: Single-phase and nucleate boiling regimes, and critical heat flux. *Int. J. Heat Mass Transfer* **115**: 1174–1205.

Lienhard, J.H. (1988). Burnout on cylinders. *J. Heat Transfer* **110**: 1271–1286.

Lienhard, J.H. and Eichhorn, R. (1976). Peak boiling heat flux on cylinders in a cross flow. *Int. J. Heat Mass Transfer* **19**: 1135–1142.

Lu, Q.W., Wen, Q., Liu, T. et al. (2014). A critical heat flux experiment with water flow at low pressures in thin rectangular channels. *Nucl. Eng. Des.* **278**: 669–678.

Ma, Y. and Chung, J.N. (2001). An experimental study of critical heat flux (CHF) in microgravity forced-convection boiling. *Int. J. Multiphase Flow* **27** (10): 1753–1767.

Ma, W., Zhang, M., and Chen, X. (1995). High-quality critical heat flux in horizontally coiled tubes. *J. Therm. Sci.* **4** (3): 205–211.

Macbeth, R.V. (1964). Burnout analysis part 5. Examination of published world data for rod bundles. Report AEEW- R 358.

Mahmoudi, S.R., Adamiak, K., and Castle, G.S.P. (2012). Two-phase cooling characteristics of a saturated free falling circular jet of HFE7100 on a heated disk: effect of jet length. *Int. J. Heat Mass Transfer* **55**: 6181–6190.

McKee, H.R. and Bell, K.J. (1969). Forced convection boiling from a cylinder normal to the flow. *Chem. Eng. Prog. Symp. Ser.* **65**: 222–230.

Merilo, M. (1979). Fluid-to-fluid modeling and correlation of flow boiling crisis in horizontal tubes. *Int. J. Multiphase Flow* **5**: 313–325.

Miropolskiy, Z.L. and Pikus, V.Y. (1969). Critical boiling heat flux in curved channels. *Heat Transfer Sov. Res.* **1** (1): 74–79.

Mishima, K. (1984). Boiling burnout at low flow rate and low pressure conditions. Dissertation thesis. Kyoto University, Japan. Quoted in Sudo et al. (1985).

Mitsutake, Y. and Monde, M. (2003). Ultra high critical heat flux during forced flow boiling heat transfer with an impinging jet. *J. Heat Transfer* **125**: 1038–1045.

Monde, M. (1980). Burnout heat flux in saturated forced convection boiling with an impinging jet. *Heat Transfer Jpn. Res.* **9**: 31–34.

Monde, M. (1985). Critical heat flux in saturated forced convective boiling on a heated disk with an impinging jet. *Warme und Stoffubergang* **19**: 205–209.

Monde, M. (1987). Critical heat flux in saturated forced convection boiling on a heated disk with an impinging jet. *J. Heat Transfer* **109**: 991–996.

Monde, M. and Mitsutake, Y. (1996). Critical heat flux in forced convective subcooled boiling with multiple impinging jets. *J. Heat Transfer* **117**: 241–243.

Monde, M., Kitajima, K., Inoue, T., and Mitsutake, Y. (1994). Critical heat flux in a forced convective subcooled boiling with an impinging jet. In: *Proceedings of the 10th International Heat Transfer Conference*, 515–520. Brighton, UK: Begell House.

Mudawar, I. (2017). Flow boiling and flow condensation in reduced gravity. *Adv. Heat Transfer* **49**: 225–306.

Mudawwar, I.A., El-Masri, M.A., Wu, C.S. et al. (1985). Boiling heat transfer and critical heat flux in high-speed rotating liquid films. *Int. J. Heat Mass Transfer* **28** (4): 795–806.

Mudawwar, I.A., Incropera, T.A., and Incropera, F.P. (1987). Boiling heat transfer and critical heat flux in liquid films falling on vertically-mounted heat sources. *Int. J. Heat Mass Transfer* **10**: 2083–2095.

Nejat, Z. (1981). Effect of density ratio on critical heat flux in closed end vertical tubes. *Int. J. Multiphase Flow* **7**: 321–327.

Nonn, T., Dagan, Z., and Jiji, L.M. (1988). Boiling jet impingement cooling of simulated microelectronic heat sources. ASME Paper no. 88-WA/EEP-3. Quoted in Hall et al. (2001).

Noyes, R.C. and Lurie, H. (1966). Boiling sodium heat transfer. In: *Proceedings of the Third International Heat Transfer Conference*, vol. **5**, 143–148. New York, NY: ASME.

Ohno, M., Mori, H., and Yoshida, S. (1999). Characteristics of critical heat flux at high pressure in an inclined tube. *Trans. Jpn. Soc. Mech. Eng. Ser. B* (in Japanese) **65** (638): 3422–3429.

Palen, J.W. and Small, W.M. (1964). A new way to design kettle and internal reboilers. *Hydrocarbon Process.* **43**: 199–208.

Palen, J.W., Yarden, A., and Taborek, J. (1972). Characteristics of boiling outside large-scale horizontal multi-tube bundles. *AIChE Symp. Ser.* **68** (118): 50–61.

Park, J.W., Baek, W.P., and Chang, S.H. (1997). Critical heat flux and flow pattern for water flow in annular geometry. *Nucl. Eng. Des.* **172**: 137–155.

Pioro, I.L., Groeneveld, D.C., Leung, L.K. et al. (2002). Comparison of CHF measurements in horizontal and vertical tubes cooled with R-134a. *Int. J. Heat Mass Transfer* **45**: 4435–4450.

Qi, S., Zhang, P., Wang, R., and Xu, L. (2007). Flow boiling of liquid nitrogen in microtubes: Part II – heat transfer characteristics and critical heat flux. *Int. J. Heat Mass Transfer* **50**: 5017–5030.

Qiu, Y.H. and Liu, Z.H. (2010). The theoretical simulation of the effect of solid–liquid contact angle on the critical heat flux of saturated water jet boiling on stagnation zone. *Int. J. Heat Mass Transfer* **53** (9–10): 1921–1926.

Qiu, L., Dubey, S., Choo, F.H., and Duan, F. (2015). Recent developments of jet impingement nucleate boiling. *Int. J. Heat Mass Transfer* **89**: 42–58.

Qu, W. and Mudawar, I. (2004). Measurement and correlation of critical heat flux in two-phase microchannel heat sinks. *Int. J. Heat Mass Transfer* **47**: 2045–2059.

Revellin, R. and Thome, J.R. (2008). A theoretical model for the prediction of the critical heat flux in heated microchannels. *Int. J. Heat Mass Transfer* **51**: 1216–1225.

Revellin, R., Dupont, V., Ursenbacher, T. et al. (2006). Characterization of diabatic two-phase flows in microchannels: flow parameter results for R-134a in a 0.5 mm channel. *Int. J. Multiphase Flow* **32**: 755–774.

Revellin, R., Mishima, K., and Thome, J.R. (2009). Status of prediction methods for critical heat fluxes in mini and microchannels. *Int. J. Heat Fluid Flow* **30**: 983–992.

Rohsenow, W.M. (1973). Boiling. In: *Handbook of Heat Transfer* (eds. W.M. Rohsenow and J.P. Hartnett), 13–64. New York, NY: McGraw-Hill.

Roy, G.M. (1966). Getting more out of BWR's. *Nucleonics* **24** (11): 41.

Sadasivan, P. and Lienhard, J.H. (1986). Burnout of cylinders in flow boiling: the role of gravity on the influences on the vapor plume. *ASME/AIChE Confernce*, Pittsburgh, PA. Qouted in Lienhard (1988).

Shah, M.M. (1976). A new correlation for heat transfer during boiling flow through pipes. *ASHRAE Transactions* **82** (2): 66–86.

Shah, M.M. (1979). A generalized graphical method for predicting CHF in uniformly heated vertical tubes. *Int. J. Heat Mass Transfer* **22**: 557–568.

Shah, M.M. (1980). A general correlation for critical heat flux in annuli. *Int. J. Heat Mass Transfer* **23**: 225–234.

Shah, M.M. (1987). Improved general correlation for critical heat flux in uniformly heated vertical tubes. *Int. J. Heat Fluid Flow* **8** (4): 326–335.

Shah, M.M. (2015a). A general correlation for CHF in horizontal channels. *Int. J. Refrig* **59**: 37–52.

Shah, M.M. (2015b). Improved general correlation for CHF in vertical annuli with upflow. *Heat Transfer Eng.* **37** (6): 557–570. https://www.tandfonline.com.

Shah, M.M. (2017). Applicability of general correlations for critical heat flux in conventional tubes to mini/micro channels. *Heat Transfer Eng.* **38** (1): 1–10. https://www.tandfonline.com.

Sharan, A. and Lienhard, J.H. (1985). On predicting burnout in the jet-disk configuration. *J. Heat Transfer* **107**: 398–401.

Stevens, G.F. & G. J. Kirby, (1964) *A quantitative comparison between burnout data for water at l000 psia and Freon-12 at 155 psia uniformly heated round tubes, vertical upflow.* United Kingdom Atomic Energy Authority, Report No. AEEW-R327 (1964).

Stoddart, R.M., Blasick, A.M., Ghiassiaan, S.M. et al. (2002). Onset of flow instability and critical heat flux in thin horizontal annuli. *Exp. Therm Fluid Sci.* **26**: 1–14.

Sturgis, J.C. and Mudawar, J. (1999). Critical heat flux in a long, rectangular channel subjected to one sided heating – II. Analysis of critical heat flux data. *Int. J. Heat Mass Transfer* **42**: 1849–1862.

Subbotin, V.I., Deev, V.I., Pridantsev, A.I. et al. (1985). Heat transfer and hydrodynamics in cooling channels of superconducting devices. *Cryogenics* **25**: 261–265.

Sudo, Y., Miyata, K., Ikawa, H. et al. (1985). Experimental study of differences in DNB heat flux between upflow and downflow in vertical rectangular channel. *J. Nucl. Sci. Technol.* **22**: 604–618.

Sun, K.H. and Lienhard, J.H. (1970). The peak pool boiling heat transfer on horizontal cylinders. *Int. J. Heat Mass Transfer* **13**: 1425–1439.

Thiagarajan, S.J., Narumanchi, S., and Yang, R. (2014). Effect of flow rate and subcooling on spray heat transfer on microporous copper surfaces. *Int. J. Heat Mass Transfer* **69** (2014): 493–505.

Tong, L.S. (1967). Heat transfer in water cooled nuclear reactors. *Nucl. Eng. Des.* **6**: 301–324.

Tong, L.S. (1968). Boundary-layer analysis of the flow boiling crisis. *Int. J. Heat Mass Transfer* **11**: 1208–1211.

Tong, L.S. and Tang, Y.S. (1997). *Boiling Heat Transfer and Two-Phase Flow*. Washington, DC: Taylor & Francis.

Tong, L.S., Currin, H.B., Larsen, P.S., and Smith, O.G. (1966). Influence of axially non-uniform heat flux on DNB. *AIChE Chem. Eng. Prog. Symp. Ser.* **62** (64): 35–40.

Ueda, T., Inoue, M., and Nagatome, S. (1981). Critical heat flux and droplet entrainment rate in boiling of falling liquid films. *Int. J. Hear Mass Transfer* **24** (7): 1257–1266.

Visaria, M. and Mudawar, I. (2008). Effects of high subcooling on two-phase spray cooling and critical heat flux. *Int. J. Heat Mass Transfer* **51**: 5269–5278.

Vliet, G.C. and Leppert, G. (1964). Critical heat flux for nearly saturated water flow normal to a cylinder. *J. Heat Transfer* **86**: 59–67.

Weisman, J. and Pei, B.S. (1983). Prediction of critical heat-flux in flow boiling at low qualities. *Int. J. Heat Mass Transfer* **26**: 1463–1477.

Weisman, J. and Ying, S.H. (1985). A theoretically based critical heat flux prediction for rod bundles at PWR conditions. *Nucl. Eng. Des.* **85**: 239–250.

Whalley, P.B. (1976). The calculation of dryout in a rod bundle – comparison of experimental and calculated results. Report AERE-R-8977.

Whalley, P.B. (1978). The calculation of dryout in a rod bundle. Report AERE-R-8319.

Whalley, P.B., Hutchinson, P., and Hewitt, G.F. (1974). The calculation of critical heat flux in forced convection boiling. *Proceedings of the Fifth International Heat Transfer Conference*, Tokyo, Japan (3–7 September).

Wojtan, L., Ursenbacher, T., and Thome, J.R. (2005). Investigation of flow boiling in horizontal tubes: Part I – A new diabatic two-phase flow pattern map. *Int. J. Heat Mass Transfer* **48**: 2955–2969.

Wojtan, L., Revellin, R., and Thome, J.R. (2006). Investigation of saturated critical heat flux in a single uniformly heated microchannel. *Exp. Therm Fluid Sci.* **30**: 765–774.

Wolf, D.H., Incropera, F.P., and Viskanta, R. (1993). Jet impingement boiling. In: *Advances in Heat Transfer* (eds. P.H. James and F.I. Thomas), 1–132. Elsevier.

Wong, Y.L., Groeneveld, D.C., and Cheng, S.C. (1990). CHF prediction for horizontal tubes. *Int. J. Multiphase Flow* **16**: 123–138.

Wu, Z., Li, W., and Ye, S. (2011). Correlations for saturated critical heat flux in microchannels. *Int. J. Heat Mass Transfer* **54**: 379–389.

Yang, J., Groeneveld, D.C., Leung, L.K.H. et al. (2006). An experimental and analytical study of the effect of axial power profile on CHF. *Nucl. Eng. Des.* **236**: 1384–1395.

Yang, B., Han, B., Liu, A., and Wang, S. (2019). Recent challenges in subchannel thermal-hydraulics-CFD modeling, subchannel analysis, CHF experiments, and CHF prediction. *Nucl. Eng. Des.* **354**: 1–19.

Yao, S.C. and Hwang, T.H. (1989). Critical heat flux on horizontal tubes in an upward crossflow of Freon-113. *Int. J. Heat Mass Transfer* **32** (I): 93–103.

Yilmaz, S. and Westwater, J.W. (1980). Effect of velocity on heat transfer to boiling Freon-113. *J. Heat Transfer* **103**: 26–31.

Zhang, H., Mudawar, I., and Hasan, M.M. (2002). Experimental assessment of the effects of body force, surface tension force, and inertia on flow boiling CHF. *Int. J. Heat Mass Transfer* **45**: 4079–4095.

Zhang, H., Mudawar, I., and Hasan, M.M. (2004). A method for assessing the importance of body force on flow boiling CHF. *J. Heat Transfer* **126**: 161–168.

Zhang, H., Mudawar, I., and Hasan, M.M. (2005). Flow boiling CHF in microgravity. *Int. J. Heat Mass Transfer* **48**: 3107–3118.

Zhang, W., Hibiki, T., Mishima, K., and Mi, Y. (2006). Correlation of critical heat flux for flow boiling of water in mini-channels. *Int. J. Heat Mass Transfer* **49**: 1058–1072.

7

Post-CHF Heat Transfer in Flow Boiling

7.1 Introduction

The various regimes in pool boiling were discussed in Section 3.1 and shown in Figure 3.1.1. The regimes in flow boiling are basically the same. In a surface temperature-controlled system, critical heat flux (CHF) is followed by transition boiling in which heat flux decreases with increasing wall temperature. This continues until the minimum film boiling temperature (MFBT) is reached. With further increase in wall temperature, film boiling starts during which heat flux increases with increasing wall temperature. In a heat flux-controlled system, film boiling starts immediately after CHF, and thus there is no transition boiling.

In this chapter, the physical phenomena related to film boiling, transition boiling, and MFBT are discussed. Methods for their predictions for various applications are presented. These include tubes, annuli, horizontal and vertical tube bundles, helical coils, etc.

7.2 Film Boiling in Vertical Tubes

7.2.1 Physical Phenomena

The flow regimes usually found after CHF at low or negative qualities are shown in Figure 7.2.1. Inverted annular flow occurs immediately downstream of the CHF location. There is a continuous liquid core in the middle of the tube surrounded by a vapor film. The vapor film may also contain some liquid droplets. This is the region of stable inverted annular film boiling (IAFB). The vapor layer becomes thicker with distance from the CHF point. The vapor–liquid interface becomes wavy and unstable. The liquid still does not touch the tube wall. Further downstream, the liquid core begins to break down into liquid chunks. Liquid begins to contact the wall at qualities higher than about 0.3. Eventually liquid chunks break down into droplets carried in a continuous vapor stream. This is the region of dispersed flow film boiling (DFFB). At high flow rates, stable inverted flow pattern does not occur. Liquid core breaks down into liquid chunks close to the CHF point and then further breaks down into droplets carried in vapor stream. Usually the part of tube occupied by stable or unstable IAFB is small, most of the tube length having DFFB.

Figure 7.2.2 shows the wall temperatures and heat transfer coefficients at moderate and high flow rates following subcooled CHF at near critical pressures.

If CHF occurs at high qualities where annular flow prevails, CHF occurs due to dryout of liquid film. Then DFFB starts immediately after CHF.

Behavior of wall temperatures lead researchers to infer that usually there is thermodynamic non-equilibrium. For example, wall temperatures at equilibrium qualities >1 were observed to differ from what will be expected for vapor-only flow. Figure 7.2.3 shows the relation between actual vapor qualities and equilibrium qualities along the tube length estimated by Laverty and Rohsenow (1964) from their tests on boiling of nitrogen at 1.2 bar in an 8.1 mm diameter tube. Presence of non-equilibrium can be expected from the fact that during film boiling, only vapor is in contact with the heated wall. Hence heat first passes into vapor and then from vapor to liquid. For heat to flow from vapor to liquid, there must be a temperature difference between the two. As the liquid is at saturation temperature, vapor has to be at a higher temperature. Hence there has to be some non-equilibrium, small or large.

Direct evidence of non-equilibrium is provided by measurements of vapor temperature. Such measurements have been reported, among others, by Annunziato et al. (1983), Evans et al. (1985), Gottula et al. (1983), and Nijhawan et al. (1980). These measurements show temperatures well above the saturation temperature and hence indicate non-equilibrium. Shah and Siddiqui (2000) found these measured temperatures to be in fair agreement with their non-equilibrium model that was shown to correctly predict wall temperatures from many studies.

Two-Phase Heat Transfer, First Edition. Mirza Mohammed Shah.

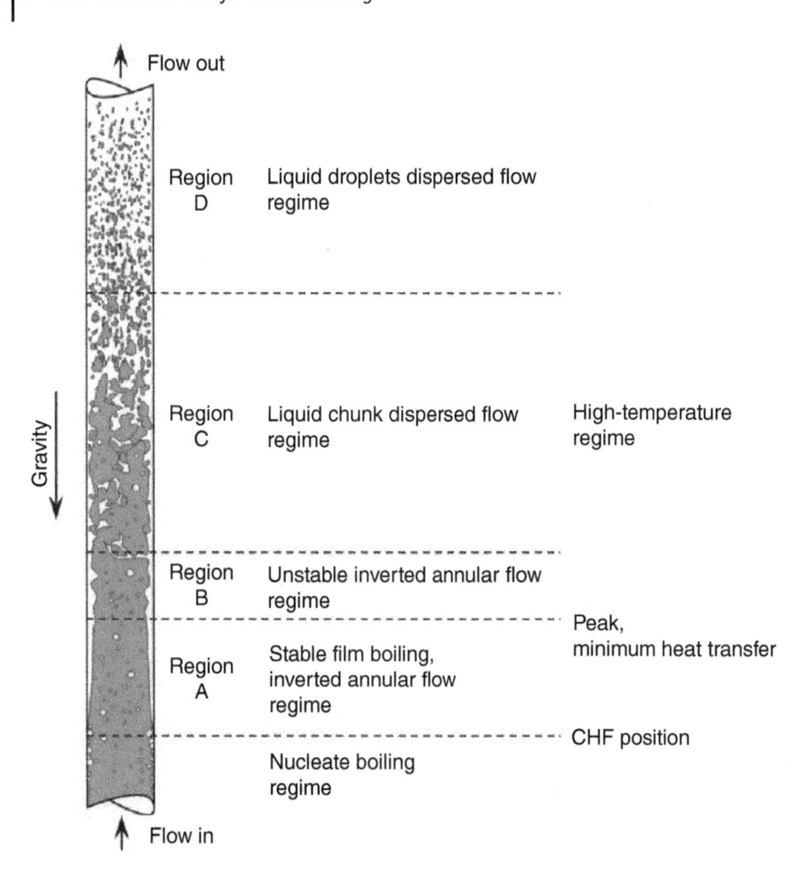

Figure 7.2.1 Film boiling regimes in a vertical tube following DNB-type boiling crisis. Source: Mawatari et al. (2014). © Begell House.

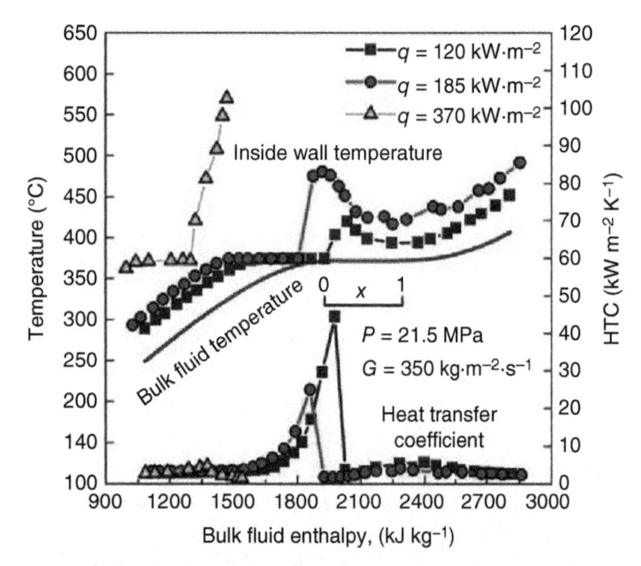

Figure 7.2.2 Wall temperatures and heat transfer coefficient before and after CHF in a vertical tube of 19 mm ID with water upflow. Source: Shen et al. (2016). © 2016 Elsevier.

7.2.2 Prediction of Dispersed Flow Film Boiling in Upflow

Many methods for prediction of heat transfer during film boiling have been proposed. These range from purely empirical formulas, which disregard non-equilibrium and other physical phenomena, to sophisticated mechanistic analyses. In between these two extremes are the phenomenological models, that take into consideration non-equilibrium and other physical phenomena. A completely different approach is that of lookup tables that list heat transfer coefficients for a wide range of conditions. These methods are discussed in the following.

7.2.2.1 Empirical Correlations

These correlations completely disregard physical phenomena such as non-equilibrium. These are modifications of single-phase heat transfer correlations. Most of the early correlations were of this type and were exclusively based on water data. Many of them have been listed in Groeneveld (1975). Among them, the one that covers the widest range of parameters is that of Groeneveld (1973) given as follows:

$$\frac{h_{\mathrm{TP}}D}{k_g} = a\left[Re_{GS}\left\{x_E + \frac{\rho_g}{\rho_f}(1 - x_E)\right\}\right]^b Pr_{g,w}^c Y^d q^e \tag{7.2.1}$$

q is in Btu h^{-1} ft^{-2}. x_E is the equilibrium quality.

$$h_{\mathrm{TP}} = \frac{q}{(T_w - T_{\mathrm{SAT}})} \tag{7.2.2}$$

$$Y = 1 - 0.1\left(\frac{\rho_f}{\rho_g} - 1\right)^{0.4}(1 - x_E)^{0.4} \tag{7.2.3}$$

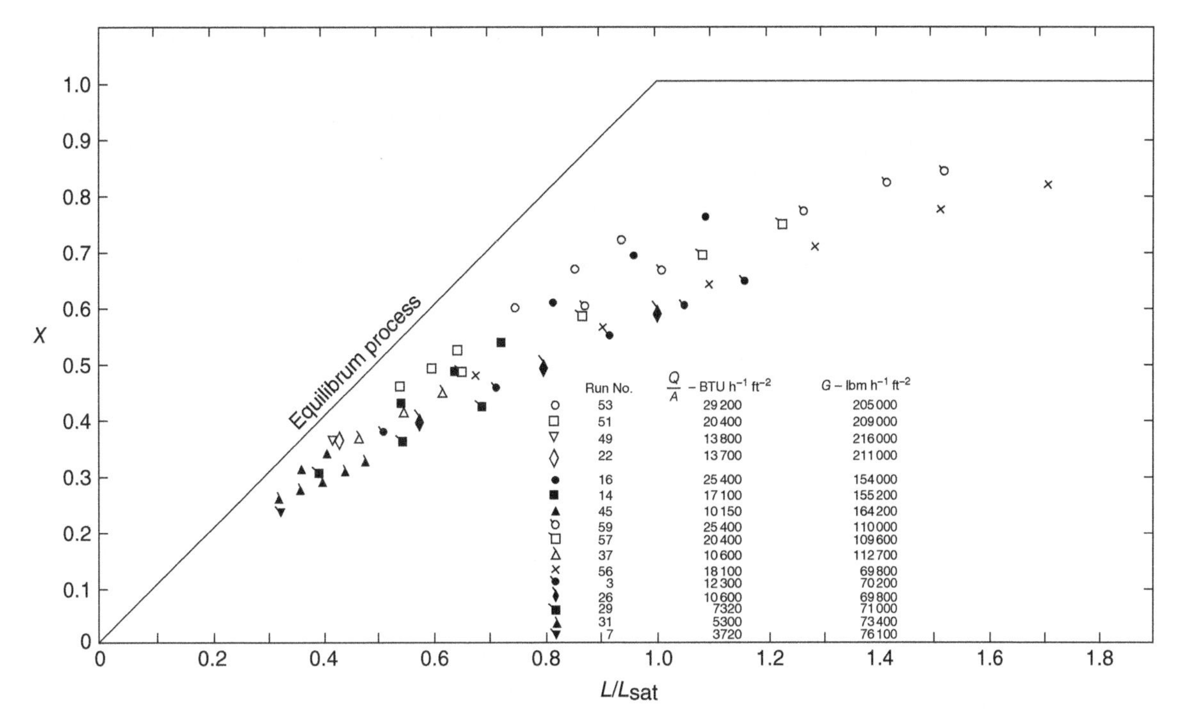

Figure 7.2.3 Estimated actual vapor quality x at various flow rates and heat fluxes. L is the location along the tube and L_{sat} is the tube length up to $x_E = 1$. Nitrogen boiling in a tube. Source: Laverty and Rohsenow (1964). © 1964 MIT.

Optimum values of the constants are the following: $a = 1.85 \times 10^{-4}$, $b = 1$, $c = 1.57$, $d = -1.12$, and $e = 0.131$. All data were for water. Their range included horizontal and vertical tubes: D = 2.5–25 mm, p = 68–215 bar, G = 700–5300 kg m^{-2} s^{-1}, and x_E = 0.1–0.9.

Another correlation that has been widely used for nuclear reactor licensing is the following by Dougall and Rohsenow (1963):

$$\frac{h_{TP}D}{k_g} = 0.023\left[Re_{GT}\left\{x_E + \frac{\rho_g}{\rho_f}(1-x_E)\right\}\right]^{0.8} Pr_g^{0.4} \tag{7.2.4}$$

All properties are at bulk fluid temperature. This correlation was found to be very erratic by Shah (2017) on comparison with a wide range of data. It mostly overpredicts heat transfer coefficients.

7.2.2.2 Mechanistic Analyses

Researchers at the Massachusetts Institute of Technology and UK Atomic Energy Establishment independently developed the two-step model, according to which heat is first transferred from tube wall to vapor and from vapor to the liquid drops suspended in the vapor stream. Dry or wet interactions are also considered possible, and radiation heat transfer is also considered in some models. Many mechanistic analysis techniques have been developed with this model. Examples are those of Bennet et al. (1967), Forslund and Rohsenow(1968), Varone and Rohsenow (1986), and Andreani and Yadigaroglu (1997). While this rational approach is highly desirable, the solution of the equations involves many assumptions and empirical factors such as droplet size distribution, effectiveness of droplet–wall interactions, effect of droplets on vapor velocity profiles, etc. Further, calculations using these models are very laborious. None of these models has been shown to agree with a wide range of data. While these analyses provide insight to the phenomena involved, these are as yet not suitable for practical design and analysis.

7.2.2.3 Phenomenological Correlations

A number of correlations have been presented, which take into consideration physical phenomena such as non-equilibrium. These may be regarded as semi-theoretical. These are discussed in this section.

There are two extreme possibilities for equilibrium following CHF, which are shown in Figure 7.2.4. One possibility is that all heat supply is utilized to evaporate liquid, and the vapor temperature stays at the saturation temperature. In other words, there is complete equilibrium. The other possibility is that no further evaporation occurs, and all the heat supply is used to superheat vapor. Thus the actual quality remains frozen at the critical quality. The actual situation is usually in between these two scenarios.

Some of the best available phenomenological correlations are described in the following text.

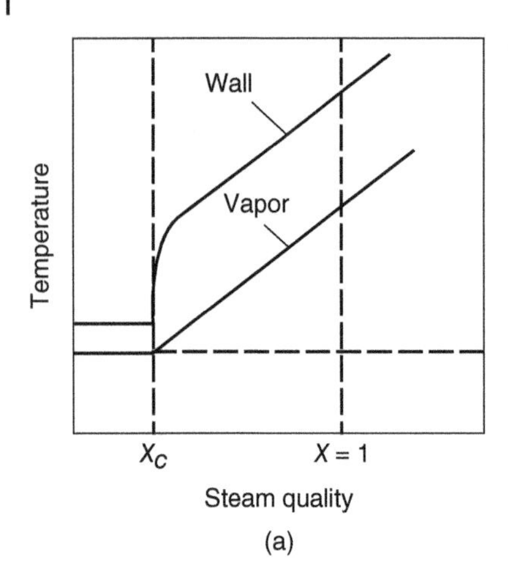

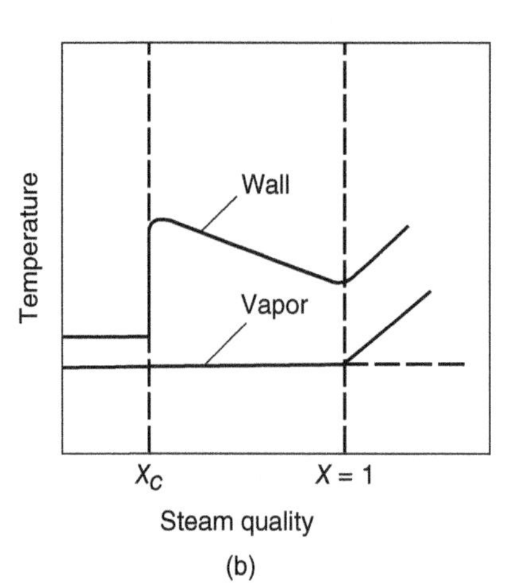

Figure 7.2.4 Extreme scenarios of equilibrium following CHF. (a) Maximum degree of non-equilibrium. (b) Complete thermal equilibrium. Source: Kohler and Hein (1986).

Groeneveld–Delorme Correlation One of this type, which has been widely used, is that of Groeneveld and Delorme (1976):

$$(i_A - i_E)/i_{fg} = \exp(-\tan\psi)\exp[-(3\alpha)^{-4}] \tag{7.2.5}$$

$$\psi = a_1 Pr_g^{a_2} Re_{TP}^{a_3} \left(\frac{qDC_{pg}}{k_g}\right)^{a4} \sum_{i=0}^{i=2} b_i (x_E)^i \tag{7.2.6}$$

If $\psi < 0$, $\psi = 0$. If $\psi > \pi/2$, $\psi = \pi/2$.

The fluid properties are at saturation temperature. Re_{TP} is defined as

$$Re_{TP} = \frac{GDx_1}{\mu_g \alpha} \tag{7.2.7}$$

Viscosity μ_g is at saturation temperature. Void fraction α is calculated based on the homogeneous model.

$$\alpha = \frac{x_1}{[x_1 + (\rho_g/\rho_f)(1 - x_1)]} \tag{7.2.8}$$

$x_1 = 0$ if $x_E < 0$. $x_1 = 1$ if $x_E > 1$. Otherwise, $x_1 = x_E$.

The constants were obtained by analysis of data for water. They are as follows: $a_1 = 0.138\,64$, $a_2 = 0.2031$, $a_3 = 0.200\,06$, $a_4 = -0.092\,32$, $b_0 = 1.3072$, $b_1 = -1.0833$, and $b_2 = 0.8455$.

Once i_A has been calculated, vapor temperature is known from thermodynamic tables. Actual vapor quality x_A is calculated by the following heat balance equation:

$$\frac{(x_E - x_A)}{x_A} = \frac{(i_A - i_E)}{i_{fg}} \tag{7.2.9}$$

Heat transfer coefficient is then calculated by the following equation of Hadaller and Bannerjee (1969), modified to include void fraction by the homogeneous model;

$$\frac{h_{TP}D}{k_g} = 0.008\,34 \left[Re_{GT}\left\{x_A + \frac{\rho_g}{\rho_f}(1 - x_A)\right\}\right]^{0.8774} Pr_g^{0.6112} \tag{7.2.10}$$

All properties are calculated at the film temperature, $(T_g + T_w)/2$. This correlation was compared to data for water from several sources, which included pressures from 7 to 215 bar, G from 130 to 4000 kg m^{-2} s^{-1}, and qualities from −0.12 to 1.6. The wall temperatures in °F were predicted with RMS error of 6.9%. Note that this correlation does not consider CHF quality x_c while experiments have shown that non-equilibrium is affected by it. For example, Hynek et al. (1969) performed tests in which all conditions (tube diameter, pressure, flow rate, etc.) were the same except that CHF occurred at different qualities. The measured wall temperatures for different critical qualities were found to be very different as seen in Figure 7.2.5.

Hein and Kohler Correlation The Hein and Kohler (1984) model assumed equilibrium at the CHF point. Downstream of the CHF point, there are three regions. In the first region, all heat is used to superheat vapor and there is no evaporation. This continues until fully developed non-equilibrium is reached. Then the second region starts in which evaporation occurs and vapor temperature remains constant. The second region ends when a limiting quality is reached. Beyond this point, wall temperature remains constant until all liquid has been evaporated. Further downstream, heat transfer occurs by single-phase convection.

The maximum superheat occurs at the end of region 1 and is expressed by

$$\Delta T_{SUP} = (T_g - T_{SAT}) = \frac{i_{fg}}{2C_{pg}}\left[\sqrt{1 + \frac{4qC_{pg}}{i_{fg}(h_{vd}A_d)}} - 1\right] \tag{7.2.11}$$

where h_{vd} is the heat transfer coefficient between vapor and droplets and A_d is the surface area of all droplets. Hein and Kohler noted that it is very difficult to calculate h_{vd} and A_d as they require the size of droplets and their distribution as well as the relative velocity between droplets and vapor. To overcome this difficulty, they developed the following correlation for $(h_{vd} \cdot A_d)$:

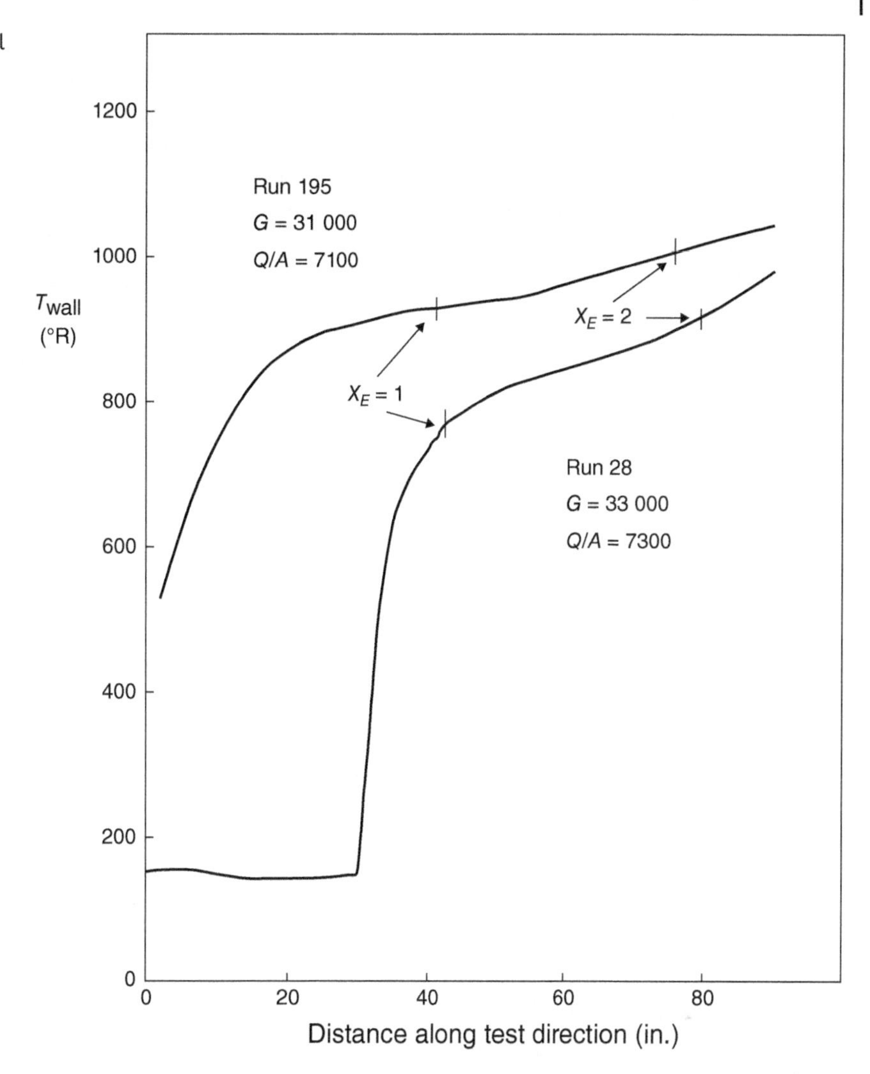

Figure 7.2.5 Effect of critical quality on wall temperature during film boiling of nitrogen in a vertical tube of 10.1 mm diameter. Source: Hynek et al. (1969). © 1969 MIT.

For $G/La \leq 1767 \times 10^3$,

$$(h_{vd}A_d) = 1.473 \times 10^{-7}(G/La)^{1.33} \tag{7.2.12}$$

For $G/La > 1767 \times 10^3$,

$$(h_{vd}A_d) = 3.078 \times 10^{4}(G/La)^{1.33} \tag{7.2.13}$$

La is the Laplace constant (also known as the capillary length) defined as

$$La = \sqrt{\frac{\sigma}{g(\rho_f - \rho_g)}} \tag{7.2.14}$$

From the above equations, the vapor temperature at the end of region 1 is known. The actual quality at this point is calculated as follows.

Equilibrium quality x_E at fully developed non-equilibrium (end of region 1), $x_{E,\min}$ is given by

$$x_{E,\min} = x_c + \frac{x_c C_{pg} \Delta T_{SUP}}{i_{fg}} \tag{7.2.15}$$

Now that the vapor temperature and equilibrium quality at the end of region 1 are known, x_A at this point is obtained from the heat balance equation, Eq. (7.2.9). Between, x_c and $x_{E,\lim}$, a linear rise in vapor temperature is assumed. Beyond this point, vapor temperature is assumed constant until a limiting quality $x_{\lim}$ is reached, which is given by

$$x_{\lim} = 0.7 + 0.002p \tag{7.2.16}$$

Pressure p is in bar. Beyond $x_{\lim}$, wall temperature is held constant until all liquid has been evaporated.

Heat transfer coefficient, h_{TP}, between vapor and wall is calculated by the Gnielinski equation (see Chapter 1) using Re_{TP} based on the assumption of homogeneous flow.

$$Re_{TP} = Re_{GT}\left\{x_A + \frac{\rho_g}{\rho_f}(1 - x_A)\right\} \tag{7.2.17}$$

All properties are calculated at the film temperature. The wall temperature T_w is

$$T_w = T_{SAT} + \Delta T_{SUP} + q/h_{TP} \tag{7.2.18}$$

This model showed good agreement with their own data for water in vertical tubes of diameter 12.5–25.4 mm. Pressure was 50–250 bar and mass flux from 300 to 2500 kg m^{-2} s^{-1}.

The above correlation is applicable only to water. Katsounis (1987) modified it to make it also applicable to other fluids. The modification is to replace Eq. (7.2.16) by the following relations:

For $\rho_f/\rho_g \leq 3$,

$$x_{\text{lim}} = 1.16 - 0.125\log_{10}(\rho_f/\rho_g) \tag{7.2.19}$$

For $3 < \rho_f/\rho_g \leq 26$,

$$x_{\text{lim}} = 1.25 - 0.32\log_{10}(\rho_f/\rho_g) \tag{7.2.20}$$

For $26 < \rho_f/\rho_g$,

$$x_{\text{lim}} = 0.97 - 0.12\log_{10}(\rho_f/\rho_g) \tag{7.2.21}$$

Katsouinis compared this modified version to data for hydrogen, water, and R-12. Agreement was satisfactory. Shah (2017) compared this correlation to a wide range of data for many fluids. Some data sets gave good agreement, while some had large deviations.

Saha Correlation A widely quoted correlation is that of Saha (1980). It also assumes equilibrium at CHF point and no interaction between liquid droplets and tube wall. Void fraction is obtained with the drift flux model of Zuber and Findlay (1965) assuming droplet distribution factor to be unity.

$$(1-\alpha) = \frac{1-x_A}{\left[1+\frac{(\rho_f-\rho_g)x_A}{\rho_g}\right]+\frac{\rho_f V_{ij}}{G}} \tag{7.2.22}$$

V_{ij} is the droplet drift velocity given by

$$V_{ij} = -1.41\left[\frac{g\sigma(\rho_f-\rho_g)}{\rho_g^2}\right] \tag{7.2.23}$$

The two-phase Reynolds number is then

$$Re_{\text{TP}} = \frac{GDx_A}{\mu_g\alpha} \tag{7.2.24}$$

Saha has given two alternative correlations. The first one is the K_1 correlation. K_1 represents the effectiveness of vapor to droplet heat transfer. It is obtained from the following equation:

$$K_1 = 6300(1-p_r)^2\left[\left(\frac{Gx_A}{\alpha}\right)^2\frac{D}{\rho_g\sigma}\right]^{0.5} \tag{7.2.25}$$

$$\frac{dx_A}{dx_E} = \frac{\Gamma_{g,A}}{\Gamma_{g,E}} = \frac{\Gamma_{g,A}}{qP_H/(A_c i_{fg})} \tag{7.2.26}$$

$\Gamma_{g,A}$ is the actual volumetric vapor generation rate and $\Gamma_{g,E}$ is that under equilibrium conditions. P_H is the heated perimeter and A_c is the cross-sectional area of tube.

$$\Gamma_{g,A} = K_1\frac{k_g(1-\alpha)(T_g-T_{\text{SAT}})}{D^2 i_{fg}} \tag{7.2.27}$$

These equations together with Eq. (7.2.9) enable the calculation of T_g and x_A. Wall-to-vapor heat transfer coefficient is calculated by the Heineman (1960) correlation as follows:

For $6 < L/D < 60$,

$$h_{\text{TP}} = 0.0157\frac{k_g}{D}Re_{\text{TP}}^{0.84}Pr_g^{0.33}\left(\frac{L}{D}\right)^{-0.04} \tag{7.2.28}$$

For $L/D > 60$,

$$h_{\text{TP}} = 0.0133\frac{k_g}{D}Re_{\text{TP}}^{0.84}Pr_g^{0.33} \tag{7.2.29}$$

L is the distance from the CHF point. All properties are to be evaluated at film temperature.

Saha's other correlation calculates the droplet diameter δ:

$$\frac{\delta}{D} = 1.47\left[\frac{\rho_g u_{g,c}^2\{\sigma/[g(\rho_f-\rho_g)]\}^{1/2}}{\sigma}\right]^{-0.675}\left(\frac{1-x_A}{1-x_c}\right)^{1/3} \tag{7.2.30}$$

where $u_{g,c}$ is the superficial vapor velocity at CHF, Gx_c/ρ_g.

$$K_1 = 6\left(\frac{h_{g-d}\delta}{k_g}\right)\left(\frac{D}{\delta}\right)^2 \tag{7.2.31}$$

where h_{g-d} is the vapor to droplet heat transfer coefficient given by the following equation:

$$h_{g-d} = \frac{k_g}{\delta}\left[2+0.459\left\{-\frac{\rho_g V_{ij}\delta}{\alpha\mu_g}\right\}^{0.55}\right]Pr_g^{0.33} \tag{7.2.32}$$

The vapor generation rate is then calculated by the following equation:

$$\Gamma_{g,A} = \left[\frac{6(1-\alpha)}{\delta}\right]\left[\frac{h_{g-d}(T_g-T_{\text{SAT}})}{i_{fg}}\right] \tag{7.2.33}$$

These together with Eq. (7.2.26) enable the calculation of x_A and T_g. Wall temperature can then be calculated using the Heineman correlation.

Saha compared this correlation with data for water in tubes (6–15 mm ID) for a wide range of pressures (29–120 bar), mass flux (393–2590 kg m^{-2} s^{-1}), equilibrium qualities (0.17–1.50), and wall superheats (167–557 K). Both versions gave good agreement with data. Shah and Siddiqui (2000) compared this correlation to data for many fluids over a wide range. They found that its void fraction model often predicted values greater than 1, which is physically incorrect. No prediction of heat transfer is possible in such cases. The same problem with the Saha correlation has been reported by Andreani and Yadigaroglu (1994) and Katsounis (1987). Webb and Chen (1986) tried the Saha correlation with void fraction by the homogeneous model; the agreement with the measured vapor generation rate was poor. For the data for which void fraction was properly calculated, Saha correlation had less deviations than the correlations of Kohler and Hein and Groeneveld and Delorme. This correlation cannot be used for subcooled CHF.

Shah Correlation Shah (1980) presented a correlation that was based on the two-step model but also considered droplet–wall interaction at high pressures. It had the functional form

$$x_A = \text{Function}\,(x_E, x_C, Bo, Fr_L) \tag{7.2.34}$$

This correlation was in graphical form. It was shown to be in agreement with data for many fluids over a wide range of data. Shah and Siddiqui (2000) converted it into equations and verified it with additional data. Finally, Shah (2017) made some more modifications and showed its agreement with additional data that included minichannels. The correlation is shown graphically in Figure 7.2.6. Equations for the correlation are the following.

The actual quality x_A is calculated as follows:

For $x_E \geq 0.4$,

$$x_A = (A_1 + A_2 x_E + A_3 x_E^2 + A_4 x_E^3) Fr_L^{0.064} \tag{7.2.35}$$

where $A_1 = -0.0347$, $A_2 = 0.9335$, $A_3 = -0.2875$, and $A_4 = 0.035$. x_A from Eq. (7.2.35) is corrected as follows: if $x_A > x_E$, then $x_A = x_E$. If $x_A > 1$, then $x_A = 1$.

For $x_E < 0.4$, the correlating curves in Figure 7.2.6 are represented by lines joining x_A at $x_E = 0.4$ from Eq. (7.2.35) and intersecting the equilibrium line ($x_A = x_E$) at

$$x_{A,\text{INT}} = x_{E,\text{INT}} = 0.19 {Fr_L}^{0.16} \tag{7.2.36}$$

Calculation is done as follows:

1. For $x_c \leq x_{E,\text{INT}}$, $x_A = x_E$ for $x_E \leq x_{E,\text{INT}}$.
2. For $x_c > x_{E,\text{INT}}$, determine the point where tangent from $x_E = x_A = x_c$ touches the curve of Eq. (7.2.35). The point of tangency ($x_{E,\text{TAN}}$, $x_{A,\text{TAN}}$) is at the intersection of Eqs. (7.2.35) and (7.2.37), obtained by simultaneous solution of the two equations.

$$x_A = x_c + (x_E - x_c) \cdot (A_2 + 2A_3 x_E + 3A_4 {x_E}^2) \cdot {Fr_L}^{0.064} \tag{7.2.37}$$

For $x_E < x_{E,\text{TAN}}$, x_A is obtained from the straight line joining the tangent point to x_c at the equilibrium line. Beyond the tangent point, x_A is given by Eq. (7.2.35). Thus for $x_E < x_{E,\text{TAN}}$,

$$x_A = \frac{(x_{A,\text{TAN}} - x_C)}{(x_{E,\text{TAN}} - x_C)} x_E + \frac{(x_{E,\text{TAN}} - x_{A,\text{TAN}})}{(x_{E,\text{TAN}} - x_C)} x_C \tag{7.2.38}$$

Having determined x_A, i_A is known from Eq. (7.2.9) and then T_g is obtained using thermodynamic tables or equations. Heat flux is then calculated as follows:

$$q = F_E F_{\text{dc}} h_g (T_w - T_g) \tag{7.2.39}$$

F_E is the entrance effect factor calculated by the following equation of McAdams (1954) for single-phase flow:

For $(z/D) < 30$,

$$h_z / h_\infty = 1 + (D/z)^{0.7} \tag{7.2.40}$$

where h_z is the heat transfer coefficient at distance z from the CHF location and h_∞ is the heat transfer coefficient at $(z/D) > 30$. For $(z/D) > 30$, $F_E = 1$.

F_{dc} is the factor for cooling due to wall–droplet interaction given by the following:

For $p_r > 0.91$,

$$F_{\text{dc}} = (0.983 - 0.015 {Fr_L}^{0.44})^{-1} \tag{7.2.41}$$

For $p_r \leq 0.91$, $F_{\text{dc}} = 1$.

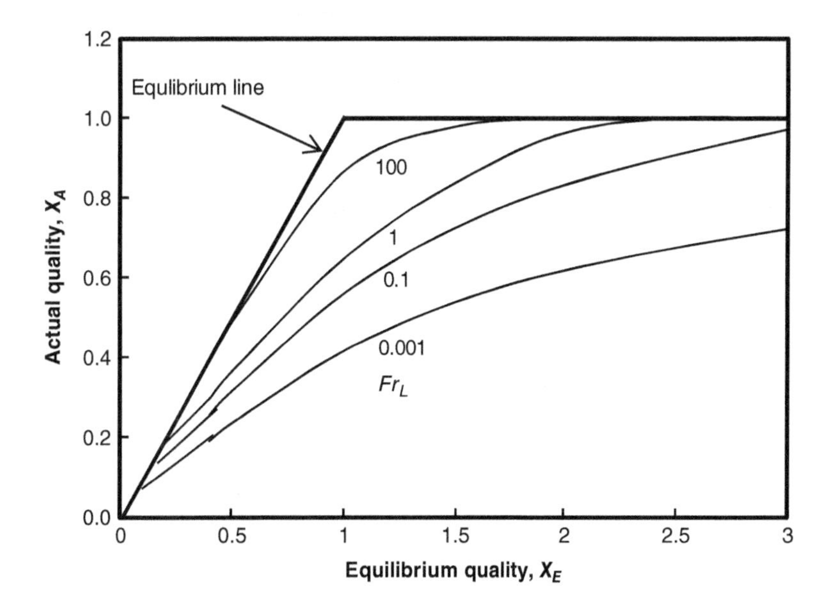

Figure 7.2.6 Shah correlation for heat transfer during film boiling in tubes. Source: Reproduced with permission from Shah (2017). © 2017 Elsevier.

Wall-to-vapor heat transfer coefficient is calculated by the following equations using properties at film temperature, which is the mean of wall and gas temperature:

For $Re_g < 10^4$,

$$\frac{h_g D}{k_g} = 0.023 Re_{TP}^{0.8} Pr_g^{0.4} \quad (7.2.42)$$

For $Re_g > 10^4$,

$$\frac{h_g D}{k_g} = 0.008\,34 Re_{TP}^{0.8774} Pr_g^{0.4} \quad (7.2.43)$$

Re_{TP} is calculated by Eq. (7.2.7) with x_1 replaced by x_A.

Void fraction α is calculated by the homogeneous model as

$$\alpha = \frac{x_A \rho_f}{(1 - x_A)\rho_g + \rho_f x_A} \quad (7.2.44)$$

This correlation was compared to data from many sources whose range is given in Table 7.2.1. All data were for uniform heat flux except those of Annuziatio et al. (1983) in which heat flux varied along the length. The mean absolute deviation (MAD) of heat transfer coefficients based on saturation temperature was 20.0%. The same data were also compared to the correlations of Dougall and Rohsenow, Groeneveld and Delorme, and Kohler and Hein (as modified by Katsounis). Their MADs were 158.3%, 28.3%, and 30.8%, respectively. Table 7.2.2 gives the details of this comparison. It is seen that the Shah correlation gives excellent agreement with data for minichannels ($D \leq 3$ mm). Figures 7.2.7–7.2.9 show comparison of some data with various correlations.

In his data analysis, Shah did not consider data for $x_E < 0.1$ as these could be for IAFB. Data for water for $(T_w - T_{SAT}) < 50$ K were excluded for $p_r < 20$ MPa so as to avoid the possibility of transition boiling. This criterion was also used by Groeneveld et al. (2003) in preparing their lookup table.

7.2.2.4 Lookup Tables

Several authors have published tables listing heat transfer coefficients during film boiling of water in tubes over a range of parameters. The most comprehensive among them

Table 7.2.1 Range of data for film boiling in vertical tubes analyzed by Shah.

Fluids	Water, helium, nitrogen, methane, propane, R-12, R-22, R-113, R-134a, and CO_2
Tube diameter, D (mm)	1.09–25.0
Reduced pressure	0.0046–0.9904
G (kg m^{-2} s^{-1})	3.7–5176
$Bo \times 10^4$	0.6–59.4
$Re_g \times 10^{-4}$	0.084–140
Fr_L	0.000 12–2518
x_C	−2.2–0.97
x_E	0.10–2.96
No. of data sources	36
No. of data points	1184

Source: Shah (2017). © 2017 Elsevier.

Table 7.2.2 Summary of results of comparison of data for film boiling in vertical tubes with various correlations.

Diameter (mm)	p_r	z/D	N	Groeneveld and Delorme	Kohler–Hein–Katsounis	Dougall and Rohsenow	Shah (2017)
All	>0.8	All	252	36.4	45.2	330.9	21.0
				−16.7	−32.5	330.7	−4.1
All	≤0.8	All	932	26.0	26.7	110.0	19.7
				2.1	2.6	106.7	0.5
All	All	≥30	1003	28.6	31.5	172.9	19.5
				1.6	−2.3	171.1	−1.9
All	All	<30	181	29.2	28.9	79.5	23.3
				−21.8	−20.2	71.6	9.0
>3	All	All	1125	27.9	30.6	152.0	20.6
				0.2	−3.4	149.2	−0.2
≤3	All	All	59	43.0	40.3	274.5	11.2
				−43.0	−35.8	274.5	1.5
All	All	All	1184	28.3	30.8	158.3	20.0
				−2.0	−5.0	155.6	−0.5

Source: Shah (2017). © 2017 Elsevier.

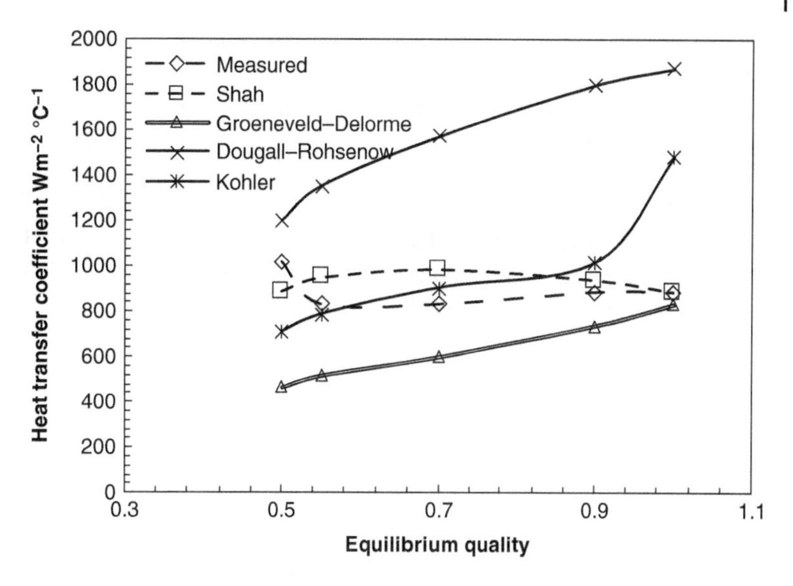

Figure 7.2.7 Comparison of the data of Ogata and Sato (1974) for helium with various correlations: $D = 1.09$ mm, $T_{SAT} = 4.35$ K, $G = 92$ kg m^{-2} s^{-1}, $q = 1.42$ kW m^{-2}, $x_C = 0.38$. Source: Shah (2017). © 2017 Elsevier.

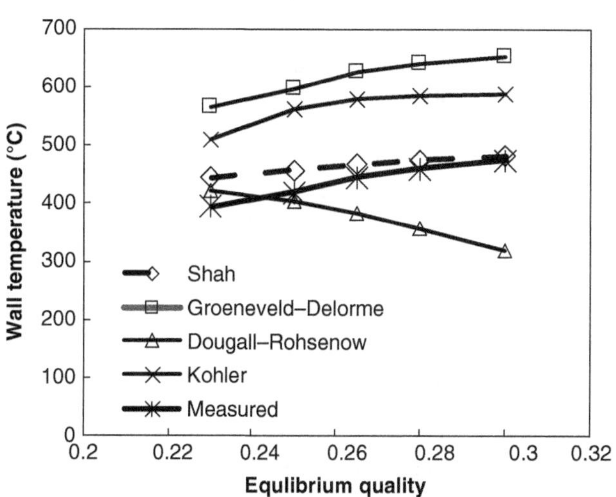

Figure 7.2.8 Comparison of the data of Nijhawan et al. (1980) for water with various correlations: $D = 14.1$ mm, $p = 2.65$ bar, $G = 42$ kg m^{-2} s^{-1}, $x_c = 0.20$. Source: Shah (2017). © Elsevier.

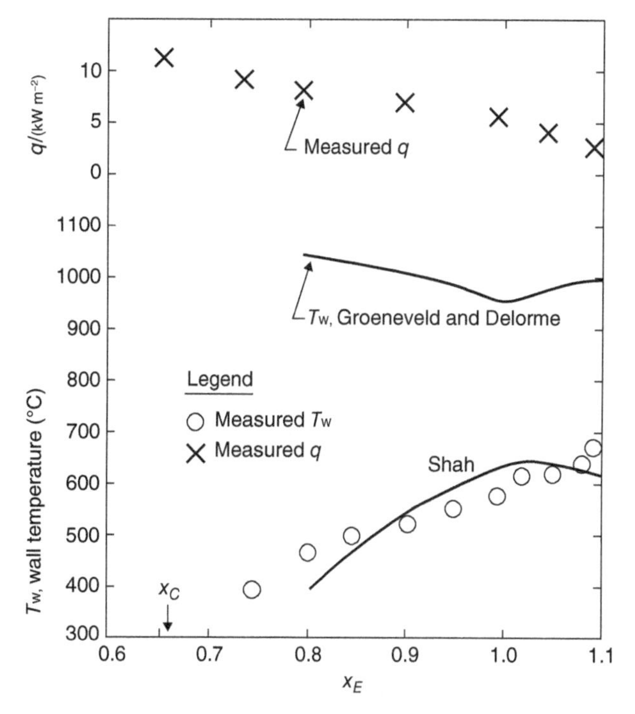

Figure 7.2.9 Data of Annunziato et al. (1983) for water in a non-uniformly heated tube compared to various correlations: $D = 12.4$ mm, $p = 1.01$ bar, $G = 4$ kg m^{-2} s^{-1}. Source: Shah and Siddiqui (2000). © 2000 Taylor & Francis.

is that of Groeneveld et al. (2003). It contains 29 744 entries over the following range: pressure from 0.1 to 20 MPa, mass flux from 0 to 7000 kg m^{-2} s^{-1}, thermodynamic equilibrium quality from −0.2 to 2.0, and wall superheat $(T_w - T_{SAT})$ from 50 to 1200 K. The listed heat transfer coefficients are for 8 mm diameter tubes. For other tube diameters, the heat transfer coefficient from the table is multiplied by $(0.008/D)^{0.2}$. The listed heat transfer coefficients are based on $(T_w - T_g)$. For $x_E < 1$, $T_g = T_{SAT}$. For $x_E > 1$, T_g is based on all mass flowing as vapor.

The table was prepared using experimental data that where available and extrapolations as well as correlations (such as Shah and Siddiqui, Kohler and Hein, and Groeneveld and Delorme) where data were not available. The entries based on test data are shown unshaded in the table and are considered the most reliable. There is less certainty about the entries shaded yellow and even less certainty about the entries shadowed red. The RMS error of predictions by the table on comparison with their database is reported to be 10.58%, while various correlations had much higher RMS error.

This prediction method does not take into account the effect of critical quality. As seen in Figure 7.2.5, test data clearly show that critical quality affects heat transfer coefficients.

The complete lookup table is available at http://www.magma.ca/~thermal.

7.2.2.5 Recommendations

The most accurate general correlation is Shah (2017). It is recommended for use for all non-metallic fluids.

The lookup table of Groeneveld et al. (2003) is recommended for water in the range of its unshaded entries. It should be kept in mind that predictions near the CHF point are likely to be inaccurate.

7.2.3 Prediction of Inverted Annular Film Boiling in Upflow

Ellion (1954) and Hsu (1977) proposed correlations for calculation of heat transfer coefficients during IAFB, which are modifications of the Bromley formula for free convection film boiling (Chapter 3). Neither of them includes the effect of forced flow. The correlation of Ellion is

$$h_{\mathrm{TP}} = h_{\mathrm{conv}} + h_{\mathrm{rad}} \tag{7.2.45}$$

The convective heat transfer coefficient h_{conv} is given by the following equation:

$$h_{\mathrm{conv}} = \frac{4}{3}\left[\frac{\rho_f \rho_g k_g^3 i_{\mathrm{fg}} g}{12\mu_g (T_w - T_{\mathrm{SAT}})z}\right]^{1/4} \tag{7.2.46}$$

$$h_{\mathrm{rad}} = \left(\frac{\sigma_{\mathrm{SB}}}{\varepsilon_w - \varepsilon_f - 1}\right)\frac{(T_w + 273)^4 - (T_{\mathrm{SAT}} + 273)^4}{(T_w - T_{\mathrm{SAT}})} \tag{7.2.47}$$

T_w is in °C, σ_{SB} is the Stefan–Boltzman constant, ε_w is the wall emissivity, and ε_f is the emissivity of liquid. z is the distance from CHF location.

Fung et al. (1979) compared these correlations with their data for water in a vertical tube of 11.9 mm ID covering a mass flux range of 50–500 kg m^{-2} s^{-1} and an inlet subcooling range of 5–70 K. Both gave satisfactory agreement with data at 100 kg m^{-2} s^{-1} but greatly underpredicted the measurements at 500 kg m^{-2} s^{-1}.

Mosaad and Johannsen (1992) developed a correlation that takes into account the effects of forced convection and subcooling. It is expressed by the following equations:

$$h_{\mathrm{TP}} = h_{\mathrm{conv}} + 0.75 h_{\mathrm{rad}} \tag{7.2.48}$$

$$h_{\mathrm{conv}} = \lambda(1 + 0.4\theta_s + 0.075\theta_s^2) h_{\mathrm{SAT}} \tag{7.2.49}$$

$$h_{\mathrm{SAT}} = \left[\frac{(\rho_f - \rho_g)\rho_g k_g^3 i'_{\mathrm{fg}} g}{16\mu_g (T_w - T_{\mathrm{SAT}})z}\right]^{1/4} \tag{7.2.50}$$

where i'_{fg} is the effective latent heat given by Eq. (3.6.2) and z is the distance from the quench front. θ_s is a local dimensionless subcooling number defined as

$$\theta_s = \frac{\Delta T_{\mathrm{SC}} h^*_{v-f}}{\Delta T_{\mathrm{SAT}} h_{\mathrm{SAT}}} \tag{7.2.51}$$

The vapor-to-liquid heat transfer coefficient at location z, h^*_{v-f} is given by

$$h^*_{v-f} = 0.8\left[1 + 3.5\left(\frac{D}{z}\right)\right] h_{v-f\infty} \tag{7.2.52}$$

$$h_{v-f\infty} = 0.06\left(\frac{k_f}{D}\right) Re_f^{0.8} Pr_f^{0.3} Pr_{\mathrm{g,film}}^{0.6}\left(\frac{\mu_g \rho_g}{\mu_f \rho_f}\right)^{0.15} \tag{7.2.53}$$

$$Pr_{\mathrm{g,film}} = Pr_g\left(\frac{i'_{\mathrm{fg}}}{C_{\mathrm{pg}}\Delta T_{\mathrm{SAT}}}\right) \tag{7.2.54}$$

The enhancement factor λ due to interfacial disturbances is given by the following equation:

$$\lambda = 1 + 1.1 Re_f^{0.45}\left(\frac{\mu_g \rho_g}{\mu_f \rho_f}\right)^{0.45}\frac{\delta}{D} \tag{7.2.55}$$

$$Re_f = \frac{(D - \delta)\rho_f u_f}{\mu_f} \tag{7.2.56}$$

u_f is the mean liquid velocity, which is calculated assuming uniform vapor layer around the liquid core.

$$u_f = \frac{G - G_g}{\rho_{f(1-\alpha)}} \tag{7.2.57}$$

The void fraction α is given by

$$\alpha = 1 - \left(1 - \frac{2\delta}{D}\right)^2 \tag{7.2.58}$$

δ is thickness of vapor film calculated as

$$\delta = \frac{\delta_s}{(1 + 0.4\theta_s + 0.075\theta_s^2)} \tag{7.2.59}$$

$$\delta_s = \frac{k_g}{h_{\mathrm{SAT}}} \tag{7.2.60}$$

This correlation is applicable for $We_g < 15$, which is defined as

$$We_g = \frac{\rho_g (u_g - u_f)^2 (D - 2\delta)}{\sigma} \tag{7.2.61}$$

The vapor velocity u_g is defined as

$$u_g = \frac{g(\rho_f - \rho_g)\delta^2}{12\mu_g} \tag{7.2.62}$$

This correlation was compared to steady-state data from five sources that covered the following range: D = 7.9–11.1 mm, p = 1–80 bar, G = 100–1010 kg m^{-2} s^{-1}, and inlet subcooling = 2–60 K. The RMS deviation for all data was 15.5%. Several other correlations were also compared with the same data. These included Sudo and Murao (1976), Groeneveld (1982) modified Berenson formula, Hsu (1975a,b), and Kaminage and Uchida (1978). All had large deviations. The correlation was also compared to reflood data. The RMS error was 22%, while the other correlations mentioned earlier had large deviations.

7.2.3.1 Recommendations

The correlation of Mosaad and Johannsen (1992) is the only one that has been verified with data from several sources. It is recommended for use in its verified range of parameters.

7.2.4 Film Boiling in Downflow

Robertshotte and Griffith (1982) measured heat transfer during flow of water down a vertical tube of 12.5 mm diameter, Mass velocity was 49–147 kg m^{-2} s^{-1}, wall temperature from 538 to 760 °C, pressure 1.3–2.6 bar, and quality 4.1–5.8%. They found that at the lowest mass flux, heat transfer approximated to the assumption of no evaporation after CHF. For the highest mass flux, heat transfer coefficients were approximated by the Dougall and Rohsenow correlation, Eq. (7.2.4).

Mawatari et al. (2014) studied film boiling heat transfer of R-134a in upflow and downflow in a vertical tube. Some of their test data is shown in Figure 7.2.10. It is seen that heat transfer coefficients during downflow are higher than those in upflow. The difference is most pronounced at the lowest flow. It becomes less with increasing mass flux, especially at higher qualities. Mawatari et al. attributed the higher heat transfer coefficients in downflow to buoyancy effects. At negative and low qualities, IAFB is likely to have occurred, while at higher qualities, DFFB would have occurred.

Shah (2017) compared the data of Mawatari et al. for upflow with his correlation for film boiling. Satisfactory agreement was found. Note that Shah compared only the data for quality >0.1 as IAFB is possible at lower qualities and that correlation is only for DFFB.

It is clear that predictive methods for upflow are applicable to downflow only above some minimum flow rate, but no method is available to determine this minimum flow rate. No prediction method is available for downflow at low flow rates.

7.3 Film Boiling in Horizontal Tubes

During DFFB in horizontal tubes, gravity tries to pull the liquid droplets down while inertia tries to keep them suspended in vapor. At high flow rates, inertia force prevails and flow remains symmetrical. Heat transfer then is the same as in vertical flow. As was described in Chapter 6, at low flow rates, gravity prevails and stratification occurs causing dryout to occur first at the top of the tube, while dryout at the bottom of the tube occurs further downstream. A typical example is shown in Figure 6.2.5. Another example is shown in Figure 7.3.1, which shows wall temperatures around the circumference at top, bottom, and side. It is seen that CHF occurs first at the top, then on the side, and then at the bottom. Wall temperature is highest at the top, lowest at the bottom, and in between the two at the side. The difference in circumferential wall

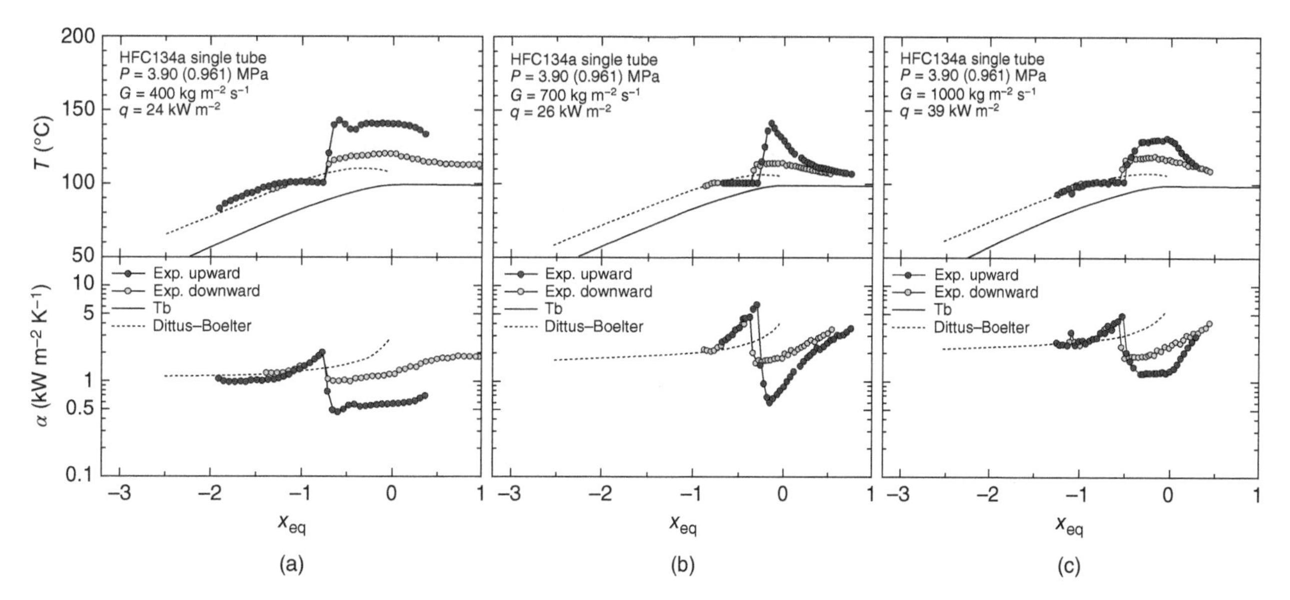

Figure 7.2.10 Comparison of heat transfer coefficient α and wall temperature T during upward and downward flow of R-134a in a vertical tube of 4.4 mm ID. Source: Mawatari et al. (2014). ©2014, with permission from Begell House, Inc.

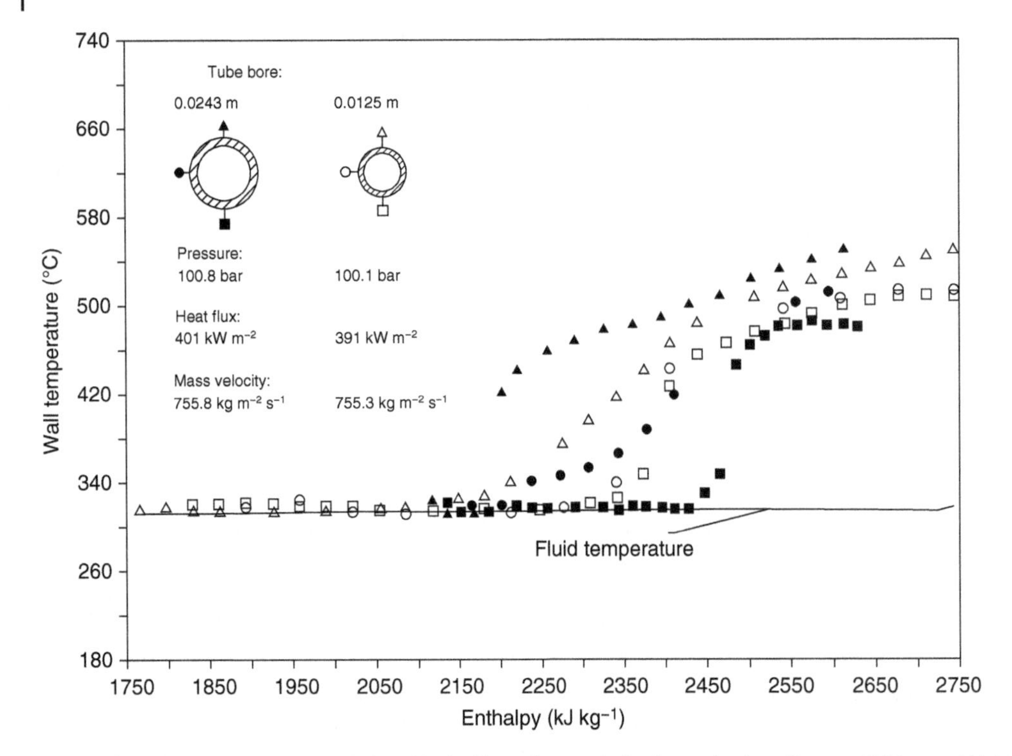

Figure 7.3.1 Wall temperatures during film boiling of water in horizontal tubes. Source: Kohler and Hein (1986).

temperature is seen to persist well beyond the dryout locations.

7.3.1 Prediction Methods

Shah and Siddiqui (2000) proposed applying their correlation for vertical tubes to the top and bottom of horizontal tubes simply by using the critical qualities at these locations and then proceeding the same way as for vertical tubes. Data from three sources were predicted with a MAD of 25.5%.

Shah (2017) took a closer look at the trends shown by test data. On the basis of the trends found, he modified the approach of Shah and Siddiqui. The following is the method for calculation when $x_{c,\mathrm{BOT}} > x_{c,\mathrm{TOP}}$:

- For the tube length between $x_{C,\mathrm{TOP}}$ and $x_{C,\mathrm{BOT}}$, calculate temperature at the top of tube assuming equilibrium. Further downstream, calculate the temperature at the top in the same way as for vertical tubes.
- For the tube length between $x_{C,\mathrm{SIDE}}$ and $x_{C,\mathrm{BOT}}$, calculate temperature at the side of tube assuming equilibrium. Further downstream, calculate the temperature at the side in the same way as for vertical tubes.
- When $(x_{C,\mathrm{BOT}} - x_{C,\mathrm{TOP}}) > 0.13$, calculate temperature of tube at bottom assuming equilibrium. When $(x_{C,\mathrm{BOT}} - x_{C,\mathrm{TOP}}) \leq 0.13$, assume non-equilibrium at the bottom for $x_E > x_{c,\mathrm{BOT}}$ and calculate x_A in the same way as for vertical tubes.

Shah compared this correlation with data from horizontal tubes from seven sources whose range is listed in Table 7.3.1. Critical quality was taken from the experimental data analyzed. The 297 data points were predicted with MAD of 17%. The same data were also compared to the correlations of Groeneveld and Delorme, Hein and Kohler, and Dougall and Rohsenow. The results are given in Table 7.3.2. The MAD of these correlations were 40.4%, 26.6%, and 100.6%, respectively. Clearly, the Shah correlation has much better accuracy. Figures 7.3.2 and 7.3.3 show comparison of the Shah correlation, and others with some test data. Note that all comparisons were made with heat transfer coefficient based on ΔT_{SAT}.

For the calculation of film boiling heat transfer with the Shah correlation, prediction of critical quality is needed. The critical quality at the top of the tube may be calculated by the Shah (2015) correlation for CHF in horizontal tubes described in Chapter 6. The quality at the bottom of tube may be obtained by using the correlation of Kohler and Hein (1986), according to which

$$(x_{C,\mathrm{BOT}} - x_{C,\mathrm{TOP}}) = \frac{16}{(2 + Fr_{\mathrm{TP}})^2} \tag{7.3.1}$$

$$Fr_{\mathrm{TP}} = \frac{x_{C,\mathrm{vert}} G}{\{gD\rho_g(\rho_f - \rho_g)\}^{0.5}} \tag{7.3.2}$$

Table 7.3.1 Range of parameters in data for film boiling in horizontal tubes analyzed by Shah (2017).

Fluids	Water, R-12, CO_2
Tube diameter, D (mm)	0.98–24.3
Reduced pressure	0.2228–0.9119
G (kg m^{-2} s^{-1})	383–2394
$Bo \times 10^4$	0.7–7.1
$Re_g \times 10^{-4}$	3.6–140
Fr_L	3.4–291
x_E	0.15–1.2
No. of data sources	7
No. of data points	297

Source: Based on Shah (2017).

The critical quality in vertical tubes $x_{c,\text{vert}}$ can be calculated by the methods given in Chapter 6. Note that the Kohler and Hein method is not applicable for subcooled burnout as Fr_{TP} becomes negative.

7.3.2 Recommendations

The Shah correlation has been verified with data from many sources and gives much better agreement than other correlations. It is therefore recommended in its verified range of parameters.

7.4 Film Boiling in Various Geometries

7.4.1 Annuli

Groeneveld (1973) analyzed data for water in internally heated annuli. Their Eq. (7.2.1) applies to annuli with the following values of the coefficients: $a = 0.013$, $b = 0.664$, $c = 1.68$, $d = -1.12$, and $e = 0.133$.

The range of data analyzed included $D_{HYD} = 1.5$–6.3 mm, $p = 34$–100 bar, $G = 800$–4100 kg m^{-2} s^{-1}, $x_E = 0.1$–0.9, and $Y = 0.61$–0.963. A total of 266 data points was predicted with RMS error of 6.1%.

Shiotsu and Hama (2000) performed tests on an internally heated annulus of 3 mm ID and 40 mm OD. R-113 flowed upwards in the annulus. Flow velocities ranged from 0 to 3 m s^{-1}, pressures from 102 to 490 kPa, liquid subcooling from 0 to 60 K, and ΔT_{SAT} up to 400 K. Tests were also done with saturated water in the same range with ΔT_{SAT} up to 500 K. They found that there was no effect of velocity up to 1 m s^{-1}. Heat transfer coefficients increased with further increase in velocity. The data for the low velocities were in agreement with the correlations of Sakurai et al. (1990a, 1990b) for natural convection pool boiling that has been given in Section 3.6.1.2. Another formula was developed for the higher velocities.

Tian et al. (2006) studied film boiling of water in bilaterally heated vertical annular channels with 1.0, 1.5, and 2.0 mm gap sizes at pressure ranging from 1.38 to 5.9 MPa

Table 7.3.2 Results of comparison of data for film boiling in horizontal tubes with various correlations done by Shah (2017).

Diameter (mm)	p_r	z/D	N	Groeneveld and Delorme	Kohler	Dougall and Rohsenow	Shah (2017)
All	>0.8	All	11	1087.0	42.4	228.4	15.6
				315.2	−42.4	228.4	3.9
All	≤0.8	All	286	41.8	25.9	90.4	17.1
				12.1	2.5	85.5	−6.3
All	All	≥30	261	45.8	25.9	109.6	15.9
				13.3	5.1	106.9	−6.8
All	All	<30	36	34.3	31.4	332.1	25.1
				5.1	−27.1	96.3	0.2
>3	All	All	264	45.3	24.9	103.0	16.3
				13.1	−3.6	96.4	−4.7
≤3	All	All	33	362.3	39.4	80.3	22.7
				105.1	36.4	78.5	−15.7
All	All	All	297	40.4	26.6	100.6	17.0
				11.8	0.9	95.7	−5.9
All	All	All	1481	30.7	29.9	146.7	19.4
				0.8	−3.8	143.6	15.8

Source: Shah (2017).© 2017 Elsevier.

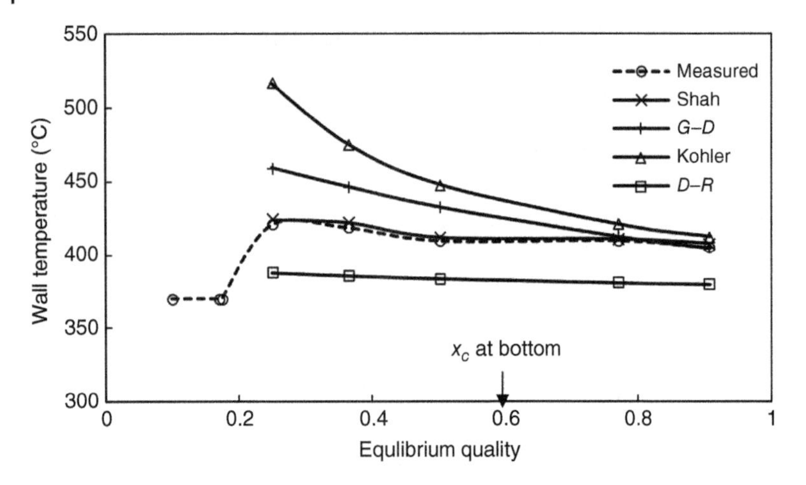

Figure 7.3.2 Comparison of the data of Kohler and Hein (1986) for top of tube with various correlations: $D = 12.5$ mm, $p = 200.7$ bar, $G = 1054$ kg m^{-2} s^{-1}, $q = 405$ kW m^{-2}, $x_{C,TOP} = 0.19$, $(z/D) = 26-271$. Source: From Shah (2017). © 2017 Elsevier..

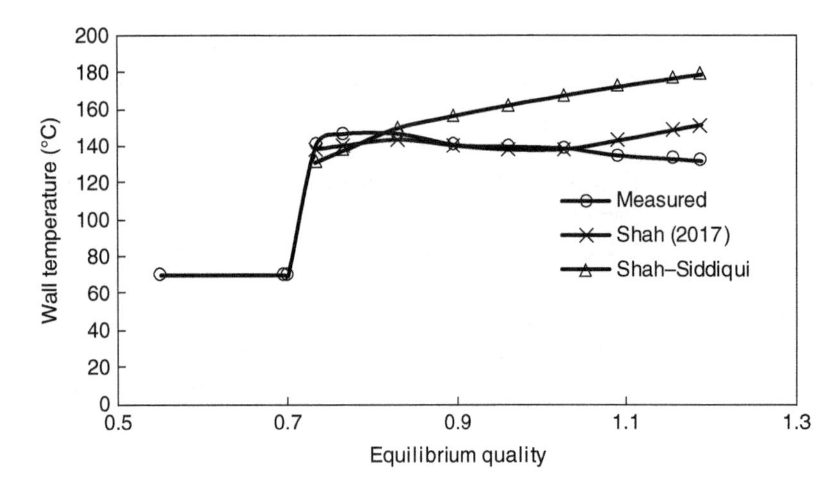

Figure 7.3.3 Comparison of correlations with data of Schnittger (1982) for bottom of tube: R-12, $D = 24.3$ mm, $p = 27.5$ bar, $G = 960$ kg m^{-2} s^{-1}, $q = 62.5$ kW m^{-2}, $x_{C,TOP} = 0.5$, $x_{C,BOT} = 0.7$, $(z/D) = 10-154$. Source: Shah (2017). © 2017 Elsevier.

and flow rate from 42.9 to 150.2 kg m^{-2} s^{-1}. The outer tube ID was fixed at 10 mm, while the inner tubes were 6, 7, and 8 mm OD. They compared their data with several empirical correlations including Groeneveld (1973), Eq. (7.2.1). All of them gave poor agreement. The flow rates in their data were much lower than in the data included in the Groeneveld correlation, and this may be the reason for its failure with these data. They performed a mechanistic analysis based on the two-step model but allowing droplet–wall interaction. It gave satisfactory agreement with their data.

In conclusion, no well-verified general predictive technique is available. The correlation of Groeneveld (1973) may be used for water in its verified range. It is possible that the phenomenological correlations for tubes may be applicable to annuli but verification is not available.

7.4.2 Vertical Tube Bundles

US Nuclear Regulatory Commission uses the following correlation in their COBRA-TF code, quoted in Sridharan (2005):

$$q = 0.62\left(\frac{D_{\mathrm{HYD}}}{\lambda_c}\right)^{0.172}\left[\frac{(\rho_f - \rho_g)\rho_g k_g^3 i'_{fg} g}{16\mu_g(T_w - T_{\mathrm{SAT}})D}\right]^{1/4}(T_w - T_{\mathrm{SAT}}) \tag{7.4.1}$$

Inverted annular film boiling is considered to end at void fraction of 0.4. λ_c is the critical wavelength due to Taylor instability given by

$$\lambda_c = 2\pi\left[\frac{\sigma}{g(\rho_f - \rho_g)}\right]^{1/2} \tag{7.4.2}$$

The TRACE code of US NRC (2007) has the following correlation for IAFB heat transfer:

$$Nu = \frac{h\delta}{k_g} = 1 + 1.3(0.268\delta^{*0.77} - 0.34) \tag{7.4.3}$$

δ^* is non-dimensional vapor film thickness, given by

$$\delta^* = \frac{\delta^3 g\rho_g(\rho_f - \rho_g)}{\mu_g^2} \tag{7.4.4}$$

δ is the thickness of vapor film, calculated with Eq. (7.2.59). IAFB is considered to end at void fraction of 0.6.

Mohanta et al. (2017) performed tests on a vertical tube bundle and developed the following correlation:

$$Nu = \frac{h_{TP}\delta}{k_g} = C(\delta^*)^{0.74} Pr_{g,film}^{0.43}(1 + 0.0016 Re_{LS,in}^{0.53}) \left(1 + 4.42 \frac{Ja_{SC}}{Ja_{SH}^{1.23}}\right) \tag{7.4.5}$$

δ is calculated with Eq. (7.2.57), $Pr_{g,film}$ by Eq. (7.2.53), $C = 0.2$ for tubes, and 0.262 for rod bundles.

$$Ja_{SC} = \frac{C_{pf}(T_{SAT} - T_f)}{i_{fg}} \tag{7.4.6}$$

$$Ja_{SH} = \frac{C_{pg}(T_w - T_{SAT})}{i_{fg}} \tag{7.4.7}$$

The coefficients and exponents in this correlation were determined from their transient reflood tests on a 7 × 7 tube bundle in a square channel of 90.2 mm. The Inconel tubes were 9.5 mm OD, the heated length was 3.657 m, and the pitch was 12.6 mm. The range of test data was as follows: pressure 138–414 kPa, reflood rate of 0.076–0.152 m s^{-1}, and inlet subcooling of 11–83 K, with void fraction of 0–0.9. As the IAFB regime is considered to end at void fraction of 0.4–0.6, this correlation also covers the inverted slug flow regime. It gave good agreement with the authors' own data using their measured void fraction to calculate the vapor film thickness δ. It also agreed well with FLECHT test data from Hochreiter et al. (2012). Good agreement was obtained with the data of Fung et al. (1979) for flow in tubes using $C = 0.2$, which was obtained by analysis of these data.

Mohanta et al. (2017) also compared their test data with other prediction methods including Sudo (1980), Mosaad and Johannsen (1989), and the correlation in the TRACE code given earlier. The RMS errors were 104%, 49%, and 31%, respectively. They did not state why they evaluated the Mosaad and Johannsen (1989) correlation instead of the improved Mosaad and Johannsen (1992) correlation that superseded it.

7.4.3 Single Horizontal Cylinder

Bromley et al. (1953) theoretically and experimentally investigated film boiling of saturated liquids on horizontal cylinders with forced flow. They concluded that the Bromley formula for natural convection film boiling, Eq. (3.6.3), can be used for forced flow provided that $Fr < 1$. Fr is defined as

$$Fr = \frac{u^2}{gD} \tag{7.4.8}$$

u is the velocity of liquid approaching the cylinder. When $Fr > 4$, the following equations apply:

$$h_{TP} = h_{conv} + 7/8 h_{rad} \tag{7.4.9}$$

$$h_{conv} = 2.7 \sqrt{\frac{u k_g \rho_g i'_{fg}}{D \Delta T_{SAT}}} \tag{7.4.10}$$

The radiation heat transfer coefficient is calculated as between two parallel infinite plates, Eq. (3.6.6).

Motte and Bromley (1957) experimentally investigated film boiling of several subcooled fluids on cylinders of diameters 9.8–16.2 mm. Velocities varied from 0.9 to 4 m s^{-1} and subcooling from 11 to 44 K. They found that subcooling greatly increased heat transfer coefficients, while cylinder diameter did not have any clear effect. Liu et al. (1992) report that this correlation does not agree with their data at higher pressure.

Epstein and Hauser (1980) performed a theoretical analysis of film boiling on spheres and cylinders neglecting heat transfer from regions other than the stagnation zone. After empirical adjustments, they arrived at the following correlation:

$$\frac{\beta Nu}{Re_f^{1/2}} = 2.5 \left[\frac{1}{24A} + \left(\frac{2}{\pi}\right)^2 \left(\frac{B}{A}\right)^4\right]^{1/4} \tag{7.4.11}$$

where

$$\beta = \left[\left(\frac{\rho_g}{\rho_f}\right)^{1/2} \frac{\nu_g}{\nu_f}\right]^{1/2} \tag{7.4.12}$$

$$A = \frac{C_{pg}(T_w - T_{SAT})}{Pr_g i_{fg}} \tag{7.4.13}$$

$$B = \beta \frac{k_f}{k_g} \frac{C_{pg}(T_{SAT} - T_\infty)}{Pr_g i_{fg}} Pr_f^{1/2} \tag{7.4.14}$$

This correlation gave satisfactory agreement with the data of Bromley et al. (1953) for saturated boiling on cylinders, the data of Motte and Bromley (1957) for subcooled boiling on cylinder, and the data of Dhir and Purohit (1978) for spheres.

Liu et al. (1992) measured heat transfer during forced convection film boiling of subcooled R-113 and water on cylinders. The range of their data included cylinder diameters 0.7–5 mm, pressures 101–490 kPa, velocities 0–1 m s^{-1}, and subcooling 9–50 K. Some of their data are shown in Figure 7.4.1. They report that the correlation of Epstein and Hauser considerably overpredicted most of their data.

Liu et al. (1992) developed the following correlation for saturated film boiling:

$$\frac{Nu_g}{\left(1 + \frac{2}{Nu_g}\right)} = HKM^{1/4} \tag{7.4.15}$$

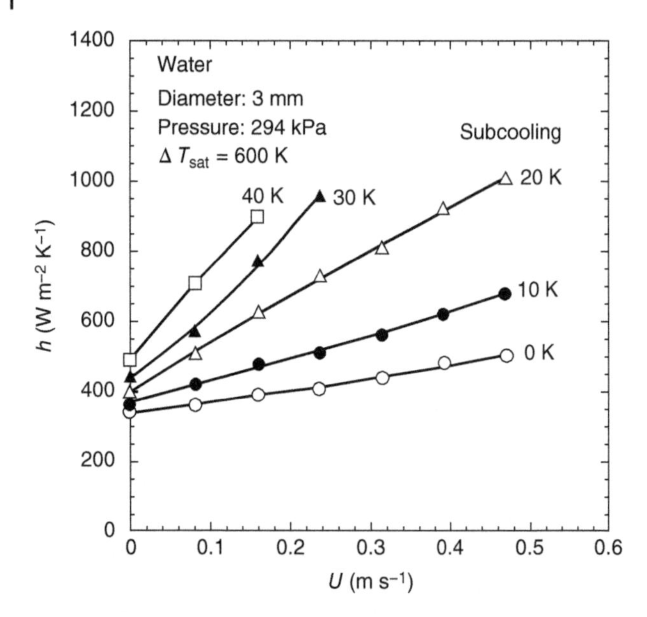

Figure 7.4.1 Effect of subcooling and velocity on heat transfer during film boiling on a horizontal cylinder. Source: Liu et al. (2009). 2009 ASME..

$$H = (1 + 0.68Fr^{2.5/4})^{1/2.5} + 0.45 \tanh\{0.04(D' - 1.3)Fr\} \quad (7.4.16)$$

$$Nu_g = h_{conv}D/k_g \quad (7.4.17)$$

$$Fr = \frac{u^2}{gD} \quad (7.4.18)$$

This correlation is a modification of the correlation of Sakurai et al. for natural convection film boiling presented in Section 3.6.1.2. The modification is the parameter H that takes into account the effect of forced convection velocity. Without H, it is the same as the natural convection correlation. The parameters K, M, and D' are the same as in the Sakurai et al. correlation.

This correlation gave good agreement with their own data, which included cylinder diameters 0.7–5 mm, pressures 101–490 kPa, and velocities 0–1 m s^{-1}. Fluids were water and R-113. Good agreement was also found with the data of Bromley et al. (1953) at atmospheric pressure, which included cylinder diameters 9.8–16.2 mm and four fluids.

Liu et al. (2009) noted that the term M takes into account the effect of subcooling in natural convection film boiling. Hence this correlation may also work for subcooled forced convection boiling. On comparison with test data, they found that the agreement was inadequate and that velocity seems to modify the effect of subcooling. Good agreement was obtained when Sc was replaced by Sc_2 defined as follows:

$$Sc_2 = [0.93Pr_f^{0.22} + 0.18(FrRe_f)^{0.22}]C_{pf}\Delta T_{SC}/i'_{fg} \quad (7.4.19)$$

where

$$Re_f = u\rho_f D/\mu_f \quad (7.4.20)$$

This correlation was compared to their own data, which included cylinder diameters 0.7–5 mm, pressures 101–490 kPa, velocities 0–1 m s^{-1}, and subcooling 9–50 K. Fluids were water and R-113. They also compared this correlation with data from several other sources including Motte and Bromley (1957) with good agreement. Figure 7.4.2 shows the comparison of this correlation with some of their data.

Liu et al. (1992, 2009) used the following method given by Sakurai et al. (1990a, 1990b) to obtain the total heat transfer coefficient h_{TP}, which includes both convection and radiation contributions:

$$h_{TP} = h_{conv} + Jh_{rad} \quad (7.4.21)$$

$$J = F + (1 - F)/(1 + 1.4h_{conv}/h_{rad}) \quad (7.4.22)$$

$$F = 1 - 0.25 \exp\left(-0.13\frac{C_{pg}\Delta T_{SAT}}{i_{fg}Pr_g}\right) \quad (7.4.23)$$

If $F < 0.19$, use $F = 0.19$.

A number of theoretical models have been developed. These include Malmazet and Berthoud (2009). Chou and Witte (1995), and Chou et al. (1995). These models were shown to agree with a limited amount of test data. The methodology involves numerical solution of differential equation sets and is difficult to use for practical design.

7.4.3.1 Recommendation

The correlations of Liu et al. (1992, 2009) for saturated and subcooled boiling, respectively, are the only prediction methods that have been verified over a wide range of data from many sources. They are therefore recommended in their verified range.

7.4.4 Spheres

Dhir and Purohit (1978) performed quenching tests on 19 and 25.4 mm diameter spheres of steel, copper, and silver by water in natural and forced flow. Water subcooling was from 0 to 50 K at velocities from 0 to 45 cm s^{-1}. They performed a theoretical analysis but the results underpredicted the data. They gave the following semi-theoretical correlation that agreed with their test data.

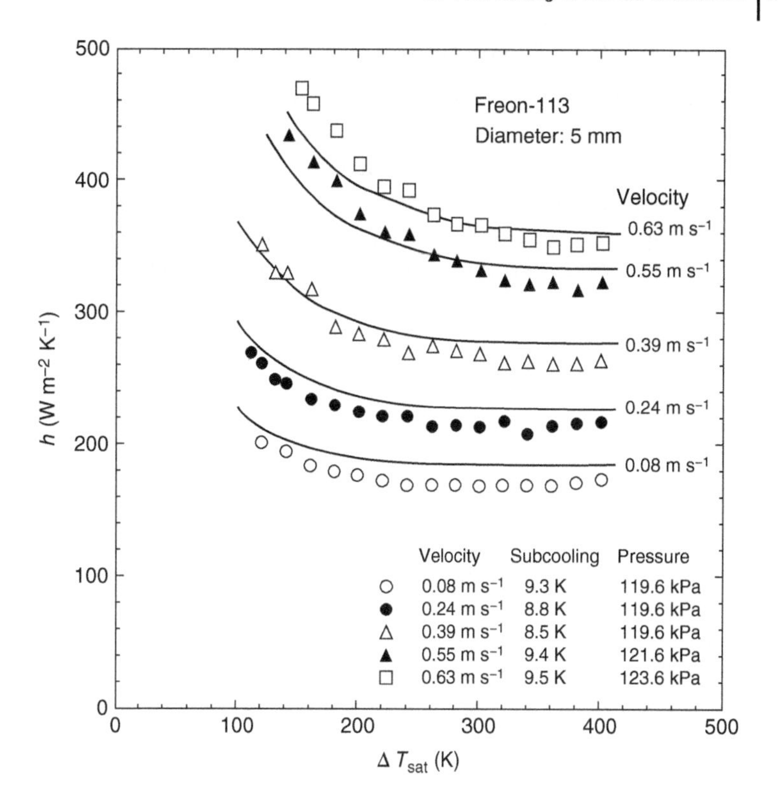

Figure 7.4.2 Comparison of the correlation of Liu et al. (2009) with some of their data for film boiling on cylinders. Source: Liu et al. (2009). © 2009 ASME..

For natural convection film boiling of saturated liquid,

$$Nu_0 = 0.8\left[\frac{g\rho_g(\rho_f - \rho_g)i_{fg}D^3}{\mu_g(T_w - T_{SAT})k_g}\right]^{1/4} \tag{7.4.24}$$

For forced convection of saturated and subcooled liquid,

$$Nu = Nu_0 + 0.8Re_f^{1/2}\left(1 + \frac{Ja_{SC}Pr_g\mu_f}{Ja_{SH}Pr_f\mu_g}\right) \tag{7.4.25}$$

Ja_{SC} and Ja_{SH} are given by Eqs. (7.4.6) and (7.4.7), respectively. Ja_{SC} varied from 0.0185 to 0.093 and Ja_{SH} varied from 0.16 to 0.45 in the data analyzed.

As stated earlier, Eq. (7.4.11) of Epstein and Hauser (1980) agrees with the data of Dhir and Purohit (1978).

Orcozo and Witte (1984) performed experiments in which R-11 flowed over a 38.1 mm copper sphere. Visual observations were made along with heat transfer measurements. The heat transfer regimes identified by them are shown in Figure 7.4.3. It is seen that there were two minimum heat flux points. The predictions of the theory of Witte and Orcozo (1984) agreed with the data in the thin wake regime that occurred after the second minimum heat flux point.

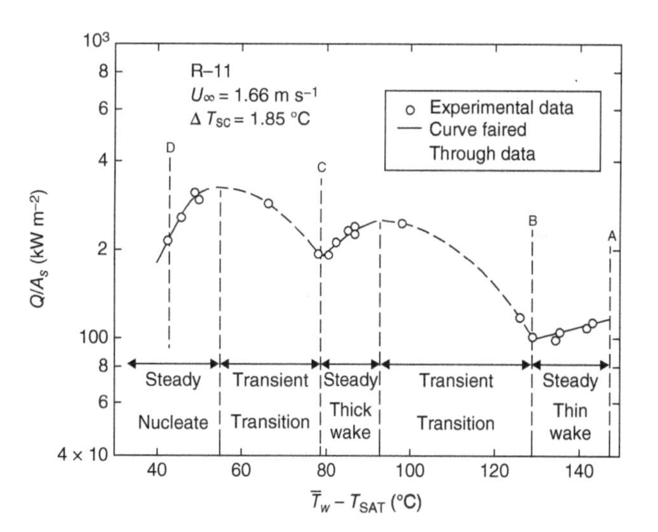

Figure 7.4.3 Heat transfer regimes during quenching of a sphere with forced flow. Source: Orcozo and Witte (1984). © ASME.

Liu and Theofanous (1996) performed experiments on film boiling on spheres with subcooled water as well as with steam–water on spheres of diameters 6–19 mm. They also performed tests with mixtures of steam–water flowing upwards and downwards. They developed a correlation that agreed with their own data. They also did tests with arrays of spheres.

Summarizing, there have been several experimental studies on film boiling on spheres, and a number of prediction methods have been proposed. However, none of the prediction methods has been verified with a wide range of data from many sources.

7.4.5 Jets Impinging on Hot Surfaces

Ruch and Holman (1975) performed tests in which vertical free stream jets of R-113 flowing upwards impinged on a copper disk of 12.9 mm diameter at 0°–45° to horizontal. Nozzles of diameter 0.21–0.433 mm were used. R-113 was at atmospheric pressure subcooled by 27 K. The data were correlated by the following dimensional equation within ±35%:

$$q = 157 u_{\mathrm{jet}}^{0.5} \Delta T_{\mathrm{SAT}}^{0.25} \tag{7.4.26}$$

The heat flux q includes both radiation and convection and is in Btu h^{-1} ft^{-2}. The jet outlet velocity u_{jet} was 14 500–81 200 ft h^{-1}, and ΔT_{SAT} ranged 110–600 °F. These units should be used in this equation. No effect of jet diameter or inclination was noticed.

Kokado et al. (1984) studied heat transfer on a stainless steel plate initially at 900 °C quenched by a circular water jet. Film boiling occurred beyond the wetted region around the point of impact that grew outwards with time. The total heat transfer due to convection and radiation in the film boiling region was correlated by the following equation:

$$h = 200 \frac{2420 - 21.7 T_f}{\Delta T_{\mathrm{SAT}}^{0.8}} \tag{7.4.27}$$

Heat transfer coefficient h is in W m^{-2} C^{-1}, and T is in °C.

Liu and Wang (2001) experimentally studied quenching of a horizontal stainless steel plate of 12 mm × 12 mm, 2 mm thick by a vertical free stream jet of water. The diameter of the jet nozzle was 10 mm, jet velocity was 1–3 m s^{-1}, and liquid subcooling was 5–80 K. By a simplified boundary layer analysis, they developed the following semi-theoretical correlation for film boiling in the stagnation zone:

$$\frac{q D_{\mathrm{jet}}}{\Delta T_{\mathrm{SAT}} k_f} = 2 Re_f^{1/2} Pr_{\mathrm{jet}}^{1/6} \left[\left(\frac{k_g}{k_f} \right) \left(\frac{\Delta T_{\mathrm{SC}}}{\Delta T_{\mathrm{SAT}}} \right) \right]^{1/2} \tag{7.4.28}$$

where

$$Re_{\mathrm{jet}} = \left(\frac{u_{\mathrm{jet}} D_{\mathrm{jet}}}{\nu_f} \right) \tag{7.4.29}$$

Vapor thermal conductivity k_g to be taken at the film temperature. This equation predicted their data within −5% and +25%. Figure 7.4.4 shows the comparison of this correlation with some of their data.

Filipovic et al. (1995a, 1995b) performed tests in which a wall jet of subcooled water quenched a heated horizontal oxygen-free copper block (25.2 mm thick, 38.1 mm wide, and 0.508 m long). Velocity was 2–4 m s^{-1} and water temperature was 25–55 °C. Using the integral solution of Filipovic et al. (1994) with the constants and exponents derived from their own test data, they arrived at the following equation:

$$\frac{h_z z}{k_g} = 0.0247 \beta (u_f^*)^{0.6} \frac{\mu_f}{\mu_g} D_2^{0.2} Re_z^{0.8} Pr_f^{1/3} \tag{7.4.30}$$

The various terms in this equation are defined as

$$u_f^* = 1 - \frac{u_i}{u_\infty} \tag{7.4.31}$$

$$\frac{u_i}{u_\infty} = \frac{1}{[1 + \beta Pr_f^{1/3}]} \tag{7.4.32}$$

$$\beta = \frac{Pr_g C_{\mathrm{pf}} (T_{\mathrm{SAT}} - T_\infty)}{Pr_f C_{\mathrm{pg}} (T_w - T_{\mathrm{SAT}})} \tag{7.4.33}$$

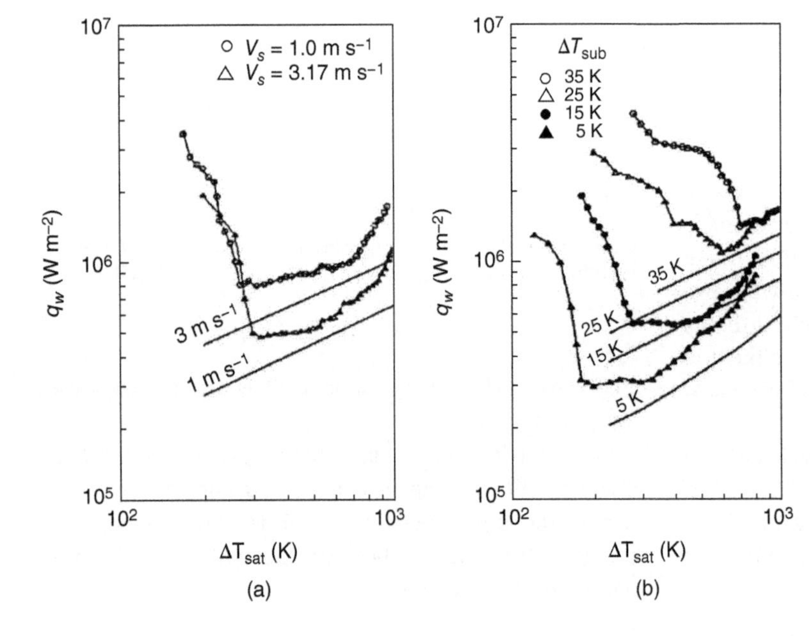

Figure 7.4.4 Effect of jet velocity V_s and subcooling on film boiling heat transfer in the stagnation zone with 6 mm diameter water jet (a) at 15 K subcooling and (b) at V_s = 2.1 m s^{-1}. The straight lines are the predictions of the Liu–Wang correlation. Source: Liu and Wang (2001). © 2001 Elsevier.

$$D_2 = 2\frac{u_i}{u_\infty} + 7 \tag{7.4.34}$$

$$Re_z = \frac{u_\infty z}{\nu_f} \tag{7.4.35}$$

In the foregoing, h_z is the local heat transfer coefficient at location z along the plate length, u_i is the velocity at liquid–vapor interface, and u_∞ is the free stream velocity. The above correlation gave good agreement with their data. Comparison was not made with any other data.

Barron and Stanely (1994) experimentally studied heat transfer to a horizontal plate during impingement of a free stream jet of liquid nitrogen flowing vertically downwards. The aluminum disk was 57 mm diameter. Jet nozzles of 1.6–4.8 mm diameter were used. Mass flow rate was 1–7 g s^{-1}. Tests were done under steady conditions as well as in transient conditions. For the steady-state tests, the mean heat transfer coefficient beyond the area directly below the jet was expressed by the following equation:

$$\frac{hD_s}{k_g} = 0.0885\left(\frac{D_s u_{\text{jet}} \rho_f}{\mu_g}\right)^{0.45} Pr_g^{1/3} \tag{7.4.36}$$

D_s is the disk diameter. The transient data were mostly higher than the steady-state data. According to these authors, this was because of the instability of the vapor layer and the waviness of the liquid layer during transient cool down.

In conclusion, there is no general prediction method for film boiling in jets. The correlations given above should be used only in their verified range.

7.4.6 Bends

Miroposkiy and Pikus (1969) measured CHF and film boiling heat transfer in bends of 45°–360°. They have given a correlation for film boiling in bends that provides factors to modify the film boiling heat transfer coefficient in straight tubes. The various factors in this correlation are given graphically.

7.4.7 Helical Coils

As was discussed in Section 6.4.6, CHF in coils occurs at higher qualities than in straight tubes. Also, CHF on the inner side of coil (that nearest to the coil axis) occurs at a lower quality than at the outer side (that furthest from the coil axis). Wall temperatures in coils are lower than in straight tubes. Wall temperatures at the outer side are lower than at the inner side of the coil. Figure 6.4.9 from the tests of Cumo et al. (1972) shows an example of this behavior. Cumo et al. (1972) also examined the effect of centrifugal acceleration a on the temperature difference between the two sides of the coil, $\Delta T_{o,s}$, during their tests with R-12 in a helical coil of 4.35 ID and 90 mm coil radius. Their results are shown in Figure 7.4.5. Centrifugal acceleration a is given by

$$a = \frac{G^2}{\rho_{\text{mix}}^2 R_{\text{coil}}} \tag{7.4.37}$$

R_{coil} is the radius of coil and ρ_{mix} is the density of vapor–liquid mixture. It is seen that the temperature

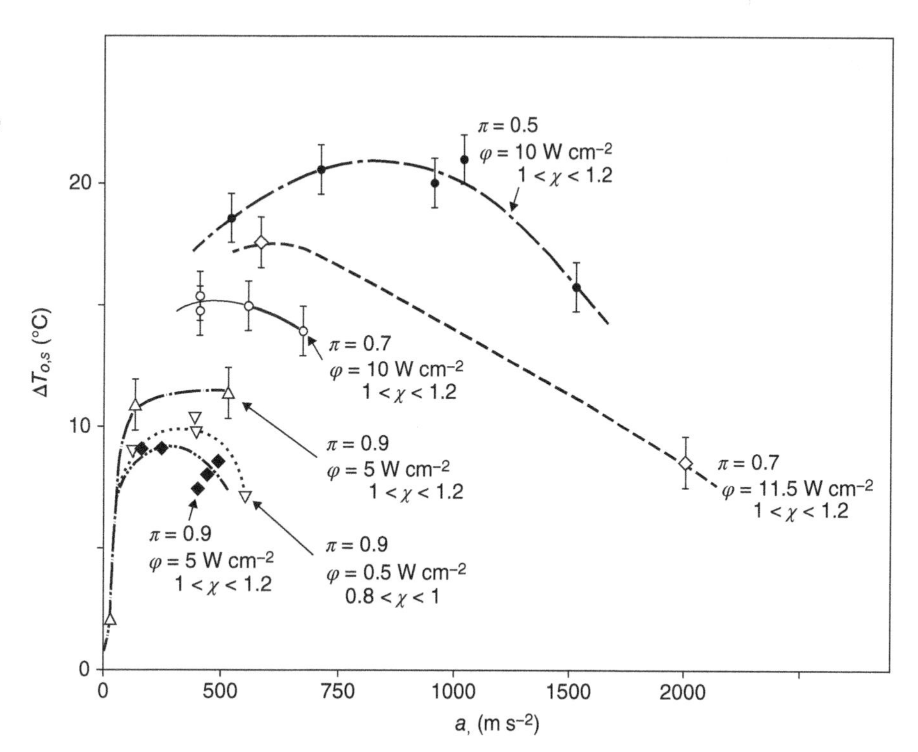

Figure 7.4.5 Post dryout temperature difference between the internal and external sides of a coil vs. the centrifugal acceleration of the mixture at various values of reduced pressure π. Source: Cumo et al. (1972). © 1972 Elsevier.

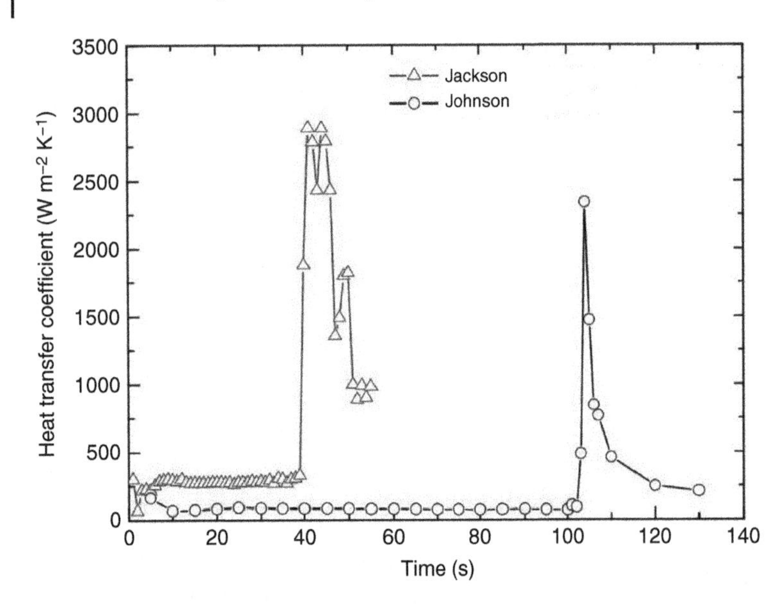

Figure 7.4.6 Heat transfer during chilldown of pipes with liquid nitrogen flowing in them: $G \approx 240$ kg m^{-2} s^{-1}. Source: Johnson and Shine (2015). © 2015 Elsevier.

difference initially increases with acceleration and then decreases with further increase in acceleration. It shows that the separating action of the centrifugal acceleration prevails against the contrary action of the secondary flow, which tends to make the mixture uniform, when acceleration (flow rate) reaches a certain value.

Chen and Zhou (1986) investigated forced convection boiling and post-dryout heat transfer in a straight tube and a vertical coil. The experiments were done in a high pressure water loop. In post-dryout region, the ratio of tube diameter to coil diameter was found to strongly affect the circumferential distribution of wall surface temperature. The post-dryout heat transfer coefficients in the coil were significantly higher than those in straight horizontal tubes. They correlated their data for the inner surface of coil with a modified form of the Miroploskiy and Pikus (1969) graphical correlation for bends.

7.4.8 Chilldown of Cryogenic Pipelines

Before transferring cryogenic liquids to tanks, the connecting pipe line and the tank have to be cooled down to cryogenic temperatures. This process is known as chilldown. It is done by passing liquid cryogens into the pipe. Nucleate, transition, and film boiling occur during the cooling process.

Johnson and Shine (2015) performed tests with liquid nitrogen entering a horizontal 20 mm diameter stainless steel pipe line. Some of their data are shown in Figure 7.4.6 together with the data of Jackson (2006) in a 12.5 mm horizontal pipe. Film boiling occurs for a considerable time; heat transfer coefficients are very low in this phase. It is followed by transition boiling during which heat transfer coefficient rises sharply. It is then followed by a short period of nucleate boiling in Jackson's test. Heat transfer coefficients then drop again and correspond to single-phase vapor flow. Johnson and Shine also performed tests in which the tube was inclined up to 15° to horizontal. Chilldown time was reduced for inclinations up to 10° but became higher at 15° inclination compared with the 10° inclination.

Hu et al. (2012) studied nitrogen flowing upward and downward in a vertical 8 mm diameter glass pipe. Mass flux was 20–80 kg m^{-2} s^{-1} and pressure was atmospheric. Figure 7.4.7 shows some of their data. As seen in it, heat transfer coefficients during transition boiling and nucleate boiling were higher during downflow, but film boiling heat transfer coefficients were higher during upflow. Chilldown time was much shorter for upflow.

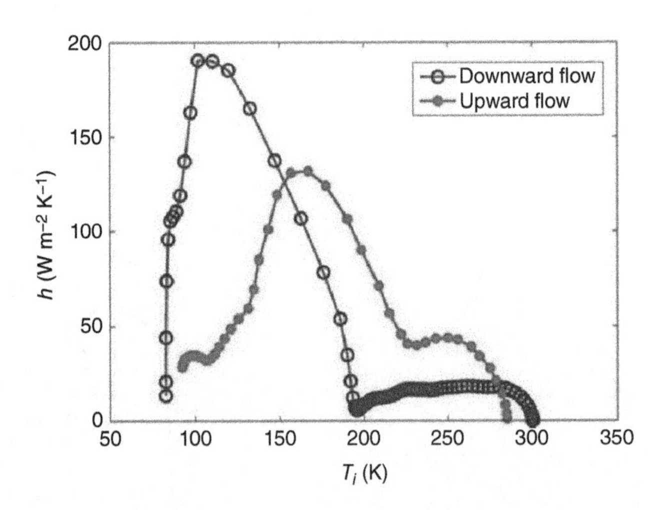

Figure 7.4.7 Heat transfer coefficients as function of wall temperature T_i during quenching of a vertical tube with upflow and downflow of nitrogen. Source: Hu et al. (2012). © Elsevier.

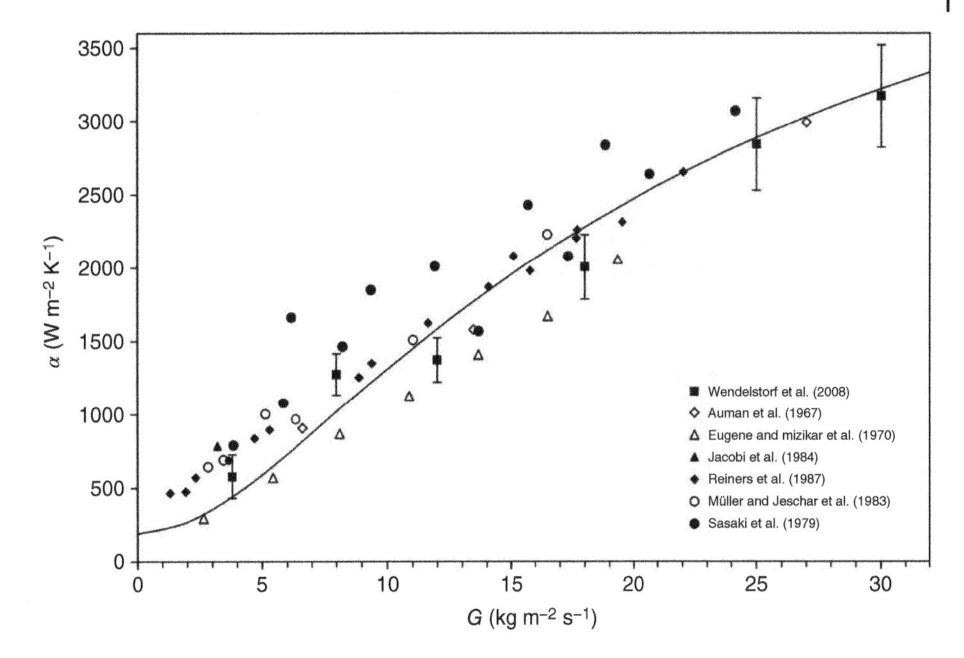

Figure 7.4.8 Film boiling heat transfer coefficients α at $\Delta T_f \approx 700$ K from several sources compared to the correlation of Wendelstorf et al. (2008). Source: Wendelstorf et al. (2008). © 2008 Elsevier.

Mercado et al. (2019) examine chilldown data from almost all published sources. They considered data from many of the studies either unreliable or not possible to analyze due to lack of needed information. The data from the remaining studies were compared with many well-known prediction methods for nucleate boiling, CHF, and film boiling. All correlations in all regimes gave poor agreement with data. Thus it appears that steady-state prediction methods are inapplicable to cryogenic chilldown. For CHF with nitrogen, their measured value was 13.8 kW m^{-2}, while the predictions of four correlations were 24–163 kW m^{-2}. For hydrogen, measured CHF was 233 kW m^{-2}. Predictions of three correlations were 87–144 kW m^{-2}, while the fourth predicted 377 kW m^{-2}. It is curious that for nitrogen all correlations grossly overpredict, while for hydrogen three of the four grossly underpredict.

Due to the interest in space exploration, there have been many studies on the effect of gravity on chilldown heat transfer.

Kawamania et al. (2007) studied chilldown in a 7 mm diameter pipe at Earth condition and at low gravity in a drop shaft. Mass flux of nitrogen was 100–300 kg m^{-2} s^{-1}. They found that heat transfer and quench front velocity under microgravity conditions increased up to 20% from those under 1 g, and the difference in the heat transfer characteristics under 1 g and μg decreased with increasing mass velocity.

Among other studies on effect of gravity are those of Verthier et al. (2009) and Yuan et al. (2008). However, these have not resulted in any reliable general methods for predicting the effects of gravity on film boiling or other regimes of boiling.

In conclusion, there are no verified general methods for predicting nucleate boiling, film boiling, and CHF during chilldown, in normal or reduced gravity.

7.4.9 Flow Parallel to a Plate

Cess and Sparrow (1961) considered laminar film boiling on a plate under the condition that effect of gravity is negligible. They developed a formula by solving the conservation equations for mass momentum and energy. No comparison was made with any test data.

7.4.10 Spray Cooling

It is generally agreed that spray flow rate is the most important parameter affecting film boiling heat transfer. There are conflicting reports about other factors such as surface temperature, nozzle type, subcooling, distance of nozzle from surface, droplet velocity, and spray orientation.

Wendelstorf et al. (2008) performed tests with water spray from full cone nozzle flowing down on a nickel plate of 70 mm diameter. Heat transfer coefficient was found to decrease with increasing ΔT_{f}, the wall-to-liquid temperature difference. The following correlation of their data was given:

$$h = 190 + \tanh\left(\frac{G}{8}\right)\left\{140G\left(1 - \frac{G\Delta T_f}{72\,000}\right)\right.$$
$$\left. +3.26(\Delta T_f)^2\left[1 - \tanh\left(\frac{\Delta T_f}{128}\right)\right]\right\} \qquad (7.4.38)$$

In these tests, water temperature was 18 °C, surface temperature is 473–1373 K, G is 3–30 kg m^{-2} s^{-1}, droplet velocity is 13–15 m s^{-1}, and droplet diameter is 300–400 μm. Test data from other sources was available for $\Delta T_f \approx 700$ K. Those were found to be in agreement with this correlation as seen in Figure 7.4.8.

Klinzing et al. (1992) did tests with water spray from one flat spray and four different full cone nozzles on a gold-plated copper disk. They gave the following correlations of their data:

For low spray flux, $0.58\times10^{-3}<Q''<3.5\times10^{-3}$ m^3 s^{-1}m^{-2},

$$q = 63.25\Delta T_f^{1.691} Q''^{0.264} D_{32}^{-0.062} \tag{7.4.39}$$

For high spray flux, 9.96×10^{-}3>Q''>3.5×10^{-3}m^{3}s^{-1}m^{-2},

$$q = 141\,345\Delta T_f^{0.461} Q''^{0.566} u_m^{0.639} \tag{7.4.40}$$

where u_m is the mean droplet velocity. The range of parameters in these tests was as follows: $Q'' = 0.58\times10^{-3}$–9.96×10^{-3} m^3 s^{-1} m^{-2}, u_m = 10.1–29.9 m s^{-1}, D_{32} = 0.137–1.350 mm, and surface temperatures up to 520 °C. In the above equations, fluid properties are calculated at the film temperature. SI units given in Nomenclature are used.

More information on this topic may be found in Liang and Mudawar (2017).

7.5 Minimum Film Boiling Temperature and Heat Flux

Very few prediction methods are available for forced convection boiling. Available information for some geometries is given in the following text.

7.5.1 Flow in Channels

Groeneveld and Stewart (1982) have given the following correlations based on water data:

For $p < 9000$ kPa,

$$T_{\mathrm{MFB}} = 284.7 + 0.0441p - 3.72\times10^{-6}p^2 + \frac{C_{\mathrm{pf}}\Delta T_{\mathrm{SC}}\times10^4}{(2.82+0.001\,22p)i_{\mathrm{fg}}} \tag{7.5.1}$$

For $p > 9000$ kPa,

$$T_{\mathrm{MFB}} = [(T_{\mathrm{MFB}} - T_{\mathrm{SAT}}) + 9000]\left(\frac{p_c - p}{p_c - 9000}\right) + T_{\mathrm{SAT}} \tag{7.5.2}$$

In these equations, p and p_c are in kPa. The applicable range is as follows: p = 0.1–9 MPa, G = 50–2750 kg m^{-2} s^{-1}, and x = −0.1 to 0.13. These are applicable when minimum film boiling (MFB) condition is reached from DFFB or from IAFB.

For flooding by water with top flow, Groeneveld and Snoek (1986) give the following correlation:

For $p < 689$ kPa,

$$T_{\mathrm{MFB}} - T_{\mathrm{SAT}} = 0.0984p + 29.1 \tag{7.5.3}$$

$689 < p \le 4200$ kPa,

$$T_{\mathrm{MFB}} - T_{\mathrm{SAT}} = 0.007\,65p + 89.1 \tag{7.5.4}$$

$p > 4200$ kPa,

$$T_{\mathrm{MFB}} = T_c = 374.15 \tag{7.5.5}$$

7.5.2 Jets Impinging on Hot Surfaces

Ishigai et al. (1978) studied quenching of a surface at about 1000 °C by a planar free surface water jet 6.2 mm wide. They found that the minimum film boiling temperature T_{MFB} increased with increasing jet velocity and subcooling. They gave the following equation to represent their data:

$$q_{\mathrm{MFB}} = 0.054\times10^6 u_{\mathrm{jet}}^{0.607}(1+0.527\Delta T_{\mathrm{SC}}) \tag{7.5.6}$$

The minimum heat flux q_{MFB}, u_{jet}, and ΔT_{SC} have the units W m^{-2}, m s^{-1}, and K, respectively. The range of parameters correlated was u_{jet} 0.65–3.5 m s^{-1} and ΔT_{SC} 5–55 K.

Liu (2003) developed the following semi-theoretical correlations for minimum heat flux and MFBT in the stagnation zone:

$$(T_{\mathrm{MFB}} - T_{\mathrm{SAT}}) = 14\Delta T_{\mathrm{SC}} Pr_f^{1/3}\frac{k_f}{k_g}\frac{\mu_g}{\mu_f} \tag{7.5.7}$$

$$\frac{q_{\mathrm{MFB}}D_{\mathrm{jet}}}{k_g} = 0.5(T_{\mathrm{MFB}} - T_{\mathrm{SAT}})Re_{\mathrm{jet}}^{1/2}\left(\frac{\mu_g}{\mu_f}\right)^{1/2} \tag{7.5.8}$$

These correlations were verified with data from three sources for free stream water jets at atmospheric pressure. Nozzle sizes were 6.2 mm × 50 mm and 10 mm diameter. Subcooling was 3–55 K. Re_{jet} was 1×10^4–6×10^4.

Sailer-Marie et al. (2004) analytically developed the following equation for MFB heat flux:

$$\frac{q_{\mathrm{MFB}}}{\rho_f u_{\mathrm{jet}} i_{\mathrm{fg}}(1+0.075Ja)} = \sqrt{\frac{\rho_g}{\rho_f}}\sqrt{\frac{La}{D_{\mathrm{jet}}}}\left(\frac{gD_{\mathrm{jet}}}{u_{\mathrm{jet}}^2}\right)^{1/4}(M^*)^{2/3} \tag{7.5.9}$$

La is the Laplace length defined as

$$La = \sqrt{\frac{\sigma D_{\mathrm{jet}}}{u_{\mathrm{jet}}^2(\rho_f - \rho_g)}} \tag{7.5.10}$$

$$Ja = \frac{C_{\mathrm{pf}}\Delta T_{\mathrm{SC}}}{i_{\mathrm{fg}}} \tag{7.5.11}$$

M^* is a kind of Mach number defined as

$$M^* = \left[\frac{\sigma u_{\mathrm{jet}}^2}{D_{\mathrm{jet}}(\rho_f - \rho_g)}\right]^{1/4}\sqrt{\frac{\rho_g}{p}} \tag{7.5.12}$$

This formula gave satisfactory agreement with the data of Robidou (2000) and Ishigai et al. (1978).

Sailer-Marie et al. (2004) also developed an analytical method to calculate the heat flux at the first MFB point and the heat flux at the start of shoulder. The first MFB point is explained in Section 7.6.2.

Ochi et al. (1984) studied quenching of a horizontal plate by vertical free surface jet of water. The plate was 210 mm × 50 mm. Nozzle diameters were 5, 10, and 15 mm, and they were 100 mm distant from the plate that was initially heated to 1100 °C. Subcooling ranged from 5 to 80 K and the jet velocity was 2–7 m s^{-1}. The following correlation was fitted to their data:

$$q_{\mathrm{MFB}} = 3.18 \times 10^5(1 + 0.383\Delta T_{\mathrm{SC}})(u_{\mathrm{jet}}/D_{\mathrm{jet}})^{0.828} \tag{7.5.13}$$

q_{MFB} is in W m^{-2}, and other units are as given in the data range.

7.5.3 Chilldown of Cryogenic Lines

Hartwig et al. (2015) performed chilldown tests with liquid hydrogen flowing in a pipe. Leidenfrost temperature was measured to be 170 K, and Berenson model, Eq. (3.5.9), predicted 144 K.

Johnson and Shine (1984) performed tests on chilldown of a 20 mm horizontal pipe with nitrogen. The minimum heat flux measured by them agreed that by the Zuber model for pool boiling using the constant from Berenson's model (see Chapter 3 for these models).

7.5.4 Spheres

Dhir and Purohit (1978) measured MFBT on a 19 mm diameter stainless steel sphere with subcooled water flowing over it. Water velocity was 2–45 cm s^{-1} and subcooling 0–50 K. They found that velocity had no effect on MFBT and that it depended only on subcooling. They also quenched spheres made of silver, copper, and stainless steel, in stagnant water pools. Data for both forced and natural convection for all materials were correlated by the following equation:

$$\Delta T_{\mathrm{MFB}} = 101 + 8\Delta T_{\mathrm{SC}} \tag{7.5.14}$$

ΔT is in K. This equation was also in good agreement with the data of Bradfield (1967) for a 63 mm diameter chrome-plated copper sphere quenched in water.

Steven and Witte (1971) studied film boiling on a 17.8 mm diameter copper sphere moving in water with a velocity of 2.9–6.1 m s^{-1}. They concluded that velocity does not affect the MFBT.

Rezakhany and Wakil (1984) performed tests on three brass spheres of diameter 3.1, 4.8, and 6.3 mm in streams of R-11 and R-113 at speeds of 0–15 cm s^{-1}, corresponding to liquid Reynolds numbers of 0–3500. Some of their data is shown in Figure 7.5.1. It is seen that MFBT decreases with increasing Reynolds number up to about 1100, rises with increasing Reynolds number and becomes nearly constant at $Re > 1800$. As noted earlier, other researchers have reported no effect of velocity on MFBT, and their measurements went down to zero velocity. The difference may be due to type of fluid as the other sources used water, or it may be because the spheres used by Rezakhany and El-Wakil were much smaller than those used by others or there may be other reasons. Hence further investigation is needed.

From the foregoing, it may be concluded that sphere material does not affect MFBT. Also, diameter has no effect if it is 17.8 mm or larger. Most of the evidence indicates

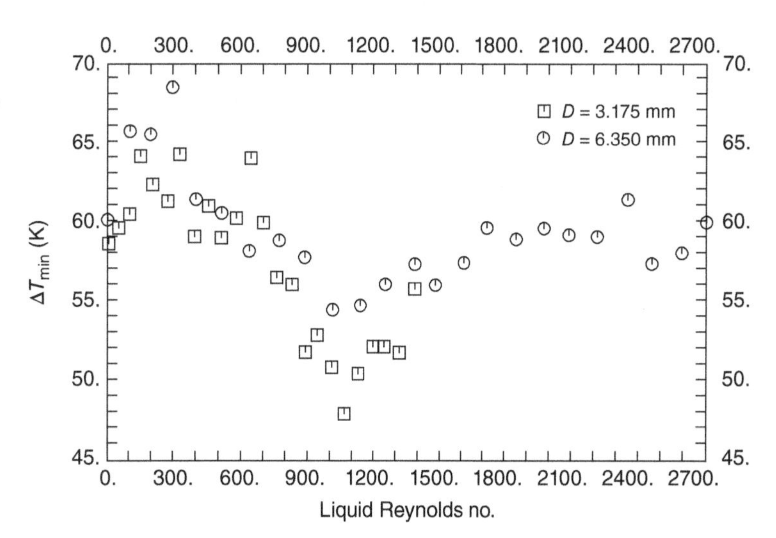

Figure 7.5.1 Effect of Reynolds number on minimum film boiling temperature difference on spheres in R-11 streams. Source: Rezakhany and El-Wakil (1984).

that velocity has no effect. Equation (7.5.13) may be used for water at atmospheric pressure for spheres of diameter 19 mm and larger.

7.5.5 Spray Cooling

Mudawar and Valentine (1989) gave the following correlation of their data for water sprays on metallic surfaces:

$$\frac{q_{MFB}}{\rho_g i_{fg} Q''} = 0.145\left(\frac{u_m}{Q''}\right)^{0.834} \tag{7.5.15}$$

The test conditions included variations in volumetric spray flux, mass mean drop diameter, and mean drop velocity of 0.6×10^{-3}–9.96×10^{-3} m^3 s^{-1} m^{-2}, 0.434–2.005 mm, and 10.6–26.5 m s^{-1}, respectively.

The tests of Klinzing et al. (1992) were described in Section 7.4.10. They gave the following correlations for MFB from the same tests.

For low spray flux, $0.58\times10^{-3}<Q''<3.5\times10^{-3}$m^{3}s^{-1}m^{-2},

$$q_{MFB} = 3.324\times10^6 Q''^{0.544} u_m^{0.324} \tag{7.5.16}$$

$$\Delta T_{MFB} = 2.049\times10^2 Q''^{0.066} u_m^{0.138} D_{32}^{-0.035} \tag{7.5.17}$$

For high spray flux, $9.96\times10^{-3}>Q''>3.5\times10^{-3}$m^{3}s^{-1}m^{-2},

$$q_{MFB} = 6.069\times10^6 Q''^{0.943} u_m^{0.864} \tag{7.5.18}$$

$$\Delta T_{MFB} = 7.99\times10^3 Q''^{-0.027} u_m^{1.033} D_{32}^{0.952} \tag{7.5.19}$$

Klinzing et al. also gave correlations that point where the slope of film boiling line begins to change. They called this point departure from film boiling (DFB). These correlations are as follows:

Low spray flux,

$$q_{DFB} = 6.1\times10^6 Q''^{0.588} u_m^{0.0.244} \tag{7.5.20}$$

$$\Delta T_{DFB} = 2.808\times10^2 Q''^{0.087} u_m^{0.11} D_{32}^{-0.035} \tag{7.5.21}$$

For high spray flux,

$$q_{DFB} = 6.536\times10^6 Q''^{0.995} u_m^{0.924} \tag{7.5.22}$$

$$\Delta T_{DFB} = 3.079\times10^4 Q''^{-0.194} u_m^{1.922} D_{32}^{1.651} \tag{7.5.23}$$

Fluid properties are calculated at the film temperature. SI units given in Nomenclature are used.

7.6 Transition Boiling

Transition boiling occurs in the region between CHF point and MFB point. In this regime, wall surface is intermittently in contact with liquid and vapor. Heat transfer is a combination of unstable nucleate and film boiling. Heat transfer coefficient decreases from the CHF point to the MFB point.

7.6.1 Flow in Channels

Many correlations have been proposed for transition boiling of water in tubes, annuli, and rod bundles. Many of these have been listed by Groeneveld and Snoek (1986) and Groeneveld and Gardiner (1977). Some of them are purely empirical, while some take into consideration the physical phenomena. A few of them are given as follows:

Ellion (1954) in an annulus, p = 110–413 kPa, G = 330–1490 kg m^{-2} s^{-1},

$$q_{TB} = 3.51\times10^8 \Delta T_{SAT}^{-2.4} \tag{7.6.1}$$

q_{TB} is the transition boiling heat flux. Bjonard and Griffith (1977) assumed that transition boiling heat transfer is composed of both nucleate boiling (wet wall) and film boiling (dry wall). Their correlation is

$$q_{TB} = \delta q_{CHF} + (1-\delta) q_{MFB} \tag{7.6.2}$$

The fraction of wall area that is wet, δ, is given by

$$\delta = \left[\frac{T_w - T_{MFB}}{T_{CHF} - T_{MFB}}\right]^2 \tag{7.6.3}$$

This correlation was verified with data for tubes and annuli from two sources that included pressure up to critical and mass flux of 310–3490 kg m^{-2} s^{-1}.

Hsu (1975b) gave the following semi-theoretical correlation:

$$h_{TB} = 0.62\left[\frac{g k_g^3 \rho_g(\rho_f-\rho_g) i_{fg}}{\Delta T_{SAT}\mu_g}\frac{1}{2\pi}\left(\frac{g(\rho_f-\rho_g)}{\sigma}\right)^{1/2}\right]^{1/4} + A\exp(-B\Delta T_{SAT}) \tag{7.6.4}$$

where

$$A = 1456 p^{0.558} \tag{7.6.5}$$

$$B = 3.758\times10^{-3} p^{0.1733} \tag{7.6.6}$$

p is in psia, ΔT_{SAT} is in °F. The expression in brackets is dimensionless. This correlation was verified with FLECHT data for reflooding a rod bundle with pressure 103–620 kPa, G = 50–250 kg m^{-2} s^{-1}, and subcooling 12–83 K.

Figure 7.6.1 shows the comparison of test data for reflooding of a vertical tube bundle with several correlations. Large differences are seen in the predictions of various correlations. Groeneveld and Snoek (1986) state that the most reasonable agreement with data was obtained by connecting experimentally determined CHF and minimum boiling points on a log–log graph of q vs. ΔT and interpolating as follows:

$$\frac{q_{TB}}{q_{MFB}} = \left(\frac{q_{CHF}}{q_{MFB}}\right)^{n_1} \tag{7.6.7}$$

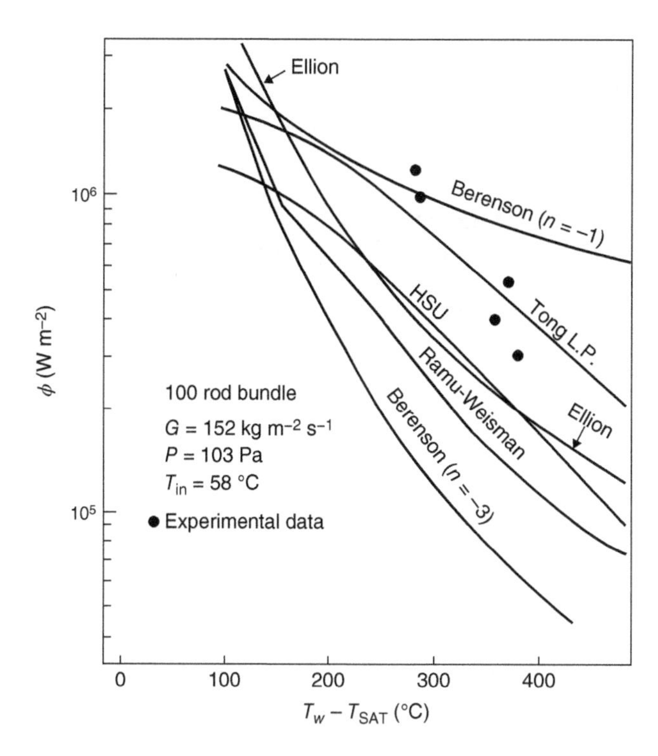

Figure 7.6.1 Heat flux φ during transition boiling in reflooding of a rod bundle compared with various correlations. Source: Groeneveld and Gardiner (1977). © ASME.

where

$$n_1 = \frac{\ln(\Delta T_{\mathrm{MFB}}/\Delta T_{\mathrm{TB}})}{\ln(\Delta T_{\mathrm{MFB}}/\Delta T_{\mathrm{CHF}})} \tag{7.6.8}$$

or

$$\frac{\Delta T_{\mathrm{TB}}}{\Delta T_{\mathrm{MFB}}} = \left(\frac{\Delta T_{\mathrm{CHF}}}{\Delta T_{\mathrm{MFB}}}\right)^{n_2} \tag{7.6.9}$$

where

$$n_2 = \frac{\ln(q_{\mathrm{TB}}/q_{\mathrm{MFB}})}{\ln(q_{\mathrm{CHF}}/q_{\mathrm{MFB}})} \tag{7.6.10}$$

Based on his pool boiling tests, Berenson (1960) had also recommended drawing a straight line between CHF and MFB points, plotted on a log–log graph.

In their model for an air conditioning system using carbon dioxide, Ayad et al. (2012) assumed linear variation of heat flux between CHF and MFB points. Their model gave satisfactory agreement with data from several sources over the entire range from nucleate boiling to film boiling.

In conclusion, the best prediction method is to interpolate between the CHF and MFB points. However, accurate prediction of MFB point can be difficult.

7.6.2 Jets on Hot Surfaces

Ishigai et al. (1978) used a planar free stream jet of water to quench a surface initially at about 1000 °C. At low subcooling (10–15 K), heat flux decrease continuously with increasing wall temperature until the MFB point was reached. At higher subcooling, heat flux initially decreased then increased with increasing wall temperature, became near constant, and then decreased with further increase in wall temperature. This near constant heat flux part is called the shoulder of flux.

Robidou (2000) and Robidou et al. (2002) studied a planar free stream jet of water on a heated surface. Tests were done in steady state. Figure 7.6.2 shows some of the data. It is seen that the heat flux in the shoulder is highest at the stagnation point and becomes lower with increasing distance from the stagnation point. Heat flux in the shoulder was found to increase with subcooling.

Sailer-Marie et al. (2004) studied the data of Robidou and developed a model to predict heat transfer in the shoulder. Heat transfer in the shoulder region was considered to be mainly related to the heating up of the liquid and not to its vaporization. Vapor bubbles/fragments are formed at the vapor–liquid interface due to Rayleigh–Taylor instability. At each bubble oscillation, a constant amount of liquid flows into the vapor and reaches the wall. This amount of liquid spreads on the wall. It is heated up by transient conduction. When the average liquid temperature nearly reaches its saturation temperature, a part of this liquid that directly touches the hot plate is evaporated. The vapor generated evicts the hot liquid from the wall. This liquid is finally evacuated to the bulk of the flow. They assumed that the heat transfer associated with vapor generation is negligible and that the heat flux is mainly controlled by periodic ejection of heated liquid. The heat flux transferred by this mechanism is the heat flux in shoulder, q_{sh}. They finally arrived at the following relation:

$$\frac{q_{\mathrm{sh}}}{\rho_f u_{\mathrm{jet}} C_{\mathrm{pf}} \Delta T_{\mathrm{SC}}} = 0.15\left(\frac{La}{D_{\mathrm{jet}}}\right)^{1/2}\left(\frac{gD_{\mathrm{jet}}}{u_{\mathrm{jet}}^2}\right)^{1/4} \tag{7.6.11}$$

This equation gave good agreement with data at the stagnation point from four sources, all for water jets. The range of data was as follows: jet diameter, 1.8–5.1 mm; jet velocity, 0.57–15 m s^{-1}; subcooling, 5–85 K.

Sakhuja et al. (1980) studied an array of circular water jets quenching a copper surface. The following expressions were given to represent the heat transfer coefficients in transition boiling:

$$h = 6.71 \times 10^4 u_{\mathrm{jet}} Pr_f^{0.6} (1.8\Delta T_{\mathrm{SAT}})^n F \tag{7.6.12}$$

$$n = -1.89 + 0.328\left(\frac{S}{D_{\mathrm{jet}}}\right) - 0.028\left(\frac{S}{D_{\mathrm{jet}}}\right)^2 \tag{7.6.13}$$

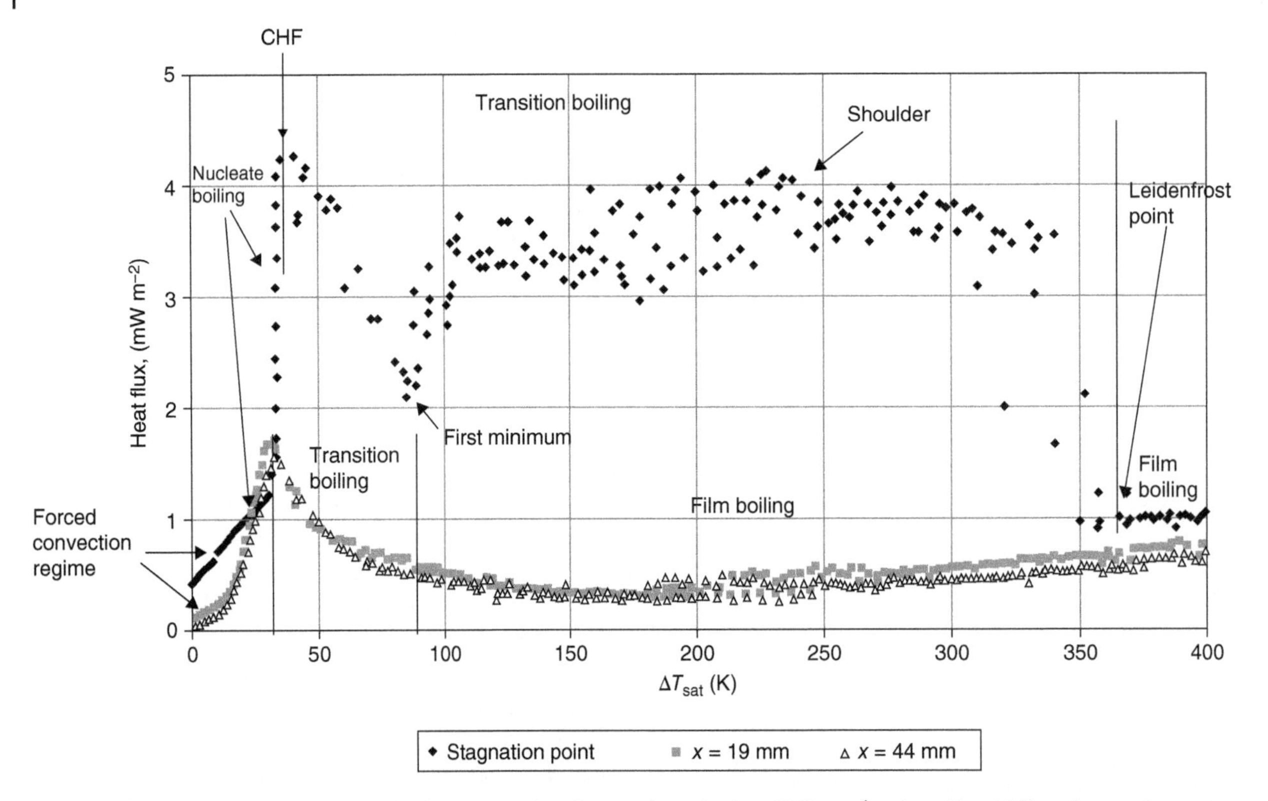

Figure 7.6.2 Quenching of a surface with free stream jet of water: jet velocity of 0.8 m s^{-1}, subcooling 16 K, and a nozzle-to-plate distance of 6 mm; x is the distance from the stagnation point. Source: Robidou et al. (2002). © 2002 Elsevier.

$$F = 27 - 0.35\left(\frac{S}{D_{\text{jet}}}\right) - 1.51\left(\frac{S}{D_{\text{jet}}}\right)^2 + 0.15\left(\frac{S}{D_{\text{jet}}}\right)^3 \tag{7.6.14}$$

The units are h in W m^{-2} K^{-1}, u_{jet} in m s^{-1}, and ΔT_{SAT} in K. S is the distance between jets. The test data included $4 \leq S/D_{\text{jet}} \leq 12$. Most of the data for $\Delta T_{\text{SAT}} > 42$ K were predicted within ±20%.

7.6.3 Spheres

Witte et al. (1968) performed tests in which a heated 12.7 mm diameter tantalum sphere was moved in subcooled sodium at atmospheric pressure. Sphere temperature was up to 1982 °C, while the sodium temperature was from 300 to 450 °C. Sphere velocity was up to 3 m s^{-1}. Measured heat transfer coefficients were correlated by the following equation:

$$q = 0.561\left[\frac{U_\infty \rho_f k_f C_{\text{pf}}}{D}\right]^{1/2}(T_w - T_f) \tag{7.6.15}$$

T_f is the bulk liquid temperature. This equation may be rearranged to the following form:

$$Nu = 0.561 Pe^{1/2} \tag{7.6.16}$$

The authors discussed the type of heat transfer that occurred during their tests. They estimated that the thickness of vapor layer required for film boiling would be smaller than the surface roughness. They therefore felt that there must have been some contact between liquid and sphere surface. Such a large temperature difference suggests the possibility of film boiling. It could be that nucleate boiling and film boiling occurred at different parts of the surface as in transition boiling. Hence these data are tentatively considered to be for transition boiling.

7.6.4 Spray Cooling

The tests of Klinzing et al. (1992) have been described in Sections 7.4.10 and 7.5.5. They gave the following correlation of their data for transition boiling:

$$q_{\text{TB}} = q_{\text{CHF}} - \frac{q_{\text{CHF}} - q_{\text{MFB}}}{(\Delta T_{\text{CHF}} - \Delta T_{\text{MFB}})^3}[\Delta T_{\text{CHF}}^3 - 3\Delta T_{\text{CHF}}^2 \Delta T_{\text{MFB}} + 6\Delta T_{\text{CHF}} \Delta T_{\text{MFB}} \Delta T_f - 3(\Delta T_{\text{CHF}} + \Delta T_{\text{MFB}})\Delta T_f^2 + 2\Delta T_f^3] \tag{7.6.17}$$

Nomenclature

Bo	boiling number = $qG^{-1}i_{fg}^{-1}$ (–)
C_p	specific heat at constant pressure (J kg^{-1} K^{-1})
D	diameter of tube or of other objects (m)
D_{HP}	equivalent diameter based on heated perimeter (m)
D_{HYD}	hydraulic equivalent diameter (m)
D_{32}	Sauter mean diameter (m)
Fr_L	Froude number for all liquid flow = $G^2\rho_f^{-2}g^{-1}D^{-1}$ (–)
g	acceleration due to gravity (m s^{-2})
h	heat transfer coefficient (W m^{-2} K^{-1})
i	enthalpy (J kg^{-1})
i_{fg}	latent heat of vaporization (J kg^{-1})
k	thermal conductivity (W m^{-1} K^{-1})
L	length or distance from CHF point (m)
La	Laplace (capillary) length (m)
M	molecular weight (–)
N	number of data points (–)
Nu	Nusselt number = hD/k (–)
Pe	Peclet number = ($Re\,Pr$) (–)
Pr	Prandtl number (–)
p	pressure (Pa)
p_c	critical pressure (Pa)
p_r	reduced pressure = p/p_c (–)
Q''	volumetric flux of spray (m^3 s^{-1} m^{-2})
q	heat flux (W m^{-2})
Re	Reynolds number = (GD/μ) (–)
T	temperature (K)
ΔT_{SAT}	$(T_w - T_{SAT})$ (K)
ΔT_{SC}	$(T_{SAT} - T_f)$ (K)
ΔT_{SUP}	$(T_g - T_{SAT})$ (K)
ΔT_f	$(T_w - T_f)$ (K)
t	time (s)
u	velocity (m s^{-1})
x	vapor quality (–)
z	length or distance (m)

Greek letters

α	void fraction, (–)
μ	dynamic viscosity (Pa s)
ν	kinematic viscosity (m^2 s^{-1})
ρ	density (kg m^{-3})
σ	surface tension (N m^{-1})

Subscripts

A	actual
BOT	bottom
CHF	CHF
c	critical, at CHF
conv	convective
E	equilibrium
FB	film boiling
FDB	fully developed boiling
f	liquid
film	at film temperature
GS	vapor phase flowing alone in the channel
GT	all mass flowing as vapor
g	vapor
in	inlet
jet	jet
LS	liquid flowing alone in the channel
LT	all mass flowing as liquid
MFB	minimum film boiling
rad	radiation
SAT	saturation
SC	subcooled
s	surface impacted by jet
TOP	top
TB	transition boiling
TP	two-phase
w	wall
∞	free stream or outside entrance length

Abbreviations

CHF	critical heat flux
DFFB	dispersed flow film boiling
IAFB	inverted annular film boiling
MAD	mean absolute deviation
MFB	minimum film boiling
MFBT	minimum film boiling temperature

References

Andreani, M. and Yadigaroglu, G. (1994). Prediction methods for dispersed flow film boiling. *Int. J. Multiphase Flow* **20** (Suppl. 1): 1–51.

Andreani, M. and Yadigaroglu, G. (1997). A 3-D Eulerian–Lagrangian model of dispersed flow film boiling including a mechanistic description of the droplet spectrum evolution – 1. The thermal hydraulic model. *Int. J. Heat Mass Transfer* **40** (8): 1753–1772.

Annuziatio, A., Cumo, M., and Pallazi, G. (1983). Post-dryout heat transfer in uncovered core accidents. In: *Thermal Hydraulics of Nuclear Reactors*, vol. **1** (ed. M. Merilo), 335–342. American Nuclear Society.

Auman, P.M., Giffiths, D.K., and Hill, D.R. (1967). Hot strip mill runout table temperature control. *Iron Steel Eng.*: volume 44, no. 9 174–179.

Ayad, F., Benelmir, R., and Souayed, A. (2012). CO_2 evaporators design for vehicle HVAC operation. *Appl. Therm. Eng.* **36**: 330–344.

Barron, R.F. and Stanley, R.S. (1994). Film boiling under an impinging cryogenic jet. In: *Advances in Cryogenic Engineering*, vol. **39** (ed. P. Kittel), 1769–1777. Boston, MA: Springer.

Bennet, A.W., Hewitt, G.F., Kearsey, H.A., and Keeys, R.K.F. (1967). Heat transfer to steam-water mixtures in uniformly heated tubes in which critical heat flux has been exceeded. Report AERE-R5373.

Berenson, P.J. (1960) Transition boiling heat transfer from a horizontal cylinder. Technical Report No. 17, M.I.T., Cambridge, MA.

Bjonard, T.A. and Griffith, P. (1977). PWR blowdown heat transfer. In: *Symposium Thermal Hydraulic Aspects of Nuclear Reactor Safety. Volume 1: Light-Water Reactors* (eds. O.C. Jones and S.G. Bankoff), 17–42. New York, NY: ASME.

Bradfield, W.S. (1967). On the effect of subcooling on wall superheat in pool boiling. *J. Heat Transfer* **89** (3): 269–270.

Bromley, L.A., Leroy, N.R., and Roberts, J.A. (1953). Heat transfer in forced convection film boiling. *Ind. Eng. Chem.* **45** (1): 2639–2646.

Cess, R.D. and Sparrow, E.M. (1961). Film boiling in a forced-convection boundary-layer flow. *J. Heat Transfer* **83** (3): 370–375.

Chen, X. and Zhou, F. (1986). Forced convection boiling and post-dryout heat transfer in helical coiled tube. In: *Proceedings of the 8th International Heat Transfer Confernce, IHTC 8*, 2221–2226. Begell House.

Chou, X.S. and Witte, L.C. (1995). Subcooled flow film boiling across a horizontal cylinder: Part 1 – Analytical model. *J. Heat Transfer* **117**: 167–174.

Chou, X.S., Sankaran, S., and Witte, L.C. (1995). Subcooled flow film boiling across a horizontal cylinder: Part 2 – Comparison to experimental data. *J. Heat Transfer* **117**: 175–178.

Cumo, M., Farello, G.E., and Ferrari, G. (1972). The influence of curvature in post-dryout heat transfer. *Int. J. Heat Mass Transfer* **15**: 2045–2062.

Dhir, V.K. and Purohit, G.P. (1978). Subcooled film boiling heat transfer from spheres. *Nucl. Eng. Des.* **47**: 49–66.

Dougall, R.S. and Rohsenow, W.M. (1963). Film boiling on the inside of vertical tubes with upward flow of the fluid at low qualities. MIT report 9076-26. Massachusetts Institute of Technology, Cambridge, MA.

Ellion, M.E. (1954). California Institute of Technology, report JPL-MEMO-20-88. Quoted in Fung et al. (1979) and Groeneveld and Snoek (1986).

Epstein, M. and Hauser, G.M. (1980). Subcooled forced-convection film boiling in the forward stagnation region of a sphere or cylinder. *Int. J. Heat Mass Transfer* **23**: 179–189.

Eugene, A. and Mizikar, A. (1970). Spray cooling investigation for continuous casting of billets and blooms. *Iron Steel Eng.* **47** (6): 53–60.

Evans, D., Webb, S.W., and Chen, J.C. (1985). Axially varying vapor superheats in convective film boiling. *J. Heat Transfer* **107**: 663–669.

Filipovic, J., Viskanta, R., and Incropera, F.P. (1994). An analysis of subcooled turbulent film boiling on a moving isothermal surface. *Int. J. Heat Mass Transfer* **37** (18): 2661–2673.

Filipovic, J., Incropera, F.P., and Viskanta, R. (1995a). Quenching phenomena associated with a water wall jet: I. transient hydrodynamic and thermal conditions. *Exp. Heat Transfer* **8** (2): 97–117.

Filipovic, J., Incropera, F.P., and Viskanta, R. (1995b). Quenching phenomena associated with a water wall jet: II. Comparison of experimental and theoretical results for the film boiling region. *Exp. Heat Transfer* **8** (2): 119–130.

Forslund, R.P. and Rohsenow, W.M. (1968). Dispersed flow boiling. *J. Heat Transfer* **90** (11): 399–407.

Fung, K., Gardiner, S., and Groeneveld, D. (1979). Subcooled and low quality flow film boiling of water at atmospheric pressure. *Nucl. Eng. Des.* **55** (1): 51–57.

Gottula, R.C., Nelson, R.A., Chen, J.C., et al. (1983). Forced convective non-equilibrium post-CHF heat transfer experiments in a vertical tube. *ASME-JSME Thermal Engineering Conference*, Honolulu, HI, 20–24 March 1983.

Groeneveld, D.C. (1973) Post-dryout heat transfer at reactor operating conditions. Atomic Energy of Canada Limited Report AECL-4513.

Groeneveld, D.C. (1975). Post-dryout heat transfer: physical mechanisms and a survey of prediction methods. *Nucl. Eng. Des.* **32** (283): 294.

Groeneveld, D.C. (1982). Prediction methods for post-CHF heat transfer and superheated steam cooling suitable for reactor accident analysis. Center d'Etudes Nuclaires de Grenoble report, TT/SETRE/82-4-E/DCGr.

Groeneveld, D.C. and Delorme, G.G.J. (1976). Prediction of thermal non-equilibrium in the post-dryout regime. *Nucl. Eng. Des.* **36**: 17–26.

Groeneveld, D.C. and Gardiner, S.R.M. (1977). Post-CHF heat transfer under forced convective condition. In: *Symposium Thermal Hydraulic Aspects of Nuclear Reactor Safety. Volume 1: Light-Water Reactors* (eds. O.C. Jones and S.G. Bankoff), 43–74. New York, NY: ASME.

Groeneveld, D.C. and Snoek, C.W. (1986). A comprehensive examination of heat transfer correlations suitable for reactor safety analysis. In: *Multiphase Science and Technology*, vol. **2** (eds. G.F. Hewitt, J.M. Delhaye and N. Zuber), 181–274. Washington, DC, USA: Hemisphere.

Groeneveld, D.C. & Stewart, J.C. (1982) The minimum film boiling temperature for water during film boiling collapse. *Proc. 7th Int. Heat Transfer Conf.*, DOI: 10.1615/IHTC7.1270, pp. 393–398.

Groeneveld, D.C., Leung, L.K.H., Vasic, A.Z. et al. (2003). A look-up table for fully developed film-boiling heat transfer. *Nuclear Engineering and Design* **225**: 83–97.

Hadaller, G. and Bannerjee, S. (1969). Heat transfer to superheated steam in round tubes. AECL unpublished report. Quoted in Groeneveld and Delorme (1976).

Hartwig, J., Hub, H., Styborski, J. et al. (2015). Comparison of cryogenic flow boiling in liquid nitrogen and liquid hydrogen chilldown experiments. *Int. J. Heat Mass Transfer* **88**: 662–673.

Hein, D. and Kohler, W. (1984). A simple-to-use post dryout heat transfer model accounting for thermal non-equilibrium. *First International Heat Transfer Workshop on Fundamental Aspects of Post-Dryout Heat Transfer*, Salt Lake City, UT (2–4 April 1984). NUREG/CP-0060.

Heineman, J.B. (1960). An experimental investigation of heat transfer to superheated steam in round and rectangular tubes. Argonne National Laboratory report ANL-6213.

Hochreiter, L.E., Cheung, F.B., Lin, T.F., et al. (2012) *RBHT reflood heat transfer experiments data and analysis*. Tech. Rep. NUREG/CR-6980, US Nuclear Regulatory Commission.

Hsu, Y.Y. (1975a) Tentative correlations for reflood heat transfer. Third Water Reactor Safety Information Meeting, Germantown, Maryland, 1975.

Hsu, Y.Y. (1975b). A tentative correlation for the regime pf transition boiling and film boiling during reflood. *Paper Presented at the Third Water Reactor Safety Review Information Meeting, US NRC*, Washington, DC. Quoted in Groeneveld and Snoek (1986).

Hsu, Y.Y. (1977). Proposed heat transfer "best estimate" packages. Quoted in Fung et al. (1979).

Hu, H., Chung, J.N., and Amber, S.H. (2012). An experimental study on flow patterns and heat transfer characteristics during cryogenic chilldown in a vertical pipe. *Cryogenics* **52** (2012): 268–277.

Hynek, S.J., Rohsenow, W.M., and Bergles, A.E. (1969). *Forced convection dispersed flow film boiling*. MIT Department of Mechanical Engineering report 70586-63.

Ishigai, S., Nakanishi, S., and Ochi, T. (1978). Boiling heat transfer for a plane water jet impinging on a hot surface. In: *Proceedings of the 6th International Heat Transfer Conference, Toronto, Canada*, vol. **1**, 445–450. Begell House, paper FB-30.

Jackson, J.K. (2006). Cryogenic two phase flow during chill down: flow transition and nucleate boiling heat transfer. PhD thesis. University of Florida.

Jacobi, H.G., Kaestle, G., and Wunnenberg, K. (1984). Heat transfer in cyclic secondary cooling during solidification of steel. *Ironmaking and Steelmaking* **11** (3): 132–145.

Johnson, J. and Shine, S.R. (2015). Transient cryogenic chill down process in horizontal and inclined pipes. *Cryogenics* **71**: 7–17.

Kaminage, K. and Uchida, H. (1978). Reflooding phenomena in a single heated rod: Part 2 – Analytical study. *Nippon kikai Gakkai Ronbushu* **44** (388): 4263–4271.

Katsounis, A. (1987). Post dryout correlations and models compared to experimental data from different fluids. In: *Heat and Mass Transfer in Refrigeration and Cryogenics* (eds. J. Bougard and N. Afgan), 152–164. Washington, DC: Hemisphere.

Kawamania, O., Azuma, H., and Ohta, H. (2007). Effect of gravity on cryogenic boiling heat transfer during tube quenching. *Int. J. Heat Mass Transfer* **50**: 3490–3497.

Klinzing, W.P., Rozzi, J.C., and Mudawar, I. (1992). Film and transition boiling correlations for quenching of hot surfaces with water sprays. *J. Heat. Treat.* **9**: 91–103.

Kohler, W. and Hein, D. (1986). Influence of the wetting state of a heated surface on heat transfer and pressure loss in an evaporator tube. US NRC report NUREG/IA-0003.

Kokado, J., Hatta, N., Tauda, H. et al. (1984). An analysis of film boiling phenomena of subcooled water spreading radially on a hot steel plate. *Arch. Eisenhuttenwes.* **55**: 113–118. Quoted in Wolf, D.H., Incropera, F.P., & Viskanta,

R. (1993) *Jet impingement boiling*. in: P.H. James, F.I. Thomas (Eds.), Advances in Heat Transfer, Elsevier, pp. 1–132.

Laverty, W.F. and Rohsenow, W.M. (1964). Film boiling of saturated liquid flowing upward through a heated tube: high vapor quality range. Technical report No. 9857–32. Massachusetts Institute of Technology, Cambridge, MA.

Liang, G. and Mudawar, I. (2017). Review of spray cooling – Part 2: High temperature boiling regimes and quenching applications. *Int. J. Heat Mass Transfer* **115**: 1206–1222.

Liu, Z.H. (2003). Prediction of minimum heat flux for water jet boiling on a hot plate. *J. Thermophys. Heat Transfer* **17** (2): 159–165.

Liu, G. and Theofanous, T.G. (1996). Film boiling on spheres in single- and two-phase flows. DOE/ER12933-3, DOE/ID-10499.

Liu, Z.H. and Wang, J. (2001). Study on film boiling heat transfer for water jet impinging on high temperature flat plate. *Int. J. Heat Mass Transfer* **44**: 2475–2481.

Liu, Q.S., Shiotsu, M., and Sakurai, A. (1992). A correlation for forced convection film boiling heat transfer from a horizontal cylinder. In: *HTD-Vol. 197, Two-Phase Flow and Heat Transfer*, 101–110. ASME.

Liu, Q., Shioutsu, M., Sakurai, A., and Fukuda, K. (2009). Forced convection film boiling heat transfer from single horizontal cylinders in saturated and subcooled liquids: Part 2 – Experimental data for subcooled liquids and its correlation. In: *ASME 2009 Heat Transfer Summer Conference July 19–23, 2009, San Francisco, California, USA*, Paper No: HT2009-88038, 39–54.

Malmazet, E. and Berthoud, G. (2009). Convection film boiling on horizontal cylinders. *Int. J. Heat Mass Transfer* **52**: 4731–4747.

Mawatari, T., Mori, H., and Kariya, K. (2014). A study on post-CHF heat transfer at near-critical pressure. *Proceedings of the 15th International Heat Transfer Conference, IHTC-15*, Kyoto, Japan (10–15 August 2014). Paper # IHTC15-9867.

McAdams, W.H. (1954). *Heat Transmission*, 3e. New York, NY: McGraw Hill.

Mercado, M., Wong, N., and Hartwig, J. (2019). Assessment of two-phase heat transfer coefficient and critical heat flux correlations for cryogenic flow boiling in pipe heating experiments. *Int. J. Heat Mass Transfer* **133**: 295–315.

Miropolskiy, Z.L. and Pikus, V.Y. (1969). Critical boiling heat fluxes in curved channels. *Heat Transfer Sov. Res.* **1** (1): 74–79.

Mohanta, L., Sohag, F.A., Cheung, F.B. et al. (2017). Heat transfer correlation for film boiling in vertical upward flow. *Int. J. Heat Mass Transfer* **107**: 112–122.

Mosaad, M. and Johannsen, K. (1989). Experimental study of steady-state film boiling heat transfer of subcooled water flowing upwards in a vertical tube. *Exp. Thermal Fluid Sci.* **2** (4): 477–493.

Mosaad, M. and Johannsen, K. (1992). An improved correlation for subcooled and low quality film boiling heat transfer of water at pressures from 0.1 to 8 MPa. *Nucl. Eng. Des.* **135**: 355–366.

Motte, E.I. and Bromley, L.A. (1957). Film boiling of flowing subcooled liquids. *Ind. Eng. Chem.* **49** (11): 1921–1928.

Mudawar, I. and Valentine, W.S. (1989). Determination of the local quench curve for spray cooled metallic surfaces. *J. Heat. Treat.* **7**: 107–121. Quoted in Liang and Mudawar (2017).

Müller, H.R. and Jeschar, R. (1983). Warmeubergang bei der Spritzwasserkuhlung von Nichteisenmetallen. *Zeitschrift fur Metallkunde* **74** (5): 257–264.

Nijhawan, S., Chen, J.C., Sundaram, N.K., and London, E.J. (1980). Measurement of vapor superheat in post-critical heat flux boiling. *J. Heat Transfer* **102**: 465–470.

Ochi, T., Nakanishi, S., Kaji, M. et al. (1984). Cooling of a hot plate with an impinging circular water jet. In: *Multi-Phase Flow and Heat Transfer III, Part A: Fundamentals* (eds. T.N. Veziroglu and A.E. Bergles), 671–681. Elsevier.

Ogata, H. and Sato, S. (1974). Forced convection heat transfer to boiling helium in a tube. *Cryogenics* **14** (7): 375–380.

Orcozo, J.A. and Witte, L.C. (1984). Flow boiling from a sphere to subcooled Freon-11. In: *Fundamentals of Phase Change: Boiling and Condensation*, HTD-Vol. 38 (eds. C.T. Avedisian and T.M. Rudy), 35–42. New York, NY: ASME.

Reiners, U. (1987). Warmeubertragung durch Spritzwasserkuhlung heiser Oberflächen im Bereich der stabilen Filmverdampfung. PhD thesis. Technische Universitat Clausthal, 1987.

Rezakhany, S. and El-Wakil, M.M. (1984). Do the forced flow and subcooling of the coolant affect the minimum flow boiling point. In: *Fundamentals of Phase Change: Boiling and Condensation,* HTD-Vol. 3 (eds. C.T. Avedisian and T.M. Rudy), 51–61. New York, NY: ASME.

Robertshotte, P. and Griffith, P. (1982). Downflow post-critical heat flux heat transfer to low pressure water. *Nucl. Technol.* **56**: 134–140.

Robidou, H. (2000). Etude expérimentale du refroidissement diphasique à haute température par jet d'eau impactant. PhD thesis. University of Henri Poincaré, Nancy 1, France.

Robidou, H., Auracher, H., Gardin, P., and Lebouché, M. (2002). Controlled cooling of a hot plate with a water jet. *Exp. Therm. Fluid Sci.* **26**: 123–129.

Ruch, M.A. and Holman, J.P. (1975). Boiling heat transfer to Freon-113 jet impinging upward onto a flat heated surface. *Int. J. Heat Mass Transfer* **18**: 51–60.

Saha, P. (1980). A non-equilibrium heat transfer model for dispersed droplet post-dryout regime. *Int. J. Heat Mass Transfer* **23**: 483–493.

Sakhuja, R.K., Lazgin, F.S., and Owen, M.J. (1980). Boiling heat transfer with arrays of impinging jets. ASME Paper 80-HT-47. Quoted in Wolf et al. (1973).

Sakurai, A., Shiotsu, M., and Hata, K. (1990a). Effect of system pressure on minimum film boiling temperature for various liquids. *Exp. Therm. Fluid Sci.* **3**: 450–457.

Sakurai, A., Shiotsu, M., and Hata, K. (1990b). A general correlation for pool film boiling heat transfer from a horizontal cylinder to subcooled liquid. Part 2: Experimental data for various liquids and its correlation. *J. Heat Transfer* **112**: 441–450.

Sasaki, K., Sugatani, Y., and Kawasaki, M. (1979). Heat transfer in spray cooling on hot surface. *Tetsu-to-Hagane (J. Iron Steel Inst. Jpn.)* **65** (January): 90–96.

Schnittger, R.B., 1982. Untersuchengen zumWarmeubergang bei vertikalen und Horizontalen Rohrstromungen in Post-Dryout Bereich, Dr. Ing. Dissertation, Technical University of Munich.

Seiler-Marie, N., Seiler, J.-M.T., and Simonin, O. (2004). Transition boiling at jet impingement. *International Journal of Heat and Mass Transfer* **47**: 5059–5070.

Shah, M.M. (1980). A general predictive technique for heat transfer during saturated film boiling in tubes. *Heat Transfer Eng.* **2**: 51–62.

Shah, M.M. (2015). A general correlation for CHF in horizontal channels. *International Journal of Refrigeration* **59**: 37–52.

Shah, M.M. (2017). Comprehensive correlation for dispersed flow film boiling heat transfer in mini/macro tubes. *Int. J. Refrig.* **78**: 32–46.

Shah, M.M. and Siddiqui, M.A. (2000). A general correlation for heat transfer during dispersed flow film boiling in tubes. *Heat Transfer Eng.* **21** (4): 1–15. https://www.tandfonline.com.

Shen, Z., Yang, D., Xie, H. et al. (2016). Flow and heat transfer characteristics of high-pressure water flowing in a vertical upward smooth tube at low mass flux conditions. *Appl. Therm. Eng.* **102**: 391–401.

Shiotsu, M. and Hama, K. (2000). Film boiling heat transfer from a vertical cylinder in forced flow of liquids under saturated and subcooled conditions at pressures. *Nucl. Eng. Des.* **2000**: 23–38.

Sridharan, A. (2005). Modeling of inverted annular film boiling using an integral method. PhD thesis. The Pennsylvania State University.

Stevens, J.W. and Witte, L.C. (1971). Transient film and transition boiling from a sphere. *Int. J. Heat Mass Transfer* **14**: 443–450.

Sudo, Y. (1980). Film boiling heat transfer during reflood phase in postulated PWR loss-of-coolant accident. *J. Nucl. Sci. Technol.* **17** (7): 516–530.

Sudo, Y. and Murao, Y. (1976). Film boiling heat transfer during reflood process. JAERl-report M6848.

Tian, W.X., Qiu, S.Z., and Jia, D.N. (2006). Investigations on post-dryout heat transfer in bilaterally heated annular channels. *Ann. Nucl. Energy* **33**: 189–197.

US NRC. (2007). TRACE V5.0 theory manual: field equations, solution methods, and physical models. Quoted in Mohanty (2015)..

Varone, A.F. and Rohsenow, W.M. (1986). Post dryout heat transfer prediction. *Nucl. Eng. Des.* **95**: 315–327.

Verthier, B., Celata, G.P., Zummo, G. et al. (2009). Effect of gravity on film boiling heat transfer and rewetting temperature during quenching. *Microgravity Sci. Technol.* https://doi.org/10.1007/s12217-009-9135-7.

Webb, S.W. and Chen, J.C. (1986). Evaluation of convective film boiling models with non-equilibrium data in tubes. *Proceedings of the 8th International Heat Transfer Conference, San Francisco*. Quoted in Andreani and Yadigaroglu (1994), 17–22 August 1986.

Wendelstorf, J., Spitzer, K.H., and Wendelstorf, R. (2008). Spray water cooling heat transfer at high temperatures and liquid mass fluxes. *Int. J. Heat Mass Transfer* **51**: 4902–4910.

Witte, L.C. and Orcozo, J.A. (1984). The effect of vapor velocity profile shape on flow film boiling from submerged bodies. *J. Heat Transfer* **186** (1): 191–197.

Witte, L.C., Baker, L., and Haworth, D.R. (1968). Heat transfer from spheres into subcooled liquid sodium during forced convection. *J. Heat Transfer* **90** (4): 394–398.

Yuan, K., Ji, Y., Chung, J.N. et al. (2008). Cryogenic boiling and two-phase flow during pipe chilldown in earth and reduced gravity. *J. Low Temp. Phys.* **150**: 101–122.

Zuber, N. and Findlay, J.A. (1965). Average volumetric concentration in two-phase flow systems. *J. Heat Transfer* **87**: 453–468.

8

Two-Component Gas–Liquid Heat Transfer

8.1 Introduction

Heat transfer to mixtures of permanent gases with liquid flowing in tubes/pipes is frequently encountered in petroleum, nuclear power, chemical, food, and pharmaceutical industries. Injection of gas into liquid is also used for enhancing heat transfer. Hence prediction of heat transfer for such applications is of considerable practical importance. Topics discussed in this chapter include heat transfer during flow of pre-mixed gas–liquid mixtures inside channels as well as heat transfer with gas injected through the surface of tubes. Also discussed are cooling of objects of various shapes by mist cooling, cooling by jets, effects of gravity, and evaporation from water pools. Sections 8.2–8.6 discuss only non-metallic fluids. Liquid metal–gas mixtures are considered in Section 8.7.

8.2 Pre-mixed Mixtures in Channels

Figure 8.2.1 shows typical experimental data for heat transfer to gas–liquid mixtures in tubes. Heat transfer coefficient increases with increasing liquid flow rate. It also increases with the ratio of superficial gas and liquid velocities till it reaches a maximum. Then it declines with further increase in u_{GS}/u_{LS}. This is similar to dryout followed by transition boiling that occurs during boiling of single-component liquids.

In some cases, somewhat different behavior is found as shown in Figure 8.2.2. It is seen that at higher liquid Reynolds numbers, heat transfer initially increases with increasing vapor Reynolds number and then becomes near constant or drops slightly before rising again. Figure 8.2.2 also shows the observed flow patterns. It is seen that in the part where heat transfer coefficient became constant or decreased with increasing gas flow rate, there was slug flow patterns. Thus, heat transfer is seen to be affected by flow pattern.

Many researchers have presented prediction methods for specific flow patterns. These are discussed in the following text.

8.2.1 Flow Pattern-Based Prediction Methods

8.2.1.1 Bubbly Flow

Collier (1972) assumed that introduction of gas into liquid serves only to increase the velocity of liquid phase, and hence,

$$\frac{h_{TP}}{h_{LS}} = \left(\frac{1}{1-\alpha}\right)^n \tag{8.2.1}$$

If liquid flow is laminar, $n = 1/3$. For turbulent flow, $n = 0.8$. For bubbly flow, homogeneous flow is likely. Calculating the void fraction α by the homogeneous model, the following relation is obtained:

$$\frac{h_{TP}}{h_{LS}} = \left(1 + \frac{u_{GS}}{u_{LS}}\right)^n \tag{8.2.2}$$

where u_{GS} and u_{LS} are the superficial velocities of the gas and liquid phases, respectively, i.e. assuming each phase is flowing alone in the channel.

Novasad (1955) studied heat transfer in a vertical 38 mm diameter tube through which liquid flowed upward and gas was introduced into it from the bottom. The tube was cooled by water. The liquids used were water, 49% glycerin solution, and n-butyl alcohol. Gases used were air and hydrogen. Heat transfer coefficient increased with increasing gas flow until a critical velocity was reached beyond which slug flow occurred. Heat transfer coefficient at velocities higher than critical velocity rose slowly and even decreased before rising again with further increase in gas velocity. The data in the bubbly flow region were correlated by the following equation:

$$Nu_{TP} = \frac{Nu_f}{1 + 30\alpha^{0.5}} + 2.28(Re'_{TP})^{0.7} Pr_f^{0.33} \tag{8.2.3}$$

The two-phase Reynolds number Re'_{TP} is defined as

$$Re'_{TP} = \frac{D\rho_f u_{GS}}{\mu_f \alpha^{0.5}} \tag{8.2.4}$$

Two-Phase Heat Transfer, First Edition. Mirza Mohammed Shah.

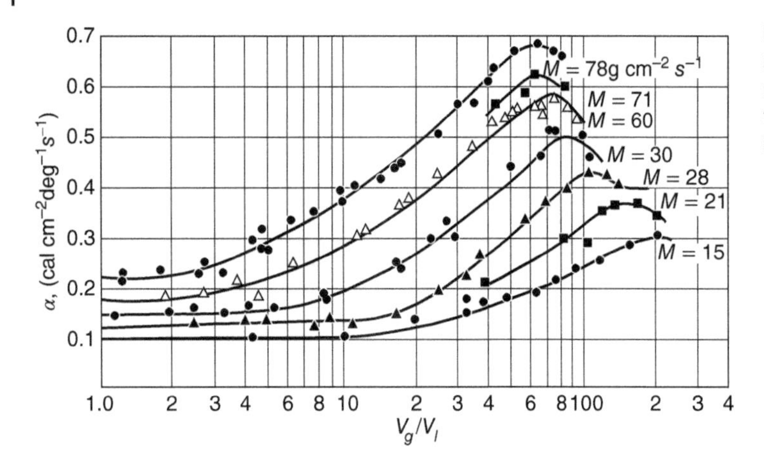

Figure 8.2.1 Effect of liquid flow rate M and ratio of superficial velocities of gas and liquid, V_g and V_l, respectively, on heat transfer coefficient α. Air–water flowing up in a heated vertical tube 14.25 mm ID. Source: Groothius and Hendal (1959). © 1959, Elsevier.

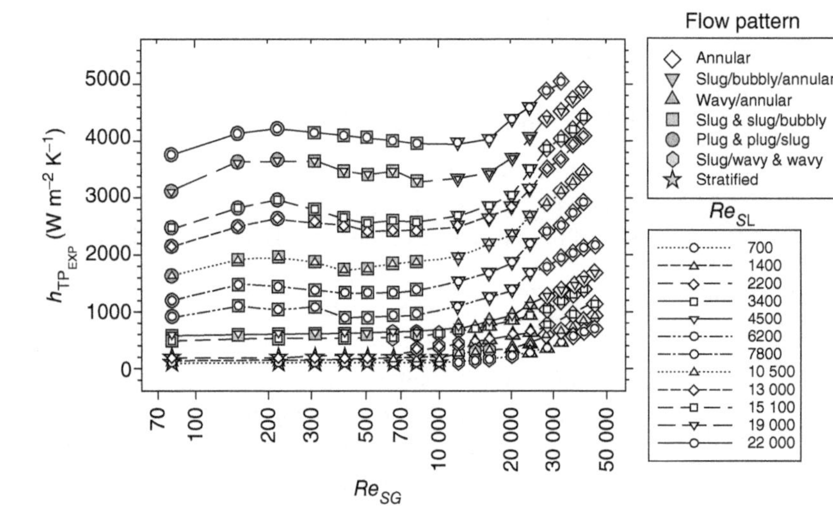

Figure 8.2.2 Effect of gas and liquid superficial Reynolds numbers and flow patterns on heat transfer during air–water flow in a 26.6 mm ID horizontal pipe. Source: Ghajar and Tang (2010). © 2010, Taylor & Francis.

Nu_f is the Nusselt number for single-phase liquid flow given by the following equation obtained from his own test data:

$$Nu_f = \frac{h_{LS}D}{k_f} = 0.3Gr_f Pr_f^{0.1} \tag{8.2.5}$$

$$Gr_f = \frac{D^3 \beta g(T_w - T_f)\rho_f^2}{\mu_f^2} \tag{8.2.6}$$

where β is the coefficient of thermal expansion of liquid. All properties are calculated at the film temperature. The void fraction α was obtained from Novasad's own test data.

Knott et al. (1959) performed tests with oil–nitrogen mixture flowing up in a vertical tube. They correlated their data by the following equation:

$$h_{TP} = h_{LS}(1 + u_{GS}/u_{LS})^{1/3} \tag{8.2.7}$$

The single-phase heat transfer coefficient was based on their own test data that agreed well with Sieder–Tate correlation. It is seen that this empirical correlation is the same as Eq. (8.2.2) derived by Collier through physical reasoning.

Zhang et al. (2006) assumed that bubbly flow and dispersed-bubble flow can be treated as pseudo single-phase flow. The fluid physical properties are adjusted on the basis of liquid holdup. With these assumptions, they derived the following expression by modifying the Pethukov correlation for single-phase flow:

$$\frac{h_{TP}D}{k_f} = \frac{0.5f_m Re_m Pr_m}{1.07 + 12.7(Pr_m^{2/3} - 1)\sqrt{0.5f_m}} \left(\frac{\mu_f}{\mu_{f,w}}\right)^{0.25} \tag{8.2.8}$$

where f_m is the Fanning friction factor for the mixture, determined using mixture density and mixture Reynolds number. Re_m is defined as

$$Re_m = \frac{(u_{GS} + u_{LS})\rho_m D}{\mu_f} \tag{8.2.9}$$

$$Pr_m = \frac{C_{pm}\mu_f}{k_f} \tag{8.2.10}$$

$$\rho_m = \alpha\rho_g + (1 - \alpha)\rho_f \tag{8.2.11}$$

$$C_{pm} = \alpha C_{pg} + (1 - \alpha)C_{pf} \tag{8.2.12}$$

This correlation was compared with the data of Manabe (2001) with satisfactory agreement.

8.2.1.2 Slug Flow

In slug flow, a succession of liquid slugs and gas bubbles are observed at a fixed location. The liquid slugs contain entrained gas. The bubbles are symmetrical in vertical flow and are usually surrounded by a liquid film. In horizontal flow, the gas bubbles have a liquid film below them while gas directly contacts the wall at the top. As the thermal conductivity of gas is much lower than that of liquid, heat transfer coefficients at the top are much lower than at the bottom. This is seen in Figure 8.2.3, which shows some of the data of Deshpande et al. (1991) for air–water flow in a horizontal pipe.

Oliver and Wright (1964) measured heat transfer coefficients during slug flow in a horizontal tube 6.35 mm ID. Liquid was 88% glycerol solution in water. They gave a correlation of their own data.

Lunde (1961) developed the following correlations:

For $Re_L < 3\times10^4$,

$$\frac{h_{TP}D}{k_f} = \frac{0.069Pr_f^{1/3}(\mu_f/\mu_{f,w})^{0.14}Re_L^{1.12}}{Re_{LS}^{0.48}(p_{actual}/p_{atmospheric})^{0.17}} \tag{8.2.13}$$

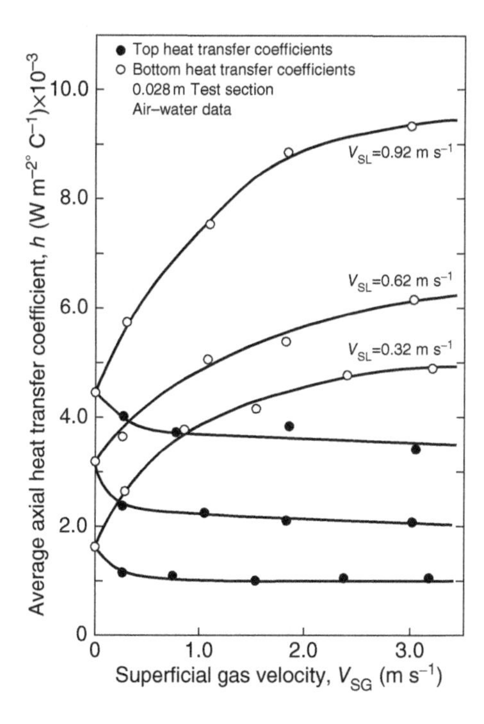

Figure 8.2.3 Heat transfer coefficients during slug flow at the top and bottom of a horizontal tube. V_{SL} is the superficial liquid velocity. Source: Deshpande et al. (1991). © 1991, American Chemical society.

For $Re_L > 3\times10^4$,

$$\frac{h_{TP}D}{k_f} = \frac{0.022Pr_f^{1/3}(\mu_f/\mu_{f,w})^{0.14}Re_L^{0.8}}{(p_{actual}/p_{atmospheric})^{0.17}} \tag{8.2.14}$$

Re_{LS} is the liquid Reynolds number based on superficial liquid velocity, while Re_L is the Reynolds number based on actual liquid velocity, $= Re_{LS}/\alpha$. It gave good agreement with horizontal tube data of Johnson (1955), Johnson and Abou-sabe (1952), and Fried (1954), as well as the vertical upflow data of Verschoor and Stemerding (1951). The data covered Re_L from 900 to 6×10^5.

A number of prediction methods have been developed through analysis of hydrodynamics of slug flow.

Deshpande et al. (1991) developed the following correlations for slug flow in horizontal pipes:

$$Nu_{bottom} = 0.023Re_L^{0.83}Pr_f^{0.4}\left(\frac{u_{LS}}{u_{NS}} - 1\right)^{0.53} \tag{8.2.15}$$

$$Nu_{top} = 1.93Re_m^{0.44}Pr_m^{0.4}\left(\frac{u_{LS}}{u_{NS}}\right)^{0.21}\left(\frac{\gamma u_{LS}}{g}\right)^{0.53} \tag{8.2.16}$$

$$Nu_{average} = 0.023Re_m^{0.83}Pr_m^{0.4}\left(\frac{u_{LS}}{u_{NS}}\right)^{0.76} \tag{8.2.17}$$

"top," "bottom," and "average" indicate top part of pipe, bottom part of pipe, and circumferential average, respectively. The various terms in the above correlation are defined as follows:

$$Re_L = \frac{u_{NS}\rho_f D}{\mu_f} \tag{8.2.18}$$

$$Re_m = \frac{u_{NS}\rho_m D}{\mu_m} \tag{8.2.19}$$

$$Pr_m = \frac{C_{pm}\mu_m}{k_m} \tag{8.2.20}$$

u_{NS} is the velocity of gas and liquid assuming no slip. The mixture properties are calculated as follows:

$$\rho_m = \rho_f R_{slug} + \rho_g(1 - R_{slug}) \tag{8.2.21}$$

$$\mu_m = \mu_f R_{slug} + \mu_g(1 - R_{slug}) \tag{8.2.22}$$

$$k_m = k_f R_{slug} + k_g(1 - R_{slug}) \tag{8.2.23}$$

$$C_{pm} = \frac{C_{pf}\rho_f R_{slug} + C_{pg}\rho_g(1 - R_{slug})}{\rho_f R_{slug} + \rho_g(1 - R_{slug})} \tag{8.2.24}$$

R_{slug} is the liquid holdup of the liquid slug. This is calculated by the correlation of Heywood and Richardson (1979). That correlation is given by Heywood and Richardson in the form of a graph. According to that graph, $R_{slug} = 1$ for $u_{GS} \le 1.4\ \mathrm{m\,s^{-1}}$. For $u_{GS} > 1.4\ \mathrm{m\,s^{-1}}$, the following equation is fitted to the graph by the present author:

$$R_{slug} = 1.11 - 0.078u_{GS} \tag{8.2.25}$$

u_{GS} is in m s^{-1}. The maximum u_{GS} in the test data shown in the graph is 6.5 m s^{-1}.

The frequency of slugs Υ is calculated by the following correlation of Heywood and Richardson (1979):

$$\gamma = 0.0434\left[\left(\frac{u_{LS}}{u_{NS}}\right)\left(\frac{2.02}{D}+\frac{u_{NS}^2}{gD}\right)\right]^{1.02} \quad (8.2.26)$$

This correlation gave satisfactory agreement with their own data for air–water mixtures in horizontal pipes of diameter 28 and 57 mm. It also gave good agreement with the tube average data of Fried (1954) and those of Shoham et al. (1982) for top and bottom of tube.

Kim and Ghajar (2002) have given an empirical flow pattern-based correlation that includes slug flow. It is presented in Section 8.2.1.3.

Franca et al. (2008) analyzed a model of slug flow in horizontal pipes. They developed equations for calculating time-averaged heat transfer coefficients that require insertion of hydrodynamic factors such as lengths of liquid slug and bubble, their frequency, void fraction, etc. Their method gave satisfactory agreement with their own test data obtained in a 25 mm diameter pipe and a 150 mm pipe.

Zhang et al. (2006) performed a mechanistic analysis to develop a formula for heat transfer during symmetrical slug flow. Their formula requires insertion of hydrodynamic parameters such as length of slug and bubble, their speeds, slug frequency, void fraction, etc. These have to be calculated by methods given in Zhang et al. (2003). This formula was compared only to the data of Manabe (2001). Agreement was satisfactory.

Bhandara et al. (2015) have reviewed the experimental and theoretical work on heat transfer and hydrodynamics during slug/Taylor flow in minichannels/micro channels that were considered to be of diameters ≤6.35 mm. They discussed a number of proposed prediction methods for heat transfer based on mechanistic analysis. On comparing them, they found wide differences, up to 500%, between their predictions. None of those methods had been compared to a wide range of data.

8.2.1.3 Annular Flow

Theoretical solutions for annular flow heat transfer have been developed by Levy (1952) and Suzuki et al. (1983). These were verified with a limited amount of test data.

Johnson and Abou-sabe (1952) performed tests on air–water mixtures in a horizontal pipe of 22.1 mm diameter. Most of their data were in the annular regime. They gave a graphical correlation of their data.

Pletcher and McManus (1968) studied heat transfer to air–water flow in a horizontal 25.4 mm diameter pipe. They have given a graphical correlation of their data, which were in annular flow.

Hughmark (1963) noted that as annular flow consists of a gas core inside a liquid film, heat transfer occurs in two steps, from wall to liquid film and from liquid film to gas core. Hence total temperature drop is the sum of temperature differences in the liquid film and in the gas core. This leads to the following equation:

$$\frac{H_{total}}{Ah_{TP}} = \frac{H_{gas}}{A_{gc}h_G} + \frac{H_{liquid}}{Ah_L} \quad (8.2.27)$$

A is the surface area of the pipe and A_{gc} is the surface area of gas core. H is the amount of heat transferred, Watt or equivalent in other units.

Heat transfer coefficient of liquid film h_L is calculated by the momentum–heat transfer analogy as

$$\frac{h_L}{C_{pf}u_L\rho_f} = \frac{\left(\frac{f_L}{2}\right)\left(\frac{\mu_f}{\mu_{f,w}}\right)^{0.14}}{\left\{1.2+11.8\left(\frac{f_L}{2}\right)^{0.5}(Pr_f-1)Pr_f^{-1/3}\right\}} \quad (8.2.28)$$

The liquid velocity u_L is

$$u_L = \frac{G_L}{\rho_f(1-\alpha)} \quad (8.2.29)$$

$$f_L = \left(\frac{\Delta p_{TP}}{\Delta L}\right)\left(\frac{D}{2\rho_f u_L^2}\right) \quad (8.2.30)$$

Heat transfer coefficient of the gas core is calculated by the method of Kropholler and Carr (1962) as

$$\frac{h_G}{C_{pg}(u_G-u_L)\rho_g} = \frac{f_G}{2\Phi} \quad (8.2.31)$$

$$u_G = \frac{G_G}{\alpha\rho_g} \quad (8.2.32)$$

$$f_G = \left(\frac{\Delta p_{TP}}{\Delta L}\right)\left(\frac{D}{2\rho_g(u_G-u_L)^2}\right) \quad (8.2.33)$$

ϕ is a function of Reynolds and Prandtl numbers expressed by the formula given in Kropholler and Carr (1962).

Hughmark compared this correlation with the data of Johnson (1955) for air–oil and air–water data of Johnson and Abou-sabe (1952), Fried (1954), and King (1952). Mean absolute deviation (MAD) was 11.8%. The values of pressure drop and void fraction used in the correlation were the measured values. Deviations are likely to be higher when these are obtained from correlations.

Zhang et al. (2006) analyzed a model similar to that of Hughmark and presented formulas for heat transfer in annular flow as well as stratified flow. These were verified only with the data of Manabe (2001).

Table 8.2.1 Constants and exponents in the flow pattern-based correlation of Kim and Ghajar (2002).

Flow pattern	C	m	n	p	q
Slug,bubbly-slug, bubbly-slug-annular	2.86	0.42	0.35	0.66	−0.72
Wavy–annular	1.58	1.4	0.54	−1.93	−0.09
Wavy	27.89	3.1	−4.44	−9.65	1.56

Source: Based on Kim and Ghajar (2002).

Kim and Ghajar (2002) gave an empirical flow pattern-based correlation expressed by the following correlation:

$$h_{\mathrm{TP}} = (1-\alpha)h_L \left[1 + C\left(\frac{x}{1-x}\right)^m \left(\frac{\alpha}{1-\alpha}\right)^n \times \left(\frac{Pr_g}{Pr_f}\right)^p \left(\frac{\mu_g}{\mu_f}\right)^q\right] \quad (8.2.34)$$

Quality x is defined as

$$x = \frac{W_g}{W_f + W_g} \quad (8.2.35)$$

It is stated that liquid phase heat transfer coefficient is to be calculated by the Sieder and Tate (1936) correlation. Sieder and Tate gave separate equations for laminar and turbulent flow. According to Mollamahmutoglu (2012), which is from the same institute, laminar equation is used when $Re_{\mathrm{SL}} < 2000$ and the turbulent equation for $Re_{\mathrm{SL}} > 2000$.

Re_{L} to be used in the Sieder–Tate correlation is defined as

$$Re_L = \frac{4W_L}{\pi(1-\alpha)^{0.5}\mu_L D} \quad (8.2.36)$$

Void fraction α is calculated by the following correlation of Chisholm (1973):

$$\alpha = \left[1 + \left(\frac{u_G}{u_L}\right)\left(\frac{1-x}{x}\right)\left(\frac{\rho_G}{\rho_L}\right)\right]^{-1} \quad (8.2.37)$$

The values of constant and exponents in Eq. (8.2.34) for various flow patterns are listed in Table 8.2.1. These were obtained by analysis of their own data for air–water as well as those of King (1952) in horizontal pipes. Flow patterns were based on their own observations. Good agreement was obtained with these data.

8.2.1.4 Post-dryout Dispersed Flow

Mastanaiah and Ganic (1981) performed tests with water drops dispersed in air flowing up in a vertical heated tube. Wall temperature was below the minimum film boiling temperature of water. Heat transfer coefficients decrease with increasing wall temperature. Analysis was performed in which heat transfer from tube surface occurred by convection with air as well as by impact of liquid drops. A calculation method was developed using a number of assumptions and approximations. It showed satisfactory agreement with their experimental data.

8.2.2 General Correlations

8.2.2.1 Horizontal Channels

Kim and Ghajar (2006) presented the following empirical correlation:

$$h_{\mathrm{TP}} = F_P h_L \left\{1 + 0.82\left[\left(\frac{x}{1-x}\right)^{0.08}\left(\frac{1-F_P}{F_P}\right)^{0.39} \times \left(\frac{Pr_G}{Pr_L}\right)^{0.03}\left(\frac{\mu_L}{\mu_G}\right)^{0.01}\right]\right\} \quad (8.2.38)$$

Void fraction is obtained by the Chisholm (1983) correlation, which is

$$\alpha = \left[1 + \left(1 - x + x\frac{\rho_L}{\rho_G}\right)^{0.5}\left(\frac{1-x}{x}\right)\left(\frac{\rho_G}{\rho_L}\right)\right]^{-1} \quad (8.2.39)$$

F_p is the flow pattern factor that takes into account the effect of flow pattern on heat transfer. It is given by the following formula:

$$F_P = (1-\alpha) + \alpha F_S^2 \quad (8.2.40)$$

F_S is a shape factor given by the following relation:

$$F_S = \frac{2}{\pi}\tan^{-1}\sqrt{\frac{\rho_g(u_G - u_L)^2}{gD(\rho_L - \rho_G)}} \quad (8.2.41)$$

The velocities of gas and liquid u_G and u_L are calculated with Eq. (8.2.32) and (8.2.29), respectively.

Liquid heat transfer coefficient h_L is calculated for liquid flowing at the in situ velocity with Reynolds number given by Eq. (8.2.36) using the Sieder–Tate correlation. They found good agreement with air–water data from their own tests as well as those from Kim and Ghajar (2002). The data included Re_{SL} from 835 to 35 503.

Shah (1981) gave the following correlation for horizontal tubes:

For $Re_{\mathrm{LS}} < 170$,

$$h_{\mathrm{TP}} = h_{\mathrm{LS}}(1 + u_{\mathrm{GS}}/u_{\mathrm{LS}})^{0.25} \quad (8.2.42)$$

For $Re_{\mathrm{LS}} > 170$, the correlation is presented in the form of a graph. It has the following functional form:

$$\psi = h_{\mathrm{TP}}/h_{\mathrm{LS}} = \mathrm{function}(u_{\mathrm{GS}}/u_{\mathrm{LS}}, Fr_L) \quad (8.2.43)$$

where

$$Fr_L = \frac{u_{\mathrm{LS}}^2}{gD} \quad (8.2.44)$$

At any particular Froude number Fr_L, ψ increases with increasing velocity ratio till a transition velocity ratio $u_{r,\mathrm{tran}}$

is reached. At velocity ratios higher than $u_{r,\text{tran}}$, ψ decreases. For $Re_{LS} < 170$, Sieder–Tate laminar equation is used, while at $Re_{LS} > 170$, the Dittus–Boelter equation is used.

Shah (1981) verified this correlation with data from seven sources that included a wide range of parameters. Good agreement of this correlation with their own data has been reported by Wang et al. (2017) for air–water and air–oil in a 26 mm diameter tube and Noville and Bannwart (2010) for air–water in a 50 mm diameter tube.

Shah (2018a) compared the Shah (1981) correlation with additional data. He made modifications to improve its accuracy and converted it into equation form so as to make it easier to use in computerized calculations. It is given in the following.

The following definitions are used:

$$u_r = \frac{u_{GS}}{u_{LS}} \tag{8.2.45}$$

With increasing u_r,ψ increases until a maximum is reached and then decreases with further increase in u_r. The value of u_r at this transition is called $u_{r,\text{tr}}$. The value of ψ at $u_{r,\text{tr}}$ is called ψ_{tr}.

For $Re_{LS} < 170$, all flow patterns, Eq. (8.2.42) applies.

For $Re_{LS} \geq 170$, when $u_r \leq u_{r,\text{tr}}$,

$$\Psi = F_{fp}F_g\frac{(414 + 89.4u_r^{0.49})}{(365 + u_r^{0.49})} \tag{8.2.46}$$

For $Re_{LS} \geq 170$, when $u_r > u_{r,\text{tr}}$

$$\Psi = F_{\text{fp}}F_g\Psi_{tr}\frac{(0.97 + 0.22u_{\text{red}})}{(1 + 0.15u_{\text{red}} + 0.038u_{\text{red}}^2)} \tag{8.2.47}$$

$$u_{\text{red}} = u_r/u_{r,\text{tr}} \tag{8.2.48}$$

The flow pattern factor $F_{\text{fp}} = 1$ except as follows:

For $D \geq 50$ mm when flow pattern is stratified, slug, or wave, $F_{\text{fp}} = 0.5$.

Further, for $28 < D < 50$ mm, $F_{\text{fp}} = 1$ for annular flow but is unknown for other flow patterns. There were no data for tube diameters between 28 and 50 mm.

As the flow pattern factor F_{fp} was developed using the map of Mandhane et al. (1974) and as definitions of flow patterns used and their predictions by other correlations may be different, only the Mandhane et al. correlation (version without fluid properties) is to be used.

The gravity correction factor $F_g = 1$ except for annular flow where it is given by

$$F_g = 1 + 0.5\left(1 - \frac{g}{g_e}\right) \tag{8.2.49}$$

The flow pattern is to be determined by a map suitable for the actual gravitational acceleration g. Flow pattern maps for low gravity are discussed in Section 8.6.3 and in Chapter 1.

The single-phase heat transfer coefficient for liquid h_{LS} is calculated by the following relations:

$Re_{LS} < 170$,

$$h_{LS} = 1.86(Re_{LS}Pr_fD/L)^{1/3}(\mu_f/\mu_{f,w})^{0.14}k_f/D \tag{8.2.50}$$

$Re_{LS} \geq 170$,

$$h_{LS} = 0.023Re_{LS}^{0.8}\,Pr_f^{0.4}k_f/D \tag{8.2.51}$$

Equation (8.2.50) is the correlation of Sieder and Tate (1936). Equation (8.2.51) is the correlation of Dittus and Boelter (1930).

The correlation as given by the above equations is shown graphically in Figure 8.2.4. Shah compared this correlation with data from many sources. Their range is given on Table 8.2.2. The same data were also compared with several other correlations. The results are given in Table 8.2.3. The MAD of the Shah (2018a,b,c) correlation is 19.2%, while

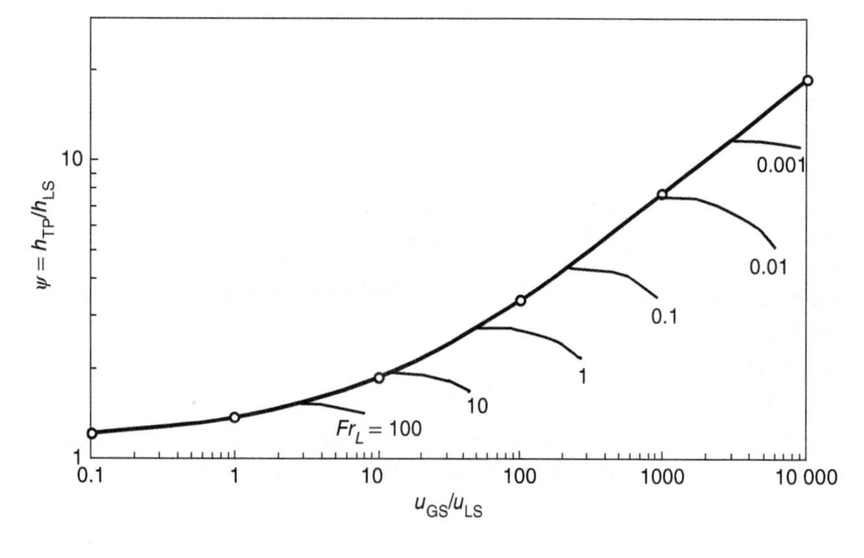

Figure 8.2.4 Shah correlation for heat transfer during gas–liquid flow in horizontal pipes at $Re_{LS} > 170$. Source: Shah (2018a). © 2018, ASME.

Table 8.2.2 Range of data with which the Shah correlation for horizontal pipes was verified.

	Range
Mixtures	Air–water, N_2–88% glycerin, air–oil
Tube diameter (L/D)	4.3–57.0 (20–244)
Heat transfer method	Cooling by liquid, heating by liquid/steam/electricity
Mixture pressure (bar)	0.48–4.1
Mixture temperature (°C)	12–62
Gravity	<0.1% Earth gravity to Earth gravity
Re_{LS}	$9-1.23E5$
Re_{GS}	$6-4.7E5$
Fr_L	0.0001–49
u_{GS}/u_{LS}	0.24–9298
Number of sources	18

Source: Shah (2018a). © 2018, ASME.

Table 8.2.3 Results of comparison of data for gas–liquid heat transfer in horizontal tubes with various correlations reported by Shah (2018a).

	Deviation (%) Mean absolute/average						
Number of data points	Kim and Ghajar	Tang and Ghajar	Knott et al.	Johnson and Abousabe	Kamin-sky	Shah Modified	Shah (2018a)
946	41.6	45.2	41.2	33.2	92.8	26.7	19.2
	−26.0	−42.1	21.1	−2.3	72.4	−9.7	−1.1

Source: Based on Shah (2018a).

those of others range from 26.7% for the modified Shah correlation to 92.8% for the Kaminsky (1999) correlation. The modified Shah correlation is Eq. (8.2.42) with h_{LS} calculated using a turbulent flow correlation. This version of Shah correlation has been used by several researchers including Mollamahmutoglu (2012) and Koviri et al. (2015). Figure 8.2.5 shows the comparison of the Shah correlations and two other correlations with the data of Nada (2017) for air–water in a 24.5 mm pipe.

While all other data analyzed by Shah were at Earth gravity, those of Witte et al. (1996) were for less than 0.1% gravity. The gravity effect factor F_g was based on these data. Its basis is described herein. Witte et al. had observed the flow patterns during the tests and provided a table giving heat transfer coefficients and the coincident flow patterns. For all the data, flow pattern was either annular or slug. When these data were compared to the Shah correlation before the development of F_g, it was found that all the data for slug flow were in satisfactory agreement, but the data for annular flow were greatly underpredicted. Most of these annular flow data were predicted to be in the slug or wave flow regimes by the Mandhane et al. map that is based on data at Earth gravity. Thus the higher heat transfer coefficients at low gravity appear to be due to the change in flow pattern caused by low gravity. In wave and slug flow, part or all of the tube surface is in contact with gas that has low thermal conductivity, while in annular flow, all the surface of the tube is in contact with liquid that has high thermal conductivity. This results in heat transfer coefficient being higher with annular flow at low gravity when comparable flow rates at Earth gravity produce slug or wave flow.

Dong and Hibiki (2018a,b) have given the following correlation.

For $Re_{LS} > 2300$,

$$\frac{h_{TP}}{h_{LS}} = (1-\alpha)^{-0.194}(1+0.687X^{-0.7}) \tag{8.2.52}$$

For $Re_{LS} < 2300$,

$$\frac{h_{TP}}{h_{LS}} = (1-\alpha)^{0.257}(1+4.27X^{-0.697}) \tag{8.2.53}$$

X is the Lockhart-Martinelli parameter. At $Re_{LS} < 2300$, Sieder–Tate equation for laminar flow is used to determine h_{LS}. At higher Re_{LS}, Gnielinski correlation is used, with calculated value multiplied by $(1 + D/L)^{2/3}$. Void fraction for slug flow is calculated by the drift flux correlation of Rassame and Hibiki (2018). The method of calculation of void fraction for other flow patterns has not been stated.

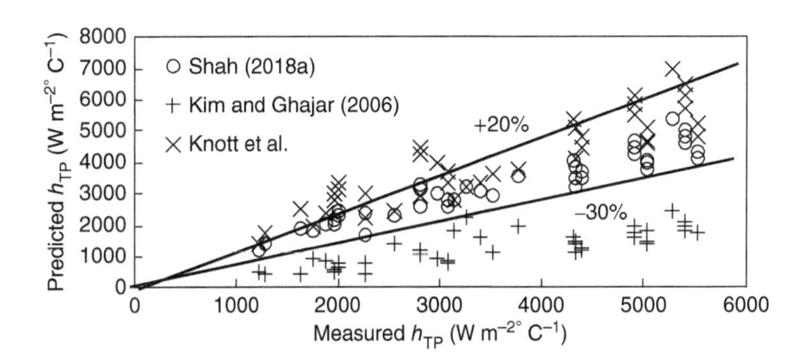

Figure 8.2.5 Comparison of data of Nada (2017) for air–water mixtures in a horizontal pipe compared with three correlations. Source: Shah (2018a). © 2018, ASME.

This correlation was compared to data for slug flow in horizontal pipes from many sources. The range of data was as follows: pipe diameters 0.87–5.15 cm, u_{GS} 0.007–2.29 m s^{-1}, and u_{LS} 0.2–20.5 m s^{-1}. Both heating and cooling conditions were included. All data were for air–water except for one set for air–oil. The data were predicted with MAD of 14.0%. They compared the same data to many other correlations. Best agreement was with the modified Shah correlation at 16.9%. All other correlations had significantly larger deviations. Correlations of Lunde (1961) and Deshpande et al. (1991) were not included in their evaluation.

While this correlation was developed primarily for slug flow, they also compared it with data for stratified, bubbly, and annular flow with air–water from four sources. Good agreement was found.

8.2.2.2 Vertical Channels

Rezkallah (1986) gave the following correlation based on his own data for air–water, air–silicone, and air–glycerin in upflow in a 11.7 mm diameter tube:

For $Re_{LS} < 2000$,

$$h_{TP} = (1 + u_r^{0.25}/Pr_f^{0.23})h_{LS} \quad (8.2.54)$$

For $Re_{LS} > 2000$,

$$h_{TP} = h_{LS}/(1-\alpha)^{0.9} \quad (8.2.55)$$

The void fraction α is calculated by the Chisholm (1973) correlation, Eq. (8.2.37).

Aggour (1978) gave the following correlation based on his data for mixtures of water with air, R-12, and helium flowing up in a 11.7 mm diameter tube:

For $Re_{LS} < 2000$,

$$h_{TP} = h_{LS}/(1-\alpha)^{0.33} \quad (8.2.56)$$

For $Re_{LS} > 2000$,

$$h_{TP} = h_{LS}/(1-\alpha)^{0.89} \quad (8.2.57)$$

In the above two correlations, h_{LS} is calculated by a laminar flow equation at $Re_{LS} < 2000$ and by a turbulent flow equation at higher Re_{LS}.

Kim et al. (2000) gave the following empirical correlation:

$$\frac{h_{TP}}{(1-\alpha)h_{LS}} = \left[1 + 0.27\left(\frac{x}{1-x}\right)^{-0.04}\left(\frac{\alpha}{1-\alpha}\right)^{1.21} \times \left(\frac{Pr_g}{Pr_f}\right)^{0.66}\left(\frac{\mu_g}{\mu_f}\right)^{-0.72}\right] \quad (8.2.58)$$

The data were from several sources, included several gas–liquid combinations, and Re_{LS} = 4000–1.26 × 10^5. Void fraction was calculated with the Chisholm (1973) correlation, while h_{LS} was calculated by the Sieder–Tate correlation for turbulent flow.

Shah (1981) gave a graphical correlation using the same parameters as his horizontal tube correlation described in Section 8.2.2.1. Shah (2018b) converted the graphical correlation to equation form, compared it to additional data, and further developed it to improve agreement with all data. The improved correlation is given as follows:

For $15 < Re_L < 175$, Eq. (8.2.42) applies.

For $Re_{LS} \le 15$,

$$h_{TP} = 0.75h_{LS}(1 + u_{GS}/u_{LS})^{0.25} \quad (8.2.59)$$

For $Re_{LS} > 175$,

$$\frac{h_{TP}}{h_{LS}} = \frac{E(414 + 89.4u_r^{0.49})}{(365 + u_r^{0.49})} \quad (8.2.60)$$

The factor E is calculated as follows:

For $Fr_L > 10$,

$$E = 1 \quad (8.2.61)$$

For $Fr_L \le 10$, E is the larger of those given by the following two equations:

$$E = 0.7\,Fr_L^{-0.36} \quad (8.2.62)$$

$$E = 1.41\,Fr_L^{-0.15} \quad (8.2.63)$$

If E given by Eqs. (8.2.62) and (8.2.63) is less than 1, use $E = 1$. The transition between Eqs. (8.2.62) and (8.2.63) occurs at $Fr_L = 0.035$. Figure 8.2.6 shows the correlation in graphical form.

Shah compared this correlation to data from 19 sources whose range of parameters is given in Table 8.2.4. The data included round pipes, annuli, and rectangular channels. For annuli and rectangular channels, hydraulic equivalent diameter was used in all equations. h_{LS} was calculated the same way as for Shah's horizontal tube correlation. Several other correlations were also compared to the same data. The results are given in Table 8.2.5 and shown in Figures 8.2.7 and 8.2.8. It is seen that for all data analyzed, the Shah correlation has MAD of 18.1%, while MAD of other correlations range from 28.6% to 45.5%. It is also seen that deviations of all correlations except the Shah correlation are high for Re_{LS} between 175 and 2300. It appears to be because these correlations used laminar flow equations to calculate liquid heat transfer coefficient. As pointed out by Shah, theoretical and experimental studies have shown that in the presence of gas, liquid films become turbulent at much lower Reynolds numbers than in single-phase flow. For example, Carpenter and Colburn (1951) examined their data for condensation in tubes and concluded that transition for laminar to turbulent liquid film occurred at Re_{LS} of 240. Rohsenow (1956) analyzed condensation of vapor flowing parallel to cooled plate. He found that the

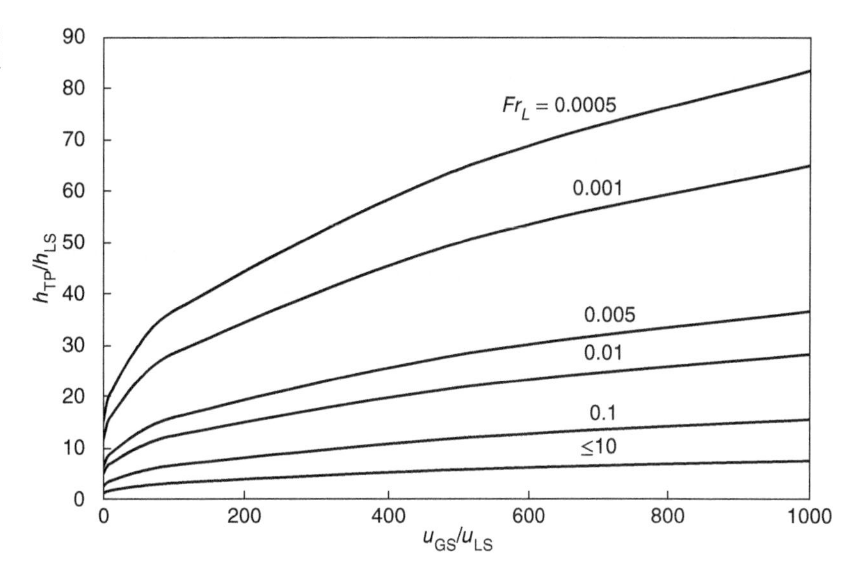

Figure 8.2.6 The correlation of Shah (2018b) for heat transfer to gas-liquid flow in vertical pipes in graphical form.

Table 8.2.4 Range of data for heat transfer to gas–liquid mixtures flowing in vertical channels analyzed by Shah (2018b).

Mixtures	Air–water, helium–water, R12–water, hydrogen–water, air–gasoil, air–toluene, air–glycerol, air–49% glycerol, air–59% glycerol, air–81.5% glycerol, air–butanol, air–methanol, nitrogen–oil
Test section types	Tubes, annuli, rectangular channel; heated or cooled
D or D_{hyd}, mm	4–70
Flow direction	Vertical up or down
Temperature (°C)	16–115
Pressure (bar)	1–6.9
Re_{LS}	2–127 231
Fr_L	0.00 017–1975
u_{GS} (m/s)	0.01–87
u_{LS} (m/s)	0.02–13
u_{GS}/u_{LS}	0.03–1630
Number of sources/data sets	19/30

Source: Shah (2018b). © 2018, ASME.

Table 8.2.5 Deviations of various correlations with data for vertical channels in various ranges of Re_{LS}.

		Deviation (%) Mean absolute/average						
Re_{LS}	**Number of data points**	**Tang and Ghajar**	**Knott et al.**	**Rezkallah**	**Kaminsky**	**Aggour**	**Shah modified**	**Shah (2018b)**
<175	124	28.0	37.1	32.3	157.3	21.9	24.1	14.7
		−18.5	29.6	32.0	157.2	−12.2	15.3	−14.6
>175	898	32.8	27.9	28.0	30.0	35.4	31.3	18.5
		−22.2	−24.8	7.5	−26.2	−2.8	−30.6	−3.2
175–2300	185	66.6	61.2	33.6	38.7	72.8	58.6	22.4
		−65.8	−61.2	−24.8	−25.5	−70.4	−58.2	−1.9
All	1022	32.3	29.0	28.6	45.5	33.8	30.4	18.1
		−21.8	−18.2	10.4	−3.9	−3.9	−20.5	−2.6

Source: From Shah (2018b). Copyright ASME, used with their permission.

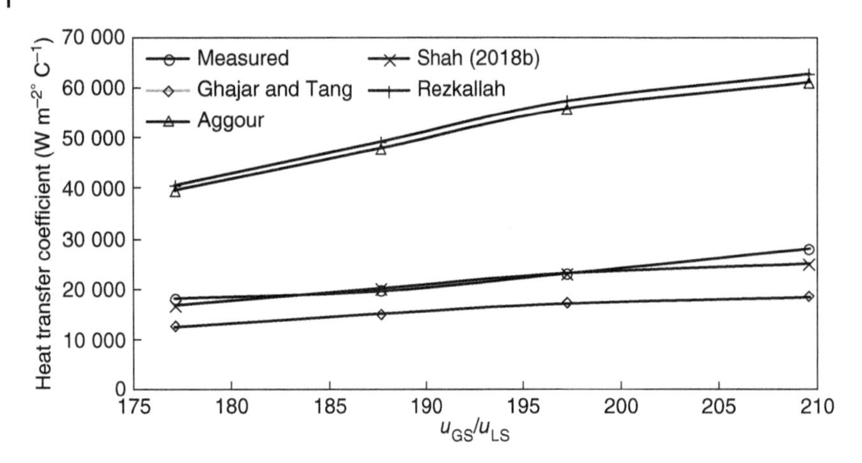

Figure 8.2.7 Data of for hydrogen–water mixture at 6.9 bar in a 4 mm diameter vertical tube compared to various correlations. $u_{LS} = 0.41\ \mathrm{m\,s^{-1}}$. Source: Shah (2018b). © 2018, ASME.

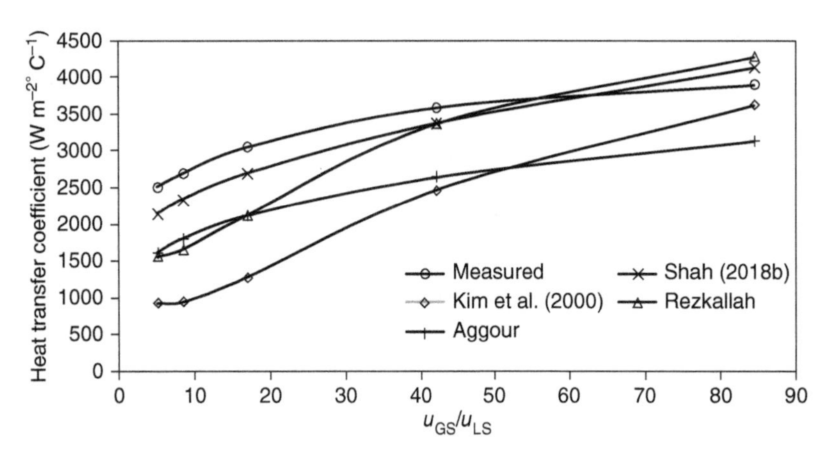

Figure 8.2.8 Comparison of various correlations with the data of Ravipudi (1976) for toluene–air mixture at atmospheric pressure in a 19 mm diameter tube. Source: Shah (2018b). © 2018, ASME.

transition Reynolds number goes down with increasing vapor shear and can be as low as 50.

All the data analyzed by Shah were for upflow except for two. Those of Dorresteijn (1970) included both upflow and downflow. While there were differences in the upflow and downflow heat transfer coefficients, both were in fairly good agreement with the Shah correlation. The data of Mollamahmutoglu (2012) were for vertical downflow and showed excellent agreement with the Shah correlation.

Dong and Hibiki (2018b) have given the following correlation:

For $Re_{LS} > 2300$,

$$\frac{h_{TP}}{h_{LS}} = (1 - \alpha)^{-0.02}(1 + 2.56X^{-0.508}) \tag{8.2.64}$$

For $Re_{LS} < 2300$,

$$\frac{h_{TP}}{h_{LS}} = (1 - \alpha)^{0.339}(1 + 4.65X^{-0.409}) \tag{8.2.65}$$

h_{LS} is calculated in the same way as in their correlation for horizontal tubes. For calculation of void fraction, correlations have been given for each flow pattern. For slug flow, data from eight sources were predicted with a MAD of 14.2%. Data for other flow patterns from three sources were predicted with MAD of 17.5%.

8.2.2.3 Horizontal and Vertical Channels

Some correlations have been given as applicable to both horizontal and vertical flow. These are discussed herein.

Kaminsky (1999) gave the following correlation as applicable to both vertical and horizontal tubes. For turbulent flow,

$$h_{TP} = h_{LS}\phi_L S^{0.5} \tag{8.2.66}$$

S is the fraction of surface that is wet and is obtained from the following implicit relation:

$$R_L = S - S\ \ln(2\pi S)/2\pi \tag{8.2.67}$$

For laminar flow,

$$h_{TP} = h_{LS}(2 - R_L)/{R_L}^{2/3} \tag{8.2.68}$$

R_L is the liquid holdup. Kaminsky compared this correlation with data for horizontal tubes from three sources and data for vertical tubes from many more sources. Satisfactory agreement was found.

Tang and Ghajar (2007) modified the Kim and Ghajar (2006) correlation for horizontal tubes to make it also applicable to other orientations. Ghajar and Tang (2010) noted that it does not take into consideration the effect of surface tension. They further modified it to arrive at the following correlation:

$$h_{\mathrm{TP}} = F_P h_{\mathrm{LS}} \left\{ 1 + 0.55 \left[\left(\frac{x}{1-x} \right)^{0.1} \left(\frac{1-F_P}{F_P} \right)^{0.4} \times \left(\frac{Pr_g}{Pr_f} \right)^{0.25} \left(\frac{\mu_f}{\mu_g} \right)^{0.25} (I)^{0.25} \right] \right. \tag{8.2.69}$$

I is a factor that depends on the tube inclination. It is expressed by the following relation:

$$I = 1 + \mathrm{Bn}\,\mathrm{Abs}(\sin\theta) \tag{8.2.70}$$

where θ is the inclination of channel to horizontal and Bn is the Bond number. Void fraction was calculated with the correlation of Woldesemayat and Ghajar (2007) and h_{SL} by the Sieder–Tate correlation. Data included horizontal and vertical tubes as well as tube inclined upwards by up to 7° to the horizontal from several sources. Good agreement was found.

8.2.2.4 Inclined Channels

As stated in Section 8.2.2.3, the Ghajar and Tang (2010) correlation showed satisfactory agreement with their own data for pipe inclined to horizontal by up to 7°.

Hosssainy (2014) measured heat transfer coefficients in a 12.5 mm diameter tube inclined 0–20° downwards to horizontal. They found that heat transfer coefficients in the stratified regime decreases with increasing inclination, the maximum decrease being 27%. The impact on heat transfer in other flow patterns was small. They compared their data to many correlations. All gave poor agreement with data in the stratified regime. For all other flow patterns, excellent agreement was found with the modified Shah correlation, which is Eq. (8.2.42) with h_{LS} calculated with a turbulent flow correlation; 93% of the data were predicted within ±30%.

John et al. (2015) made a study similar to Hossainy (2014) in which inclinations varied from 0° to 90° downwards. Some of their data are shown in Figure 8.2.9. It is seen that heat transfer drops considerably between 0° and 30° only at the lowest gas and liquid Reynolds numbers. At higher Reynolds numbers, the effect is comparatively small. They found best agreement with the modified Shah correlation. The correlation of Tang and Ghajar (2007), which has a factor for pipe inclination, had considerably higher deviations except in the annular flow regime,

Kalapatapu (2012) studied the effect of upward inclination on heat transfer during flow of air–water mixtures in a 12.5 mm ID tube. Pipe inclinations were 0–20° upwards. It was found that the two-phase heat transfer coefficient increases from 0° to +5° and + 10° and then decreases at +20°. However, the variations with inclinations were comparatively small. They compared their data with many correlations including the Tang and Ghajar (2007) correlation. Best results were with the modified Shah correlation, 97% of the data being predicted within ±30%. As this correlation does not have any factor for the effect of inclination, it may be concluded that reliable correlations for horizontal tubes can be used for tubes inclined up to 20° upwards.

8.2.3 Recommendations

Among the flow pattern specific correlations, only those of Dong and Hibiki (2018a,b) for slug flow in horizontal and vertical channels have been verified with a wide range of data. These are therefore recommended for slug flow. For other flow patterns, they have had only limited verification.

It should be noted that prediction of flow patterns can be in considerable error. This significantly reduces the accuracy of flow pattern-based correlations. It is therefore preferable to use well-verified correlations that do not require determination of flow patterns.

The general correlations of Shah (2018a,2018b) have been shown to be reliable over a very wide range of data that include all flow patterns. Therefore, the recommendation is to use the Shah correlations for horizontal and vertical (up and down) orientations.

There is no well-verified method for calculation of heat transfer to inclined tubes. The experimental data examined here indicates that except for very low gas and liquid flow

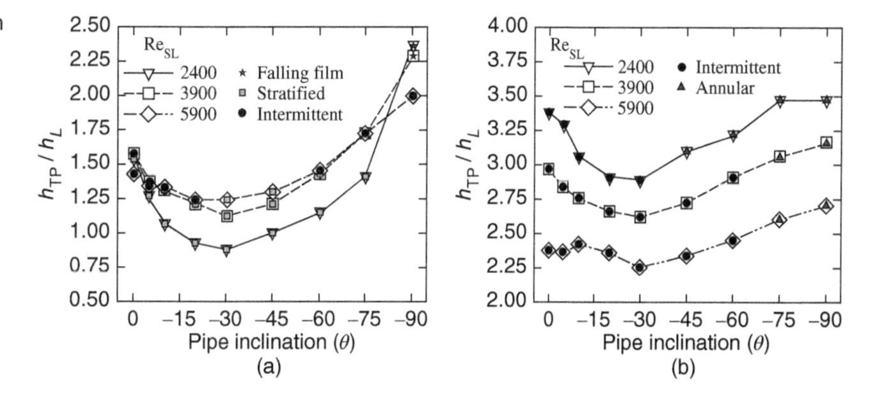

Figure 8.2.9 Effect of downward inclination on ratio of two-phase to single-phase heat transfer. (a) Re_{SG} = 2800 and (b) Re_{SG} = 14 000. Source: John et al. (2015). © 2015 with permission from Begell House, Inc.

rates, the correlations for vertical and horizontal flow can be used for changes of inclination up to about 20°.

8.3 Gas Flow through Channel Walls

8.3.1 Experimental Studies

Pioneering research in this field was done by Gose et al. (1957). They used an electrically heated sintered bronze tube through which water flowed. The porosity of the tube material was about 25%, and the diameter of the pores was between 7 and 13 μm. Nitrogen was injected through the surface of tube. The measured heat transfer coefficients are shown in Figure 8.3.1. It is seen that heat transfer coefficient rises rapidly with injection of only a small amount of gas. It increases slowly with further increase in gas injection rate. Enhancement in heat transfer is greater at lower water flow rates. Gose et al. hypothesized that the increase in heat transfer was due to increase in turbulence in the boundary layer caused by gas injection.

Kudirka (1964) performed tests in a sintered 15.8 mm ID vertical tube. Mixtures of air with water or ethylene glycol entered at the bottom of tube and flowed upwards. Air was also bubbled through the tube surface into the flowing mixture. Test data are shown in Figures 8.3.2 and 8.3.3. These show heat transfer coefficients vs. ratio of superficial gas and liquid velocities of pre-mixed mixture

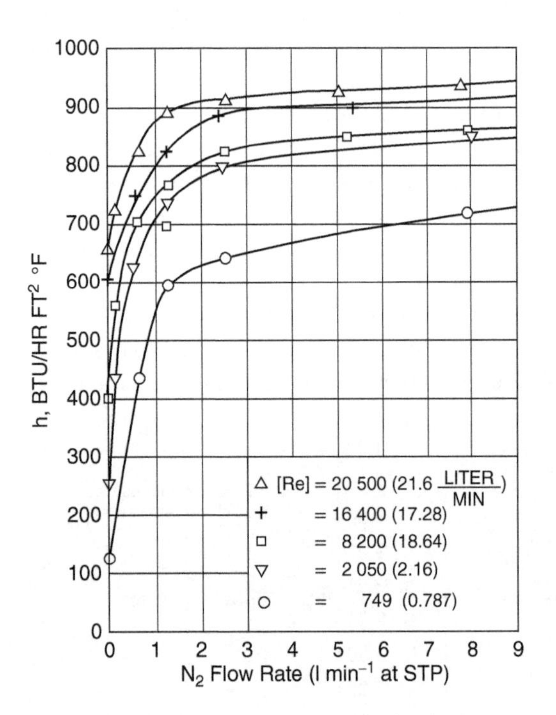

Figure 8.3.1 Effect of nitrogen injection through tube surface into water at various Reynolds numbers. Source: Gose et al. (1957). © 1957, AIP publishing.

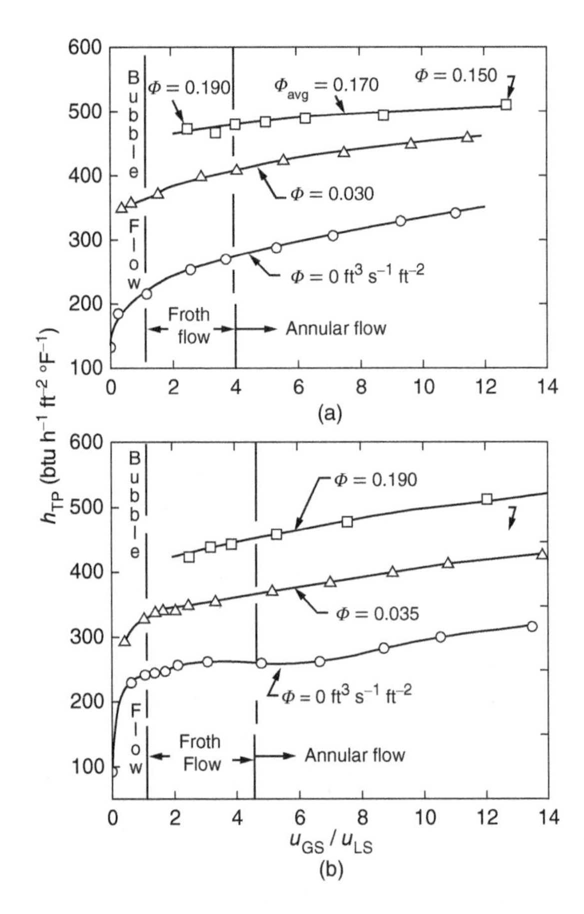

Figure 8.3.2 Effect of air injection through tube wall into air–ethylene glycol mixtures flowing in a vertical tube. Panel (a) is at superficial liquid velocity u_{LS} = 4.5 ft s^{-1}, and panel (b) is at u_{LS} = 1 ft s^{-1}. Source: Kudirka (1964). © 1964, US Department of Energy.

entering through the bottom of tube. For air–ethylene glycol, it is seen that heat transfer coefficient always increased with increasing surface air injection rate ϕ. The situation with air–water mixtures was different. At superficial water velocity of 0.3 m s^{-1}, heat transfer coefficient increased with increasing ϕ. At water velocity of 1.4 m s^{-1}, heat transfer coefficient decreased with increasing ϕ. The enhancement in heat transfer by gas injection appears to be dependent on liquid Reynolds number. If the liquid has low or moderate Reynolds number, heat transfer increases. If it is high, heat transfer decreases. The viscosity of ethylene glycol is about 18 times that of water. Hence its Reynolds numbers at the same velocity are much lower than those of water. Under the test conditions, superficial Reynolds number of ethylene glycol at 0.3 m s^{-1} velocity was about 270 and at 1.4 m s^{-1} about 1200. On the other hand, superficial Reynolds number of water at these velocities would have been 4800 and 22 000. The intensity of turbulence increases with Reynolds number. At low and medium Reynolds numbers, gas injection increases turbulence

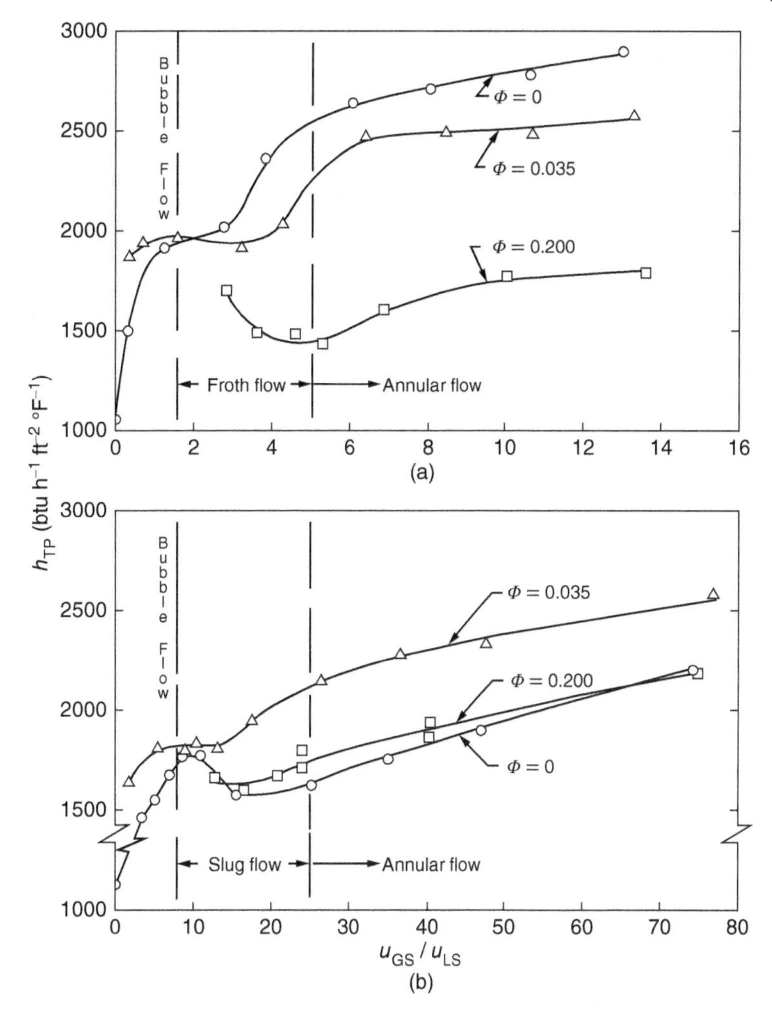

Figure 8.3.3 Effect of air injection through tube wall into air–water mixtures flowing in a vertical tube. ϕ is the rate of air injection through tube surface, ft^3 s^{-1} ft^{-2}. Panel (a) is at superficial liquid velocity $u_{LS} = 4.5$ ft s^{-1} and panel (b) is at $u_{LS} = 1$ ft s^{-1}. Source: Kudirka (1964). © 1964, US Department of Energy.

in the boundary layer and thus increases heat transfer coefficients. At high Reynolds numbers, liquid is already highly turbulent and gas injection does not increase it further. Instead, bubbles prevent contact of liquid with the wall in their vicinity and thus reduce the area available for convection by liquid causing reduction of heat transfer coefficients.

Martin and Sims (1971) performed tests in a horizontal rectangular channel 13.2 mm × 6.5 mm cross section. The bottom of the channel was heated and was made of porous material. Tests were done with only water at entrance and also when air–water mixture entered the channel. Air was injected through the bottom of the channel in both types of tests. Figure 8.3.4 shows the results for zero air at inlet. It is seen that heat transfer coefficient initially rose sharply with a small amount of gas injection. In this region, observed flow pattern was mostly bubbly. With further increase of gas injection rate, heat transfer coefficient reached a maximum and then remained constant or became slightly lower. The gas injection velocity at which maximum heat transfer occurred was called the critical velocity.

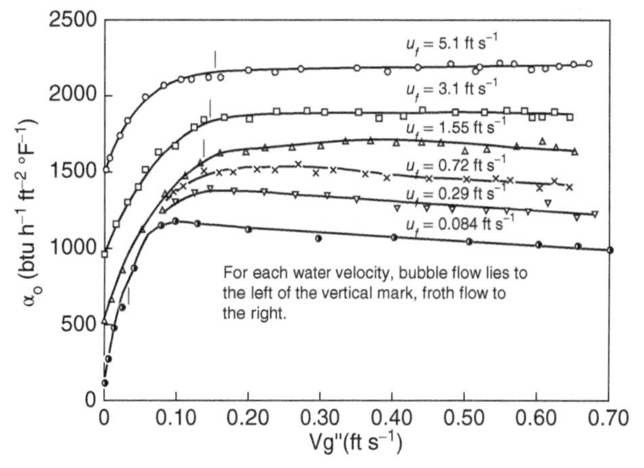

Figure 8.3.4 Effect of nominal velocity of air bubbled through the wall V_g'' on heat transfer coefficient α_o at various water superficial velocities U_f. No air mixed with water at entrance to channel. Source: Martin and Sims (1971). © 1971, Elsevier.

The tests with pre-mixed air–water entering the channel were done with entering superficial gas velocity 0–18 m s^{-1} and varying liquid superficial velocities. With liquid entering the channel at 0.088 m s^{-1} ($Re_{LS} \approx 730$), heat transfer coefficient almost always increased with increasing air injection through wall. With liquid entering at 1.55 m s^{-1} ($Re_{LS} \approx 12\,800$), air injection always decreased heat transfer. With liquid entering at 0.47 m s^{-1} ($Re_{LS} \approx 3900$), heat transfer coefficients increased at lower channel entrance gas velocities and decreased at higher entrance velocities.

Michiyoshi (1978) studied heat transfer in an annulus with air–water mixtures. In some tests, air and water were pre-mixed before entering the test section. In other tests, only water entered the annulus and air was injected through the wall. The quality (volumetric gas flow rate divided by total gas and liquid flow rate) varied from 0.03 to 0.38. For the same quality, heat transfer coefficients in the two cases were almost the same. The exception was that at the highest liquid flow rate, heat transfer coefficients were higher with wall injection than with pre-mixing at quality below 0.15. Shah (2018b) had found their data for pre-mixed condition to be in excellent agreement with his correlation. However, these data had a very limited range and therefore cannot be the basis for any general conclusion.

8.3.2 Heat Transfer Prediction

For gas injection rates lower than those that caused reduction in heat transfer coefficient, Martin and Sims (1971) considered the effect of gas injection to be similar to that of bubble nucleation in boiling. Accordingly, they postulated the following relation:

$$h_{TP} = Fh_{LS} + h_{bub}\psi \tag{8.3.1}$$

where h_{bub} is the heat transfer coefficient due to bubbling of gas through the wall and ψ is the effectiveness of bubbling. F is the enhancement due to air entering together with liquid into the channel. It is given by

$$F = 1 + 0.64\left(\frac{u_{GS,inlet}}{u_{LS}}\right)^{0.5} \tag{8.3.2}$$

The following equations were developed for h_{bub}:

$0.009 \le K_T \le 0.15$

$$\frac{h_{bub}}{k_f}\left[\frac{\sigma}{g(\rho_f - \rho_g)}\right]^{0.5} Pr_f^{0.2} = 167K_T^{0.75} \tag{8.3.3}$$

$0.15 \le K_T \le 1.27$

$$\frac{h_{bub}}{k_f}\left[\frac{\sigma}{g(\rho_f - \rho_g)}\right]^{0.5} Pr_f^{0.2} = 48.7K_T^{0.1} \tag{8.3.4}$$

$$K_T = \frac{V_g\rho_f^{0.5}}{[\sigma g(\rho_f - \rho_g)]^{0.25}} \tag{8.3.5}$$

V_g is the nominal velocity of gas flow through channel wall, volumetric flow rate of gas divided by wall surface area.

The bubbling effectiveness factor, ψ is given by

$$\psi = \psi_1\psi_0 \tag{8.3.6}$$

ψ_0 is the effectiveness factor when there is no pre-mixed gas entering the channel. For $Re_{SL} < 2000$, $\psi_0 = 1$. For $Re_{SL} > 2000$,

$$\psi_0 = 9.78Re_{SL}^{-0.3} \tag{8.3.7}$$

ψ_1 represents the effect of pre-mixed gas and is given by

$$\psi_1 = \exp\left(-\frac{0.68}{10^6}\frac{V_g}{u_{SL}}Re_{SL}^{1.4}\right) \tag{8.3.8}$$

This correlation gave good agreement with their own data but did not agree with the data of Gose et al. (1957). They did not give any method to determine the critical velocity beyond which gas injection decreases heat transfer, nor have they given any explanation of why heat transfer deteriorates beyond the critical gas injection velocity.

8.3.3 Conclusions

The following conclusions may be reached from the experimental work described earlier:

1. With zero quality at channel entrance, wall gas injection increases heat transfer coefficient rapidly at low gas flow rate. The flow pattern under these conditions is mainly bubbly. With further increase in wall injection rate, heat transfer reaches a maximum and then remains constant or decreases.
2. With finite quality at inlet, gas injection through wall increases heat transfer when liquid superficial Reynolds number is low, roughly less than 10 000. At high liquid Reynolds number, wall injection decreases heat transfer. The increase in heat transfer appears to be due to the increase in turbulence in the liquid boundary layer caused by gas injection. The decrease appears to be due to gas bubbles preventing liquid from contacting the wall in their vicinity.
3. The correlation of Martin and Sims is applicable only to their experimental conditions and only up to the critical gas injection rate. No other prediction method appears to be available.

8.4 Cooling by Air–Water Mist

8.4.1 Single Cylinders in Crossflow

Acrivos et al. (1964) conducted the first heat transfer investigation from a cylinder in air–water spray flow. They used a 38.1 mm diameter cylinder positioned vertically in a horizontal wind tunnel. They found as much as an eightfold increase in the local heat transfer coefficients when compared with dry air flow.

Early studies on heat transfer with mist flow across cylinders were also done by Hoelscher (1965), Saterbak (1967), Takahara (1965), and Smith(1966).

Finlay and McMillan (1967–1968) studied heat transfer to a heated horizontal cylinder located in a horizontal duct with air–water mixture flowing over it. The duct was 200 mm square and the cylinder was 10 mm diameter. Velocity of air approaching the cylinder was 20–75 m s^{-1} and mixture qualities were 0–9% of water by weight. Heat transfer coefficients were found to be strongly dependent on mixture quality. The average values of the surface heat transfer coefficient were up to 20 times those for dry air. Heat transfer coefficient was highest at the forward stagnation point, fell down to an angle of about 100° from the stagnation point, and then became nearly constant. Most of the heat transfer occurred in the front half of the cylinder. They observed the presence of a continuous liquid film in the front part of cylinder up to the separation point. In the region of separation, 80–130° from stagnation point, this film gradually thickened to form an unstable wavy ridge from which the liquid was re-entrained in the mainstream. Figure 8.4.1 shows some of their data.

Heat transfer coefficients of air flowing alone were in close agreement with the correlation of Hilpert (1933). Heat transfer coefficient at the forward stagnation point was correlated by the following equation:

$$\ln\left[\frac{Nu_s}{Re_{TP}^{1-6(1-x)}}\right] + 4.1 = 56(1-x)^{0.8} \tag{8.4.1}$$

Subscript "*s*" indicates at stagnation point. For the average heat transfer coefficient for the cylinder h_{avg}, the following equation was given:

$$\ln\left[\left(\frac{h_s}{h_{avg}}\right)/Re_{TP}^{0.1+2(1-x)}\right] + 0.58 = -17.6(1-x) \tag{8.4.2}$$

In Eqs. (8.4.1) and (8.4.2), h_s is the heat transfer coefficient at stagnation point and h_{avg} is the surface average heat transfer coefficient. Other definitions are

$$Nu_s = \frac{h_s D}{k_{mix}} \tag{8.4.3}$$

$$Re_{TP} = \frac{(G_{GS} + G_{LS})D}{\mu_{mix}} \tag{8.4.4}$$

$$x = \frac{W_G}{W_G + W_L} \tag{8.4.5}$$

The mixture properties k_{mix} and μ_{mix} are calculated assuming homogeneous flow. Re_{TP} varied from 25 000 to 95 000.

Hodgson et al. (1968) measured heat transfer to a 76.2 mm diameter horizontal brass cylinder with air–water mist flowing vertically down over it. The distribution of heat transfer coefficients around the cylinder was similar to that shown in Figure 8.4.1 except that the minimum occurred at 90°. Figure 8.4.2 shows the effect of air flow rate on the overall heat transfer coefficient of the entire circumference. It is seen that it increases with increasing air Reynolds number.

Mednick (1967) investigated heat transfer to an electrically heated vertical cylinder in a horizontal duct through which flowed a mixture of air and water. Cylin-

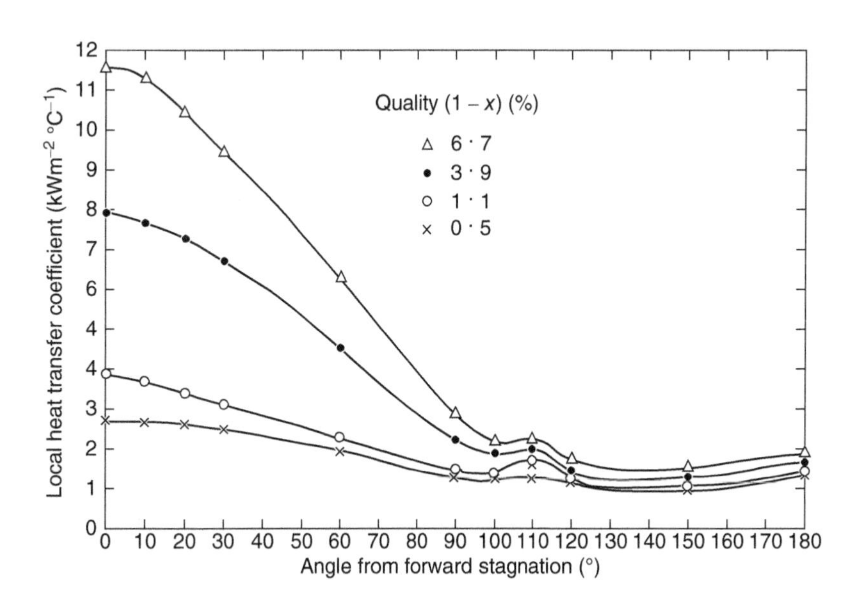

Figure 8.4.1 Effect of water content in air flowing over a cylinder on heat transfer coefficients at various angles. Re_{TP} = 95 000. Source: Finlay and McMillan (1967–1968).

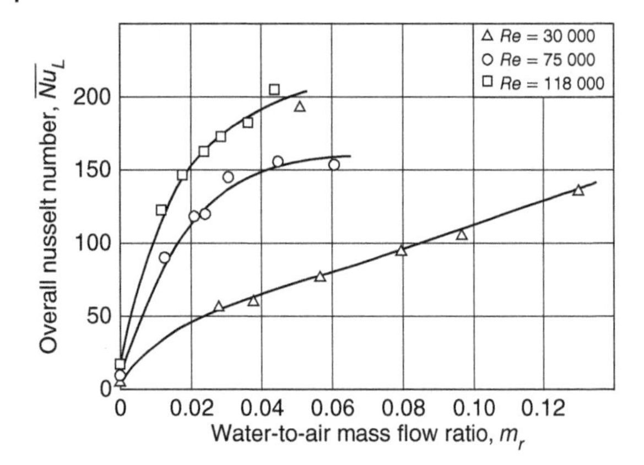

Figure 8.4.2 Effect of air Reynolds number and water to air mass flow ratio on circumferentially averaged Nusselt number for a cylinder in mist flow. Source: Hodgson et al. (1968). © 1968, ASME.

ders of diameter 25.4 and 38.1 mm were used. Air velocity approaching the cylinder was 18–43 m s^{-1}. A wide range of water spray densities, 12–95 kg m^{-2} s^{-1}, were used. Heat transfer coefficients at the stagnation point with spray were up to 40 times those with air alone. Heat transfer coefficients decreased with increasing angle from the stagnation point, similar to Figure 8.4.2. The following correlation was obtained for the average heat transfer coefficient for the entire cylinder surface:

$$\frac{h_{\text{avg}}D}{k_f} = 0.44\left(\frac{G_{\text{LS}}D}{\mu_f}\right)^{0.63}\left(\frac{DG_{\text{GS}}}{\mu_g}\right)^{0.264} \tag{8.4.6}$$

Fluid properties were evaluated at the temperature of the front half of the cylinder.

Aihara et al. (1990a) performed a comprehensive numerical analysis of heat transfer during downflow of air–water mist on a horizontal cylinder. The results were presented in the form of graphs.

Aihara et al. (1990b) studied heat transfer to a horizontal heated cylinder 50 mm diameter in a 125 mm×200 mm duct with air–water mist flowing vertically down on it. Average heat transfer coefficient at the rear half of the cylinder h_{rear} was correlated by the following equation;

$$\frac{h_{\text{rear}}}{h_s} = 0.093Re_{\text{GS}}^{1/6}\lambda^{-1/3} \tag{8.4.7}$$

$$\lambda = \left(\frac{G_{\text{LS}}}{u_{g,c}\rho_g}\right)Pr_g Re_{\text{GS}}^{1/2}\frac{C_{\text{pf}}}{C_{\text{pg}}} \tag{8.4.8}$$

where $u_{g,c}$ is the velocity of air at the center of duct without the presence of the cylinder. This equation is valid for D/b =0.1–0.4, Re_{GS} =4×10^5–1.2×10^5, and λ =2–70, where b is the width of duct in which the cylinder is located. The heat transfer coefficient at the front stagnation point h_s is obtained from the numerical solution in Aihara et al. (1990a).

8.4.2 Flow over Tube Banks

Elperin (1961) measured heat transfer in tube banks with air–water flow over them. Up to 20 times increase in heat transfer compared to air alone was found.

Finlay and Harris (1984) reviewed available test data and correlations for rating of various arrangements for spray cooling of coils. They have made recommendations for methods of calculation.

Nakayama et al. (1988) experimentally investigated enhancement of cooling with mist flow through banks of smooth, micro-finned, and finned tubes. Air velocity was 1–3 m s^{-1} and water spray in the range of 50–390 kg m^{-2} s^{-1}. Water injection increased average heat transfer coefficients four to five times those without water injection. They concluded that evaporation of water from the wetted surface was the primary mechanism of enhancement of heat transfer. They have presented a method for the design of mist cooled tube banks.

8.4.3 Flow Parallel to Plates

Hishida et al. (1980) did tests in which air–water mist flowed down parallel to a vertical heated plate 200 mm high and 100 mm wide. Air velocity varied from 2.7 to 9.8 m s^{-1}, plate temperature 50–80 °C, and water mass content was up to 1.67% of air mass. Visual observations showed that there was no water film on the plate. They therefore assumed that the water drops either evaporated within the boundary layer or evaporated immediately on contact with the plate. Heat transfer coefficients increased with air flow rate and water flow rate, and decreased with increasing plate temperature. Enhancement ratio up to 6 was obtained. They postulated that the total heat flux q is the sum of those by convection to air, sensible heat gain by evaporating water drops, and the latent heat absorbed by the evaporating water. Thus,

$$q = h_0(T_w - T_{g,\infty}) + C_{\text{pf}}G_{\text{ev}}(T_v - T_{g,\infty}) + G_{\text{ev}}i_{fg} \tag{8.4.9}$$

G_{ev} is the mass flux of drops that evaporate in the boundary layer, and h_0 is the heat transfer coefficient of air without water injection. Approximately, the temperature of evaporating drops T_v may be considered to equal T_w. Equation (8.4.9) can then be put in the following form:

$$\frac{h_{\text{TP}}}{h_0} = 1 + \frac{G_{\text{ev}}}{h_0(T_w - T_{g,\infty})}[i_{\text{fg}} + C_{\text{pf}}(T_w - T_{g,\infty})] \tag{8.4.10}$$

This equation gave satisfactory agreement with their data and the trend of decreasing heat transfer coefficients with increasing wall temperature. This indicates that their physical model is correct. However, they have not given any method to predict G_{ev}. Hence this equation cannot be used for design calculations.

Trela (1981) analyzed laminar flow over a plate that is fully or partially covered with a liquid film. The equations developed showed agreement with some data for low flow conditions.

8.4.4 Wedges

Thomas and Sunderland (1970) performed tests in which air–water mist flowed downwards over a wedge with its edge facing up. The rate of heat transfer was increased over that for air alone as a result of evaporation and sensible heating of the continuous liquid film that formed on the solid surface. Typically, heat transfer rates were increased about 20 times by adding 5% liquid water to the air stream. The liquid film formed was 0–0.2 mm thick. Analytical solutions for heat transfer coefficients were developed by solving the integral energy equation. These agreed with their own test data.

Aihara et al. (1979) studied heat transfer to a wedge with a horizontal jet of air–water mist flowing over it. Heat transfer coefficients with mist were up to 14 times those with air alone. Analytical formulas were developed, which agreed with their own data.

8.4.5 Jets

Jets of mists are often used for cooling of thin metal sheets and for tempering of glass at high temperature film boiling conditions. Mist jets are also being considered for cooling of electronic equipment.

Graham and Ramadhyani (1996) used vertical jets of air–water and air–methanol to cool a flat surface whose temperature was lower than the boiling point of water and methanol. Thus, cooling occurred without any nucleate boiling. The nozzle diameter was 2 mm and the plate 6.35 mm square. Single-phase heat transfer coefficient $h_{g,s}$ at the stagnation point for air without liquid was correlated by the following equation:

$$\frac{h_{g,s}D_{\text{noz}}}{k_g} = \left(\frac{u_{\text{noz}}D_{\text{noz}}}{\nu_g}\right)^{0.5} Pr_g^{0.4} \tag{8.4.11}$$

D_{noz} is the nozzle diameter and u_{noz} is the velocity at nozzle exit. Heat transfer coefficients increased with increasing velocity, with increasing amount of liquid, and with increasing wall temperature. They developed an analytical model that agreed with their data for air–methanol but did not agree quantitatively with their air–water data.

Quinn et al. (2017) studied a dilute air–water vertical jet impinging on a horizontal flat surface. Heat transfer coefficients were measured, and shadowgraph imaging was done in an attempt to understand the mechanism of heat transfer. Surface temperature was low so that no boiling occurred. Figure 8.4.3 shows some of their data. It is seen that heat transfer coefficients increase with mist loading fraction and decrease with distance from the stagnation point. The trends were similar when distance of nozzle from the surface was doubled though heat transfer coefficients were lower. They have described in detail the observed behavior of liquid on the surface. Briefly, isolated liquid pools that they call slugs are formed at the surface at low mist loading. These grow with increasing mist loading and eventually coalesce together to form a continuous film. They proposed that intermittent thermal disturbance caused by the impingement of the mist jet droplets is the dominant heat transfer mechanism when mist loading is very low. With increasing mist loading fraction, evaporation from the surface liquid film and convection due to film flow become more significant.

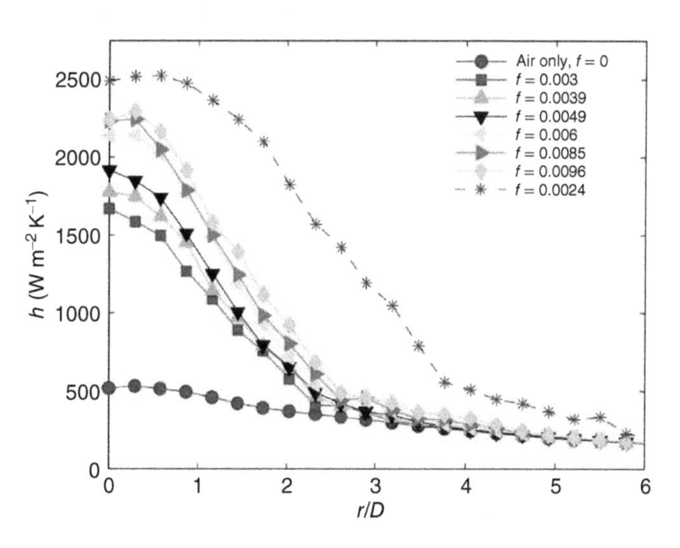

Figure 8.4.3 Heat transfer to a horizontal plate with vertical air–water jet impingement. D is the nozzle diameter, r the distance from stagnation point, and f is the water-to-air mass flow ratio. Reynolds number 4500, $H/D = 5$, H being the distance between nozzle and plate. Source: Quinn et al. (2017).

Sozbir et al. (2003) studied heat transfer during film boiling. They used a vertical jet of air–water mist flowing down on a heated stainless steel plate 102 mm diameter that was heated to a temperature of 815–870 °C before turning on the jet. The diameter of jet nozzle was 55 mm and it was 40 mm from the plate. The total spray angle was about 13°. Single-phase heat transfer coefficient $h_{g,s}$ at the stagnation point for air without water was in satisfactory agreement with Eq. (8.4.11). Heat transfer coefficients were found to increase with increasing water injection rates that were up to 7.6 kg m^{-2} s^{-1}. The plate loses heat by radiation and by convection. The convective heat transfer coefficient h_{conv} at the stagnation point is expressed as follows:

$$h_{\text{conv}} = h_{g,s} + h_{\text{mist}} \tag{8.4.12}$$

The following correlation was developed for h_{mist}, with G being the mass flux of water:

$$\frac{h_{\text{mist}} D_{\text{noz}}}{k_f} = 194 \left(\frac{G D_{\text{noz}}}{\mu_f} \right)^{0.8} \tag{8.4.13}$$

In developing the above correlation, only the data in film boiling were considered. The Leidenfrost temperature was found to increase with air velocity as well as with the amount of water injected.

Khangembam and Singh (2019) studied cooling of a cylinder with a vertical air–water mist jet. Cylinder diameter was 50.65 mm. Jet nozzle was annular, 2.5 mm OD, and 1.9 mm ID with a 0.5 mm water nozzle in the middle. Reynolds number varied from 8820 to 17 106, mist loading fraction from 0.25% to 1.0%, and nozzle-to-surface spacing, H/D_{noz}, from 30 to 60. H is the distance between nozzle and cylinder. Heat transfer coefficients increased with air flow and mist loading and decreased with distance from the stagnation point, axially as well as circumferentially. The latter were similar to those observed with cylinders in air–water crossflow as seen in Figure 8.4.2. They satisfactorily correlated their data for stagnation point by the following equation:

$$\frac{h_{\text{TP}}}{h_0} = 54.43 \left(\frac{H}{d} \right)^{-0.2942} \left(\frac{W_{\text{air}}}{W_{\text{water}}} \right)^{-0.3831} \tag{8.4.14}$$

where d is the outer diameter of the annular nozzle and h_0 is the heat transfer coefficient of air alone.

According to Lee et al. (2005), there are three regimes of heat transfer. In Regime I, heat transfer is dominated by radiation. All liquid drops are evaporated due to radiation before reaching the boundary layer. Hence heat transfer in this regime is by radiation and single-phase convection. In Regime II, liquid drops are evaporated in the boundary layer and the surface does not have a liquid film. Heat transfer is dominated by film boiling. In Regime III, surface is wetted by liquid drops and heat transfer is by film cooling. The heat transfer coefficient in Regime II, h_{II}, is given by

$$h_{\text{II}} = 0.75 h_{\text{rad}} + h_{\text{spc}} + h_{fb} \tag{8.4.15}$$

where h_{spc} is the heat transfer coefficient due to single-phase convection and h_{fb} is the heat transfer coefficient due to film boiling.

$$h_{\text{rad}} = \frac{\varepsilon \sigma_{sb} (T_w^4 - T_\infty^4)}{(T_w - T_\infty)} \tag{8.4.16}$$

where ε is the surface emissivity and σ_{sb} is the Stefan–Boltzman constant. h_{spc} is obtained by the following equation:

$$\frac{h_{\text{spc}} D}{k_{\text{mix}}} = 0.3 + \frac{0.62 Re_{\text{mix}}^{1/2} Pr_{\text{mix}}^{1/3}}{[1 + (0.4/Pr_{\text{mix}})^{2/3}]^{1/4}} \times \left[1 + \left(\frac{Re_{\text{mix}}}{282\,000} \right)^{5/8} \right]^{4/5} \tag{8.4.17}$$

D is the diameter of cylinder and $Re_{\text{mix}} = G_{\text{mix}} D / \mu_{\text{mix}}$. The subscript "mix" indicates the air–water mixture.

The film boiling heat transfer coefficient is calculated by the following equation:

$$h_{\text{fb}} = 0.62 \left[\frac{\rho_g k_g^3 (\rho_{\text{water}} - \rho_g) i_{fg} g}{\mu_g (T_w - T_{\text{SAT}}) D} \right]^{1/4} \frac{(T_w - T_{\text{SAT}})}{(T_w - T_\infty)} \tag{8.4.18}$$

Properties are calculated at the film temperature.

In Regime III, it is assumed that essentially the entire heat transfer occurs due to evaporation of liquid film on the cylinder surface. Therefore, heat flux in Regime III, q_{III} is given by

$$q_{\text{III}} = \frac{W_{L,\text{proj}} i_{\text{fg}}}{A_s} \tag{8.4.19}$$

$$h_{\text{III}} = \frac{q_{\text{III}}}{(T_w - T_\infty)} \tag{8.4.20}$$

$W_{L,\text{proj}}$ is mass of liquid drops passing through the projected area of cylinder (width × length) and A_s is the surface area of the cylinder. Thus it is assumed that all drops approaching the cylinder are captured into the liquid film and all cooling is due to their evaporation. Lee et al. (2005) showed that their test data for a 10 mm diameter cylinder were in good agreement with these equations. The cylinder was heated to 800 °C before exposure to air–water jet.

Buckingham and Haji-Sheikh (1995) performed tests on a 50.8 mm diameter cylinder heated to 1000 °C quenched by an air–water mist jet. They have discussed calculations in Regime I. It amounts to using a single-phase convection equation with mixture mass flux and properties to calculate the convective contribution.

8.4.6 Sphere

Abed et al. (2019) performed experiments in which heat transfer to a 34 mm diameter sphere by natural convection was enhanced by water mist generated by an ultrasonic humidifier located well below the sphere. Water droplet diameter was 2–10 μm. Tests were done with the sphere in open space as well as with the sphere in a cylindrical channel of diameter 46.6 mm. Heat transfer coefficients without mist were in close agreement with the following equation of Churchill (1983) for laminar free convection:

$$Nu = 2 + \frac{0.589Ra^{1/4}}{[1 + (0.469/Pr)^{9/16}]^{4/8}} \tag{8.4.21}$$

The Rayleigh number Ra is given by the following equation:

$$Ra = \frac{(T_w - T_g)\beta g D_{sp}^2 \rho_g^2}{\mu_g^2} Pr_g \tag{8.4.22}$$

D_{sp} is the diameter of sphere and β is the coefficient of thermal expansion of air.

Test data for heat transfer by air–water mixture are shown in Figure 8.4.4. They gave the following correlation of the data:

$$\frac{Nu_{mist}}{Nu_0} = 1 + Ra^{0.34}We^{0.5} \tag{8.4.23}$$

Nu_{mist} is the Nusselt number with mist flow and Nu_0 is that with air only, Weber number We is given by

$$We = \frac{j^2 D_p}{2\rho_f \sigma} \tag{8.4.24}$$

D_p is the average diameter of water drops in the mist. Particle diameter was 5–10 μm in their tests. j is the mass flux of water spray. Figure 8.4.4 shows the data for sphere in open space. Data taken when the sphere was located inside cylindrical duct were not too different.

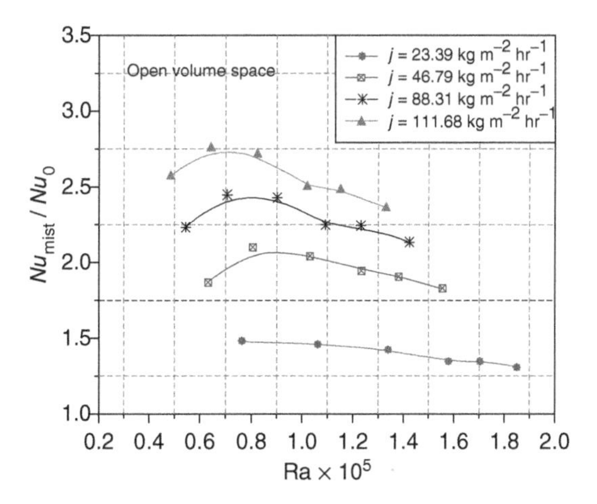

Figure 8.4.4 Ratio of heat transfer to a sphere in mist flow to that in air alone. j is the mass flux of water spray. Natural convection in open space. Source: Abed et al. (2019). Licensed under CCBY 3.0.

8.5 Evaporation from Water Pools

8.5.1 Introduction

Calculation of evaporation from water pools is needed for many applications including swimming pools, water reservoirs, pools containing spent nuclear fuel, process tanks, and water spills. Air flow over the pools may be by natural convection as in most indoor pools or by forced convection as in outdoor swimming pools and lakes. Heat transfer between air and water surface occurs by single-phase convection while evaporation is occurring. Many methods empirical and theoretical have been proposed for the calculation of evaporation rate. These are discussed in the following.

8.5.2 Empirical Correlations

Many authors have proposed equations of the following form to their own data from laboratory tests on evaporation into quiet air:

$$E = c(p_w - p_r)^m \tag{8.5.1}$$

In most of them, $m = 1$.

For evaporation with forced air flow, many equations of the following type have been proposed:

$$E = (b + cu^n)(p_w - p_r)^m \tag{8.5.2}$$

where u is the velocity of air. All of them are based on their authors' own data. For $b = 0$ and $n = 0$, Eq. (8.5.2) is the same as Eq. (8.5.1). Quite often, formulas based on forced convection evaporation data have been applied to natural convection conditions by inserting $u = 0$. Such formulas do not take into account natural convection, which depends on air density difference and hence cannot be expected to be generally applicable to situations in which natural convection effects are prevalent, for example, indoor swimming pools. This was pointed out by several early researchers. For example, Himus and Hinchley (1924) performed tests on evaporation by natural convection as well as with forced air flow and gave formulas for both cases. They found that their forced convection formula extrapolated to zero velocity predicted three times the prediction of their formula for natural convection. Lurie and Michailoff (1936) and Baturin (1972) stated that forced convection equations should not be extrapolated to zero velocity as they will give erroneous results. Nevertheless, such use continues.

Table 8.5.1 Constants and exponents for Eq. (8.5.1) for evaporation from calm water surfaces in various sources.

Source	*b*	*c*	*n*	*m*	Basis (notes)
Carrier (1918)	1.33*E* − 4	1.17 *E* − 4	1	1	Pool with forced air flow data $u = 0–6.3\ \mathrm{m\ s^{-1}}$
Smith et al. (1993)	9.84 *E* − 5	8.66 *E* − 5	1	1	Swimming pool data.
					Water temperature 28.9 °C, air temperature 14.4–27.8 °C, RH 27–65%, $u = 03\ \mathrm{m\ s^{-1}}$
ASHRAE (2015)	6.65 *E* − 5	5.85 *E* − 5	1	1	0.5 × Carrier formula
VDI (1994)	4 *E* − 5	0	1	1	
Hens (2009)	4.086 *E* − 5	0	1	1	Swimming pool data
					Water temperature 27–31 °C
Baturin (1972)	4.72*E* − 5	1.05*E* − 4	1	1	Wind tunnel tests
Lurie and Michailoff (1936)	1.65 *E* − 4	1.26 *E* − 4	1	1	Wind tunnel tests
					Air temperature 40–225 °C, $u = 1–7.5\ \mathrm{m\ s^{-1}}$
Rohwer (1931)	1.37 *E* − 4	8.25 *E* − 5	1	1	Wind tunnel tests
Meyer (1942)	1.56 *E* − 4	3.51 *E* − 5	1	1	Data for small shallow lakes
Boelter et al. (1946)	1.62 *E* − 5	0	0	1.22	Small vessel in quiet air data
Box (1876)	7.78 *E* − 5	0	0	1	
Himus and Hinchley (1924)	2.58 *E* − 5	0	0	1.2	Small vessel in quiet air data

Source: Shah (2018c) © 2018, Taylor & Francis.

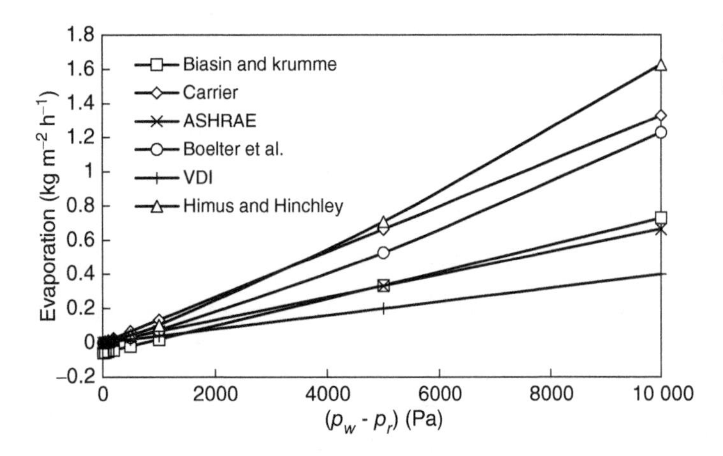

Figure 8.5.1 Comparison of predicted evaporation by some empirical correlations from water pools without forced convection. Source: Shah (2018c). © 2018, Taylor & Francis.

A notable example is the use of the Carrier (1918) formula for indoor swimming pools by inserting $u = 0$.

In Table 8.5.1 are listed values of b, c, m, and n in some of the formulas that have been published. Figure 8.5.1 shows some of the formulas for natural convection. Large differences between the predictions of the various formulas are seen. It clearly shows that these types of correlations are inadequate. Figure 8.5.2 compares the formulas for forced convection. Large differences in the predictions by various correlations are seen.

Raimundo et al. (2014) gave the following formula based on their own data for forced convection in a wind tunnel:

$$E = (3.78 + 37.15u)(y_w - y_r) \tag{8.5.3}$$

where y_w is the mass fraction of water vapor in air at the water surface and y_r is that outside the boundary layer.

8.5.3 Analytical Models

8.5.3.1 Shah Model

Shah (1992, 2008, 2012a, b) presented analytical formulas based on the analogy between heat and mass transfer. According to Shah, the physical phenomena are as follows. A very thin layer of air that is in contact with water quickly gets saturated due to molecular movement at the air–water

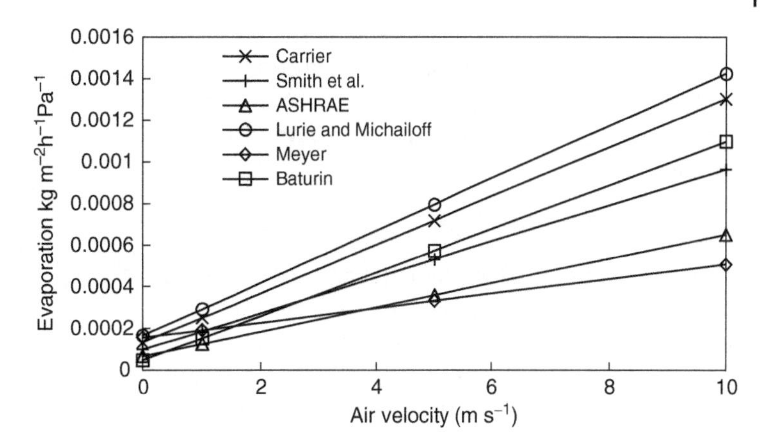

Figure 8.5.2 Comparison of some empirical correlations for evaporation from water pools with forced air flow over them. Source: Shah (2018c). © 2018, Taylor & Francis.

interface. If there is no air movement at all, further evaporation proceeds entirely by molecular diffusion, which is a very slow process. If there is air movement, this thin layer of saturated air is carried away and is replaced by the comparatively dry room air and evaporation proceeds rapidly. Thus, for any significant amount of evaporation to occur, air movement is essential. Air movement can occur due the following mechanisms:

1. Air currents caused by natural convection (buoyancy effect). Room air in contact with the water surface gets saturated and thus becomes lighter compared with the room air and moves upwards. The heavier and drier room air moves downwards to replace it.
2. For indoor pools, air currents caused by the building ventilation system or infiltration/exfiltration.
3. For outdoor pools, air flow by wind.

The rate of evaporation is given by the following relation, Eckert and Drake (1972):

$$E = h_M \rho_w (W_w - W_r) \tag{8.5.4}$$

W is the specific humidity of air, mass of moisture divided by mass of dry air. The air density ρ is evaluated at the water surface temperature as recommended by Kusuda (1965).

Heat transfer during turbulent natural convection to a heated plate facing upwards is given by the following relation, McAdams (1954):

$$Nu = 0.14(Gr_H Pr)^{1/3} \tag{8.5.5}$$

Using the analogy between heat and mass transfer, the corresponding mass transfer relation is

$$Sh = 0.14(Gr_M Sc)^{1/3} \tag{8.5.6}$$

Equations (8.5.5) and (8.5.6) are for turbulent conditions. They were used as analysis of wide-ranging data for pools that showed (Gr_M Pr) to be greater than $2 \times 10^7$and hence in the turbulent range

Gr_M is defined as, Eckert and Drake (1972):

$$Gr_M = \frac{\gamma g (W_w - W_r) L^3 \rho^2}{\mu^2} \tag{8.5.7}$$

where

$$\gamma = -\frac{1}{\rho}\left(\frac{\partial \rho}{\partial W}\right) \tag{8.5.8}$$

Equation (8.5.8) was approximated as

$$\gamma = \frac{\rho_r - \rho_w}{\rho (W_w - W_r)} \tag{8.5.9}$$

Combining Eqs. (8.5.7) and (8.5.8),

$$Gr_M = \frac{g(\rho_r - \rho_w) L^3 \rho^2}{\mu^2} \tag{8.5.10}$$

Substituting Gr_M from Eq. (8.5.10) into Eq. (8.5.6) and expanding Sc and Sh,

$$\frac{h_M L}{D} = 0.14\left[\frac{(\rho_r - \rho_w) g L^3}{\mu D}\right]^{1/3} \tag{8.5.11}$$

D is the coefficient of molecular diffusion. Combining Eqs. (8.5.4) and (8.5.11),

$$E = (0.14 g^{1/3} D^{2/3} \mu^{-1/3}) \rho_w (\rho_r - \rho_w)^{1/3} (W_w - W_r) \tag{8.5.12}$$

The value of $(D^{2/3}\ \mu^{-1/3})$ does not vary much over the typical range of room air conditions. Inserting a mean value, Eq. (8.5.12) becomes

$$E = 35 \rho_w (\rho_r - \rho_w)^{1/3} (W_w - W_r) \tag{8.5.13}$$

The above derivation was first given in Shah (1992). It is noteworthy that Eqs. (8.5.12) and (8.5.13) have been derived from theory without any empirical adjustments. Also note that the air density ρ is defined as mass of dry air per unit volume of moist air. This is in accordance with the methodology of ASHRAE (2017).

When density of air at water surface is higher than density of room air, natural convection essentially ceases, and air movement needed to remove saturated air from water surface is entirely due to the air currents caused by building ventilation system or infiltration/exfiltration. By analyzing data for $\rho_r < \rho_w$, Shah (2012a, 2012b) obtained the following formula for evaporation due to these air currents:

$$E = 0.00\,005\,(p_w - p_r) \tag{8.5.14}$$

Air conditioning systems for building are designed such that air velocity at floor level does not exceed $0.15\,\text{m}\,\text{s}^{-1}$. The computational fluid dynamics (CFD) simulation of Li and Heisenberg (2005) of a large indoor swimming pool also showed that velocity above pool did not exceed $0.15\,\text{m}\,\text{s}^{-1}$. Therefore $0.15\,\text{m}\,\text{s}^{-1}$ was proposed as the changeover velocity in Shah (2012a). Data analysis in Shah (2018c) showed that deviations are minimized using changeover velocity as $0.12\,\text{m}\,\text{s}^{-1}$. Therefore Eq. (8.5.14) is to be used when air velocity over the pool does not exceed this value.

Evaporation rate is the higher of those given by Eq. (8.5.13) and (8.5.14).

The following formula was developed for air velocities over the pool surface greater than $0.12\,\text{m}\,\text{s}^{-1}$ Shah (2018c):

$$E = 0.00\,005(u/0.12)^{0.8}(p_w - p_r) \tag{8.5.15}$$

u is in $\text{m}\,\text{s}^{-1}$. For air velocity over the pool surface $>0.12\,\text{m}\,\text{s}^{-1}$, evaporation from the pool surface is the greater of Eq. (8.5.13) and (8.5.15). This air velocity should be beyond the boundary layer over the pool surface. Smith et al. (1993,1999) had measured air velocity 0.3 m above water surface in their tests on indoor and outdoor pools. This height is expected to be beyond the boundary layer. Use of air velocity at 0.3 m height is, therefore, recommended.

The range of data with which this model has been verified is listed in Table 8.5.2. The data from 21 studies were predicted with MAD of 13.9% giving equal weight to each data set. The same data were also compared to a number of other prediction methods. All had much higher deviations. Figures 8.5.3 and 8.5.4 show the comparison of the Shah model as well as some other prediction methods with test data.

8.5.3.2 Other Models

Hugo (2015) proposed the following equation based on a diffusion model:

$$E = 9.24(1 + u^{1.35})^{0.67} \ln\left(\frac{p_{at} - p_w}{p_{at} - p_r}\right) \tag{8.5.16}$$

where p_{at} is the atmospheric pressure. Hugo recommends this formula for natural convection as well as for forced convection. For natural convection, $u = 0$ is inserted into

Table 8.5.2 Range of data for which the Shah model for evaporation from water pools was verified.

	Range of data
Types of pools	Laboratory vessels, swimming pools, spent fuel pool
Pool area (m²)	0.022–425
Water temperature (°C)	7–94
Air temperature (°C)	6–50
Air relative humidity (%)	28–98
Air velocity (m/s)	0–1.9
$(p_w - p_r)$ (Pa)	210–80 156
$(\rho_r - \rho_w)$ (kg/m³)	−0.004 to +1.002
Number of data sources	21

Source: Shah (2019). © 2019, ASME.

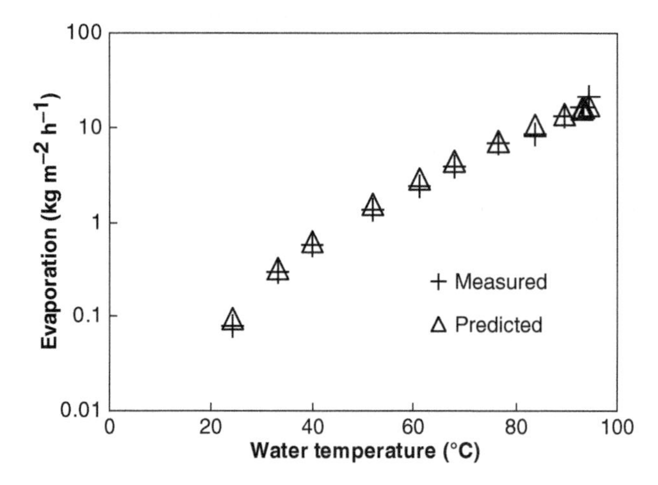

Figure 8.5.3 Comparison of the predictions of Shah model with data of Boelter et al. (1946) for natural convection from a small calm water pool. Source: Shah (2014). © 2014, ASHRAE.

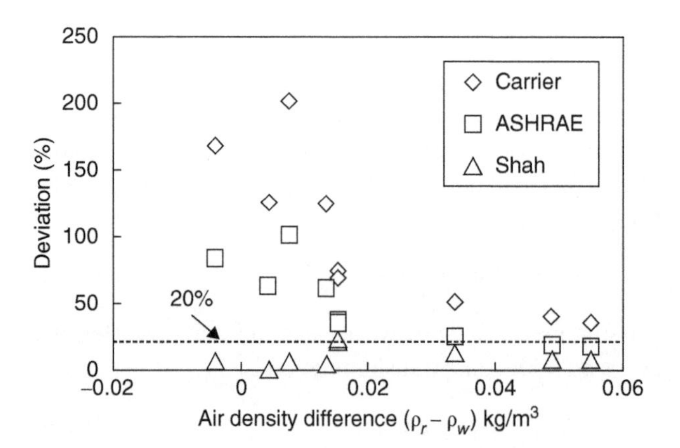

Figure 8.5.4 Deviations of various prediction methods for evaporation with test data for natural convection. Source: Reprinted from Shah (2012b). © 2012, with permission from Elsevier..

the formula. For spent nuclear fuel pools, he recommends using air velocity of 0.51 m s^{-1} in the above formula. Shah (2018c) compared this correlation with the database whose range is listed in Table 8.5.2. It had large deviations with data for natural convection. It had a MAD of 21.3% with data for forced convection; that of the Shah model was 15.6%.

Wang et al. (2011), Yanagi et al. (2015), and Brewester (2018) gave analytical models for calculation of evaporation from nuclear fuel pools under accident conditions. These were compared to a limited amount of such data.

8.5.4 CFD Models

Several researchers have developed CFD models to predict evaporation from water pools. For example, Raimundo et al. (2014) developed a model that was shown to be in agreement with their own tests data for forced convection evaporation at air velocities of 0.1–0.7 m s^{-1}. Blazquez et al. (2017) presented a CFD model whose predictions were reported to be in agreement with the experimental data of Raimundo et al. and some other sources.

CFD modeling is a promising technique, but more thorough verification with a wide range of data is needed to establish its reliability.

8.5.5 Occupied Swimming Pools

All discussions till now are for the situation in which water is undisturbed. In occupied swimming pools, water is disturbed by the activities of the occupants, which enhances the increase of evaporation. Further, the pool deck is wetted by water dripping from occupants as they move in and out of the pool. Shah (2013) developed the following correlation for evaporation from occupied swimming pools:

For $N/A \geq 0.05$,

$$\frac{E}{E_0} = 1.9 - 21(\rho_r - \rho_w) + 5.3N/A \tag{8.5.17}$$

For $N/A < 0.05$, perform linear interpolation between E_0 and E at $N/A = 0.05$. N is the number of occupants and A is the surface area of pool in m^2. This correlation has been shown to be in good agreement with test data for five swimming pools, Shah (2018c).

8.5.6 Conclusions and Recommendations

While there are many published methods for calculation of evaporation from water pools, the only one that has been shown to agree with a wide range of data from many sources is the Shah model. It is therefore recommended for all undisturbed water pools.

8.6 Various Topics

8.6.1 Jets Impinging on Hot Surfaces

There have been attempts to enhance the cooling capacity of liquid jets by adding gas to them. While mist jets discussed in Section 8.4.5 consist of air containing a small amount of water, the jets discussed here consist of water containing a small amount of air.

Zumbrunnen and Balasubramanian (1995) used a planar free surface vertical jet flowing down to impact a horizontal plate. The jet was 5.08 mm wide and 50.8 mm long. Liquid was water and air was injected near the outlet of the nozzle with capillary tubes. The plate surface was always at a temperature below the saturation temperature of water and hence no boiling occurred. Tests were done with and without air injection. Heat transfer coefficients without air injection were correlated by the following equations:

For $0 < z/w < 1.18$,

$$\frac{h_L w}{k_f}\left(\frac{u_{\text{jet}} w}{\nu_f}\right)^{-0.551} Pr_f^{-0.4} = 0.3261 - 0.0552\left(\frac{z}{w}\right)^2 \tag{8.6.1}$$

For $1.18 < z/w < 5.0$,

$$\frac{h_L w}{k_f}\left(\frac{u_{\text{jet}} w}{\nu_f}\right)^{-0.551} Pr_f^{-0.4} = 0.3823 - \frac{z}{w} \times \left(0.1438 - 0.0286\frac{z}{w} + 0.0019\left(\frac{z}{w}\right)^2\right) \tag{8.6.2}$$

where u_{jet} is the jet velocity at nozzle outlet, w is the jet width, and z is the distance from the stagnation point.

Heat transfer coefficients at the stagnation points with air were up to 2.2 times those without air. The enhancement ratio was 1.6 at a distance five times the nozzle width. The test data at the stagnation point were correlated by the following equation:

$$\frac{h_{\text{TP}}}{h_L} = 1 + 4.433X_{\text{tt}}^{-0.7385} \tag{8.6.3}$$

X_{tt} is the Martinelli parameter. If there is no slip between gas and liquid (homogeneous model), it can be written in terms of void fraction as follows:

$$X_{\text{tt}} = \left(\frac{1-\alpha}{\alpha}\right)^{0.9}\left(\frac{\rho_f}{\rho_g}\right)^{0.4}\left(\frac{\mu_f}{\mu_g}\right)^{0.1} \tag{8.6.4}$$

Hall et al. (2001) quenched a horizontal hot plate with a vertical jet of air–water mixture flowing downwards. The nozzle diameter was 5.1 mm and it was at a distance of 110 mm from the plate. Air was introduced into the nozzle well-ahead of the discharge point. The void fraction of air–water mixture in the nozzle varied from 0 to 0.4, calculated assuming homogeneous flow. Liquid-only velocities in jet were 2–4 m s^{-1}, and subcooling was constant at 75 K. Figure 8.6.1 shows the boiling curves at various distances from the stagnation point. It is seen that maximum heat flux was unaffected by void fraction. Minimum film boiling heat flux decreased with increasing void fraction. Minimum film boiling temperature increased with increasing void fraction and with distance from stagnation point. Non-boiling heat transfer occurred in the stagnation zone. Heat transfer coefficients increased with increasing jet velocity and void fraction. At a velocity of 4 m s^{-1}, heat transfer coefficient doubled as void fraction rose from 0 to 0.2.

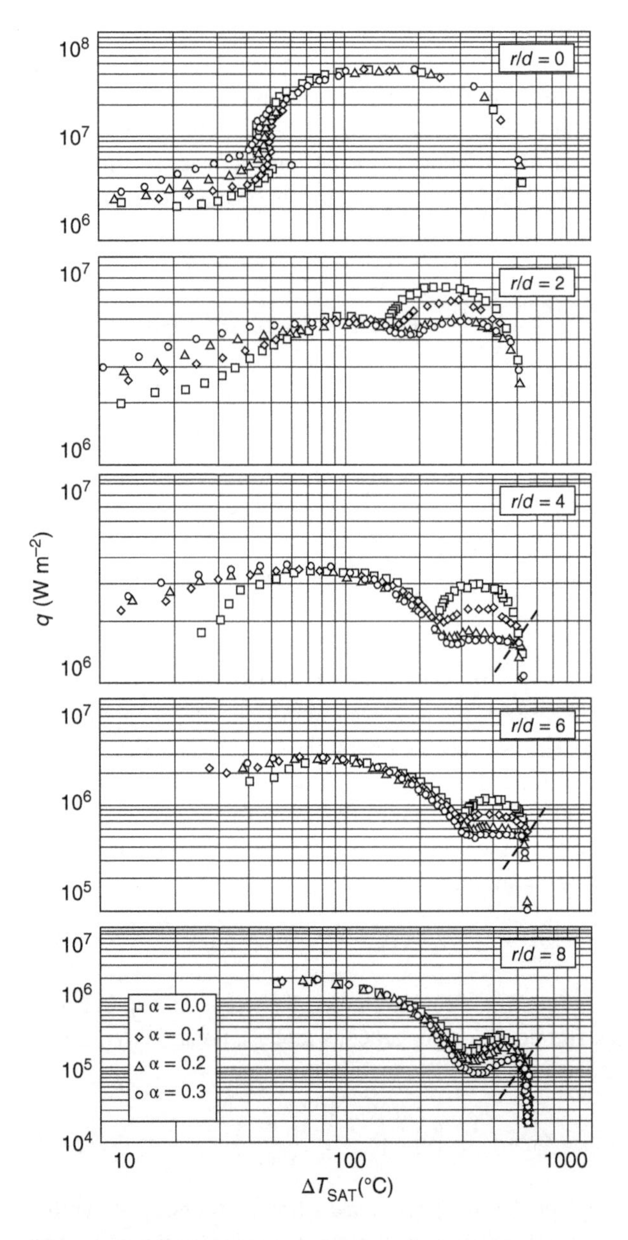

Figure 8.6.1 Boiling curves for quenching of a plate by an air–water free surface vertical jet, r is the distance from the stagnation point and d is the nozzle diameter. Source: Hall et al. (2001). © 2001, ASME.

8.6.2 Vertical Tube Bundle

Drucker et al. (1982) performed tests on a vertical tube bundle with a mixture of water and nitrogen flowing up through it. The bundle had four tubes of 11.1 mm diameter with a pitch of 14.1 mm arranged in a square grid. These were placed in a plexiglass tube of 50.8 mm diameter. Hydraulic diameter of the bundle was 22 mm. The data were correlated by the following equation:

$$\frac{h_{\mathrm{TP}}}{h_{\mathrm{LS}}} = 1 + 3.25\left(\frac{\alpha Gr}{Re_{\mathrm{SL}}^2}\right)^{0.5} \tag{8.6.5}$$

Gr is the Grashof number given by

$$Gr = \frac{(\rho_f - \rho_g) g D_{\mathrm{HYD}}^3}{\rho_f v_f^2} \tag{8.6.6}$$

The heat transfer coefficient of water flowing alone was correlated by the following equation:

$$h_{\mathrm{LS}} = 0.159 Re_{\mathrm{SL}}^{0.6} Pr_f^{1/3} \frac{k_f}{D_{\mathrm{HYD}}} \tag{8.6.7}$$

Re_{SL} ranged from 1840 to 6900. The void fraction was up to 0.15 and flow was bubbly.

8.6.3 Effect of Gravity

While there have been many studies on boiling and condensation heat transfer in low gravity, very little work has been done on gas–liquid heat transfer.

Witte et al. (1996) performed low gravity tests (gravity <0.01% Earth gravity) during parabolic flights with air–water and air–glycerin in a 25.4 mm horizontal tube. The flow patterns were bubble, slug, and annular. As was discussed in Section 8.2.2.1, Shah (2018a) found that the heat transfer coefficients in annular flow were higher than those in Earth gravity for the same parameters. Therefore, he introduced a gravity effect factor in his correlation. To use his correlation, one has to be able to predict the occurrence of annular flow pattern in low gravity. Witte et al. presented their flow pattern data in graphical form. The following correlation is fitted by the present author to their data for the 12.7 and 25.4 mm diameter tubes.

Annular flow occurs when $u_{\mathrm{GS}} \geq 5$ m s^{-1} and

$$u_{LS} \leq 0.066 u_{\mathrm{GS}} \tag{8.6.8}$$

Fore et al. (1997) compared the low gravity data of Witte et al. (1996) for slug flow with several correlations based on data under terrestrial conditions. Those correlations

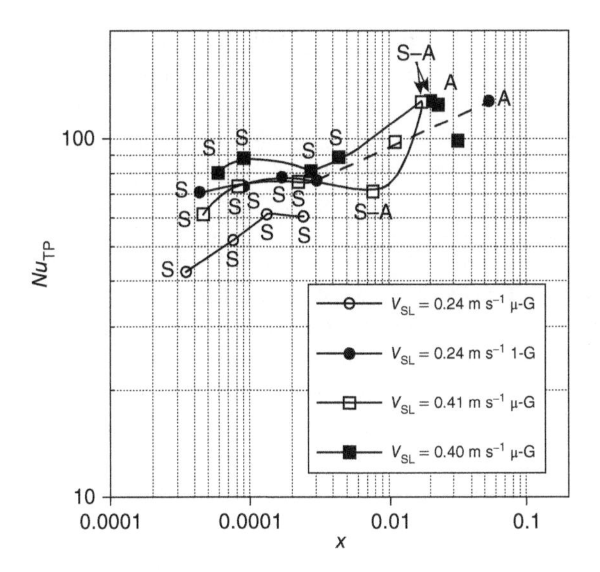

Figure 8.6.2 Comparison of heat transfer at microgravity and Earth gravity during air–water flow in a vertical tube. V_{SL} is the superficial liquid velocity, and S, A, S–A are slug flow, annular, and slug–annular flow patterns. Source: Rite and Rezkallah (1997). © 1997, Elsevier.

overpredicted the data. Fore et al. concluded that low gravity reduces heat transfer. The correlations with which Fore et al. compared the Witte et al. data were based only on their own data and are not reliable beyond their range. As discussed in Section 8.2.2.1, Shah (2018a) found that the data of Witte et al. for slug flow are in agreement with his correlation, which has been verified with data for Earth gravity from many sources.

Rite and Rezkallah (1994, 1997) performed microgravity tests (gravity about 0.04% of Earth gravity) during parabolic flights, with air-water mixtures flowing up in a vertical 9.53 mm ID tube. Tests were also done at Earth gravity. They found that at two-phase Reynolds numbers less than 10^4, heat transfer coefficients in microgravity at low qualities were as much as 50% lower than those at Earth gravity. Figure 8.6.2 shows some of their test data. At higher Reynolds numbers, heat transfer coefficients at low and normal gravity levels were within 10–15% of each other, which is about the level of uncertainty in the data. Two-phase Reynolds number was defined as

$$Re_{TP} = \frac{Re_{SL}}{(1-\alpha)} \tag{8.6.9}$$

where α is the void fraction.

8.7 Liquid Metal–Gas in Channels

Effect of presence of gas on heat transfer to liquid metals has been studied mainly for liquid metal fuel breeder reactor (LMFBR)-type nuclear reactors as these use inert gases as cover. These studies are discussed in the following.

8.7.1 Mercury

Mizushina et al. (1964) studied heat transfer to mercury–gas mixtures in horizontal and vertical tubes. The gases were helium and nitrogen. The horizontal tube was 26 mm ID, while the vertical tube was 14 mm ID. Both were made of stainless steel and were heated by steam. Their single-phase heat transfer coefficients for mercury were lower than theoretical values. In horizontal tube, addition of a small amount of gas sharply decreased heat transfer coefficient but did not decrease further on addition of more gas. In vertical tube, heat transfer coefficient decreased sharply on addition of a small amount of gas but increased slowly with further addition of gas. To explain this behavior, they hypothesized that addition of gas has two opposing effects, decrease in heat transfer due to the lowering of thermal conductivity as gas thermal conductivity is low and acceleration of flow that increases heat transfer. At low gas flow rates, the first effect dominates in both horizontal and vertical tubes. In the case of horizontal tube, the two effects canceled each other at higher gas rates. In the vertical tube at higher gas flow, enhancement due to increased velocity overcomes the reduction due to lowering of thermal conductivity. There was no difference between the results with helium and nitrogen though the thermal conductivity of helium is nine times that of nitrogen. In these tests, gas-to-liquid volumetric ratio varied from 0% to 5%.

Michiyoshi et al. (1982) studied heat transfer to mercury–argon mixtures flowing up in a vertical annulus, with and without transverse magnetic field. The annulus consisted of a 19 mm ID Lucite tube with a 6 mm OD stainless steel type 304 heater in the middle. Heat transfer coefficients of mercury flowing alone were in satisfactory agreement with formulas for turbulent flow. Measurements for mercury–argon flow without magnetic field are shown in Figure 8.7.1. It is seen that heat transfer increases with increasing quality. These results are quite similar to those for non-metallic liquids, for example, Michiyoshi (1978) for air–water in an annulus. On the other hand, they differ completely from the results of Mizushina et al. (1964) discussed earlier. Michiyoshi et al. fitted the following equation to their test data at $L/D_{HYD} = 70$:

$$\frac{h_{TP}}{h_{LT}} = 1 + 118X_{tt}^{-0.911} \tag{8.7.1}$$

The data included Peclet number 400–900.

Application of magnetic field reduced heat transfer coefficient both in single-phase and two-phase flow. Some of their data is shown in Figure 8.7.2. They fitted the following formula to their data:

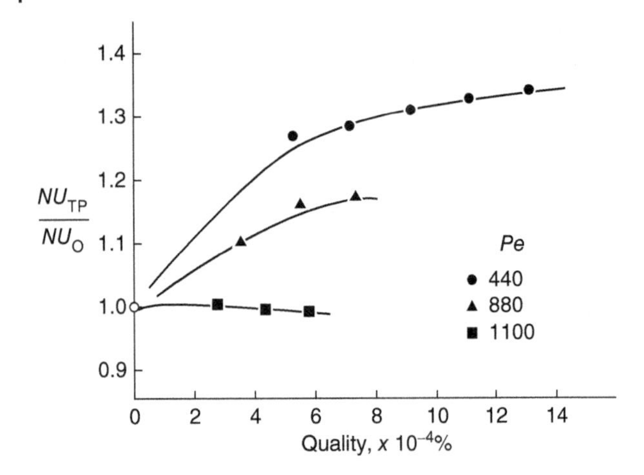

Figure 8.7.1 Heat transfer to mercury–argon mixtures flowing up in an annulus. No magnetic field. Source: Michiyoshi et al. (1982). © 1982, Elsevier.

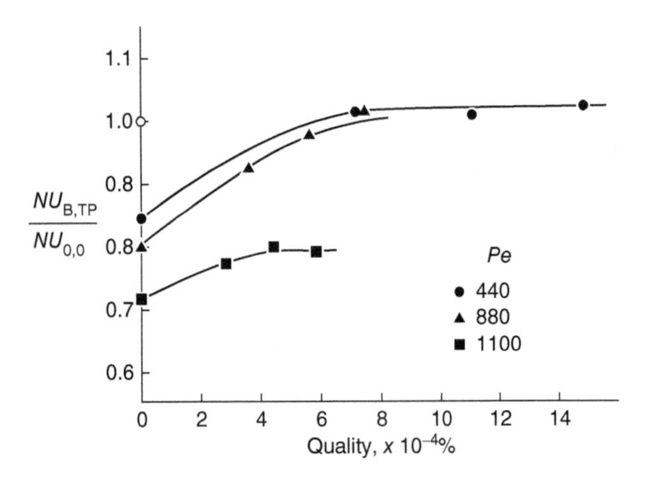

Figure 8.7.2 Effect of transverse magnetic field on heat transfer to mercury–argon mixture in a vertical annulus. Hartmann number = 110, L/D_{HYD} = 70. Source: Michiyoshi et al. (1982). © 1982, Elsevier.

$1075 < X_{tt} < 2100$; $H < 83$,

$$Nu_{B,TP} = Nu_{0,0} \exp\left(-0.23\frac{H}{120}\right) \tag{8.7.2}$$

$Nu_{B,TP}$ is the two-phase Nusselt number in the presence of magnetic field, and $Nu_{0,0}$ is the single-phase Nusselt number in the absence of magnetic field. H is the Hartmann number defined as

$$H = 0.5D_{HYD}B(\sigma_f/\mu_f)^{1/2} \tag{8.7.3}$$

B is the magnetic flux density, T.

8.7.2 Various Liquid Metals

Hori and Friedland (1970) performed an analysis to determine the effect of gas entrainment on heat transfer to sodium flowing in a tube. For bubble flow, they assumed that gas and liquid form a homogeneous mixture with mixture mean properties and that gas thermal conductivity, specific heat, and density are negligible compared to those of liquid sodium. They arrived at the following relation:

$$\frac{h_{TP}}{h_{LS}} = \frac{1-\alpha}{(1+0.5\alpha)^{0.6}} \tag{8.7.4}$$

This equation predicts a gradual and continuous decrease in heat transfer coefficient with increasing void fraction. This trend is quite different from the measurements of Mizushina et al. (1964) on air–mercury bubble flow discussed earlier that show a sharp decline in heat transfer coefficient with a small amount of gas and then constant or slowly rising heat transfer coefficient with further increase in gas flow.

For annular flow, Hori and Friedland arrived at the following relation:

$$\frac{h_{TP}}{h_{LS}} = (1-\alpha)^{-0.4} \tag{8.7.5}$$

Ochai et al. (1971) measured heat transfer in a 22 mm ID vertical tube with sodium–argon flow. Heat transfer coefficients were 0.9–0.65 times the single-phase values.

Winterton (1974) assumed that the wall of the pipe gets covered with a layer of gas bubbles of spherical shape while the liquid flows in the middle. This gas layer reduces the heat transfer. The reduction in heat transfer depends on the arrangement of the spheres, how closely they are packed, and the contact angles (static, receding, and advancing). They showed that if certain assumptions are made about these factors, their theory was in agreement with the data of Macdonald and Quittenton (1974) for sodium containing entrained gas.

Bishop et al. (1979) studied test data from several sources and proceeded to develop the following correlation for sodium with entrained gas:

$$\frac{h_{TP}}{h_{LS}} = 1 - 15\left(\frac{T}{T_{cw}}\right)\left(\frac{4\alpha}{1+3\alpha}\right) \times \left[\frac{1-(Re_{SL}/Re_{limit})^{1.22}}{1+10(Re_{SL}/Re_{limit})^{1.22}}\right]^{0.9} \tag{8.7.6}$$

T is the system temperature, while T_{cw} is the temperature for complete wetting of heating surface. Re_{limit} is the limiting value of Reynolds number beyond which there is no reduction in heat transfer due to presence of gas. The limiting Reynolds number was given as 10^5. Thus, there is no decrease in heat transfer if there is complete wetting of surface or if Reynolds number equals or exceeds the limiting Reynolds number. This equation was developed using the data of Ochiai et al. (1971) and was shown to be in good agreement with it. For the calculation of void fraction, they note that slip velocity causes the volumetric inert gas flow rate to be greater than the void fraction by a factor of approximately 1.6. They recommend 532 °C as the temperature for complete wetting in sodium–stainless steel systems and 300 °C as the temperature for completely unwetted condition.

Kalish and Dwyer (1967) studied heat transfer to NaK amalgam flowing up in a 19 rod unbaffled bundle. In some of the tests, they found heat transfer coefficients to be up to 50% lower than theoretical values. The discrepancy increased with decreasing Peclet number. On investigation, it was concluded that this was due to the presence of argon that was used as cover gas as well as due to incomplete wetting of rod surface by sodium.

8.7.3 Discussion

The behavior in mercury–gas systems studied by Mizushina et al. (1964) and Michiyoshi et al. (1982) was very different. The trend found by the latter is similar to that in non-metallic liquid–gas mixtures. The reason for the Mizushina et al. data showing sharp reduction in heat transfer with a small amount of gas is unknown. It could be that there was incomplete wetting of surface that allowed gas to form insulating layer at the unwetted parts of the wall. The correlation of Michiyoshi et al. is recommended under full wetted condition.

For sodium–gas systems, the correlation of Bishop et al. seems much more reasonable than the other models discussed earlier. Considering the results of Michiyoshi et al. (1982) with mercury–gas, one would expect similar increase in heat transfer with gas content when the surface is fully wetted. However, this cannot be established from available data. The use of Bishop et al. correlation is recommended that predicts that presence of gas reduces heat transfer at low flow rates when surface is not fully wetted.

Nomenclature

- A area of water pool (m^2)
- C_p specific heat at constant pressure (J kg^{-1} K^{-1})
- D diameter (m)
- D coefficient of molecular diffusion (–)
- E rate of evaporation (kg m^{-2} h^{-1})
- Fr_L Froude number of liquid phase = $G_L^2 \rho_L^{-2} g^{-1} D^{-1}$ (–)
- f friction factor (–)
- G mass flux (kg m^{-2} s^{-1})
- Gr Grashof number (–)
- g acceleration due to gravity (ms^{-2})
- h heat transfer coefficient (W m^{-2} K^{-1})
- h_M mass transfer coefficient (m h^{-1})
- i_{fg} latent heat of vaporization (J kg^{-1})
- k thermal conductivity (W m^{-1} K^{-1})
- L characteristic length of water pool (m)
- L length of channel (m)
- Nu Nusselt number hDk^{-1} (–)
- P total pressure of air (Pa)
- Pr Prandtl number (–)
- p pressure (Pa)
- p_w partial pressure of water vapor in air saturated at water temperature (Pa)
- p_r partial pressure of water vapor in air far from water surface (Pa)
- R_L liquid holdup (–)
- Re Reynolds number = $(GD\mu^{-1})$ (–)
- Sc Schmidt number = $\mu\rho^{-1} D^{-1}$ (–)
- Sh Sherwood number = $h_M DL^{-1}$ (–)
- T temperature (K)
- u velocity (ms^{-1})
- u_r velocity ratio, = $u_{GS} u_{LS}^{-1}$ (–)
- W specific humidity of air, kg of moisture/kg of dry air (–)
- W mass flow rate (kg s^{-1})
- X Lockhart–Martinelli parameter (–)
- X_{tt} X for turbulent–turbulent flow (–)
- x quality, mass of gas/mass of gas–liquid mixture (–)

Greek Symbols

- α void fraction
- ν kinematic viscosity (m^2 s^{-1})
- μ dynamic viscosity (kg m^{-1}s^{-1})
- ϕ_L = $[(\Delta p)_{LS}/(\Delta p)_{TP}]^{0.5}$ (–)
- ρ Density, (kg m^{-3})
- σ Stefan–Boltzman constant = 5.67 × 0^{-8} W m^{-2} K^{-4}

Subscripts

- f liquid
- H heat transfer
- HYD hydraulic
- G gas, vapor
- GS superficial gas, gas flowing alone in channel
- g gas, vapor

L	liquid
LS	superficial liquid, liquid flowing alone in channel
M	mass transfer
m	mixture
mix	mixture
noz	nozzle
r	air far away from water surface
rad	radiation
s	stagnation point
TP	two-phase
w	air saturated at water temperature
w	wall
∞	far from the surface, outside boundary layer

References

Abed, A.H., Klimova, V.A., Shcheklein, S.E., and Pakhaluev, V.M. (2019). On the possibility to improve heat transfer of a sphere by natural convection and water mist. *J. Phys.*: Conference Series 1382, 012124, IOP Publishing, doi:https://doi.org/10.1088/1742-6596/1382/1/012124.

Acrivos, A., Ahem, J.E., and Nagy, A.R. (1964). Research Investigation of Two-component Heat Transfer. The Marquardt Corporation, *ARL Report ARL 64-116*, Wright-Patterson AFB, OH.

Aggour, M.A. (1978). Hydrodynamics and heat transfer in two-phase two-component flow. PhD dissertation. University of Manitoba, Winnipeg, Canada.

Aihara, T., Tagas, M., and Haraguchi, T. (1979). Heat transfer from a uniform heat flux wedge in air-water mist flows. Ins. J. *Heat Mass Transf.* **22**: 51–60.

Aihara, T., Fu, W.S., and Suzuki, Y. (1990a). Numerical analysis of heat and mass transfer from honzontal cylinders in downward flow of art-water mist. *J. Heat Transf.* **112**: 472–478.

Aihara, T., Fu, W.S., Hongoh, M., and Shimoyama, T. (1990b). Experimental study of heat and mass transfer from a horizontal cylinder in downward air-water mist flow with blockage effect. *Exp. Thermal Fluid Sci.* **3**: 623–631.

ASHRAE (2015). *ASHRAE Handbook – HVAC Applications*, Chapters 5. Atlanta: American Society of Heating, Refrigerating and Air-Conditioning Engineers, Inc.

ASHRAE (2017). *ASHRAE Handbook—Fundamentals*, Chapters 6. Atlanta: American Society of Heating, Refrigerating and Air-Conditioning Engineers, Inc.

Baturin, V.V. (1972). *Principles of Industrial Ventilation*, 3e, Chapter 9. Oxford: Pergamon Press.

Bhandara, T., Nguyen, N., and Rosengarten, G. (2015). Slug flow heat transfer without phase change in microchannels: a review. *Chem. Eng. Sci.* **126**: 283–295.

Bishop, A.A., Engel, F.C., and Markley, R.A. (1979). Heat transfer effect of entrained gas in liquid sodium systems. *Nucl. Eng. Des.* **52**: 1–13.

Blazquez, J.L.F., Maestere, A.R., Gallero, F.J.G., and Gomez, P.A. (2017). A new practical CFD-based methodology to calculate the evaporation rate in indoor swimming pools. *Energ. Buildings* **149**: 133–141.

Boelter, L.M.K., Gordon, H.S., and Griffin, J.R. (1946). Free evaporation into air of water from a free horizontal quiet surface. *Ind. Eng. Chem.* **38** (6): 596–600.

Box, T. 1876. *A Practical Treatise on Heat*. Quoted in Boelter et al. (1946).

Brewester, M.Q. (2018). Evaporation of water at high mass-transfer rates by natural convection air flow with application to spent-fuel pools. *Int. J. Heat Mass Transf.* **116**: 703–714.

Buckingham, F.P. and Haji-Sheikh, A. (1995). Cooling of high-temperature cylindrical surfaces using a water-air spray. *J. Heat Transf.* **117**: 1018–1026.

Carpenter, F.G. and Colburn, A.P. (1951). The effect of vapor velocity on condensation inside tubes. In: *General Discussion on Heat Transfer*, 20–26. Proceedings of the Institution of Mechanical Engineers.

Carrier, W.H. (1918). The temperature of evaporation. *ASHVE Trans.* **24**: 25–50.

Chisholm, D. (1973). Void fraction during two-phase flow. *J. Mech. Eng. Sci.* **15**: 235–236. Quoted in Kim and Ghajar (2002).

Chisholm, D. (1983). *Two-phase Flow in Pipelines and Heat Exchangers* (ed. G. Godwin). London and New York: Institution of Chemical Engineers. Quoted in Kim and Ghajar (2006).

Churchill, S.W. (1983). Comprehensive, theoretically based, correlating equations for free convection from isothermal spheres. *Chem. Eng. Commun.* **24** (4–6): 339–352. Quoted in Abed et al. (2019).

Collier, J.G. (1972). *Convective Boiling and Condensation*. London: McGraw-Hill.

Deshpande, S.D., Bishop, A.A., and Karandikar, B.M. (1991). Heat transfer to air-water plug-slug flow in horizontal pipes. *Ind. Eng. Chem. Res.* **30**: 2172–2180.

Dittus, F.W. and Boelter, L.M.K. (1930). *Heat Transfer in Automobile Radiators of the Tubular Type*, vol. **2**, 443–461. University of California Publications in Engineering.

Dong, C. and Hibiki, T. (2018a). Heat transfer correlation for two-component two-phase slug flow in horizontal pipes. *Appl. Therm. Eng.* **141**: 866–876.

Dong, C. and Hibiki, T. (2018b). Correlation of heat transfer coefficient for two-component two-phase slug flow in a vertical pipe. *Int. J. Multiph. Flow* **108**: 124–139.

Dorresteijn, W.R. (1970). Experimental study of heat transfer in upward and downward two-phase flow of air and oil through 70 mm tubes. Proceedings of the 4th International Heat Transfer Conference, 31 August–5 September, Paris-Versailles, France.

Drucker, M.I., Dhir, V.K., Duffey, R.B., et al. (1982). Two-phase heat transfer in tubes rod bundles and blockages. ASME Paper 82-WA/HT-47.

Eckert, E.R.G. and Drake, R.M. (1972). *Analysis of Heat and Mass Transfer*. New York: McGraw-Hill.

Elperin, T. (1961). Heat transfer between a two-phase flow and a cluster of pipes. *Inz. Fiz. Zh.* **4** (8): 30–35. Quoted in Trela (1981).

Finlay, I.C. and Harris, D. (1984). Evaporative cooling of tube banks. *Int. J. Refrig.* **7**: 214–224.

Finlay, I.C. and McMillan, T. (1967-1968). Heat transfer during two-component mist flow across a heated cylinder. *Proc. Instn. Mech. Eng.* **182**, part 3H: 27–287.

Fore, L.B., Witte, L.C., and McQuillen, J.B. (1997). Heat transfer to two-phase slug flows under reduced-gravity conditions. *Int. J. Multiphase Flow* **23** (2): 301–311.

Franca, F.A., Bannwart, A.C., Ricardo, M.T. et al. (2008). Mechanistic modeling of the convective heat transfer coefficient in gas-liquid intermittent flows. *Heat Transfer Eng.* **29** (12): 984–998.

Fried, L. (1954). Pressure drop and heat transfer for two-phase, two-component flow. *AIChe. Symp. Ser.* **50** (9): 47–51.

Ghajar, A.G. and Tang, C.C. (2010). Importance of non-boiling two-phase flow heat transfer in pipes for industrial applications. *Heat Transfer Eng.* **31** (9): 711–732.

Gose, E.E., Peterson, E.E., and Acrivos, A. (1957). On the rate of heat transfer in in liquids with gas injection through the boundary layer. *J. Appl. Phys.* **28**: 1509. https://doi.org/10.1063/1.323870.

Graham, K.M. and Ramadhyani, S. (1996). Experimental and theoretical studies of mist jet impingement cooling. *J. Heat Trans.-ASME* **118**: 343–349.

Groothius, H. and Hendal, W.P. (1959). Heat transfer in two-phase flow. *Chem. Eng. Sci.* **11**: 212–220.

Hall, D.E., Incropera, F.P., and Viskanta, R. (2001). Jet impingement boiling from a circular jet during quenching: part 2 – two-phase jet. *J. Heat Transf.* **123**: 911–917.

Hens, H. (2009). Indoor climate and building envelope performance in indoor swimming pools. In: *Energy Efficiency and New Approaches* (eds. N. Bayazit, G. Manioglu, G. Oral and Z. Yilmaz), 543–552. Istanbul Technical University.

Heywood, N.I. and Richardson, J.F. (1979). Slug flow of air-water mixtures in a horizontal pipe: determination of liquid holdup by X-ray absorption. *Chem. Eng. Sci.* **34**: 17–30.

Hilpert, R. (1933). Experimental study of heat dissipation of heated wires and pipes in current of air. *Forsch. Arb. Geb. Ing. Wes.* **4** (5): 215.

Himus, G.W. and Hinchley, J.W. (1924). The effect of a current of air on the rate of evaporation of water below the boiling point. *Chem. Ind.* **22**: 840–845.

Hishida, K., Maeda, M., and Ikai, S. (1980). Heat transfer from a flat plate in two-component mist flow. *J. Heat Transf.* **102** (3): 513–518.

Hodgson, W., Staterbak, R.T., and Sunderland, I.E. (1968). An experimental investigation of heat transfer from a spray cooled isothermal cylinder. *J. Heat Transf.* **9OC**: 457–463.

Hoelscher, J. F. (1965). Study of heat transfer from a heated cylinder in two-phase, water-air flow. M.S. thesis. Air Force Institute of Technology.

Hori, M. and Friedland, A.J. (1970). Effect of gas entrainment on thermal-hydraulic performance of sodium cooled reactor core. *J. Nucl. Sci. Technol.* **7**: 256–263.

Hossainy, T.A. (2014). Non-boiling heat transfer in horizontal and near horizontal downward inclined gas-liquid two phase flow. MSc thesis. University of Oklahoma, Norman, OK.

Hughmark, G.A. (1963). Heat transfer in horizontal annular gas-liquid flow. *Chem. Eng. Prog.* **59** (7): 176–178.

Hugo, B.R. (2015). Modeling evaporation from spent nuclear fuel storage pools: a diffusion approach. PhD thesis. Washington State University.

John, T.J., Bhagwat, S.M., Ghajar, A.J. (2015). Heat transfer measurements and correlations assessment for downward inclined gas-liquid two phase flow. Proceedings of the 1st Thermal and Fluid Engineering Summer Conference, TFESC New York, USA (9–12 August 2015).

Johnson, H.A. (1955). Heat transfer and pressure drop for viscous-turbulent flow of oil-air mixtures in a horizontal pipe. *ASME Trans.* **77**: 1257–1264.

Johnson, H.A. and Abou-Sabe, A.H. (1952). Heat transfer and pressure drop for turbulent flow of air-water mixtures in a horizontal pipe. *Trans. ASME* **74**: 977–987.

Kalapatapu, S.B. (2012). Non-boiling heat transfer in horizontal and near horizontal upward inclined gas-liquid two phase flow. MSc thesis. University of Oklahoma, Norman, OK.

Kalish, S. and Dwyer, O.E. (1967). Heat transfer to NaK flowing through unbaffled rod bundles. *Int. J. Heat Mass Transf.* **10**: 1533–1558.

Kaminsky, R.D. (1999). Estimation of two-phase flow heat transfer in pipes. *J. Energy Resour. Technol.* **121** (2): 75–80.

Khangembam, C. and Singh, D. (2019). Experimental investigation of air–water mist jet impingement cooling over a heated cylinder. *J. Heat Transf.* **141**: 082201-1–082201-12.

Kim, D. and Ghajar, A.J. (2002). Heat transfer measurements and correlations for air-water flow of different flow patterns in a horizontal pipe. *Exp. Thermal Fluid Sci.* **25**: 659–676.

Kim, J. and Ghajar, A. (2006). A general heat transfer correlation for non-boiling gas-liquid flow with different flow patterns in horizontal pipes. *Int. J. Multiphase Flow* **32**: 447–465.

Kim, D., Ghajar, A.J., and Dougherty, R.L. (2000). Robust heat transfer correlation for turbulent gas-liquid flow in vertical pipes. *J. Thermophys. Heat Transf.* **14**: 574–578.

King, C.D.G. (1952). Heat transfer and pressure drop for an air-water mixture flowing in a 0.737 inch i.d. horizontal tube. M.S. thesis. University of California, Berkeley, CA.

Knott, R.F., Anderson, R.N., Acrivos, A., and Petersen, E.E. (1959). An experimental study of heat transfer to nitrogen-oil mixtures. *Ind. Eng. Chem.* **51** (11): 1369–1372.

Koviri, R.N.K., Bhagwat, S.M., and Ghajar, A.J. (2015) Heat transfer measurements and correlations assessment for upward inclined gas-liquid two phase flow. Proceedings of the 1st Thermal and Fluid Engineering Summer Conference, TFESC, New York, USA (9–12 August 2015).

Kropholler, H.W. and Carr, A.D. (1962). The prediction of heat and mass transfer coefficients for turbulent flow in pipes at all values of the Prandtl or Schmidt number. *Int. J. Heat Mass Transf.* **5**: 1191–1205.

Kudirka, A.A. (1964). Two-phase heat transfer with gas injection through a porous boundary surface. *Argonne National Laboratory Report ANL-6862*.

Kusuda, T. (1965). Calculation of the temperature of a flat plate wet surface under adiabatic conditions with respect to the Lewis relation. In: *Humidity and Moisture*, vol. **1** (ed. R.E. Ruskin), 16–32. Rheinhold.

Lee, S., Park, J., Lee, P., and Kim, M. (2005). Heat transfer characteristics during mist cooling on a heated cylinder. *Heat Transfer Eng.* **26** (8): 24.

Levy, S. (1952). *Proceedings of Second Midwestern Conference on Fluid Mechanics*, 337. Quoted in Collier (1972).

Li, Z. and Heisenberg, P. (2005). *CFD Simulations for Water Evaporation and Airflow Movement in Swimming Baths*. Aalborg, Denmark: Department of Building Technology, University of Aalborg.

Lunde, K.E. (1961). Heat transfer and pressure drop in two-phase flow. *Chem. Eng. Prog. Symp. Ser.* **57** (22): 104–110.

Lurie, M. and Michailoff, N. (1936). Evaporation from free water surfaces. *Ind. Eng. Chem.* **28** (3): 345–349.

MacDonald, W.C. and Quittenton, R.C. (1974). Critical analysis of metal wetting and gas entrainment in heat transfer to molten metal. *Chem. Eng. Prog. Symp. Ser.* **50** (9): 59–67.

Manabe, R. (2001). A comprehensive mechanistic heat transfer model for two-phase flow with high-pressure flow pattern validation. PhD dissertation. University of Tulsa, Tulsa.

Mandhane, J.M., Gregory, G.A., and Aziz, K. (1974). A flow pattern map for gas-liquid flow in horizontal pipes. *Int. J. Multiphase Flow* **1**: 537–553.

Martin, B.W. and Sims, G.E. (1971). Forced convection heat transfer to water with air injection in a rectangular duct. *Int. J. Heat Mass Transf.* **14**: 1115–1134.

Mastanaiah, K. and Ganic, E.N. (1981). Heat transfer in two-component dispersed flow. *J. Heat Transf.* **103**: 300–306.

McAdams, W.H. (1954). *Heat Transmission*. New York: McGraw Hill.

Mednick, R.L. (1967). Air-water spray flow heat transfer across a cylinder dissertation. PhD thesis. University of Oklahoma, Norman, OK.

Meyer, A.F. (1942). *Evaporation from Lakes and Reservoirs*. 0Minnesota Resources Commission.

Michiyoshi, I. (1978). Heat transfer in air-water two-phase flow in a concentric annulus. Proceedings of the Sixth International Heat Transfer Conference, Toronto, Canada (7–11 August 1978).

Michiyoshi, I., Tanaka, M., and Takahashi, O. (1982). Mercury-argon two-phase heat transfer in a vertical annulus under transverse magnetic field. *Int. J. Heat Mass Transf.* **25** (10): 1481–1487.

Mizushina, T., Sasano, T., Hirayama, M. et al. (1964). Effect of gas entrainment on liquid metal heat transfer. *Int. J. Heat Mass Transf.* **7**: 1419–1425.

Mollamahmutoglu, M. (2012). Study of isothermal pressure drop and non-boiling heat transfer in vertical downward two-phase flow. MSc thesis. University of Oklahoma, Tulsa, OK.

Nada, S.A. (2017). Experimental investigation and empirical correlations of heat transfer in different regimes of air–water two-phase flow in a horizontal tube. *J. Therm. Sci. Eng. Appl.* **9**: 021004-1–021004-9.

Nakayama, W., Kuwahara, H., and Hirasawa, S. (1988). Heat transfer from tube banks to air/water mist flow. *Int. J. Heat Mass Transf.* **31** (2): 449–462.

Novasad, Z. (1955). Heat transfer in two-phase gas-liquid systems. *Collect. Czechoslov. Chem. Commun.* **20** (2): 477–498.

Noville, I. and Bannwart, A.C. (2010). Experimental study of heat transfer in horizontal gas-liquid intermittent flow.

Proceedings of the 14th International Heat Transfer Conference, Washington, DC, 7–13 August 2010.

Ochiai, M., Kuroyanagi, T., Kobayashi, K., and Furukawa, K. (1971). Void fraction and heat transfer coefficient of 13. Sodium-argon two-phase flow in vertical channel. *J. Atom. Energy Soc. Jpn* **13**: 566–573. Quoted in Bishop et al. (1979).

Oliver, D.R. and Wright, S.J. (1964). Pressure drop and heat transfer in gas-liquid slug flow in horizontal tubes. *British Chem. Eng.* **9** (9): 590–596.

Perroud, P. and de La Harpe, A. (1960). Transfert de Chaleur par Liquedes Entraines Daus Encoulementgazeux Turbulent. Centre d' Etudes Nucleaires de Grenoble, *Report CEA No. 1422.*

Pletcher, P.H. and McManus, H.N. (1968). Heat transfer and pressure drop in horizontal annular two-phase, two-component flow. *Int. J. Heat Mass Transf.* **11**: 1087–1104.

Quinn, C., Murray, D.B., and Persoons, T. (2017). Heat transfer behaviour of a dilute impinging air-water mist jet at low wall temperatures. *Int. J. Heat Mass Transf.* **111**: 1234–1249.

Raimundo, A.M., Gaspar, A.R., Virgílio, A. et al. (2014). Wind tunnel measurements and numerical simulations of water evaporation in forced convection airflow. *Int. J. Therm. Sci.* **86**: 28–40.

Rassame, S. and Hibiki, T. (2018). Drift-flux correlation for gas-liquid two-phase flow in a horizontal pipe. *Int. J. Heat Fluid Flow* **69**: 33–42.

Ravipudi, S.R. (1976). The effect of mass transfer on heat transfer rates for two-phase flow in a vertical pipe. Ph.D. thesis. Chemical Eengineering, Vanderbilt University. Nashville, TN. Quoted in Rezkallah (1986).

Rezkallah, K.S. (1986). Heat transfer and hydrodynamics in two-phase two-component flow in a vertical tube. PhD Dissertation. University of Manitoba, Winnipeg, Canada.

Rite, R.W. and Rezkallah, K.S. (1994). Heat transfer in two-phase flow through a circular tube at reduced gravity. *J. Thermophys. Heat Transf.* **8** (4): 702–708.

Rite, R.W. and Rezkallah, K.S. (1997). Local and mean heat transfer coefficients in bubbly and slug flows under microgravity conditions. *Int. J. Multiphase Flow* **23** (1): 37–54.

Rohsenow, W.M. (1956). Effect of vapor velocity on laminar and turbulent condensation. *Trans. ASME*: 1637–1643.

Rohwer, C. (1931). Evaporation from free water surfaces. United States Department of Agriculture. Bulletin No. 271.

Saterbak, R.T. (1967). An experimental investigation of heat transfer from an isothermal cylinder exposed to two-component crossflow. MSc thesis. Georgia Institute of Technology, Atlanta, GA.

Shah, M.M. (1981). Generalized prediction of heat transfer during two component gas-liquid flow in tubes and other channels. *AIChE Symp. Ser.* **77** (208): 40–151.

Shah, M.M. (1992). Calculating evaporation from pools and tanks. Heating/Piping/Air Conditioning (April 1992, pp. 69–71).

Shah, M.M. (2008). Analytical formulas for calculating water evaporation from pools. *ASHRAE Trans.* **114** (2).

Shah, M.M. (2012a). Calculation of evaporation from indoor swimming pools: further development of formulas. *ASHRAE Trans.* **118** (2).

Shah, M.M. (2012b). Improved method for calculating evaporation from indoor water pools. *Energ. Buildings* **49** (6): 306–309.

Shah, M.M. (2013). New correlation for prediction of evaporation from occupied swimming pools. *ASHRAE Trans.* **119** (2): 450–456.

Shah, M.M. (2014). Methods for calculation of evaporation from swimming pools and other water surfaces. *ASHRAE Trans.* **120** (2).

Shah, M.M. (2018a). Improved general correlation for heat transfer during gas-liquid flow in horizontal tubes. *J. Therm. Sci. Eng. Appl.* **10**: 1–7.

Shah, M.M. (2018b). General correlation for heat transfer to gas-liquid flow in vertical channels. *J. Therm. Sci. Engi. Appli.* **10**: 1–9.

Shah, M.M. (2018c). Improved model for calculation of evaporation from water pools. *Sci. Technol. Built Env.* **24** (10): 1064–1074.

Shah, M.M. (2019). Calculation of evaporation from Fukushima NPP spent fuel pools. *J. Nucl. Eng. Rad. Sci.* **5**: 1–6.

Shoham, O., Dukler, A.E., and Taitel, Y. (1982). Heat transfer during intermittent/slug flow in horizontal tubes. *Ind. Eng. Chem. Fundam.* **21**: 312–318.

Sieder, E.N. and Tate, G.E. (1936). Heat transfer and pressure drop of liquids in tubes. *Ind. Eng. Chem.* **28** (12): 1429–1435.

Smith, J.E. (1966). Heat transfer studies of water-spray flows. Northern Research and Engineering Corporation, *ARL Report ARL 66-0091*, Wright-Patterson AFB, OH.

Smith, C.C., Jones, R.W., and Lof, G.O.G. (1993). Energy requirements and potential savings for heated indoor swimming pools. *ASHRAE Trans.* **99** (2): 864–874.

Smith, C.C., Lof, G.O.G., and Jones, R.W. (1999). Rates of evaporation from swimming pools in active use. *ASHRAE Trans.* **104** (1A): 514–523.

Sozbir, N., Chang, Y.W., and Yao, S.C. (2003). Heat transfer of impacting water mist on high temperature metal surfaces. *J. Heat Transf.* **125**: 70–74.

Suzuki, K., Hagiwara, Y., and Sato, T. (1983). Heat transfer and flow characteristics of two-phase two-component annular flow. *Int. J. Heat Mass Transf.* **26** (4): 597–605.

Takahara, E. W. (1965). Experimental study of heat transfer from a heated circular cylinder in two-phase, water-air flow. M.S. thesis. Air Force Institute of Technology.

Tang C.C. and Ghajar A.J. (2007). Validation of a general heat transfer correlation for non-boiling two-phase flow with different flow patterns and pipe inclination angles. Proceedings of the 2007 ASME-JSME Thermal Engineering Heat Transfer Conference, Vancouver, Canada (8–12 July).

Thomas, W.C. and Sunderland, J.E. (1970). Heat transfer between a plane surface and air containing suspended water droplets. *Ind. Eng. Chem.* **9** (3): 363–374.

Trela, M. (1981). An approximate calculation of heat transfer during flow of an air-mist along a heated flat plate. *Int. J. Heat Mass Transf.* **24** (4): 749–755.

VDI (1994). *Wärme, Raumlufttechnik, Wasserer- und -entsorgung in Hallen und Freibädern*, VDI 2089.

Verschoor, H. and Stemerding, S. (1951). Heat transfer in two-phase flow. In: *Proceedings of the General Discussion on Heat Transfer*, Section 2, 27. London: Institute of Mechanical Engineers.

Wang, D., Gauld, J.C., Yoder, G. et al. (2011). Study of Fukushima Daiichi nuclear power station unit 4 spent-fuel pool. *Nucl. Technol.* **180**: 205–215.

Wang, X., Wang, Z., Zhang, X., and Limin, H.E. (2017). Heat transfer of oil-gas slug flow in horizontal pipe. *CISE J.* **68** (6): 2306–2314.

Winterton, R.H.S. (1974). Effect of gas bubbles on liquid metal heat transfer. *Int. J. Heat Mass Transf.* **17**: 549–554.

Witte, L.C., Bousman, W.S., and Fore, L.B. (1996). Studies of two-phase flow dynamics and heat transfer at reduced gravity conditions. *NASA Contractor Report 198459*, NASA Lewis Research Center, Cleveland, OH.

Woldesemayat, M.A. and Ghajar, A.J. (2007). Comparison of void fraction correlations for different flow patterns in horizontal and upward inclined pipes. *Int. J. Multiphase Flow*: 337–370.

Yanagi, C., Murase, M., Yoshida, Y., and Kusunoki, T. (2015). Prediction of temperature and water level in a spent fuel pit during loss of all AC power supplies. *J. Nucl. Sci. Technol.* **52** (2): 193–203.

Zhang, H.Q., Wang, Q., and Sarica, C. (2003). Unified model for gas-liquid pipe flow via slug dynamics – part 1: model development. *J. Energy Resour. Technol.* **125** (4): 266–273.

Zhang, H.Q., Wang, Q., Sarica, C., and Brill, J.P. (2006). Unified model of heat transfer in gas-liquid pipe flow. *SPE: Prod. Oper.* **21** (1): 114–122.

Zumbrunnen, D.A. and Balasubramanium (1995). Convective heat transfer enhancement due to gas injection into an impinging liquid jet. *J. Heat Transf.* **117**: 1011–1017.

9

Gas-Fluidized Beds

9.1 Introduction

Fluidized beds are widely used in many commercial applications, such as catalytic reactions, drying, coating, combustion, power plants, and ore roasting.

Fluidized beds consist of solid particles suspended in gas. The conventional fluidized beds operate below the carryover velocity, and hence there is no significant outflow of solids. The fast or recirculating beds operate above the particle carryover velocity; solid particles are carried out of the bed. In this chapter, discussions are mostly confined to conventional beds though heat transfer to gas–solid suspensions flowing in pipes is also addressed.

While this chapter is mainly concerned with heat transfer, the hydrodynamics related to heat transfer is also discussed.

9.2 Regimes of Fluidization

Consider a bed of solid particles resting on a perforated plate and a gas is injected at the bottom. At low gas velocity, the gas just passes through the spaces between the particles that remain stationary. With increasing velocity, pressure drop and frictional drag on the particles increase. With further increase in velocity, a point is reached where the frictional drag just equals the weight of the particles and the particles get suspended in the gas stream. The particles have now been fluidized, and the superficial gas velocity at which this occurs is known as the minimum fluidizing velocity u_{mf}. Pressure drop does not increase when velocity increases beyond u_{mf}.

With further increase in velocity, particles remain suspended in the gas. If gas velocity keeps increasing, the particles are carried along by the gas, and the bed will empty unless particles are added to make up for the loss. This velocity is called the carryover or transport velocity u_{tr}. Conventional fluidized beds operate below the conveying velocity. Many beds operate beyond the carryover velocity. These are called recirculating or fast-fluidized beds. Pneumatic conveyers also operate at velocities higher than u_{tr}.

As the velocity increases from u_{mf} to u_{tr}, a number of fluidization regimes are possible. These are shown in Figure 9.2.1. As velocity increases beyond minimum fluidization, the particles are uniformly distributed and the operation is without pressure fluctuations. This is the particulate or homogeneous regime. The bubbly regime starts when the velocity reaches u_{mb}. Bubbles form near the distributor and move up, usually growing in size as they progress upwards. Bubbles break through the top surface of the bed, which looks like a boiling liquid. Significant pressure fluctuations occur. Further increase in velocity to u_{ms} starts the slug regime in which bubbles join together to form large voids. There are heavy pulsations and regular pressure fluctuations. The voids can be large enough to cover the entire bed cross section. This is called the plug regime. With further increase in velocity, pressure fluctuations decrease. This is the start of the transitional or intermediate turbulent regime and this velocity is denoted by u_c. With further increase in velocity to u_k, pressure fluctuations become steady at a much lower level than at u_c. This is the fully turbulent regime. Small clusters of particles and voids move randomly. Top surface of the bed is not seen clearly. Beyond u_{tr} is the fast fluidization regime in which particle move out continuously from the top and are recirculated into the bed near the bottom. The higher the velocity, the lower is the concentration of particles in gas.

Fluidization in all these regimes except the particulate regime is considered heterogeneous, the concentration of solids varying from point to point and at different instants. In particulate regime, fluidization is called uniform or homogeneous as the distribution of particles is uniform throughout the bed.

Two-Phase Heat Transfer, First Edition. Mirza Mohammed Shah.

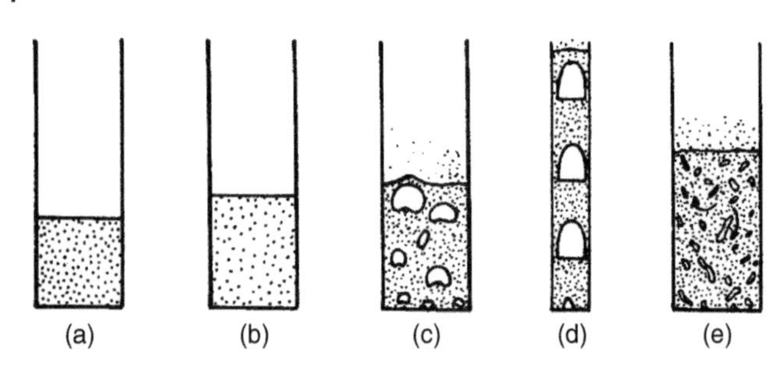

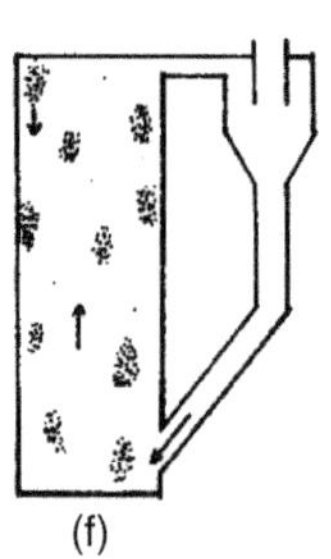

Figure 9.2.1 Regimes of fluidization and their appearance. The regimes are (a) fixed bed, (b) particulate, (c) bubbly, (d) slug, (e) turbulent, and (f) fast. Source: Grace (1982a).

9.2.1 Regime Transition Velocities

9.2.1.1 Minimum Fluidization Velocity

Many correlations have been proposed for the calculation of minimum fluidizing velocity. The following equation is obtained using the Ergun (1952) equation for pressure drop, Botterill et al. (1982):

$$Ar = \frac{150\left(1-\varepsilon_{\text{mf}}\right)}{\varphi_s^2\varepsilon_{\text{mf}}^3}Re_{\text{mf}} + \frac{1.75}{\varphi_s\varepsilon_{\text{mf}}^3}Re_{\text{mf}}^2 \tag{9.2.1}$$

where φ_s is the shape factor of particle.

Goroshko et al. (1958) have the following equation:

$$Re_{\text{mf}} = \frac{Ar}{\frac{150(1-\varepsilon_{\text{mf}})}{\varepsilon_{\text{mf}}^3} + \sqrt{\frac{1.75Ar}{\varepsilon_{\text{mf}}^3}}} \tag{9.2.2}$$

With the void fraction at minimum fluidization $\varepsilon_{\text{mf}} = 0.4$, this becomes

$$Re_{\text{mf}} = \frac{Ar}{1400 + 5.22\sqrt{Ar}} \tag{9.2.3}$$

Ar is the Archimedes number, also called Galileo number by some authors, defined as

$$Ar = \frac{\left(\rho_s-\rho_g\right)\rho_g g D_p^3}{\mu_g^2} \tag{9.2.4}$$

$$\text{Re}_{\text{mf}} = \frac{u_{\text{mf}}\rho_g D_p}{\mu_g} \tag{9.2.5}$$

Several researchers have given correlations of the following form:

$$Re_{\text{mf}} = \sqrt{\left(C_1^2 + C_2 Ar\right)} - C_1 \tag{9.2.6}$$

Wen and Yu (1966a, b) gave $C_1 = 33.7$ and $C_2 = 0.0408$. Richardson (1971) gives $C_1 = 25.7$ and $C_2 = 0.0365$. Grace (1982a) recommends $C_1 = 27.2$ and $C_2 = 0.0408$.

Rabinovich and Kalman (2011) have given the following correlation that was based on the analysis of data from many sources:

For Geldart group A ($1 < Ar < 80$),

$$Re_{\text{mf}} = 0.0008Ar \tag{9.2.7}$$

For Geldart group B particles ($80 < Ar < 30\,000$),

$$Re_{\text{mf}} = 0.000\,955Ar^{0.96} \tag{9.2.8}$$

For group D particles ($30\,000 < Ar$),

$$Re_{\text{mf}} = 0.059Ar^{0.56} \tag{9.2.9}$$

9.2.1.2 Various Regime Transition Velocities

Geldart and Abrahamsen (1978) developed the following dimensional equation for minimum bubbling velocity:

$$\frac{u_{\text{mb}}}{u_{\text{mf}}} = \frac{4.125\times10^4\mu_g^{0.9}\rho_g^{0.1}}{\left(\rho_s-\rho_g\right)gD_p} \tag{9.2.10}$$

If predicted $u_{\text{mb}} < u_{\text{mf}}$, $u_{\text{mb}} = u_{\text{mf}}$. The SI units listed in Nomenclature are to be used.

For the minimum slugging velocity, Rabinovich and Kalman (2011) have given the following formula applicable to groups A, B, and D particles:

$$Re_{\text{ms}} = 0.059Ar^{0.56} \tag{9.2.11}$$

It will be noticed that Re_{ms} from this equation equals the minimum fluidizing velocity for group D particles given by Eq. (9.2.9).

Arnaldos and Casal (1996) evaluated numerous available correlations for various regime transition velocities by comparison with test data. The correlations that they found to be the best are as follows.

For the velocity u_c at the start of the intermediate/transitional turbulent regime:

For Geldart group A particles, $54 < D_p < 2600\,\mu\text{m}$,

$$Re_c = 0.936Ar^{0.472} \tag{9.2.12}$$

For Geldart group B particles, the correlation of Nakajima et al. (1991),

$$Re_c = 0.633Ar^{0.467} \tag{9.2.13}$$

For the velocity at the start of the fully turbulent fluidization regime, u_k:

Group A particles:

$$Re_k = 1.41Ar^{0.562} \tag{9.2.14}$$

Group B particles, correlation of Perales et al. (1991):

$$Re_k = 1.95Ar^{0.0.453} \tag{9.2.15}$$

For the transport velocity, u_{tr}:

Group A particle, correlation of Adanez et al. (1993):

$$Re_{tr} = 2.078Ar^{0.0.463} \quad (9.2.16)$$

For group B particles, correlation of Perales et al. (1991):

$$Re_{tr} = 1.45Ar^{0.0.484} \quad (9.2.17)$$

Krugell-Emden and Vollmari (2016) measured transition velocities for complex-shaped particles and compared them to correlations for spherical particles. The particles included shapes such as elongated plates, cylinders, and cuboids. They developed factors to adjust the predictions of correlations for spherical particles to fit their data.

9.2.2 Void Fraction and Bed Expansion

Void fraction is the fraction of total bed volume occupied by gas. The relation between bed height H and void fraction ε is expressed by the following relation:

$$\frac{H}{H_{mf}} = \frac{1-\varepsilon_{mf}}{1-\varepsilon} \quad (9.2.18)$$

Wen and Yu (1966a) have given the following correlation for ε_{mf}, the void fraction at minimum fluidization:

$$\varepsilon_{mf} = \left(14\varphi_s\right)^{-1/3} \quad (9.2.19)$$

The void fraction at minimum fluidization for spherical particles is near 0.4. For non-spherical particles, it is higher. King and Harrison (1982) found no change in void fraction over the pressure range of 1–25 bar. Pattipati and Wen (1981) found no effect of temperature on ε_{mf} in the range 20–850 °C.

Broadhurst and Becker (1975) proposed the following correlation for void fraction at minimum bubbling:

$$\varepsilon_{mb} = 0.586\varphi_s^{-0.72}Ar^{-0.029}\left(\rho_g/\rho_s\right)^{0.021} \quad (9.2.20)$$

Its verified range is $\varphi_s = 0.85$–1, $Ar = 1$–10^5, and $\rho_s/\rho_g = 500$–$50\,000$.

Babu et al. (1978) analyzed data for void fraction in beds of various types of coal and related materials such as dolomite and limestone.

For $D_{bed} \leq 63.5$ mm,

$$\frac{H}{H_{mf}} = 1 + \frac{0.762\left(u-u_{mf}\right)^{0.57}\rho_g^{0.083}}{\rho_s^{0.166}u_{mf}^{0.063}D_{bed}^{0.445}} \quad (9.2.21)$$

For $D_{bed} > 63.5$ mm,

$$\frac{H}{H_{mf}} = 1 + \frac{10.978\left(u-u_{mf}\right)^{0.738}\rho_s^{0.376}D_p^{1.006}}{\rho_g^{0.126}u_{mf}^{0.937}} \quad (9.2.22)$$

The maximum bed diameter D_{bed} in the data analyzed was 305 mm and u/u_{mf} was up to 40. Pressure was 1–69 bar. This correlation is dimensional: u in ft s^{-1}, D_{bed} and D_p in ft, and ρ in lb ft^{-3}.

Staub and Canada (1978) have given the following method for calculating bed expansion in bubbling beds. The voidage of the whole bed ε_{bed} is given by

$$\varepsilon_{bed} = \frac{u\varepsilon_{mf}}{1.05u\varepsilon_{mf} + \left(1-\varepsilon_{mf}\right)u_{mf}} \quad (9.2.23)$$

The fraction of bed volume occupied by bubble ε_{bub} is given by

$$\varepsilon_{bub} = \frac{\varepsilon_{bed}-\varepsilon_{mf}}{\left(1-\varepsilon_{mf}\right)} \quad (9.2.24)$$

According to Grace (1982a, b), this procedure is inaccurate at velocities near u_{mf} but gives satisfactory predictions at higher velocities approaching the turbulent regime.

For preliminary calculation of void fraction of bed at any velocity u between u_{mf} and u_{tr}, Zabrodsky (1966) recommends the following formula of Goroshko et al. (1958):

$$\varepsilon_{bed} = \left(\frac{18Re + 0.36Re^2}{Ar}\right)^{0.21} \quad (9.2.25)$$

where

$$Re = \frac{u\rho_g D_p}{\mu_g} \quad (9.2.26)$$

9.3 Properties of Solid Particles

9.3.1 Density

The absolute density of the particle ρ_s is the mass of particle divided by its volume. The bulk density ρ_b is the mass of particles in the bed divided by the volume of bed. The two are related as follows:

$$\rho_b = \rho_s(1-\varepsilon) \quad (9.3.1)$$

The mean density of a mixture of particles is given by

$$\rho_{s,mean} = \sum_{1}^{N}\left(x_i\rho_{s,i}\right) \quad (9.3.2)$$

where x_i and $\rho_{s,I}$ are the mass fraction and density of component "i" of the mixture and N is the number of components.

9.3.2 Particle Diameter

For non-spherical particles, equivalent diameter is defined as the diameter of a sphere that has the same volume as the particle. Thus,

$$D_p = \left(\frac{6V_p}{\pi}\right)^{1/3} \quad (9.3.3)$$

V_p is the volume of the particle. Another definition used by some authors, for example, Krugell-Emden and Vollmari

(2016), is the diameter of a sphere that has the same surface area to volume ratio as the particle. This is expressed by

$$D_{SV} = \frac{6V_p}{A_p} \tag{9.3.4}$$

A_p is surface area of particle. This is also known as Sauter mean diameter or d_{32}. The definition commonly used is according to Eq. (9.3.3). This is the definition used here except where stated otherwise.

When a mixture of particles of different sizes are involved, average particle diameter is calculated as follows:

$$\frac{1}{D_{p,\text{mean}}} = \sum \frac{x_i}{D_{p,i}} \tag{9.3.5}$$

where x_i is the mass fraction of particles of diameter $D_{p,i}$. It is generally determined by sieve analysis as the mass fraction collected between sieves of mean aperture equal to $D_{p,i}$.

Another definition that has been used by some authors is given as follows:

$$D_{p,\text{mean}} = \sum x_i D_{p,i} \tag{9.3.6}$$

9.3.3 Particle Shape Factor

Shape factor φ_s of a particle is defined as the surface area of a sphere of equivalent volume divided by the particle surface area S_p. Thus,

$$\varphi_s = \frac{\pi\left(6V_p/\pi\right)^{2/3}}{S_p} \tag{9.3.7}$$

V_p is the particle volume. Shape factor of spherical particles is 1. That of particles of other shapes is <1. Shape factor of particles of a particular material type can vary over a wide range. Shape factor is also known as sphericity.

9.3.4 Classification of Particles

Many schemes for classification of particles have been proposed. The best known among them is that of Geldart (1973). He divided particles into four groups called A, B, C, and D.

Group A particles are aeratable. These deaerate slowly after air is shutdown. Bed expansion is high and they have particulate fluidization at velocities near minimum fluidization velocity. Examples are fluid cracking catalysts.

Group B particles bubble readily. An example is sand. They do not have particulate fluidization. Bubbling regime starts immediately at start of fluidization.

Group C particles are cohesive and hence are difficult to fluidize. An example is flour.

Group D particles are spoutable, that is, they can form spouted beds. Wheat is an example of this type of particle.

Group B and D particles are called coarse particles, while A and C particles are regarded as fine particles.

The boundary between group A and B particles is where minimum fluidization and minimum bubbling velocities are equal. The equations given in Section 9.2.2 for u_{mf} and u_{mb} may be used for finding this boundary. Based on their own correlations for u_{mf} and u_{mb}, Geldart and Abrahamsen (1978) gave the following formula, which may be used for determining the boundary between A and B particles:

$$\frac{u_{mb}}{u_{mf}} = \frac{4.125 \times 10^4 \mu_g^{0.9} \rho_g^{0.1}}{\left(\rho_s - \rho_g\right) g D_p} \tag{9.3.8}$$

With $u_{mb} = u_{mf}$ at the boundary between groups A and B, a particle is in group B if

$$D_p \geq \frac{4.125 \times 10^4 \mu_g^{0.9} \rho_g^{0.1}}{\left(\rho_s - \rho_g\right) g} \tag{9.3.9}$$

For the boundary between *B* and *D* particles, Geldart (1973) gave the following relation:

$$D_p = \left(\frac{0.001}{\rho_s - \rho_g}\right)^{0.5} \tag{9.3.10}$$

Equations (9.3.8) and (9.3.9) are dimensional. The units listed in the Nomenclature should be used.

The boundaries according to the Geldart classification for air at normal atmospheric pressure and temperature are plotted in Figure 9.3.1.

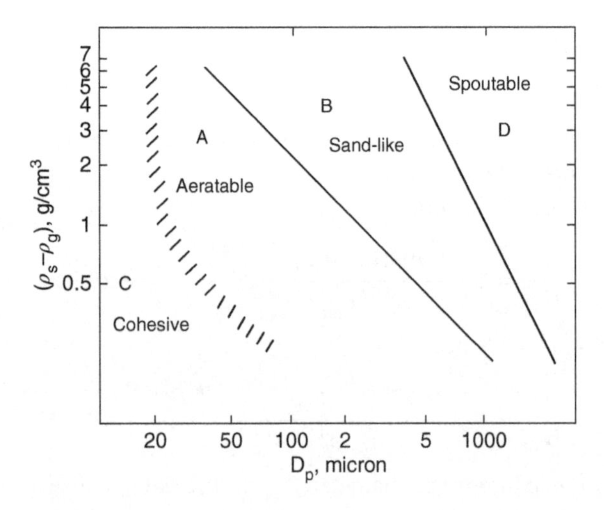

Figure 9.3.1 Representation of Geldart classification of particles for air at normal temperature and pressure. Source: Geldart and Abrahamsen (1978). © 1978, Elsevier.

9.4 Parameters Affecting Heat Transfer to Surfaces

In this section, the parameters affecting heat transfer between fluidized beds and surface immersed in them are discussed.

Heat transfer coefficients are generally not constant but fluctuate with time. The heat transfer coefficients reported in literature are time-averaged values. Heat transfer coefficients around a horizontal tube in the bed vary around the circumference. The reported values are the integrated mean around the circumference. For vertical tubes, heat transfer coefficients are integrated average for the height of the tube. For a sphere, heat transfer is integrated mean for its entire surface. In the following discussions, heat transfer coefficients referred to are the time-averaged mean values for the surfaces.

9.4.1 Gas Velocity

By gas velocity is meant the superficial velocity, which is the velocity of gas assuming that the bed is empty, that is, it is volumetric gas flow rate divided by the total cross-sectional area of the bed.

Heat transfer coefficients typically increase sharply as velocity goes above the minimum fluidization, reaches a maximum, and then decreases. The velocity at the point of maximum heat transfer is called the optimum velocity. For small particles, heat transfer coefficients drop fairly quickly as velocity goes beyond the optimum velocity. For larger particles, the decrease in heat transfer coefficients beyond optimum velocity is gradual. Figure 9.4.1 shows typical behavior.

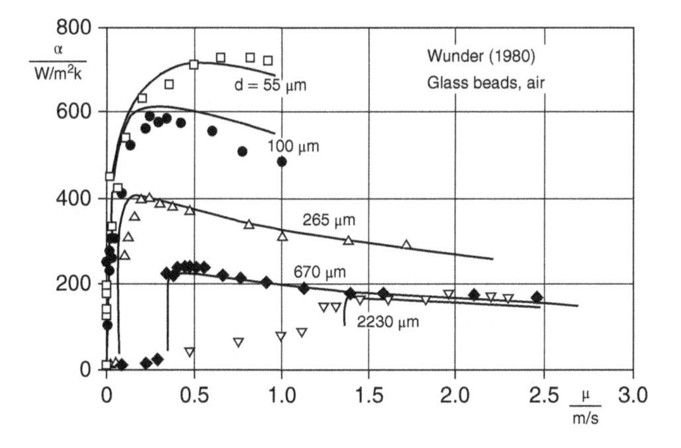

Figure 9.4.1 Effect of particle size and air velocity on heat transfer to vertical tubes in a fluidized bed at room temperature. Data of Wunder (1980). Solid lines are predictions of Martin model. Source: Martin (1984b). © 1984, Elsevier.

The explanation of this behavior is as follows. As gas velocity increases, there are two opposing effects. Higher velocity increases mixing and turbulence in the bed. This tends to increase heat transfer. But increasing gas flow rate also increases the bed voidage and thus the density of the bed. This tends to decrease heat transfer. During the initial increase in velocity, the first effect dominates, while at high velocities the second effect is stronger.

9.4.2 Particle Size and Shape

In Figure 9.4.1, it is seen that heat transfer coefficient decreased as particle diameter increased from 55 to 22 230 μm. This is the typical tendency for smaller particle sizes. As the diameter increases, a point is reached where heat transfer coefficient begins to increase with increasing particle size. This is illustrated by Figure 9.4.2, which shows the maximum heat transfer coefficients for corundum particles of diameters up to 10 mm. The heat transfer coefficients decrease as particle diameter increases to about 2 mm and then begins to increase with increasing particle diameter.

These variations of heat transfer with particle size may be explained in terms of the penetration theory described in Section 9.5.2. With increasing particle size, the area of tube in contact with particles decreases. This reduces the heat transfer due to conduction. At the same time, area of tube in contact with gas increases, which increases the convective heat transfer. The first effect dominates for the smaller particles and the second effect for the larger particles. For the largest particles, almost all heat transfer is by convection.

Very small particles fall into Geldart group C as seen in Figure 9.3.1. These are cohesive and do not fluidize well. Their poor fluidization results in poor heat transfer. This is the likely explanation for the very low heat transfer coefficient of the 0.01 mm particle seen in Figure 9.4.2. Baerns (1967) reported that addition of coarse particles to fine particles improved fluidization and heat transfer.

Addition of small particles to larger particles has been found to increase heat transfer. Figiola et al. (1986) performed tests in which glass beads of diameter 256 and 509 μm were mixed in several proportions to provide mixtures of different mean diameters. Tests were performed in the bubbly regime. Measured heat transfer coefficients were compared to several correlations for uniform size particles and were found to be higher than predictions using the mixture's particle mean diameter.

Baerg et al. (1950) performed tests that included spherical and non-spherical particles. They found that heat transfer coefficients with spherical particles were higher than with non-spherical particles. They considered it to be due to

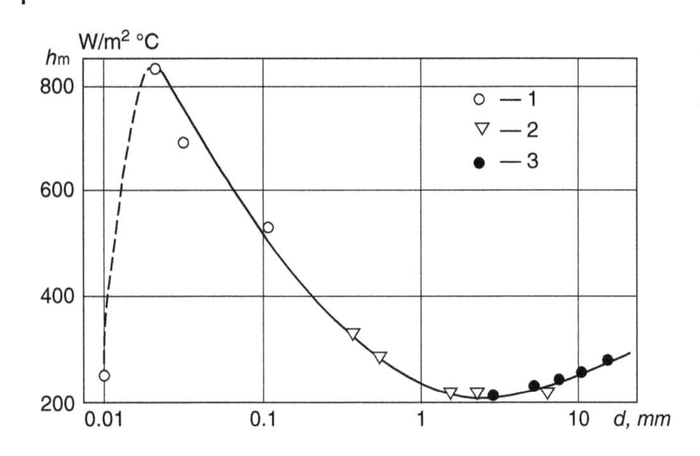

Figure 9.4.2 Effect of particle diameter d on maximum heat transfer coefficient h_m to objects in a fluidized bed of corundum particles. (1) Horizontal cylinder 6 mm diameter. (2) Vertical plate 220 mm × 160 mm. (3) Vertical cylinder 40 mm diameter. Source: Baskakov et al. (1973). © 1973, Elsevier.

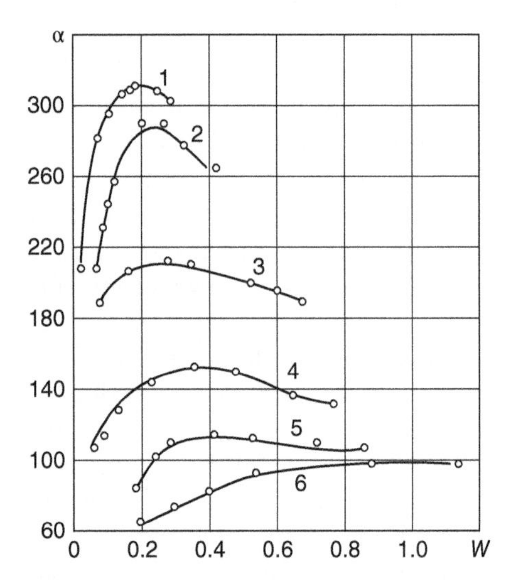

Figure 9.4.3 Effect of low pressure on heat transfer coefficient α (W m^{-2} s^{-1}) at superficial air velocity w (m s^{-1}) on vertical cylinders in a bed of 0.2 mm sand. Pressure, Pa: (1) 10^5, (2) 26 600, (3) 13 300, (4) 931, (5) 400, and (6) 266. Source: Shlapkova (1969). © 1969, Springer nature.

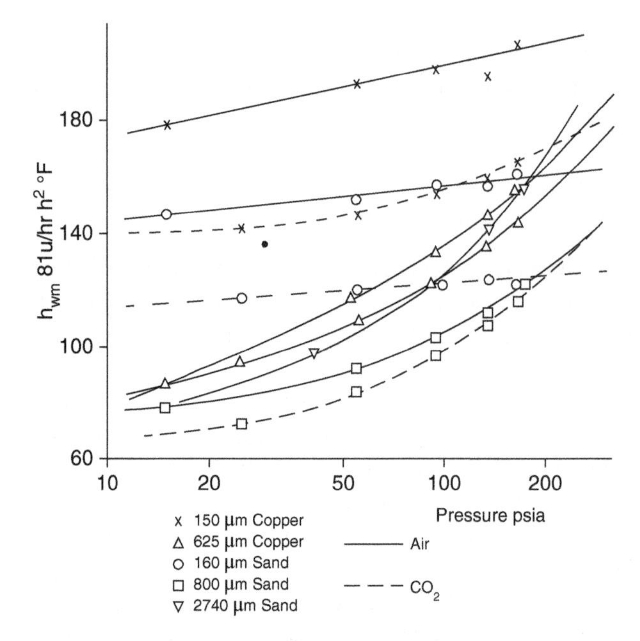

Figure 9.4.4 Effect of increase in pressure on maximum heat transfer coefficient in fluidized beds. Source: Botterill and Desai (1972). © 1972, Elsevier.

the greater mobility of spherical particles. Shah (1983) analyzed data for maximum heat transfer from many sources. He concluded that maximum heat transfer coefficient of spherical particles is higher than that of non-spherical particles by 24%. With spherical particles, bed voidage is lower and hence the bed density is higher. This may be a factor in their having higher heat transfer.

9.4.3 Pressure and Temperature

Shlapkova (1969) experimentally investigated the effect of low pressure on heat transfer to vertical cylinders in a fluidized bed. Cylinder diameters were 3.4 and 6.8 mm. The results are shown in Figure 9.4.3. It is seen that heat transfer coefficients became lower with decreasing pressure.

Bhatt and Whitehead (1963) measured heat transfer to a vertical cylinder in a fluidized bed of −52 + 100 mesh silica sand over a pressure range of 0.1–1 bar. They found no effect of pressure on heat transfer coefficients. Thus their results disagree with those of Shlapkova.

Botterill and Desai (1972) studied the effect of pressure on heat transfer to the wall of fluidized beds. A variety of particle types and sizes were used. Figure 9.4.4 shows their results for the maximum heat transfer coefficients. It is seen that there is large increase in heat transfer coefficient with increasing pressure for the larger particles. The effect of pressure is much less for the smaller particles.

Heat transfer coefficients have been found to increase with temperature. Thermal conductivity of gases increases with temperature. This tends to increase convective heat transfer between gas and surface as well as the particle convection heat transfer as conduction to particles occurs through a thin layer of gas at the surface. At high temperatures, heat transfer also increases due to radiation.

9.4.4 Heat Transfer Surface Diameter

Different opinions have been expressed about the effect of diameter. On examining data from a number of sources, Gelperin and Einstein (1971) concluded that heat transfer decreases with increasing cylinder diameter until it reaches 10 mm. Beyond that, there is no effect of diameter. The widely used correlation of Zabrodsky (1958) for maximum heat transfer does not include diameter of surface. In the correlations of Grewal and Saxena (1981) and Grewal (1982), maximum heat transfer decreases with increasing surface diameter. At smaller diameters, heat transfer coefficient increases with decreasing diameter. Lechner et al. (2014) performed tests with horizontal cylinders of diameter 12–200 mm in fluidized bed of lignite particles of diameter 73 μm. Bed was 390 mm square. They concluded that maximum heat transfer coefficient is proportional to $D_t^{-0.3}$.

Shah (2018a, b) examined data for maximum heat transfer from many sources for surface diameters from 0.05 to 220 mm. He concluded that diameter has no effect if it is greater than 3 mm.

9.4.5 Properties of Gas and Solid

Heat transfer in fluidized bed occurs by both gas convection and conduction to particles across a thin gas layer. The properties that affect single-phase heat transfer also affect heat transfer in fluidized bed. Among these, the most important is thermal conductivity. Jacob and Osberg (1957) used several gases in their tests with wires in fluidized beds and found heat transfer coefficients to increase with gas thermal conductivity. Almost all correlations show increasing heat transfer with increasing thermal conductivity.

In his extensive data analysis, Shah (2018a, b) found that there was no effect of particle thermal conductivity when it was greater than about $0.5\,\mathrm{W\,m^{-1}\,K^{-1}}$. At lower thermal conductivities, heat transfer coefficients decreased with decreasing thermal conductivity. His explanation was that it is because low thermal conductivity of particles severely reduces heat transfer due to conduction. A particle exchanges heat by conduction with the surface through direct contact or by conduction through a very thin gas film adjacent to the surface and then moves away into the gas stream. If the particle thermal conductivity is very low, heat exchange by conduction will become very small and the total heat transfer coefficient will also become low.

Similar effect of particle thermal conductivity was also noted by Martin (1984a, b). He compared the data of Wunder (1980) for styropor ($k \approx 0.035\,\mathrm{W\,m^{-1}\,K^{-1}}$) with his analytical model. The measurement was much lower than his model, which neglected effect of thermal conductivity of particles but agreed with his model, which took into account the particle thermal conductivity. When particle thermal conductivities were higher than $1\,\mathrm{W\,m^{-1}\,K^{-1}}$, it had no effect on the predictions of his model.

9.4.6 Gas Distribution

Frankel and Kondukov (1978) studied the effect of gas velocity distribution on heat transfer. They found that flat velocity profile (uniform velocity across bed cross section) gave much higher heat transfer than convex or jet-like profile.

9.4.7 Length and Location of Tube

The effect of length of vertical tubes has been investigated in a number of studies. A few show decrease in mean heat transfer coefficient with increasing height while many other show no effect of height. Gelperin and Einstein (1971) reviewed experimental evidence and noted that the measurements that show effect of tube length were done close to the gas distributor where heat transfer coefficients are high. Away from the distributor, length of tube has no effect.

Vreedenberg (1952) measured heat transfer coefficients to a vertical tube placed at different radial locations. He found that heat transfer coefficient was higher at locations away from the axis, up to 1.75 times that at the axis of bed. Gelperin et al. (1958) found that heat transfer coefficient decreased when tube was moved away from the bed axis. Their results for the effect of tube location may be expressed as

$$h \propto \left(1 - \frac{r}{R_{\mathrm{bed}}}\right)^{0.36} \quad (9.4.1)$$

where r is the distance of tube axis from the bed axis and R_{bed} is the radius of bed.

However, there have been other studies in which little or no effect of location was found. For example, Morooka et al. (1979) found no effect of tube location on heat transfer except close to the bed wall. Stefanova et al. (2011) found no effect of location for (r/R_{bed}) from 0 to 0.8 except at the lowest gas velocities.

Regarding the location of horizontal tubes, most studies indicate that there is no effect of height above the distributor as long as it is not too close to it. Heat transfer coefficients usually remain unchanged at locations above one bed diameter. An example of such studies is that of Abubakar et al. (1980).

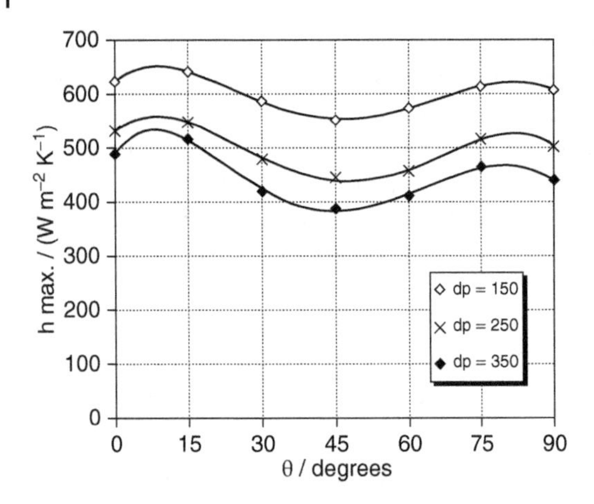

Figure 9.4.5 Effect of tube inclination on maximum heat transfer coefficient in fluidized beds of sand of particle diameters 150–350 μm. Source: Abid et al. (2011) © 2011 Elsevier.

9.4.8 Bed Diameter and Height

Most test data show no effect of bed diameter or height. For example, Kobayashi et al. (1970) experimented with beds of height 100–300 mm. Heat transfer coefficients were the same for all bed heights. Gelperin and Einstein (1971) reviewed data from many sources and concluded that bed height and diameter do not affect heat transfer. Data that show effect of height are taken close to the gas distributor or at heights where density of bed was low.

9.4.9 Tube Inclination

Genetti et al. (1971) performed tests with tubes at various inclinations from horizontal to vertical. Heat transfer coefficients for horizontal and vertical orientations were the same but were lower at in between orientations. Gelperin and Einstein (1971) report their own experiments with 160 μm quartz sand particles with 20 and 80 mm diameter tubes. No significant difference in heat transfer coefficient was found as tube inclination varied from horizontal to vertical. Abid et al. (2011) performed tests with a 20 mm diameter cylinder in beds of 150–350 μm sand particles fluidized by air. They found heat transfer coefficients to be highest at an inclination of 15° to horizontal and lowest at 45°. Heat transfer coefficients at 0° and 90° were about the same. Their maximum heat transfer results are shown in Figure 9.4.5. The trends at other velocities were similar.

Shah (2018a, b) analyzed a very wide range of data for maximum heat transfer to horizontal and vertical tubes. His correlation agrees well with data for both orientations even though it has no factor for tube orientation.

For maximum heat transfer coefficients, the conclusion is that they are the same for vertical and horizontal tubes. The situation at in-between orientations is not clear as the available data are conflicting.

The situation regarding heat transfer coefficients over a wide range of velocities is not clear. The experimental studies mentioned earlier suggest that heat transfer coefficients in horizontal and vertical orientations are about the same. The analytical model of Martin (1984b) was shown to agree with data for both horizontal and vertical tubes while it has no factor for tube orientation. However, there has been no comprehensive examination of applicability of correlations for horizontal tubes to vertical tubes. Saxena (1989) reviewed data and correlations for all orientations over the entire range of velocities. He gives separate correlations for horizontal and vertical tubes.

Baskakov et al. (1973) performed tests with a plate in a fluidized bed. They found that heat transfer is lowered if the plate orientation is changed from vertical. This is because voids are formed at the lower face of the plate, while solid particles accumulate over the upper surface of plate.

9.5 Theories of Heat Transfer

Many theories have been put forward about mechanism of heat transfer in fluidized beds. They can be put in two broad categories, namely, the film theory and the penetration theory. These are briefly discussed herein. Details of various theories can be found, among others, in Grace (1982b), Gelperin and Einstein (1971), Saxena (1989), Yusuf et al. (2005), and Abdelmotalib et al. (2015).

9.5.1 Film Theory

The film theory was first put forward by Dow and Jakob (1951). According to it, heat transfer occurs only by convection between the gas and the heat transfer surface. The solid particles enhance the convective heat transfer by thinning out the boundary layer. Thermophysical properties of solid particles have no effect on heat transfer.

9.5.2 Penetration Theory

In this theory, heat transfer occurs by convection between gas and the heater surface as well as by transient conduction to solid particles. In some analyses, particles actually touch the surface, while in others, there is a thin gas layer between the particles and surface. The conduction mechanism dominates at low velocities with small particles, while the gas convection mechanism dominates for large particles and high gas velocities. There are two basic versions of the penetration theory. In one version, conduction to distinct particles is analyzed. In the other version, conduction to groups or packets of particles is analyzed. In both versions, conduction, convection, and radiation are considered to be independent of each other and additive. Thus,

$$h_w = h_{gc} + h_{pc} + h_{rad} \tag{9.5.1}$$

$$q = h_w \left(T_{bed} - T_w\right) \tag{9.5.2}$$

where h_{gc} is the heat transfer coefficient for gas convection, h_{pc} is the heat transfer due to conduction (usually called particle convection), and h_{rad} is the heat transfer coefficient due to radiation.

In the film theory, $h_{pc} = 0$.

9.5.2.1 Particle Theory

In this version of the penetration theory, heat conduction to individual particles is considered. In some analyses, conduction to a single particle in contact with the surface or close to it with a gas gap in between is considered, and conduction from this particle to other particles is neglected. In other analyses, chains of particles are considered. Conduction first occurs to the particle near the surface and then from that particle to the next particle in the chain and so on. Models with two, three, or infinite number of particles have been proposed. Various gap sizes between the particle and surface and in between the particles have been proposed.

9.5.2.2 Packet Theory

The packet theory was first put forward by Mickley and Fairbanks (1955) and has been further developed by many others. In this theory, conduction heat transfer occurs between packets of particles that contact the heater surface, gain heat from it, and then move back into the bulk of the bed where they exchange heat with the material there. The packets have the properties of the quiescent bed (bed at minimum fluidization) such as void fraction, thermal conductivity, and specific heat. The packets are not permanent; they keep on forming and breaking up. Calculations with this model involve factors such as frequency and duration of contacts of the packets with surface that are difficult to predict. A number of empirical factors have to be used for calculating heat transfer.

9.6 Prediction Methods for Single Tubes and Spheres

In this section, prediction of heat transfer between bed and single tubes and spheres immersed in it is discussed. Early work on this subject has been reviewed among others by Zabrodsky (1966), Gelperin and Einstein (1971) and Saxena et al. (1978).

9.6.1 Analytical Models

9.6.1.1 Particle Models

The first model of this type appears to have been proposed by Jacob and Osberg (1957). It was verified only with their own data. A number of more verified models of this type have been published. Some of them are discussed herein.

Natale et al. (2008) developed a model in which the surface-to-bed heat flow due to particle convection was calculated as the sum of the heat flows between the surface and each particle, weighted by means of the particle residence time distribution, and multiplied by the number of particles simultaneously in contact with the surface. The following relations were developed for particle convection heat transfer coefficient:

$$h_{pc} = \rho_s C_s \lambda \left(1 - \varepsilon_w\right) \frac{f}{1 + f\tau} \tag{9.6.1}$$

ε_w is the void fraction near the surface given by the following relation:

$$\varepsilon_w = \varepsilon_{mf} + 0.045 \log(Ar) \tag{9.6.2}$$

f is the renewal frequency (inverse of particle residence time) obtained by the following formula:

$$f = 2.045 \times 10^{-4} Re_p^{0.287} D_p^{-1} \tag{9.6.3}$$

Re_p is defined as

$$Re_p = \frac{\left(u - u_{mf}\right) \rho_g D_p}{\mu_g} \tag{9.6.4}$$

τ is the characteristic time for particle heat exchange given by the following relation:

$$\tau = \frac{\delta D_p C_s \rho_s}{6 k_g} \tag{9.6.5}$$

λ is the thickness of thermal gradient between the bed and the surface. Based on the study of published experimental data, they assigned the following values to λ:

For $D_p < 300\ \mu m$, $\lambda = D_p$ and for $D_p > 300\ \mu m$, $\lambda = 500\ \mu m$

δ is the thickness of gas layer between the particle and surface. Based on Botterill (1973), the following values were assigned to it:

$\delta = D_p/10$ for spherical particles. $\delta = D_p/24$ for non-spherical particles.

Gas convection heat transfer coefficient is obtained by the following relation:

$$h_{gc} = \left(1 - \varepsilon_{bub}\right) h_{mf} + h_{bub} \varepsilon_{bub} \tag{9.6.6}$$

The heat transfer coefficient at minimum fluidization is calculated by the following dimensional correlation of Botterill and Denloye (1978):

$$\frac{h_{mf} D_p^{1/2}}{k_g} = 0.863 Ar^{0.39} \tag{9.6.7}$$

This equation is applicable to Ar from 10^3 to 2×10^6. Units listed in Nomenclature should be used. The bubble fraction within the fluidized bed, ε_{bub}, is calculated by the classical two-phase theory for aggregative fluidization given in Kunii and Levenspiel (1991). h_{bub} is the heat transfer coefficient due to rising bubbles. They recommend neglecting its contribution. This model was found in satisfactory agreement with a wide range of data from many sources that included particle diameter 115 μm to 10 mm and pressures from 1 to 25 bar. All data were for room temperature.

Another single particle model that was verified with a wide range of data is by Martin (1981, 1984a,b). The model is based on concepts derived from the kinetic theory of gases. Single particles with more or less random motion exchange heat with a cooling or heating surface by transient conduction through a thin air gap then move back into the bulk of the bed. The air gap is a modified mean free path of the gas molecule calculated using the kinetic theory of gases. This gap near the contact point of particle is always less than the mean free path. They developed the following equations to predict the particle convection heat transfer coefficient:

$$\frac{h_{\text{pc}}D_p/k_g}{1-\varepsilon_{\text{mf}}} = \left[1-\left(\varepsilon_{\text{bed}}-\varepsilon_{\text{mf}}\right)\right] Z\left(1-e^{-N}\right) \quad (9.6.8)$$

$$N = \frac{Nu_{\text{WP}}}{CZ} \quad (9.6.9)$$

$$\frac{1}{Nu_{\text{WP}}} = \frac{1}{Nu_{\text{WP(max)}}} + \frac{k_g/k_s}{Nu_{\text{ip}}} \quad (9.6.10)$$

$$Nu_{\text{WP(max)}} = 4\left[(1+Kn)\ln\left(1+\frac{1}{Kn}\right)-1\right] \quad (9.6.11)$$

$$Nu_{\text{ip}} = 4\left(1+\sqrt{\frac{1.5Ck_gZ}{\pi k_s}}\right) \quad (9.6.12)$$

$$Z = \frac{C_s\rho_s}{6k_g}\sqrt{\frac{gD_p^3}{5\left(1-\varepsilon_{\text{mf}}\right)}}\sqrt{\frac{\left(\varepsilon-\varepsilon_{\text{mf}}\right)}{1-\left(\varepsilon-\varepsilon_{\text{mf}}\right)}} \quad (9.6.13)$$

$$Kn = \frac{4}{D_p}\left(\frac{2}{\gamma}-1\right)\frac{k_g\sqrt{2\pi RT/M}}{p\left(2C_{\text{pg}}-R/M\right)} \quad (9.6.14)$$

C is the only adjustable constant in this model and it was given the value of 2.6. Kn is the Knudsen number; R is the universal gas constant; and M the molecular weight of gas. Υ is the accommodation coefficient representing the incompleteness of heat exchange between particle and wall during contact. It varies between 0 and 1. Values of Υ at 25 °C for various gases are given in Table 9.6.1. Calculation of Υ at temperatures other than 25 °C is done as follows:

$$\log\left(\frac{1}{\gamma}-1\right) = 0.6 - \frac{\frac{1000}{T}+1}{C_A} \quad (9.6.15)$$

Table 9.6.1 Values of accommodation coefficient Υ at 25 °C in the Martin model.

Gas	Υ	Gas	Υ
Air, CO_2	0.9	H_2O	0.8
Ar	0.876	Kr	0.933
CH_4	0.7	Ne	0.573
H_2	0.2	NH_3	0.9
He	0.235	Xe	0.956

T is in K and C_A is given by

$$C_A = \frac{(1000/298.5)-1}{0.6-\log\left(1/\gamma_{25}-1\right)} \quad (9.6.16)$$

Gas convection heat transfer coefficient was calculated by the following equation of Baskakov et al. (1973):

$$\frac{h_{\text{gc}}D_p}{k_g} = 0.009\text{Pr}_g^{1/3}Ar^{0.5}\left(\frac{u}{u_{\text{opt}}}\right)^n \quad (9.6.17)$$

The exponent $n=0.3$ for $u < u_{\text{opt}}$ and $n=0$ for $u > u_{\text{opt}}$ where u_{opt} is the superficial velocity at which maximum heat transfer coefficient occurs.

Radiation heat transfer coefficient is calculated as

$$h_{\text{rad}} = 4E\sigma\left(\frac{T_w+T_{\text{bed}}}{2}\right)^3 \quad (9.6.18)$$

E is effectivity emissivity and σ is the Stefan–Boltzmann constant. They used $E=0.5$ in comparing with a data set.

Martin (1984b) reported good agreement with data from several sources, which included particle diameters 400 μm to 10 mm, pressures 0.3–40 bar, temperatures up to 830 °C, and particle thermal conductivity down to 0.035 W m^{-1} K^{-1}. Data included horizontal and vertical tubes as well as spheres. Grewal and Saxena (1983) found it to be in good agreement with their data in horizontal tube banks.

Botterill (1981) considered Martin's physical model to be incorrect and reported that it gave increasing underprediction of data from tests in his laboratory as bed temperature rose from 220 °C to 850 °C, even after taking into consideration the contribution of radiation. As mentioned previously, Martin showed agreement of his correlation with data up to 830 °C. An independent evaluation of the accuracy of Martin's correlation is desirable.

9.6.1.2 Packet Models

Mickley and Fairbanks (1955) were the first to propose the packet model, according to which heat transfer occurs between packets (clusters) of particle in intermittent contact with the heating/cooling surface. Their equation is

$$h_w = \sqrt{k_c\rho_cC_cS} \quad (9.6.19)$$

where the subscript "c" indicates for the cluster/packet of particles. S is a stirring factor depending on the motion of the packets and geometry of the bed.

Many variations of the packet model have been proposed. Baskakov et al. (1973) included the resistance of a thin gas gap between the packet and heater surface. Their model also includes heat transfer by gas convection. They express the particle convection heat transfer coefficient by the following equation:

$$h_{gc} = (1 - f_0)\, h_I \tag{9.6.20}$$

where f_0 is the fraction of time the heater surface is covered by gas bubbles (lean phase) and h_I is the heat transfer coefficient between the packet and heater surface during the fraction of time they are in contact. It is approximated by the following expression:

$$h_I = \frac{1}{R_{gap} + R_c} \tag{9.6.21}$$

R_{gap} is the resistance of the gas gap between the packet and surface and R_c is the resistance of the packet. R_c is obtained from the Mickley–Fairbanks model as

$$R_c = \left(\frac{k_c \rho_c C_c (1 - \varepsilon_{bed})}{\pi \tau} \right)^{-1/2} \tag{9.6.22}$$

where τ is the contact time between packet and surface. It is seen that as τ approaches zero, R_c tends to zero and therefore heat transfer coefficient tends to infinity. As this does not agree with test data, the resistance of gas gap is introduced into this model and other packet models.

R_{gap} is given by the following equation:

$$R_{gap} = \frac{D_p}{2k_c} \tag{9.6.23}$$

By analyzing their own data for vertical 15 and 30 mm diameter tubes in beds of corundum (0.12–0.5 mm diameter) and 0.65 mm slag beads, they obtained the following correlations for τ and f_0:

$$\tau = 0.44 \left[\frac{g D_p}{u_{mf}^2 (u_r - \eta)^2} \right]^{0.14} \left(\frac{D_p}{D_t} \right)^{0.225} \tag{9.6.24}$$

$$f_0 = 0.33 \left[\frac{u_{mf}^2 (u_r - \eta)^2}{g D_p} \right]^{0.14} \tag{9.6.25}$$

D_t is the diameter of tube and $u_r = u/u_{mf}$. η is an empirical factor. It was found to decrease with increasing tube diameters and with increasing particle shape factor. For corundum particles, it was 0.8 and 0.9 for the 30 and 15 mm diameter tubes, respectively. For slag beads, the values were 0.7 and 0.75.

Baskakov et al. compared this methodology with their own test data as well as data from several other sources that included particles from 0.02 to 14 mm with good agreement. The gas convection heat transfer coefficient was calculated with Eq. (9.6.17). For radiation heat transfer contribution, they used their own experimental data.

Karimipour et al. (2007) developed a model in which heat transfer to the packets was divided into two time periods. The time t_0 for heat to completely penetrate a packet/cluster is given by

$$t_0 = \frac{D_c^2}{6\alpha_c} \tag{9.6.26}$$

D_c is the diameter of cluster and α_c is its thermal diffusivity. Separate expressions were developed for heat transfer for $t < t_0$ and $t > t_0$, which used correlations for properties of clusters from other researchers. Approximate agreement was shown with a limited amount of experimental data.

Masoumifard et al. (2008) noted that the model of Karimipour et al. (2007) gives very high heat transfer coefficients as time approaches zero, which is incorrect. They modified the Karimipour et al. model by introducing the resistance of a gas layer between the cluster and surface. Their model showed good agreement with some data for particles of diameter 205–750 μm.

Stefanova et al. (2019) have given a probabilistic packet model. It showed reasonable agreement with some data for heat transfer between a tube and FCC and alumina particles of diameter 0.29 and 1.56 mm.

In conclusion, none of the packet models has been shown to agree with a wide range of test data.

9.6.2 Empirical Correlations

9.6.2.1 Maximum Heat Transfer

Many empirical correlations have been proposed. Among them, the most verified is that of Shah (2018a). The other correlations are first described and then the Shah correlation is discussed.

Various Correlations The ones that have had considerable verification include the following:

A very well-known correlation is the dimensional formula of Zabrodsky (1958):

$$h_{max} = 35.7 k_g^{0.6} D_p^{-0.36} \rho_s^{0.2} \tag{9.6.27}$$

Zabrodsky et al. (1976) gave the following dimensionless correlation:

$$Nu_p = 0.21\, Ar^{0.213} \tag{9.6.28}$$

Maskaev and Baskakov (1973) proposed the following correlation:

$$Nu_p = 0.21\, Ar^{0.32} \tag{9.6.29}$$

Grewal and Saxena (1981) proposed the following correlation:

$$Nu_p = 2.0(1-\varepsilon)Ar^{0.21}(0.0127/D_t)^{0.21}(C_s/C_g)^{45.5Ar^{-0.7}} \tag{9.6.30}$$

A formula was given for calculating the void fraction ε. It includes particle shape factor.

Grewal (1982) modified Eq. 9.30 to the following:

$$Nu_p = 2.0(1-\varepsilon)Ar^{0.21}(0.0127/D_t)^{0.21}(C_s/C_g)^{0.2} \tag{9.6.31}$$

Chen and Pie (1985) correlated data from several sources by the following relations:

For Ar $20 < Ar < 2\times10^4$,

$$Nu_p = 0.074Ar^{0.2}\mathrm{Pr}_g^{1/3}\left(\frac{\rho_s C_s}{\rho_g C_g}\right)^{1/3} \tag{9.6.32}$$

For $2\times10^4 < Ar < 10^7$,

$$Nu_p = 0.013Ar^{0.37}\mathrm{Pr}_g^{1/3}\left(\frac{\rho_s C_s}{\rho_g C_g}\right)^{1/3} \tag{9.6.33}$$

Shah (1983) gave the following correlation:

$$\text{For } Re_{\text{opt}} < 170,\ \frac{h_{\max}}{h_g} = 315B\left(\frac{C_s}{C_g}\right)^{0.18} Re_{\text{opt}}^{-0.647} \tag{9.6.34}$$

$$\text{For } Re_{\text{opt}} > 170,\ \frac{h_{\max}}{h_g} = 19.3BRe_{\text{opt}}^{-0.11} \tag{9.6.35}$$

where h_g is the heat transfer coefficient of gas flowing alone calculated by

$$h_g = 0.027Re_t^{0.805}\mathrm{Pr}_g^{0.33}k_g/D_t \tag{9.6.36}$$

$B = 1.24$ for spherical particles and $B = 1$ for non-spherical particles. Using the measured optimum velocities, excellent agreement was found with a very wide range of data that included tube diameters from 0.13 to 220 mm, particle diameters from 60 to 15 000 μm, several gases, and a wide range of pressures and temperatures. Using the measured optimum velocity, data from 33 studies were predicted with mean absolute deviation (MAD) of only 17%.

For optimum Reynolds number Re_{opt}, the Todes (1965) equation below was found by Shah (1983) to be in good agreement with data for horizontal cylinders and spheres in air-fluidized beds:

$$Re_{\text{opt}} = \frac{Ar}{18 + 5.22Ar^{0.5}} \tag{9.6.37}$$

Kim and Kim (2013) developed the following correlation based on their own as well as data from other sources:

$$Nu_p = 2.044Ar^{0.12}\mathrm{Pr}_g^{0.33}\left(\frac{u_{\text{mf}}^2}{D_p g}\right)^{0.12} \tag{9.6.38}$$

The data correlated included copper, glass, and sand with pressure 1–11.2 bar and D_p 115–475 μm.

Shah (2018a) Correlation Shah (2018a) noted that accurate prediction of optimum velocity is difficult, especially as it is not well defined in some cases as, for example, is seen in Figure 9.4.1 for the largest particles. Therefore, he developed a new correlation that does not involve it. This correlation is given as follows:

$$Nu_p = F_t F_k \left(13.1 - 13.8\exp(-0.103Ar^{0.202})\right)(C_s/C_g)^{0.18} \tag{9.6.39}$$

$$Nu_p = 0.053F_t F_k Ar^{0.402} \tag{9.6.40}$$

The predicted Nu_p is the larger of that given by the above equations. The transition between them occurs at $Ar = 5\times10^5$. All properties are calculated at the bed temperature.

F_t corrects for the effect of small tube diameters and is given by the following equation:

$$F_t = 3.35 - 2.35\left[1 - \exp\left(-1604D_t\right)\right] \tag{9.6.41}$$

D_t in Eq. (9.6.41) is in meters. According to Eq. (9.6.41), $F_t = 1$ for $D_t > 3$ mm. Figure 9.6.1 shows F_t graphically.

F_k takes into account low particle conduction heat transfer of particles with very low thermal conductivity. It is given by

$$F_k = 0.51 + 0.49\left[1 - \exp\left(-29.7k_p^{\ 5}\right)\right] \tag{9.6.42}$$

The particle thermal conductivity k_p is in W m^{-1} K^{-1}. F_k is shown in Figure 9.6.2.

This correlation was verified with the range of data in Table 9.6.2. These include spheres and vertical and

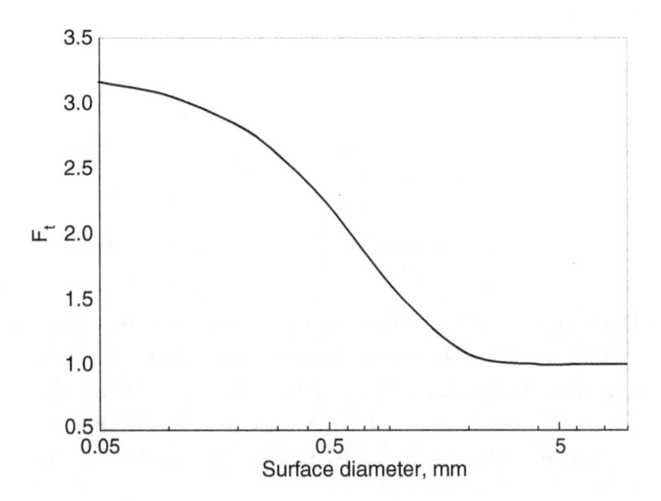

Figure 9.6.1 The factor F_t for the effect of heat transfer surface in the Shah correlation. Source: Shah (2018b). © 2018, ASME.

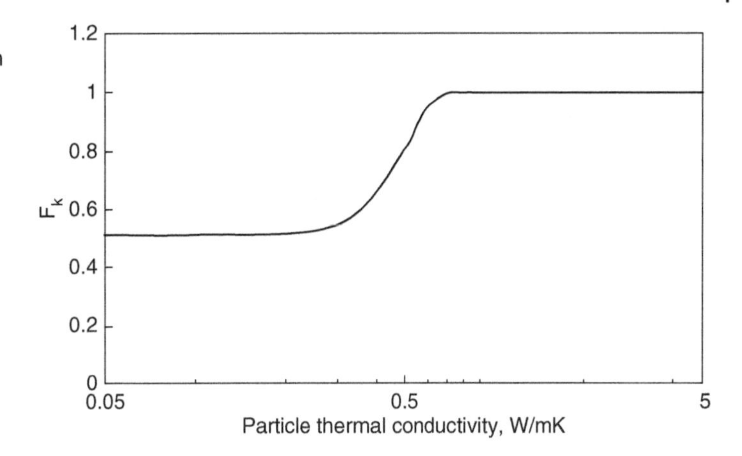

Figure 9.6.2 The factor F_k for the effect of particle thermal conductivity in the Shah correlation. Source: Shah (2018b). © 2018, ASME.

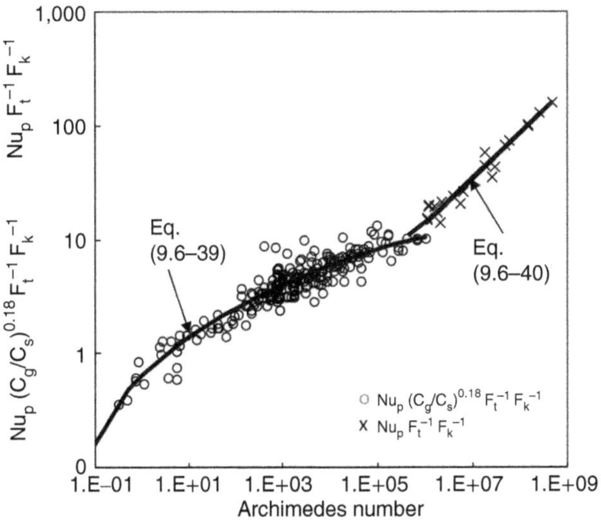

Figure 9.6.3 Comparison of the Shah correlation with test data over the entire range of Archimedes number. Source: Reprinted from Shah (2018a). © 2018, with permission from Elsevier.

Table 9.6.2 Range of parameters in data for maximum heat transfer with which the Shah correlation was verified.

Type of heat transfer surfaces	Horizontal and vertical cylinders (tubes and wires), spheres
Particle materials	Glass, quartz, various types of sands, zircon, shale, alumina, SiC, limestone, dolomite, corundum, carborundum, graphite, silica gel, mulite, FCC, polystyrene, styropor, polymers, sponge iron, steel, copper, lead, solder, bronze, and nickel
Particle density ($kg\,m^{-3}$)	26–11 340
Particle specific heat ($J\,kg^{-1}{}^{\circ}C^{-1}$)	129–2100
Thermal conductivity of particles ($W\,m^{-1}\,K^{-1}$)	0.035–440
Gases	Air, carbon dioxide, helium, hydrogen, nitrogen, methane, R-12, ammonia, and three mixtures of hydrogen and nitrogen
Particle diameter (μm)	31–1500
Diameter of heat transfer surface (mm)	0.05–220
Pressure (MPa)	0.03–0.95
Temperature (°C)	13–1028
Archimedes number	0.3–4.6*E*8

Source: Reprinted from Shah (2018a). © 2018, with permission from Elsevier.

horizontal cylinders (tubes and wires), surface diameters from 0.05 to 220 mm, particle diameters from 31 to 15 000 μm, many gases, and extreme ranges of bed pressure and temperature. The 363 data points from 53 studies are predicted with MAD of 16.2%. Figure 9.6.3 shows the comparison of this correlation with data over the entire range of Archimedes number.

Shah also compared the entire database with many of the other correlations listed previously. The results are shown in Table 9.6.3. It is seen that all have much larger deviations than the Shah correlation. Among these, the correlation that gives the least deviation is that of Grewal with MAD of 24.1%.

The trends for the effect of particle diameter shown by the Shah, Zabrodsky, and Grewal correlations are shown in Figure 9.6.4. Shah correlation predicts reduction in heat transfer coefficient with increasing particle diameter up to a minimum and then an increase with further increase in particle diameter. The other correlations predict continuous decrease in heat transfer with increasing particle diameter. That the trend predicted by the Shah correlation is correct is seen in Figure 9.6.5, which shows its agreement

Table 9.6.3 Deviations of various correlations for maximum heat transfer with test data.

			Deviation (%) Mean absolute/average						
Type	Ar	Number of data points	Zabrodsky (1958)	Zabrodsky et al. (1976)	Grewal and Saxena (1981)	Grewal (1982)	Maskaev and Baskakov (1973)	Chen and Pie (1985)	Shah (2018a)
Cylinder, horizontal	All	176	36.0	29.8	38.6	24.5	53.7	41.0	17.7
			−19.0	−13.9	−7.2	6.3	−50.4	−21.6	2.5
Cylinder, vertical	All	134	19.7	25.8	36.2	24.6	39.1	31.2	15.6
			−0.8	7.8	−15.5	−8.2	−33.8	6.0	−0.1
Sphere	All	53	20.2	22.3	6.5*E*4	20.7	49.0	21.9	12.8
			8.8	8.4	6.5*E*4	−11.1	−48.9	0.9	−6.3
All	<5*E*5	338	27.6	27.1	1.01*E*4	23.5	49.5	31.7	16.4
			−7.7	−2.2	1.01*E*4	−0.7	−48.1	−14.2	0.2
All	>5*E*5	25	29.2	29.7	30.6	31.6	23.0	74.0	13.5
			−14.7	−8.9	−5.4	−13.2	11.5	73.5	0.9
All	All	363	27.7	27.2	9599	24.1	47.6	34.6	16.2
			−8.2	−2.6	9551	−2.2	−44.0	−8.1	0.3

Source: Reprinted from Shah (2018a). © 2018, with permission from Elsevier.

with data for particles from 100 to 1500 μm. The mechanisms causing this trend may be explained as follows: fluidizing velocity increases with increasing particle size and hence convective heat transfer contribution increases with increasing particle size. According to the packet theory of Baskakov et al. (1973), the contact time and the thermal resistance between the packet and surface increase with increasing particle diameter. These result in reduction in contribution due to conduction with increasing particle diameter. Due to these opposing trends of convective and conductive components, initially total heat transfer decreases with increasing particle size, becomes near constant as the two components become comparable, and then increases with increasing particle size because convection contribution far exceeds conduction contribution.

According to the Shah correlation, diameter of surface has no effect if it is greater than 3 mm. For D_t less than 3 mm, heat transfer coefficient increases with decreasing diameter of surface. This was verified with data from four sources. An example is seen in Figure 9.6.6. The Shah correlation is in excellent agreement with data, while the Zabrodsky and Grewal correlations show large deviations. The theoretical model of Martin does not include the diameter of surface as a parameter.

The data analyzed by Shah included several data sets for high temperatures. They were analyzed without any correction for radiation heat transfer. An example is seen in Figure 9.6.7, which shows comparison with the data of Li et al. (1993) in which temperatures were up to 1028 °C. Good agreement is seen with the Shah correlation, while that of Zabrodsky et al. (1976) overpredicted the data.

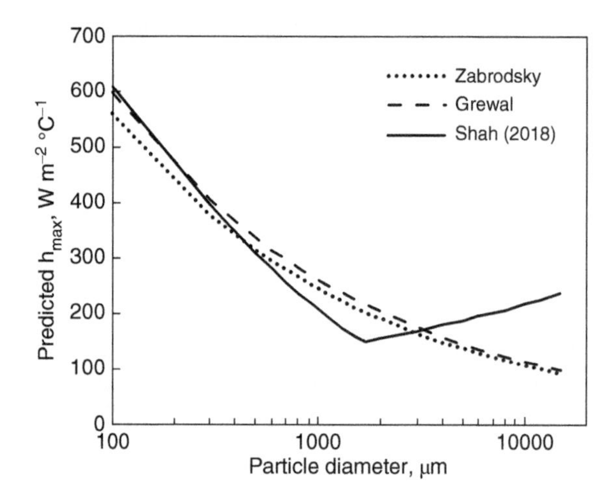

Figure 9.6.4 Effect of particle diameter on maximum heat transfer coefficient according to various correlations for silica sand fluidized by air at room conditions. $D_t \geq 3$ mm. Source: Reprinted from Shah (2018a). © 2018, with permission from Elsevier.

9.6.2.2 Correlations for the Entire Range

Correlations applicable over the velocity range u_{mf} to u_{tr} are considered in this section.

Grewal and Saxena (1980) developed the following correlation for horizontal tubes in fluidized beds:

$$\frac{h_w D_t}{k_g} = 47\left(1 - \varepsilon_{bed}\right)\left(\frac{GD_t\rho_s}{\rho_f\mu_g}\frac{\mu_g^2}{gD_p^3\rho_s^2}\right)^{0.325}\left(\frac{\rho_s C_s D_t^{3/2} g^{1/2}}{k_g}\right)^{0.23}\mathrm{Pr}_g^{0.3} \quad (9.6.43)$$

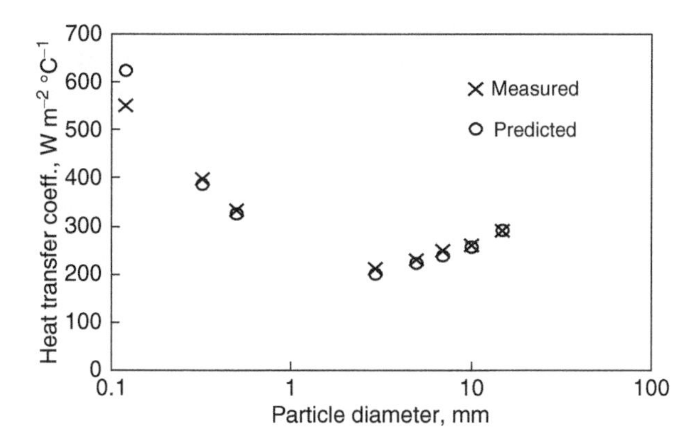

Figure 9.6.5 Comparison of the data of Baskakov et al. (1973) for corundum particles in air-fluidized beds with the Shah correlation. Source: Reprinted from Shah (2018a). © 2018, with permission from Elsevier.

If the particle shape factor is known, bed void fraction is calculated by the following equation:

$$\varepsilon = \frac{1}{2.1}\left[0.4 + \left\{4\left(\frac{\mu_g G}{D_p^2 \rho_g (\rho_s - \rho_g) g \varphi_s^2}\right)^{0.43}\right\}^{1/3}\right] \tag{9.6.44}$$

If particle shape factor is not known, void fraction is calculated by the correlation of Goroshko et al. (1958), Eq. 9.25. All properties are calculated at the mean of surface and bed temperature.

This correlation was compared to data from many sources that included minerals and metals, particle diameter 163–1450 μm, tube diameters 12.7–50.8 mm, and bed temperatures up to 1100 °C. All data were predicted within ±25%. They have not mentioned any adjustments for radiation heat transfer.

Grewal and Saxena (1980) also compared the same database to seven other correlations, but none of them was found satisfactory. Those included correlations by Vreedenberg (1958), Genetti et al. (1971), Ternovskaya and Korenberg (1971), and Petrie et al. (1968).

Mathur and Saxena (1986) gave the following correlations for large particles, $Ar > 130\,000$:

$$Nu_p = 5.95(1-\varepsilon)^{2/3} + 0.055 Ar^{0.3} Re_p^{0.2} \mathrm{Pr}_g^{1/3} \tag{9.6.45}$$

The data correlated included particle diameters 0.62–4 mm, pressure 1–81 bar, and $\mathrm{Ar} \times 10^{-6} = 0.2$–231. Both horizontal and vertical tubes were included. Prediction accuracy was ±35%. The same data were compared to several other correlations. All gave large deviations with most of the data.

Saxena (1989) reviewed the correlations for vertical tubes in fluidized bed of small particles with $Ar < 130\,000$. He concluded that none of them was satisfactory. The best among them is the following correlation by Wender and Cooper (1958):

$$\frac{h_w D_t}{(1-\varepsilon) k_g}\left(\frac{k_g}{C_{pg}\rho_g}\right)^{0.43} = 0.033 C_R Re_p^{0.23}\left(\frac{C_s}{C_{pg}}\right)^{0.8}\left(\frac{\rho_s}{\rho_g}\right)^{0.66} \tag{9.6.46}$$

C_R is a function of $r^* = (r/R_{bed})$ where r is the distance of tube from the bed axis and R_{bed} is the radius of the bed. It is given in the form of a graph based on the test data of Vreedenberg (1952). The following equation has been fitted to that graph by the present author:

$$C_R = \frac{1 + 10.88r^*}{1 + 3.64r^* + 3.9(r^*)^2} \tag{9.6.47}$$

Note that $k_g/(C_{pg}\rho_g)$ is dimensional. The following units apply: k_g in Btu h^{-1} ft^{-1} F^{-1}), C_{pg} in Btu lb^{-1} F^{-1}), and ρ_g in lb ft^{-3}.

The data on which the Wender–Cooper correlation was based included particle diameter 40–880 μm, tube diameter 7.3–19.3 mm, and particle density 795–2870 kg m^{-3}. Fluidizing gases included air, R-12, ammonia, helium, argon, and methane. Vreedenberg (1960) stated that when he tried this correlation with data other than those used in its development, large deviations were found. Grewal and Gupta (1989) found this correlation to be in good agreement with their data for a 26.2 mm tube with air-fluidized 0.237–0.896 mm diameter particles of silica sand.

There have been many experimental studies on finned tubes of various types, and correlations for researchers' own data have been given. These have been reviewed by Saxena (1989), among others. No generally applicable correlation is available.

9.6.3 Recommendations

The following prediction methods are recommended in their verified range for heat transfer to plain surfaces:

1) For maximum heat transfer to horizontal and vertical tubes as well as spheres, Shah (2018a) correlation is recommended.
2) For heat transfer to horizontal tubes at all gas velocities with $Ar < 130\,000$, Grewal and Saxena (1981) correlation is recommended.
3) For heat transfer to vertical tubes at all gas velocities, Wender and Cooper (1958) correlation in its verified range is recommended.
4) For $Ar > 130\,000$ at all gas velocities, Mathur and Saxena (1986) correlation for both horizontal and vertical tubes is recommended.

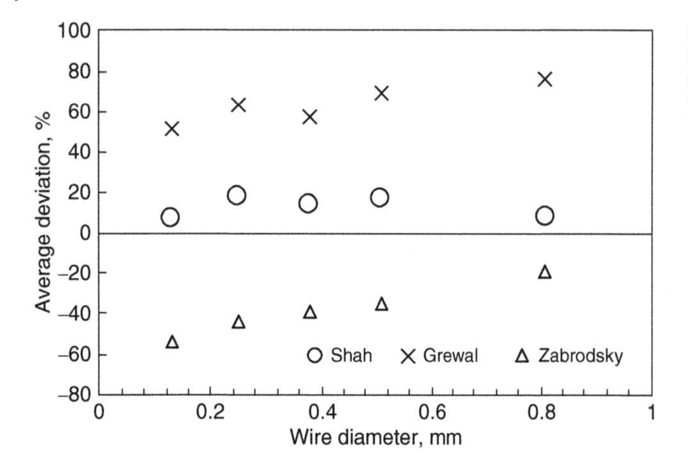

Figure 9.6.6 Predictions of some correlations for maximum heat transfer to wires in fluidized beds. Data of Turton (1986) for polythene particles. Source: Reprinted from Shah (2018a). © 2018, with permission from Elsevier.

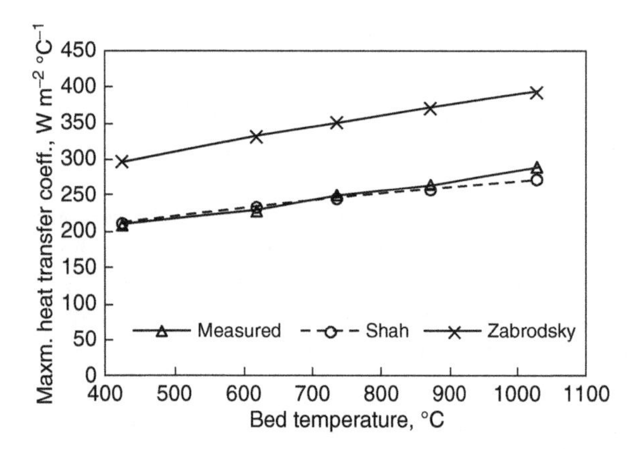

Figure 9.6.7 Comparison of the Shah (2018a) and Zabrodsky et al. (1976) correlations with the data of Li et al. (1993). Silica sand, $D_p = 1815\,\mu m$, $D_t = 40\,mm$, and air at room pressure. Source: Reprinted from Shah (2018a). © 2018, with permission from Elsevier.

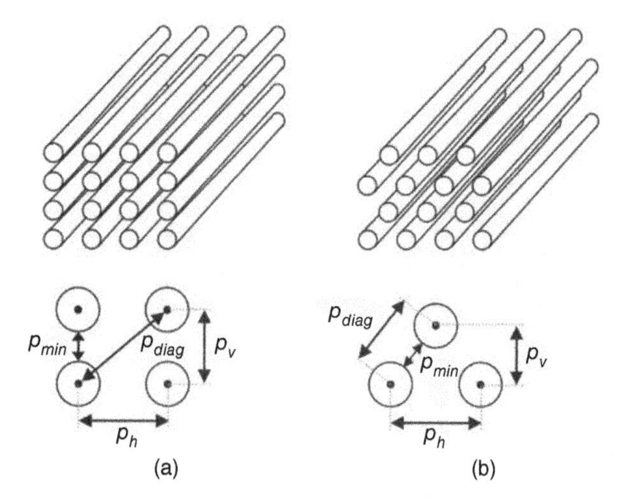

Figure 9.7.1 Common types of horizontal tube bundles: (a) in-line bundle and (b) staggered bundle. Source: Lechner et al. (2014). © 2014, Elsevier.

Outside the verified range of the previously mentioned correlations, the analytical model of Martin is recommended.

9.7 Tube Bundles

The heat exchangers used in industry are usually bundles of horizontal or vertical tubes. Heat transfer coefficients for them can be different from those for single tubes.

9.7.1 Horizontal Tube Bundles

The tubes in horizontal tube bundles are arranged either in-line or staggered as shown in Figure 9.7.1.

Gelperin et al. (1968a, b) performed tests on in-line and staggered bundles of 20 mm diameter tubes with P_h/D_t from 2 to 9 and P_v/D_t 1–8 (in-line bundles) and 0–10 (staggered bundles). Quartz sand particles of 160–350 μm diameter were used.

For the in-line bundles, they found that vertical pitch has no effect except that there is a slight deterioration when the tubes almost touch each other. Decrease in horizontal pitch decreased maximum heat transfer coefficient by 25% as P_h/D_t decreased from 6 to 2. In staggered bundles, both horizontal and vertical pitches affected heat transfer. Maximum heat transfer coefficient decreased with decreasing horizontal and vertical pitches.

Gelperin et al. (1968a) gave the following correlation based on their data for in-line bundle described above for P_h/D_t 2–9:

$$Nu_{p,\max} = 0.79Ar^{0.22}\left[1 - \left(D_t/P_h\right)^{0.25}\right] \tag{9.7.1}$$

For staggered bundles of 20 mm diameter tubes, Gelperin et al. (1969) gave the following correlation:

$$Nu_{p,\max} = 0.74Ar^{0.22}\left[1 - \left(D_t/P_h\right)\left(1 + \left(\frac{D_t}{P_v + D_t}\right)\right)\right]^{0.25} \tag{9.7.2}$$

This is applicable to P_h/D_t 2–9 and P_v/D_t 0–10.

Saxena (1979) measured heat transfer coefficients in tube banks of 12.7-mm outside diameters in fluidized beds of silica sands with average diameters of 167 and 504 pm. He found that the heat transfer coefficient from a single tube to the fluidized bed was not influenced by adding two or four more rows as long as the horizontal pitch $\geq 4D_t$, and vertical pitch $\geq 2D_t$ for staggered tube bundles, and the pitches were $\geq 4D_t$, for in-line tube bundles.

Grewal and Saxena (1983) performed tests on horizontal tube bundles with tubes located at the vertices of equilateral triangle. P_h varied from 1.75 to 9 times the tube diameter. Particles of silica sand $D_p = (167$ and $504\,\mu\text{m})$ and alumina $(D_p = 259\,\mu\text{m})$ were used, fluidized by air. At $P_h/D_t \geq 5.5$, heat transfer coefficients were the same as for a single tube. At $P_h/D_t = 2.25$, heat transfer coefficients were lower than for single tubes. Figure 9.7.2 shows some of their data. They developed the following correlation based on the data from their own tests:

$$\frac{h_{\text{max,bund}}}{h_{\text{max,st}}} = 1 - 0.21\left(\frac{P}{D_t}\right)^{-1.75} \tag{9.7.3}$$

$h_{\text{max,st}}$ is the maximum heat transfer coefficient of the single tube, calculated by the Grewal correlation, Eq. 9.31, and $h_{\text{max,bund}}$ is the maximum heat transfer coefficient of the bundle. For in-line bundles, $P = P_h$. For staggered bundles, $P = P_{\text{diag}}$. The range of applicability is stated to be $1.75 \leq P/D_t \leq 9$; $75 \leq Ar \leq 20\,000$. They also compared this correlation to data from 11 other sources. These included in-line and staggered bundles. Those data were also compared to six other prediction methods. Radiation contribution was calculated from the data of Baskakov et al. (1973). All correlation gave large deviations with data from one source. Data from all other sources were well predicted by the correlations of Grewal and Saxena and Gelperin et al. given previously. The Martin (1984a) model also gave good agreement. This is interesting as the Martin model was verified only with data for single tubes and had no factor for bundle geometry. This indicates that bundle geometry has only a small effect over a wide range of horizontal and vertical pitches. This is also shown by Figure 9.7.2.

Lechner et al. (2014) performed tests in which measurements were done on a single tube located in the middle of in-line and staggered bundles of tubes. Tubes of 12, 22, and 35 mm diameter were used. P_h/D_t was 1.54–4.0 and P_v/D_t was 1–3.45. Tests were done with lignite particles of mean diameter of 73 μm. Bundle maximum heat transfer coefficients were as low as half of those for single tubes. They

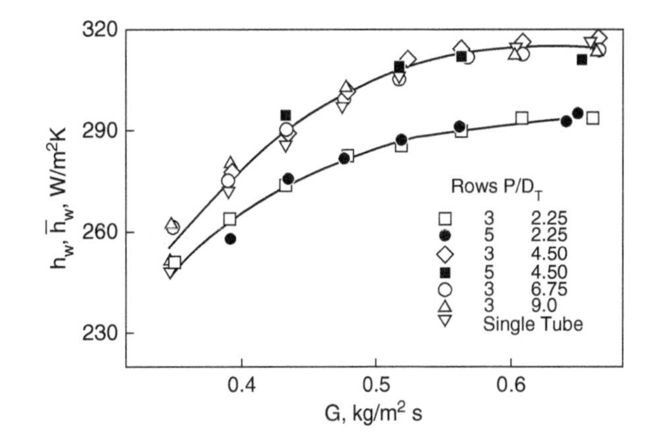

Figure 9.7.2 Effect of P/D_t on heat transfer coefficients in a horizontal bundles of 12.7 mm diameter tubes arranged in equilateral triangle grid. Silica sand particles $D_p = 504\,\mu\text{m}$. Source: Grewal and Saxena (1983). © 1983, American Chemical Society.

correlated their data by the following equation:

$$\frac{h_{\text{max,bund}}}{h_{\text{max,st}}} = \left(1 - \frac{D_t}{P_h}\right)^{0.36}\left(1 - \frac{D_t}{P_{\text{diag}}}\right)^{0.24} \left(1 - \frac{D_p}{P_{\text{min}}}\right)^{4}\left(\frac{D_t}{D_{22\text{mm}}}\right)^{0.09} \tag{9.7.4}$$

This equation closely represented their data, while the correlations of Gelperin et al. and Grewal and Saxena given earlier overpredicted the data.

Lechner et al. also performed tests with 3.5 mm diameter plastic beads. Bundle heat transfer coefficients were about 10–15% lower than those for a single tube.

Lechner et al. noted that for small particles (Geldart type A), heat transfer occurs almost entirely by particle convection. For large particles (Geldart type D), heat transfer occurs mostly by gas convection. Therefore, they recommend that Eq. 9.4 be applied only to the particle convection term in the Martin model or other similar models.

Grewal (1981) did tests on a staggered tube bundle with equilateral triangular arrangement. Tubes of 12.7 and 28.6 mm were used with pitch 1.75–9 times the tube diameter. The test data were correlated by the following equation:

$$\frac{h_{\text{bund}}}{h_{\text{st}}} = 1 - 0.21\left(\frac{P}{D_t}\right)^{-1.75} \tag{9.7.5}$$

The heat transfer coefficient for single tube h_{st} is obtained from the correlation of Grewal and Saxena (1980), Eq. (9.6.43). They also compared this correlation with data from 11 other sources that included in-line and staggered bundles with P/D_t 1.17–8.5 and bed temperature up to 1100 K. Almost all data were predicted within ±25% by Eq. (9.7.5). Note that for in-line bundles, $P = P_h$, while for staggered bundles, it is the minimum distance between centers of adjacent tubes.

Grewal and Menart (1987) tested a staggered tube bundle in a bed of silica sand with average diameter ranging from 0.544 to 2.335 mm. Bed temperature was from 587 to 1205 K. Superficial gas velocity was 0.73–2.58 m s^{-1}. Their test data was satisfactorily correlated by Eq. (9.7.5) after deducting the contribution by radiation. Their estimated contribution by radiation did not exceed 13% of the total heat transfer.

Goshayeshi et al. (1986) experimented with bundles of tubes arranged in equilateral triangles with $P/D_t = 3$, $D_t = 50.8$ mm and particles of 2.14 and 3.33 mm. Heat transfer coefficients were about the same as on a single tube.

Shah (1983) compared his correlation for maximum heat transfer to single tubes with data for staggered and in-line bundles from five sources. He concluded that correlations for single tubes can be used for bundles with $P/D_t > 3$.

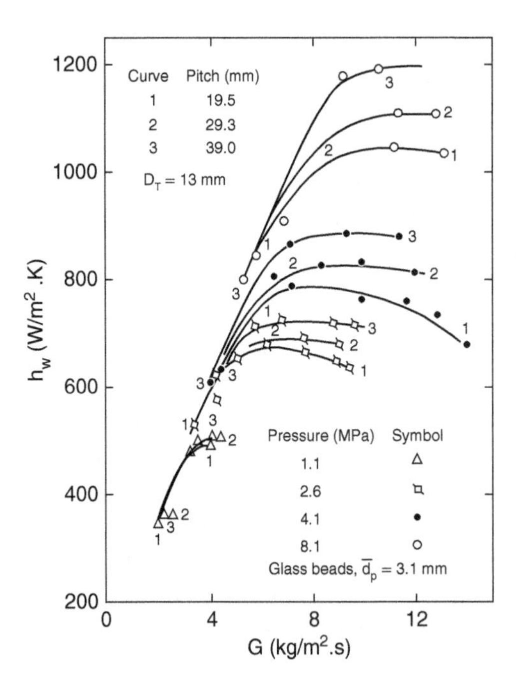

Figure 9.7.3 Heat transfer coefficients measured on vertical tube bundles in fluidized beds. Source: Borodulya et al. (1983). © 1983, Elsevier.

9.7.2 Vertical Tube Bundles

There has been much less research on vertical tube bundles than on horizontal bundles.

Gelperin and Einstein (1971) have described their tests on bundles of vertical tubes arranged in triangular pattern. Tube diameter was 20–40 mm with P/D_t 1.25–5. Quartz sand particles of various sizes were used. When P/D_t decreased from 5 to 2, maximum heat transfer coefficient decreased up to 7%. When P/D_t decreased to 1.25, maximum heat transfer coefficient fell by 15–20%. These results are represented by the following equation of Gelperin et al. (1963):

$$Nu_{p,\max} = 0.64\,Ar^{0.22}\left(P/D_t\right)^{0.09} \tag{9.7.6}$$

They state that on the rising branch of u–h curve, heat transfer coefficients in closely packed bundles could fall by 35–50%, especially near the minimum fluidization velocity.

Borodulya et al. (1983) measured heat transfer to vertical tube bundles in fluidized beds of glass beads ($D_p = 1.25$ and 3.10 mm) and sands ($D_p = 0.794$ and 1.225 mm) at pressures of 1.1–8.1 MPa at ambient temperature. Tubes of 13 mm diameter were arranged in equilateral triangles with pitch of 19.5–39 mm. Figure 9.7.3 shows some of their data. It is seen that heat transfer coefficient decreases with decreasing pitch at higher velocities. The decrease is greater at higher pressures. However, the decrease is limited to about 15%. They compared their maximum heat transfer data with the correlation of Maskaev and Baskakov for single tubes, Eq. (9.6.29) with satisfactory agreement. Considering all velocities, the data were in fairly good agreement with the semi-theoretical correlation of Ganzha et al. (1982) for large particles. This correlation has no factor for bundle geometry. It was verified with data for single horizontal tubes and tube bundles. These results suggest that there is no difference in heat transfer between horizontal and vertical orientations, and that bundle effects are negligible for the tube pitch range in these data ($P/D_t = 1.5$–3).

Noe and Knudsen (1968) studied heat transfer to a vertical tube in a bundle of vertical tubes. Glass particles of 0.145 and 0.477 mm were used with superficial air mass velocity 4.8–240 kg m^{-2} s^{-1}. The bundle average heat transfer coefficients were compared to the Wender and Cooper correlation, Eq. (9.6.46). Using $C_R = 2$, most of the data were correlated within ±50%.

9.7.3 Recommendations

1) If P/D_t is above a certain limit, maximum heat transfer coefficient of bundles can be calculated using methods for single tubes. For vertical bundles, this limit is about 1.5. For horizontal staggered bundles, $P_h/D_t \geq 4$, and $P_v/D_t \geq 2$. For in-line tube bundles, $P_v/D_t > 1.1$, $P_h/D_t \geq 4$.
2) For maximum heat transfer, the recommended correlations are as follows: horizontal bundles, Eq. (9.7.3) by Grewal and Saxena (1983) and vertical bundles, Eq. (9.7.6) of Gelperin et al. (1963).

3) For heat transfer coefficients at all velocities in horizontal bundles, the recommended correlation is Eq. (9.7.5) by Grewal (1981).
4) For heat transfer coefficients at all velocities in vertical bundles with Geldart D particles, use prediction methods for single horizontal tubes if $P/D_t \geq 1.5$.

9.8 Radiation Heat Transfer

9.8.1 Radiation Heat Transfer Coefficient and Effective Emissivity

The radiation heat flux between bed and heat transfer surface may be expressed by the following equation:

$$q_{\text{rad}} = \sigma E_{\text{eff}} \left(T_{\text{bed}}^4 - T_w^4 \right) \tag{9.8.1}$$

where E_{eff} is the effective emissivity between bed and surface. It includes the effects of emissivities of bed and surface as well as view factor. Further, the particles near the bed may be cooler than the bulk of the bed due to particle convection; E_{eff} has to also include this effect unless T_{bed} is replaced by the actual temperature of the particles facing the surface.

The radiation heat transfer coefficient is defined as

$$h_{\text{rad}} = \frac{q_{\text{rad}}}{T_{\text{bed}} - T_w} = \frac{\sigma E_{\text{eff}}(T_{\text{bed}}^4 - T_{w)}^4}{T_{\text{bed}} - T_w} \tag{9.8.2}$$

The bed may be regarded to be an impervious cylinder surrounding the cooling tube. Effective emissivity is then given by the following relation:

$$\frac{1}{E_{\text{eff}}} = \frac{1}{E_w} + \left(A_w / A_{\text{bed}} \right) \left(\frac{1}{E_{\text{bed}}} - 1 \right) \tag{9.8.3}$$

As the air gap between wall and particles is very small, $A_w/A_{\text{bed}} \approx 1$. The relation for E_{eff} then becomes.

$$\frac{1}{E_{\text{eff}}} = \frac{1}{E_w} + \frac{1}{E_{\text{bed}}} - 1 \tag{9.8.4}$$

Using algebraic relations, Eq. 9.2 may be written as

$$h_{\text{rad}} = \sigma E_{\text{eff}} \left(T_{\text{bed}}^2 - T_w^2 \right) \left(\text{T}_{\text{bed}} + T_w \right) \tag{9.8.5}$$

In the range of temperatures used in fluidized beds, this is approximated by

$$h_{\text{rad}} = 4\sigma E_{\text{eff}} T_m^3 \tag{9.8.6}$$

T_m is the mean of bed and surface temperature. This is the form used by Martin (1984b).

Zabrodsky (1966) recommends $E_{\text{eff}} = E_w$. Grace (1982b) examined the experimental data of Botterill (1975) that included several minerals, particle diameter 0.25–1.5 mm, a range of velocities, and bed temperature from 500 to 1200 °C. He recommends.

$$E_{\text{eff}} = 0.5 \left(1 + E_p \right) \tag{9.8.7}$$

Baskakov (1985) has given the following relation applicable when $0.3 < E_p < 0.6$:

$$h_{\text{rad}} = 7.3 E_p E_w T_w^3 \tag{9.8.8}$$

This equation gave good agreement with a wide range of data for tubes, spheres, and small vertical plates. However, it overpredicted data for tubes 90–220 mm diameter. Yamada et al. (1991) applied this formula to their data for alumina particles of diameter 183–353 μm with $E_p = 0.8$ and $E_w = 0.8$. Good agreement was found.

9.8.2 Temperature for Significant Radiation Contribution

Many experimental and theoretical studies on radiation heat transfer in fluidized beds have been reported. All studies conclude that radiation contribution is negligible at lower temperatures and it becomes increasingly significant at higher temperatures. However, there is considerable disagreement about the temperature above which it becomes significant.

Yoshida et al. (1974) performed experiments on a vertical tube in a bed of catalyst particles of 149–210 μm diameter fluidized by nitrogen. They found that effect of radiation heat transfer was negligible at bed temperatures up to 1000 °C. They developed a theoretical model that agreed closely with their test data. Their theoretical model showed contribution of radiation to be 6% at a bed temperature of 1500 °C, with tube emissivity of 0.8. Szekeley and Fisher (1968) had reached the same conclusion from their tests and theoretical analyses.

Kharchenko and Makhorin (1964) performed tests on a 60 mm diameter sphere in beds of quartz sand and chamotte particles of diameter 0.34–1.66 mm. Bed temperature was up to 1050 °C. They argued that radiation heat transfer was insignificant as the measured maximum heat transfer coefficients varied linearly with wall to bed temperature difference. Shah (2018a) reports these data to be in close agreement with his own data as well as that of Zabrodsky (1958) without any adjustment for radiation contribution. He also reports good agreement with other data for temperature up to 1028 °C without any adjustment for radiation heat transfer.

As noted in Section 9.6.2.2, Grewal and Saxena (1980) report good agreement of their general correlation with data for temperatures up to 1100 °C without any accounting for radiation contribution.

Based on their study of experimental data from many sources, Gelperin and Einstein (1971) concluded that for temperatures up to 1000 °C, radiation heat transfer is less than 5% of the total heat transfer. At temperature of

1500 °C, radiation contribution may be 25–30% of total heat transfer.

Botterill et al. (1984) performed tests on a tube immersed in beds of alumina and sand of diameter 0.37–3 mm. Bed temperature was up to 980 °C. They estimate that the contribution of radiation did not exceed 10%.

While the previously quoted studies indicate negligible effect of radiation up to temperatures of 1000–1100 °C, there are many studies indicating significant contribution of radiation at lower temperatures. One of them is the experimental study of Baskakov et al. (1973) whose results are shown in Figure 9.8.1. It is seen that they found significant contributions of radiation at wall temperatures above 500 °C and radiation heat transfer increased with increasing particle diameter.

The packet theory model of Vedamurthy and Sastri (1974) shows that the contribution of radiation to total heat transfer increases with increasing gas velocity. At higher velocities, it becomes equal to or larger than the conduction heat transfer. The contribution of radiation heat transfer was found to be significant at temperatures of 800 °C and higher.

From their tests on a single 35 mm diameter horizontal tube, Renzhang et al. (1987) estimated that the contribution of radiation to total heat transfer was 3%, 24%, and 27% at bed temperatures of 600, 950, and 1000 °C, for particles of 0.802 mm diameter. The contribution was 17% for particles of 0.452 mm at a bed temperature of 950 °C.

Blaszczuk et al. (2017) analyzed test data from a large circulating fluidized bed furnace. For bed temperatures of 764–936 °C, they estimated contribution of radiation to be up to 60% of total. Particle diameter was 0.219–0.411 mm.

9.8.3 Conclusions and Recommendations

There are considerable differences regarding the temperature at which radiation heat transfer becomes significant. Many of them indicate negligible contribution of radiation at temperatures up to 1100 °C, while there are others that show significant effect of radiation starts at a temperature of about 600 °C.

When calculations of heat transfer by conduction and convection are done by the maximum heat transfer correlation of Shah (2018a) or the correlation of Grewal and Saxena (1980) for a range of velocities, consider radiation heat transfer to be negligible for temperatures up to about 1100 °C. These have been validated with data close to this temperature.

When using the Martin model (1984a), calculate h_{rad} by Eq. (9.8.6) with E_{eff} by Eq. (9.8.7).

9.9 Heat Transfer to Bed Walls

9.9.1 Prediction Methods

Toomey and Johnstone (1953) gave the following correlation for heat transfer to bed wall:

$$\frac{h_{bw}D_p}{k_g} = 3.75\left(\frac{D_p u_{mf}\rho_g}{\mu_g}\log\frac{u}{u_{mf}}\right)^{0.47} \tag{9.9.1}$$

This was based on their own test data in a bed of 115 mm diameter as well as those of Mickley and Trilling (1949). The solids were glass beads of diameter 55–848 μm. At very low velocities, heat transfer coefficients were low due to channeling. Excluding those data, average deviation of data from this equation was ±5%.

Wender and Cooper (1958) analyzed data from several sources for heat transfer between fluidized bed and its external walls. They developed the graphical correlation shown in Figure 9.9.1. The abscissa is particle Reynolds number Re_p and the ordinate y that is defined as

$$y = \frac{h_{bw}D_p/\left[k_g(1-\varepsilon)\left(\rho_s C_s/C_{pg}\rho_g\right)\right]}{1+7.5e^{-0.44}\left(L_H/D_{bed}\right)\left(C_{pg}/C_s\right)} \times 10^4 \tag{9.9.2}$$

L_H is the heated length of bed and D_{bed} is the diameter of bed. The present author has fitted the following equation to the curve in Figure 9.9.1:

$$y = \frac{-0.253 + 40.2Re_p}{4.05 + Re_p} \tag{9.9.3}$$

As seen in Figure 9.9.1, it predicts quite close to the Wender–Cooper curve. This correlation was developed using data from five sources. The MAD of the 425 data points was 22.1%. As seen in the figure in their paper, some data have large deviations. The data analyzed included particle diameter 49–847 μm, bed diameters 25.3–120 mm and heated length 43–1015 mm. Particles were of glass, sand, calcium carbonate, aluminum, and iron catalysts. Gases were air, helium, and carbon dioxide.

Gunn and Hilal (1994) studied heat transfer to walls of fluidized beds of diameter 90 and 290 mm. The 290 mm bed consisted of five sections 120 mm high, each heated by a steam jacket. The bed of 90 mm diameter was also jacketed and was provided as two bed depths of 95 and 192 mm. Particles were glass, daikon, and nickel of diameter 100–1000 μm. Figure 9.9.2 shows some of their test data. They developed the following correlation of their data:

$$\frac{h_{bw}D_p}{k_g} = 0.005\,15Ar^{0.184}\left(\frac{\rho_s C_s}{\rho_g C_{pg}}\right)^{0.6}\left(\frac{D_{bed}}{D_{ref}}\right)^{-0.12} \tag{9.9.4}$$

D_{ref} = 90 mm. This equation is valid for $Ar = 10^2$–10^5. All data were predicted with MAD of 12.8%.

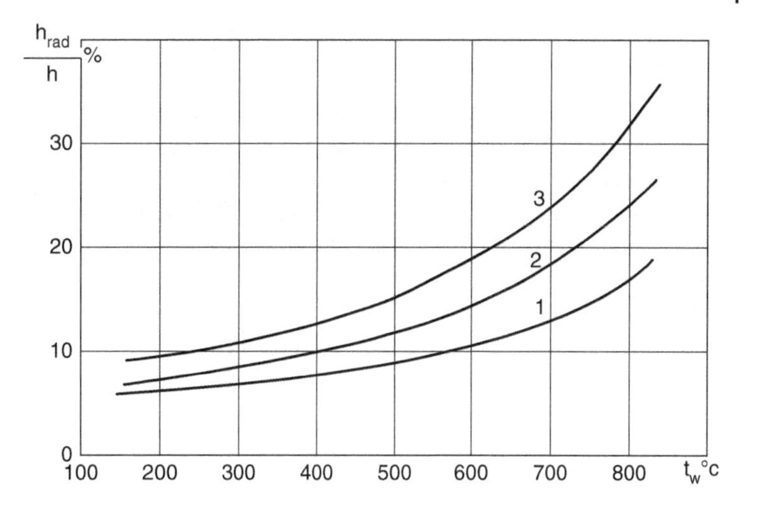

Figure 9.8.1 Fraction of total heat transfer provided by radiation as a function of wall temperature t_w. Chamotte particles in a bed at 850 °C Particle diameters 1–0.35 mm, 2–0.63 mm, and 3–1.25 mm. Source: Baskakov et al. (1973). © 1973 Elsevier.

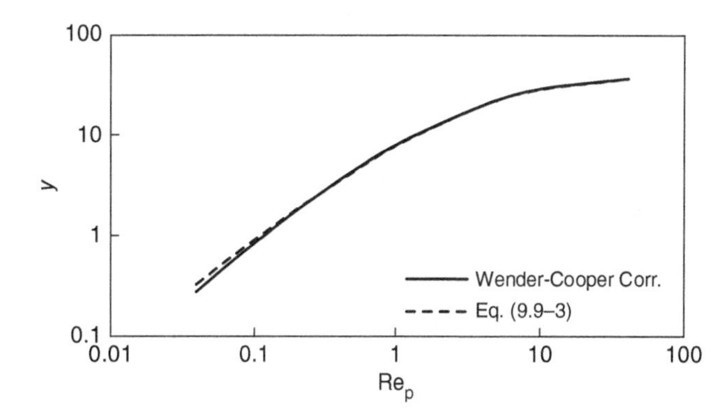

Figure 9.9.1 The graphical correlation of Wender and Cooper (1958) for heat transfer to external walls of fluidized beds together with the equation fitted to it.

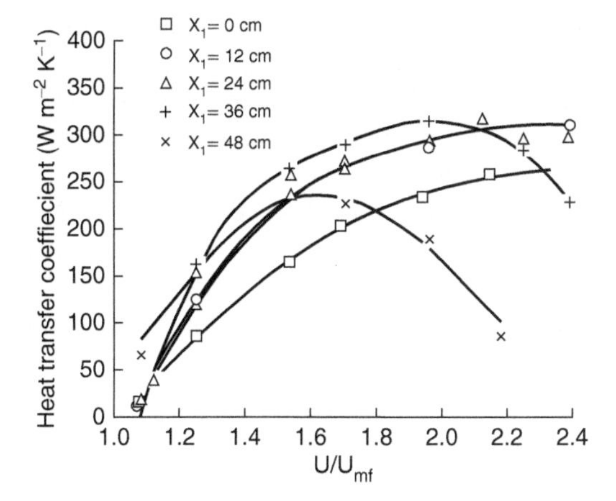

Figure 9.9.2 Variation of heat transfer with distance X_1 from the distributor. Bed diameter 290 mm, 500 μm glass ballotini fluidized by air. Source: Gunn and Hilal (1994). © 1994, Elsevier.

The previously mentioned correlations are for conventional fluidized beds operating at velocities below the transport velocity. Heat transfer in circulating fluidized beds has been reviewed, among others, by Basu and Nag (1996) and Abdelmotalib et al. (2015).

9.9.2 Conclusions and Recommendations

There is no well-verified general prediction method. Various correlations should be used only in their verified range.

9.10 Heat Transfer in Freeboard Region

The freeboard is the region from the top of the expanded bed to the top of the apparatus. In bubbling beds, particles are ejected from the bed surface due to bubbles erupting from the top of bed. In the freeboard, movement of particles is determined by a balance of drag, inertia, and gravity forces. The heavier particles slow down with distance and eventually fall back into the bed. The smaller particles continue to move upwards. These have to be separated and reintroduced into the bed. The height above which the rate of decrease in carryover reaches a minimum is called transport disengagement height (TDH). Some authors consider freeboard to be the region above the TDH and call the region below TDH as the splash zone. As the gases and solids in the freeboard contain considerable heat, heat exchangers are installed in the freeboard to recover it. Hence prediction of heat transfer in this region is of considerable interest.

9.10.1 Experimental Studies and Prediction Methods

Biyikli et al. (1987) measured heat transfer to a horizontal 38 mm diameter tube located in freeboard region of air-fluidized beds. Bed temperature was 300–750 °C and pressure was atmospheric. Silica sand and limestone particles with 465–1400 μm mean diameter were used. Figure 9.10.1 shows some of their data. It is seen that heat transfer coefficient increases with superficial gas velocity and decreases with distance from expanded bed surface. The data of Byam et al. (1981) are seen to be in agreement with their data. They correlated their data by the following relation:

$$h_n = \left[1 + 4.1 \times 10^{-9}\left\{\frac{HD_p(\rho_s - \rho_g)^2 g}{(u - u_{\mathrm{mf}})\,\rho_g \mu_g}\right\}^{2.58}\right]^{-1} \tag{9.10.1}$$

H is the distance above the expanded bed surface, and h_n is the normalized particle convection heat transfer coefficient defined as

$$h_n = \frac{h_{\mathrm{fb}} - (h_{\mathrm{rad}} + h_{\mathrm{gc}})}{h_{\mathrm{ib}} - (h_{\mathrm{rad}} + h_{\mathrm{gc}})} \tag{9.10.2}$$

where h_{fb} is the heat transfer coefficient in the freeboard and h_{ib} is the heat transfer coefficient when immersed in the bed. h_{ib} was obtained from their own tests. h_{rad} was calculated using their own experimental data and h_{gc} was calculated by the following correlation of Knudsen and Katz (1958):

$$h_g = 0.68 Re_t^{0.47} \mathrm{Pr}_g^{0.33} k_g / D_t \tag{9.10.3}$$

All data were predicted with average deviation of 29%.

Biyikli et al. (1989) developed a phenomenological model for heat transfer to tubes located in freeboard. The correlation assumes that tubes in freeboard are intermittently splashed with dense emulsions from bursting bubbles and are in contact with lean phase in between these bursts. It is shown to agree with some test data, but it involves factors such as contact time of lean and dense phase that are difficult to predict.

Dyrness et al. (1992) tested six horizontal tube bundles placed in the freeboard region of bubbling beds of iron grit, 0.2 and 0.25 mm diameter. The bundles were of 12.7 mm tubes and included in-line and staggered types with horizontal pitch 19–38.1 mm and vertical pitch 13.7–38.1 mm. Heat transfer coefficients were found to decrease with distance from the expanded bed surface; Figure 9.10.2 is an example. They developed the following formula to correlate

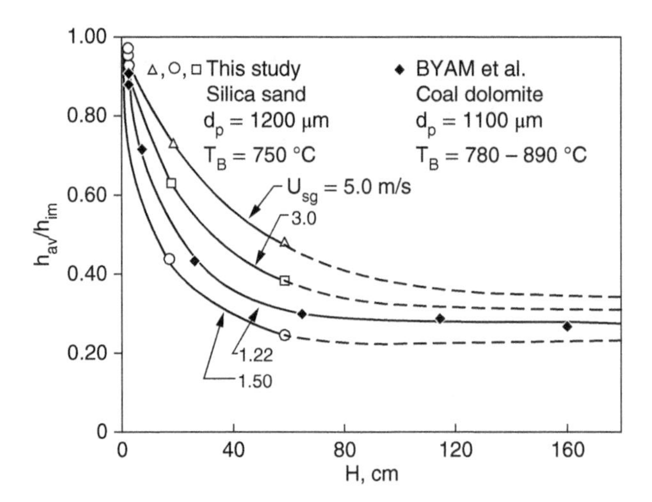

Figure 9.10.1 Effect of superficial gas velocity U_{sg} and height H above bed surface on ratio of heat transfer coefficients in freeboard (h_{avg}) and inside the bed (h_{im}). Source: Biyikli et al. (1987). © 1987, Elsevier.

their data:

$$h_{\mathrm{fb}} = h_{\mathrm{ib}} \exp\left[-\frac{1.5 H u_{\mathrm{mf}}}{P_h (u - u_{\mathrm{mf}})}\right] \tag{9.10.4}$$

Xavier and Davidson (1981) studied heat transfer to a cylinder and a rod in the freeboard region well above fluidized beds of sand particles of diameter 1.34 and 0.885 mm. Heat transfer coefficients were found to be the same as for gas flowing alone. They concluded that there was no particle carryover during their tests.

Farag and Tsai (1992) have reported tests on plates located in the freeboard region. The plates were placed in a variety of orientations and radial locations.

Some numerical/computational fluid dynamics (CFD) models have been reported, but they were validated with very limited range of data. Hence their general applicability is unknown. An example is the model of Dounit et al. (2008).

9.10.2 Recommendation

There is no adequately verified general predictive technique. Available methods should be used only in their verified range.

9.11 Heat Transfer Between Gas and Particles

A well-known correlation for single particles is the following by Ranz and Marshall (1952):

$$Nu_p = 2 + 0.6 Re_p^{1/2} \mathrm{Pr}_g^{1/3} \tag{9.11.1}$$

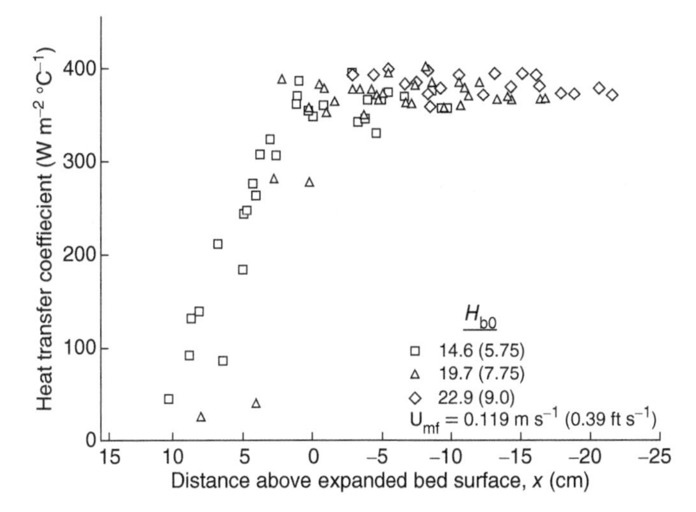

Figure 9.10.2 Heat transfer coefficients of a tube bundle as a function of distance from expanded bed surface. Staggered bundle with 1.9 cm horizontal and vertical pitch. Iron grit particles 0.2 mm diameter. H_{bo} is the bed height at minimum fluidization, cm (inch). Source: Dyrness et al. (1992). © 1992, Elsevier.

Whitaker (1972) has given the following correlation for single spheres:

$$Nu_p = 2 + \left(0.4Re_p^{1/2} + 0.06Re_p^{2/3}\right) \mathrm{Pr}_g^{0.4}\left(\mu_g/\mu_{g,w}\right)^{1/4} \tag{9.11.2}$$

The data included Re_p from 4 to 10^5.

Many correlations have been proposed for gas to particle heat transfer in fluidized beds. These have been reviewed, among others, by Zabrodsky (1966), Gelperin and Einstein (1971), and Xavier and Davidson (1985). There are very large differences in the predictions of various correlations. Among the reasons for the differences are difficulties in temperature measurements, different definitions of driving temperature difference, and differences in the heat transfer models assumed in interpreting data.

As noted by Grace (1982a, b), the temperature is so uniform in most fluidized beds due to rapid solid mixing and other reasons that gas to particle heat transfer can be neglected beyond 1–2 cm above the distributor grid. There are exceptions such as beds with exothermic reactions and beds with large particles and closely spaced tubes that severely impede circulation.

9.12 Gas–Solid Flow in Pipes

Heat transfer to flowing gas–solid mixtures in pipes occurs in many applications including chemical processing, pneumatic transport, and nuclear reactors. Use in concentrating solar power plants is being investigated. Many experimental studies have been done, and many prediction methods, theoretical and empirical, have been proposed. Before going into them, the regimes of gas–solid flow are discussed.

9.12.1 Regimes of Gas–Solid Flow

There are two basic regimes of gas–solid flow, namely, dense phase and dilute phase. If gas flow rate is decreased at a constant solid rate, pressure drop initially decreases. With further decrease in gas flow rate, a point is reached where pressure drop increases sharply with further decrease in gas flow rate. The gas velocity at this point is called the critical or choking velocity. Systems operating at velocities higher than the critical velocity are called dilute phase or lean phase, and those at velocities lower than the critical velocity are called dense phase.

Many formulas for prediction of choking velocity have been proposed. Yang (1975, 1983) gave the following semi-theoretical correlation for upflow in vertical pipes:

$$\frac{2gD_t\left(\varepsilon_{ch}^{-4.7} - 1\right)}{\left(u_{ch} - u_T\right)^2} = 6.8 \times 10^5 \tag{9.12.1}$$

where u_{ch} is the gas velocity at choking, u_T is the single particle terminal velocity, and ε_{ch} is the void fraction at the choking point. By assuming that the slip velocity equals u_T, mass balance yields

$$G_s = \left(u_{ch} - u_T\right)\rho_s\left(1 - \varepsilon_{ch}\right) \tag{9.12.2}$$

G_s is the mass velocity of solid particles. For the desired solids transport rate G_s, u_{ch} and ε_{ch} can be calculated by simultaneous solution of the above two equations. This correlation was verified with data covering the following range: D_p 40–3400 μm, u_T 0.12–22.9 m s^{-1}, D_t 2.54–7.5 cm, and ρ_s 910–7700 kg m^{-3}.

The terminal velocity u_T is the constant velocity reached by a particle falling in a viscous medium. The particle is accelerated by gravity and retarded by viscous drag. These opposing forces equalize at the terminal velocity. Thus,

$$\frac{1}{2}C_d\left(\frac{\pi}{4}D_p^2\rho_g\right)u_T^2 = \frac{\pi}{6}D_p^3\left(\rho_s - \rho_g\right)g \tag{9.12.3}$$

C_d is the drag coefficient. Rearrangement of Eq. (9.12.3) yields the following equation:

$$\frac{u_T \rho_g D_p}{\mu_g} = \left(\frac{4Ar}{3C_d}\right) \tag{9.12.4}$$

Many expressions for C_d have been proposed. Kunii and Levenspiel (1969) give the following relations for spherical particles:

$$Re_p < 0.4, C_d = 24/Re_p \tag{9.12.5}$$

$$0.4 < Re_p < 500, C_d = 10/Re_p^{0.5} \tag{9.12.6}$$

$$500 < Re_p < 2x10^5, C_d = 0.43 \tag{9.12.7}$$

Drag coefficients for non-spherical particles are larger than for spherical particles, and therefore their terminal velocities are lower. Kunii and Levenspiel (1969) have given a graphical correlation that includes the effect of particle shape.

Various flow regimes that occur in vertical dilute and dense phase flow are shown in Figure 9.12.1. These regimes and their prediction have been discussed by Rabinovich and Kalman (2011).

9.12.2 Experimental Studies of Heat Transfer

Numerous experimental studies have been done on heat transfer to flowing gas–solid mixtures. These were most recently reviewed by Shah (2020) and are discussed in the following.

Farbar and Morley (1957) studied heat transfer to air–catalyst mixtures flowing upwards in a vertical heated pipe of 17.8 mm diameter. Average particle diameter was 50 μm. They found that addition of solid catalyst particles to air always increased heat transfer coefficient though the increase was less at higher Reynolds numbers. Danziger (1963) studied upflow of 50 μm catalyst particles with air in cooled vertical pipes of 17.5–38.1 mm diameter. Solids loading ratio was up to 446. He found similar trends with solids loading and Reynolds number as Farbar and Morley. Schluderberg et al. (1961) experimented with graphite particles of 1–5 μm in air and various gases flowing up in heated vertical pipes of 8–22.2 mm diameter. Solids loading ratio was up to 90. They found that heat transfer coefficient increased with increasing solids loading as well as increasing ratio of specific heats of particles to those of gases. They found no effect of Reynolds number.

While the previously mentioned studies showed that addition of solids always increased heat transfer coefficient, many others showed different trends. Boothroyd and Haque (1970a, b) studied mixtures of air with 15 μm zinc particles in heated pipes of diameter 25.4, 50.8, and 76.2 mm. Flow direction was vertically upwards. Solids loading ratio was up to 12. In all cases, addition of solids initially lowered heat transfer coefficient followed by rise with further increase in solids loading. Further, ψ decreased with decreasing pipe diameter, where $\psi = (h_{TP}/h_g)$ is the ratio of mixture heat transfer coefficient to that of gas flowing alone. This behavior was attributed to modification of turbulence in gas flow caused by addition of solids. Other studies showing a minimum in ψ with addition of solids include Brotz et al. (1958), Wahi (1977), Sukomel et al. (1967), Hasegawa et al. (1983), Jepson et al. (1963), Depew and Cramer (1970), and Bowen (1969).

Sukomel et al. (1967) performed tests with vertical upflow as well as with horizontal flow. They found no effect of orientation. Tube diameter was 8.1 mm. Solids were aluminum oxide and graphite, particle diameters from 65 to 290 μm. Gases were air, helium, and nitrogen.

Tien and Quan (1962) studied the effect of 30 and 200 μm glass and lead particles on air flow in a heated 18 mm diameter vertical tube. Solids loading ratio was up to 3.3. Figure 9.12.2 shows their data. They found the effect of solids loading to be similar to that reported by Boothroyd and Haque discussed earlier. While the heat transfer coefficients for the 200 μm and lead particles were comparable, those for 30 μm lead particles were much lower than those for 30 μm glass particles. These data are shown in Figure 9.12.1. As the thermal conductivity of lead is much higher than that of glass, it indicates that thermal conductivity becomes a factor for small particles.

Most studies, such as those by Sukomel et al. (1967), Boothroyd and Haque (1970a), and Depew and Cramer (1970) show that entrance lengths in suspensions are much longer than for gas alone. The entrance lengths in these studies were up to 90 diameters, while for gas alone, entrance length does not exceed 30 diameters. Variations of entrance length are illustrated in Figure 9.12.3. Boothroyd and Haque (1970a) attribute the greater entrance length of suspensions to the higher thermal capacity of the mixture, and delay in transfer of heat from gas to particles.

9.12.3 Prediction of Heat Transfer

Many methods for prediction of heat transfer have been proposed. These were most recently reviewed by Shah (2020) as described herein.

9.12.3.1 Various Methods

Tien (1961) developed a theoretical solution using many assumptions. It predicts that entrance length is longer than for gas alone, local heat transfer coefficient of suspension in the entrance length is higher than that of gas alone, while the asymptotic heat transfer coefficient of suspension is

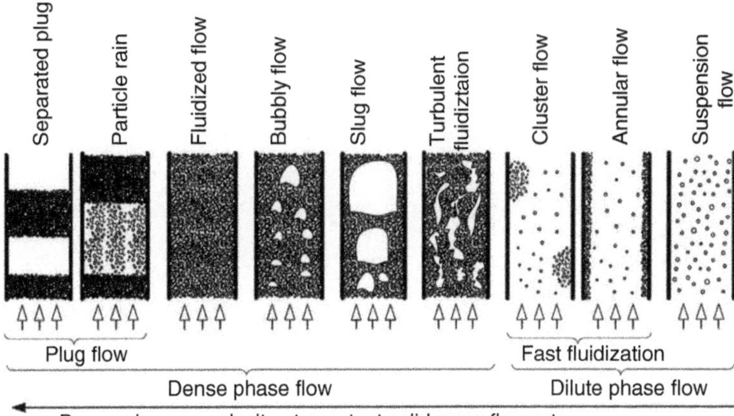

Figure 9.12.1 Flow regimes during gas–solid upward flow in vertical pipes. Source: Rabinovich and Kalman (2011). © Elsevier.

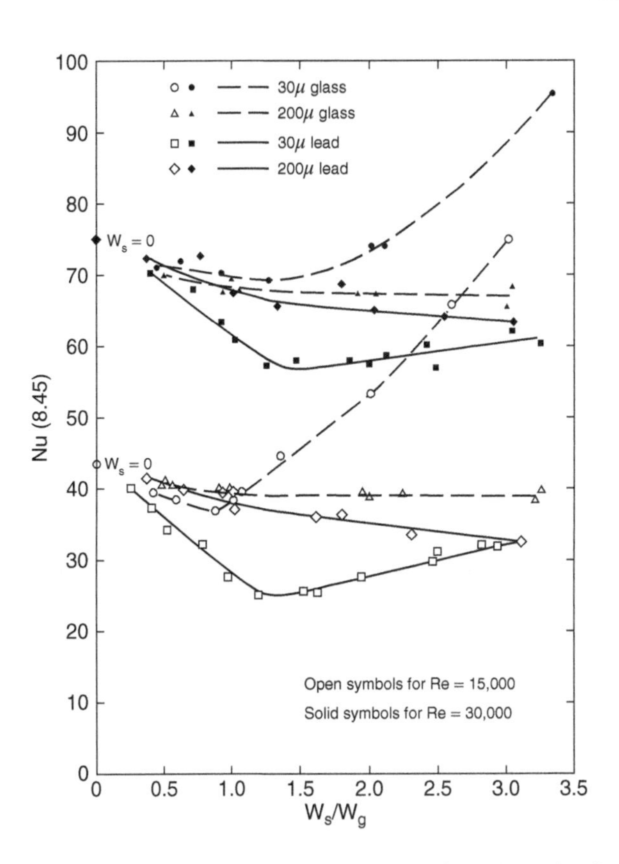

Figure 9.12.2 Effect of solid-to-gas ratio on Nusselt number for glass and lead particles flowing with air in a vertical tube at $x/D = 8.45$. Source: Tien and Quan (1962). © ASME.

equal to that of gas alone. Tien and Quan (1962) found it to disagree with their data for glass and lead suspensions. Indeed, it contradicts most other studies qualitatively and quantitatively as seen in the discussion in Section 9.12.2. Tien and Quan attributed the discrepancies to incorrect assumptions in the Tien analysis.

Many other theoretical models have been presented for gas–solid flows. Examples are Michaelides (1986), Han et al. (1991), Louge and Yusof (1993), Li and Mason (2000), Bertoli (2000), and El-Behery et al. (2011). Each of these was shown to agree with a few data points from one or two studies. None of the theoretical solutions have been demonstrated to agree with a wide range of data. Hence none of them can be considered to be generally applicable. Furthermore, they are difficult to use.

Numerous empirical correlations have been proposed. Almost all of them are based on authors' own data. These are given in the following text.

Farbar and Morley (1957) gave the following correlation based on their data for silica alumina catalyst in air:

$$Nu_{\mathrm{TP}} = 0.14 Re_t^{0.6} \Gamma^{0.45} \tag{9.12.8}$$

where $\Gamma = (W_s/W_g)$ and $Nu_{\mathrm{TP}} = h_{\mathrm{TP}} D_t / k_g$.

Danziger (1963) gave the following correlation based on his data for 60 μm particles of silica alumina catalyst flowing through banks of vertical tubes:

$$Nu_{\mathrm{TP}} = 0.0784 Re_t^{0.66} \Gamma^{0.45} \tag{9.12.9}$$

Schluderberg et al. (1961) analyzed their data for heating of graphite particles of 1–5 μm diameter fluidized by several gases and came up with the following correlation:

$$Nu_{\mathrm{TP}} = 0.02 Re_t^{0.8} (1 + \eta\Gamma)^{0.45} \tag{9.12.10}$$

where $\eta = (C_s/C_{\mathrm{pg}})$. Pipe diameters were 8–22.2 mm. Gases used were carbon dioxide, helium, nitrogen, and carbon tetrafluoride. Pressure was 3–10 bar and temperature 32–593 °C. Solids loading ratio was 0–90. L/D was 100–389.

Gorbis and Bakhtiozin (1962) proposed the following correlation:

$$\psi = 1 + \left(6.3 Re_t^{-0.3} Re_p^{-0.33}\right) \eta\Gamma \tag{9.12.11}$$

where $\psi = h_{\mathrm{TP}}/h_{\mathrm{GS}}$. This formula gave reasonable agreement with their data for mixtures of air with graphite particles of 150–2080 μm heated in vertical pipes of diameter 12–33 mm. L/D was 50. Solids loading ratio was up to 32.

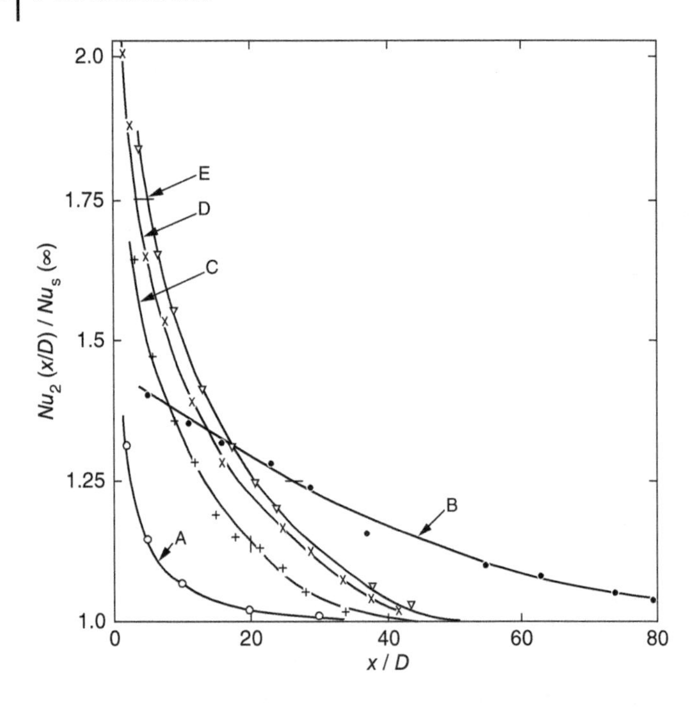

Figure 9.12.3 Effect of solid-to-gas ratio on entrance length. W_s/W_g for the curves are $A = 0$, $B = 6.9$ (graphite), $C = 24.2$ (graphite), $D = 3.35$ (glass), and $E = 1.17$ (glass). $Re_t = 10^4$ to 3×10^4. Source: Boothroyd and Haque (1970a). © SAGE.

Pfeffer et al. (1966) analyzed data from several sources and gave the following correlation:

$$\psi = 1 + \left(4Re_t^{-0.32}\right)\eta\Gamma \qquad (9.12.12)$$

They recommend this correlation in the following range: Re_t from 5000 to 100 000, $(\Gamma\eta)$ between 2 and 10, and particle diameters between 5 and 50 μm.

Sukomel et al. (1967) studied graphite and aluminum oxide particles flowing with air, helium, and nitrogen and gave the following correlation:

$$\psi = F_{\text{ent}}\left(0.55\eta^{0.1}\Gamma^{a}\right) \qquad (9.12.13)$$

where $F_{\text{ent}} = h_{\text{TP},x}/h_{\text{TP},\infty}$, $a = 0.33$ for particles of 65 μm and 0.25 for the larger particles. Their expression for entrance effect factor is

$$F_{\text{ent}} = 1 + be^{-c\left(\frac{x}{D_t}\right)} \qquad (9.12.14)$$

where $b = 0.7$ and $c = 0.045$ for $D_p = 65$ μm; $b = 0.35$, and $c = 0.04$ for larger particles.

9.12.3.2 Shah Correlation

Shah (2020) has given a general correlation that was shown to agree with data covering a very wide range. It is expressed by the following relation:

$$\psi = \frac{h_{\text{TP}}}{h_g} = F_{\text{ent}}\psi_\infty \qquad (9.12.15)$$

where ψ_∞ is the asymptotic value of ψ, i.e. beyond the thermal entrance length. It is given by the following equation:

$$\psi_\infty = 382\Gamma^{n}Re_t^{-0.67} \qquad (9.12.16)$$

If $Re_t < 10^4$, use $Re_t = 10^4$ in Eq. (9.12.16).

If ψ_∞ given by Eq. (9.12.16) < 0.8, $\psi_\infty = 0.8$

If $Re_t > 15\,000$ together with $(D_t/D_p) < 170$ and Eq. 9.16 gives $\psi_\infty > 1$, use $\psi_\infty = 1$.

The exponent n in Eq. (9.12.16) is given by the following equation:

$$n = 0.027\left(D_t/D_p\right)^{0.48} \qquad (9.12.17)$$

If n given by Eq. (9.12.17) < 0.25, use n = 0.25. If n given by Eq. (9.12.17) > 0.6, use $n = 0.6$.

Gas only heat transfer coefficient is calculated by the following equation, generally known as the Dittus–Boelter equation:

$$h_g = 0.023Re_t^{0.8}\text{Pr}_g^{0.4}k_g/D_t \qquad (9.12.18)$$

Properties of gas are calculated at the temperature of suspension.

The entrance effect factor F_{ent} is determined as follows:

$$F_{\text{ent}} = 1 + B/y^{0.7} \qquad (9.12.19)$$

$B = 0.3$ for local heat transfer coefficient and $B = 1$ for mean heat transfer coefficient for the length.

The factor y is calculated as follows; x is the distance from entrance:

$$\text{For } Re_t > 10^5, y = x/D_t \qquad (9.12.20)$$

$$\text{For } Re_t \leq 10^5, y = \left(\frac{x}{D_t}\right)\psi_\infty/(Z+1) \qquad (9.12.21)$$

$$\text{For } (\eta\Gamma) > 7, Z = 7 \qquad (9.12.22)$$

$$\text{For } (\eta\Gamma) \leq 7, Z = \eta\Gamma \qquad (9.12.23)$$

Table 9.12.1 The range of data for which the Shah (2020) correlation for heat transfer during gas–solid flow in pipes was verified.

	Range
Solids	Glass, coal, sand, clay, aluminum oxide, alumina silica catalyst, quartz sand, zinc, copper, lead, and graphite
Gases	Air, nitrogen, and helium
Particle diameter (μm)	13–1130
Tube diameter (mm)	5.1–77
Flow direction	Vertical up and down, horizontal
L/D_t, x/D_t	5–288
Re_t	60–130 000
C_s/C_{pg}	0.13–1.15
k_s ($W\,m^{-1}\,K^{-1}$)	0.12–385
W_s/W_g	0.02–520
Number of data sources	20

Source: Shah (2020)©2020, ASME.

The above correlation is not applicable if all the following three conditions are met:

$$D_t < 26\text{ mm}, k_s > 5\text{ W m}^{-2}\text{ K}^{-1}, D_p < 40\ \mu\text{m}$$

The reason for this exclusion is that data in this range were found to be much lower than this correlation. Note that the correlation is applicable if only one or two of the previously mentioned conditions are met.

The test data analyzed included temperatures up to 550 °C. These were satisfactorily predicted without considering radiation heat transfer. At higher temperatures, addition of radiation contribution should be considered. Methods for calculating radiation heat transfer are discussed in Section 9.8.

This correlation was compared to test data whose range is given in Table 9.12.1. As seen therein, the data include a wide variety of particles of 13–1130 μm in pipes of 5.1–77 mm diameter in all orientations and solids loading ratios from 0 to 520. Thus, both lean phase and dense phase regimes are included. The MAD of this correlation with the 630 data points from 20 sources was 18.9%.

Six published correlation were compared to the same data. None of them was satisfactory, their MAD ranging from 35.3% to 57%. Table 9.12.2 shows the deviations of all correlations with data in various ranges.

Figures 9.12.4–9.12.6 show the comparison of various correlations with a wide range of data. Figure 9.12.4 shows that the Shah correlation is able to correctly predict the effect of entrance length. Figures 9.12.5 and 9.12.6 show its agreement with both lean and dense phase data with solids to gas ratio from near zero to near 500. It should be noted, as seen in Table 9.12.1, that MAD of the Shah correlation for solids loading Γ up to 10 is 18% while that for higher loading is 24%. This suggests that its accuracy for dense phase flow is comparatively lower than for lean phase flow. However, this is not conclusive as the data for dense phase were much fewer and from fewer sources.

9.12.4 Recommendation

The Shah (2020) correlation is the only prediction method that has been verified over a wide range of data. It is therefore recommended. Care should be taken not to use it beyond the limit stated by Shah. Beyond that limit, guidance should be taken from test data.

9.13 Solar Collectors with Particle Suspensions

Solar power plants using concentrating collectors offer the possibility of high temperatures and hence high cycle efficiency. Until now, mostly molten salts have been used. However, they have an upper temperature limit of about 560 °C and a lower temperature limit of 140–230 °C because they solidify below these temperatures. This limits cycle efficiency and causes problems in operation and thermal storage. Gas-solid suspensions have no limits on operating temperatures and have large thermal capacity. They have therefore been under consideration as the heat transfer fluid in concentrating power plants since around 1980. Zhang et al. (2016) have reviewed the research in this field that goes as far back as 1980 though most of the work has been done in recent years. Most researchers used receivers of diameter less than 100 mm. The flowing suspensions in receiver were either lean with very low solid content or a real fluidized bed with solids volume fraction between 25% and 35%.

Zhang et al. (2016) performed tests on a solar collector that consisted of a 43.4 mm OD/36 mm ID vertical stainless steel tube 50 cm long placed at the focus of a solar furnace. Solar radiation was received through a 0.10 m × 0.50 m slot at the focus plane, with aperture angle 126°. An air-silicon carbide suspension flowed through the tube. The particle mean Sauter diameter was 63.9 μm. The superficial particle mass flux was 9.3–45.1 $kg\,m^{-2}\,s^{-1}$; the air superficial velocity was 0.032–0.135 $m\,s^{-1}$; and mean bed temperature was 402–808 K. The suspensions were in the bubbly regime. Measured mean heat transfer coefficients were 431–1116 $W\,m^{-2}\,K^{-1}$. They developed the following

Table 9.12.2 Results of comparison of test data for heat transfer to gas–solid flowing in pipes with various correlations.

				Deviation (%) Mean absolute/average							
Flow direction	W_s/W_g	Local or mean heat trans. coefficient	Number of data points	Shah (2018a)	Pfeffer et al. (1966)	Gorbis and Bakhtiazin (1962)	Yousfi et al. (1974)	Sukomel et al. (1967)	Schuldberg et al. (1961)	Farbar and Morley (1957)	Danziger (1963)
H	<10	*Both*	50	15.7	59.9	33.7	38.1	22.5	48.9	70.0	64.8
				9.4	59.9	33.5	37.1	−14.1	48.9	70.0	64.8
	>10	*Both*	11	13.7	153.8	39.2	61.2	7.1	109.5	203.1	198.7
				9.4	153.8	38.8	61.2	6.0	109.5	203.1	198.7
	All	*Both*	61	15.3	76.8	34.6	42.2	19.7	66.4	94.5	93.9
				9.4	76.8	34.6	41.4	−10.5	65.0	91.7	91.7
VD	All	*Both*	27	26.6	44.1	29.9	28.8	43.3	44.3	51.5	55.3
				−5.6	31.7	16.6	28.3	−40.1	16.1	24.9	18.8
VU	<10	*Both*	451	17.7	32.7	21.6	24.4	35.2	27.7	39.5	38.5
				−2.1	27.0	11.1	11.3	−33.5	17.3	19.8	18.9
VU	>10	*Both*	91	25.2	167.8	148.7	222.4	50.5	49.5	65.4	60.9
				−6.2	167.8	148.7	222.4	−49.5	12.5	54.4	48.0
VU	All	*Both*	542	19.0	55.4	43.0	57.7	37.7	31.3	43.8	42.3
				−2.8	49.3	32.3	41.0	−36.2	16.5	25.6	23.8
All	All	*Mean*	384	17.6	61.6	49.2	68.5	41.9	30.3	42.8	40.9
				−5.9	57.6	40.3	52.8	−40.9	14.9	23.2	20.4
All	All	*Local*	158	22.2	40.2	27.9	31.2	27.5	33.8	46.2	45.7
				4.8	29.1	12.8	12.1	−24.1	20.3	31.5	31.9
All	<10	*Both*	528	18.0	35.8	23.2	25.9	34.4	31.3	43.0	42.5
				−1.2	30.4	13.5	14.6	−32.0	20.8	24.5	23.5
All	>10	*Both*	102	24.0	166.3	136.9	205.0	45.8	56.0	80.2	75.8
				−4.5	158.9	126.4	174.4	−43.5	23.0	70.4	64.3
All	All	*Both*	630	18.9	57.0	41.6	54.9	36.2	35.3	49.1	47.8
				−1.7	51.2	31.8	40.5	−33.9	21.2	32.0	30.1

Source: Shah (2020). ©2020, ASME.

correlation to fit their data:

$$\frac{h_{TP}}{h_g} = 1 + 0.1\,(\Gamma\eta) \tag{9.13.1}$$

Zhang et al. (2017) reported further tests on the same facility using receivers made of 46 and 50 mm plain tubes as well as finned tubes of 50 mm diameter. Their measured heat transfer coefficients are shown in Figure 9.13.1. Some of the tests were done using silicon carbide of 64 μm diameter, and some were done using cristobalite of diameter 58 μm.

Flamant et al. (2013) have reported a study on a solar receiver using a suspension of silicon carbide particles. Heat transfer coefficients up to 500 kW/m^2 K were obtained.

Nomenclature

Ar	Archimedes number defined by Eq. (9.2.4) (−)
C_{pg}	specific heat of gas at constant pressure (J kg^{-1} K^{-1})
C	specific heat (J kg^{-1} K^{-1})
D	diameter (m)
E	emissivity (−)
F_{ent}	entrance effect factor (−)
G	superficial mass flux (kg m^{-2} s^{-1})
g	acceleration due to gravity (ms^{-2})
h	heat transfer coefficient (W m^{-2} K^{-1})

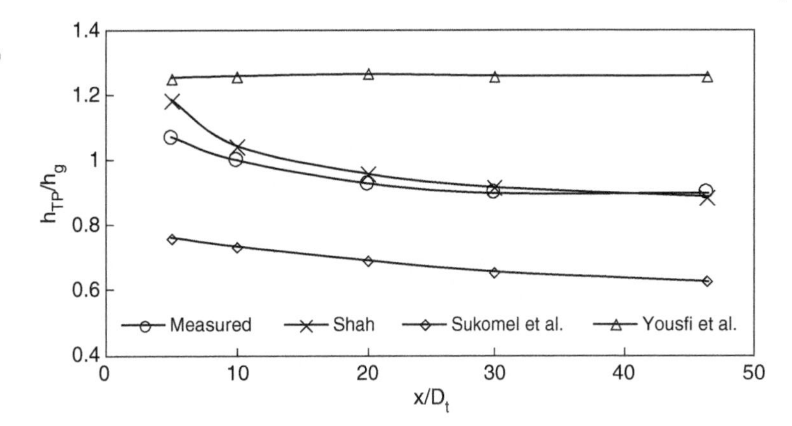

Figure 9.12.4 Data of Tien and Quan (1962) for 200 μm lead particles with air in a pipe compared to various correlations. Re_t = 15 000. Source: Shah (2020). © 2020, ASME.

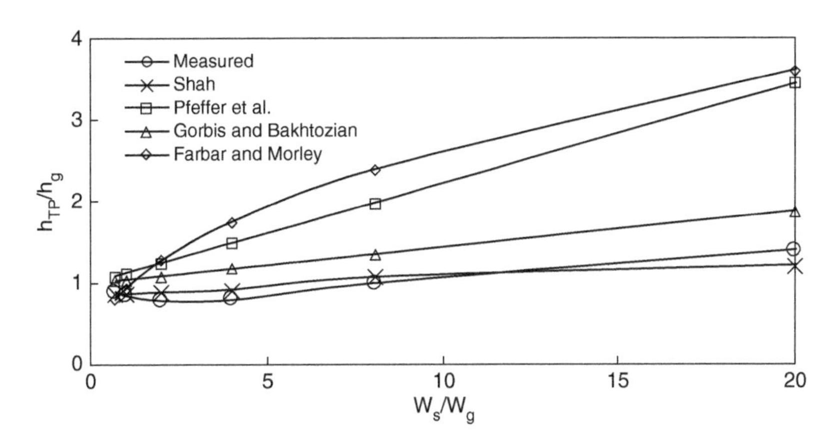

Figure 9.12.5 Data of Sukomel et al. (1967) for air–graphite mixture in horizontal pipe compared to various correlations. D_p = 65 μm. Source: Shah (2020). © 2020, ASME.

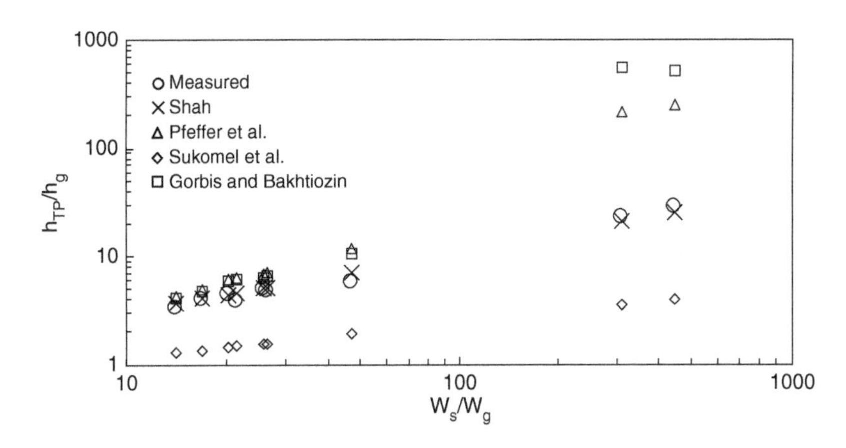

Figure 9.12.6 Data of Danziger (1963) for catalyst with air flowing up in a 28.5 mm diameter vertical tube compared to various correlations. Source: Shah (2020). © 2020, ASME.

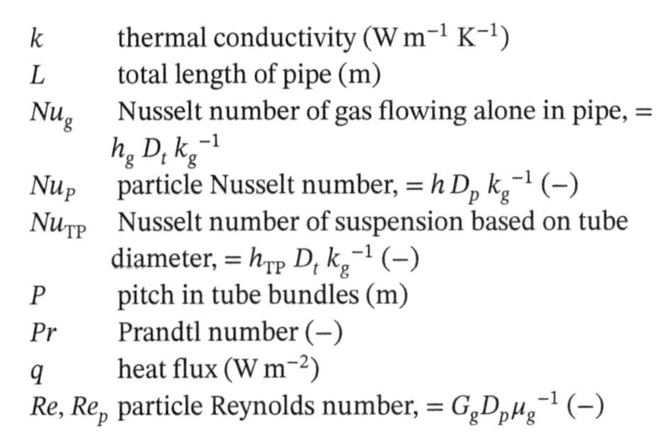

k	thermal conductivity (W m^{-1} K^{-1})
L	total length of pipe (m)
Nu_g	Nusselt number of gas flowing alone in pipe, = $h_g\, D_t\, k_g^{-1}$
Nu_P	particle Nusselt number, = $h\, D_p\, k_g^{-1}$ (–)
Nu_{TP}	Nusselt number of suspension based on tube diameter, = $h_{TP}\, D_t\, k_g^{-1}$ (–)
P	pitch in tube bundles (m)
Pr	Prandtl number (–)
q	heat flux (W m^{-2})
Re, Re_p	particle Reynolds number, = $G_g D_p \mu_g^{-1}$ (–)
Re_t	Reynolds number of gas flowing alone in pipe, = $G_g D_t \mu_g^{-1}$ (–)
T	temperature (°C or K)
u	superficial velocity assuming gas flowing alone (ms^{-1})
u_{ms}	minimum slugging velocity (ms^{-1})
u_{opt}	superficial gas velocity at which maximum heat transfer occurs (ms^{-1})
u_{tr}	transport or carryover velocity (ms^{-1})
W	mass flow rate (kg s^{-1})
x	distance from pipe entrance (m)

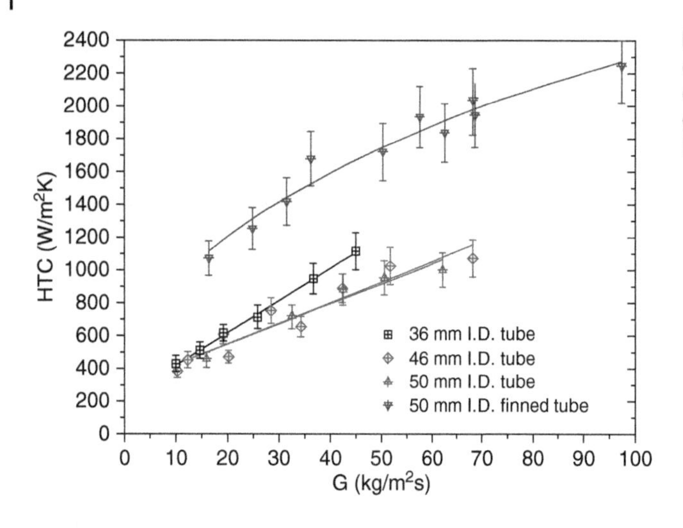

Figure 9.13.1 Effect of particle mass flux G and receiver tube diameter on measured heat transfer coefficients in a concentrating solar collector with flowing dense phase gas–solid suspension. Source: Zhang et al. (2017). © 2017, Elsevier.

Greek letters

ε void fraction (–)
φ_s particle shape factor (–)
Γ solids loading ratio, $= W_s W_g^{-1}$ (–)
μ dynamic viscosity (Ns m^{-2})
η ratio of specific heats, $= C_s C_{pg}^{-1}$ (–)
ρ density (kg m^{-3})
σ Stefan–Boltzmann constant, 5.67×10^{-8} W m^{-2} K^{-4}
ψ $h_{TP}\, h_g^{-1}$ (–)

Subscripts

bed bed
bub bubble
bw between bed and bed wall
c start of the intermediate/transitional turbulent regime
c cluster/packet
ch choking
fb freeboard
g gas
gc gas convection
ib immersed in bed
k start of fully turbulent regime
max maximum
mb minimum bubbling
mf minimum fluidization
p particle
pc particle convection
rad radiation
s solid particle
T terminal
TP two-phase
t tube, surface
w wall
x distance from beginning of heated length
∞ beyond thermal entrance length

References

Abdelmotalib, H.M., Youssef, M.M.A., Hassan, A.A. et al. (2015). Heat transfer process in gas-solid fluidized bed combustors: a review. *Int. J. Heat Mass Transf.* **89**: 567–575.

Abid, B.A., Ali, J.M., and Alzubaidi, A.A. (2011). Heat transfer in gas-solid fluidized bed with various heater inclinations. *Int. J. Heat Mass Transf.* **54**: 2228–2233.

Abubakar, M.Y., Bergougnou, M.A., Tarasuk, J.D. et al. (1980). Local heat transfer coefficients around a horizontal tube in a shallow fluidized bed. *J. Powder Bulk Solids Technol.* **4** (4): 11–18.

Adanez, J., Diego, L.F., and Gayan, P. (1993). Transport velocities of coal and sand particles. *Powder Technol.* **77**: 61–68.

Arnaldos, J. and Casal, J. (1996). Prediction of transition velocities and hydrodynamical regimes in fluidized beds. *Powder Technol.* **86**: 285–298.

Babu, S.P., Shah, B., and Talwalkar, A. (1978). Fluidization correlations for coal gasification materials – minimum fluidization velocity and fluidized bed expansion ratio. *AIChe. Symp. Ser.* **176** (74): 176–185.

Baerg, A., Klassen, J., and Gishler, P.E. (1950). Het transfer in a fluidized solids bed. *Can. J. Res.* **F28**: 287–307.

Baerns, M. (1967). Fluidization of fine particles. In: *Proc. Int. Symp. Fluidization*, 403–416. Eindhoven, Amsterdam: Netherland University Press.

Baskakov, A.P. (1985). Heat transfer in fluidized beds. In: *Fluidization*, 2e (eds. J.F. Davidson, R. Clift and D. Harrison), 465–471. London: Academic Press.

Baskakov, A.P., Berg, B.V., Fillipovsky, N.F. et al. (1973). Heat transfer to objects immersed in fluidized beds. *Powder Technol.* **8**: 273–282.

Basu, P. and Nag, P.K. (1996). Heat transfer to walls of a circulating fluidized-bed furnace. *Chem. Eng. Sci.* **51** (1): 1–26.

Bertoli, S.L. (2000). Radiant and convective heat transfer on pneumatic transport of particles: an analytical study. *Int. J. Heat Mass Transf.* **43**: 2345–2363.

Bhatt, G.N. and Whitehead, A.B. (1963). Heat transfer in subatmospheric fluidized bed. *Aust. J. Basic Appl. Sci.* **14**: 198–203.

Biyikli, S., Tuzla, K., and Chen, J.C. (1987). Freeboard heat transfer in high-temperature fluidized beds. *Powder Technol.* **53**: 187–194.

Biyikli, S., Tuzla, K., and Chen, J.C. (1989). A phenomenological model for heat transfer in freeboard of fluidized beds. *Can. J. Chem. Eng.* **67** (2): 230–236.

Blaszczuk, A., Nowak, W., and Kryzwanski, J. (2017). Effect of bed particle size on heat transfer between fluidized bed of group B particles and vertical rifled tubes. *Powder Technol.* **316**: 111–122.

Boothroyd, R.G. and Haque, H. (1970a). Experimental investigation of heat transfer in the entrance region of a heated duct conveying fine particles. *Trans. Inst. Chem. Eng.* **48**: T109–T120.

Boothroyd, R.G. and Haque, H. (1970b). Fully developed heat transfer to a gaseous suspension of particles flowing turbulently in ducts of different sizes. *J. Mech. Eng. Sci.* **12** (3): 191–200.

Borodulya, V.A., Ganzha, V.L., Podberezsky, A.I. et al. (1983). High pressure heat transfer investigations for fluidized beds of large particles and immersed vertical tube bundles. *Int. J. Heat Mass Transf.* **26** (11): 1571–1584.

Botterill, J.S.M. (1973). Bed-to-surface heat transfer. *AIChE. Symp. Ser.* **69** (128): 26–27.

Botterill, J.S.M. (1975). *Fluid Bed Heat Transfer*. New York: Academic Press.

Botterill, J.S.M. (1981). Comments on the paper "Fluid-bed heat exchangers – A new model for particle convective energy transfer" by H. Martin. *Chem. Eng. Commun.* **13** (1–3): 17–19.

Botterill, J.S.M. and Denloye, A.O.O. (1978). Gas convective heat transfer to packed and fluidized beds. *AIChE. Symp. Ser.* **176**: 194–202.

Botterill, J.S.M. and Desai, M. (1972). Limiting factors in gas-fluidized bed heat transfer. *Powder Technol.* **6**: 231–238.

Botterill, J.S.M., Yeoman, Y., and Yuregir, K.R. (1982). The effect of operating temperature on the velocity of minimum fluidization, bed voidage and general behavior. *Powder Technol.* **31**: 101–110.

Botterill, J.S.M., Yeoman, Y., and Yuregir, K.R. (1984). Factors affecting heat transfer between gas-fluidized surfaces. *Powder Technol.* **39**: 177–189.

Bowen, W. (1969). Heat transfer to stratified suspension flows with 30, 62, and 200 micron particles. MS thesis. University of Washington, Seattle. Quoted in Depew and Cramer (1970).

Broadhurst, T.E. and Becker, H.A. (1975). Onset of fluidization and slugging in beds of uniform particles. *AICHE J.* **21**: 238.

Brotz, W., Hiby, J.W., and Muller, K.G. (1958). Warmeubergang auf eine Flugstaubstomung im Senkrechten Rohr. *Chem. Ing. Tech.* **30** (3): 138–143.

Byam, J., Pillai, K.K., and Roberts, R.G. (1981). Heat transfer to cooling coils in the "splash" zone of a pressurised fluidized bed combustor. *AIChe. Symp. Ser.* **77** (208): 351–358.

Chen, P. and Pei, D.C.T. (1985). A model of heat transfer between fluidized beds and immersed surfaces. *Int. J. Heat Mass Transf.* **28** (3): 675–682.

Danziger, W.J. (1963). Heat transfer to fluidized gas-solids mixtures in vertical transport. *Ind. Eng. Chem. Process. Des. Dev.* **2** (4): 269–276.

Depew, C.A. and Cramer, E.R. (1970). Heat transfer to horizontal gas-solid suspensions. *J. Heat Transf.* **92**: 77–82.

Dounit, S., Hemati, M., and Andreux, R. (2008). Modelling and experimental validation of a fluidized-bed reactor freeboard region: application to natural gas combustion. *Chem. Eng. J.* **140**: 457–465.

Dow, W.M. and Jakob, M. (1951). Heat transfer between a vertical tube and a fluidized air-solid mixture. *Chem. Eng. Prog.* **47** (12): 637–648.

Dyrness, A., Glicksman, L.R., and Yule, T. (1992). Heat transfer in the splash zone of a bubbling fluidized bed. *Int. J. Heat Mass Transf.* **35** (4): 847–860.

El-Behery, S.M., El-Askary, W.A., Hamed, M.H., and Ibrahim, K.A. (2011). Hydrodynamic and thermal fields analysis in gas-solid two-phase flow. *Int. J. Heat Fluid Flow* **32**: 740–754.

Ergun, S. (1952). Fluid flow through packed columns. *Chem. Eng. Prog.* **48**: 89–94.

Farag, I.H. and Tsai, K. (1992). Fluidized bed freeboard measurements of heat transfer. *Can. J. Chem. Eng.* **70** (4): 664–673.

Farbar, L. and Morley, M.J. (1957). Heat transfer to flowing gas-solids mixtures in a circular tube. *Ind. Eng. Chem.* **49** (7): 1143–1150.

Figiola, R.S., Suarez, E.G., and Pitts, D.R. (1986). Mixed particle size distribution effects on heat transfer in a fluidized bed. *J. Heat Transf.* **108**: 913–915.

Flamant, G., Gauthier, D., Benoit, H. et al. (2013). Dense suspension of solid particles as a new transfer fluid for concentrated solar thermal plants: on-sun proof of concept. *Chem. Eng. Sci.* **102**: 567–576.

Frankel, L.L. and Kondukov, N.B. (1978). Intensification of heat transfer of surface to bed heat transfer in a fluidized bed at controlled gas distribution. In: *Proceedings of Sixth International Heat Transfer for Conference*, vol. **6**, 37–42. Washington, DC: Hemisphere Publishing Corporation.

Ganzha, V.L., Upadhyay, S.N., and Saxena, S.C. (1982). A mechanistic theory for heat transfer between fluidized beds of large particles and immersed surfaces. *Int. J. Heat Mass Transf.* **25**: 1531–1540.

Geldart, D. (1973). Types of gas fluidization. *Powder Technol.* **7**: 285–292.

Geldart, D. and Abrahamsen, A.R. (1978). Homogeneous fluidization of fine powders using various gases and pressures. *Powder Technol.* **19**: 133–136.

Gelperin, N.I. and Einstein, V.G. (1971). Heat transfer in fluidized beds. In: *Fluidization* (eds. J.F. Davidson and D. Harrison). London: Academic Press.

Gelperin, N.I., Kruglikov, V.Y., and Ainshtein, V.G. (1958). Heat transfer between a fluidized bed and the surface of a single tube in longitudinal and transverse gas flow. *Khim. Prom.* **6**: 358–363. Quoted in Zabrodsky (1966).

Gelperin, N.I., Einstein, V.G., and Romanova, N.A. (1963). *Khim. Prom.* **1**: 823. Quoted in Saxena et al. (1978).

Gelperin, N.I., Einstein, V.G., and Zaikpvski, V.G. (1968a). *Khim. Mashinostr.* **3**: 17. Quoted in Gelperin and Einstein (1971). Also in Grewal and Saxena (1983).

Gelperin, N.I., Einstein, V.G., and Korontjanskaya, L.A. (1968b). *Khim. Prom.* **6**: 427. Quoted in Gelperin and Einstein (1971).

Gelperin, N.I., Ainshtein, V.G., and Korontjanskaya, L.A. (1969). Heat transfer between a fluidized bed and staggered bundles of horizontal tubes. *Int. Chem. Eng.* **9**: 137–142. Quoted in Grewal and Saxena (1983).

Genetti, W.E., Schmall, R.A., and Grimmet, E.S. (1971). The effect of tube orientation on heat transfer with bare and finned tubes in a fluidized bed. *Chem. Eng. Prog. Symp. Ser.* **67** (116): 90–96.

Gorbis, Z.R. and Bakhtiozin, R.A. (1962). Investigation of convection heat transfer to a gas-graphite suspension under conditions of internal flow in vertical channels. *Sov. At. Energy* **12** (5): 402–409.

Goroshko, V.D., Rozenbaum, R.B., and Todes, O.M. (1958). Approximate hydraulic relationships for suspended beds and hindered fall. *Izvestiya Vuzov, Neft'i Gaz* **1**: 125–131. Quoted in Zabrodsky (1966).

Goshayeshi, A., Welty, J.R., Adams, R.L., and Alavizadeh, N. (1986). Local heat transfer coefficients for horizontal tube arrays in high temperature large particle fluidized beds. *J. Heat Transf.* **108**: 907–915.

Grace, J.R. (1982a). Hydrodynamics of fluidization. In: *Handbook of Multiphase Flow* (ed. G. Hetsroni), 8-5–8-64. Washington,DC: Hemisphere Publishing Corporation.

Grace, J.R. (1982b). Fluidized bed heat transfer. In: *Handbook of Multiphase Flow* (ed. G. Hetsroni), 8-65–8-83. Washington, DC: Hemisphere Publishing Corporation.

Grewal, N.S. (1981). A generalized correlation for heat transfer between a gas-solid fluidized bed of small particles and an immersed staggered array of horizontal tubes. *Powder Technol.* **30** (1981): 145–154.

Grewal, N.S. (1982). A correlation for maximum heat transfer between a horizontal tube and a gas-solid fluidized bed of small particles. *Lett. J. Heat Mass Tran.* **9**: 377–384.

Grewal, N.S. and Gupta, A. (1989). Total and gas convective heat transfer from a vertical tube to a mixed particle gas-solid fluidized bed. *Powder Technol.* **57**: 27–38.

Grewal, N.S. and Menart, J. (1987). Heat transfer to horizontal tubes immersed in a fluidized-bed combustor. *Powder Technol.* **52**: 149–159.

Grewal, N.S. and Saxena, S.C. (1980). Heat transfer between a horizontal tube and a gas-solid fluidized bed. *Int. J. Heat Mass Transf.* **23**: 1505–1519.

Grewal, N.S. and Saxena, S.C. (1981). Maximum heat transfer coefficient between a horizontal tube and a gas-solid fluidized bed. *Ind. Eng. Chem. Process. Des. Dev.* **20**: 108–116.

Grewal, N.S. and Saxena, S.C. (1983). Experimental studies of heat transfer between a bundle of horizontal tubes and a gas-solid fluidized bed of small particles. *Ind. Eng. Chem. Process. Des. Dev.* **22**: 367–376.

Gunn, D.J. and Hilal, N. (1994). Heat transfer from vertical surfaces to dense gas-fluidized beds. *Int. J. Heat Mass Transf.* **37** (16): 2465–2473.

Han, K.S., Sung, H.J., and Chung, M.K. (1991). Analysis of heat transfer in a pipe carrying two-phase gas-particle suspension. *Int. J. Heat Mass Transf.* **34** (1): 69–78.

Hasegawa, S., Echigo, R., Kanemaru, K. et al. (1983). Experimental study on forced convective heat transfer of flowing gaseous solid suspension at high temperature. *Int. J. Multiphase Flow* **9** (2): 131–145.

Jacob, A. and Osberg, G.B. (1957). The effect of thermal conductivity on local heat transfer in a fluidized bed. *Can. J. Chem. Eng.* **35** (June): 5–9.

Jepson, G.A., Poll, A., and Smith, W. (1963). Heat transfer from gas to wall in a gas/solid transport line. *Trans. Inst. Chem. Eng.* **41**: 207–211.

Karimipour, S., Zarghami, R., Mostoufi, N., and Sotudeh-Gharebagh, R. (2007). Evaluation of heat transfer coefficient in gas-solid fluidized beds using cluster-based approach. *Powder Technol.* **172**: 19–26.

Kharchenko, N.V. and Makhorin, K.F. (1964). The rale of heat transfer between a fluidized bed and an immersed body at high temperatures. *Int. Chem. Eng.* **4**: 650–654.

Kim, W.K. and Kim, S.D. (2013). Heat transfer characteristics in a pressurized fluidized bed of fine particles with immersed horizontal tube bundle. *Int. J. Heat Mass Transf.* **64**: 269–277.

King, D.F. and Harrison, D. (1982). The dense phase of a fluidized bed at elevated pressure. *Trans. Inst. Chem. Eng.* **60**: 26–30.

Knudsen, J.D. and Katz, D.L. (1958). *Fluid Dynamics and Heat Transfer*. New York: McGraw-Hill. Quoted in Biyikli et al. (1987).

Kobayashi, M., Ramaswamy, D., and Brazelton, W.T. (1970). Heat transfer from an internal surface to a pulsed bed. *Chem. Eng. Prog. Symp. Ser.* **66** (105): 58–76.

Krugell-Emden, H. and Vollmari, K. (2016). Flow-regime transitions in fluidized beds of non-spherical particles. *Particuology* **29**: 1–15.

Kunii, D. and Levenspiel, O. (1969). *Fluidization Engineering*. New York: Wiley.

Kunii, D. and Levenspiel, O. (1991). *Fluidization Engineering*, 2e, 325. Newton, MA: Butterworth-Heinemann. Quoted in Natale et al. (2008).

Lechner, S., Merzsch, M., and Krautz, K.J. (2014). Heat-transfer from horizontal tube bundles into fluidized beds with Geldart A lignite particles. *Powder Technol.* **253**: 14–21.

Li, J. and Mason, D.J. (2000). A computational investigation of transient heat transfer in pneumatic transport of granular particles. *Powder Technol.* **112**: 273–282.

Li, H.S., Qian, R.Z., Huang, W.D., and Bi, K.J. (1993). Investigation on instantaneous local heat transfer coefficients in high-temperature fluidized beds-l. Experimental results. *Int. J. Heat Mass Transf.* **36** (18): 4389–4395.

Louge, M. and Yusof, J.M. (1993). Heat transfer in the pneumatic transport of massive particles. *Int. J. Heat Mass Transf.* **36** (2): 265–275.

Martin, H. (1981). Fluid-bed heat exchangers – a new model for particle convective energy transfer. *Chem. Eng. Commun.* **13** (1–3): 1–16.

Martin, H. (1984a). Heat transfer between gas fluidized beds of solid particles and the surfaces of immersed heat exchanger elements, part I. *Chem. Eng. Process.* **18**: 157–169.

Martin, H. (1984b). Heat transfer between gas fluidized beds of solid particles and the surfaces of immersed heat exchanger elements, part II. *Chem. Eng. Process.* **18**: 199–223.

Maskaev, V.K. and Baskakov, A.P. (1973). Characteristics of heat transfer in fluidized bed of coarse particles. *J. Eng. Phys.* **6**: 589–593.

Masoumifard, N., Mostoufi, N.A., Hamidi, A., and Sotudeh-Gharebagh, R. (2008). Investigation of heat transfer between a horizontal tube and gas-solid fluidized bed. *Int. J. Heat Fluid Flow* **29**: 1504–1511.

Mathur, A. and Saxena, S.C. (1986). A correlation for heat transfer between immersed surfaces and gas-fluidized beds of large particles. *Energy* **11** (9): 843–852.

Michaelides, E.E. (1986). Heat transfer in particulate flows. *Int. J. Heat Mass Transf.* **29** (2): 265–273.

Mickley, H.S. and Fairbanks, D.F. (1955). Mechanism of heat transfer to fluidized beds. *AICHE J.* **1** (3): 374–384.

Mickley, H.S. and Trilling, C.A. (1949). Heat transfer characteristics of fluidized beds. *Ind. Eng. Chem.* **41**: 1135–1147.

Morooka, S., Maruyama, Y., Kawazuishi, K. et al. (1979). Heat transfer coefficient between bed and vertically inserted tube wall, and holdup of solid particles, in the dense and dilute zones of a fluidized bed. *Heat Trans. Jpn. Res.* **8** (1): 61–70.

Nakajima, M., Harada, M., Asai, M. et al. (1991). *Circulating Fluidized Bed Technology Ill* (eds. P. Basu, M. Horio and M. Hasatani), 79–84. Oxford: Pergamon. Quoted in Arnoldas & Casal (1996).

Natale, F.D., Lancia, A., and Nigro, R. (2008). A single particle model for surface-to-bed heat transfer. *Powder Technol.* **187**: 68–78.

Noe, A.R. and Knudsen, J.G. (1968). *Chem. Eng. Symp. Ser.* **64** (82): 202–211. Quoted in Saxena et al. (1978).

Pattipati, R.R. and Wen, C.Y. (1981). Minimum fluidization velocity at high temperatures. *Ind. Eng. Chem. Process. Des. Dev.* **20** (4): 705–707.

Perales, J.F., Coll, T., Llop, M.F. et al. (1991). *Circulating Fluidized Bed Technology Ill* (eds. P. Basu, M. Horio and M. Hasatani), 73–78. Oxford: Pergamon. Quoted in Arnoldas & Casal (1996).

Petrie, J.C., Freeby, A., and Buckham, J.A. (1968). In-bed heat exchangers. *Chem. Eng. Prog. Symp. Ser.* **64** (67): 45–51.

Pfeffer, R., Rosetti, S., and Lieblein, S., (1966). Analysis and correlation of heat transfer coefficient and friction factor data for dilute gas-solid suspensions. NASA TN D3603.

Rabinovich, E. and Kalman, H. (2011). Flow regime diagram for vertical pneumatic conveying and fluidized bed systems. *Powder Technol.* **207**: 119–133.

Ranz, W.E. and Marshall, W.R. (1952). Evaporation from drops. *Chem. Eng. Prog.* **48** (3): 141–146.

Renzhang, Q., Wendi, H., Yunsheng, X., and Dechang, L. (1987). Experimental research of radiative heat transfer in fluidized beds. *Int. J. Heat Mass Transf.* **30** (5): 827–831.

Richardson, J.F. (1971). Incipient fluidization and particulate systems. In: *Fluidization* (eds. J.F. Davidson and D. Harrison), 26–64. New York: Academic Press.

Saxena, S.C. (1979). Heat transfer from a bank of immersed horizontal smooth tubes in a fluidized bed. *Lett. Heat Mass Transfer* **6**: 225–229.

Saxena, S.C. (1989). Heat transfer between immersed surfaces and gas-fluidized beds. *Adv. Heat Tran.* **19**: 97–190.

Saxena, S.C., Grewal, N.S., Gabor, J.D. et al. (1978). Heat transfer between a gas fluidized bed and immersed tubes. *Adv. Heat Tran.* **14**: 149–246.

Schluderberg, D.C., Whitelaw, R.L., and Carlson, R.W. (1961). Gaseous suspensions – a new reactor coolant. *Nucleonics* **19** (8): 67–76. Quoted in Pfeffer et al. (1966).

Shah, M.M. (1983). Generalized prediction of maximum heat transfer to single cylinders and spheres in gas fluidized beds. *Heat Transfer Eng.* **4** (3–4): 107–122.

Shah, M.M. (2018a). General correlation for maximum heat transfer to surfaces submerged in gas-fluidized beds. *Chem. Eng. Sci.* **185**: 127–140.

Shah, M.M. (2018b). A correlation for maximum heat transfer to cylinders and spheres in gas-fluidized beds. Paper IMECE2018-86586.

Shah, M.M. (2020). A correlation for heat transfer to gas-solid suspensions flowing in pipes. *J. Therm. Sci. Eng. Appl.* **12**: 021009-1–021009-8.

Shlapkova, Y.P. (1969). Heat transfer to a cylindrical surface immersed in a fluidized bed at low pressure. *J. Eng. Phys.* **10** (3): 187–188.

Staub, F.W. and Canada, G.S. (1978). Effect of tube bank and gas density on flow behavior and heat transfer in fluidized beds. In: *Fluidization* (eds. J.F. Davidson and D.L. Keairns), 339–344. Cambridge: Cambridge University Press.

Stefanova, A., Bi, H.T., Lim, J.C., and Grace, J.R. (2011). Local hydrodynamics and heat transfer in fluidized beds of different diameter. *Powder Technol.* **212**: 57–63.

Stefanova, A., Xiaotao, T.B., Lim, C.J., and Grace, J.R. (2019). A probabilistic heat transfer model for turbulent fluidized beds. *Powder Technol.* https://doi.org/10.1016/j.powtec.2019.01.066.

Sukomel, A.S., Tavetkov, F.F., and Kerimov, R.V. (1967). A study of local heat transfer from a tube wall to a turbulent flow of gas bearing suspended solids. *Therm. Eng.* **14**: 116–121.

Szekeley, J. and Fisher, R.J. (1968). Bed to wall radiation heat transfer in a gas-solid fluidized bed. *Chem. Eng. Sci.* **24**: 833–849.

Ternovskaya, A.N. and Korenberg, Y.G. (1971). *Pyrite Kilning in a Fluidized Bed*. Moscow: Izd. Khimiya. Quoted in Grewal and Saxena (1980).

Tien, C.L. (1961). Heat transfer by a turbulently flowing fluid-solids mixture in a pipe. *J. Heat Transf.* **83** (2): 183–188.

Tien, C. L. and Quan, V. (1962). Local heat transfer characteristics of air-glass and air-lead mixtures in turbulent pipe flow. ASME Paper No. 62-HT-15.

Todes, O.M. (1965). *Applications of Fluidized Beds in the Chemical Industry*, part II, 4–27. Izd. Znanie, Leningrad. Quoted in Gelperin and Einstein (1971).

Toomey, R.D. and Johnstone, H.F. (1953). Heat transfer between beds of fluidized solids and the walls of the container. *Chem. Eng. Symp. Ser.* **49** (5): 51–63.

Turton, R., (1986) Heat Transfer Studies in Fine Particle Fluidized Beds. PhD Thesis, Oregon State University.

Vedamurthy, V.N. and Sastri, V.M.K. (1974). An analysis of the conductive and radiative heat transfer to the walls of fluidized bed combustors. *Int. J. Heat Mass Transf.* **17**: 1–9.

Vreedenberg, H.A. (1952). Heat transfer between fluidized beds and vertically inserted tubes. *J. Appl. Chem.* **1**: S26–S33.

Vreedenberg, H.A. (1958). Heat transfer between a fluidized bed and a horizontal tube. *Chem. Eng. Sci.* **9**: 52–60.

Vreedenberg, H.A. (1960). Heat transfer between a fluidized bed and a vertical tube. *Chem. Eng. Sci.* **11**: 274–285. Quoted in Zabrodsky (1966).

Wahi, M.K. (1977). Heat transfer to flowing gas-solid mixtures. *J. Heat Transf.* **99**: 145–148.

Wen, C.Y. and Yu, Y.H. (1966a). Mechanics of fluidization. *Chem. Eng. Prog. Symp. Ser.* **62** (62): 100–111.

Wen, C.Y. and Yu, Y.H. (1966b). Generalized method for prediction of minimum fluidization velocity. *AICHE J.* **12**: 610–612.

Wender, L. and Cooper, G.T. (1958). Heat transfer between fluidized-solids beds and boundary surfaces – correlation of data. AIChE J. **4** (1): 15–23.

Whitaker, S. (1972). Forced convection heat transfer correlations for flow in pipes, past flat plates, single cylinders, single spheres, and for flow in packed beds and tube bundles. *AIChE J.* **18** (2): 361–371.

Wunder, R. (1980). Waermeubergang an vertikalen Wiirmetauscherflachen in Gaswirbelschichten. Doctorate Dissertation, TU Munich, F.R.G. Quoted in Martin (1984b).

Xavier, A.M. and Davidson, J.F. (1981). Heat transfer to surfaces immersed in fluidized beds and in the freeboard region. *AIChe. Symp. Ser.* **77** (208): 368–373.

Xavier, A.M. and Davidson, J.F. (1985). Heat transfer in fluidized beds: convective heat transfer in fluidized beds. In: *Fluidization*, 2e (eds. J.F. Davidson, R. Clift and D. Harrison), 437–459. London: Academic Press.

Yamada, Y., Takahashi, S., and Maki, H. (1991). Study of gas combustion fluidized beds. In: *ASME/JSME Thermal Science Proceedings*, vol. **4**, 499–506. ASME.

Yang, W.C. (1975). A mathematical definition of choking phenomenon and a mathematical model for predicting choking velocity and choking voidage. AIChE J. **21**: 1013–1015.

Yang, W.C. (1983). Criteria for choking in vertical pneumatic conveying lines. *Powder Technol.* **35**: 143–150.

Yoshida, K., Ueno, T., and Kuni, D. (1974). Mechanism of bed-wall heat transfer in a fluidized bed at high temperatures. *Chem. Eng. Sci.* **29**: 77–82.

Yousfi, G., Gau, G., and Goff, P.L. (1974). Heat transfer to an air-solid suspension at high concentration and low velocity. In: *Proceedings of the Fifth International Heat Transfer Conference, 3–7 September*, vol. **5**, 218–222. Tokyo, Japan: Begell House.

Yusuf, R., Melaaen, M.C., and Mathiesen, V. (2005). Convective heat and mass transfer modeling in gas-fluidized beds. *Chem. Eng. Technol.* **28**: 13–24.

Zabrodsky, S.S. (1958). The fundamental laws of heat transfer in fluidized beds. *Inzh. Fiz. Zhurn.* **1** (3): 40–51. Quoted in Zabrodsky (1966).

Zabrodsky, S.S. (1966). *Hydrodynamics and Heat Transfer in Fluidized Beds*. Cambridge, MA: M.I.T. Press. Translated by F.A. Zenz.

Zabrodsky, S.S., Antonishin, N.V., Vasiliev, G.M., and Paranas, A.L. (1976). On fluidized bed to surface heat transfer. *Can. J. Chem. Eng.* **54** (1–2): 52–58.

Zhang, H., Benoit, H., Gauthier, D. et al. (2016). Particle circulation loops in solar energy capture and storage: gas-solid flow and heat transfer considerations. *Appl. Energy* **161**: 206–224.

Zhang, H., Benoit, H., Perez-Lopez, I. et al. (2017). High-efficiency solar power towers using particle suspensions as heat carrier in the receiver and in the thermal energy storage. *Renew. Energy* **111**: 438–446.

Appendix

Table A.1 Unit conversion factors.

Quantity	To convert from	To	Multiply by	Quantity	To convert from	To	Multiply by
Length	m	Foot	3.2808	Specific heat	$kJ\,kg^{-1}\,K^{-1}$	$Btu\,lb^{-1}\,°F^{-1}$	0.2388
		Inch	39.37			$kcal\,kg^{-1}\,K^{-1}$	0.2388
		Yard	1.0936	Heat flux	$W\,m^{-2}$	$Btu\,h^{-1}\,ft^{-2}$	0.317
		Mile	6.21×10^{-4}			$kcal\,h^{-1}\,m^{-2}$	0.86
Mass	kg	lb	2.2046			$cal\,s^{-1}\,cm^{-2}$	2.39×10^{-5}
		Slug	0.0685	Heat transfer coefficient	$W\,m^{-2}\,K^{-1}$	$Btu\,h^{-1}\,ft^{-2}\,°F^{-1}$	0.1761
Force	Newton ($kg\,m\,s^{-2}$)	lbf	0.2248			$kcal\,h^{-1}\,m^{-2}\,K^{-1}$	0.86
		kp	0.10197			$cal\,s^{-1}\,cm^{-2}\,K^{-1}$	2.39×10^{-5}
Density	$kg\,m^{-3}$	$lb\,ft^{-3}$	0.06242	Thermal conductivity	$W\,m^{-2}\,K^{-1}\,m^{-1}$	$Btu\,h^{-1}\,ft^{-2}\,°F^{-1}\,ft^{-1}$	0.5779
Pressure	Pa, $N\,m^{-2}$	psi	0.000145			$Btu\,h^{-1}\,ft^{-2}\,°F^{-1}\,in^{-1}$	6.938
		in. Hg	0.000295			$kcal\,h^{-1}\,m^{2}\,K\,m^{-1}$	0.86
		mm Hg	0.0075			$cal\,s^{-1}\,cm^{-2}\,K^{-1}\,cm^{-1}$	0.002388
		torr	0.0075	Dynamic viscosity	Pa s, $Ns\,m^{-2}$ $kg\,m^{-1}\,s^{-1}$	centipoise	1000
		bar	10^{-5}			$lb\,s^{-1}\,ft^{-1}$	0.672
		atm	0.987×10^{-5}			$lb\,h^{-1}\,ft^{-1}$	2420
		$kg\,cm^{-2}$	1.0197×10^{-5}			$lbf\,s^{-1}\,ft^{-2}$	0.0209
		$dyne\,cm^{-1}$	10	Kinematic viscosity, thermal diffusivity	$m^2\,s^{-1}$	$ft^2\,h^{-1}$	38 750
Mass flux	$kg\,m^{-2}\,s^{-1}$	$lb\,h^{-1}\,ft^{-2}$	738.3			$ft^2\,s^{-1}$	10.764
Heat, energy	J (W s)	Btu	9.48×10^{-4}			Stoke	10^4
		kcal	2.39×10^{-4}	Surface tension	$N\,m^{-1}$	$dyne\,cm^{-1}$, $erg\,cm^{-2}$	10^3
Enthalpy, latent heat	$kJ\,kg^{-1}$	$Btu\,lb^{-1}$	0.4299			$lbf\,in^{-1}$	5.71×10^{-3}
		$kcal\,kg^{-1}$	0.2388			g force cm^{-1}	1.0197

Two-Phase Heat Transfer, First Edition. Mirza Mohammed Shah.

Table A.2 Physical properties of some fluids.

Fluid	Molecular weight	Boiling temperature at 1.013 Bar (°C)	T_c (°C)	p_c (bar)
Air	28.959	−194.25	−140.59	37.89
Ammonia	17.03	−33.6	132.25	113.33
Argon	39.95	−185.85	−122.46	48.63
CO_2	44.01	−78.4	30.98	7.38
Ethane	30.07	−88.58	32.72	48.72
Isobutane	58.122	−11.75	134.66	36.29
Methane	16.04	−161.48	−82.59	45.99
Methanol	32.0	64.70	240.0	79.5
Oxygen	32	−182.96	−118.57	50.43
Propane	44.1	−42.11	96.74	42.51
Propylene	42.08	47.62	91.06	45.55
R-22	86.47	−40.81	96.14	49.9
R-134a	102.03	−26.07	101.06	40.59
R-410A	72.58	−51.45	71.35	49.03
R-1234yf	114.04	−29.45	94.7	33.82
Water	18.01	99.97	373.95	220.64

Table A.3 Properties of dry air at room temperature.

Temperature (°C)	Density (kg m^{-3})	C_p (kJ kg^{-1} K^{-1})	Thermal conductivity (mW m^{-1} K^{-1})	Viscosity (μPa s)
−40	1.5261	1.0057	21.225	15.152
−35	1.4939	1.0057	21.626	15.417
−30	1.463	1.0056	22.024	15.681
−25	1.4334	1.0056	22.419	15.942
−20	1.405	1.0056	22.812	16.201
−15	1.3776	1.0056	23.203	16.459
−10	1.3513	1.0056	23.591	16.714
−5	1.326	1.0056	23.977	16.967
0	1.3017	1.0057	24.361	17.219
5	1.2782	1.0058	24.742	17.468
10	1.2556	1.0059	25.122	17.716
15	1.2337	1.006	25.499	17.962
20	1.2126	1.0062	25.874	18.206
25	1.1922	1.0063	26.247	18.448
30	1.1725	1.0065	26.618	18.689
35	1.1534	1.0067	26.987	18.928
40	1.135	1.0069	27.354	19.165
45	1.1171	1.0072	27.72	19.401
50	1.0998	1.0074	28.083	19.635
55	1.083	1.0077	28.445	19.868
60	1.0667	1.008	28.804	20.099

Table A.4 Properties of saturated ammonia.

Temperature (°C)	Pressure (MPa)	Liquid density (kg m^{-3})	Vapor density (kg m^{-3})	Liquid enthalpy (kJ kg^{-1})	Vapor enthalpy (kJ kg^{-1})	Liquid, C_p (kJ kg^{-1} K^{-1})	Vapor, C_p (kJ kg^{-1} K^{-1})	Liquid thermal conductivity (mW m^{-1} K^{-1})	Vapor thermal conductivity (mW m^{-1} K^{-1})	Liquid viscosity (μPa s^{-1})	Vapor viscosity (μPa s^{-1})	Surface tension (mN m^{-1})
−70	0.01094	724.72	0.1110	32.343	1498.7	4.245	2.085	792.06	19.729	475.03	7.0318	42.342
−60	0.02189	713.62	0.2125	75.093	1516.9	4.3031	2.125	757	19.935	391.29	7.2963	40.184
−50	0.04083	702.09	0.38055	118.43	1534.3	4.359	2.177	722.28	20.238	328.87	7.5729	37.953
−40	0.07169	690.15	0.6438	162.32	1550.9	4.413	2.244	688.11	20.641	281.24	7.8588	35.668
−30	0.11943	677.83	1.0374	206.76	1566.5	4.464	2.325	654.63	21.149	244.07	8.1516	33.346
−20	0.19008	665.14	1.6033	251.71	1580.8	4.5138]	2.424	621.96	21.768	214.41	8.4495	31.001
−10	0.29071	652.06	2.3906	297.16	1593.9	4.5636	2.541	590.14	22.503	190.22	8.7511	28.647
0	0.42938	638.57	3.4567	343.15	1605.4	4.6165	2.679	559.2	23.365	170.09	9.0558	26.295
10	0.61505	624.64	4.8679	389.72	1615.3	4.6757	2.841	529.12	24.365	153.03	9.3638	23.955
20	0.85748	610.2	6.7025	436.94	1623.3	4.7448	3.029	499.86	25.519	138.32	9.6761	21.636
30	1.1672	595.17	9.0533	484.91	1629.3	4.8282	3.25	471.35	26.846	125.45	9.9953	19.346
40	1.5554	579.44	12.034	533.79	1633.1	4.9318	3.510	443.54	28.379	114.04	10.325	17.092
50	2.034	562.86	15.785	583.77	1634.2	5.0635	3.823	416.32	30.16	103.79	10.673	14.883
60	2.6156	545.24	20.493	635.12	1632.4	5.2351	4.208	389.59	32.262	94.483	11.049	12.726
70	3.3135	526.31	26.407	688.2	1627.1	5.4648	4.699	363.24	34.803	85.933	11.468	10.628
80	4.142	505.67	33.888	743.5	1617.5	5.7837	5.354	337.11	37.995	77.979	11.954	8.598
90	5.1167	482.75	43.484	801.76	1602.3	6.2501	6.290	311.02	42.24	70.468	12.549	6.6482
100	6.2553	456.63	56.117	864.16	1579.8	6.9912	7.762	284.77	48.363	63.231	13.322	4.7937
110	7.5783	425.61	73.55	932.84	1546.2	8.3621	10.46	258.13	58.329	56.028	14.42	3.0582
120	9.1125	385.49	100.07	1013.1	1493.4	11.94	17.21	231.24	78.402	48.34	16.212	1.4845
130	10.898	312.29	156.77	1135.2	1382.5	54.21	76.49	221.87	160.39	37.29	20.63	0.19069

Table A.5 Properties of saturated carbon dioxide.

Temperature (°C)	Pressure (MPa)	Liquid density (kg m^{-3})	Vapor density (kg m^{-3})	Liquid enthalpy (kJ kg^{-1})	Vapor enthalpy (kJ kg^{-1})	Liquid C_p (kJ kg^{-1} K^{-1})	Vapor C_p (kJ kg^{-1} K^{-1})	Liquid thermal conductivity (mW m^{-1} K^{-1})	Vapor thermal conductivity (mW m^{-1} K^{-1})	Liquid viscosity (μPa s^{-1})	Vapor viscosity (μPa s^{-1})	Surface tension (mN m^{-1})
−48	0.73949	1147.1	19.373	96.905	433.29	1.9779	0.96657	169.48	11.759	221.63	11.418	14.498
−44	0.86445	1132	22.547	104.87	434.39	1.9933	0.99817	164.36	12.138	207.16	11.641	13.583
−40	1.0045	1116.4	26.121	112.9	435.32	2.0117	1.0333	159.3	12.54	193.75	11.869	12.681
−36	1.1607	1100.5	30.137	121.01	436.07	2.0333	1.0725	154.29	12.971	181.3	12.102	11.791
−32	1.3342	1084.1	34.644	129.2	436.62	2.0587	1.1165	149.32	13.434	169.71	12.341	10.915
−30	1.4278	1075.7	37.098	133.34	436.82	2.0731	1.1406	146.86	13.68	164.22	12.464	10.482
−28	1.5261	1067.2	39.696	137.5	436.96	2.0886	1.1663	144.4	13.937	158.9	12.589	10.053
−26	1.6293	1058.6	42.445	141.69	437.04	2.1055	1.1938	141.95	14.205	153.77	12.716	9.6271
−24	1.7375	1049.8	45.356	145.91	437.06	2.1238	1.2234	139.51	14.486	148.8	12.846	9.2052
−22	1.8509	1040.8	48.437	150.16	437.01	2.1437	1.2551	137.07	14.782	143.99	12.979	8.7873
−20	1.9696	1031.7	51.7	154.45	436.89	2.1653	1.2893	134.64	15.093	139.33	13.115	8.3733
−18	2.0938	1022.3	55.155	158.77	436.7	2.1889	1.3263	132.22	15.422	134.81	13.255	7.9634
−16	2.2237	1012.8	58.816	163.14	436.44	2.2146	1.3664	129.8	15.77	130.43	13.4	7.5578
−14	2.3593	1003.1	62.697	167.55	436.09	2.2426	1.4099	127.38	16.141	126.17	13.549	7.1565
−12	2.501	993.13	66.814	172.01	435.66	2.2734	1.4572	124.96	16.537	122.04	13.703	6.7597
−10	2.6487	982.93	71.185	176.52	435.14	2.3072	1.5091	122.54	16.96	118.02	13.863	6.3676
−8	2.8027	972.46	75.829	181.09	434.51	2.3446	1.566	120.13	17.417	114.11	14.03	5.9803
−6	2.9632	961.7	80.77	185.71	433.79	2.386	1.6288	117.71	17.909	110.3	14.205	5.5981
−4	3.1303	950.63	86.032	190.4	432.95	2.4322	1.6986	115.29	18.445	106.58	14.388	5.221
−2	3.3042	939.22	91.647	195.16	431.99	2.4839	1.7767	112.86	19.029	102.95	14.581	4.8494
0	3.4851	927.43	97.647	200	430.89	2.5423	1.8648	110.43	19.671	99.394	14.786	4.4835
2	3.6733	915.23	104.07	204.93	429.65	2.6086	1.9649	107.99	20.381	95.916	15.004	4.1236
4	3.8688	902.56	110.98	209.95	428.25	2.6846	2.0799	105.54	21.171	92.502	15.237	3.7699
6	4.072	889.36	118.41	215.08	426.67	2.7724	2.2134	103.08	22.057	89.144	15.489	3.4228
8	4.2831	875.58	126.44	220.34	424.89	2.8753	2.3704	100.6	23.06	85.833	15.761	3.0827
10	4.5022	861.12	135.16	225.73	422.88	2.9976	2.5578	98.119	24.206	82.557	16.059	2.75
12	4.7297	845.87	144.67	231.29	420.62	3.1454	2.7856	95.622	25.53	79.304	16.387	2.4253
14	4.9658	829.7	155.11	237.03	418.05	3.3278	3.0684	93.116	27.081	76.059	16.752	2.1092
16	5.2108	812.41	166.66	243.01	415.12	3.5583	3.429	90.608	28.926	72.802	17.165	1.8024
18	5.4651	793.76	179.57	249.26	411.76	3.8581	3.9046	88.118	31.162	69.509	17.637	1.5059
20	5.7291	773.39	194.2	255.87	407.87	4.2637	4.5599	85.683	33.943	66.148	18.187	1.2208
22	6.0031	750.77	211.08	262.93	403.26	4.8464	5.5186	83.395	37.519	62.67	18.847	0.94871
24	6.2877	725.02	231.1	270.61	397.7	5.7674	7.0487	81.469	42.351	58.998	19.663	0.69165
26	6.5837	694.46	255.86	279.26	390.71	7.4604	9.862	80.451	49.439	54.984	20.731	0.45286
28	6.8918	655.28	289.11	289.62	381.2	11.549	16.691	81.885	61.728	50.302	22.272	0.23778
30	7.2137	593.31	345.1	304.55	365.13	35.338	55.822	95.356	98.023	43.768	25.17	0.058862

Table A.6 Properties of saturated helium-4.

Temperature (K)	Pressure (MPa)	Liquid density (kg m^{-3})	Vapor density (kg m^{-3})	Liquid enthalpy (kJ kg^{-1})	Vapor enthalpy (kJ kg^{-1})	Liquid C_p (kJ kg^{-1} K^{-1})	Vapor C_p (kJ kg^{-1} K^{-1})	Liquid thermal conductivity (mW m^{-1} K^{-1})	Vapor thermal conductivity (mW m^{-1} K^{-1})	Liquid viscosity (μPa^{-1} s^{-1})	Vapor viscosity (μPa^{-1} s^{-1})	Surface tension (mN m^{-1})
2.2	0.005 326	145.98	1.2321	−6.7459	16.019	3.0871	5.5498	13.607	4.0395	3.5939	0.54486	0.28449
2.4	0.008 348	145.37	1.8029	−6.1786	16.818	2.4983	5.6781	14.488	4.5337	3.713	0.60721	0.26802
2.6	0.0123 75	144.29	2.5184	−5.6734	17.562	2.3514	5.831	15.252	4.9938	3.7454	0.66948	0.25047
2.8	0.017 562	142.89	3.396	−5.169	18.245	2.4062	6.0144	15.931	5.4434	3.7297	0.73232	0.232
3	0.0240 62	141.24	4.4554	−4.6359	18.863	2.5656	6.2367	16.538	5.8939	3.6856	0.79626	0.21278
3.2	0.0320 27	139.35	5.7198	−4.0578	19.408	2.7901	6.5107	17.075	6.3513	3.6238	0.86182	0.19296
3.4	0.0416 07	137.22	7.2175	−3.4237	19.871	3.0683	6.8556	17.542	6.8196	3.5502	0.92954	0.17271
3.6	0.0529 56	134.81	8.9852	−2.7244	20.243	3.4049	7.3003	17.936	7.3081	3.4679	1	0.15221
3.8	0.0662 27	132.09	11.072	−1.9495	20.511	3.8182	7.8898	18.251	7.8185	3.3786	1.0739	0.1316
4	0.081 581	128.99	13.544	−1.0862	20.655	4.3431	8.7001	18.486	8.3661	3.2829	1.1522	0.11108
4.2	0.099 188	125.41	16.502	−0.11703	20.652	5.0426	9.8711	18.643	8.9752	3.1804	1.236	0.090789
4.4	0.119 23	121.23	20.093	0.98282	20.464	6.0396	11.689	18.732	9.6919	3.0699	1.3272	0.070915
4.6	0.141 91	116.2	24.555	2.2514	20.032	7.6162	14.832	18.783	10.608	2.9489	1.4286	0.05163
4.8	0.167 45	109.9	30.322	3.7551	19.255	10.657	21.376	18.836	11.921	2.8121	1.5453	0.033118
5	0.196 1	101.23	38.492	5.6581	17.865	20.341	42.082	19.008	14.226	2.6439	1.6905	0.015601

Table A.7 Properties of saturated nitrogen.

Temperature (K)	Pressure (MPa)	Liquid density (kg m^{-3})	Vapor density (kg m^{-3})	Liquid enthalpy (kJ kg^{-1})	Vapor enthalpy (kJ kg^{-1})	Liquid C_p (kJ kg^{-1} K^{-1})	Vapor C_p (kJ kg^{-1} K^{-1})	Liquid thermal condition (mW m^{-1} K^{-1})	Vapor thermal condition (mW m^{-1} K^{-1})	Liquid viscosity (μPa s^{-1})	Vapor viscosity (μPa s^{-1})	Surface tension (mN m^{-1})
64	0.014602	863.73	0.77689	−149.03	65.589	2.0016	1.0605	171.53	5.7104	297.5	4.4386	12.001
66	0.020623	855.44	1.0673	−145.02	67.472	2.0053	1.0664	167.5	5.9228	267.79	4.5861	11.522
68	0.028481	847.03	1.4359	−141	69.31	2.0095	1.0734	163.48	6.1374	242.27	4.7343	11.047
70	0.038545	838.51	1.896	−136.97	71.098	2.0145	1.0816	159.47	6.3547	220.22	4.8835	10.576
72	0.051213	829.88	2.4616	−132.93	72.832	2.0204	1.0911	155.46	6.5752	201.07	5.0339	10.109
74	0.066914	821.11	3.1475	−128.87	74.504	2.0273	1.102	151.47	6.7998	184.33	5.1856	9.646
76	0.086102	812.2	3.9695	−124.79	76.11	2.0353	1.1145	147.47	7.0291	169.63	5.339	9.1876
78	0.10926	803.15	4.9442	−120.7	77.644	2.0447	1.1287	143.49	7.264	156.64	5.4943	8.7337
80	0.13687	793.94	6.0894	−116.58	79.099	2.0555	1.1449	139.5	7.5057	145.11	5.6517	8.2844
82	0.16947	784.56	7.4237	−112.43	80.47	2.0681	1.1633	135.58	7.7554	134.81	5.8115	7.8399
84	0.20757	774.99	8.9672	−108.26	81.749	2.0826	1.1842	131.64	8.0146	125.58	5.9741	7.4003
86	0.25174	765.23	10.742	−104.05	82.93	2.0993	1.2079	127.69	8.285	117.25	6.1397	6.9658
88	0.30251	755.24	12.77	−99.805	84.006	2.1185	1.2349	123.75	8.5688	109.7	6.3088	6.5366
90	0.36046	745.02	15.079	−95.517	84.97	2.1407	1.2655	119.8	8.8682	102.82	6.4818	6.1129
92	0.42616	734.54	17.696	−91.181	85.812	2.1664	1.3005	115.86	9.1862	96.522	6.6593	5.6949
94	0.5002	723.77	20.654	−86.789	86.524	2.196	1.3406	111.92	9.5263	90.731	6.8419	5.2828
96	0.58316	712.67	23.989	−82.336	87.095	2.2305	1.3867	107.99	9.8927	85.378	7.0303	4.8771
98	0.67565	701.22	27.742	−77.813	87.514	2.2707	1.4402	104.05	10.29	80.403	7.2255	4.4779
100	0.77827	689.35	31.961	−73.209	87.766	2.318	1.5026	100.11	10.726	75.758	7.4285	4.0856
102	0.89166	677.03	36.705	−68.514	87.837	2.3739	1.5762	96.176	11.207	71.395	7.6408	3.7006
104	1.0164	664.17	42.042	−63.713	87.706	2.4406	1.6637	92.24	11.743	67.274	7.864	3.3233
106	1.1533	650.7	48.057	−58.79	87.35	2.5212	1.7692	88.305	12.347	63.357	8.1005	2.9544
108	1.3028	636.5	54.857	−53.723	86.739	2.6199	1.8986	84.37	13.036	59.609	8.3531	2.5943
110	1.4658	621.45	62.579	−48.486	85.835	2.7433	2.0618	80.437	13.834	55.993	8.626	2.2439
112	1.643	605.36	71.405	−43.044	84.587	2.901	2.2746	76.507	14.775	52.475	8.9246	1.904
114	1.8351	587.98	81.587	−37.35	82.925	3.1091	2.5633	72.583	15.913	49.014	9.257	1.5757
116	2.0431	568.96	93.489	−31.337	80.753	3.3964	2.9733	68.673	17.331	45.563	9.6354	1.2604
118	2.2678	547.73	107.67	−24.902	77.919	3.8192	3.5937	64.795	19.172	42.064	10.079	0.96006
120	2.5106	523.36	125.09	−17.87	74.173	4.5076	4.6309	61.006	21.715	38.425	10.624	0.67738
122	2.7727	493.97	147.62	−9.9021	69.021	5.8453	6.6981	57.518	25.616	34.489	11.34	0.41663
124	3.0562	454.65	180.13	−0.11039	61.211	9.6811	12.731	55.372	33.113	29.873	12.426	0.18574
126	3.3645	372.04	255.22	17.576	42.656	112.02	161.38	75.013	80.151	22.247	15.357	0.008937

Table A.8 Properties of saturated propane.

Temperature (°C)	Pressure (MPa)	Liquid density (kg m^{-3})	Vapor density (kg m^{-3})	Liquid enthalpy (kJ kg^{-1})	Vapor enthalpy (kJ kg^{-1})	Liquid, C_p (kJ kg^{-1} K^{-1})	Vapor, C_p (kJ kg^{-1} K^{-1})	Liquid thermal condition. (mW m^{-1} K^{-1})	Vapor thermal condition (mW m^{-1} K^{-1})	Liquid viscosity (μPa s^{-1})	Vapor viscosity (μPa s^{-1})	Surface tension (mN m^{-1})
−150	5.38E-06	694.61	0.000232	−123.78	402.06	1.9616	1.0205	192.88	3.681	1342.7	3.545	29.413
−140	2.9E-05	684.51	0.001157	−104.09	412.43	1.977	1.0541	187.68	4.2752	985.44	3.7962	28.291
−130	0.000121	674.4	0.004474	−84.234	423.12	1.9937	1.0866	182.17	4.8984	761.71	4.0503	27.119
−120	0.000408	664.26	0.014127	−64.207	434.11	2.0119	1.1186	176.4	5.5502	611.59	4.3067	25.905
−110	0.001164	654.05	0.037899	−43.988	445.38	2.0318	1.151	170.45	6.2304	504.98	4.5645	24.655
−100	0.002899	643.74	0.089039	−23.56	456.88	2.0538	1.1845	164.37	6.9382	425.69	4.8229	23.375
−90	0.006448	633.32	0.18762	−2.8974	468.58	2.0783	1.22	158.23	7.673	364.51	5.0813	22.071
−80	0.013049	622.76	0.36132	18.028	480.44	2.1059	1.2583	152.06	8.4343	315.9	5.3389	20.75
−70	0.024404	612.02	0.6457	39.251	492.41	2.1369	1.3003	145.9	9.222	276.36	5.5955	19.416
−60	0.042693	601.08	1.084	60.811	504.44	2.172	1.3465	139.81	10.037	243.59	5.8512	18.074
−50	0.070569	589.9	1.727	82.753	516.48	2.2115	1.3971	133.83	10.88	216.03	6.1067	16.73
−40	0.11112	578.43	2.6326	105.12	528.48	2.2558	1.4526	127.95	11.756	192.55	6.363	15.388
−35	0.13723	572.58	3.2042	116.49	534.45	2.2799	1.4822	125.06	12.208	182.07	6.4921	14.72
−30	0.16783	566.64	3.8669	127.97	540.38	2.3054	1.5133	122.21	12.67	172.32	6.6222	14.053
−25	0.20343	560.6	4.6302	139.6	546.28	2.3323	1.546	119.4	13.144	163.23	6.7536	13.39
−20	0.24452	554.45	5.5046	151.36	552.13	2.3608	1.5803	116.62	13.631	154.73	6.8868	12.729
−15	0.29162	548.19	6.5012	163.28	557.93	2.391	1.6165	113.89	14.132	146.76	7.0222	12.073
−10	0.34528	541.8	7.6321	175.35	563.65	2.423	1.6548	111.21	14.649	139.28	7.1603	11.421
−5	0.40604	535.27	8.9103	187.59	569.3	2.457	1.6954	108.57	15.185	132.23	7.3019	10.773
0	0.47446	528.59	10.351	200	574.87	2.4932	1.7387	105.97	15.742	125.59	7.4475	10.131
5	0.55112	521.75	11.969	212.6	580.33	2.5318	1.7852	103.43	16.323	119.3	7.598	9.4956
10	0.6366	514.73	13.783	225.4	585.67	2.5733	1.8353	100.93	16.93	113.35	7.7544	8.8664
15	0.73151	507.5	15.813	238.4	590.89	2.6179	1.8897	98.476	17.569	107.68	7.9176	8.2442
20	0.83646	500.06	18.082	251.64	595.95	2.6662	1.9492	96.073	18.244	102.29	8.0891	7.6297
25	0.95207	492.36	20.618	265.11	600.84	2.7189	2.0147	93.718	18.96	97.132	8.2702	7.0236
30	1.079	484.39	23.451	278.83	605.54	2.7767	2.0877	91.409	19.724	92.188	8.4628	6.4265
35	1.2179	476.1	26.618	292.84	610.01	2.8408	2.1697	89.145	20.545	87.433	8.6691	5.8391
40	1.3694	467.46	30.165	307.15	614.21	2.9127	2.2632	86.923	21.432	82.844	8.8918	5.2621
45	1.5343	458.4	34.146	321.79	618.12	2.9946	2.3714	84.742	22.4	78.396	9.1343	4.6966
50	1.7133	448.87	38.63	336.8	621.66	3.0893	2.4987	82.598	23.466	74.066	9.401	4.1433
55	1.9072	438.76	43.706	352.23	624.77	3.2013	2.6519	80.485	24.654	69.828	9.6977	3.6035
60	2.1168	427.97	49.493	368.14	627.36	3.3375	2.8414	78.398	26	65.654	10.032	3.0785
65	2.343	416.34	56.152	384.6	629.29	3.5089	3.0863	76.331	27.557	61.513	10.415	2.5699
70	2.5868	403.62	63.916	401.75	630.37	3.735	3.4214	74.277	29.412	57.364	10.864	2.0797
75	2.8493	389.47	73.14	419.76	630.33	4.0529	3.914	72.233	31.714	53.15	11.403	1.6106
80	3.1319	373.29	84.406	438.93	628.73	4.5445	4.7067	70.213	34.746	48.787	12.075	1.1663
90	3.7641	328.83	119	483.71	616.47	7.6233	9.8876	67.139	46.659	38.819	14.282	0.37921

Table A.9 Properties of saturated R1234yf.

Temperature (°C)	Pressure (MPa)	Liquid density (kg m^{-3})	Vapor density (kg m^{-3})	Liquid enthalpy (kJ kg^{-1})	Vapor enthalpy (kJ kg^{-1})	Liquid, C_p (kJ kg^{-1} K^{-1})	Vapor, C_p (kJ kg^{-1} K^{-1})	Liquid thermal condition (mW m^{-1} K^{-1})	Vapor thermal condition (mW m^{-1} K^{-1})	Liquid viscosity (μPa s^{-1})	Vapor viscosity (μPa s^{-1})	Surface tension (mN m^{-1})
−50	0.037423	1318.4	2.3545	139.63	329.85	1.1278	0.74633	88.952	7.6375	419.34	8.2689	17.088
−45	0.048624	1305.2	3.0067	145.31	333.21	1.1425	0.76181	87.122	8.0349	386.53	8.449	16.271
−40	0.062367	1291.9	3.7945	151.07	336.58	1.1575	0.77774	85.306	8.4314	357.54	8.6282	15.465
−35	0.079039	1278.3	4.7372	156.9	339.95	1.1728	0.79416	83.506	8.8273	331.73	8.8067	14.67
−30	0.099056	1264.5	5.8553	162.81	343.32	1.1883	0.81108	81.724	9.2228	308.62	8.9846	13.888
−25	0.12286	1250.5	7.1712	168.8	346.69	1.2042	0.82856	79.962	9.6186	287.8	9.1624	13.118
−20	0.15092	1236.3	8.7093	174.87	350.05	1.2204	0.84663	78.219	10.015	268.94	9.3404	12.36
−15	0.18372	1221.8	10.496	181.02	353.4	1.2369	0.86535	76.497	10.414	251.78	9.5192	11.616
−10	0.22178	1207	12.559	187.26	356.72	1.2539	0.8848	74.797	10.815	236.08	9.6995	10.884
−5	0.26563	1191.8	14.931	193.59	360.02	1.2713	0.90504	73.12	11.221	221.65	9.8822	10.167
0	0.31582	1176.3	17.647	200	363.29	1.2893	0.92618	71.465	11.632	208.33	10.068	9.4631
5	0.37292	1160.4	20.744	206.5	366.52	1.308	0.94835	69.833	12.051	195.99	10.259	8.7739
10	0.43753	1144	24.267	213.1	369.7	1.3274	0.97177	68.225	12.481	184.51	10.455	8.0997
15	0.51025	1127.2	28.266	219.8	372.83	1.3478	0.99674	66.639	12.925	173.79	10.66	7.441
20	0.59172	1109.9	32.796	226.6	375.89	1.3693	1.0237	65.076	13.385	163.73	10.874	6.7984
25	0.68258	1091.9	37.925	233.5	378.87	1.3921	1.0533	63.535	13.868	154.26	11.102	6.1726
30	0.78351	1073.3	43.729	240.51	381.75	1.4166	1.0864	62.015	14.379	145.32	11.345	5.5642
35	0.89521	1054	50.301	247.64	384.52	1.4434	1.1239	60.517	14.925	136.83	11.609	4.974
40	1.0184	1033.8	57.753	254.9	387.17	1.4732	1.167	59.039	15.517	128.75	11.898	4.4031
45	1.1538	1012.6	66.223	262.3	389.66	1.5074	1.2175	57.578	16.167	121.02	12.219	3.8523
50	1.3023	990.38	75.884	269.85	391.98	1.5476	1.2775	56.13	16.891	113.59	12.581	3.323
60	1.6419	941.34	99.754	285.53	395.93	1.6564	1.4417	53.249	18.668	99.407	13.477	2.3346
70	2.0445	883.23	132.33	302.22	398.57	1.8373	1.7235	50.337	21.186	85.677	14.744	1.4535
80	2.5194	808.98	180.33	320.54	398.9	2.2269	2.3636	47.43	25.333	71.601	16.76	0.70508
90	3.0803	694.07	269.1	342.79	393.32	4.1862	5.688	46.096	35.463	54.872	21.049	0.14385

Table A.10 Properties of saturated R-134a.

Temperature (°C)	Pressure (MPa)	Liquid density (kg m^{-3})	Vapor density (kg m^{-3})	Liquid enthalpy (kJ kg^{-1})	Vapor enthalpy (kJ kg^{-1})	Liquid, C_p (kJ kg^{-1} K^{-1})	Vapor, C_p (kJ kg^{-1} K^{-1})	Liquid thermal condition (mW m^{-1} K^{-1})	Vapor thermal condition (mW m^{-1} K^{-1})	Liquid viscosity (μPa s^{-1})	Vapor viscosity (μPa s^{-1})	Surface tension (mN m^{-1})
−100	0.000559	1582.4	0.039694	75.362	336.85	1.1842	0.59322	143.23	3.3445	1882.4	6.9606	26.835
−90	0.001524	1555.8	0.10236	87.226	342.76	1.1892	0.6173	137.27	4.1459	1341	7.3562	25.188
−80	0.003672	1529	0.23429	99.161	348.83	1.1981	0.64165	131.54	4.9479	1020.3	7.7484	23.563
−70	0.007981	1501.9	0.48568	111.2	355.02	1.2096	0.66654	126.03	5.7509	809.1	8.1364	21.959
−60	0.015906	1474.3	0.92676	123.36	361.31	1.223	0.69239	120.71	6.5554	660.51	8.5194	20.377
−50	0.029451	1446.3	1.6496	135.67	367.65	1.2381	0.71969	115.57	7.3625	550.89	8.897	18.819
−40	0.051209	1417.7	2.7695	148.14	374	1.2546	0.749	110.59	8.1736	467.03	9.269	17.285
−34	0.069512	1400.2	3.689	155.71	377.8	1.2654	0.76778	107.68	8.6629	425.62	9.4898	16.378
−28	0.092703	1382.4	4.8356	163.34	381.57	1.2767	0.78759	104.82	9.1549	389.39	9.709	15.48
−22	0.12165	1364.4	6.2477	171.05	385.32	1.2888	0.80854	102	9.6504	357.42	9.927	14.592
−16	0.15728	1345.9	7.9673	178.83	389.02	1.3017	0.83075	99.225	10.15	329	10.144	13.714
−10	0.2006	1327.1	10.041	186.7	392.66	1.3156	0.85435	96.491	10.655	303.55	10.362	12.847
−4	0.25268	1307.9	12.521	194.65	396.25	1.3304	0.8795	93.794	11.168	280.61	10.58	11.991
2	0.31462	1288.1	15.465	202.69	399.77	1.3466	0.90641	91.128	11.689	259.81	10.8	11.147
8	0.38761	1267.9	18.938	210.84	403.2	1.3641	0.93532	88.491	12.221	240.83	11.023	10.316
14	0.47288	1246.9	23.015	219.09	406.53	1.3835	0.96659	85.878	12.769	223.42	11.252	9.4969
20	0.57171	1225.3	27.78	227.47	409.75	1.4049	1.0007	83.284	13.335	207.37	11.488	8.6915
26	0.68543	1202.9	33.335	235.97	412.84	1.4288	1.0382	80.705	13.925	192.48	11.735	7.9004
32	0.81543	1179.6	39.799	244.62	415.78	1.4559	1.08	78.136	14.548	178.61	11.995	7.1244
38	0.96315	1155.1	47.316	253.43	418.55	1.487	1.1272	75.571	15.213	165.61	12.275	6.3645
44	1.1301	1129.5	56.064	262.43	421.11	1.5232	1.1818	73.003	15.935	153.37	12.579	5.6218
50	1.3179	1102.3	66.272	271.62	423.44	1.5661	1.2461	70.427	16.734	141.77	12.917	4.8977
60	1.6818	1052.9	87.379	287.5	426.63	1.6602	1.3868	66.091	18.326	123.61	13.587	3.737
70	2.1168	996.25	115.57	304.28	428.65	1.8039	1.6051	61.672	20.471	106.51	14.475	2.6429
80	2.6332	928.24	155.08	322.39	428.81	2.0648	2.0122	57.147	23.735	89.846	15.773	1.6318
82	2.7473	912.56	165.05	326.24	428.51	2.147	2.1426	56.235	24.627	86.487	16.116	1.4418
84	2.8653	895.91	175.97	330.2	428.05	2.2473	2.3026	55.329	25.642	83.09	16.5	1.2565
86	2.9874	878.1	188.05	334.28	427.42	2.373	2.504	54.435	26.811	79.637	16.936	1.0763
88	3.1136	858.86	201.52	338.51	426.55	2.5358	2.7658	53.568	28.18	76.102	17.437	0.90189
90	3.2442	837.83	216.76	342.93	425.42	2.7559	3.1207	52.755	29.819	72.45	18.023	0.73378
92	3.3793	814.43	234.31	347.59	423.92	3.0719	3.6303	52.047	31.837	68.628	18.725	0.57288
94	3.5193	787.75	255.08	352.58	421.92	3.5669	4.4258	51.551	34.44	64.551	19.593	0.42037
96	3.6645	756.09	280.73	358.07	419.18	4.4602	5.8478	51.514	38.058	60.062	20.727	0.27805
98	3.8152	715.51	315.13	364.47	415.14	6.5735	9.1402	52.609	43.917	54.801	22.359	0.14895
100	3.9724	651.18	373.01	373.3	407.68	17.592	25.35	58.884	58.976	47.429	25.42	0.039964

Table A.11 Properties of saturated water.

Temperature (°C)	Pressure (MPa)	Liquid density (kg m^{-3})	Vapor density (kg m^{-3})	Liquid enthalpy (kJ kg^{-1})	Vapor enthalpy (kJ kg^{-1})	Liquid, C_p (kJ kg^{-1} K^{-1})	Vapor, C_p (kJ kg^{-1} K^{-1})	Liquid thermal condition (mW m^{-1} K^{-1})	Vapor thermal condition (mW m^{-1} K^{-1})	Liquid viscosity (μPa s^{-1})	Vapor viscosity (μPa s^{-1})	Surface tension (mN m^{-1})
10	0.001228	999.65	0.009407	42.021	2519.2	4.1955	1.8947	578.71	17.412	1306	9.2384	74.221
20	0.002339	998.16	0.017314	83.914	2537.4	4.1844	1.9059	597.95	18.087	1001.6	9.5441	72.736
30	0.004247	995.61	0.030415	125.73	2555.5	4.1801	1.918	614.34	18.786	797.22	9.8602	71.194
40	0.007385	992.18	0.051242	167.53	2573.5	4.1796	1.9314	628.44	19.509	652.72	10.185	69.596
50	0.012352	988	0.083147	209.34	2591.3	4.1815	1.9468	640.57	20.261	546.5	10.516	67.944
60	0.019946	983.16	0.13043	251.18	2608.8	4.1851	1.9648	650.96	21.043	466.02	10.854	66.238
70	0.031201	977.73	0.19843	293.07	2626.1	4.1902	1.9862	659.72	21.86	403.53	11.195	64.481
80	0.047414	971.77	0.29367	335.01	2643	4.1969	2.012	666.97	22.717	354.04	11.539	62.673
90	0.070182	965.3	0.4239	377.04	2659.5	4.2053	2.0429	672.77	23.618	314.17	11.885	60.816
100	0.10142	958.35	0.59817	419.17	2675.6	4.2157	2.08	677.21	24.57	281.58	12.232	58.912
110	0.14338	950.95	0.82693	461.42	2691.1	4.2283	2.1244	680.35	25.579	254.61	12.58	56.962
120	0.19867	943.11	1.1221	503.81	2705.9	4.2435	2.177	682.24	26.652	232.03	12.927	54.968
130	0.27028	934.83	1.497	546.38	2720.1	4.2615	2.2389	682.95	27.795	212.94	13.273	52.932
140	0.36154	926.13	1.9667	589.16	2733.4	4.2826	2.3109	682.53	29.016	196.64	13.618	50.856
150	0.47616	917.01	2.5481	632.18	2745.9	4.3071	2.3939	681.02	30.321	182.61	13.961	48.741
160	0.61823	907.45	3.2596	675.47	2757.4	4.3354	2.4883	678.73	31.721	170.43	14.304	46.591
170	0.79219	897.45	4.1222	719.08	2767.9	4.3678	2.5944	675.52	33.221	159.77	14.645	44.406
180	1.0028	887	5.1588	763.05	2777.2	4.405	2.7129	671.28	34.832	150.38	14.985	42.19
190	1.2552	876.08	6.3954	807.43	2785.3	4.4474	2.8443	666.09	36.563	142.04	15.325	39.945
200	1.5549	864.66	7.861	852.27	2792	4.4958	2.9895	660.01	38.426	134.58	15.666	37.675
210	1.9077	852.72	9.5885	897.63	2797.3	4.5512	3.1503	653.06	40.435	127.87	16.009	35.381
220	2.3196	840.22	11.615	943.58	2800.9	4.6146	3.3289	645.26	42.606	121.77	16.354	33.067
230	2.7971	827.12	13.985	990.19	2802.9	4.6876	3.5285	636.63	44.96	116.19	16.705	30.736
240	3.3469	813.37	16.749	1037.6	2803	4.7719	3.7537	627.17	47.525	111.06	17.062	28.394
250	3.9762	798.89	19.967	1085.8	2800.9	4.8701	4.0105	616.89	50.336	106.28	17.429	26.043
260	4.6923	783.63	23.712	1135	2796.6	4.9856	4.3075	605.78	53.446	101.81	17.81	23.689
270	5.503	767.46	28.073	1185.3	2789.7	5.123	4.6563	593.83	56.926	97.585	18.208	21.337
280	6.4166	750.28	33.165	1236.9	2779.9	5.2889	5.0731	581.03	60.878	93.55	18.63	18.993
290	7.4418	731.91	39.132	1290	2766.7	5.4931	5.5821	567.32	65.456	89.657	19.083	16.664
300	8.5879	712.14	46.168	1345	2749.6	5.7504	6.2197	552.65	70.9	85.855	19.58	14.36
310	9.8651	690.67	54.541	1402.2	2727.9	6.0848	7.0449	536.92	77.587	82.094	20.135	12.089
320	11.284	667.09	64.638	1462.2	2700.6	6.5373	8.1589	520.02	86.156	78.31	20.773	9.8644
330	12.858	640.77	77.05	1525.9	2666	7.1863	9.7526	501.76	97.74	74.428	21.532	7.7026
340	14.601	610.67	92.759	1594.5	2621.8	8.208	12.236	481.93	114.52	70.331	22.477	5.6255
350	16.529	574.71	113.61	1670.9	2563.6	10.116	16.692	460.47	141.27	65.803	23.739	3.6654
360	18.666	527.59	143.9	1761.7	2481.5	15.004	27.356	439.16	191.44	60.306	25.638	1.8772
370	21.044	451.43	201.84	1890.7	2334.5	45.155	96.598	445.42	349.46	52.263	29.659	0.38822

Index

a

Air-water mist flow
  flow across cylinder 292
    Aihara et al. correlation 294
    Finlay–McMillan correlation 293
    Mednick correlation 294
  flow across sphere 297
  flow across tube banks 294
  flow parallel to plates 294
    Hishida et al. correlation 294
  jets 295
    Graham & Ramadhyani correlation 295
  wedges 295
Aladyev et al. correlation for potassium boiling in tubes 185
Amalfi et al. correlation boiling in PHE 163
Ayub et al. correlation boiling in PHE 164

b

Baker flow pattern map 6
Barnea flow pattern map 8
Barnett correlation CHF in annuli 216
Basu et al. correlation inception of boiling 124
Beaty and Katz model condensation 44
Bell–Ghaly model mixture condensation 60
Bends
  boiling heat transfer 139, 170
  critical heat flux 233
  film boiling 265
Bergles–Rohsenow model boiling inception 123
Bertsch et al. correl. boiling in minichannels 158
Boiling
  critical heat flux
    Annuli 216
    inside tubes 201
    liquid metals 95
    pool boiling 90
    subcooled boiling 93
  tube bundles 224, 227
  waiting period 80
Boundary mini - macro channels
  Kandlikar classification 3
  Kew–Cornwell classification 4
  Li and Wang classification 3
  Ong–Thome criterion 4
  Shah criteria 4
  Ullman–Brauner classification 5
Bubble departure point. *See* Onset of significant void (OSV)
Bubble ebullition cycle
  bubble departure diameter 79
  bubble frequency 80
  bubble growth rate 80
    Cole–Shulman equation 80
    Fritz–Ende formula 80
    Kim and Kim formula 80
  liquid metals 86
  waiting period 80

c

CFD simulation 18, 30
Chen correlation boiling in tubes 152
CHF during flow in channels
  analytical models 201
    Tong correlation 201
  annuli 216
    Barnett correlation 216
    Doerfferet al. method 217
    eccentric annuli 221
    Katto correlation 217
    Shah correlation 218
    Stoddart et al. correlation 221
  Bertoletti et al. correl. *See* CISE corr.
  Biasi et al. correlation 203
  Bowring correlation 203
  CISE correlation 217
  critical quality correlations
    Kim–Mudawarr correlation 212
    Wojtan et al. correlation 212
  fluid to fluid modelling 213
  Groeneveld et al. CHF lookup table 202
  horizontal channels 206, 216
  inclined tubes 210
  Katto–Ohno correlation vertical tubes 203
  Kefer et al. correlation horizontal tubes 208
  liquid metals 207
  lookup tables 202
  mechanisms 201
  Merilo correlation horizontal tubes 208
  minichannels 212
    Wu et al. correlation 212
    Zhang et al. correlation 212
  non-uniform heat flux 214
    boiling length average (BLA) method 215
    F-factor correlation 214
    local condition hypothesis 214
    overall power hypothesis 214
  Shah correlation for horizontal tubes 209
  Shah correlation for vertical tubes 204
  Sudo et al. correlation 211
  vertical channels 203
  vertical tube with closed bottom 210
CHF flow parallel to plate 230
  Haramura & Katto correlation 230
  Katto & Kurata correlation 230

CHF horizontal tube bundle
  Cumo et al. correlation 224
  Hasan et al. correlation 225
  Jensen & Tang correlation 224
  Palen & Small correlation 226
CHF in crossflow on cylinders 222
  Haramura–Katto correlation 224
  Hasan et al. correlation 223
  Hwang & Yao correlation 224
  Lienhard and Eichorn analysis 222
CHF in pool boiling
  bubble interference model 90
  cryogenic fluids 93
  dry spot model 91
  effect of heater thickness 94
  effect of subcooling 93
  effect of surface roughness 94
  Haramura–Katto model 91
  hydrodynamic instability model 90
  inclined surfaces 92
  interfacial lift off model 92
  Kandlikar model 92
  Kutateladze correlation 91
  liquid metals 95
  macrolayer dryout model 91
  multicomponent mixture 95
  Vishnev correlation for inclined surfaces 92
  Yagov model 91
  Zuber model 90
CHF liquid metals
  boiling instability 97
  correlations for pool boiling 98
  effect of cover gases 98
  effect of surface wetting 98
  mechanisms 97
CHF vertical tube bundles 227
  mixed flow analysis 227
  sub-channel analysis 228
  phenomenological analysis 228
Chilldown cryogenic pipelines 366
Chun–Seban correlation falling films 171
CISE correlation for CHF 217
Cole–Shulman correlation bubble growth 80
Colebrook correlation for friction factor 14
Condensation heat transfer
  Akers et al. correlation 33
  Ananiev et al. correlation 33
  annular flow models 29
  Cavallini et al. correlation 30
  Dobson–Chato correlation 33
  Dorao–Fernandino correlation 34
  finned tubes 44
  flow pattern-based models 32, 33
  on horizontal tubes 41
  inclined channels 39
  inside tubes 28, 47
  Kim and Mudawar correlation 33
  liquid metals 61
  multicomponent mixture 59
  Nusselt analysis 25, 41
  on plates 25, 27
  Shah correlation 31, 34
  superheated vapor 49
  Thome et al. correlation 33
  on tube bundles 42, 46
  underside of plates 27
Condensation in plate type condensers 50
  Longo et al. correlation 51
  superheated vapor 51
Condensation inside enhanced tubes 47
  Cavallini et al. model 47
Condensation of liquid metals 61
  effect of non-condensable gases 62
  heat transfer correlations 62
  interfacial resistance 62
Condensation of mixtures 59
  azeotropic mixtures 59
  Bell–Ghaly correlation 60
  bubble point 59
  Del Col et al. model 60
  dew point 59
  glide 59
  McNaught model 60
  mechanism 59
  zeotropic mixtures 59
Condensation on cones
  Rohsenow analysis 51
  rotating cone 52
    Sparrow–Hartnett analysis 52
  stationary cone 51
Condensation on enhanced tubes
  Beatty and Katz model 44
  Kumar et al. model 45
Condensation on rotating disk
  Rohsenow analysis 51
  Sparrow–Gregg analysis 51
  visual observations 52
Condensation on rotating tubes 52
Condensation superheated vapor 49
  forced flow external 49
  forced flow in tubes 50
  plate type heat exchanger 51
  Sparrow–Minkowycz analysis 49
  stagnant vapor 49
  Webb model 50
Cooper correlation pool boiling 82
Critical heat flux (CHF) types 90
Critical (choking) velocity
  gas-solid flow in pipes 333

d

Dittus–Boelter correlation 13
Dorao–Fernandino correlation 34
Drift flux
  definition 2
  Rouhani–Axelsson void fraction model 128
  Zuber–Findlay model void fraction 128
Drift velocity 2
Dropwise condensation 63
  effect of non-condensables 64
  effect of surface orientation 64
  heat transfer prediction 64
  mechanisms 63
  promoters 63
  resistances to heat transfer 64
  theories 63
  transition dropwise to filmwise 63

e

Effect of gravity
  boiling with crossflow 188
  CHF flow boiling 237
  CHF in pool boiling 112
  criteria for countering effect of gravity 237
  condensation 55
  experimental studies condensation 55
  gas-liquid flow heat transfer 302
  onset of nucleate boiling 112
  pool boiling heat transfer 112
  pool boiling of mixtures 112
  saturated boiling in tubes 188
  scaling method of Raj et al. 110
Effect of non-condensables on condensation
  Colburn–Hougen analysis 56
  degradation factor 57
  Ge et al. correlation 56
  Lee and Kim correlation 57
  Meisenberg correlation 56

Effect of oil in refrigerants
ammonia condensers 54
ammonia evaporators 189
boiling in tubes 189
boiling in tube bundles 113
condensation 54
enhanced tubes 54, 113
mechanisms 54, 190
miscible oils 54, 190
non-miscible oils 54. 189
halogenated refrigerants 54, 113, 190
pool boiling 113
Evaporation from water pools 297
CFD models 301
empirical correlations 297
Hugo model 300
occupied swimming pools 301
Shah model 298

f

Falling film evaporators
flow modes/patterns 177
heat transfer prediction 178
Chyu et al. correlation 178
Zhao et al. correlation 178
Falling thin films on vertical surfaces
CHF 229
Chun–Seban correlation for heat transfer 171
Fujita & Ueda heat transfer correlation 171
Mudawwar et al. CHF correlation 171
Ueda CHF correlation 229
Film boiling flow across sphere 262
Film boiling flow parallel to plate 267
Film boiling in annuli 259
Film boiling in flow across cylinders
Bromley correlation 261
Epstein and Hauser correlation 261
Liu et al. correlation 261
Film boiling in pool boiling
Berenson model 107
Breen & Westwater correlation 107
Bromley model 106, 107
cylinder 107
horizontal plate 108
Klimenko correlation horizontal plate 108
liquid metals 109
radiation heat transfer 107
Sakurai model 107
spheres 109
vertical plate 106
effective latent heat 106
Film boiling in tubes
dispersed flow 248
Dougal–Rohsenow correlation 249
Downflow 257
Groeneveld correlation 248
Groeneveld–Delorme correlation 250
Hein–Kohler correlation 250
horizontal tubes 257
inverted annular 256
Ellion correlation 256
Mosaad & Johannsen correlation 256
lookup tables 254
mechanistic analyses 249
physical phenomena 247
Saha correlation 252
Shah correlation horizontal channels 258
Shah correlation vertical upflow 252
Film boiling in vertical tube bundles
COBRA code 260
Mohanta et al. correlation 261
TRACE code 260
Flooding
during condensation in tubes 57
in upflow 57
McQuillan–Whalley correlation 57
mechanisms 57
types of 57
Wallis correlation 57
Flooding angle finned tubes 45
Honda et al. correlation 45
Flow boiling heat transfer, saturated 151
correlations for tubes
Chen correlation 152
Gungor–Winterton correlations 156
Kandlikar correlation 155
Liu–Winterton correlation 157
Shah correlation 154
Steiner Taborek correlation 157
tube bundles 172
Flow boiling, subcooled
annuli 131
fully developed boiling regime 123
Gungor–Winterton correlation 132
high subcooling regime 123
Kandlikar correlation 134
Liu–Winterton correlation 133
low subcooling regime 123
partial boiling regime 123
onset of significant void (OSV) 126
Saha-Zuber model for OSV 126
Shah correlation for heat transfer 130
Flow patterns
annuli 8
Baker map 7
Barnea map 8
effect of gravity 10
Dukler map 10
Zhao–Rezkallah map 11
horizontal tube bundles 9
Kanizawa–Ribatski map 9
inclined channels 7
Mandhane et al. map 8
McQuillan and Whalley map 10
minichannels 8
Taitel–Dukler map 6
vertical tube bundles 10
Liu–Hibiki map 10
Venkateswararao map 10
Fluidized beds
carryover/transport velocity
Arnaldas and Casal correlation 312
classification of particles 314
Geldart classification 314
heat transfer between gas and particles 332
heat transfer in freeboard region
Biyikli et al. correlation 332
Dyrness et al. correlation 332
heat transfer to bed walls
Gunn–Hilal correlation 330
Toomey–Johnstone correlation 330
Wender–Cooper correlation 330
heat transfer to objects inside beds
Chen- Pie correl. max.heat transfer 322
Gelperin et al. correl. tube bundles 326
Grewal correl. maximum heat transfer 322
Grewal tube bundles 327

Fluidized beds (*contd.*)
Grewal–Saxena correlation 322, 327
Lechner et al. correl. tube bundles 327
Martin model 320
Mathur&Saxena correlation 325
parameters affecting heat transfer 315
Saxena correlation tube bundle 327
Shah correlation max.heat transfer 322
Wender & Cooper correlation tubes 325
Zabrodsky correlation 321
minimum bubbling velocity 312
Geldart–Abrahamson correlation 312
minimum fluidization velocity
Goroshko et al. correlation 312
Rabinovich–Kalman correlation 312
minimum slugging velocity
Rabinovich–Kalman correlation 312
optimum velocity 322
regimes of fluidization 311
properties of solid particles
density 313
diameter 313
shape factor 314
theories of heat transfer
film theory 318
packet theory 318
Baskaov et al. model 321
Karimipouretal.model 321
Masoumifard et al. model 321
Mickley–Fairbank model 320
Stefanova et al. model 321
particle theory 319
Martin model 320
Natale et al. model 319
penetration theory 318
radiation heat transfer 329
void fraction
Babu et al. correlation 313
Broadhurst–Becker minm.Bubbling 313
Goroshko et al. correlation 313
Staub & Canada correlation 313
Wen & Yu minimum fluidization 313
Fluid to Fluid Modelling 213
Ahmad model 213
Fritz–Ende formula for bubble growth 80
Frost–Dzakovic correlation for ONB 79

g
Gas-liquid flow in tubes
annular flow correlations 282
Hughmark correlation 282
Kim–Ghajar correlation 283
bubbly flow correlations 279
Novasad correlation 279
Zhang et al. correlation 280
gas-injection through channel walls
experimental studies 290
Martin & Sims correlation 291
general correlations
Aggour correlation 286
Dong–Hibiki correlation 285
Kaminsky correlation 288
Kim et al. correlation 286
Kim–Ghajar correlation 283
Rezkallah correlation 286
Shah correlation horizontal tubes 283
Shah correlation vertical tubes 286
Tang–Ghajar correlation 288
inclined channels 289
slug flow correlations 281
Deshpande correlation 281
Lunde correlation 281
Gas-solid flow in pipes
choking velocity 333
experimental studies 334
flow regimes 334
heat transfer prediction 334
Danziger correlation 335
Farbar–Morley correlation 335
Gorbis & Bakhtiozin correlation 335
Pfeffer et al. correlation 336
Schluderberg et al. correlation 335
Shah correlation 336
Sukomel et al. correlation 336
solar collector with particle suspensions 337
Gerstmann & Griffith correlation 27
Gnielinski correlation 13
Gorenflo correlation pool boiling 82
Gungor–Winterton correlation 132, 156

h
Han–Griffith model for waiting period 80
Heat pipes, thermosiphons
boiling 115
condensation 58
Heat transfer single-phase
cylinder with crossflow 14
Dittus–Boelter correlation 13
Gnielinski correlation 13
liquid metals 14
Pethukov correlation 13
plate 14
sphere 14
tube bundles 13
Hein–Kohler correlation 250
Helical coils
boiling heat transfer correlations 166
condensation heat transfer correlations 58
critical heat flux
Hardik–Prabhu correlation 231
Jensen–Bergles correlation 231
Kaji et al. correlation 231
Ma et al. correlation 231
film boiling 265
Hibiki–Ishii model active nucleation sites 81
Holdup 1
liquid holdup 1
Homogeneous flow model 2

i
Inception of nucleate boiling. *See* onset of nucleate boiling
Inclined channels
CHF 209
condensation 39
film boiling 257
flow patterns 7
Shah correlation for condensation 41
Wurfel et al. correlation for condensation 41
Ivey–Morris correlation subcooled CHF 94

j
Jens–Lottes correlation subcooled boiling 130

Jets of mist 295
Jets of liquid impinging on hot surfaces
critical heat flux 234
effect of subcooling 235
Katto–Yokoya correlation 234
Li et al. correlation 234
Monde correlation 234
Qiuand Liu correlation 235
film boiling
Barron & Stanley correlation 265
Fillipovic et al. correlation 264
Kokada et al. correlation 264
Liu & Wang correlation 264
Ruch–Holman correlation 264
flow regions 141
nucleate boiling heat transfer
array of jets 145
effect of contact angle 144
effect of orientation 144
effect of jet diameter 144
inception of boiling 144
Ma-Bergles correlation 142
types of jet 141

k

Kandlikar correlation
boiling in tubes 134, 155
onset of significant void 127
pool boiling CHF 92
Kern correlation condensation in bundles 42
Kim and Mudawar correlation
boiling in minichannels 158
condensation in minichannels 33
critical quality in CHF 212
pressure drop in minichannels 16
Katto–Yokoya correlation jet CHF 234
Kolokosta & Yanniotis correlation 168
Kutateladze CHF correlation 91

l

Laminar film condensation
cone 51, 52
horizontal single tube 41
horizontal tube bundle 42
Nusselt theory
horizontal tube 41
horizontal tube bundle 42
vertical plate 25
vertical tube 26
Lazarek and Black correlation 158
Li and Wu correlation 158
Liquid metals
boiling in tubes 182
boiling in tube bundles 183
boiling stability 86
critical heat flux 95
ebullition cycle 86
effect of magnetic field 87
gas-liquid flow heat transfer 303
Bishop et al. correlation 304
Hori–Friedland correlation 304
inception of boiling 172
pool boiling 86
Shah correlation pool boiling 87
single-phase heat transfer 14
Subbotin correlation pool boiling 87
waiting period 86
Leidenfrost temperature 104, 268
Liu–Winterton correlation 157
Longo et al. correlation plate HX 163

m

Minichannels
boiling heat transfer 158
condensation heat transfer 34
criteria mini-macro boundary 3
critical heat flux 203, 210
flow patterns 8
Minm. film boiling temp.in flow boiling
chilldown of cryogenic lines 269
flow in tubes 268
Groeneveld & Snoek correlation 268
Groeneveld–Stewart correlation 268
jets
Ishigai et al. correlation 268
Liu et al. correlation 268
Ochi et al. correlation 269
Sailer–Marie correlation 268
spheres with crossflow 269
spray cooling 270
Klinzing et al. correlation 270
Mudawar & Valentine correlation 270
Minm.film boiling temp. in pool boiling
Baumeister and Simon correlation 105
Berenson model 104
effect of pressure 106
Henry correlation 105
Olek et al. correlation 106
Sakurai et al. correlation 105
Zuber model 104
Monde correlation jet CHF 234
Mostinski correlation pool boiling 81
Multicomponent mixtures
boiling in tubes, heat transfer prediction
Shah method 180
Thome method 180
boiling in plate heat exchangers 182
bubble point 59
condensation 59
Bell–Ghaly correlation 60
dew point 59
Fujita and Tsutsui correlation 86
Palen–Small correl. CHF tube bundles 226
physical phenomena 59, 83
plate type heat exchangers 182
Schlunder correlation 85
Stephan–Korner correlation 84
temperature glide 59
Thome correlation 84
Thome–Shakir correlation 85

n

Nucleation site density
active nucleation sites 81
Hibiki–Ishii model 81
Nucleate boiling mechanisms
bubble agitation 77
evaporative mechanism 78
vapor-liquid exchange 77
Nusselt analysis for condensation
horizontal tube 41
horizontal tube bundle 42
vertical plate 25

o

Onset of nucleate boiling (ONB)
Basu et al. correlation 124
Bergles–Rohsenow model 79, 123
Davis & Anderson model 124
effect of dissolved gases 126
flow in tubes 123
Frost–Dzakovic correlation 79
Han Griffith model 78
Hsu–Graham model 79
liquid metals 87, 183
pool boiling 78
Sato & Matsumura model 124

Onset of significant void (OSV)
  Kandlikar correlation 127
  Saha–Zuber correlation 126
  Shah correlation 127
Onset of turbulence in condensation 28

p

Palen–Small correlation tube bundle CHF 226
Pethukov correlation 13
Plate type heat exchangers
  boiling
    Amalfi et al. correlation 163
    Ayub et al. correlation 164
    Longo et al. correlation 163
    multicomponent mixtures 182
  condensation
    Longo et al. correlation 53
  types 162–165
Pool boiling
  critical heat flux
    Kutateladze formula 91
    Zuber model 90
  film boiling 106
  heat transfer in pool boiling
    Cooper correlation 82
    Gorenflo correlation 82
    Mostinski correlation 81
    Stephan–Abdelsalam correlation 82
  liquid metals
    Shah correlation 87, 99
    Subbotin et al. correlation 87, 98
  mechanisms 77
  in low gravity 110
  on plates 81, 108
Pressure drop calculation
  Chisolm correlation 15
  Colebrook equation 14
  Darcy friction factor 14
  Fanning friction factor 14
  homogeneous model 15
  Kanizawa–Ribatski correlation 17
  Kim–Mudawar correlation minichannels 16
  Lockhart–Martinelli correlation 15
  Muller–Steinhagen & Heck correlation 16
  Shannak correlation 15
  pipes 14
  single-phase 14
  tube bundles 17
  two-phase 15
  Zigrang and Sylvester correlation 14

q

Quality
  actual
    in film boiling 250
    in subcooled boiling 128
  critical for CHF 201
  definition 2
  equilibrium 2

r

Radiation heat transfer
  film boiling 107
  fluidized beds 329
  jet cooling 296
Raj et al. scaling for gravity effect 110
Rotating disk
  boiling 168
    Kolokosta & Yanniotis correlation 168
  condensation 51
Rotating liquid film CHF 232
Rotating tube
  boiling 169
  condensation 52

s

Sakurai correlation
  film boiling heat transfer 107
  minimum film boiling temperature 105
Saturated boiling in channels
  annuli 153
  correlations for macro channels 152
    Bennet & Chen correlation 154
    Chen correlation 152
    Gungor–Winterton correlation 156
    Liu–Winterton correlation 157
    Kandlikar correlation 155
    Shah correlation 154
    Steiner–Taborek correlation 157
  correlations for minichannels 158
  effect of oil in refrigerants 189
  effect of stratification 154
  effect of various parameters 151
Schlunder correlation mixture boiling 85
Subcooled boiling in channels 123
Stephan–Korner correlation mixture boiling 84
Separated flow model 2
Shah correlation
  boiling in tubes 154
  boiling in tube bundles 175
  condensation in tubes 31, 34, 41
  condensation in inclined tubes 41
  critical heat flux
    annuli 218
    horizontal tubes 209
    liquid metal pool boiling 99
    vertical tubes 204
  evaporation from water pools 298
  film boiling in tubes
    horizontal tubes 258
    vertical tube 253
  flow pattern-based condensation correlation 32
  fluidized bed heat transfer 322
  gas-liquid flow in horizontal channels 284
  gas-liquid flow in vertical channels 286
  gas-solid flow in pipes 336
  mixtures boiling in tubes 181
  onset of significant void 127
  pool boiling of liquid metals 86
  saturated boiling in tubes 154
  subcooled boiling in channels 130
  subcooled boiling flow across cylinders 135
Shekriladze–Gomelauri condensation analysis 42
Sieder–Tate correlation 12
Slip ratio 2
Slip velocity 2
Slug flow 6
Spiral plate heat exchangers 172
Spiral wound heat exchanger 170
Spray cooling
  critical heat flux
    Thiagarajan et al. correlation 237
    Visaria and Mudawar correlation 237
  film boiling
    Klinzing et al. correlation 268
    Wendelstorf et al. correlation 267
  nucleate boiling 145
Spray evaporator. *See* falling film evaporator
Steiner–Taborek correlation 157

Stephan–Abdelsalam correlation 82
Subcooled boiling in channels
mechanism of heat transfer 129
prediction of heat transfer
Gungor–Winterton correlation 132
Jens–Lottes correlation 130
Kandlikar correlation 134
Liu–Winterton correlation 133
Moles & Shaw correlation 134
Shah correlation 130
Subcooled boiling in flow across a cylinder
Lemmert–Chawla correlation 138
McKee correlation 138
regimes of boiling 125
Shah correlation 135
Subcooled boiling regimes
fully developed boiling 123
highsubcooling 123
lowsubcooling 123
partial boiling 123
prediction of regimes 126
Sun and Mishima correlation 158
Superficial gas/liquid velocity 2

t

Taitel–Dukler flow pattern map 6
Taylor flow. *See* slug flow
Thermosiphons
boiling 115
condensation 58
Jouhara–Robinson correlation 115
Thom et al. correlation boiling water 130
Tong correlation for CHF 214
Transition boiling
flow in channels
Bjonard & Griffith correlation 270
Ellion correlation 270
Groeneveld & Snoek method 270
Hsu correlation 270
jet on hot surface
Sailer–Marie et al. correlation 271
Sakhuja et al. correlation 271
pool boiling
Kalinin et al. correlation 102
Ramilieson & Lienhard model 103
Wang et al. method 102
Tube bundles
boiling in horizontal bundles
ammonia evaporator 172
Brisbane et al. model 173
Polley et al. correlation 174
Shah correlation 175
boiling in vertical bundles 172
condensation
Kern model 42
Nusselt analysis 42
Zeinelabdeen et al correlation 44
vapor entry from side 44
gas-liquid flow 294
single-phase heat transfer 13
single-phase pressure drop 17
Two-fluid model 3

v

Vishnev correlation CHF inclined surfaces 92
Void fraction
drift flux model 128
flow in tubes 127
fluidized beds 313
homogeneous model 2
Rouhani–Axelsson model 128
subcooled boiling in tubes
Delhaye et al. correlation 128
Levy model 127
Zuber–Findlay model 128

w

Waiting period in bubble nucleation
Han–Griffith model 80
liquid metals 86
ordinary liquids 79
Webb model superheated condensation 50

y

Yagov model pool boiling CHF 91

z

Zhao et al. correl. falling film evaporator 178
Zuber model
CHF in pool boiling 90
minimum film boiling heat flux104